Lyman®

# Pistol and Revolver Handbook

C. Kenneth Ramage, editor

The Pistol & Revolver Handbook



ISBN: 0-912412-12-7
Library of Congress: 78-55731

Published by Lyman Publications
Rt. 147
Middlefield, Ct. 06455
J. Mace Thompson, president
Printed in the U.S.A.

# Centennial

*During the year 1878, William Lyman began limited manufacture of his #1 aperture tang sight. Initially, he didn't have his own factory building so these first sights were machined and assembled in a corner of the Metropolitan Washing Machine Company facility in Middlefield, Ct. A patent was granted on the sight in 1879, the following year.*

*Within a very few years, Lyman's sight business grew to the point where he had to move into his own building; still a part of the much-expanded Lyman plant today. In the ensuing years, Lyman refined his basic tang sight and added other types to the line, including sights for shotguns and revolvers.*

*New product lines were to follow—metallic reloading, bullet casting and reloading handbooks via the purchase of the Ideal Tool Company in 1925. Two years later Lyman entered the telescopic sight market by purchasing rights to 'scopes from Winchester Repeating Arms Co. and the J. Stevens Arms and Tool Co.*

*Business picked up and the company prospered, generally, through times that were rather lean, nationally. In fact, the shotgun choke line was expanded by the addition of the Cutts Compensator in 1932.*

*During WW II, much of Lyman's production went into support of the national war effort while "civilian" production took a back seat. WW II is something of a milestone for Lyman in that a number of its early competitors did not resume production after the war—while current competitors (in the main) started their businesses in the latter 1940's and early 50's.*

*Today we can say quite correctly—and proudly—that Lyman is the oldest current manufacturer of metallic aperture sights, telescopic rifle sights, bullet casting moulds and related equipment, metallic reloading dies and presses, reloading handbooks and add on shotgun chokes—in the United States.*

*One hundred years in business—that's an accomplishment which would not have been possible without the interest and support of the American shooter. The staff and management at Lyman would like to take this opportunity to say "thank you".*

# Table of Contents

# Table of Contents
(continued)

# Acknowledgements

**Lyman would like to thank the following firms whose contributions of materials and/or information implemented the production of this handbook:**

**E.I. Dupont De Nemours & Co.**
Explosives Dept. Chemical Sales Div.
10th and Market St.
Wilmington, Delaware 19898
Attn: Mr. L.W. Werner
(Manufacturer of Dupont Powders)

**Federal Cartridge Corporation**
Dept. L
2700 Foshay Tower
Minneapolis, Minn. 55402
Attn: Mike Bussard

**Hercules, Inc.**
910 Market St.
Wilmington, Delaware 19899
Attn: Mr. John C. Delaney
(Manufacturer of Hercules Powders)

**Hodgdon Powder Co.**
7710 W. 50 Highway
Shawnee Mission, Kansas 66202
Attn: Bob Hodgdon

**Hornady Manufacturing Co.**
Grand Island, Nebraska 68801

**Omark-C.C.I., Inc.**
Box 660
Lewiston, Idaho 83501
Attn: Mr. Arlen Chaney
(Manufacturer of C.C.I. Primers & Speer Bullets)

**Remington Arms Company, Inc.**
939 Barnum Ave.
Bridgeport, Conn. 06602
Attn: Mr. Joe Callahan

**Sierra Bullets**
10532 South Painter Ave.
Santa Fe Springs, Ca. 90670

**Smith & Wesson**
2100 Roosevelt Ave.
Springfield, Mass.01101
(firearms)

**Thompson-Center Arms Co.**
Rochester, New Hampshire 03867

**Winchester-Western**
275 Winchester Ave.
New Haven, Conn. 06504

# Lyman's Handbook Staff

**Editor:**

**C. Kenneth Ramage**
**Manager, Publications**

**Associates:**

**Raymond I. Cowles**
**Manager, Ballistics Laboratory**

**Edward R. Bryant**
**Assistant Editor**

**Walter M. Florczak**
**Technical Assistant**

**Peter N. Thomas**
**Technical Assistant**

**George F. Marx**
**Technical Assistant**

RAYMOND I. COWLES, LAB MGR.

C. KENNETH RAMAGE, EDITOR

EDWARD R. BRYANT,
ASST. EDITOR

WALTER M. FLORCZAK, TECHNICAL ASST.

GEORGE F. MARX
TECHNICAL ASST.

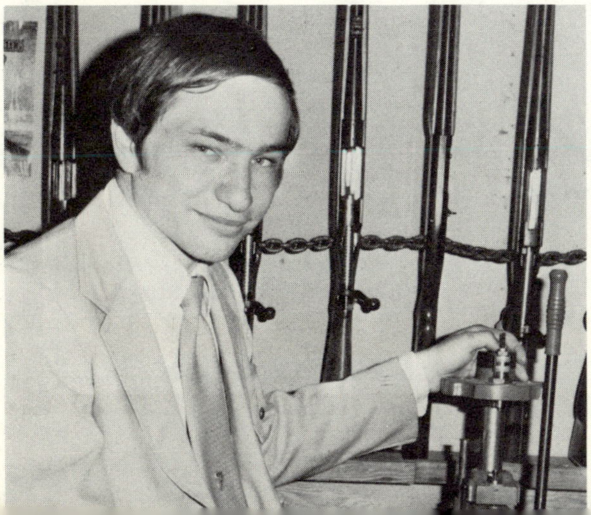

PETER N. THOMAS,
TECHNICAL ASST.

# Editor's Foreword

*We are pleased to introduce to you Lyman's third specialized technical title, the PISTOL & REVOLVER HANDBOOK. This book contains, without a doubt, the most extensive reference of load and trajectory data ever compiled for the handgun.*

*The PISTOL & REVOLVER HANDBOOK was done, as you see, in a larger format than, say, the Lyman RELOADING HANDBOOK, 45th Edition. This increase in page size allows the use of larger, easier-to-read type in the data sections and permits greater ease of illustration. An all-around easier to read book, in short. Along the spine you'll notice what looks like holes for a three-ring notebook binder . . . and that is exactly what they are. This book can be placed into any standard three-ring binder by simply taking the book apart page by page and placing it into the binder. Lyman, by the way, also offers one and two inch binders plus a packet of load note sheets.*

*As is the case in any book—regardless of the subject—we couldn't cover everything. However, we did spend over a year in solid ballistic research and data collection to generate the data you'll see further back in this book. We're confident you will find some useful information for your particular interests.*

*Since this book is for the novice as well as the advanced reloader, the front of the book covers most of the points necessary to start reloading safely.*

*Our technical staff stands ready to answer your reloading questions and we are always pleased to receive constructive comments from shooters who use Lyman products. We hope you enjoy this book.*

*Sincerely,*

*C. Kenneth Ramage*

*C. Kenneth Ramage, Editor*

# 1

# Why Reload?

# Why Reload?

**WHY RELOAD?**

Shooters reload for two primary reasons: to save money and tailor their ammuntion to a particular gun or purpose. On the way to reaching the foregoing goals, they find a third benefit—that pleasure derived from making something with one's own hands.

Reloading is the only way most of us will be able to enjoy our shooting sport to the greatest extent. Just a few hours spent one evening during the week can produce enough well-made ammunition to get through a pretty active weekend at the range.

Whether your're a hunter, plinker, target shooter or one of the growing number of silhouette shooters, there is information in this Handbook which will both save your money and improve your handgun performance.

Handgunning is an art and, like all arts, requires considerable practice. If you should ask any top shooter the secret of his success, he'd probably reply: "Plenty of practice and ten thousand dollars worth of ammunition"! If you wish to become a really good handgunner and your budget is limited, then you must eventually resort to reloading.

This book is about reloading for handguns. And it can quite possibly provide the solutions to your shooting problems, too. The text and illustrations will lead you through the basic mechanics of reloading, guide you in the slightly more complex areas, and leave you with a safe and sound knowledge of a most fascinating hobby. By stretching your ammunition dollars, you'll be able to afford a great deal more shooting. And in the process, you'll become a more skillful pistol shot. Who knows? . . . maybe even a national champion!

**Sure I can save money by reloading—but what about the worry, the danger, the skill required?**

After hearing the reloading lingo around the local gun shop or reading conflicting statements that have appeared in shooting publications, it is little wonder many shooters think of metallic reloading as some sort of mystic art. It's easy for the novice to become bewildered when the so-called experts disagree, and when information and misinformation are blended together in print and discussion.

Is reloading a highly technical hobby that requires a touch of genius? Or can anyone participate? Can an average Joe Novice, who possesses little technical ability, actually assemble good, safe ammo without endangering his life or limb?

Of course, he can! But let's not try to oversimplify the problems of reloading because they do require serious consideration. From strictly a mechanical aspect, reloading is a fairly simple hobby. Certainly, it does not require any genius to operate the handle of a reloading press or move a cartridge case through a normal loading cycle. It does, however, require a good deal of common sense on the part of the reloader if he is to remain out of trouble. The reloader should realize that his job is one of assembling ammunition; not experimenting with untested loads.

A knowledgeable reloader selects his data carefully from a reliable source and he uses this data in an intelligent manner. In other words, he follows instructions to the letter, and does not alter or substitute components in any way. Equally important, he avoids unreliable sources of data like the plague! Simply because a certain load is mentioned in an article on shooting, this does not necessarily mean that it is a safe load.

For example, a sports columnist may write something like: "Here's a whiz of a load sent in to me by a reader from Texas." The reloader should immediately question the validity of such opinion. Who tested the load? Is that reader qualified? Is the load endorsed by a reputable powder or component manufacturer? If these questions have not been answered satisfactorily, then the reloader should back off—he could be led along a dangerous road.

The safe route (and there is a safe route) to reloading is also the easiest one. It consists of selecting a reputable source of data, such as this Handbook, and abiding by the rules. Even the rules are easy for they require only that you employ common sense and do not experiment by altering components. It's as simple and as unexciting as that. But, who needs the excitement when so much is at stake?

***Safe Reloading***
***is very much like safe gun handling.***
***Both require an intelligent mature approach.***

***Caution to Consumer***
***Reloading can be hazardous if not done properly.***
***In all cases, follow instructions.***

# 2

# Getting Started

# Getting Started

In view of your present interest in the reloading subject, it is safe to assume that you have already had considerable shooting experience. Certainly, you have purchased and fired a good many rounds of factory loaded ammunition. Hence you are equipped with a working knowledge of the various caliber designations and cartridge shapes, and you have learned to recognize the correct cartridge for your particular pistol. Probably many readers will have an overall understanding of shooting that extends beyond these simple basics. However, for the benefit of the uninitiated, we'll begin at this point.

Each factory loaded cartridge we purchase and fire represents extensive thought and care on the part of its manufacturer. If we are to reload a fired cartridge and duplicate if possible, the original factory loading, we must first learn to appreciate some of the intricacies of ammunition design. For instance, the fired cartridge case is the most important reloading component. The function of the case, how it is constructed, and its **condition after firing** are important considerations to reloaders.

## Equipment Selection

You'll need to equip a reloading bench. While there are much less sophisticated reloading systems on the market, starting with the best gear you can afford quickly pays for itself by turning out excellent reloads on a volume basis.

We've listed our thoughts on basic equipment below and, for reloaders needing complete sets of matched equipment, Lyman now provides the two reloading kits illustrated which contain all needed items—plus an accessory collection for the reloader already owning a press.

### EQUIPMENT LIST:

A reloading press
A shell holder
A priming arm
A ram
Complete set of dies
Powder scale
Powder measure
Powder funnel
Case lubricant with pad
Case blocks
Trimmer
Deburring Tool
Vernier caliper or case length gauge
Reloading manual

### OPTIONAL EQUIPMENT:

Automatic primer feed
Case neck brush
Inside neck reamers
Outside neck turners
Drop tube funnel
Primer pocket cleaner
Primer pocket reamer
Powder Trickler
Bullet puller

Buying a full set of reloading equipment is a fairly complicated task—especially if your dealer isn't deeply committed to reloading products. He may have a "Brand A" press, "Brand B" measure, "Brand C" shell holder head and so forth. It's possible for the novice reloader to stagger from the shop laden with two hundred dollars worth of reloading gear and still have doubts about whether he has all he needs and whether what he has will fit together.

The foregoing is a very real problem and prevents many people from enjoying the reloading hobby.

Lyman has developed three reloading kits which contain everything a person needs to perform reloading operations within certain areas. Two kits, the Expert and Sharpshooter, contain reloading presses and needed hardware to reload a given rifle or pistol cartridge. The third kit, the Marksman, is intended for the person who already owns a press and dies but needs to cover his powder handling and case lubrication requirements.

Of course, Lyman sells reloading equipment in other than kit form. Currently a reloader has a choice of three types of metallic bench presses and a full array of accessories. Lyman also offers a full line of bullet moulds and casting equipment. The best way to see all this is to write for your copy of the current annual catalog.

## Universal Trimmer ™

Simplicity and precision clearly describe our trimmer. Its patented chuck head accepts all metallic rifle or pistol cases, regardless of rim thickness. To change calibers, simply change the inexpensive case head pilot.

You get additional quality features like coarse and fine cutter adjustments and an oil impregnated bronze bearing. A rugged cast base assures precision alignment and years of service.

## Sharpshooter Reloading Kit

**The Sharpshooter Kit contains:**

1. Spartan Press
2. Spartan Ram
3. Spartan Auto-Primer Feed
4. Spartan Shell Holder Head
5. Spartan Primer Arm
6. Spartan Primer Catcher
7. Universal Trimmer & Pilot
8. D-7 Scale
9. No. 55 Powder Measure
10. Deburring Tool
11. Case Lube Kit
12. Powder Funnel
13. 45th Edition—Reloading Handbook
14. Reloading Die Set & Die Wrench
15. Allen Wrench
16. Extra Decapping Pins
17. Instruction Manual for Easy Set-Up

## Expert Reloading Kit

**Nothing Else To Buy!**

Here's a kit that offers the new reloader a complete set of reloading equipment featuring the famous high-speed turret loading Spar-T press.

**The Expert Kit Contains:**

1. Spar-T Press
2. Spar-T Auto Primer Feed
3. Spartan Primer Arm
4. Spartan Ram
5. Spartan Shell Holder Head
6. Primer Catcher
7. Universal Trimmer & Pilot
8. D-7 Scale
9. No. 55 Powder Measure
10. Deburring Tool
11. Case Lube Kit
12. Powder Funnel
13. 45th Edition-Reloading Handbook
14. Reloading Die Set & Die Wrench
15. Allen Wrench
16. Extra Decapping Pins
17. ⅞'' x 14'' Adapter to mount Powder Measure
18. Instruction Manual for Easy Set-Up

Both the Expert and Sharpshooter Kit are available in the 12 most-used Lyman die calibers:

1. 9mm Luger
2. .38 Spl.
3. .357 Mag.
4. .44 Mag.
5. .45 Automatic
6. .222 Remington
7. .22-250 Remington
8. .243 Winchester
9. .270 Winchester
10. .30-30 Winchester
11. .308 Winchester
12. .30-06

## The Marksman Kit

The Marksman Kit is for the reloader who has a metallic reloading press but needs the accessories for it. The Marksman Kit contains Lyman's popular case lubrication and powder handling components. This all-in-one package is the answer to the reloader's case-sizing and powder charging problems!

**The Lyman Marksman Kit includes:**

1. No. 55 Powder Measure
2. D-7 Scale
3. Deburring Tool
4. Case Lube Kit
5. Powder Funnel
6. Lyman's new Reloading Basics

## The #55 Powder Measure

This powder measure dispenses charge after charge of smokeless powder with consistent accuracy to a fraction of a grain. The key to this unfailing precision is the 55's unique three-slide micrometer, adjustable cavity. Extra-fine adjustments of width and depth provide consistent charges with almost any powder. A 2400 grain reservoir of clear PVC plastic resists chemical action and protects your powder from harmful light rays. An attached knocker assures a complete charge every time. Mount the 55 on your bench or any turret press. No funnel is required.

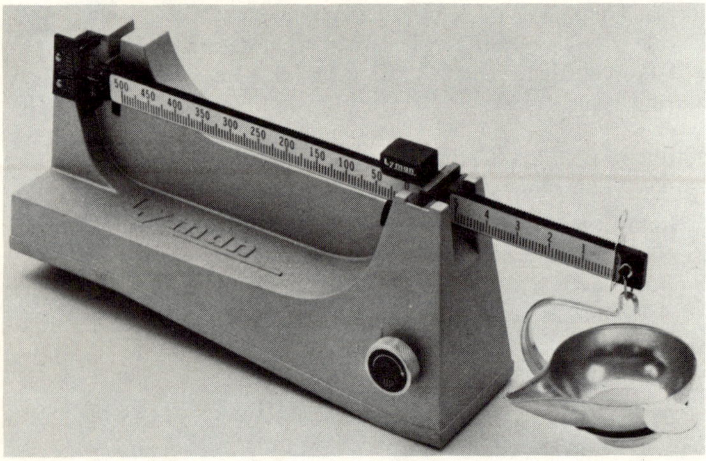


## Lyman's D-7 Scale

Our D-7 represents the ultimate reloader's scale. These eight features add up to consistent, long-lasting accuracy:

- 505 grain capacity . . . enough for even the heaviest charges.
- Black on white markings . . . faster and easier readings.
- Magnetic damping . . . assures unsurpassed ease and speed.
- Precision-ground knife edge on agate bearings . . . assure continued accuracy after years of service.
- New, unique beam lifter . . . guarantees less wear and tear when not in use or moving.
- Positive pan positioning . . . this unique feature assures constant accuracy during use.
- Heavy base casting . . . reduces vibration for increased accuracy.
- ± 1/10 grain accuracy . . . each scale is factory calibrated before shipment.

## All-American® Reloading Die Sets

These dies are designed for rapid, accurate reloading of metallic cartridges on our Spartan, Spar-T and AA Turret presses, as well as all others with ⅞'' x 14 thread die stations. Carefully constructed of ''first quality'' steel, AA dies are first hardened and then polished to a mirror smooth interior finish to produce a perfectly finished and sized case.

To assure the reloader years of service, the outer surfaces of the dies are chrome-plated for maximum protection. Precision rolled threads and hexagonal steel lock-rings securely hold die in precise adjustment during reloading.

For best reloads, sizing dies for bottleneck cartridges are carefully vented to prevent air traps and bullet seating dies adjust to crimp or not crimp case neck, plus all three die sets include our Deluxe Two-Step ''M'' Neck-Expanding Die.

**Deluxe Pistol Set**

The pistol marksman can really appreciate the features that this set offers. The special Lyman T-C (Tungsten Carbide) Full-Length Resizing and Decapping Die assures extra long life, while the Two-Step ''M'' Neck-Expanding Die prepares the case for perfect bullet seating every time. A Bullet Seating and Crimping Die is also included.

# 3

# Components

Cases
Powder
Projectiles
Primers

# Components

## CARTRIDGE CASES

The chief function of a cartridge is to seal off the breech at the time of firing. To accomplish this, the case walls must expand freely so that they are tight against the sides of the chamber. This sealing action prevents the hot powder gases from leaking back around the cartridge and out through the action. Along with this, the cartridge case must withstand the chamber pressure that is built up during firing. To achieve this, the case requires a structural strength of its own—plus the additional supporting strength supplied by the breech face and chamber walls. In essence, the case functions as an intrinsic part of the pistol. **A pistol is no stronger than the case that is used in it—nor is the case stronger than the pistol.**

Cartridge brass is carefully tempered in its final manufacture. The head of the case is thick and tough which gives it the strength and rigidity necessary to resist the force of the chamber pressure. The forward section of the case (neck, shoulder and body) is considerably thinner than the head section. In manufacture, these portions are given an anneal which leaves them soft and ductile. The obvious advantage is that the case walls and neck will now expand freely to release the bullet and seal the chamber while the cartridge is fired.

As shooters, we may have been rather casual in our regard for empty brass cases, but as reloaders we soon come to think differently. Without a quantity of strong and serviceable cases, we would not get far in reloading ammunition. The most usual way for a reloader to obtain serviceable cases is to purchase factory loaded ammunition. After this ''store bought'' ammo has been fired, the empty cases are retained for future reloading. The reloader may also purchase new cases from his component dealer.

**To make sure your cases are in prime condition, start with either new or once-fired cases. Never use brass of unknown origin such as that found on a shooting range.**

Each firing and resizing has an influence on the serviceability of the case. The battering of chamber pressure and the forces applied by the resizing die eventually work-harden the forward portion of the case and destroy its usefulness.

**Cartridge Case Nomenclature**

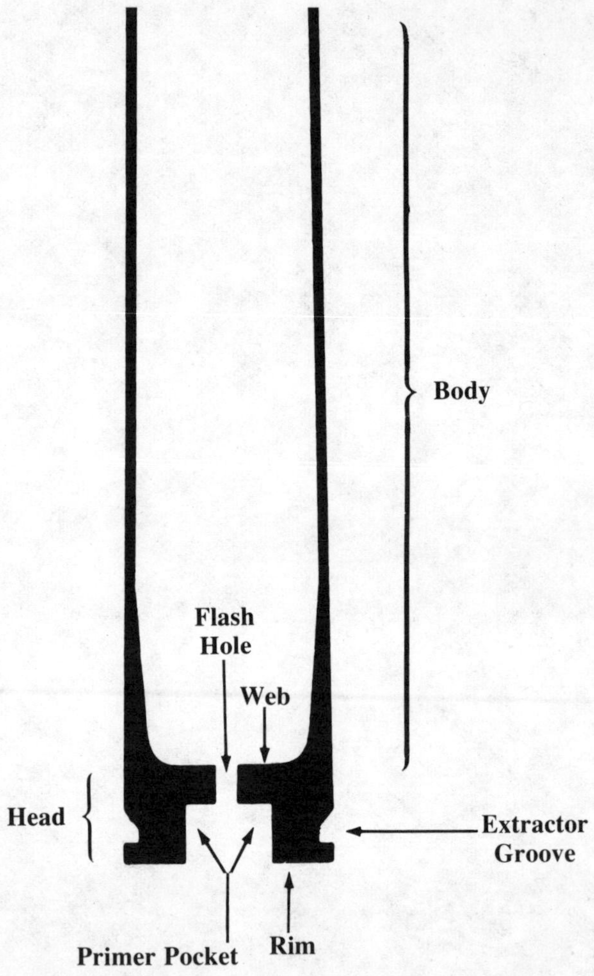


**Nomenclature of a pistol case. While cases vary in shape and size, the same basic nomenclature applies to all center fire cartridges.**

**Carefully inspect your cases before each reloading.** If your cases are new, or once fired, they will not reveal fatigue at the first reloading. However, fatigue signs will show up in subsequent loading, so you must learn to look for them. **Check your cases for splits or cracks in the neck, shoulder or body.** Reject all cases that show signs of defects, but flatten them with a pair of pliers before discarding to prevent their reuse. We suggest you separate your cartridge cases into lots and keep a record of their history.

**Trimming** is necessary when your cases have lengthened after numerous firings. Check your cases after resizing and never allow them to exceed the maximum listed measurement. How often you will need to trim the cases depends largely on the type of case you are using and the pressure of the load. Bottle-neck cases take more abuse from pressure than straight sided cases and require trimming more often.

### CASE STYLES

*Function is basically the same as Rimless Type.
**Straight walled cases have a slight taper which is not visibly apparent.

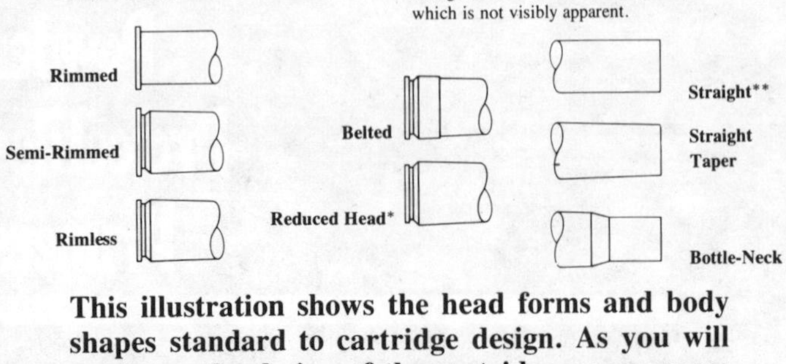


**This illustration shows the head forms and body shapes standard to cartridge design. As you will see later, the design of the cartridge case governs the headspace dimensions. Case examples for almost any combination of the above shapes may be encountered.**

**Case length** is of vital importance to the auto pistol reloader since most of those cartridges headspace on the case mouth; rather than a projecting rim at the base.

Typically, rimmed straight-wall cases require little or no trimming. Case length directly affects overall cartridge length and is of particular importance to crimped cartridges. Magnum revolver loadings should be crimped into a cannelure or groove to ensure the bullets will not back out under recoil and bind the cylinder.

Cases which are notably short will require the bullet to make a longer ''jump'' to reach the rifling. Usually this affects accuracy adversely.

The trimming of any case more than four times is not recommended. After this amount of trimming, it may be assumed that the case walls are now too thin, and the case should be discarded.

## Case Diagnosis

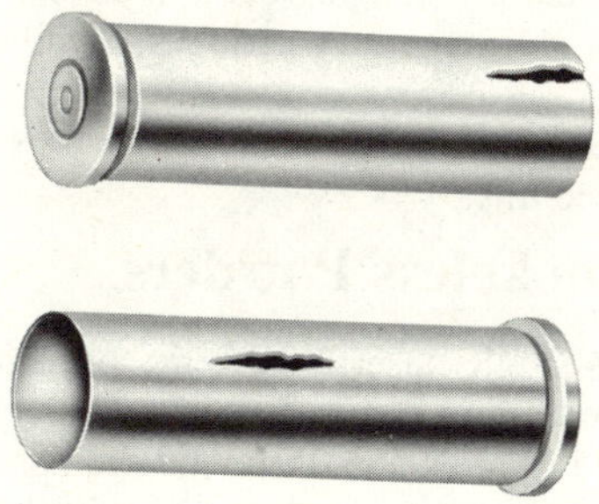

Fatigue signs show up in the form of splits or cracks in the case mouth or body. Discard cases that show signs of defects.

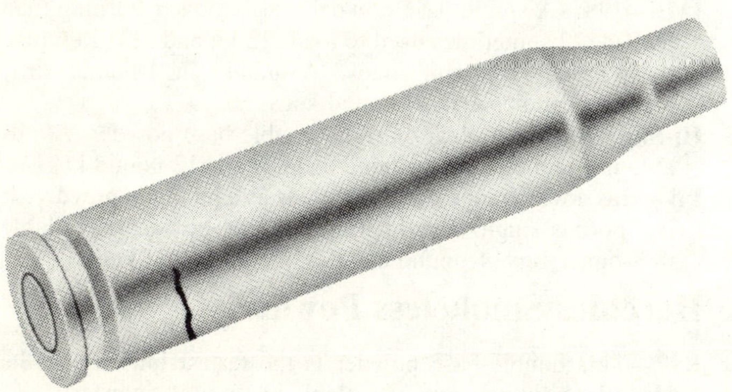

**CONDITION ''B'' EXCESSIVE HEAD SPACE**
The condition pictured in this example is quite different from case fatigue. This crack runs around the circumference of the case and indicates an excessive headspace condition. Never use any pistol which shows signs of excessive headspace.

To properly check the length of your cases, you will require a good vernier caliper available at most hardware stores.

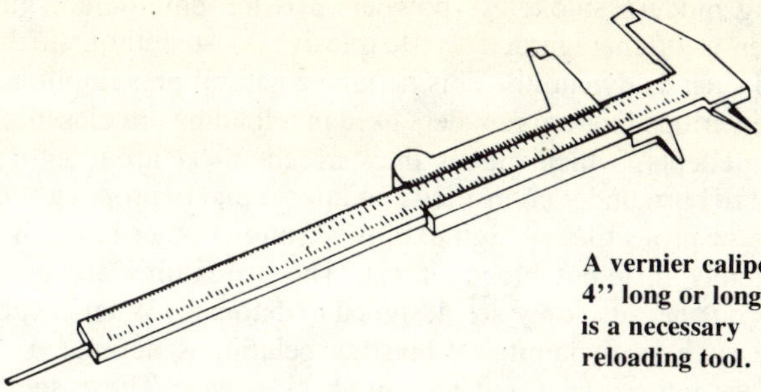

**A vernier caliper, 4'' long or longer, is a necessary reloading tool.**

In addition, the reloader needs an accurate means of case trimming. The Lyman Universal Trimmer (pictured below) is actually a miniature lathe featuring a unique chuck head which accepts all metallic cartridge cases, rifle or pistol. The tool is supplied with complete instructions and will accomplish the trimming operation quickly and easily.

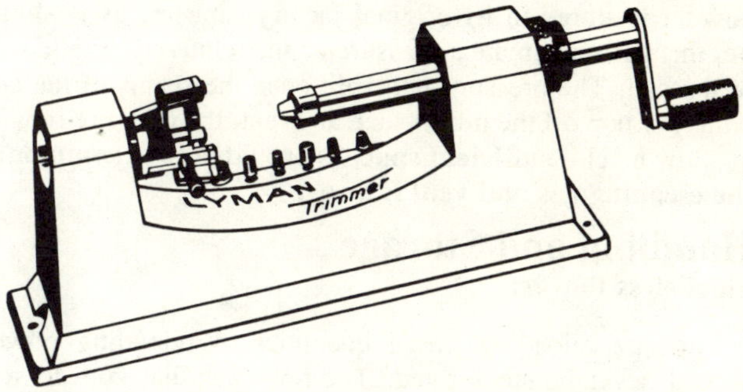


**LYMAN Universal Trimmer**

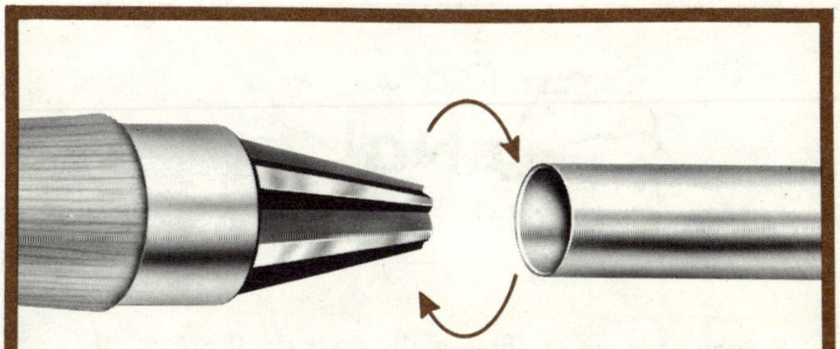

When chamfering turn the reamer lightly, removing very little case material.

When loading new or once-fired cases, it is necessary to remove the sharp inside edges of the case mouth. This operation is called **chamfering** and its purpose is to ease insertion of the new bullet. Chamfering is required only for the first reloading of a new or once-fired case. The inexpensive Lyman hand reamer chamfers a case easily and with uniformity. Hold the case in one hand, while you lightly turn the reamer in the case mouth with the other hand. Remove very little material and **do not** cut a sharp knife edge on the case.

# POWDER

Many people are concerned with the dangers of handling and loading modern smokeless powders. To the uninitiated, gun powder is often regarded as "explosive"—something to be feared such as dynamite. This is quite a natural presumption.

The various modern powders used in reloading are classified as propellents, which means they are chemical mixtures designed to burn under controlled conditions, and to propel a shot charge or projectile. A high explosive, on the other hand, is a completely different breed of cat. These mixtures are quite dangerous because they are designed to detonate. When a substance such as dynamite or blasting gelatin, is detonated, it produces intense heat and violent shock waves. These shock waves exert tremendous pressure on anything they contact, which makes it almost impossible to vent away the effects of detonation involving any appreciable quantity of explosive material.

While not to be compared to explosives, modern reloading powders are nonetheless highly flammable. They not only burn, they burn vigorously. In case of accidental ignition, a great amount of gas at high temperature will be formed. If the powder is stored in its original factory canister, **as it should be,** this gas will create a pressure on the relatively fragile sides of the can. The pressure will split open the seams of the container, or pop off the lid. In such an event, the pressure remains at a low level **if sufficient space is provided to accommodate the escaping gas and vent it away.**

## Handling and Storage

### Smokeless Powder

For safe reloading, large quantities of reloading powder should never be stock-piled. The few cans that you do store must be handled with the care and caution due a flammable substance. In this way you control the situation so that it cannot get out of hand. Even in the case of a total house fire arising from other sources, **smokeless powders will not produce the effect of an explosion** if the proper storage precautions have been taken.

**Strong storage cabinet walls restrain the gas, allowing it to build up to a violent release.**

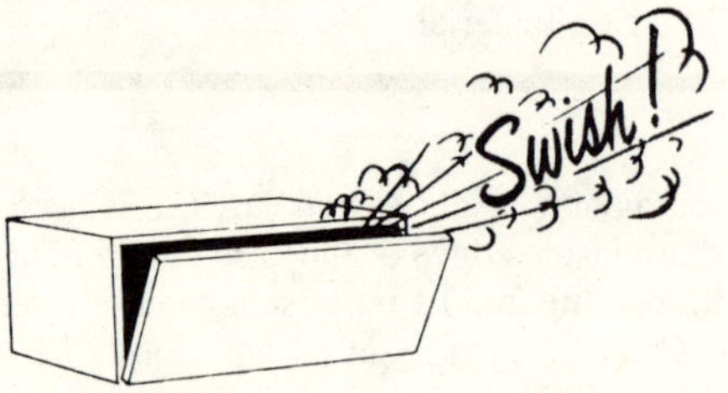


**Weak storage cabinet wall, or walls, push out to allow easy venting of gas before appreciable build up of force.**

### Black Powder

Unlike smokeless powder, black powder is classified as a Class "A" Explosive. Also, black powder is a good deal touchier than smokeless powder. Its ignition does not necessarily require a direct flame. It can be ignited by percussion (dropping container, etc.) or by static electricity. We recommend that black powder be handled with extreme care. Quantities should be kept to a minimum and should not be stored in the home or with other reloading powders. A clean storage area in an out-building (away from the home) should be used. This area should comply with the regulations set for smokeless powder. The out-building should be **locked** to protect children.

**Never smoke—or have an open flame—in the presence of smokeless or black powder.**

## Alcan Smokeless Powders

**AL-5**—A single-base, progressive-burning powder. Available in 8-ounce tin.

**AL-7**—A dense, very progressive-burning powder. Available in 8-ounce tins.

**AL-8**—This coarse-grained powder proved an excellent choice when used in .44 magnum cartridges. Available in 8-ounce tins.

**AL-120**—A clean-burning powder that ignites easily. Available in 8-ounce tins.

## DuPont Smokeless Powders

**SR-4756**—This single-base powder has been used with good success in medium to heavy pistol loads. Packed in 8-ounce cans, 4-pound caddies and 25-pound drums.

**SR-4759**—A single base powder used with several handgun loads.

**SR-7625**—Suitable for a wide variety of light to medium center-fire handgun loads, this single-base powder produced excellent results throughout the testing program. Available in 8-ounce tins, 4-pound caddies and 25-pound canisters.

**IMR-4227**—A single base powder that is the fastest burning of the IMR series. Well suited for magnum handgun loads. Available in 1-pound tins, 8-pound caddies and 20-pound kegs.

**IMR-4198**—A single-base powder that is slower burning than IMR-4227 sometimes used to load .22 Jet and .221 Remington Fireball handgun loads. Available in 1-pound tins, 8-pound caddies and 20-pound kegs.

**Hi-Skor 700X**—A clean-burning double-base powder. Available in 8-ounce tins, 5-pound caddies and 12-pound kegs.

**PB**—Has a wide range of applications as a handgun powder. It is a porous single-base medium-burning powder. Available in 8-ounce tins, 4-pound caddies and 25-pound kegs.

## Hercules Smokeless Powders

**Rx7**—This double-base powder is the fastest burning of the Hercules rifle powders. Available in 1-pound canister.

**Red Dot**—This double-base, medium fast-burning powder is useful for light handgun loads. Available in 1-pound canisters, 4-pound canisters, 8- and 15-pound kegs.

**Green Dot**—This double-base powder offers uniform ignition, minimum blast and it's non-hygroscopic. Suited to a wide range of pistol and revolver loads. Comes in 1-pound tins, 4-pound canisters, 8- and 15-pound kegs.

**Blue Dot**—A shotshell powder with some limited application to large caliber handgun loads. Available in 1-pound tins, and 5-pound canisters.

**Herco**—A double-base medium slow-burning powder suited to heavy handgun loads. Available in 1- and 4-pound canisters and 8- and 15-pound kegs.

**Unique**—A double-base medium-burning shotgun powder used in magnum handgun loads and for automatic pistol loads. Available in 1- and 4-pound canisters and 8- and 15-pound kegs.

**2400**—A medium slow-burning powder with a double-base that is well suited for heavy handgun loads. Available in 1- and 4-pound canisters and 8- and 15-pound kegs.

**Bullseye**—One of the most widely used pistol powders in the world, this powder is capable of excellent accuracy in a wide range of pistol cartridges, particularly the small calibers. Available in 1- and 4-pound canisters and 8- and 15-pound kegs.

## Hodgdon Smokeless Powders

**Trap 100**—A spherical powder that is suited for target loads in centerfire handgun cartridges. Availablein 1-pound tins.

**HS-5**—A spherical shotshell powder. Available in 1-pound tins.

**HS-6**—A dense, spherical shotshell powder that ignites easily. Available in 1-pound tins.

**HS-7**—This is the slowest of the Hodgdon spherical powders. Comes in 1-pound tin.

**H-110**—A spherical double-base powder that can be used for heavy pistol loads. Available in 1-pound tins.

**H-322**—An extruded powder that performs best in small to medium cases. Available in 1-pound tin.

**HP-38**—A fast-burning pistol powder suited for many handgun loads but particularly applicable to .38 Special. Available in 12-ounce can.

**Spherical BL-C2**—A double-base ball-type powder. Available in 1-pound can.

**Pyrodex P**—For black powder pistol cartridges. Will duplicate black powder loads. Available in 1-pound canisters.

## Winchester Smokeless Powders

**231**—A very fast and high energy pistol powder designed for target and standard velocity loads in handguns. This powder is popular with reloaders and police departments. Available in 1-, 3-, 8- and 12-pound tins.

**296**—A .410 bore powder designed to produce the optimum in case volume. This is the factory powder in calibers such as the .357 magnum, .44 magnum and .30M1 Carbine. Available in 1-, 3- and 8-pound tins. Use only as directed by Winchester: 1) A high bullet pull must be used. A heavy crimp is appropriate; 2) A magnum pistol primer must be used in handgun calibers; 3) At least 90 per cent loading density must be maintained; 4) Only heavy bullets should be used.

**630**—A moderately slow but high energy pistol powder appropriate for cartridges such as the .45 A.C.P., .38 Super and a wide variety of other pistol calibers. Available in 1- and 8-pound units.

**680**—A very fast rifle powder designed specifically for small cases like the .30 Carbine and .22 Hornet. Available in 1-pound canisters.

**748**—A center fire rifle powder that has a wide range of applications. Available in 1- and 8-pound canisters.

## Projectiles

Reloaders of pistol cartridges have, in the past, tended to use more jacketed bullets—on the average. Only target shooters and those who sought the most economical form of projectile used lead bullets to any great degree.

Today, the trend is shifting. The retail price of a box of jacketed bullets has reached the point where more and more reloaders (always economy-minded) are becoming bullet casters. Today we know much more about various alloys than we did 30 years ago. New lubricants, like Alox/beeswax, have permitted maximum pistol velocities without leading. Now the handgunner can efficiently use cast bullets for most of his needs without sacrificing efficiency.

Basically, bullet shapes are similar for both cast lead and jacketed pistol bullets; an exception being the jacketed rifle projectiles useable in pistol cartridges like the .221 Fireball and .256 Winchester.

There are several factors to consider when selecting a bullet besides having carefully tested load data for a given projectile weight:

**1. Weight**—For optimum potential accuracy the bullet must fall within a range peculiar to a given caliber. Also, the velocity at which that bullet will be driven is a factor. The whole idea is to properly blend the three (weight, rate of twist and velocity) to produce the results satisfying you, the reloader. Weight is important to the target shooters for its effect upon sectional density; hunters are concerned with weight for its implications to penetration potential on game.

**2. Base Design**—By far the most common base design for lead or jacketed pistol bullets is the flat base. Variations include hollow bases in lead bullets and a boattail on some of those .224" and .257" rifle bullets mentioned above. Considering that the practical pistol range does not exceed that at which a boattail base starts to pay for itself, the handgunner is not at any disadvantage with flat base bullets.

**3. Point Design**—There are various configurations within a given caliber ranging from aerodynamically sleek spitzer points for minimum drag to flat noses particularly suited for short range hunting loads. These point shapes can also incorporate an exposed lead nose, hollow point or full coverage jacket.

One last thing to remember about jacketed bullets is that the actual jacket material and projectile bearing surface will vary between manufacturers for bullets of identical weight and similar design.

All our data lists the specific bullet for which the loads were developed. The following is a table of explanation for symbols used in the Data Section to describe bullet designs:

| Key: | (Bullet design) |
|---|---|
| MC | Metal Case |
| SP | Soft Point |
| SPP | Spire Point |
| HP | Hollow Point |
| PSP | Pointed Soft Point |
| JHP | Jacketed Hollow Point |
| HJHP | Half-Jacketed Hollow Point |
| HC | Hollow Cavity |
| HJSP | Half-Jacketed Soft Point |
| SJHP | Semi-Jacketed Hollow Point |
| FP | Flat Point |
| MCWC | Metal Case Wad Center |
| Keys: (other) | |
| OAL | Overall Length With Bullet |

# PRIMERS

For a cartridge to function properly, the successful operation of each component is required. Primarily from a safety standpoint, we have stated that the cartridge case is the most important component. However, if a primer does not ignite, we will have a misfire and all of our concern for case inspection, powder selection, etc., will be for naught. Therefore, every component, including the primer, is critical to the performance of the load.

Actually, the primer is responsible for only a small percentage of the accuracy of a given load, but its burning characteristics will add or detract from the overall pressure. How much these burning characteristics affect pressure is relative. It depends primarily upon the application of a specific primer to a specific set of conditions. Generally, the primer's influence on pressure is minimal when compared to the possible effects of other conditions and components. This influence, however, is not so slight that it can be overlooked.

In testing metallic cartridges we noted pressure variations (due to primer change) of approximately 2,000 C.U.P. Further testing could indicate still greater variations. The point is that while metallic cartridge primers do not seem to vary greatly, enough variation exists to require us to re-work a load when changing a primer type or brand. The reloader should stick with the same brand and type of primer when working up and using a load. If he changes the primer brand or type, he must then go back to the "starting load" and begin again.

If you examine a "Boxer Type" primer carefully, you will note that the anvil protrudes slightly beyond the end of the primer cup. This is not an oversight in manufacture! The anvil is supposed to protrude so that it will seat solidly against the bottom of the primer pocket and firmly resist the force of the firing pin blow. If the primer is seated incorrectly (leaving space between anvil and bottom of pocket) then the entire primer will move forward when struck by the firing pin. Such a condition retards the blow of the firing pin and causes inconsistent ignition.

When seating primers, make sure that the primer bottoms in the primer pocket. However, the primer should not be crushed. Normally, when a primer is seated correctly, it will be flush with the head of the case, or perhaps a few thousandths below. Under no circumstances should the primer stick out beyond the case head. Such a condition can prove dangerous particularly in auto loading firearms, for the primer is in an exposed position. If such an improperly seated primer were to receive a blow from the bolt face, it could fire before the round was safely chambered.

Boxer primers having either round or flat dome construction may be encountered and the correctly shaped priming punch is available for either contour. Use the flat punch for the flat primer and the concave punch for the round primer. Before seating a primer it is wise to examine it visually to make sure the anvil is not missing. Although this rarely occurs, the resultant misfire could be embarrassing if it should come at the wrong moment (such as when that trophy buck shows himself from behind the hemlocks.).

### Handling Primers

A word of caution on handling primers. Primers are detonated by percussion (a sharp blow). As packaged by the manufacturer, primers are quite safe, but they should not be tossed about or handled in a careless manner. Keep your primers in the factory container until they are to be used. Never store primers in a makeshift container. An old glass jar full of primers could become lethal if dropped or accidentally knocked onto the floor.

### Crimped Primers

Some military cases are loaded in a manner which employs a slight crimp over the primer to hold it snugly in place. These cases can be decapped without difficulty, but the crimp must be removed before the case can be primed again. Various commercial tools, which include the Lyman primer pocket reamer, are available to do this job. Primer pocket reamers come in two sizes, large and small.

One last thought—never decap live primers. If for some reason live ammunition must disassembled, pull the bullet, pour out the powder, chamber the primed case in a firearm, and snap the primer before decapping.

**NOMENCLATURE OF "BOXER TYPE" PRIMER**

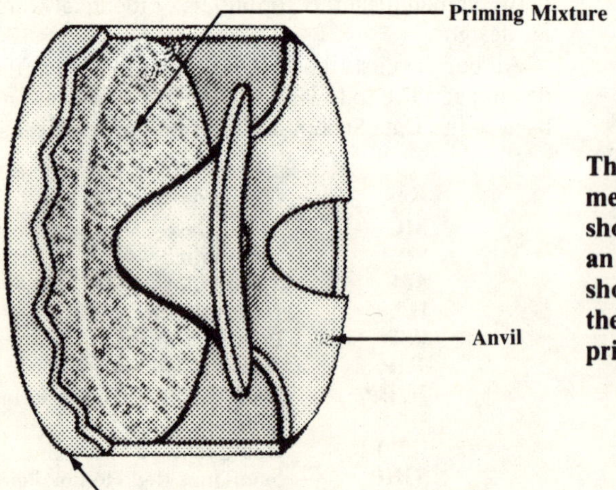


**The boxer type primer is used in all American metallic cartridges. The battery cup primer used in shotshells is of the same basic design, but contains an extra cup to facilitate its use in the less rigid shotshell base. The Berdan type primer functions in the same manner as the Boxer type, but with this primer the anvil is actually part of the case.**

NOTE: This chapter refers only to rifle and pistol cartridge primers. For information pertaining to shotshell primers, see the Lyman Shotshell Handbook, 2nd Edition.

# 4

# Reloading

# Reloading

## Choosing A Load

To select the proper components, refer to the data page of the Handbook that lists your cartridge. Make a note of the **bullet diameter** and **primer size** specified for this cartridge. This information will be needed when you purchase components. For example, if your cartridge was a .38 Special, you would look under the specifications for that cartridge and take note of the following data:

**EXAMPLE (38 SPECIAL)**

**SPECIFICATIONS:**

Bullet Dia. Jacketed & Cast ......... .354'' to .360''
Maximum Case Length ......................... 1.155''
Trim-to Length ..................................1.149''
Maximum Overall Length (w/Bullet) ........ 1.550''
Primer Size ................................ Small Pistol
Lyman Shell Holder Number ......................... 1

*The correct bullet diameter is related to the groove diameter of your particular handgun. With handguns, this dimension can vary from one handgun to another. The only way to be certain of your exact groove diameter is to **slug the barrel.**

**EXAMPLE (38 SPECIAL)**

| 158 gr. Hornady HP .357'' dia./1.480'' OAL | | | | | | |
|---|---|---|---|---|---|---|
| Powder | Sug. Starting Grains | Velocity F.P.S. | Pressure C.U.P. | Max. Grains | Velocity F.P.S. | Pressure C.U.P. |
| SR-4756 | 4.9 | 611 | 13,800 | 5.4 | 705 | 16,800 |
| SR-4759 | 9.0 | 685 | 14,800 | 9.5 | 756 | 16,700 |

***Use Starting Load.***

***Jacketed Bullets may be purchased from your dealer. Their dia. need not correspond exactly to the groove dia. of your pistol.***

| 141 gr. Cast Bullet #358495 (Linotype) 1.310'' OAL | | | | | | |
|---|---|---|---|---|---|---|
| Powder | Sug. Starting Grains | Velocity F.P.S. | Pressure C.U.P. | Max. Grains | Velocity F.P.S. | Pressure C.U.P. |
| 700X | 2.2 | 636 | 8,300 | 4.0 | 994 | 16,600 |
| SR-7625 | 3.1 | 683 | 9,100 | 4.3 | 905 | 15,600 |

***Use Starting Load.***

***Exact bullet number and alloy must be used. Size to groove dia. or larger molds available from Lyman.***

Now, you must decide on a bullet weight and a type of powder. You will note, in most pistol calibers, that a selection of **bullet types** (Cast, Jacketed and Half-Jacketed) is available.

If you should decide to use either a Jacketed or a Half-Jacketed bullet, then these bullets may be purchased from your component dealer. Jacketed and Half-Jacketed bullets are manufactured to a standard diameter which may vary from the groove diameter of your pistol. The **starting loads** for these bullet types take into consideration these possible variations, and they will be safe to use in your handgun.

If you should decide to use a cast bullet, however, then **you must purchase a mould for the bullet specified.** The loads shown for cast bullets are quite a specific and the correct bullet number and alloy must be used. These bullets should be sized to the groove diameter (or larger) of your handgun for which a sizing die (G.H. & I.) is available.

**Note:**

1. **Starting loads may not function automatic pistols. Increase loads gradually; do not exceed the maximum safe charge level.**
2. **Half-jacketed bullets can prove dangerous when used with reduced loads. The jackets sometimes lodge in the bore while the core exits. Never load below listed starting loads.**

Heavy magnum type loads shorten case life tremendously, while with light mid-range loads, cases may last almost indefinitely.

## Basic Mechanics of Loading

Reloading a pistol cartridge requires a series of SIX basic mechanical operations. So that you will understand the fundamentals of each operation, we will first treat them graphically and then explain why each operation is necessary. Four out of the six operations are performed by the reloading dies.

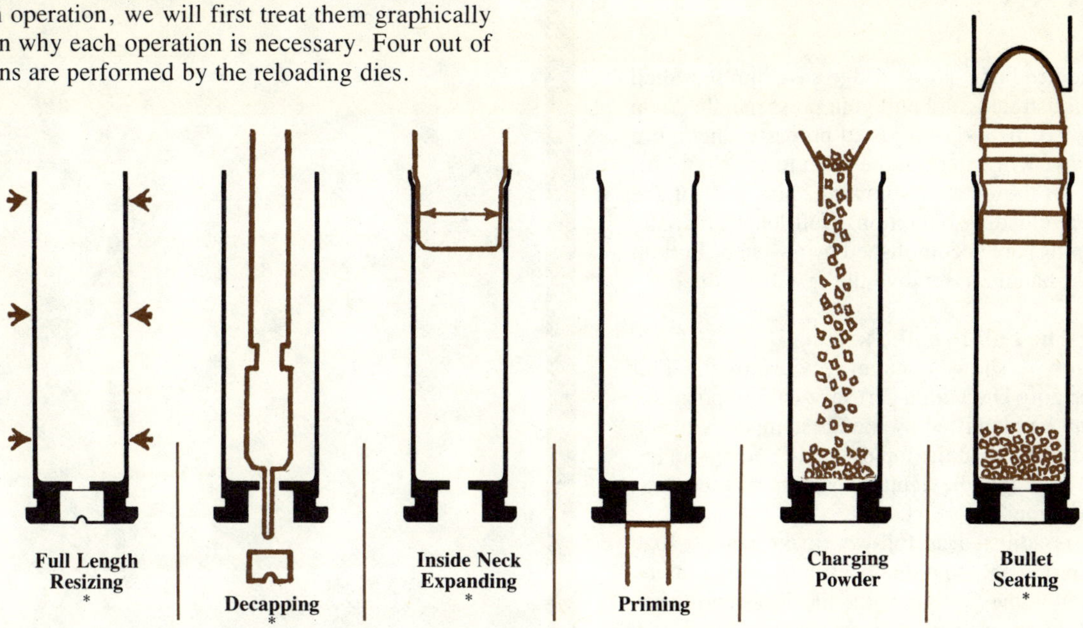


*Operations 1, 2, 3, and 6 are performed by the reloading dies.

1. **FULL LENGTH RESIZING:** When a cartridge is fired, the side walls of the case expand to the chamber size of the handgun. This is necessary for the case to function properly and seal the chamber. These walls remain pretty much at their expanded size and do not snap back to original dimensions. Since all pistol chambers are not identical, cases fired in one pistol may not chamber in another. Even the chambers of a revolver cylinder will, in fact, vary from one to another. For this reason, it is necessary to compress the walls of the case to a standard diameter that is acceptable to all handguns. This operation is called resizing.
2. **DECAPPING:** This operation consists of simply removing the old or fired primer.
3. **INSIDE NECK EXPANDING:** After the case has been resized, the inside diameter of the neck will be too small to accept the bullet. Inside neck expanding enlarges the inside diameter of the neck to a size which will receive and hold the bullet securely. For pistol cases, a two-step expanding plug is used to open up the inside of the case neck. The first step on this plug is slightly smaller than bullet diameter, while the second step is a few thousandths larger. The idea behind this is to allow the bullet to enter the case freely without shaving lead. The actual difference between the two steps is not visually apparent. The illustrations have been exaggerated for purposes of clarification.
4. **PRIMING:** This operation consists of inserting a new primer into the primer pocket.
5. **CHARGING POWDER:** This operation consists of carefully weighing out and pouring the appropriate powder charge into the case.
6. **BULLET SEATING:** The last operation in the reloading process is seating a new bullet into the case.

In the preceding text we covered reloading graphically and have given the reader a general idea of what is required. Now by employing photographs we will explain the actual reloading of a cartridge. You will note that we are using a set of three reloading dies to perform four of the six operations. Further along in reloading you will hear of two-die sets and even four-die sets. The difference is that two and three die sets combine some of the operations, while a four die set performs each operation separately. Due to their shape, most pistol cartridges require the use of a three die set.

## Making Cartridges

We now assume that your cases have been properly inspected, that you have selected a load and purchased the necessary components. Your reloading press should be assembled and mounted according to the instructions supplied with it. Many reloading presses may be assembled to function on either the up stroke or down-stroke of the handle. The press illustrated here is operating on the down-stroke. First, lubricate your cases by wiping them with a cloth sparingly greased with Lyman Sizing Lubricant. This special lubricant will cut friction to a minimum and ease the sizing operation. Apply a very thin coat, for too much grease will trap air in the die and cause ''lube dents.'' Although cases dented in this manner may be used for reloading, as the dents are ironed out in firing, it is not considered good reloading practice. Screw your Full Length Resizing Die into the head of the press, adjust it according to the instructions furnished, and you are ready to commence loading.

# RELOADING ON THE SPARTAN PRESS

Slide the head of your cartridge case into the Shell Holder as illustrated, and pull your press handle down all the way. If the die is adjusted properly, the entire cartridge case will enter the die flush to the shell holder. Note, in the cutaway drawing, how two of the original six reloading operations (full-length resizing and decapping) are accomplished by this step. Pull up on the press handle to remove the case from the die.

**Adjusting The Full Length Resizing Die**

Illustration A shows a cutaway view of the Full Length Resizing Die with a cartridge case in position. This die can be identified by the FL stamped between the knurled surfaces at the top of the die. Note—in the illustration how this die actually performs two operations at one time.

The die is adjusted as follows: make certain that decapping pin is tight and that does not protrude more than ⅛'' below the bottom of the die. Loosen the set screw in the lock ring on the outside of the die and screw the ring toward the knurled end of the die. This will give you room to screw the die into the Press. Lower the Press Handle until the Shell Holder is at its highest point. Screw the die into the Press until the bottom edge of the die makes firm contact with the Shell Holder. The die is now adjusted for full length resizing and decapping. Screw down the lock ring and tighten the set screw.

Apply a **thin** film of Size-Ezy Lubricant to each case before inserting them into the die. Excessive lubricant should be avoided as it traps air in the die and dents the cartridge case.

**Note—The Entire Cartridge Case Enters This Die, Flush To The Shell Holder.**

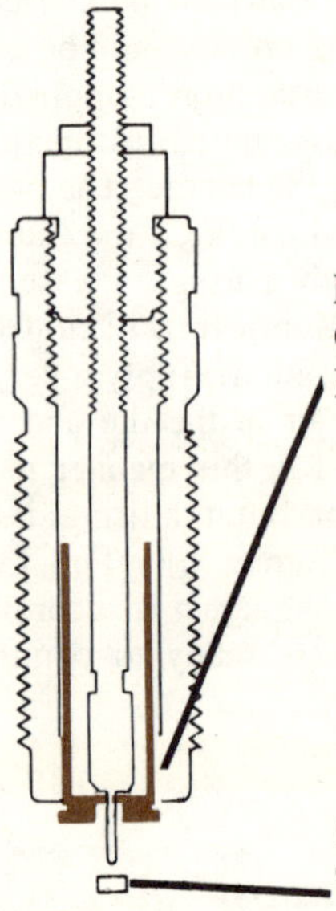

**STEP ONE (Full-Length Resizing and Decapping)**

Screw the Neck Expanding Die into your press and adjust it according to the instructions supplied with the die. Place the resized cartridge case into the shell holder and pull down on the press handle. Note, in the drawing, how the two-step plug enters and expands the case neck. Actually, there is only a few thousandths difference in diameter between the first and second steps on the plug. This difference is so slight that it is not visually apparent. The illustration has been exaggerated for clarification.

**Adjusting The 2-Step Neck Expanding Die**

The die is adjusted as follows; Remove the expanding plug stem from the die body. Loosen the set screw in the lock ring on the outside of the die and screw the ring toward the knurled end of the die. This will give you room to screw the die body into the Press. Lower the press handle until the shell holder is at its highest point and screw the die into the Press until the bottom edge of the die makes firm contact with the shell holder. The die body is adjusted at this point. Screw down the lock ring and tighten the set screw.

Your expanding button must now be adjusted to the proper depth. Start the expanding plug into the die body. Place an empty sized cartridge case in the Shell Holder and pull down on the press handle until your shell holder is again at its highest point. Screw down on the expanding plug until you feel the button contact the case. This means that the first (and narrower) stage of the button has entered the case. Lift the handle slightly, screw in the plug a bit more, then pull down the handle again. Do this a number of times until increased resistance on the handle tells you that the second stage of the button has reached the mouth of the case. Once the second step of the expanding button enters the case about 1/16'', lock up tight on the lock ring and your die is adjusted.

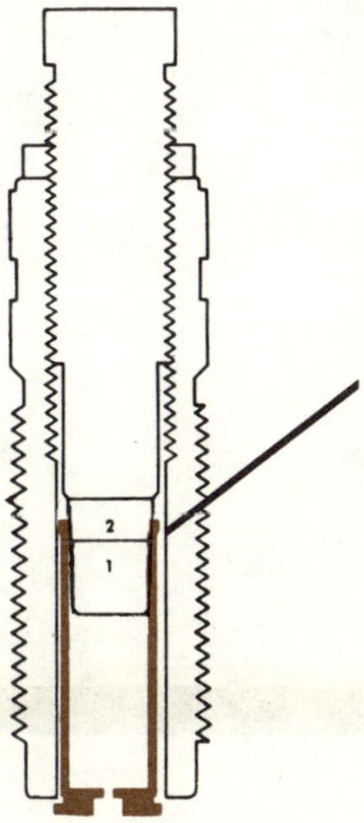


**Second Step Enters Case about 1/16'' to Allow Bullet To Start Freely.**

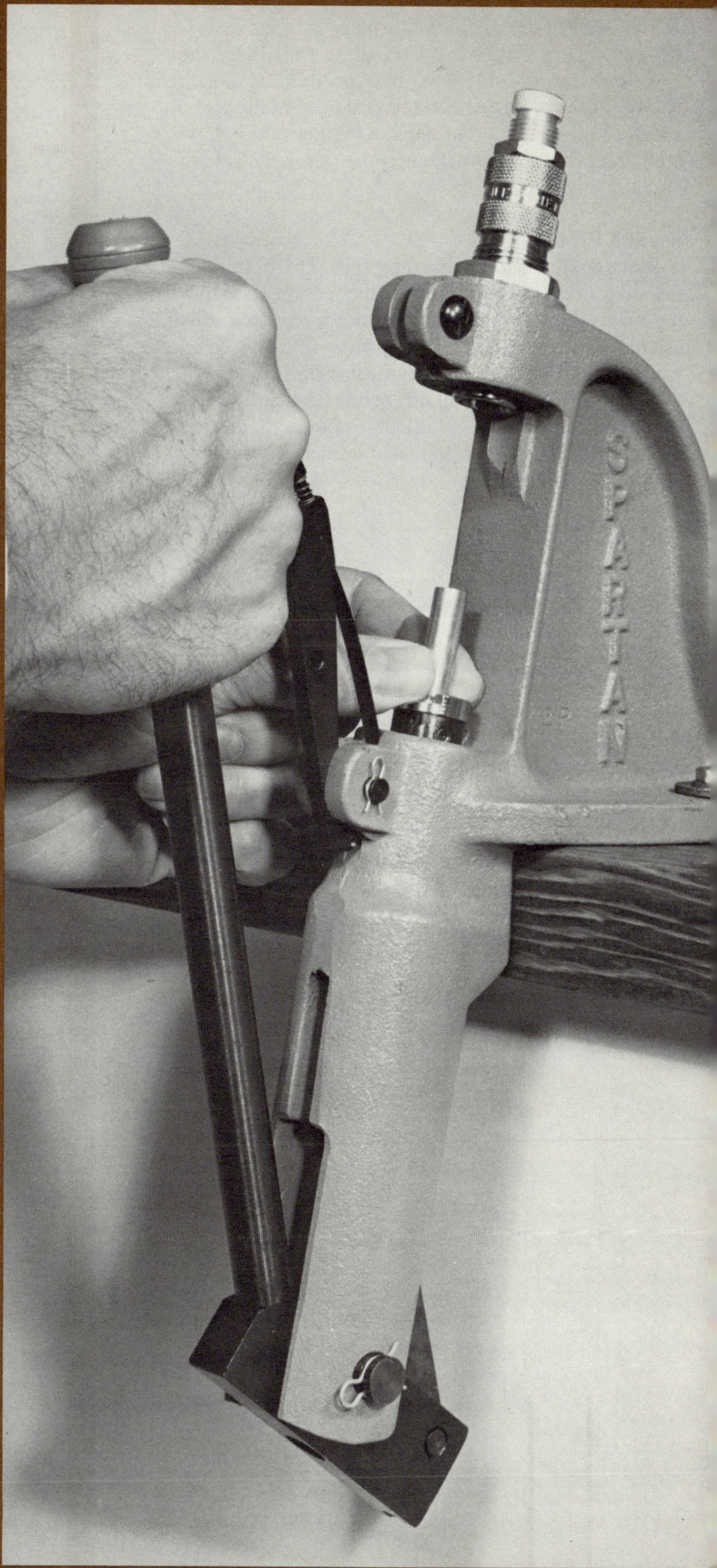


**STEP TWO (Inside Neck Expanding)**

The priming operation takes place as your case is withdrawn from the Neck Expanding Die. Place the new primer (cup side up) into the priming punch sleeve. Push the primer arm forward (toward the press) and pull up on the press handle. As the ram is lowered, the priming arm will enter the slot in the side of the ram and seat the primer.

**Seating Primers:**

Primers are seated mainly by feel. The bottom of the anvil **must bottom** in the primer pocket. Depending on the brand of case and primer being used, this usually works out so that the primer is fully seated when the top of the primer is flush with the head of the case, or a few thousandths below the head. Under no circumstances should primers protrude. Use care not to crush the primer. Crushed primers give erratic ignition, or fail to fire.

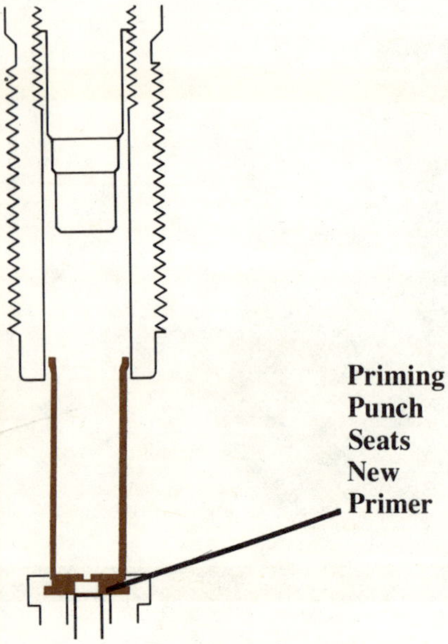


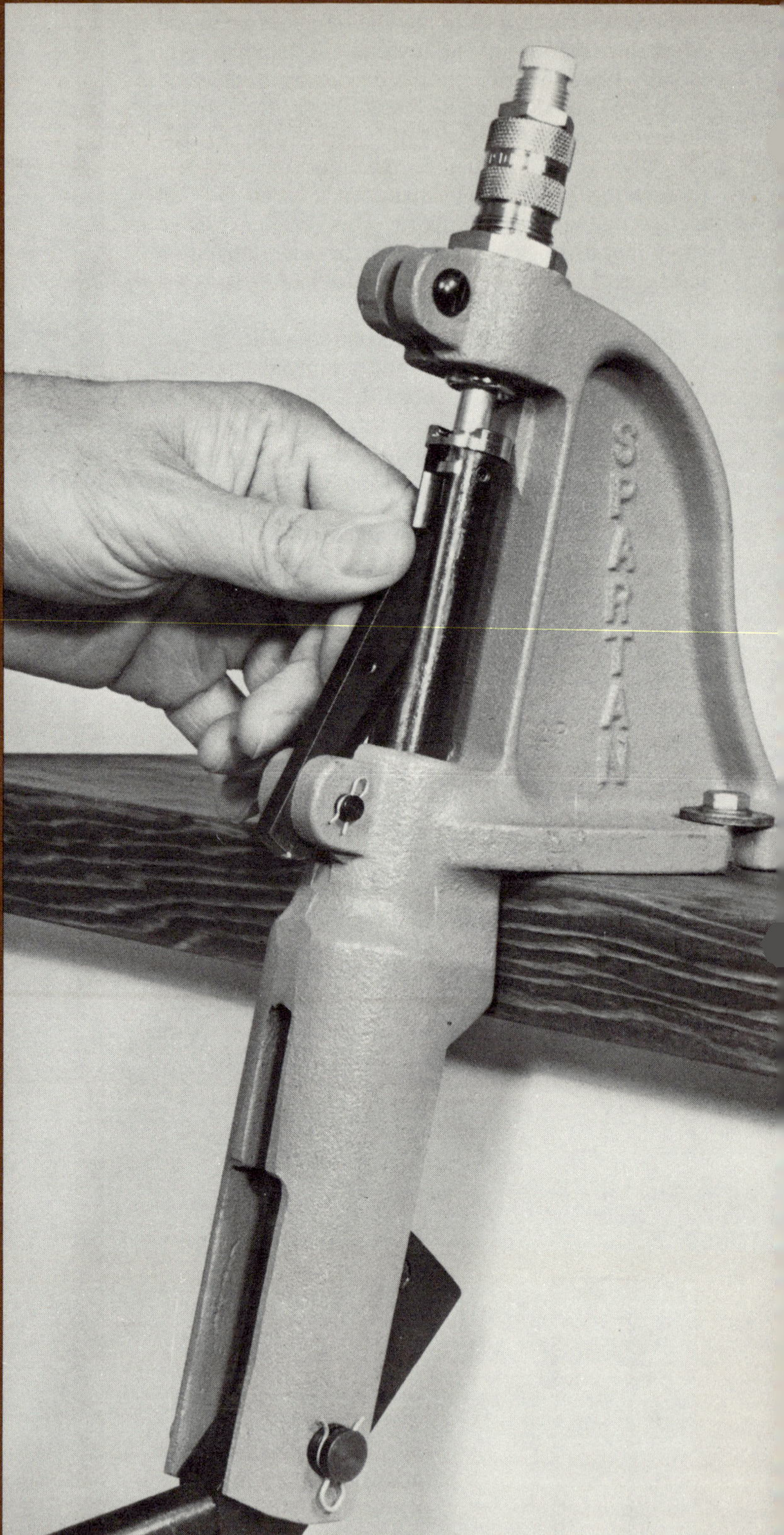


**STEP THREE** ***(Priming)***

For the weighing of powder, you require an accurate powder scale such as the Lyman D-7 shown. The Data Section of the Handbook specifies the powders appropriate for your particular cartridge. It also lists a suggested weight of the charge in grains and in fractions of grains. For example, 2.2 would be read as TWO and TWO TENTHS grains. 3.0 would be read as THREE grains. We recommend that the novice restrict himself, at least temporarily, to the suggested starting load. Carefully level the powder scale as explained in the instructions and set it to weigh your required charge. See the accompanying illustration for scale adjustment.

Slowly trickle small amounts of powder into the scale pan until the beam comes into balance. The beam is in balance when the pointed end (extreme left) is exactly on the zero mark.

Carefully remove the pan and pour its contents into the cartridge case. Use a powder funnel to make sure all the powder enters the case. Because pistol powders are comparatively fast burning, most normal charges take up very little room in the cartridge case. In other words, it is possible to **accidentally double charge,** or even triple charge many pistol cases. This, of course, would prove extremely dangerous and a foolproof system of loading must be developed. A suggested method is to place all the uncharged cartridge cases on your left. As you pick up each case for charging, turn it up-side-down and shake it. This will insure that the case is empty. Turn the case right-side-up, charge it and place it carefully on your right. Take care, when removing or replacing the scale pan that the poise are not accidentally moved.

## CAUTION

Technicians in the Lyman Lab have observed a potentially serious phenomena involving powder scales and plastic loading blocks and/or styrofoam packaging. It seems these substances sometimes retain a certain amount of static electricity.

This electrostatic condition has proven capable of radical deflection of uncharged and "zeroed" scales of all brands (available to us at the time). Of course, powder in the pan will tend to dampen the deflection but some still occurs depending on the charging level. Generally, the heavier the charge the less error... assuming the scale was first "zeroed" correctly.

We suggest you clear the surface of your reloading bench and make very sure the scale is set up accurately. Then move your equipment back piece by piece, paying particular attention to the plastic or styrofoam mentioned earlier. Please note this caution applies to all forms of reloading, not just handgun.

**Each Graduation On This Side Is Equal To 5 FULL GRAINS**

**Each Graduation On This Side Is Equal To 1/10 GRAIN**

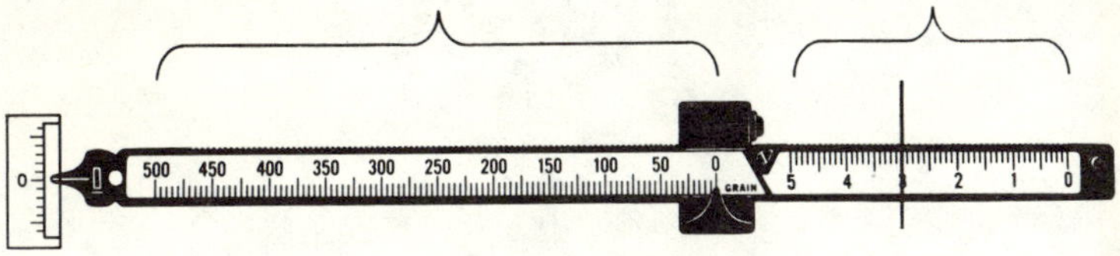


The illustration shows the beam of a modern reloading scale. Note how it graduated on both sides of the pivot point. The scale is set by moving the two weights (poise) to the proper graduations. The large poise (on the left) is used to obtain multiples of FIVE GRAINS, while the small poise (on the right) is used for FRACTIONS of a grain or SINGLE grains from one to five.

EXAMPLE: The illustration shows a setting of 3.0 grains. If you wish to increase this to 8.0 grains, simply move the large poise one notch to the left.

## STEP FOUR *(Charging Powder)*

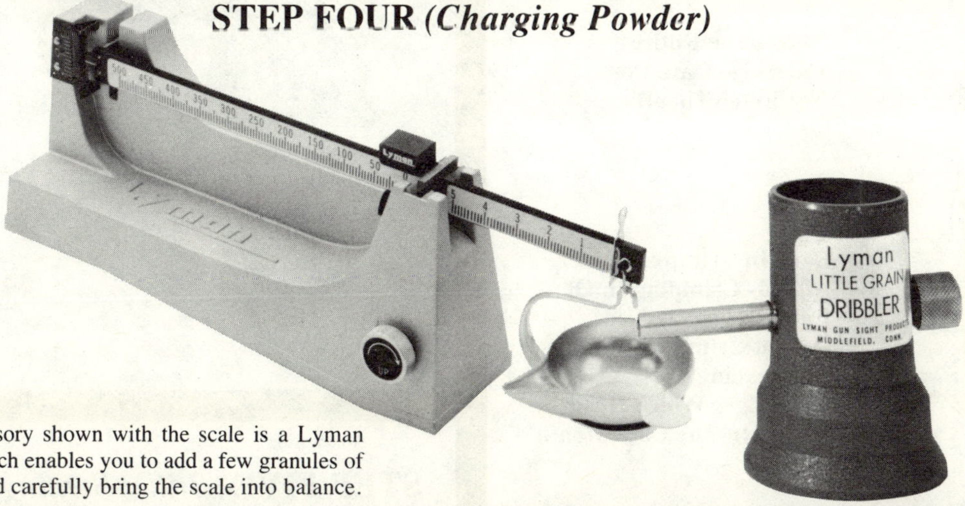


The reloading accessory shown with the scale is a Lyman Powder Dribbler which enables you to add a few granules of powder at a time, and carefully bring the scale into balance.

The last operation in reloading a cartridge case is seating the new bullet. Be sure the overall length of the finished round is not longer than the MAXIMUM OVERALL LENGTH listed in the Data Section. Adhering to this measurement will make certain that the finished round will function properly in your magazine or cylinder. Also, a bullet not seated to the proper depth can engage the rifling in an autoloading pistol, and build up pressure upon firing.

This illustration shows how a bullet is seated. Screw the bullet seating die into the head of the press and adjust it according to the instructions supplied. Place a primed, charged cartridge case in the shell holder and start a bullet in the mouth of the case. Pull your press handle all the way down, so when the case enters the die, the bullet will be pushed firmly into the neck of the case. Adjusting the seating screw controls the depth to which the bullet is seated. Adjusting the die body controls the crimp.

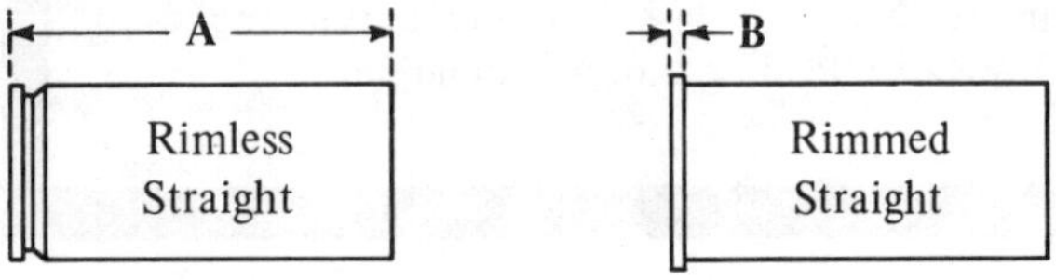


**Letter "A" indicates headspace in the first example. If a case of this type is crimped, it can shorten the overall length of the case and create an excessive headspace condition. Rimmed cases of the "B" type may be crimped if desired. Magnum cartridges, due to their heavy recoil, require the use of a crimp to hold the bullet securely.**

**CAUTION: AFTER RELOADING AND BEFORE FIRING, WIPE YOUR CASES TO REMOVE ALL SIZING LUBRICANT. THE PRESENCE OF OIL OR GREASE ON A CARTRIDGE MAY DANGEROUSLY INCREASE THRUST ON THE BOLT FACE.**

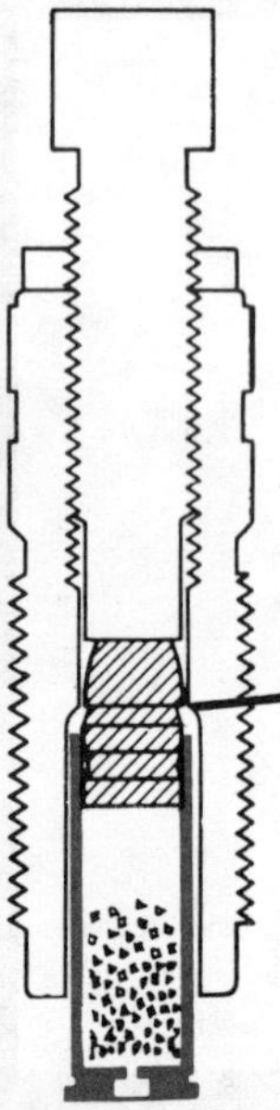

**Note—The Entire Cartridge Case Does Not Enter This Die.**

**Built-In Crimp Shoulder Affords Crimping-In Of Bullet When Desired. Do Not Crimp On Rimless Autoloading Cartridges Which Head-Space On The Case Mouth.**

**STEP FIVE** ***(Bullet Seating)***

This section of the Handbook has concerned itself with the basics of pistol cartridge reloading. For the benefit of the novice, the text and illustrations have been intentionally simplified, but not over-simplified. Still, there is considerably more to learn, for reloading is that kind of a hobby. Stay with the recommended "starting load" until you are familiar with the fundamentals of reloading. Progress slowly and never load over your head. If you don't understand something, ask. Seek the advice of knowledgeable reloading friends, or your local component dealer. If you have further questions, write the Lyman Technical Staff who gladly offer assistance.

**Adjusting The Bullet Seating Die:**

The illustration shows a cutaway view of the Bullet Seating Die. This die can be identified by the DA stamped between the knurled surfaces at the top of the die. Note carefully that the entire cartridge does not enter this die. Lyman Bullet Seating dies are equipped with a built-in crimping shoulder which may, or may not, be used depending upon the reloader's preference. The cartridge case in the illustration is shown in the non-crimp position. DO NOT CRIMP rimless auto-loading cartridges such as 45 ACP, 9M/M Luger, and 380 ACP. These cartridges headspace on the case mouth and should not be crimped. On all other pistol cases, a crimp is allowed if desired.

The die is adjusted as follows; back off several turns on the die lock ring and start the die into the Press. Put an empty sized and neck expanded cartridge case into the shell holder and pull down the press handle until the shell holder is at its highest point. Screw the die down into the Press until you "feel" the mouth of the case contact the crimp shoulder in the die. When you feel the end of the case touch the inside of the die, back off ¼ turn on the die and screw down finger tight on the lock ring. Your case is now in the non-crimp position shown in the illustration.

Your bullet must now be seated to the proper depth. Back off on the bullet seating screw about ½". Place a cartridge case into the shell holder and start a bullet in the mouth of the case. Pull down on your press handle until your shell holder is again at its highest point and turn down on the seating screw until it comes into contact with the bullet.

Alternately raise the handle and turn down on the bullet seating screw a little at a time until the bullet is seated to the desired depth. If you **do not** wish to crimp, your die is adjusted at this point and you should tighten the lock rings and set screw.

If you wish to crimp, continue as follows; make certain that your bullet is seated exactly to the crimping groove. Back off ¼ of a turn on the bullet seating screw. Screw the die body down ¾ of a turn and tighten the lock rings set screw. Your die is now properly adjusted.

## Reloading with the Lyman #310 Tool

Reloading Dies are pictured in their loading order.

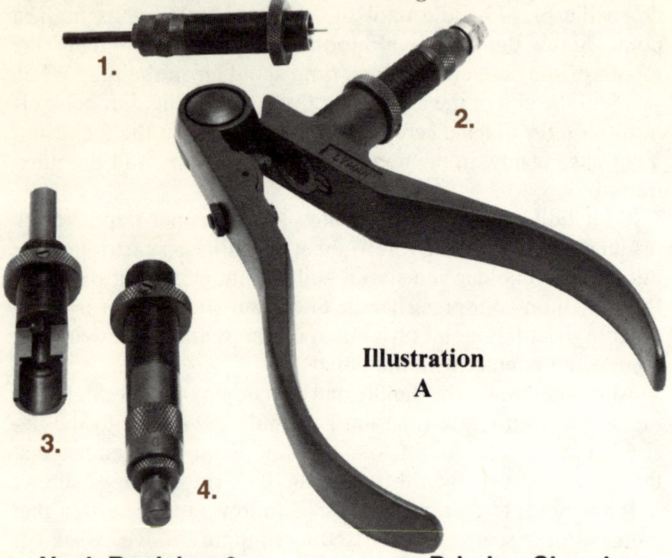


Illustration A

1. Neck Resizing & Decapping Die
2. Neck Expanding Die
3. Priming Chamber
4. Bullet Seating Die

## Pre-Loading Instructions

Carefully inspect each fired case before reloading. Reject all cases which show signs of cracks or splits about the necks or bodies. The fired cases which pass inspection should then be tried into the chamber of your rifle or handgun to insure that they fit. Cases which do not fit into the gun chamber will require FULL LENGTH RESIZING and for these cases you will need an additional Full Length (hand type) Resizing Die (see Lyman Catalog). The remaining cases may be reloaded with your 310 Tool.

The cases which are to be loaded should be lubricated. This is done by wiping the outside of each case neck with an oily cloth, or a cloth sparingly greased with Lyman Size-Ezy Lubricant. After reloading and before firing, always wipe your cases to remove this sizing lubricant.

## Step 1. Neck Resizing & Decapping

Select the Neck Resizing Die which is pictured as number one in illustration A. Make certain that the Decapping Pin is screwed tightly to the end of the Decapping Rod. Now, turn the Decapping Rod so that the Decapping Pin is a short ways up inside the die. Screw the die body into the threaded portion of the tool handles.

The neck of the case need only be resized to the depth that the bullet will be seated. To do this, screw the die body into the tool handle just far enough that the neck of the case will be forced into the die the desired distance. By checking the sizing marks on the case neck and making slight adjustments on the die body, you can quickly bring it into the proper adjustment. When the adjustment is correct, turn up on the lock nut and tighten the set-screw.

Now you must adjust your Decapping Rod. Turn in on the threaded stem until the pin protrudes slightly above the opening in the handles (about ⅛"). Be certain that the Decapping Pin does not contact the handles when closing. Lock the Decapping Rod tightly in this position.

## Step 2. Inside Neck Expanding

After the neck diameter of the case is reduced by resizing, it will be found too small for the bullet to enter. It must now be opened up to a uniform diameter that will receive and hold the bullet securely. This operation is accomplished with the Neck Expanding Die.

Select the Neck Expanding Die which is pictured as number two in illustration A. Screw the die body into the threaded portion of the tool handles. This die is equipped with a 2-step Expanding Plug. The first step on the plug expands the neck of the case to slightly under bullet diameter. The second step expands the first 1/32" of the neck to slightly over bullet diameter, allowing the bullet to enter the case freely.

By adjusting both the die body and the plug stem, you can position the expanding plug exactly so that the second step on the plug is barely (about 1/32") entering the case. Once your adjustments are made, tighten up on the lock nuts and set screw.

## Step 3. Priming

To seat the primers, screw the Priming Chamber into the tool handles until the Priming Punch protrudes slightly beyond the flat surface of the Shell Holder when the tool handles are closed.

Then place an unprimed cartridge case in the Shell Holder part of the Priming Chamber. Place a primer in the opening, close the tool handles and the primer will be seated. The primer should be seated to the bottom of the pocket and flush with the case head. The depth can be regulated by screwing the Priming Chamber in or out of the threaded portion of the tool handles. Once your adjustment is correct, tighten up on the lock nut and set screw.

## Step 4. Bullet Seating

Select the Bullet Seating Die which is pictured as number four in illustration A. Back off the bullet seating screw and screw the die body into the threaded portion of the tool handles.

To adjust this die it is best to make up a dummy round (less primer and powder). Slip a neck resized and neck expanded cartridge case into the tool handles and close the handle. Holding the handles in this closed position, screw in the die body until you feel it bear lightly on the mouth of the case. When you feel it touch the case tighten up on the lock nut to hold the die body in this position, but do not tighten the set screw.

Now you must adjust the seating screw to the proper depth. Place a bullet into the mouth of your dummy case and insert it back into the tool handles. Gradually turn in on the seating screw and check your seating depth a little at a time, until the bullet reaches the proper depth. Lock your seating screw in this position.

If you wish to crimp-in the bullet, one last adjustment is necessary. Back off slightly on the body lock nut and turn in the die body ¼ turn. Tighten the lock nut and set screw and die is adjusted.

When using the Bullet Seating Die, it is necessary to hold the tool handles so that the die is on top of the handles as shown in the illustration. Holding the handles in this manner prevents powder from spilling into the die and insures accurate alignment of bullet.

# 5

# Bullet Casting

# Bullet Casting

## Cast Bullets

Cast lead bullets can do most things for the handgunner that jacketed bullets can do. By varying the hardness of the alloy, using proper lubricants, sizing correctly and loading to a needed velocity level, the bullet casting reloader can truly customize his loads while saving a fair amount of money.

For handgunners, the matter of a plain base versus gas check is much less critical—except in the specialty .22's and .25's—than it is for riflemen. Handgun velocities seldom exceed the 1,500—1,700 fps level at which use of a gas check usually proves beneficial.

Properly alloyed and lubricated, cast bullets can be driven just as hard as jacketed bullets. Experimentation with the foregoing—plus sizing diameter—will no doubt be necessary. But it can be done; besides, experimenting within safe reloading limits is fun!

## Bullet Metal

For years, Lyman has used #2 Alloy as the standard for its centerfire bullet moulds. This alloy is quite hard—but not as hard as Linotype metal. Shotgun slugs and all muzzleloading projectiles are cast in pure lead, a very soft substance.

Over the years, shooters have logically combined the hardest practical alloy, Linotype metal, with the finest high-speed lubricant, Alox/beeswax, to achieve high velocity with little or no sacrifice in accuracy.

Linotype metal is an excellent choice for maximum handgun velocities. Happily, it is very similar to #2 Alloy as far as shrinkage goes. Chances are, there will be little dimensional variance between the two.

Over the past few years we have noticed an interesting situation unfold concerning accuracy of bullet metal alloys. The Lyman Technical Department buys its bullet metal—pure lead,

#2 Alloy and Linotype metal—from a foundry.

Contents of the metals are clearly specified and the metal should be of "scientific" quality.

To shorten a long story, we found, via independent analysis, a significant variation from specification in the #2 Alloy and Linotype metals in lots purchased over a span of several years. More recently, we have tried to define the contents of wheelweights—and found not only a great deal of variation in the formula across the country but a universal decrease in the antimony content. We have adjusted our do-it-yourself #2 Alloy formulas accordingly.

We report the foregoing simply to alert you, the bullet caster, to the fact that all might not be as it is labeled. So if one alloy fails where you think it shouldn't, perhaps it is a bit out of spec.

## Number 2 Alloy

As previously stated, #2 Alloy is a rather hard bullet metal which contains three ingredients: lead, tin and antimony.

While we have recommended that Lyman #2 Alloy be used for the cast bullet loads in this Handbook, this does not necessarily mean the bullet caster has to purchase this specific metal in order to cast good bullets. Lyman used this bullet metal **to furnish a known standard for bullet weights, loads, etc.** For example, a bullet will vary in weight (and performance) in relation to the blend of the alloy. If a standard alloy of a specific blend is not specified and followed, consistent performance will not be possible.

The reloader can, and should, make his own bullet material by blending the required metals to the proper proportions. The metal required can be gathered from junk yards, gas stations, automotive supply houses, plumbing supply houses, hardware stores and printing shops.

| Composition & Hardness of Common Bullet Metals | | | | |
|---|---|---|---|---|
| Alloy | Lead | Tin | Antimony | BHN |
| Linotype | 86% | 3% | 11% | 22 |
| #2 Alloy | 90% | 5% | 5% | 15 |
| 16-1 | 94% | 6% | 0% | N/A |
| 10-1 | 91% | 9% | 0% | 11.5 |
| Wheelweights | 95.5% | 0.5% | 4% | 9 |
| Pure Lead | 100% | — | — | 5 |

### Recipes For Mixing Your Own # 2 Alloy

| Ingredients | Yield |
|---|---|
| 9 pounds wheelweights<br>1 pound 50/50 bar solder | Makes 10 pounds # 2 Alloy |
| 4 pounds Linotype<br>1 pound 50/50 solder<br>5 pounds pure lead | Makes 10 pounds # 2 Alloy |

Some of the materials gathered from junk yards and such will contain small traces of other metals, but for the purpose of making bullets they may be considered as pure. Bar solder containing 50% tin and 50% lead is available from plumbing houses and is an excellent source for tin. Linotype is available from printing shops and while it is a little too hard in its existing form (11% antimony, 3% tin, 86% lead), it can be cut back by blending the proper proportions of tin to the mix. Wheelweights contain about 4% antimony and make an excellent base material to blend with the tin. Pure lead can be purchased either in ingot form or salvaged from a junk yard in the form of pipe, etc. *Caution: Do not salvage metal from auto or marine batteries. There has been a change in the material which can be hazardous to your health. Avoid batteries, period.*

To more clearly illustrate the effect of alloy variation on as-cast diameters and sized-to diameters in given moulds and sizing dies, we have prepared the chart on the next page.

While this data is based on extensive research which took place over the last two years, the table title carries the word "predicted" since minor variances—even in measuring from one person to the next—will always be with us.

## Sizing Cast Bullets

Whenever we form an object by the casting process (using a liquid or molten material which possesses the ability to solidify within a mould) we must consider a **shrinkage factor.** This shrinkage may be great or small depending upon the specific properties of the particular material being used. As they solidify, lead alloys will shrink in size and the freshly cast bullet will be somewhat smaller than the mould cavity in which it was cast. The more lead content in the alloy, the more noticeable is the shrinkage. Depending on the type of alloy used, the size of the particular bullet, the heat of the mould, and the mixture, shrinkage can be as much as .003″.

Our control over this kind of shrinkage is somewhat limited. Who can say with any certainty that a bullet will shrink this or that amount, or that all sides of the projectile will be exactly the same? Naturally, if our projectile is to function properly, it must fit the groove diameter of the firearm and it has to be as nearly perfect in shape as possible.

To compensate for shrinkage, the mould cavity is made considerably larger than the finished bullet diameter. For instance, a 30 caliber bullet that requires a finished diameter of .308″ drops from the mould at approximately .311″ diameter. This leaves enough metal (even after shrinkage) so that the bullet can be sized down to the standard groove diameter of the firearm.

The sizing process is merely a method of swaging cast bullets to a standard size that corresponds to or slightly above the groove diameter of the pistol. Bullet sizing also insures that each and every bullet in the group is perfectly round in shape. As no metal is removed from the bullet (bullet is swaged to shape and size), sizing does not alter the bullet's cast weight.

## Melting And Fluxing Bullet Metal

When working with bullet metal, one of the first requirements is a suitable heat source capable of heating the metal to about 750° or 800° Fahrenheit. A simple cast iron pot to hold the metal and most any heat source (kitchen stove, etc.) will suffice. Improved equipment, such as the Lyman Electric Furnace, is more desirable. The electric furnace is cleaner, safer and more convenient and its adjustable thermostat allows best control of the melt temperature. Whether you are blending the various metals into an alloy or actually casting bullets, the same melting and fluxing procedure is followed.

Heat the metal for about twenty minutes until it becomes liquified and flows freely. It is then ready for fluxing. As the metal melts, a gray scum will rise to the surface, contrasting sharply with the silver brightness of the molten lead. **DO NOT REMOVE THIS SCUM.** This contains tin, the most valuable component of the bullet metal. Fluxing will recom-

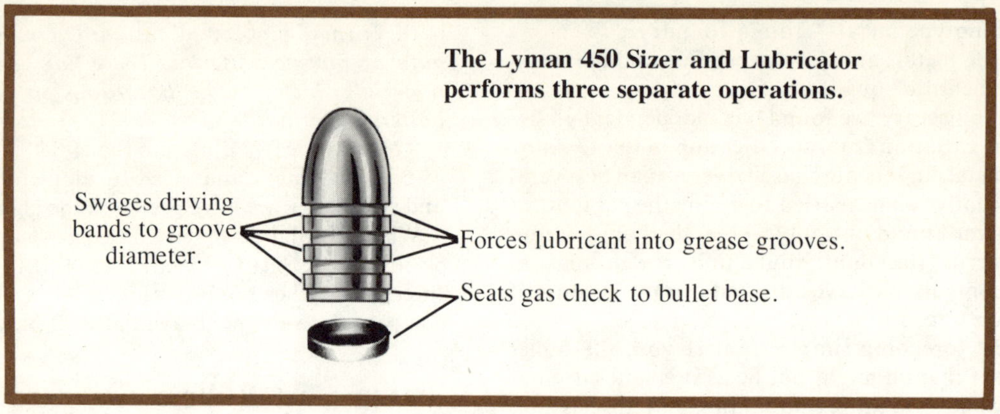


bine the tin-lead-antimony mixture. This operation is extremely important and should be done carefully.

To flux the metal merely drop a small bit of tallow, beeswax, or bullet lubricant into the mixture. A smokey gas will rise from the top of the pot and this gas should be immediately ignited with a match. This will eliminate the smoke.

A more modern-and much more pleasant-fluxing procedure is to use a dry substance, such as the product called MARVELUX. Smoke and greasy fumes are eliminated—an important benefit to those casters with wives and/or mothers in residence—and a good flux is obtained. This method is much preferable to the foregoing traditional technique.

Whichever fluxing substance you choose, be sure to stir the mixture with the dipper. As you stir, hold the dipper so the cup side is down and raise it out of the metal with each stirring stroke. This scoops air into the mixture and helps the flux.

Metal that has been properly fluxed will leave the surface almost mirror bright and flecked with small particles of black and brown impurities. Skim off and discard these impurities. Flux the metal whenever, by its appearance, it seems to need it. While the dipper is not in use, it should be left in the molten metal to keep it hot.

## Casting Bullets

After the metal has been fluxed and is hot enough to pour easily through the dipper, it is ready for casting. In addition to the mould, you should have on hand a hardwood stick (about 10" long) to be used for opening the mould. Also, pad a small area of your bench with an old piece of cloth material. This will soften the fall of the hot bullets as they drop from the mould, and prevent them from being damaged.

While there is no set way to cast good bullets, we do offer the following as a suggested method. Fill the dipper half-full of metal and place the spout of the dipper against the pouring hole as shown. Holding the mould and dipper together, slowly turn them to a vertical position with the dipper on the mould as shown.

The extra metal that runs over the top of the mould is called sprue. When it hardens, which takes only several seconds, pick up the hardwood stick and tap the sprue cutter sharply. This will separate the sprue from the base of the bullet. Drop the sprue into a cardboard box, or other receptacle. Open the mould and let the bullet fall to the pad. If the bullet does not drop out readily, use the stick to rap the hinge rivet sharply, as shown in the illustration. Use only wood for this purpose and never strike the mould blocks themselves.

As the mould will be cool, your first bullets will be imperfect. Casting bullets, one after the other, will bring the mould to the proper temperature. If you wish, the mould can be preheated by placing it on the rim of your furnace, or along side your lead pot on the stove. Never, under any circumstances, dunk the mould in the molten metal!

Wrinkled bullets indicate that the mould, and/or metal, is too cool. Frosted bullets indicate that the mould, and/or metal, is too hot. Good bullets should be clean, sharp, and fill the mould. Imperfect bullets should be collected and, along with the sprue, returned to the pot.

**Dirty bullet metal before fluxing.**

**Stir metal thoroughly while fluxing.**

**Clean, well-fluxed bullet metal.**

Bullets selected for accuracy shooting should be carefully weighed on the reloading scale. This reveals air pockets that may have formed in the bullet, lightening or unbalancing it. The actual weight of your bullets will depend pretty much on the consistency of your alloy material, and it may vary slightly from lot to lot. Weigh about twenty or thirty bullets out of each group and determine an average bullet weight. Reject any bullets that vary more than ½ grain (+ or −) of this weight. Rejected bullets may be saved and recast at a future date.

**The pouring spout on the ladle and the pouring hole on the sprue cutter are designed to fit closely together.**

**After the spout and mould are snugly together pour the lead into the mould by slowly rotating the block/ladle unit with the ladle on top. Experiment with your rate of pour to obtain the best results.**

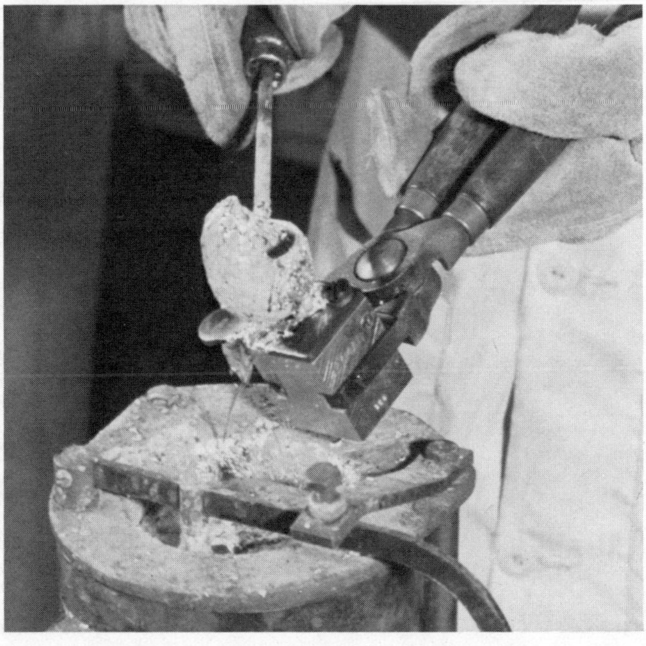

**Hitting the hinge pin is the best way to shake loose a sticking bullet. Don't land random blows on the blocks or sprue cutter since this could cause deformation and affect the quality of bullets.**

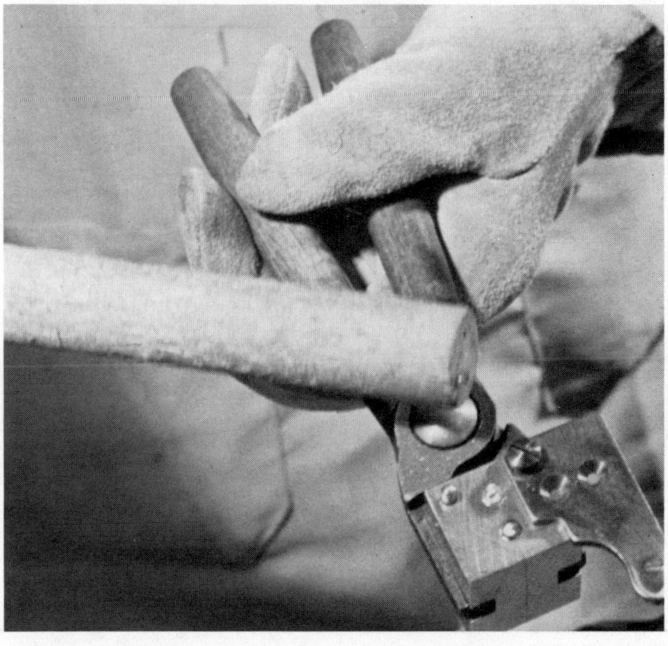

## Sizing Bullets

As stated previously, the cast bullet must be properly sized to or above the groove diameter of the handgun. To determine the exact groove diameter, the barrel should be slugged. Lyman "H & I" sizing dies corresponding to the groove diameter of the firearm are available. The Lyman 450 Bullet Sizer and Lubricator is designed to hold these dies and to supply the necessary leverage to complete the sizing operation.

### COMPLETE LIST OF H & I DIAMETERS

| | | | | | | | | | | | | |
|---|---|---|---|---|---|---|---|---|---|---|---|---|
| .222 | .223 | .224 | .225 | .228 | .243 | .244 | .245 | .257 | .258 | .259 | .263 | .264 |
| .266 | .277 | .278 | .280 | .284 | .285 | .287 | .299 | .301 | .308 | .309 | .310 | .311 |
| .312 | .313 | .314 | .315 | .316 | .318 | .319 | .320 | .321 | .322 | .323 | .325 | .338 |
| .340 | .348 | .350 | .352 | .354 | .355 | .356 | .357 | .358 | .359 | .360 | .366 | .375 |
| .377 | .378 | .379 | .380 | .386 | .400 | .401 | .403 | .406 | .410 | .412 | .414 | .419 |
| .424 | .425 | .427 | .428 | .429 | .430 | .431 | .434 | .439 | .446 | .450 | .451 | .452 |
| .454 | .456 | .457 | .459 | .509 | .512 | .515 | .580 (lub. only) | | | | | |

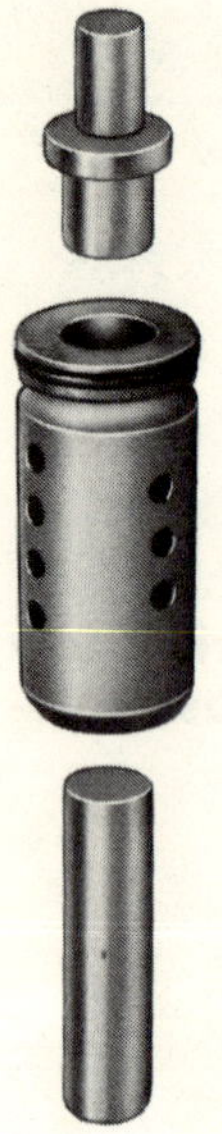

"G" Top Punch
Designed to fit the contour of the bullet nose. Order by bullet number.

"H & I" Sizing Assembly
Designed to swage the cast bullet to a specific diameter. Order by groove diameter.

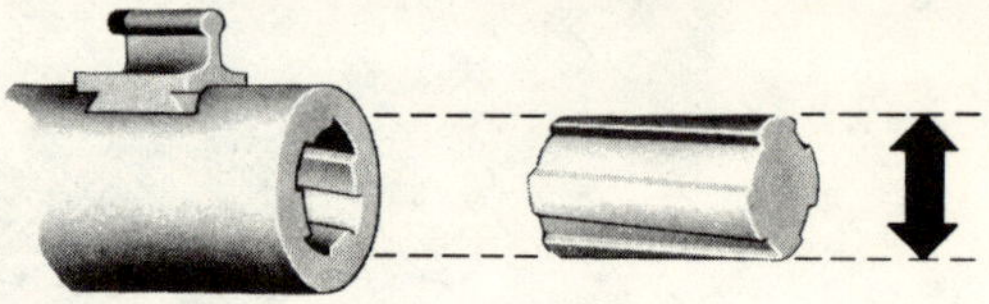

To determine your groove diameter drive SOFT lead slug through barrel and measure the widest point of this slug with micrometer.

### Pictured Is The Lyman 450 Bullet Sizer And Lubricator With "G, H & I" Dies Locked Into Position

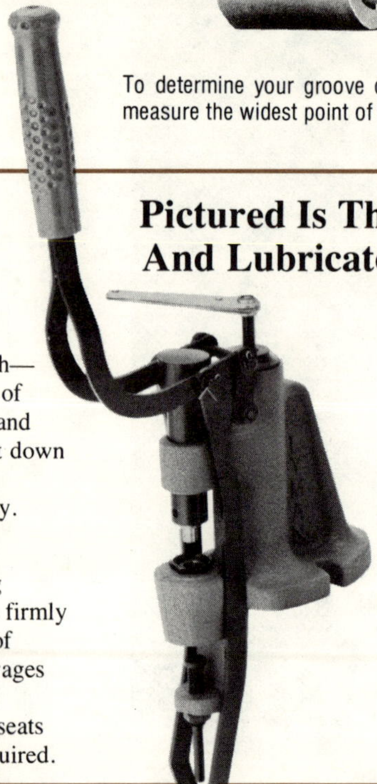

"G" Top Punch—fits the contour of the bullet nose and forces the bullet down into "H & I" Sizing Assembly.

"H & I" Sizing Assembly locks firmly into front boss of machine and swages bullet to correct diameter. Also seats gas check if required.

Grease in reservoir is forced under presssure into the lubricating grooves of the bullet.

### LYMAN CAST BULLETS USED IN THIS HANDBOOK

| Lyman Cast Bullet No. | Lyman Seating Screw No. | Lyman Top Punch No. | Lyman Cast Bullet No. | Lyman Seating Screw No. | Lyman Top Punch No. |
|---|---|---|---|---|---|
| 225107 | 415 | 415 | 356404 | 495 | 495 |
| 225438 | 438 | 438 | 357446 | 429 | 429 |
| 225415 | 415 | 415 | 40143 | 43 | 43 |
| 257420 | 415 | 420 | 41028 | 43 | 43 |
| 257312 | 415 | 420 | 410610 | 429 | 610 |
| 311359 | 359 | 359 | 410459 | 429 | 610 |
| 311576 | 465 | 465 | 410426 | 263 | 263 |
| 311410 | 467 | 467 | 42798 | 98 | 98 |
| 313249 | 226 | 226 | 429360 | 421 | 421 |
| 313226 | 226 | 226 | 429215 | 360 | 360 |
| 311252 | 465 | 465 | 429421 | 421 | 421 |
| 3118 | 8 | 8 | 429625 | 421 | 421 |
| 356402 | 402 | 402 | 429244 | 421 | 421 |
| 358242 | 311 | 311 | 429348 | 348 | 348 |
| 358345 | 429 | 429 | 429303 | 303 | 303 |
| 358480 | 429 | 429 | 452389 | 374 | 374 |
| 358495 | 495 | 495 | 452488 | 374 | 374 |
| 358477 | 429 | 429 | 452460 | 460 | 460 |
| 358156 | 429 | 429 | 452374 | 374 | 374 |
| 358311 | 311 | 311 | 452423 | 424 | 424 |
| 358429 & 624 | 429 | 429 | 452626 | 424 | 424 |
| 358212 | 311 | 311 | 45468 | 460 | 460 |
| 35863 | 344 | 344 | 454190 | 190 | 190 |
| 35891 | 495 | 495 | 454424 | 424 | 424 |

# 6

# Composite Cast Pistol Bullets

# Composite Cast Pistol Bullets

**Lab manager Ray Cowles holds targets of interest to revolver shooters. Composite #358624 was being fired for accuracy from a newly-established Ransom Rest fixture. Loads were identical for both targets. Why the difference? Target on the left was fired using all chambers while the target on the right was fired using only one chamber (selected at random).**

Lyman's new composite cast pistol bullets are a combination of an old idea, new technique and technology. The desire to create an expanding lead hunting bullet which can be driven at maximum velocity without leading is not a new one.

Composite bullets, of varying sorts for various purposes, have been used by American shooters for about 150 years. One type of projectile, used in muzzleloading slug guns, may be seen even today at any of the larger muzzleloading shoots. These slug gun bullets consist of a soft lead jacket and hard lead nose/core. Usually both pieces are cast and assembled within 24 hours to prevent a film of oxidation from affecting the fit.

To assemble, a special swaging die is used and the two lead parts are loosely joined before a precisely-fitted nose punch unit is set in place. Then the head of the punch is struck sharply with a hammer a certain number of times and the finished bullet is ready for inspection and use.

The slug gun bullet works this way: the soft exterior of the bullet is perfect considering the bullet needs to engrave deeply on loading. A shooter could not load a hard alloy bullet of proper diameter in one of these guns. As it is, a levered bullet starter is used which offers considerable mechanical advantage.

Paper-patch bullets, although not made of two different lead alloys, sought the same goal; to allow soft lead bullets to be driven at maximum possible velocity (without leading the bore) which would mushroom upon striking game. Paper-patch bullet systems were typically used on the old blackpowder cartridges; passing from the scene with the introduction of smokeless powder. The patching system was almost always applied to rifle bullets.

Some years back, Lyman offered a special mould line which would cast a nose/core of soft lead. Design of the nose portion was a direct ''lift'' from given popular rifle bullets and the resultant soft nose was intended to be identical to the nose of the same design cast full length in hard alloy. The mould system worked this way: After preparing an ample supply of the soft lead noses, the caster turned to the full length mould and a pot of harder alloy. Before each cast, a soft lead nose was set into the blocks and hard alloy poured in on top of it.

The first of these bullets was introduced in Ideal Handbook #11. The soft nose feature was made only for #457124 and #457193—both intended for the .45/70 Govt. round. Ideal Handbooks #17 through #31 show an expanded composite rifle bullet offering to include designs for .30, .32 and .38 caliber—as well as the .45/70 bullet, now called #457194. It is interesting to note that after Ideal Handbook #11, the design of the smaller bullets featured a rather short stem and the accompanying copy touted the advantages of the gas check feature (just introduced) and said not a word about terminal performance on game!

Reports of actual performance have eluded us but we may speculate about the system itself. In all probability, these bullets suffered from fabrication problems which, in the extreme, caused rejection; in a lesser state (assuming the caster didn't weigh each projectile) caused poor accuracy.

In the first place, the nose/core mould had to produce a near-perfectly fitting piece. If too small, the hard alloy would flow around and/or over its sides; if too small, the hard alloy would flow around and over its sides; if too large, the mould would not close completely and molten metal could flow between the blocks, causing a severe ''flashing'' problem at very least.

**Accuracy tests were fired in 10 shot groups with a variety of propellants. Technical assistant George Marx is replacing a just-fired #358624 target.**

An outside Ransom Rest fixture was rigged which worked fairly well. We used it primarily for the time of flight measurements from which this book's trajectory tables were derived. It was pressed into service for some composite accuracy tests too. Our fixture wasn't without its problems, keeping technical assistant Pete Thomas busy.

Penetration and expansion tests were always accuracy tests, too. Here your editor's old Super Blackhawk sends #429625 into a box of DucSeal 50 yards downrange.

In either case, the chance of a nose/core misalignment was fairly good. However, these two-piece bullets were offered in the larger bore chamberings like the .38-55 Winchester, a rather low intensity cartridge which operated at modest velocity and pressure levels. The .38-55 was, and still is, an excellent cast bullet cartridge which gracefully made the transition from blackpowder to smokeless loading.

A problem almost certainly encountered simultaneously with the foregoing was that of casting the hard alloy down over a cool lead stem and having the mould fill properly with no voids. Again, specific information documenting performance of this mould system has not been available. However, anyone who has ever cast a bullet with a hollow point or hollow base knows a cold nose or base plug is not conducive to the production of void-free bullets. Those plugs must be brought up to temperature, just like the mould blocks. The caster of those early composite bullets was faced with what amounted to a cold plug on each cast.

The current Lyman composite system for pistol bullets has, we believe, circumvented the inherent problems of its predecessor. Furthermore, most bullet casters will have the necessary lubricator/sizer and need only buy the composite blocks set from their dealer and two-part epoxy from a local store to be in business! There is no swaging involved which, as we know, requires expensive dies.

**Performance of the New Lyman Bullets**

Lyman does not make any spectacular claims regarding composite bullet performance. We have satisfied ourselves that these projectiles, with proper load development, are as accurate as most other equivalent cast or jacketed pistol bullets driven at maximum safe velocity.

Externally these bullets are designed to be virtually identical to three proven plain-base pistol bullets designed by Elmer Keith. We did slightly alter the location of the crimping groove on #358624 to permit a practical crimping depth when loaded in .357 Magnum brass. Here's the list of comparisons:

#358429 (168 gr.) = Composite #358624 (170 gr.) Both use same top punch & seating screw.

#429421 (245 gr.) = Composite # 429625 (232 gr.) Both use same top punch & seating screw.

#454424 (255 gr.) = Composite #452626 (245 gr.) Both use same top punch & seating screw.

**NOTE:** Load data for the composites is the same as that for the appropriate one-piece bullet. The nominal weights are, practically speaking, the same and reflect the kind of variation a cast bullet shooter experiences when shifting from one lot of alloy to another; or even additional ingots within the same lot.

**Terminal Ballistics**

Terminal performance is better than that of a hard alloy one-piece bullet of the same design. The hard bullet typically punches right through and (probably) exits on the far side, imparting only a portion of its energy to the target.

The composite bullets' soft nose begins to mushroom on impact. This nose expansion increases the frontal area, slowing the bullet much faster than its hard alloy counterpart. Part of the soft lead nose might break free; sheared off as it "washes" down over the the Linotype metal jacket mouth. When this occured in the Lyman tests, the fragments were found right with

**Test goals were to capture several projectiles in the DucSeal expansion medium—as was done here. This 50 yard collection was deliberately spread to avoid collisions within the medium.**

the spent bullet. There was never a trail of fragments through the Duc Seal expansion medium. Practically speaking, the bullets did their job.

However, should a bullet's nose mushroom, wash down and be sheared off, the shooter is left with a flat-nosed slug of at least 70% of original weight. This flat nose, coupled with a lighter overall weight, will cause continued rapid deceleration and transmission of energy to the target. Probability of the composite bullet expending its energy and remaining in the target, compared to its hard-cast counterpart, is much greater.

Thus, for the same basic reasons, the Lyman composite pistol bullet system will yield terminal performance approaching that of a jacketed bullet of the same weight. Interestingly, a broad systemic comparison can be made between the Lyman composite system and that of the famous Nosler Partition Rifle Bullet. If the Nosler's nose should be lost, a specially-designed jacket keeps a lead-filled base unit intact, ensuring continued penetration and energy transmission.

**Lyman Tests**

These three projectiles (#358624, #429625 and #452626) were tested for accuracy and expansion at 50 and 100 yards on the Lyman factory range. ''Full-house'' loads were used in all tests to duplicate probable hunting conditions. We learned two things: a) The composite pistol bullets are more comparable to a jacketed bullet than a hard alloy cast bullet in terms of accuracy/expansion. b) The proper choice of propellant can halve, literally, your group size! Be sure to try several different powders before settling on a load.

**After recovery, each bullet and any fragments were cleaned, weighted, individually packaged and labeled with pertinent data.**

## Assembly, Sizing and Lubrication

Separate moulds (sold as a set) are used to cast the nose/core and the jacket/base. These moulds are intended for pure lead and Linotype alloy, respectively. #2 Alloy may be substituted for Lino if loads are reduced somewhat and no tolerance problem develops with the nose/core piece. The shooter can experiment, as well, with slightly harder nose/core metal but should be aware of two potential problem areas:

1. Tolerances for each mould cavity are based on a known shrinkage factor for a given alloy. Shrinkage differs notably between pure lead and Lino; much less difference occurs between #2 Alloy and Lino. Changing the composition of metal for either part could impair fit of the two parts through increase/decrease in part dimensions.
2. We did, in fact, experiment with a stiffer nose/core metal but found it caused some fragmentation in the base/jacket of brittle Lino alloy. In short, staying with the pure lead nose/core seems a practical choice to us. Now, on to actual assembly.

- After casting an adequate quantity of each component, set up the lubricator/sizer with the proper sizing (H&I) die and the special "G" punch furnished in each composite mould set.
- Back off pressure on the lubricant reservoir so that no lube will flow into the sizing die.
- Size the jacket/base unit (only) without application of lubricant. The special "G" punch closely fits the jacket cavity and serves to ensure accurate alignment during this phase of the operation.
- Mix the two-part epoxy according to product directions. Use of this type bonding agent allows plenty of time to process bullets, correct early mistakes and keep equipment clean and free from epoxy.

**Sizing the jacket without the nose takes care of the hard sizing work. Special "G" punch keeps jacket aligned as it starts in the sizing die.**

**Prepare clean hard surface upon which to mix your epoxy. Most two-part formulas call for a 50-50 mix of resin and hardener.**

A small flat applicator of some sort is handy for spreading the epoxy on the core sidewalls.

Assemble the two pieces. Twist the nose within the jacket to coat all internal surfaces with epoxy.

- Dip the core end of the nose into the epoxy and evenly coat the base and sides. Set the nose/core into the sized jacket and put both aside. Using the same technique, quickly process the rest of the components. Note: Don't apply excessive epoxy. Clean off any overflow before going on to the next bullet.
- Allow the epoxy in the loosely-assembled bullets to begin to stiffen. This could be a matter of 30 to 45 minutes; perhaps more or less. While you are waiting, install the appropriate regular "G" punch and load the reservoir with an Alox-type lubricant such as that sold by Lyman.
- After the foregoing time has elapsed, try one of the bullets in the lube/sizer. Don't crank in lubricant until the bullet parts stay together as they come out of the sizer. If the parts separate, leaving the nose in the top punch, reassemble that one bullet and let the whole batch sit another ten minutes or so—then repeat the test.
- Once the epoxy has adequately set up, crank in the lubricant and process the bullets as you would a regular cast bullet. The lube/sizer positively aligns the two parts by centering the nose in the jacket. Keep the sizer and bullet free of excess epoxy.
- Set aside the finished bullets to cure completely; usually overnight will do the job. See directions from manufacturer of the epoxy. Afterwards, load as you would any cast pistol bullet.

We would appreciate hearing from shooters who use the bullets in the field. While these three bullets have been pretty well tested to the point where there'll be no surprises, there is no real substitute for information based on actual field use on game. Please write to the attention of the editor. Thanks . . . and good shooting.

**During last stage of assembly, some additional pressure can be exerted on the lube/sizer downstroke to firmly press the nose into the jacket. A firm fit is all that's needed—don't overdo it. Neither the bullet nor your lubricator/sizer can handle much swaging-type pressure.**

# 7

# How Bullet Moulds Are Made

# How Bullet Moulds Are Made

1. Cast malleable iron blocks of a special formula are purchased by Lyman Products for its complete line of moulds.

2. The rough blocks are milled so that the sides and faces are squared and correctly dimensioned.

3. The second cut is the groove which eventually will accommodate one of the pincer-type handles. At the same time the ends of the blocks are squared and milled to dimension.

4. The facing surfaces of the blocks are ground uniformly to a perfect match. They are then paired, numbered and air-vent lines milled on the faces.

5. Holes are drilled and tapped for dowel pins and setscrews.

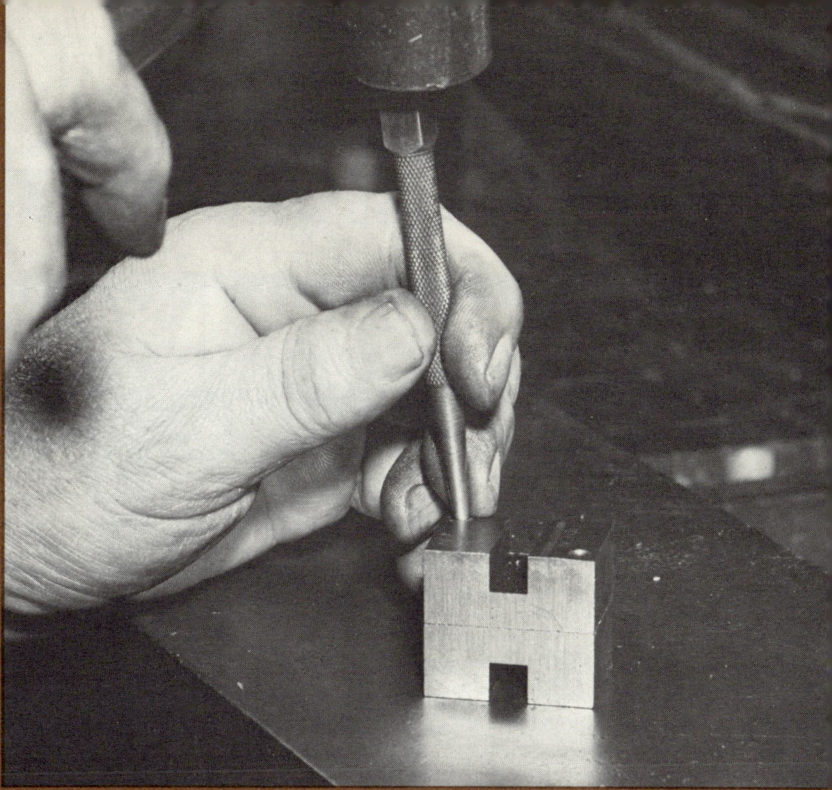

6. At this stage the dowel pins are driven home and the mould blocks checked for alignment.

7. Cherrying drills, prepared by a toolmaker, cut the mould cavities. Hundreds of these drills, the "cherrys", are needed to supply the tremendous variety of calibers and bullet designs within the Lyman line.

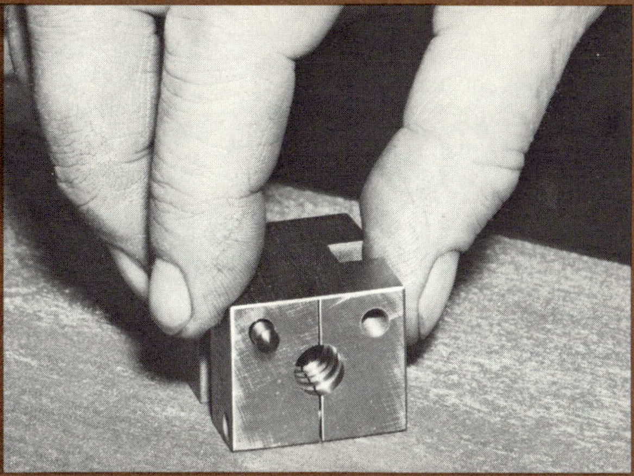

8. At several stages each block is strapped to remove any burrs that may have resulted from earlier handling and sharp edges are slightly broken.

10. A screw fastens the sprue cutter in place and the blocks are ready for final inspection, packaging and shipment.

9. The blocks are heat-cured to oxidize the metal. This "blues" the mould and minimizes the time required for the breaking-in process.

# CASTING EQUIPMENT

## LYMAN BULLET CASTING KITS MAKE IT EASY TO GET STARTED.

Lyman's Bullet Casting Kits make bullet casting easy and convenient. All three kits offer a choice of mould blocks from 15 of the most popular designs and diameters to service a wide spectrum of shooting requirements. Lyman kits are economical, too—save money on the individual component prices by purchasing the kit. Bullet Casting Kits are available exclusively from Lyman.

### The Basic

Start bullet casting with Lyman's Basic Kit. Contents include a single cavity mould with handles, a lead pot and dipper, a sizing die and punch, lube cutter, 4'' lube pan, Lyman's Ideal Lubricant, and Lyman's Cast Bullet Handbook.

### The Custom

Lyman's Custom Bullet Casting Kit features the #450 Lube/Sizer, which fully prepares almost any cast bullet for high accuracy reloading. Plus the Custom Kit offers an ingot mould, G-H&I dies, Lyman's Cast Bullet Handbook and Ideal Lubricant, a single cavity mould with handles and a lead pot with dipper.

### The Master

Lyman's new Mould Master XX casting furnace goes with our finest Bullet Casting Kit, *plus* the #450 Lube/Sizer! The kit also features a double cavity mould with handles, Ideal bullet lube, Lyman's Cast Bullet Handbook, G-H&I dies and an ingot mould.

Lyman bullet casting kits currently feature only the following mould designs. Check the current Lyman catalog for revisions. Kit components are available individually; again, check the current Lyman catalog for details.

1. #358242
2. #356402
3. #358495
4. #35891
5. #358477
6. #358156*
7. #358311
8. #357446
9. #358429
10. #429215*
11. #429244*
12. #429421
13. #452460
14. #452374
15. #454424

*with 100 gas checks

# 8

# Reference

# Reference

## PRESSURE

Every shooter has a general idea of what happens when a cartridge is fired in the chamber of a firearm. The powder charge is ignited by the primer flash. As the powder burns, it generates an enormous amount of gas which creates a force in all directions. This force, which we identify as chamber pressure, presses against the case head and drives the cartridge back against the bolt face. It pushes the case walls out tight against the sides of the chamber. Seeking to escape the confines of the case, the pressure force builds to the point where it unseats the bullet and drives it with increasing speed down the barrel. What concerns the reloader is how high this pressure builds and how safely it is controlled during firing.

## VARIABLES AFFECTING PRESSURE

The pressure for a given pistol load varies considerably with the tolerance of the individual firearm. The condition of the throat, bore diameter, groove diameter, and chamber dimensions, etc., all have a substantial bearing on the pressure generated in a handgun. When we add to this list such other variables as primer brand and type, case capacity, bullet type and hardness, bullet weight, etc., we see there is more to this business of pressure than simply pouring the proper amount of powder into the case.

## CONDITIONS THAT CAN RAISE PRESSURE

**Powder:** When we add more powder to the case, we naturally increase the pressure of the load. All other conditions being equal, fast burning powders create higher pressures than slow burning powders when used in equal amounts. The powders shown in this Handbook were carefully chosen for the applications and loads listed. Use these powders as recommended and never exceed the maximum load listed.

Under certain conditions, reduced loads can also cause trouble. This phenomenon, popularly termed "detonation", has never been satisfactorily explained. Yet, evidence does exist that under certain conditions reduced loads can jump pressures enormously. The reader is, therefore, cautioned against reducing the suggested listed starting loads.

**Bullets:** As previously indicated, slight variations in bullet style can influence the pressure of the load. All else being equal, a heavier or a harder bullet will boost pressure due to its increased weight mass and/or its greater resistance to the rifling. Oversize bullets and bullets with more bearing surface will also up pressure. Your bullet diameter should conform to the groove diameter of your rifle or handgun. When working up a load, use a specific brand, style and weight of bullet. Do not assume that you can change from one bullet brand, style, etc., to another without altering the pressure of the load.

**Cases:** The inside capacity (volume) of the cartridge case has a direct relationship to the pressure of the load. The experienced reloader sticks with one specific brand of case when working up his loads. He realizes that assorted brand cases, when used with the same powder charge, will give variations in pressure.

**Primers:** Primer characteristics will affect the pressure of the load. When working up a load, use the same brand and type of primer.

**Chamber and Bore Dimensions:** The dimensions of your particular handgun do have a direct relationship to the pressure generated by your reloaded cartridge. The tolerance range for domestic firearms has been taken into consideration when determining the suggested loads. If your handgun is imported, (sporter or military), we suggest you slug the barrel before attempting to reload. Your bullet diameter should correspond to the groove diameter of the firearm. Cast bullets can vary .002'' over the groove diameter. In all instances, the load should be worked up gradually.

**Overall Length:** To conform to chamber dimensions, the total measurement of the seated bullet and case should not exceed the listed overall length in the Data Section. This accomplishes two things: (1) it ensures the loaded round will feed properly through the magazine and (2) it keeps the bearing surface of the bullet back away from the rifling. A bullet that is seated too far forward may bump the rifling and receive greater resistance at the time of ignition. Such a condition tends to increase the chamber pressure of the load.

The whole thing is, of course, somewhat relative. When the bullet is in this forward position, case volume is increased which, in itself, tends to decrease the pressure. Chances are, however, that the increased resistance on the bullet has a greater effect than the slight increase in case volume. The only safe practice is not to exceed the overall cartridge length listed with your data. This, of course, does not mean that the cartridge cannot be shorter than this length. Lighter and shorter bullets of the same caliber are, naturally, seated to a shorter overall length.

**Case Length:** Repeated firing and resizing affects the overall length and thickness of the cartridge brass. This combination of forces constantly pushes the case material forward, gradually making the case neck longer and somewhat thicker than it was originally. When the case has exceeded the maximum tolerance allowed by the chamber, an increase in pressure results. The overall case length which is listed in the Data Section will inform you as to the proper length for your cartridge brass. A suggested "Trim-to-Length" is also specified.

**Crimp:** Crimping the case does have some influence on pressure. If a crimp is to be used, then all loads should be crimped in a uniform manner.

**Deep Seating Bullets:** Our testing indicates that excessive (below the junction of the neck and shoulder) seating of the bullet does increase pressure. Most shorter bullets, in a given caliber, will conform to maximum overall length without excessive seating. Longer bullets should be seated exactly to overall length and in all cases, the load should be worked up gradually.

# VISUAL SIGNS OF EXCESS PRESSURE

Lacking pressure testing equipment, the experienced reloader confines himself to the guidelines set down in his reloading manual. In other words, he begins with the suggested starting load and works slowly upwards until he finds his best performing load. When working up a load, he holds the variables that affect pressure to a minimum by using components of the same brand and style. His cases are either new or once fired.

After each firing he inspects the cases carefully for signs of excess pressure. If all is in order, he continues by adding another increment of powder to the total charge. He never exceeds the maximum load, for he knows that the moment he does, he is eliminating all margins of safety. He concerns himself with accuracy and performance rather than ''block-busting energy''. In most cases, if he is concentrating on accuracy and barrel life, his loads will probably remain a little on the mild side.

The following text deals with pressure signs as they might appear when working with new cases:

**Excessive Headspace:** When this condition is present, the bolt face is not properly supporting the case head. Before reloading, we must be sure that our firearm is completely free from headspace problems.

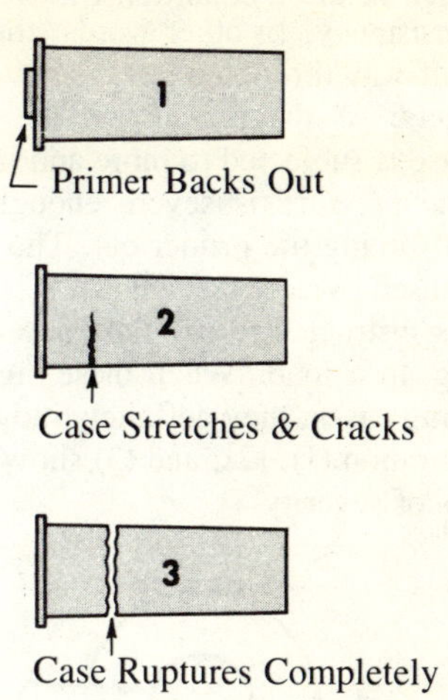


The above illustrations are indications of excessive headspace. Do not use any firearm which shows such signs of headspace problems.

At low pressure (the starting load) the first sign of excessive headspace is usually the backing out of primers. As the pressure increases slightly, the case moves back in the chamber and reseats the primer. At this stage, the case appears quite normal. As the pressure of the load builds, the case will no longer move backward but will tend, instead, to cling to the chamber wall. Lacking necessary support in the head area, the case will stretch and eventually crack around its circumference.

**Primers:** We feel that flattened or cratered primers, if interpreted correctly, **are a positive indication of extremely high pressure.** Working with pressure guns in our laboratory, we have never experienced cratered primers when the pressures were reasonable. When the pressure climbs the primer starts to show it.

The following illustration shows several changes in primer contour which may be interpreted as signs of excess pressure. The reloader should stick with the same brand and type of primer when developing his load. If all is going well and he then suddenly experiences one of these changes in contour, **he should question the safety of his load.**

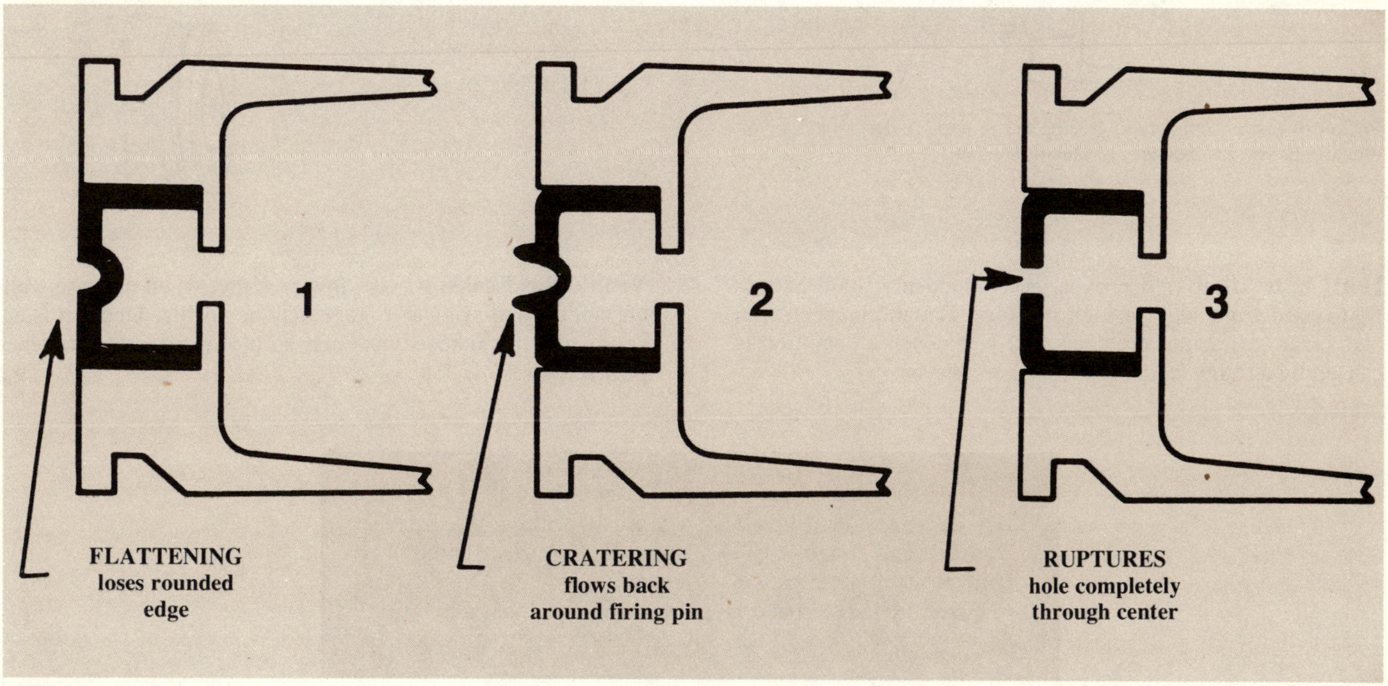

To interpret these signs correctly, common sense must be used. For example, soft (or thin) primers will flatten more than hard (or thick) primers with the same amount of pressure. Also, a long firing pin can cause the condition shown in illustration (3). Each of these conditions, however, would be obvi ous by its consistency. In other words, they would not occur abruptly or half-way through a test.

**Primer Pockets:** As the pressure of the load increases, the head of the case is subjected to more and more stress. Eventually, when the pressure is severe enough, it will open the primer pocket forcing the primer out. The signs shown below can also be caused by case fatigue and it is for this reason that the reloader is instructed to use only new or once fired cases when working up a load. When these signs are experienced with **new strong cases,** they definitely indicate extremely high pressure. Illustration (1), (2), and (3) show the same condition in three stages of severity.

1
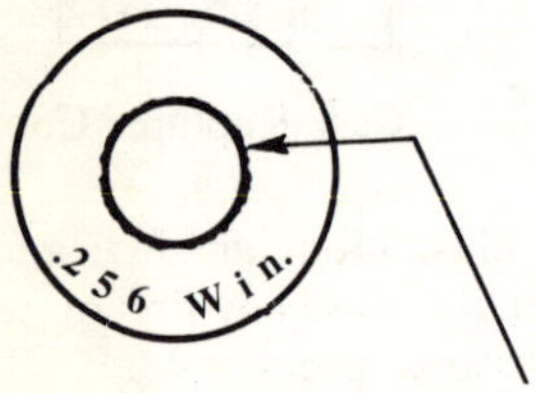


Dark smudge line around primer indicates gas leakage.

2
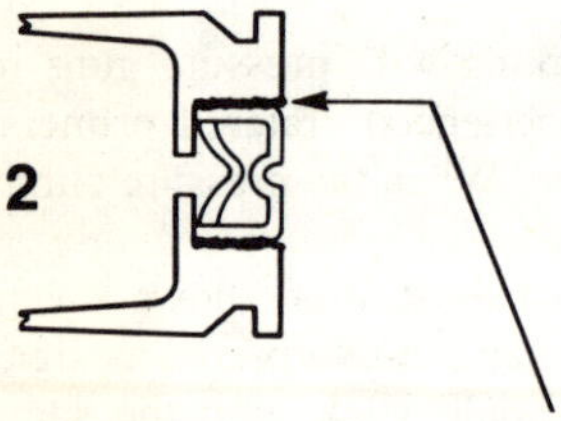

Pocket opens, leaving primer loose. A dark smudge, indicating gas leakage, is usually present.

3
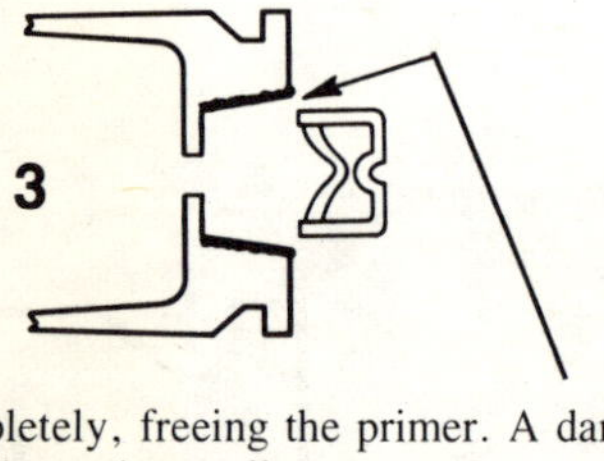

Pocket of opens completely, freeing the primer. A dark smudge, indicating heavy gas leakage, is usually present.

**Hard Extraction:** When experienced suddenly, hard extraction is a definite sign of high pressure. Rough chambers can also create extraction problems, but when such a condition is caused by a rough chamber it remains constant.

**Flowing of Brass in Head Area:** When the cartridge brass flows back into the extractor port, or the head stamping flattens, or the head takes on the irregularities of the bolt face, we have positive indications of very high pressure. Indications so severe, in fact, that they are usually accompanied by other signs of pressure such as blown primers, stretched pockets, visible gas leaks, etc.

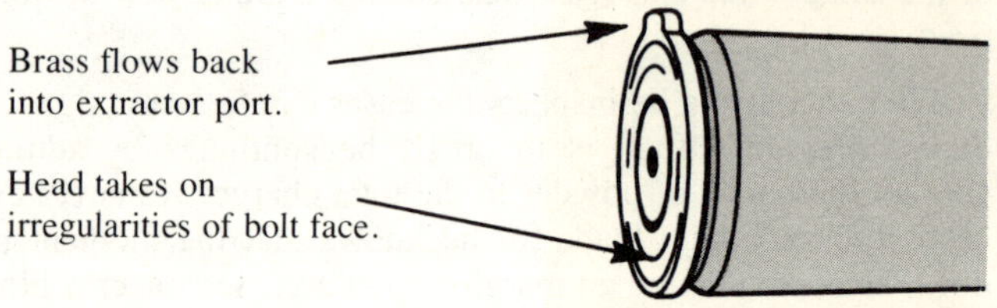


**Head Expansion:** The side walls and neck of the cartridge case are designed to expand freely and to seal the chamber at the time of firing. This is necessary to prevent gas from leaking back through the action. When a cartridge case is badly smudged by gas on its outer walls and neck (without other indications) it is actually a sign that the pressure is too low, and that the case walls are not expanding properly.

The head of the case, designed to resist pressure and expansion may be viewed for signs of pressure. However, it is one of the most difficult signs to interpret correctly and a good micrometer is required. If the chamber dimensions of your firearm are slightly on the large side, you will always experience some head expansion (on the first firing) with new brass or with factory loaded rounds. The best method is to fire a few factory loaded rounds in your pistol and measure the expansion of the brass. This measurement can then be accepted as normal and any expansion over it should be considered excessive. Normally, expansion of .001'' is considered as evidence of very high pressure.

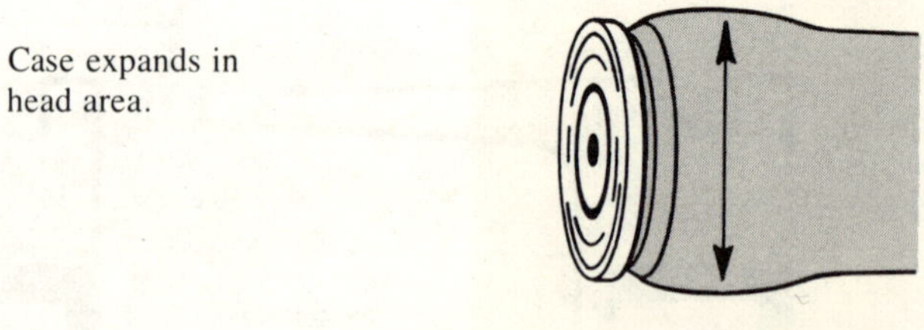


**Visible Gas Leaks:** At very low pressure, when the case walls are not expanding, gas leakage may be visible. Under it is accompanied by other signs such as blown primers, stretched pockets, etc.

**CAUTION:** A Severe Pressure Condition Can Alter The Headspace Of The Firearm. If The Reader Should Experience One Of The Conditions Discussed Herein, He Should Not Only Reduce His Load At Once, But Also Have His Firearm Checked For Headspace By A Competent Gunsmith Before Continuing!

## Rifling Twist–Handguns

| | |
|---|---|
| .22 Jet - 1 in 15″: | Smith & Wesson. |
| .22 Hornet - 1 in 14″: | Thompson-Contender. |
| .221 Fireball - 1 in 12″: | Remington XP100 |
| .256 Winchester - 1 in 14″: | Ruger; Thompson-Contender. |
| .30 Carbine - 1 in 12″: | Thompson-Contender. |
| .30 Carbine - 1 in 20″: | Ruger. |
| .32 Colt New Police - 1 in 14″: | Colt. |
| .32 ACP - 1 in 16″: | Mauser. |
| .32 Smith & Wesson, Smith & Wesson Long - 1 in 18¾″: | Smith & Wesson. |
| .357 Magnum - 1 in 14″: | Colt. |
| .357 Magnum - 1 in 16″: | Ruger. |
| .357 Magnum - 1 in 18″: | Thompson-Contender. |
| .357 Magnum - 1 in 18¾″: | Smith & Wesson. |
| .38 Smith & Wesson - 1 in 18¾″: | Smith & Wesson. |
| .38 Special - 1 in 14″: | Colt. |
| .38 Special - 1 in 17″: | Charter. |
| .38 Special - 1 in 18¾″: | Smith & Wesson. |
| .38 Super - 1 in 16″: | Colt. |
| 9mm Parabellum - 1 in 10″: | Smith & Wesson. |
| 9mm Parabellum - 1 in 16″: | Colt. |
| .41 Magnum - 1 in 18¾″: | Smith & Wesson. |
| .41 Magnum - 1 in 20″: | Ruger. |
| .44 Special - 1 in 16″: | Colt. |
| .44 Magnum - 1 in 20″: | Ruger; Smith & Wesson. |
| .44 Magnum - 1 in 22″: | Thompson-Contender. |
| .45 Auto Rim - 1 in 15″: | Smith & Wesson. |
| .45 Auto - 1 in 16″: | Colt. |
| .45 Colt - 1 in 16″: | Colt. |
| .45 Colt - 1 in 24″: | Thompson-Contender. |

## Conversion Factors

| Multiply | By | To Obtain |
|---|---|---|
| Atmospheres | 14.70 | Pounds per square inch |
| Kilograms per square centimeter | 14.23 | Pounds per square inch |
| Pounds per square inch | 0.07032 | Kilograms per $cm^2$ |
| Drams | 1.772 | Grams |
| Drams | 0.0625 | Ounces |
| Grains (Troy) | 1 | Grains (Avoirdupois) |
| Grains | 0.0648 | Grams |
| Grams | 15.43 | Grains |
| Grams | 0.03527 | Ounces |
| Kilograms | 1000 | Grams |
| Kilograms | 2.205 | Pounds (Avoirdupois) |
| Ounces | 16 | Drams |
| Ounces | 437.5 | Grains |
| Ounces | 0.0625 | Pounds (Avoirdupois) |
| Ounces | 28.35 | Grams |
| Pounds | 7000 | Grains |
| Pounds | 453.6 | Grams |
| Pounds | 16 | Ounces |
| Centimeters | 0.3937 | Inches |
| Centimeters | 0.01 | Meters |
| Centimeters | 10 | Millimeters |
| Inches | 2.540 | Centimeters |
| Meters | 100 | Centimeters |
| Meters | 3.281 | Feet |
| Meters | 39.37 | Inches |
| Meters | 1.094 | Yards |
| Feet per Second | 0.3048 | Meters per Second |
| Feet per Second | 0.6818 | Miles per Hour |
| Meters per Second | 3.281 | Feet per Second |
| Miles per Hour | 88 | Feet per Minute |
| Miles per Hour | 1.467 | Feet per Second |
| Liters | 0.2642 | Gallons |
| Liters | 1.057 | Quarts |
| Liters | 1000 | Milliliters |
| Square Centimeters | 0.1550 | Square Inches |
| Square Inches | 6.452 | Square Centimeters |

# Decimal-Fraction Tables

| Drill Size | Decimal | Drill Size | Demical | Drill Size | Decimal | Drill Size | Decimal |
|---|---|---|---|---|---|---|---|
| 80 | .0135 | 42 | .0935 | 13/64 | .2031 | X | .3970 |
| 79 | .0145 | 3/32 | .0938 | 6 | .2040 | Y | .4040 |
| 1/64 | .0156 | 41 | .0960 | 5 | .2055 | 13/32 | .4062 |
| 78 | .0160 | 40 | .0980 | 4 | .2090 | Z | .4130 |
| 77 | .0180 | 39 | .0995 | 3 | .2130 | 27/64 | .4219 |
| 76 | .0200 | 38 | .1015 | 7/32 | .2188 | 7/16 | .4375 |
| 75 | .0210 | 37 | .1040 | 2 | .2210 | 29/64 | .4531 |
| 74 | .0225 | 36 | .1065 | 1 | .2280 | 15/32 | .4688 |
| 73 | .0240 | 7/64 | .1094 | A | .2340 | 31/64 | .4844 |
| 72 | .0250 | 35 | .1100 | 15/64 | .2344 | 1/2 | .5000 |
| 71 | .0260 | 34 | .1110 | B | .2380 | 33/64 | .5156 |
| 70 | .0280 | 33 | .1130 | C | .2420 | 17/32 | 5312 |
| 69 | .0292 | 32 | .1160 | D | .2460 | 35/64 | .5469 |
| 68 | .0310 | 31 | .1200 | 1/4 | .2500 | 9/16 | .5625 |
| 1/32 | .0312 | 1/8 | .1250 | E | .2500 | 37/64 | .5781 |
| 67 | .0320 | 30 | .1285 | F | .2570 | 19/32 | .5938 |
| 66 | .0330 | 29 | .1360 | G | .2610 | 39/64 | .6094 |
| 65 | .0350 | 28 | .1405 | 17/64 | .2656 | 5/8 | .6250 |
| 64 | .0360 | 9/64 | .1406 | H | .2660 | 41/64 | .6406 |
| 63 | .0370 | 27 | .1440 | I | .2720 | 21/32 | .6562 |
| 62 | .0380 | 26 | .1470 | J | .2770 | 43/64 | .6719 |
| 61 | .0390 | 25 | .1495 | K | .2810 | 11/16 | .6875 |
| 60 | .0400 | 24 | .1520 | 9/32 | .2812 | 45/64 | .7031 |
| 59 | .0410 | 23 | .1540 | L | .2900 | 23/32 | .7188 |
| 58 | .0420 | 5/32 | .1562 | M | .2950 | 47/64 | .7344 |
| 57 | .0430 | 22 | .1570 | 19/64 | .2969 | 3/4 | .7500 |
| 56 | .0465 | 21 | .1590 | N | .3020 | 49/64 | .7656 |
| 3/64 | .0469 | 20 | .1610 | 5/16 | .3125 | 25/32 | .7812 |
| 55 | .0520 | 19 | .1660 | O | .3160 | 51/64 | .7969 |
| 54 | .0550 | 18 | .1695 | P | .3230 | 13/16 | .8125 |
| 53 | .0595 | 11/64 | .1719 | 21/64 | .3281 | 53/64 | .8281 |
| 1/16 | .0625 | 17 | .1730 | Q | .3320 | 27/32 | .8438 |
| 52 | .0635 | 16 | .1770 | R | .3390 | 55/64 | .8594 |
| 51 | .0670 | 15 | .1800 | 11/32 | .3438 | 7/8 | .8750 |
| 50 | .0700 | 14 | .1820 | S | .3480 | 57/64 | .8906 |
| 49 | .0730 | 13 | .1850 | T | .3580 | 29/32 | .9062 |
| 48 | .0760 | 3/16 | .1875 | 23/64 | .3594 | 59/64 | .9219 |
| 5/64 | .0781 | 12 | .1890 | U | .3680 | 15/16 | .9375 |
| 47 | .0785 | 11 | .1910 | 3/8 | .3750 | 61/64 | .9531 |
| 46 | .0810 | 10 | .1935 | V | .3770 | 31/32 | .9688 |
| 45 | .0820 | 9 | .1960 | W | .3860 | 63/64 | .9844 |
| 44 | .0860 | 8 | .1990 | 25/64 | .3906 | 1 | 1.0000 |
| 43 | .0890 | 7 | .2010 | | | | |

# U.S. Boxer Primer Chart

| | Large Rifle | Large Rifle Magnum | Small Rifle | Small Rifle Magnum | Large Pistol | Large Pistol Magnum | Small Pistol | Small Pistol Magnum |
|---|---|---|---|---|---|---|---|---|
| Remington | 9½ | 9½M | 6½ | 7½ | 2½ | | 1½ | 5½ |
| Winchester Western | 8½-120 | | 6½-116 | | 7-111 | 7M-111F | 1½-108 | 1½M-108 |
| Federal | 210 | 215 | 200 | | 150 | | 100 | |
| | | | *205 | | | | | |
| Alcan | Large Rifle Max-Fire | | Small Rifle Max-Fire | | Large Pistol Max-Fire | | Small Pistol Max-Fire | |
| CCI | 200 | 250 | 400 | 450 | 300 | 350 | 500 | 550 |

* Thick cup design, especially for .17 Rem. and .22 centerfire loads.

# 9

# Reloading Data Introduction

The Ballistic Pendulum as invented by Benjamin Robbins in 1742. This device allowed, for the first time, reasonably accurate measurement of projectile velocity.

# Reloading Data Introduction

The data listed in this section has been tested by our technicians and found to be safe when loaded with our test components and fired (under our laboratory controlled conditions) in our testing equipment. Since Lyman Products Corporation has no control over the manufacture of the various components listed, nor over the actual loading, choice or condition of the firearms and components used, no responsibility for use of this data is implied or assumed.

**COMPONENTS:** The reader should bear in mind that the components listed are not of Lyman manufacture. Therefore, it is possible that production changes affecting ballistic performance can occur at any time without our knowledge. If there is ever a question as to the correctness of the component specified, write to its manufacturer whose address is listed in the front of this publication.

**STARTING LOAD:** It is essential that the reader begin with the suggested weight of powder listed in this bracket and work up slowly (following load development precautions) to his best performing load. The novice should use only the ''starting load'' for a period of time until he builds confidence and experience. Never decrease this charge as an increase in pressure could be encountered.

**MAXIMUM LOAD:** All loads which are listed as maximum were tested and classified as maximum by our technicians in accordance with our laboratory standards. **Under no circumstances should these loads be exceeded,** nor should they be quickly accepted by the reader as a safe working maximum for his particular handgun.

Many reloaders misinterpret the meaning of the ''maximum load''. They wrongly assume that if a high pressure load proved safe in a test laboratory then it is equally safe under any and all conditions. This is not true. The reader must start with the ''starting load'' and work up his loads carefully. Working with his particular firearm and component combination, he may encounter signs of excess pressure before he reaches the maximum charge listed.

The technician classifies a load as maximum after carefully considering many aspects of its ballistic performance. **The maximum average pressure of the load is not the only criteria.** Often a load having an acceptable maximum average pressure will be rejected (or reduced) due to its erratic performance. Accuracy must also be considered, particularly when dealing with cast lead alloy bullets. In all instances, the maximum listing represents what our technicians consider to be the maximum working combination for the bullet, powder and caliber listed. These loads do not exceed the ammunition manufacturer's standards.

**ACCURACY LOAD:** As accuracy is not consistently the same with all firearms, this load requires some interpretation by the reader. Where an accuracy load is listed, it merely indicates our most accurate test results with this particular bullet weight and caliber. Slight variations of the load may be necessary to produce optimum accuracy in your handgun.

**FACTORY DUPLICATION LOAD:** Where possible, we have listed the load which duplicated factory velocity for the bullet weight tested. It should not be assumed that this load duplicates factory pressure. Where this load is printed in a tinted panel, it indicates that the load was at, or near, a maximum pressure level. Such a load requires that the reader work up to it slowly.

**The Universal Bond Receiver operates very much like a falling-block rifle action. These receivers can be used for any centerfire cartridge or shotshell testing; requiring an appropriately chambered and collared pressure barrel.**

**VELOCITIES:** The velocities printed in this Handbook are instrumental (not corrected to the muzzle) and were recorded at a distance of fifteen feet from the muzzle of the test firearm.

**POWDERS:** We have limited our testing to those powders which are manufactured in the United States and which are readily available to the consumer. The following brands are listed: Dupont, Winchester, Hercules, Alcan, Hodgdon and Gearhart-Owen.

**BULLETS:** The brand, weight and style of the various jacketed bullets used in our tests are listed with the specifications under each cartridge heading. Cast bullet designs are identified with each block of cast bullet data. Here again, it would be impossible to test every known bullet. We have restricted our testing to major domestic brands. Sierra, Winchester, Remington, Hornady and Speer—plus proven Lyman cast bullet designs.

**COMPRESSED LOADS:** All compressed loads are indicated with an asterisk. Depending upon the volume of the specific cartridge case used by the reader, he may, or may not, have difficulty starting bullets in such loads. If the bullet will not start, reduce the load sufficiently so that 1/10'' of space remains in the case neck. Start the bullet into the case and use whatever additional pressure is required to fully seat the bullet. Failure to comply could result in a bulged case.

For viewing clarity, tweezers place the .146'' x .400'' copper crusher atop the pressure piston head. Next, the overhead anvil will be aligned above it and the two side screws tightened. Finally, the large screw in the center will be turned in by hand until a firm—but not excessive—pressure is exerted upon the fresh crusher.

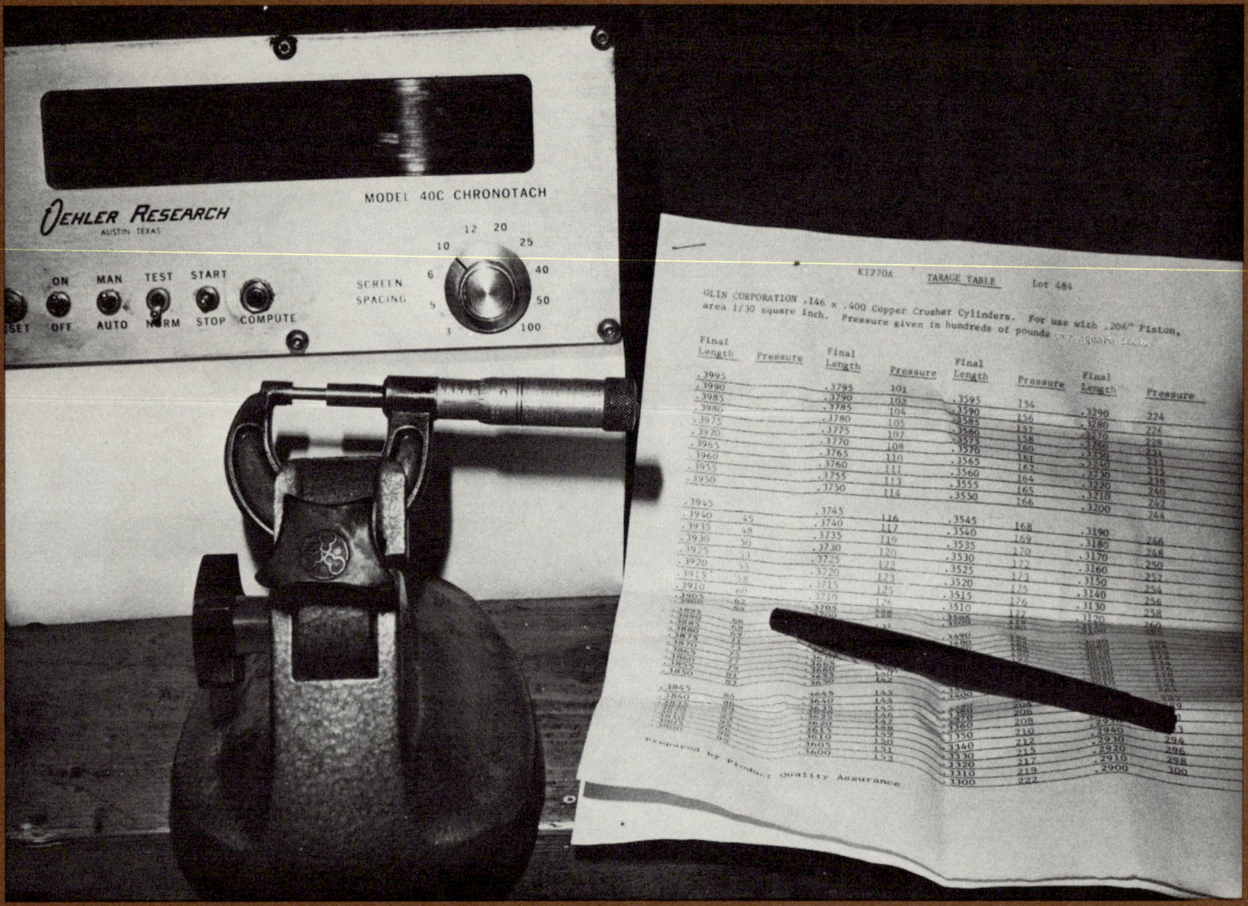


**Fired crushers must be measured by micrometer and that measurement applied to a "Tarage Table" for translation into Units of Pressure; not Pounds per Square Inch. CUP means Copper Units of Pressure. Each lot of crushers comes with a specially regulated Tarage Table which links amount of crusher compression with chamber pressure.**

# 10

# Metallic Data

# .22 REMINGTON JET

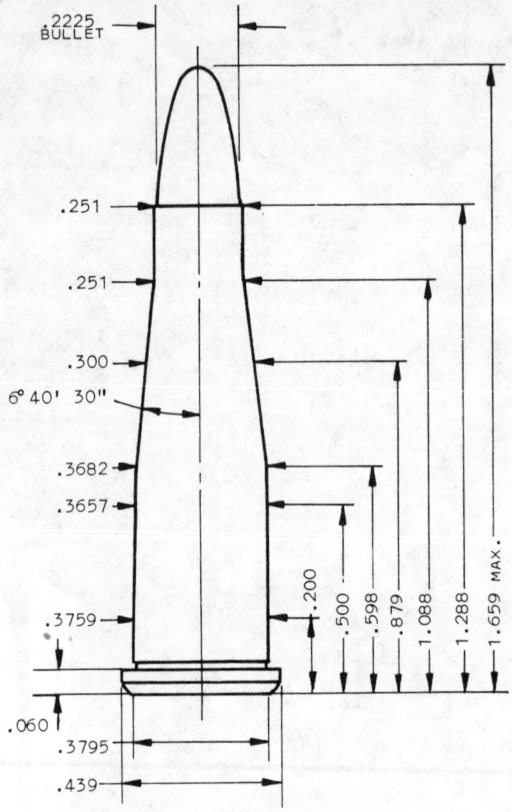


## TEST COMPONENTS

Cases ................................................Remington
Trim-to Length ..........................................1.283''
Primers ..........................................Remington 6½
Primer Size ..........................................Small Rifle
Lyman Shell Holder .....................................No. 1
Jacketed Bullets Used ..............................Remington
SP#B22990, 40 gr.
Cast Bullets Used ........................(size to .223'' dia.)
*Gas Check *#225107, 38 gr.
*#225438, 45 gr.

## TEST SPECIFICATIONS (Velocity Only)

Firearm Used ..................Thompson/Center Contender
Barrel Length ................................................10''
Twist ........................................................ 1-15''
Groove Diameter ....................................... .223''

## COMMENTS

Remove all traces of lubricant from both the cylinder and cases before firing. This will insure the proper functioning of the cylinder. Do not use .224'' diameter bullets in this cartridge. Use .222'' diameter or .223'' diameter bullets only.

**40 gr. Remington SP**
#B22990 (.222'' dia.) 1.652'' OAL

| Powder | Sug. Starting Grains | Velocity F.P.S. | Pressure C.U.P. | Max. Grains | Velocity F.P.S. | Pressure C.U.P. |
|---|---|---|---|---|---|---|
| SR-4759 | 6.0 | 921 | — | 10.0 | 1879 | — |
| IMR-4227 | 6.0 | 896 | — | 13.5 | 2316 | — |
| 2400 | 6.0 | 995 | — | 12.0 | 2332 | — |
| 630 | 4.3 | 959 | — | 12.0 | 2431 | — |
| H110 | 6.0 | 989 | — | 11.0 | 1839 | — |
| AL-8 | 5.0 | 1032 | — | 10.0 | 2182 | — |

**38 gr. Cast Bullet**
#225107 (Linotype) 1.480'' OAL

| Powder | Sug. Starting Grains | Velocity F.P.S. | Pressure C.U.P. | Max. Grains | Velocity F.P.S. | Pressure C.U.P. |
|---|---|---|---|---|---|---|
| 2400 | 8.5 | 1848 | — | 12.0 | 2400 | — |
| IMR-4227 | 9.0 | 1827 | — | 13.5 | 2409 | — |
| IMR-4198 | 12.0 | 1734 | — | 14.5+ | 2125 | — |
| 630 | 7.5 | 1712 | — | 11.0 | 2370 | — |
| HS-7 | 7.0 | 1797 | — | 10.0 | 2434 | — |
| AL-8 | 7.5 | 1730 | — | 10.0 | 2346 | — |

**Note:** Loads shown in shaded panels are maximum.
+Designates a compressed powder charge.

# .22 REMINGTON JET
(Continued)

### 45 gr. Cast Bullet
#225438 (Linotype) 1.495'' OAL

| Powder | Sug. Starting Grains | Velocity F.P.S. | Pressure C.U.P. | Max. Grains | Velocity F.P.S. | Pressure C.U.P. |
|---|---|---|---|---|---|---|
| 2400 | 8.5 | 1704 | — | 11.5 | 2211 | — |
| IMR-4227 | 9.5 | 1807 | — | 13.0 | 2256 | — |
| IMR-4198 | 12.0 | 1708 | — | 14.5+ | 2126 | — |
| 630 | 7.5 | 1702 | — | 11.5 | 2446 | — |
| HS-7 | *7.5 | 1769 | — | 10.5 | 2370 | — |
| AL-8 | 8.0 | 1838 | — | 10.5 | 2304 | — |

**Note:** Loads shown in shaded panels are maximum.
* Designates potentially most accurate load.
\+ Designates a compressed powder charge.

# .221 REMINGTON FIREBALL

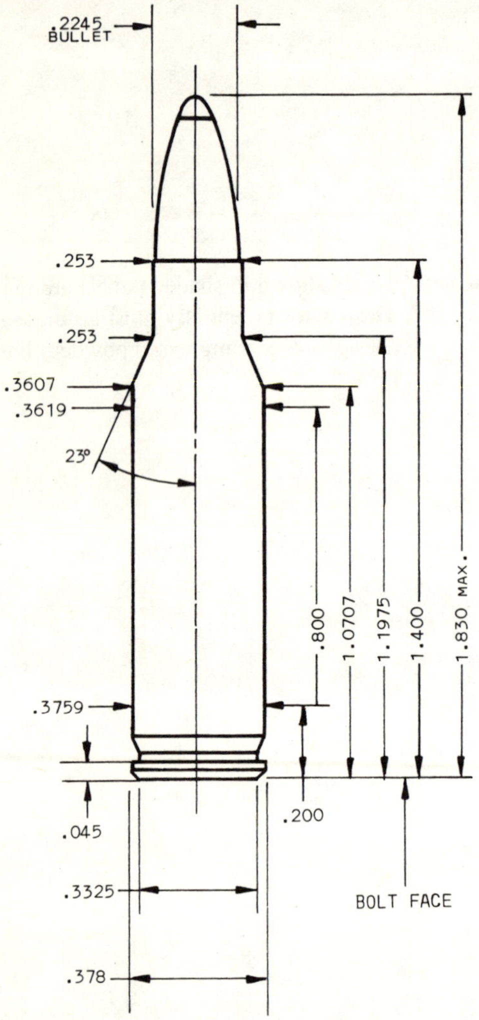


## TEST COMPONENTS

Cases ............................................. Remington
Trim-to Length ........................................ 1.395''
Primers ........................................ Remington 7½
Primer Size ........................................ Small Rifle
Lyman Shell Holder .................................... No. 26
Jacketed Bullets Used ........... Speer SPP #1017, 40 gr.
Remington HP #B22702, 45 gr.
Remington PSP #B22710, 50 gr.
Cast Bullets Used ....................... (size to .224'' dia.)
*Gas Check *#225438, 45 gr.
*#225415, 50 gr.

## TEST SPECIFICATIONS (Velocity Only)

Firearm Used ............................ Remington XP-100
Barrel Length ........................................... 10½''
Twist ........................................................ 1-12''
Groove Diameter ........................................ .224''

**40 gr. Speer SPP**
#1017 (.224'' dia.) 1.780'' OAL

| Powder | Sug. Starting Grains | Velocity F.P.S. | Pressure C.U.P. | Max. Grains | Velocity F.P.S. | Pressure C.U.P. |
|---|---|---|---|---|---|---|
| RX7 | 15.0 | 1921 | — | 18.0 | 2520 | — |
| IMR-4227 | 13.0 | 2141 | — | 16.5 | 2739 | — |
| IMR-4198 | 16.0 | 2232 | — | 18.0+ | 2680 | — |
| IMR-3031 | 16.0 | 1890 | — | 18.5+ | 2237 | — |
| H110 | 8.5 | 1710 | — | 14.0 | 2580 | — |

**45 gr. Remington HP**
#B22702 (.224'' dia.) 1.780'' OAL

| Powder | Sug. Starting Grains | Velocity F.P.S. | Pressure C.U.P. | Max. Grains | Velocity F.P.S. | Pressure C.U.P. |
|---|---|---|---|---|---|---|
| IMR-4227 | 13.0 | 2074 | — | 16.0 | 2631 | — |
| IMR-4198 | 16.0 | 2197 | — | 18.0+ | 2666 | — |
| IMR-3031 | 16.0 | 1862 | — | 18.5+ | 2232 | — |
| RX7 | 7.0 | 923 | — | 17.0 | 2246 | — |
| 748 | 9.0 | 960 | — | 20.0+ | 2107 | — |
| 2400 | 5.5 | 1129 | — | 14.7 | 2465 | — |

**Note:** Loads shown in shaded panels are maximum.
\+ Designates a compressed powder charge.

# .221 REMINGTON FIREBALL
## (Continued)

### 50 gr. Remington PSP
#22710 (.224'' dia.) 1.827'' OAL

| Powder | Sug. Starting Grains | Velocity F.P.S. | Pressure C.U.P. | Max. Grains | Velocity F.P.S. | Pressure C.U.P. |
|---|---|---|---|---|---|---|
| IMR-4227 | 12.0 | 2028 | — | 15.5 | 2512 | — |
| IMR-4198 | 16.0 | 2227 | — | 17.8 | 2610 | — |
| IMR-3031 | 16.0 | 1831 | — | 18.5 | 2202 | — |
| RX7 | 6.0 | 988 | — | 16.5 | 2234 | — |
| 748 | 9.0 | 1044 | — | 19.5 | 1990 | — |
| H322 | 6.0 | 814 | — | 17.5 | 2062 | — |

### 45 gr. Cast Bullet
#225438 (#2 Alloy) 1.675'' OAL

| Powder | Sug. Starting Grains | Velocity F.P.S. | Pressure C.U.P. | Max. Grains | Velocity F.P.S. | Pressure C.U.P. |
|---|---|---|---|---|---|---|
| Red Dot | 2.5 | 1050 | — | 5.0 | 1785 | — |
| Green Dot | 3.0 | 1135 | — | 5.5 | 1800 | — |
| Unique | 3.5 | 1180 | — | 6.5 | 1910 | — |
| Blue Dot | 7.0 | 1752 | — | 9.0 | 2183 | — |
| 2400 | 12.0 | 2173 | — | 15.0 | 2732 | — |
| 700X | 2.5 | 1050 | — | 4.8 | 1570 | — |
| PB | 3.0 | 1090 | — | 5.5 | 1770 | — |
| SR-7625 | 3.5 | 1180 | — | 5.5 | 1775 | — |
| SR-4759 | 9.5 | 1725 | — | *12.5 | 2153 | — |
| IMR-4227 | 13.0 | 2237 | — | 16.0 | 2590 | — |
| 630 | 8.0 | 1745 | — | 11.0 | 2210 | — |
| HS-7 | 7.5 | 1708 | — | 10.0 | 2144 | — |
| AL-8 | 8.5 | 1776 | — | 11.0 | 2171 | — |

### 50 gr. Cast Bullet
#225415 (#2 Alloy) 1.665'' OAL

| Powder | Sug. Starting Grains | Velocity F.P.S. | Pressure C.U.P. | Max. Grains | Velocity F.P.S. | Pressure C.U.P. |
|---|---|---|---|---|---|---|
| Red Dot | 2.5 | 1015 | — | 4.5 | 1440 | — |
| Green Dot | 3.0 | 1105 | — | 5.0 | 1665 | — |
| Unique | 3.5 | 1110 | — | 6.0 | 1755 | — |
| Blue Dot | 7.5 | 1731 | — | 9.7 | 2184 | — |
| 2400 | 12.0 | 2207 | — | 14.5 | 2512 | — |
| 700X | 2.5 | 1005 | — | 4.5 | 1460 | — |
| PB | 3.0 | 1050 | — | 5.3 | 1685 | — |
| SR-7625 | 3.5 | 1115 | — | 5.3 | 1675 | — |
| SR-4759 | 10.5 | 1706 | — | 13.5 | 2181 | — |
| IMR-4227 | 13.0 | 2272 | — | 15.5 | 2475 | — |
| 630 | 8.0 | 1738 | — | 10.5 | 2157 | — |
| HS-7 | 8.0 | 1742 | — | 10.5 | 2212 | — |
| AL-8 | *8.5 | 1699 | — | 11.0 | 2181 | — |

**Note:** Loads shown in shaded panels are maximum.
* Designates potentially most accurate load.
\+ Designates a compressed powder charge.

# .25 A.C.P.

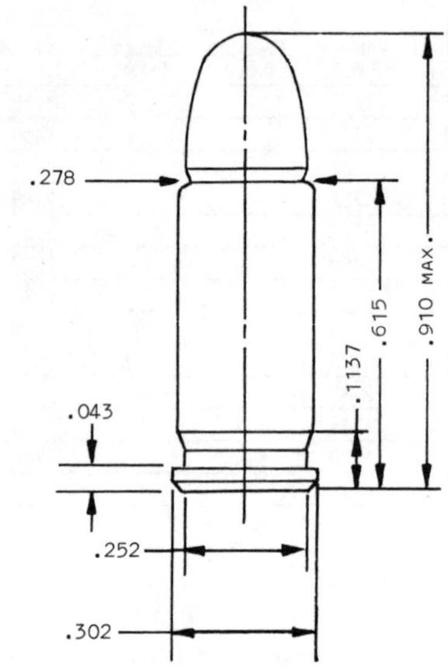


## TEST COMPONENTS

Cases .......................................... Remington-Peters
Trim-to Length .......................................... .610''
Primers ................................... Winchester 1½-108
Primer Size ........................................ Small Pistol
Shell Holder ........................................ RCBS #29
Cast Bullets Used ....................... (size to .251'' dia.)
#252435, 51 gr.

## TEST SPECIFICATIONS (Velocity Only)

Firearm Used ............................... Sterling .25 Auto
Barrel Length ................................................ 2''
Twist ...................................................... 1-16''
Groove Diameter ....................................... .251''

**51 gr. Cast Bullet**
#252435 (#2 Alloy) .910'' OAL

| Powder | Sug. Starting Grains | Velocity F.P.S. | Pressure C.U.P. | Max. Grains | Velocity F.P.S. | Pressure C.U.P. |
|---|---|---|---|---|---|---|
| Bullseye | 0.8 | 460 | — | 1.1 | 560 | — |
| Red Dot | 0.8 | 470 | — | 1.1 | 615 | — |
| Green Dot | 1.0 | 500 | — | 1.3 | 625 | — |
| Unique | 1.2 | 470 | — | 1.6 | 660 | — |
| 700X | 0.8 | 470 | — | 1.1 | 625 | — |
| PB | 1.0 | 495 | — | 1.4 | 680 | — |
| 630 | 2.8 | 595 | — | 3.2 | 695 | — |

**Note:** Loads shown in shaded panels are maximum.

# .256 WINCHESTER

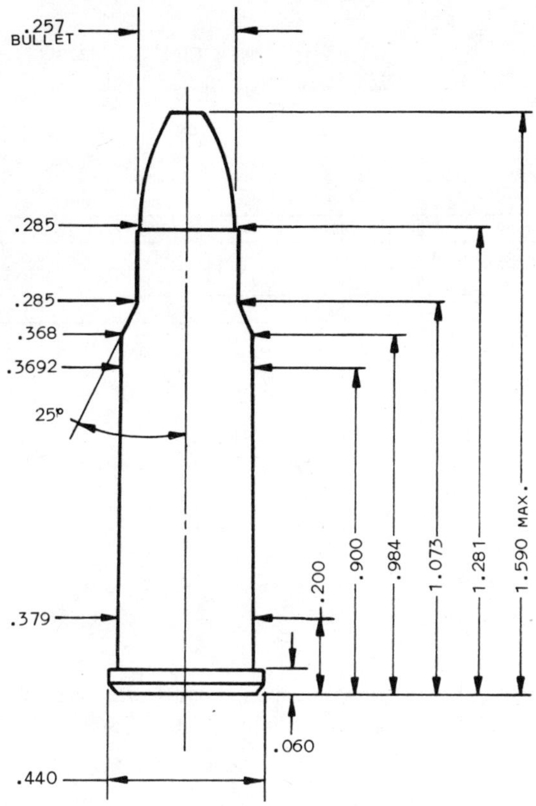


## TEST COMPONENTS

Cases ........ Winchester
Trim-to Length ........ 1.281''
Primers ........ CCI 400
Primer Size ........ Small Rifle
Lyman Shell Holder ........ No. 1
Jacketed Bullets Used ........ Hornady SP, 60 gr.
Sierra HP #1600, 75 gr.
Hornady SPP, 87 gr.
Cast Bullets Used ........ (size to .257'' dia.)
*Gas Check ........ *#257420, 65 gr.
*#257312, 89 gr.

## TEST SPECIFICATIONS (Velocity Only)

Firearm Used ........ Thompson/Center Contender
Barrel Length ........ 10''
Twist ........ 1-14''
Groove Diameter ........ .257''

## COMMENTS

The small case capacity of the cartridge limits suitable reloading powders to a very few. Ballistics and acccuracy, however, are good.

**60 gr. Hornady SP**
.257'' dia./1.563'' OAL

| Powder | Sug. Starting Grains | Velocity F.P.S. | Pressure C.U.P. | Max. Grains | Velocity F.P.S. | Pressure C.U.P. |
|---|---|---|---|---|---|---|
| IMR-4227 | 8.0 | 973 | — | 15.8 | 2108 | — |
| IMR-4198 | 9.3 | 902 | — | 17.8 | 2147 | — |
| 2400 | 6.8 | 873 | — | 13.8 | 2106 | — |
| 680 | 5.0 | 922 | — | 16.5 | 2076 | — |
| H110 | 5.0 | 853 | — | 15.5 | 2258 | — |

**75 gr. Sierra HP**
#1600 (.257'' dia.) 1.730'' OAL

| Powder | Sug. Starting Grains | Velocity F.P.S. | Pressure C.U.P. | Max. Grains | Velocity F.P.S. | Pressure C.U.P. |
|---|---|---|---|---|---|---|
| IMR-4227 | 8.0 | 968 | — | *14.5 | 2019 | — |
| IMR-4198 | 9.5 | 926 | — | 18.0+ | 2133 | — |
| 2400 | 7.0 | 893 | — | 13.7 | 2014 | — |
| Rx7 | 10.5 | 912 | — | 17.5 | 1886 | — |
| 680 | 5.0 | 805 | — | 16.0 | 2061 | — |
| H110 | 6.0 | 761 | — | 15.0 | 2108 | — |

**Note:** Loads shown in shaded panels are maximum.
* Designates potentially most accurate load.
\+ Designates a compressed powder charge.

### 87 gr. Hornady SPP
.257" dia./1.820" OAL

| Powder | Sug. Starting Grains | Velocity F.P.S. | Pressure C.U.P. | Max. Grains | Velocity F.P.S. | Pressure C.U.P. |
|---|---|---|---|---|---|---|
| IMR-4227 | 8.0 | 855 | — | 14.0 | 1804 | — |
| IMR-4198 | 10.0 | 929 | — | 16.0 | 1787 | — |
| 2400 | 7.5 | 932 | — | 12.5 | 1802 | — |
| Rx7 | 9.0 | 818 | — | 16.0 | 1663 | — |
| 680 | 6.0 | 893 | — | 15.5 | 1956 | — |
| H110 | 7.0 | 860 | — | 13.5 | 1894 | — |

### 65 gr. Cast Bullet
#257420 (Linotype) 1.525" OAL

| Powder | Sug. Starting Grains | Velocity F.P.S. | Pressure C.U.P. | Max. Grains | Velocity F.P.S. | Pressure C.U.P. |
|---|---|---|---|---|---|---|
| SR-4756 | 4.0 | 962 | — | 8.0 | 1876 | — |
| SR-4759 | 7.0 | 1031 | — | 10.0 | 1528 | — |
| Blue Dot | 4.3 | 897 | — | 9.0 | 1864 | — |
| HS-7 | 4.5 | 920 | — | 10.0 | 1857 | — |
| AL-8 | 4.3 | 857 | — | 9.0 | 1718 | — |

### 89 gr. Cast Bullet
#257312 (Linotype) 1.600" OAL

| Powder | Sug. Starting Grains | Velocity F.P.S. | Pressure C.U.P. | Max. Grains | Velocity F.P.S. | Pressure C.U.P. |
|---|---|---|---|---|---|---|
| SR-4756 | 3.8 | 874 | — | *7.5 | 1624 | — |
| SR-4759 | 6.8 | 949 | — | 9.5 | 1420 | — |
| Blue Dot | 4.0 | 820 | — | 9.5 | 1805 | — |
| HS-7 | 4.2 | 829 | — | 9.5 | 1657 | — |
| AL-8 | 4.3 | 802 | — | 8.5 | 1539 | — |

**Note:** Loads shown in shaded panels are maximum.
* Designates potentially most accurate load.

# .30 LUGER

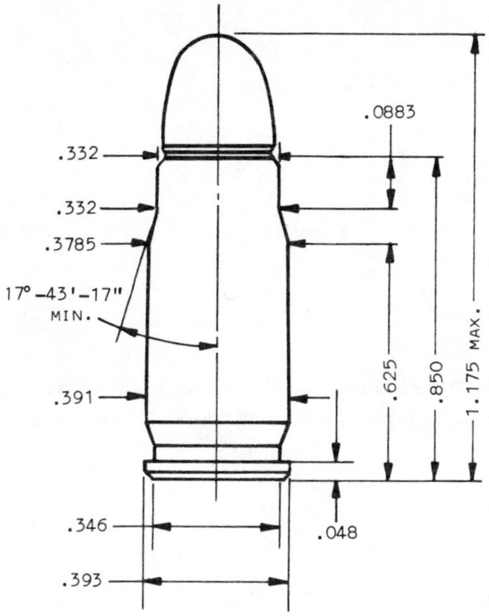


## TEST COMPONENTS

Cases .............................................. Remington
Trim-to Length .......................................... .845''
Primers ....................................... Remington 1½
Primer Size ....................................... Small Pistol
Lyman Shell Holder ................................... No. 12
Jacketed Bullets Used ............................ Remington
MC #B22764, 93 gr.
Cast Bullets Used ....................... (size to .309'' dia.)
#313249, 84 gr.
#313226, 93 gr.

## TEST SPECIFICATIONS (Velocity Only)

Firearm Used .................................... Walther P-38
Barrel Length ......................................... 4 15/16''
Twist .................................................... 1-9.85''
Groove Diameter ....................................... .3085''

## COMMENTS

Variations in groove diameters will be encountered with pistols chambered for this cartridge. For best results, it is recommended that you slug your barrel. Size cast bullets as near to the exact groove diameter as possible. When using cast bullet No. 313249, the case should be crimped on the leading edge of the first driving band.

With some pistols, the starting loads may not function the action. In such an instance, the load should be increased, but with caution.

**93 gr. Remington MC**
#B22764 (.308'' dia.) 1.175'' OAL

| Powder | Sug. Starting Grains | Velocity F.P.S. | Pressure C.U.P. | Max. Grains | Velocity F.P.S. | Pressure C.U.P. |
|---|---|---|---|---|---|---|
| 700X | 3.6 | 1065 | — | *4.1 | 1171 | — |
| SR-4756 | 5.0 | 1038 | — | 5.5 | 1161 | — |
| Bullseye | 3.5 | 968 | — | 4.0 | 1067 | — |
| Unique | 4.5 | 1003 | — | 5.0 | 1115 | — |
| 231 | 3.7 | 935 | — | 4.2 | 1046 | — |
| HP38 | 3.5 | 1028 | — | 4.0 | 1115 | — |
| AL-5 | 4.5 | 837 | — | 5.4 | 1004 | — |

**84 gr. Cast Bullet**
#313249 (#2 Alloy) 1.125'' OAL

| Powder | Sug. Starting Grains | Velocity F.P.S. | Pressure C.U.P. | Max. Grains | Velocity F.P.S. | Pressure C.U.P. |
|---|---|---|---|---|---|---|
| 700X | 3.7 | 1143 | — | 4.2 | 1204 | — |
| PB | 4.0 | 1105 | — | 4.5 | 1177 | — |
| SR-4756 | 4.7 | 966 | — | 5.6 | 1182 | — |
| Bullseye | 3.8 | 1106 | — | 4.5 | 1244 | — |
| Red Dot | 3.8 | 1116 | — | 4.3 | 1201 | — |
| Green Dot | 4.2 | 1111 | — | 4.7 | 1210 | — |
| Unique | 4.6 | 1110 | — | 5.2 | 1226 | — |
| 231 | 3.9 | 1054 | — | 4.4 | 1144 | — |
| HS-5 | 5.4 | 1044 | — | *6.0 | 1146 | — |

**Note:** Loads shown in shaded panels are maximum.
* Designates potentially most accurate load.

# .30 LUGER
(Continued)

### 93 gr. Cast Bullet
#313226 (#2 Alloy) 1.150'' OAL

| Powder | Sug. Starting Grains | Velocity F.P.S. | Pressure C.U.P. | Max. Grains | Velocity F.P.S. | Pressure C.U.P. |
|---|---|---|---|---|---|---|
| Z00X | 3.5 | 1037 | — | 4.0 | 1128 | — |
| PB | 3.9 | 1047 | — | 4.4 | 1126 | — |
| SR-4756 | 4.6 | 960 | — | 5.1 | 1094 | — |
| Bullseye | 3.6 | 1026 | — | 4.4 | 1159 | — |
| Red Dot | 3.7 | 1065 | — | 4.2 | 1152 | — |
| Green Dot | 4.1 | 1065 | — | 4.6 | 1150 | — |
| Unique | 4.5 | 1046 | — | 5.1 | 1163 | — |
| 630 | 7.2 | 1091 | — | 8.2 | 1205 | — |
| HS-5 | 5.3 | 1062 | — | *5.8 | 1134 | — |

**Note:** Loads shown in shaded panels are maximum.
* Designates potentially most accurate load.

# .30 MAUSER
## (7.63mm MAUSER)

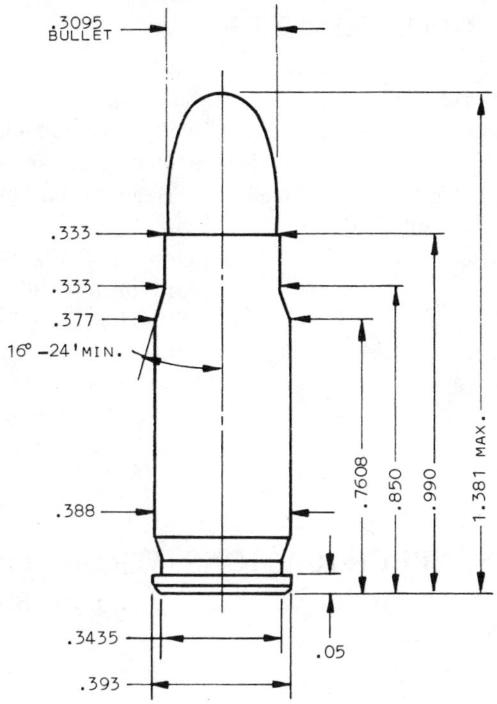


### TEST COMPONENTS

Cases .......... Remington
Trim-to Length .......... .985''
Primers .......... Remington 1½
Primer Size .......... Small Pistol
Lyman Shell Holder .......... No. 12
Cast Bullets Used .......... (size to .311'' dia.) #313249, 84 gr.

### TEST SPECIFICATIONS (Velocity Only)

Firearm Used .......... Military Mauser
Barrel Length .......... 5½''
Twist .......... 1-8''
Groove diameter .......... .311''

### COMMENTS

Variations in groove diameter may be encountered. Slug your barrel for best results. If your barrel is larger than .309'', we suggest that you use cast bullets sized as close to exact groove diameter as possible.

**84 gr. Cast Bullet**
#313249 (#2 Alloy) 1.381'' OAL

| Powder | Sug. Starting Grains | Velocity F.P.S. | Pressure C.U.P. | Max. Grains | Velocity F.P.S. | Pressure C.U.P. |
|---|---|---|---|---|---|---|
| Bullseye | 4.2 | 1128 | — | 5.0 | 1243 | — |
| Red Dot | 5.0 | 1065 | — | 5.6 | 1115 | — |
| Unique | 5.2 | 1018 | — | 6.8 | 1193 | — |

**Note:** Loads shown in shaded panels are maximum.

# .30 M1 CARBINE (REVOLVER)

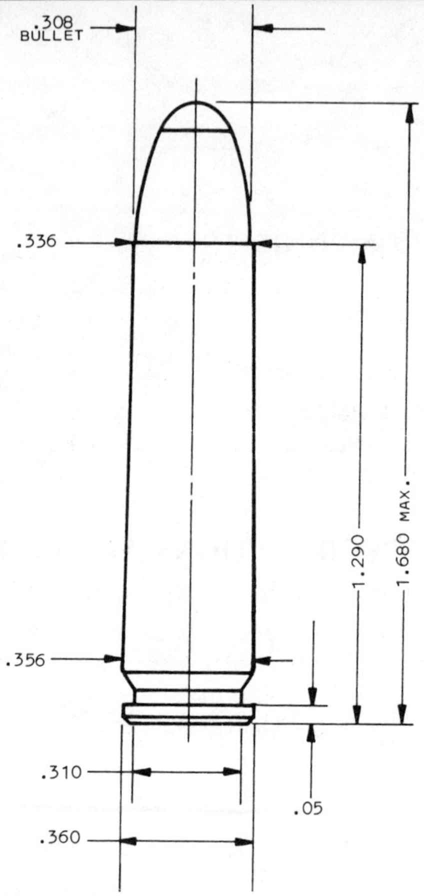


## TEST COMPONENTS

Cases .................................................... Federal
Trim-to Length ......................................... 1.286''
Primers ..................... CCI 450 (jacketed bullets only)
Remington 6½ (cast bullets only)
Primer Size ......... Small Rifle Magnum and Small Rifle
Lyman Shell Holder .................................... No. 19
Jacketed Bullets Used .. Remington MC #B22764, 93 gr.
Speer Plinker #1805, 100 gr.
Hornady HP, 110 gr.
Cast Bullets Used ....................... (size to .308'' dia.)
*Gas Check *#311359, 120 gr.
*#311576, 123 gr.
#311410, 130 gr.

## TEST SPECIFICATIONS (Velocity Only)

Firearm Used ............................... Ruger Blackhawk
Barrel Length ............................................. 7½''
Twist ..................................................... 1-20''
Groove Diameter ........................................ .308''

## COMMENTS

This cartridge headspaces on the mouth and case length is, therefore, critical. Never trim cases shorter than the trim-to length shown and never crimp bullets. Refer to Opening Remarks when using Winchester 296 powder.

**93 gr. Remington MC**
#B22764 (.308'' dia.) 1.595'' OAL

| Powder | Sug. Starting Grains | Velocity F.P.S. | Pressure C.U.P. | Max. Grains | Velocity F.P.S. | Pressure C.U.P. |
|---|---|---|---|---|---|---|
| SR-4759 | 9.5 | 948 | — | 12.5 | 1216 | — |
| IMR-4227 | 10.0 | 1076 | — | 15.0 | 1491 | — |
| Blue Dot | 6.5 | 922 | — | 8.5 | 1284 | — |
| 2400 | 9.5 | 1078 | — | 12.5 | 1390 | — |
| **296 | — | — | — | 15.0 | 1460 | — |
| H110 | 10.0 | 1176 | — | 15.5 | 1456 | — |
| AL-8 | 8.5 | 1227 | — | 12.0 | 1510 | — |

**100 gr. Speer Plinker**
#1805 (.308'' dia.) 1.600'' OAL

| Powder | Sug. Starting Grains | Velocity F.P.S. | Pressure C.U.P. | Max. Grains | Velocity F.P.S. | Pressure C.U.P. |
|---|---|---|---|---|---|---|
| 2400 | 9.5 | 1147 | — | 12.5 | 1412 | — |
| SR-4759 | 9.0 | 918 | — | 12.0+ | 1248 | — |
| IMR-4227 | 10.0 | 1112 | — | 15.0+ | 1438 | — |
| 630 | *8.5 | 1115 | — | 12.0 | 1414 | — |
| H110 | 10.0 | 1124 | — | 14.5 | 1416 | — |

**Note:** Loads shown in shaded panels are maximum.
* Designates potentially most accurate load.
\+ Designates a compressed powder charge.
** Reduced loads not recommended by Winchester.

# .30 M1 CARBINE (REVOLVER)
## (Continued)

### 110 gr. Hornady HP
.308" dia./1.660" OAL

| Powder | Sug. Starting Grains | Velocity F.P.S. | Pressure C.U.P. | Max. Grains | Velocity F.P.S. | Pressure C.U.P. |
|---|---|---|---|---|---|---|
| 2400 | 10.0 | 1161 | — | 12.0 | 1348 | — |
| SR-4759 | 9.0 | 851 | — | 12.0+ | 1172 | — |
| IMR-4227 | 11.0 | 1188 | — | 14.0 | 1377 | — |
| **296 | — | — | — | 14.5 | 1428 | — |
| H110 | 9.5 | 1054 | — | 14.0 | 1369 | — |

### 120 gr. Cast Bullet
#311359 (#2 Alloy) 1.680" OAL

| Powder | Sug. Starting Grains | Velocity F.P.S. | Pressure C.U.P. | Max. Grains | Velocity F.P.S. | Pressure C.U.P. |
|---|---|---|---|---|---|---|
| Unique | 5.5 | 1138 | — | 6.5 | 1335 | — |
| Blue Dot | 6.8 | 1066 | — | 8.8 | 1339 | — |
| 2400 | 11.0 | 1236 | — | 12.5+ | 1436 | — |
| SR-7625 | 4.6 | 1027 | — | 5.6 | 1162 | — |
| SR-4756 | 5.5 | 1050 | — | 6.5 | 1220 | — |
| IMR-4227 | 11.5 | 1169 | — | 13.3+ | 1388 | — |
| **296 | — | — | — | 13.5 | 1415 | — |
| H110 | 10.0 | 1055 | — | 13.5 | 1358 | — |
| AL-8 | 9.5 | 1239 | — | 11.0+ | 1408 | — |

### 123 gr. Cast Bullet
#311576 (#2 Alloy) 1.610" OAL

| Powder | Sug. Starting Grains | Velocity F.P.S. | Pressure C.U.P. | Max. Grains | Velocity F.P.S. | Pressure C.U.P. |
|---|---|---|---|---|---|---|
| Unique | 5.0 | 1017 | — | 6.3 | 1316 | — |
| Blue Dot | 6.5 | 997 | — | 8.8 | 1290 | — |
| 2400 | 10.8 | 1271 | — | 12.3+ | 1445 | — |
| SR-4756 | 5.1 | 923 | — | 6.1 | 1172 | — |
| IMR-4227 | 11.0 | 1124 | — | 13.1+ | 1342 | — |
| **296 | — | — | — | 13.5 | 1431 | — |
| H110 | 10.0 | 1037 | — | 13.5 | 1420 | — |
| AL-8 | 9.2 | 1196 | — | 10.8+ | 1384 | — |

### 130 gr. Cast Bullet
#311410 (#2 Alloy) 1.680" OAL

| Powder | Sug. Starting Grains | Velocity F.P.S. | Pressure C.U.P. | Max. Grains | Velocity F.P.S. | Pressure C.U.P. |
|---|---|---|---|---|---|---|
| Unique | 4.9 | 1064 | — | 6.1 | 1270 | — |
| Blue Dot | 6.3 | 971 | — | 8.2 | 1249 | — |
| 2400 | 10.5 | 1282 | — | 12.0+ | 1414 | — |
| SR-4756 | 5.0 | 1011 | — | 6.0 | 1182 | — |
| IMR-4227 | 10.6 | 1116 | — | 12.7+ | 1387 | — |
| **296 | — | — | — | 12.5 | 1356 | — |
| H110 | 9.0 | 979 | — | 12.5 | 1350 | — |
| AL-8 | 9.0 | 1225 | — | 10.4+ | 1346 | — |

**Note:** Loads shown in shaded panels are maximum.
* Designates potentially most accurate load.
+ Designates a compressed powder charge.
** Reduced loads not recommended by Winchester.

# .32 A.C.P.
## (7.65mm BROWNING)

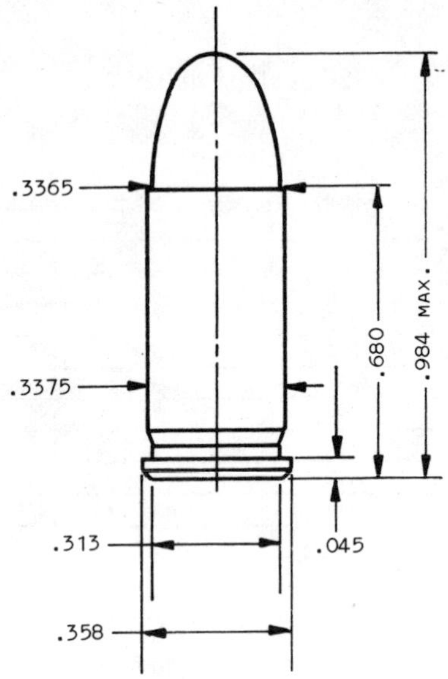


## TEST COMPONENTS

Cases ................................ Remington; Winchester
Trim-to Length ........................................ .672''
Primers ................ Remington 1½; Winchester 1½-108
Primer Size ........................................ Small Pistol
Lyman Shell Holder ..................................... No. 9
Jacketed Bullets Used ........................ Remington MC #B22816, 71 gr.
Cast Bullets Used ........................ (size to .309'' dia.) #311252, 77 gr. #313249, 84 gr.

## TEST SPECIFICATIONS (Velocity and Pressure)

Firearm Used .................................... Mauser HSc; also Special Pressure Test Barrel
Barrel Length ..................................... 3''(Mauser); 3¼''(Special Pressure Barrel)
Twist ...................................................... 1-16''
Groove Diameter ........................................ .309''

## COMMENTS

A wide variation in groove diameters is present in handguns chambered for this cartridge. The loads listed may be used with .311'' jacketed bullets in pistols which have a groove diameter as small as .309''.

**71 gr. Remington MC**
#B22816 (.311'' dia.) .984'' OAL

| Powder | Sug. Starting Grains | Velocity F.P.S. | Pressure C.U.P. | Max. Grains | Velocity F.P.S. | Pressure C.U.P. |
|---|---|---|---|---|---|---|
| **700X | 1.7 | 706 | 9,500 | ‡2.2 | 902 | 14,700 |
| ++Bullseye | 1.5 | 603 | — | 2.2 | 825 | — |
| ++Red Dot | 1.7 | 705 | — | 2.5 | 937 | — |
| ++Unique | 2.0 | 571 | — | 3.1 | 945 | — |
| **630 | 3.6 | 617 | 6,900 | 5.2 | 897 | 14,000 |
| **HP-38 | 1.7 | 733 | 11,100 | 2.2 | 883 | 14,900 |

**77 gr. Cast Bullet**
#311252 (#2 Alloy) .975'' OAL

| Powder | Sug. Starting Grains | Velocity F.P.S. | Pressure C.U.P. | Max. Grains | Velocity F.P.S. | Pressure C.U.P. |
|---|---|---|---|---|---|---|
| 700X | 1.5 | 750 | — | 1.9 | 895 | — |
| PB | 1.6 | 665 | — | 2.2 | 870 | — |
| SR-7625 | 1.7 | 675 | — | 2.3 | 860 | — |
| Bullseye | 1.5 | 685 | — | 2.0 | 830 | — |
| Red Dot | 1.5 | 695 | — | 2.3 | 940 | — |
| Green Dot | 1.6 | 705 | — | 2.4 | 895 | — |
| Unique | 1.8 | 625 | — | 2.6 | 860 | — |
| **231 | 1.5 | 620 | 7,200 | 2.5 | 888 | 15,000 |
| **HP38 | 1.5 | 688 | 9,200 | *2.0 | 824 | 13,800 |

**84 gr. Cast Bullet**
#313249 (#2 Alloy) .984'' OAL

| Powder | Sug. Starting Grains | Velocity F.P.S. | Pressure C.U.P. | Max. Grains | Velocity F.P.S. | Pressure C.U.P. |
|---|---|---|---|---|---|---|
| 700X | 1.3 | 685 | — | 1.7 | 830 | — |
| PB | 1.5 | 665 | — | 2.1 | 830 | — |
| SR-7625 | 1.7 | 675 | — | 2.2 | 860 | — |
| Bullseye | 1.5 | 660 | — | 1.9 | 795 | — |
| Red Dot | 1.5 | 665 | — | 2.1 | 895 | — |
| Green Dot | 1.6 | 680 | — | 2.3 | 845 | — |
| Unique | 1.8 | 650 | — | 2.5 | 860 | — |
| **630 | 3.5 | 649 | 7,700 | 5.1 | 882 | 14,400 |
| **Trap 100 | 2.0 | 664 | 8,700 | *2.8 | 814 | 13,700 |

**Note:** Loads shown in shaded panels are maximum.
* Designates potentially most accurate load.
‡ Designates factory velocity duplication load.
** Designates Winchester cases, primers, and special pressure test barrel.
++ Designates Remington cases and primers.

# .32 SMITH & WESSON

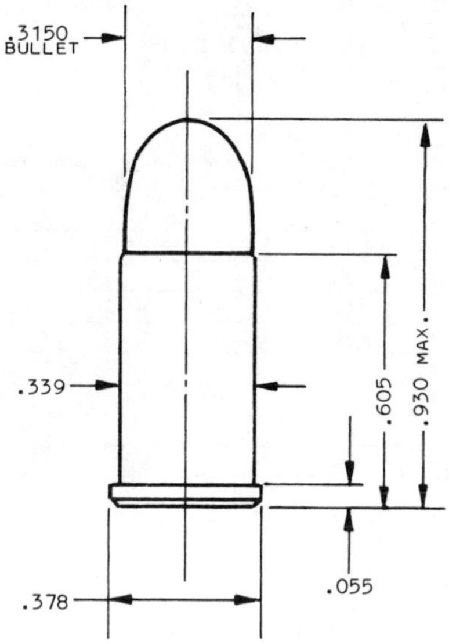


## TEST COMPONENTS

Cases ................ Remington; Norma (cast bullets only)
Trim-to Length ........................................... .600''
Primers ......................................... Remington 1½; Winchester 1½-108 (cast bullets only)
Primer Size ....................................... Small Pistol
Lyman Shell Holder ..................................... No. 9
Jacketed Bullets Used ... Remington MC #B22816, 71 gr. Remington SP #B22810, 100 gr.
Cast Bullets Used ........................ (size to .310'' dia.) #311252, 77 gr. #313249, 84 gr.

## TEST SPECIFICATIONS (Velocity Only)

Firearm Used .................... Smith & Wesson Model 31
Barrel Length ................................................. 4''
Twist .................................................... 1-18¾''
Groove Diameter ........................................ .312''

## COMMENTS

Use maximum loads only in solid frame revolvers. For top-break-models (in good condition) use only the suggested starting loads. Do not go higher as these pistols are of a relatively weak design.
Variations in groove diameter may be encountered. Slug your barrel before reloading and size your cast bullets to as near groove diameter as possible.

**71 gr. Remington MC**
#B22816 (.311'' dia.) .897'' OAL

| Powder | Sug. Starting Grains | Velocity F.P.S. | Pressure C.U.P. | Max. Grains | Velocity F.P.S. | Pressure C.U.P. |
|---|---|---|---|---|---|---|
| 700X | 2.0 | 562 | — | 2.5 | 722 | — |
| SR-7625 | 2.4 | 522 | — | 2.9 | 674 | — |
| Bullseye | 2.0 | 515 | — | 2.5 | 671 | — |
| Unique | 2.7 | 482 | — | 3.2+ | 650 | — |
| 231 | 2.4 | 568 | — | 2.9 | 711 | — |
| HP38 | 2.0 | 562 | — | 2.5 | 710 | — |

**100 gr. Remington SP**
#B22810 (.310'' dia.) .897'' OAL

| Powder | Sug. Starting Grains | Velocity F.P.S. | Pressure C.U.P. | Max. Grains | Velocity F.P.S. | Pressure C.U.P. |
|---|---|---|---|---|---|---|
| 700X | 2.0 | 625 | — | 2.5 | 787 | — |
| SR-7625 | 2.0 | 536 | — | 2.5 | 682 | — |
| Bullseye | 1.9 | 578 | — | 2.5 | 736 | — |
| Unique | 2.5 | 556 | — | 3.0 | 712 | — |
| 630 | 4.0 | 562 | — | 4.7 | 667 | — |
| Trap 100 | 2.9 | 595 | — | 3.5+ | 695 | — |

**Note:** Loads shown in shaded panels are maximum.
* Designates potentially most accurate load.
\+ Designates a compressed powder charge.
‡ Designates factory velocity duplication load.
** Designates the use of Remington cases and primers.

# .32 SMITH & WESSON
## (Continued)

**77 gr. Cast Bullet**
**#311252 (#2 Alloy) .860" OAL**

| Powder | Sug. Starting Grains | Velocity F.P.S. | Pressure C.U.P. | Max. Grains | Velocity F.P.S. | Pressure C.U.P. |
|---|---|---|---|---|---|---|
| 700X | 1.3 | 598 | — | 1.8 | 758 | — |
| PB | 1.4 | 523 | — | 2.1 | 803 | — |
| Bullseye | 1.1 | 478 | — | 1.5 | 578 | — |
| Red Dot | 1.4 | 578 | — | 1.9 | 763 | — |
| Green Dot | 1.5 | 553 | — | 2.0 | 723 | — |
| Unique | 1.6 | 493 | — | 2.5 | 743 | — |
| **231 | 1.9 | 641 | — | *2.5 | 800 | — |
| **HP38 | 1.4 | 642 | — | 2.2 | 821 | — |
| **AL-120 | 1.3 | 537 | — | 1.8 | 721 | — |

**84 gr. Cast Bullet**
**#313249 (#2 Alloy) .860" OAL**

| Powder | Sug. Starting Grains | Velocity F.P.S. | Pressure C.U.P. | Max. Grains | Velocity F.P.S. | Pressure C.U.P. |
|---|---|---|---|---|---|---|
| 700X | 1.2 | 613 | — | 1.7 | 798 | — |
| PB | 1.3 | 543 | — | 1.9 | 768 | — |
| **SR-7625 | 1.6 | 625 | — | 2.1 | 760 | — |
| Bullseye | 1.0 | 468 | — | 1.4 | 578 | — |
| Red Dot | 1.3 | 588 | — | 1.7 | 703 | — |
| Green Dot | ‡1.4 | 528 | — | 1.9 | 728 | — |
| Unique | 1.5 | 493 | — | 2.4 | 753 | — |
| **630 | 2.5 | 485 | — | 4.0 | 726 | — |
| **Trap 100 | 1.6 | 558 | — | 2.0 | 645 | — |

**Note:** Loads shown in shaded panels are maximum.
* Designates potentially most accurate load.
‡ Designates factory velocity duplication load.
** Designates the use of Remington cases and primers.

# .32 SMITH & WESSON LONG
## (.32 COLT NEW POLICE)

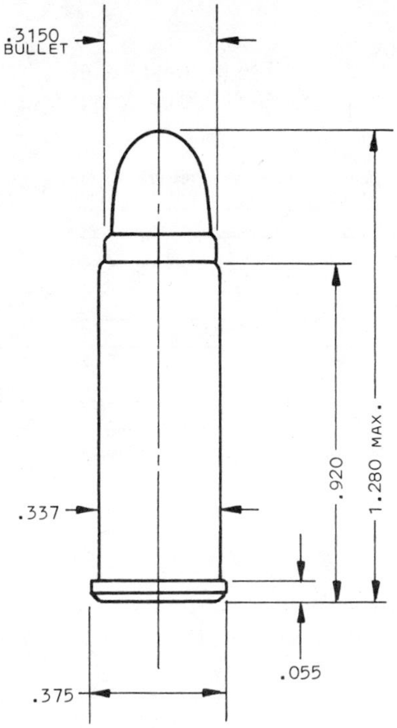


### TEST COMPONENTS

Cases .................................................. Remington
Trim-to Length ........................................... .915''
Primers ......................................... Remington 1½
Primer Size ........................................ Small Pistol
Lyman Shell Holder ..................................... No. 9
Jacketed Bullets Used .......................... Remington SP
#B22810, 100 gr.
Cast Bullets Used ......................... (size to .312'' dia.)
#311252, 77 gr.
#313226, 93 gr.
#3118, 115 gr.

### TEST SPECIFICATIONS (Velocity Only)

Firearm Used .................... Smith & Wesson Model 31
Barrel Length ................................................. 4''
Twist ................................................... 1-18¾''
Groove Diameter ........................................ .312''

### COMMENTS

Use maximum loads only in solid frame revolvers. For top break models (in good condition) use only the suggested starting loads. Do not go higher because these pistols are of a relatively weak design.

Variations in groove diameter may be encountered. Slug your barrel before reloading and size your cast bullets to as near groove diameter as possible.

**100 gr. Remington SP**
#B22810 (.310" dia.) 1.280" OAL

| Powder | Sug. Starting Grains | Velocity F.P.S. | Pressure C.U.P. | Max. Grains | Velocity F.P.S. | Pressure C.U.P. |
|---|---|---|---|---|---|---|
| PB | Do not use reduced load | | | 3.3 | 733 | — |
| SR-4756 | 3.7 | 677 | — | 4.1 | 761 | — |
| Bullseye | 2.4 | 624 | — | 3.0 | 781 | — |
| Unique | 3.5 | 674 | — | 4.0 | 779 | — |
| 630 | 4.9 | 626 | — | 5.5 | 737 | — |
| HS-5 | 4.0 | 680 | — | 4.8 | 838 | — |
| AL-5 | 3.8 | 649 | — | 4.5 | 784 | — |

**77 gr. Cast Bullet**
#311252 (#2 Alloy) 1.115" OAL

| Powder | Sug. Starting Grains | Velocity F.P.S. | Pressure C.U.P. | Max. Grains | Velocity F.P.S. | Pressure C.U.P. |
|---|---|---|---|---|---|---|
| Bullseye | 1.5 | 513 | — | 3.0 | 958 | — |
| Red Dot | 1.8 | 588 | — | 3.3 | 1028 | — |
| Green Dot | 1.9 | 563 | — | 3.4 | 993 | — |
| Unique | 2.5 | 608 | — | 4.0 | 943 | — |
| 700X | 1.8 | 598 | — | 3.3 | 1098 | — |
| SR-7625 | 2.4 | 583 | — | 3.6 | 988 | — |
| 231 | 2.0 | 677 | — | 3.5 | 1043 | — |
| Trap 100 | 1.7 | 645 | — | 3.5 | 1003 | — |
| AL-120 | 1.9 | 635 | — | 3.5 | 1071 | — |

**Note:** Loads shown in shaded panels are maximum.
* Designates potentially most accurate load.

# .32 SMITH & WESSON LONG
(Continued)

**93 gr. Cast Bullet**
#313226 (#2 Alloy) 1.185'' OAL

| Powder | Sug. Starting Grains | Velocity F.P.S. | Pressure C.U.P. | Max. Grains | Velocity F.P.S. | Pressure C.U.P. |
|---|---|---|---|---|---|---|
| Bullseye | 1.5 | 518 | — | 2.9 | 898 | — |
| Red Dot | 1.7 | 548 | — | 3.0 | 943 | — |
| Green Dot | 1.8 | 528 | — | 3.1 | 893 | — |
| Unique | 2.4 | 543 | — | 3.7 | 883 | — |
| 700X | 1.6 | 558 | — | 3.0 | 983 | — |
| SR-7625 | 2.2 | 553 | — | 3.3 | 953 | — |
| 231 | 2.0 | 639 | — | 3.7 | 1064 | — |
| HP38 | 1.9 | 738 | — | 3.3 | 1038 | — |
| AL-120 | 2.0 | 657 | — | 3.5 | 1070 | — |

**115 gr. Cast Bullet**
#3118 (#2 Alloy) 1.180'' OAL

| Powder | Sug. Starting Grains | Velocity F.P.S. | Pressure C.U.P. | Max. Grains | Velocity F.P.S. | Pressure C.U.P. |
|---|---|---|---|---|---|---|
| Bullseye | 1.4 | 528 | — | 2.0 | 683 | — |
| Red Dot | 1.6 | 553 | — | 2.5 | 783 | — |
| Green Dot | 1.7 | 538 | — | 2.6 | 783 | — |
| Unique | 2.1 | 558 | — | 3.5 | 863 | — |
| 700X | 1.5 | 548 | — | 2.4 | 818 | — |
| SR-7625 | 1.9 | 538 | — | 3.0 | 863 | — |
| 630 | 4.0 | 611 | — | *7.0 | 1046 | — |
| HS-5 | 3.0 | 671 | — | 5.3 | 1102 | — |
| AL-5 | 3.0 | 644 | — | 5.1 | 1003 | — |

**Note:** Loads shown in shaded panels are maximum.
* Designates potentially most accurate load.

# 9mm LUGER
## (9mm PARABELLUM)

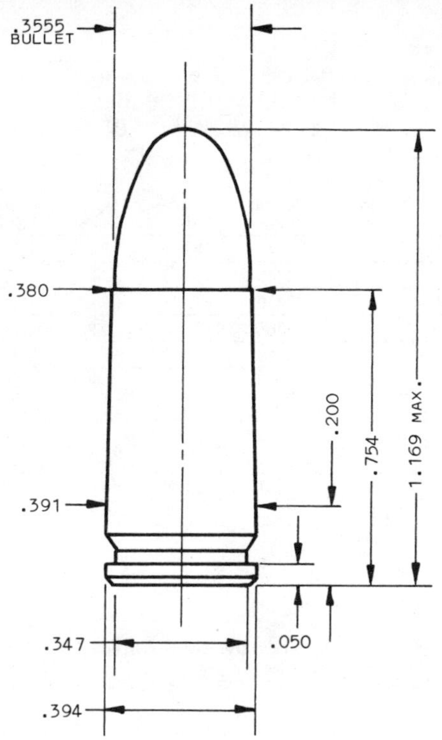


### TEST COMPONENTS

Cases ..................................... Federal; Winchester
Trim-to Length ............................................ .751''
Primers ................................................ CCI 500; Winchester 1½-108
Primer Size ....................................... Small Pistol
Lyman Shell Holder .................................... No. 12
Jacketed Bullets Used ... Remington MC #B22844, 95 gr. Hornady HP, 115 gr. Speer SP #4005, 125 gr.
Cast Bullets Used ......................... (size to .358'' dia.) #358242, 92 gr. #358345, 115 gr. #356402, 121 gr.

### TEST SPECIFICATIONS (Velocity and Pressure)

Firearm Used ............................. Universal Receiver
Barrel Length .................................................. 4''
Twist ........................................................ 1-10''
Groove Diameter ........................................ .358''

### COMMENTS

Variations in groove diameter are often encountered with pistols chambered for this cartridge. However, the most common diameter is .354''. Jacketed bullets of .355'' may be used safely in all barrels ranging from .354'' to .356''. This particular cartridge headspaces on the case mouth and, therefore, the case should not be crimped. Also, the trim-to length should be adhered to closely. A short case or a crimp on the mouth can cause problems with headspace. Our most popular cast bullet for this caliber is No. 356402.

**95 gr. Remington MC**
#B22844 (.355'' dia.) 1.075'' OAL

| Powder | Sug. Starting Grains | Velocity F.P.S. | Pressure C.U.P. | Max. Grains | Velocity F.P.S. | Pressure C.U.P. |
|---|---|---|---|---|---|---|
| 700X | 3.5 | 1013 | 15,900 | 4.5 | 1259 | 31,000 |
| SR-4756 | 4.5 | 1001 | 18,000 | 6.4+ | 1303 | 33,000 |
| Bullseye | 3.8 | 1046 | 17,100 | 5.3 | 1318 | 31,200 |
| **Unique | 5.0 | 1068 | 19,200 | 6.2 | 1345 | 31,700 |
| Blue Dot | 6.5 | 1022 | 19,200 | 8.9+ | 1326 | 30,800 |
| 630 | 8.0 | 1055 | 18,100 | *11.0+ | 1345 | 30,100 |
| HS-6 | 6.2 | 1035 | 16,700 | 7.9 | 1365 | 31,900 |

**115 gr. Hornady HP**
.355'' dia./1.090'' OAL

| Powder | Sug. Starting Grains | Velocity F.P.S. | Pressure C.U.P. | Max. Grains | Velocity F.P.S. | Pressure C.U.P. |
|---|---|---|---|---|---|---|
| 700X | 3.1 | 918 | 20,200 | *4.5 | 1171 | 31,700 |
| SR-4756 | 4.5 | 946 | 18,600 | ‡6.2+ | 1147 | 28,100 |
| Bullseye | 3.5 | 945 | 18,700 | 4.8 | 1184 | 31,700 |
| Unique | 4.4 | 996 | 20,600 | 5.8 | 1233 | 30,700 |
| Blue Dot | 6.3 | 910 | 17,100 | 8.7+ | 1211 | 29,200 |
| 630 | 7.5 | 942 | 16,700 | 10.2+ | 1219 | 30,200 |
| HS-6 | 5.7 | 956 | 17,400 | 7.2 | 1193 | 30,400 |

**Note:** Loads shown in shaded panels are maximum.
* Designates potentially most accurate load.
\+ Designates a compressed powder charge.
‡ Designates factory velocity duplication load.
** Designates use of Winchester cases and primers.

# 9mm LUGER
(Continued)

### 125 gr. Speer SP
#4005 (.355" dia.) 1.120" OAL

| Powder | Sug. Starting Grains | Velocity F.P.S. | Pressure C.U.P. | Max. Grains | Velocity F.P.S. | Pressure C.U.P. |
|---|---|---|---|---|---|---|
| 700X | 3.0 | 835 | 19,300 | 4.4 | 1095 | 32,000 |
| Bullseye | 3.3 | 876 | 18,700 | 4.8 | 1142 | 32,500 |
| Unique | 4.1 | 859 | 18,200 | 5.3+ | 1118 | 31,100 |
| Blue Dot | 6.1 | 895 | 18,200 | 8.0+ | 1190 | 32,600 |
| 630 | 7.3 | 900 | 18,500 | 10.1 | 1191 | 32,700 |
| HS-6 | *5.4 | 901 | 19,300 | 6.8 | 1144 | 31,100 |
| AL-5 | 5.0 | 879 | 19,200 | 6.9 | 1149 | 31,600 |

### 92 gr. Cast Bullet
#358242 (#2 Alloy) 1.045" OAL

| Powder | Sug. Starting Grains | Velocity F.P.S. | Pressure C.U.P. | Max. Grains | Velocity F.P.S. | Pressure C.U.P. |
|---|---|---|---|---|---|---|
| 700X | 4.1 | 1125 | 22,800 | 5.1 | 1285 | 31,000 |
| PB | 4.8 | 1115 | 19,800 | 6.0 | 1360 | 33,000 |
| SR-7625 | 5.0 | 1180 | 23,400 | 5.9 | 1310 | 30,500 |
| SR-4756 | 5.3 | 1168 | 22,400 | 7.2 | 1380 | 30,500 |
| Bullseye | 4.0 | 1060 | 19,800 | 5.2 | 1315 | 32,000 |
| Red Dot | 4.1 | 1120 | 22,800 | 5.5 | 1340 | 32,500 |
| Green Dot | 4.3 | 1095 | 20,400 | 6.0 | 1325 | 31,000 |
| Unique | 5.0 | 1065 | 19,800 | 6.8 | 1325 | 31,500 |
| 630 | 8.5 | 1121 | 20,200 | 12.0 | 1405 | 32,100 |
| HS-5 | 5.9 | 1101 | 19,300 | *7.6 | 1376 | 32,100 |
| AL-5 | 5.9 | 1096 | 19,500 | 9.0 | 1405 | 31,500 |

### 115 gr. Cast Bullet
#358345 (#2 Alloy) .997" OAL

| Powder | Sug. Starting Grains | Velocity F.P.S. | Pressure C.U.P. | Max. Grains | Velocity F.P.S. | Pressure C.U.P. |
|---|---|---|---|---|---|---|
| **700X | 3.7 | 1045 | 25,200 | 4.4 | 1160 | 32,500 |
| **PB | 3.7 | 995 | 23,400 | 4.5 | 1130 | 32,000 |
| **SR-7625 | 3.8 | 1005 | 24,000 | 4.5 | 1130 | 31,000 |
| SR-4756 | 4.2 | 1005 | 18,800 | 6.1 | 1284 | 32,000 |
| **Bullseye | 3.8 | 1035 | 22,800 | 4.9 | 1200 | 31,000 |
| **Red Dot | 3.9 | 1055 | 24,600 | 4.9 | 1200 | 32,000 |
| **Green Dot | 4.0 | 1030 | 24,000 | 5.0 | 1200 | 32,500 |
| **Unique | 4.5 | 1025 | 22,200 | 5.7 | 1210 | 31,500 |
| Blue Dot | 5.8 | 983 | 16,800 | 8.4 | 1339 | 32,300 |
| 630 | 7.4 | 1028 | 18,700 | *10.8+ | 1322 | 29,700 |
| HS-5 | 5.0 | 1000 | 18,100 | 7.2 | 1356 | 32,800 |
| AL-5 | 5.0 | 991 | 18,900 | 7.2 | 1269 | 28,700 |

### 121 gr. Cast Bullet
#356402 (#2 Alloy) 1.110" OAL

| Powder | Sug. Starting Grains | Velocity F.P.S. | Pressure C.U.P. | Max. Grains | Velocity F.P.S. | Pressure C.U.P. |
|---|---|---|---|---|---|---|
| **700X | 3.3 | 975 | 25,200 | 4.0 | 1080 | 33,000 |
| **PB | 3.4 | 915 | 20,400 | 4.3 | 1095 | 31,500 |
| **SR-7625 | 3.5 | 935 | 21,000 | 4.3 | 1080 | 31,500 |
| SR-4756 | 4.0 | 925 | 12,500 | 5.5 | 1198 | 32,900 |
| **Bullseye | 3.7 | 950 | 21,600 | 4.4 | 1125 | 32,000 |
| **Red Dot | 3.8 | 1020 | 26,400 | 4.5 | 1130 | 32,500 |
| **Green Dot | 3.9 | 970 | 22,200 | 4.8 | 1165 | 32,500 |
| **Unique | 4.3 | 940 | 21,000 | 5.3 | 1135 | 31,000 |
| Blue Dot | 5.9 | 973 | 18,200 | 8.0 | 1253 | 31,600 |
| 630 | 7.2 | 985 | 18,800 | *10.3 | 1259 | 32,400 |
| HS-5 | 5.3 | 1001 | 20,500 | 6.4 | 1176 | 30,200 |
| AL-7 | 5.4 | 958 | 17,900 | 8.4+ | 1260 | 32,900 |

**Note:** Loads shown in shaded panels are maximum.
* Designates potentially most accurate load.
\+ Designates a compressed powder charge.
‡ Designates factory velocity duplication load.
** Designates use of Winchester cases and primers.

# .38 SMITH & WESSON
## (COLT NEW POLICE)

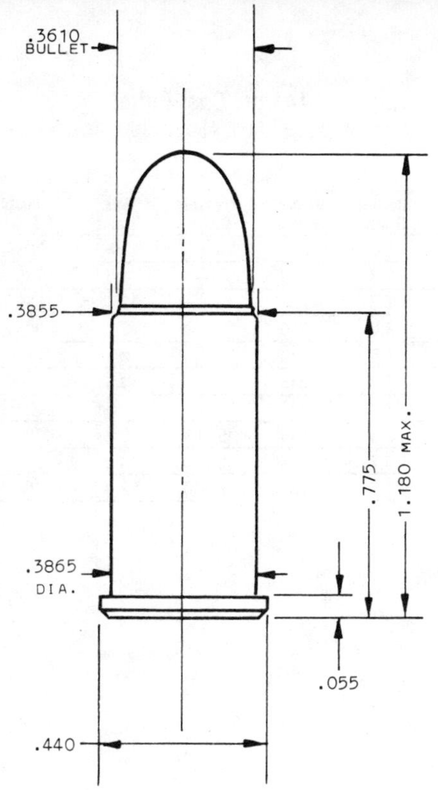


### TEST COMPONENTS

Cases ................................ Remington; Winchester
Trim-to Length ........................................ .765''
Primers .................................. Winchester 1½-108
Primer Size ...................................... Small Pistol
Lyman Shell Holder .................................. No. 21
Cast Bullets Used ...................... (size to .360'' dia.)
#358242, 92 gr.
#358242, 121 gr.
#358480, 133 gr.
#358495, 141 gr.
#358311, 158 gr.

### TEST SPECIFICATIONS
### (Velocity and Pressure)

Firearm Used ................. Special Pressure Test Barrel; Smith & Wesson Model 33
Barrel Length ........................................ Both 4''
Twist ................................................ 1-18¾''
Groove Diameter ...................................... .361''

### COMMENTS

The loads listed are intended for solid frame revolvers which are in good condition.

Wide variations in groove diameter may be encountered. We suggest that you slug your barrel before reloading and size cast bullets as near to actual groove diameter as possible.

**92 gr. Cast Bullet**
#358242 (#2 Alloy) 1.125'' OAL

| Powder | Sug. Starting Grains | Velocity F.P.S. | Pressure C.U.P. | Max. Grains | Velocity F.P.S. | Pressure C.U.P. |
|---|---|---|---|---|---|---|
| **Bullseye | 2.4 | 718 | 6,500 | 3.4 | 965 | 12,800 |
| **Unique | 3.5 | 748 | 7,000 | 4.6 | 941 | 11,400 |
| **Blue Dot | 5.4 | 778 | 6,600 | 6.7 | 958 | 10,800 |
| **700X | 2.3 | 720 | 5,600 | 3.2 | 972 | 12,200 |
| **SR-4756 | *4.0 | 758 | 7,300 | 4.9 | 968 | 12,500 |
| **231 | 2.7 | 719 | 6,500 | 3.6 | 944 | 12,100 |
| **HS-5 | 4.5 | 782 | 6,600 | 5.3 | 946 | 10,800 |
| **AL-5 | 4.3 | 798 | 7,200 | 5.5 | 980 | 12,300 |

**121 gr. Cast Bullet**
#358242 (#2 Alloy) 1.125'' OAL

| Powder | Sug. Starting Grains | Velocity F.P.S. | Pressure C.U.P. | Max. Grains | Velocity F.P.S. | Pressure C.U.P. |
|---|---|---|---|---|---|---|
| Bullseye | 2.0 | 560 | — | 3.0 | 805 | — |
| Unique | 3.5 | 700 | — | 4.5 | 945 | — |
| 700X | 2.3 | 680 | — | 3.1 | 895 | — |
| PB | 2.6 | 575 | — | 3.4 | 825 | — |
| SR-7625 | 2.7 | 580 | — | 3.5 | 830 | — |
| **AL-5 | 3.5 | 702 | 9,700 | 4.3 | 827 | 13,000 |
| **HS-5 | 3.6 | 676 | 8,700 | *4.5 | 831 | 12,100 |
| **231 | 2.0 | 610 | 6,900 | 2.8 | 798 | 12,400 |
| Red Dot | 2.4 | 650 | — | 3.2 | 855 | — |
| Green Dot | 2.5 | 615 | — | 3.3 | 815 | — |

**Note:** Loads shown in shaded panels are maximum.
* Designates potentially most accurate load.
‡ Designates factory velocity duplication load.
** Designates use of Special Pressure Test Barrel.

# .38 SMITH & WESSON
(Continued)

### 133 gr. Cast Bullet
#358480 (#2 Alloy) 1.062" OAL

| Powder | Sug. Starting Grains | Velocity F.P.S. | Pressure C.U.P. | Max. Grains | Velocity F.P.S. | Pressure C.U.P. |
|---|---|---|---|---|---|---|
| Bullseye | 2.0 | 595 | — | 2.9 | 820 | — |
| Red Dot | 2.4 | 705 | — | 3.1 | 865 | — |
| Green Dot | 2.5 | 680 | — | 3.2 | 830 | — |
| Unique | 3.5 | 790 | — | 4.2 | 920 | — |
| 700X | 2.3 | 735 | — | 2.9 | 875 | — |
| PB | 2.6 | 670 | — | 3.2 | 830 | — |
| SR-7625 | 2.7 | 675 | — | 3.3 | 850 | — |
| **AL-5 | 3.0 | 644 | 8,300 | 3.9 | 799 | 12,400 |
| **HS-5 | 3.0 | 620 | 7,500 | *3.8 | 766 | 12,300 |
| **231 | 1.9 | 568 | 6,400 | 2.8 | 786 | 12,300 |

### 141 gr. Cast Bullet
#358495 (#2 Alloy) .990" OAL

| Powder | Sug. Starting Grains | Velocity F.P.S. | Pressure C.U.P. | Max. Grains | Velocity F.P.S. | Pressure C.U.P. |
|---|---|---|---|---|---|---|
| Bullseye | 1.6 | 510 | — | 2.4 | 720 | — |
| Red Dot | 1.7 | 560 | — | 2.3 | 715 | — |
| Green Dot | 1.8 | 530 | — | ‡2.4 | 680 | — |
| Unique | 3.0 | 710 | — | 3.7 | 885 | — |
| 700X | 1.5 | 550 | — | 2.1 | 715 | — |
| PB | 1.8 | 545 | — | 2.4 | 710 | — |
| SR-7625 | 1.9 | 535 | — | 2.5 | 725 | — |
| **AL-5 | 2.5 | 546 | 6,100 | 3.2 | 702 | 11,500 |
| **HS-7 | 3.7 | 619 | 8,200 | 4.5 | 737 | 11,600 |
| **231 | 1.7 | 537 | 6,900 | *2.3 | 705 | 11,700 |

### 158 gr. Cast Bullet
#358311 (#2 Alloy) 1.150" OAL

| Powder | Sug. Starting Grains | Velocity F.P.S. | Pressure C.U.P. | Max. Grains | Velocity F.P.S. | Pressure C.U.P. |
|---|---|---|---|---|---|---|
| Bullseye | 1.8 | 545 | — | 2.6 | 730 | — |
| Red Dot | 2.1 | 605 | — | 2.8 | 770 | — |
| Green Dot | 2.2 | 595 | — | 2.9 | 745 | — |
| Unique | 3.2 | 715 | — | 3.7 | 805 | — |
| 700X | 2.1 | 665 | — | 2.6 | 785 | — |
| PB | 2.3 | 620 | — | 2.8 | 730 | — |
| SR-7625 | 2.4 | 660 | — | 2.9 | 770 | — |
| **AL-7 | 2.7 | 516 | 5,200 | 4.0 | 754 | 12,200 |
| **HS-7 | *3.6 | 619 | 8,100 | 4.5 | 749 | 11,900 |
| **231 | 1.5 | 486 | 4,600 | 2.5 | 727 | 12,400 |

**Note:** Loads shown in shaded panels are maximum.
* Designates potentially most accurate load.
‡ Designates factory velocity duplication load.
** Designates use of Special Pressure Test Barrel.

# .38 SUPER AUTO

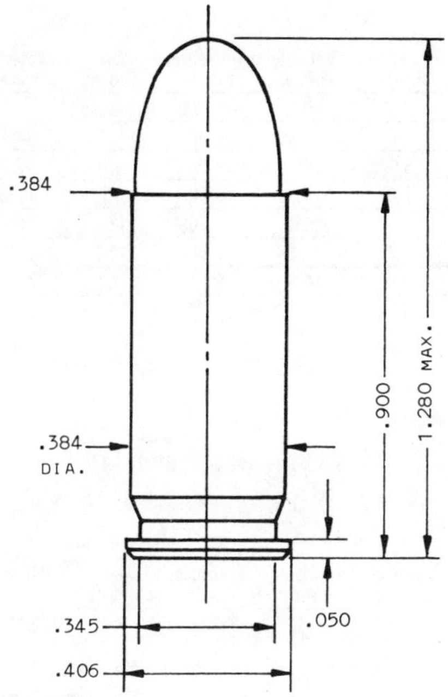


## TEST COMPONENTS

Cases .......... Remington
Trim-to Length .......... .895''
Primers .......... CCI 500; CCI 550; Winchester 1½-108 (cast bullets only)
Primer Size .......... Small Pistol; Small Pistol Magnum;
Lyman Shell Holder .......... No. 12
Jacketed Bullets Used .......... Sierra HC #8100, 90 gr.
Speer HP #3983, 100 gr.
Hornady HP, 115 gr.
Remington MC #B22842, 124 gr.
Speer SP #4005, 125 gr.
Remington MC #B22848, 130 gr.
Cast Bullets Used .......... (size to .355'' dia.)
#358242, 92 gr.
#356404, 95 gr.
#356402, 121 gr.
#358242, 121 gr.
#358480, 133 gr.
#358311, 158 gr.

## TEST SPECIFICATIONS (Velocity and Pressure)

Firearm Used .......... Universal Receiver
Barrel Length .......... 5''
Twist .......... 1-16''
Groove Diameter .......... .355''

## COMMENTS

These loads are intended only for those pistols which are chambered for the ''Super Auto'' cartridge.

For the ''Standard .38 Auto'' cartridge, it is recommended that you use a bullet of 133 grains or less. Reduce the starting load by ½ grain and work up to the starting load. Do not exceed this load with the ''Standard Auto'' cartridge.

Because this particular auto cartridge headspaces on the case rim, it may be crimped if desired.

**90 gr. Sierra HC**
#8100 (.355'' dia.) 1.125'' OAL

| Powder | Sug. Starting Grains | Velocity F.P.S. | Pressure C.U.P. | Max. Grains | Velocity F.P.S. | Pressure C.U.P. |
|---|---|---|---|---|---|---|
| Bullseye | 5.0 | 1222 | 17,600 | 6.8 | 1472 | 30,600 |
| Unique | 6.0 | 1221 | 19,400 | 7.1 | 1409 | 30,000 |
| Herco | 6.0 | 1193 | 17,500 | 8.2 | 1544 | 30,500 |
| PB | 5.5 | 1193 | 20,400 | 7.3 | 1482 | 29,900 |
| SR-7625 | 6.0 | 1203 | 21,200 | 7.8 | 1515 | 30,600 |
| 231 | 5.5 | 1198 | 20,500 | 6.9 | 1383 | 30,400 |
| HS-6 | 7.5 | 1210 | 20,000 | 9.4 | 1550 | 30,500 |
| AL-5 | 7.5 | 1220 | 20,500 | 10.0+ | 1543 | 29,300 |

**100 gr. Speer HP**
#3983 (.355'' dia.) 1.185'' OAL

| Powder | Sug. Starting Grains | Velocity F.P.S. | Pressure C.U.P. | Max. Grains | Velocity F.P.S. | Pressure C.U.P. |
|---|---|---|---|---|---|---|
| Bullseye | 5.3 | 1158 | 21,100 | 6.5 | 1374 | 30,800 |
| Unique | 5.8 | 1126 | 18,800 | 7.1 | 1343 | 31,000 |
| Herco | 6.1 | 1141 | 19,600 | 7.9 | 1397 | 31,100 |
| 700X | 4.6 | 1136 | 21,800 | 6.5 | 1358 | 31,400 |
| PB | 5.3 | 1153 | 21,100 | 7.0+ | 1333 | 30,400 |
| 231 | 5.5 | 1101 | 18,700 | 6.5 | 1303 | 28,900 |
| HS-6 | 7.5 | 1172 | 17,800 | 9.3 | 1433 | 30,800 |
| AL-120 | 5.3 | 1131 | 20,200 | 7.2+ | 1388 | 31,200 |

**Note:** Loads shown in shaded panels are maximum.
+ Designates a compressed powder charge.

# .38 SUPER AUTO
(Continued)

### 115 gr. Hornady HP
.355" dia./1.265" OAL

| Powder | Sug. Starting Grains | Velocity F.P.S. | Pressure C.U.P. | Max. Grains | Velocity F.P.S. | Pressure C.U.P. |
|---|---|---|---|---|---|---|
| Bullseye | 5.0 | 1084 | 19,100 | 6.2 | 1182 | 31,300 |
| Unique | *5.4 | 1039 | 19,900 | 6.9 | 1186 | 30,700 |
| 700X | 4.4 | 1032 | 20,200 | 6.0 | 1202 | 32,600 |
| SR-7625 | 5.3 | 1042 | 19,200 | 6.6 | 1195 | 31,900 |
| 231 | 5.1 | 1026 | 20,100 | 6.6 | 1173 | 31,200 |
| HS-6 | 7.5 | 1114 | 20,500 | 9.0 | 1316 | 31,400 |
| AL-120 | 4.5 | 986 | 19,400 | 6.2 | 1152 | 31,300 |
| Herco | 5.6 | 1039 | 19,600 | 6.9 | 1200 | 30,600 |
| AL-5 | 6.6 | 1097 | 20,600 | 8.6 | 1308 | 31,000 |

### 124 gr. Remington MC
#B22842 (.355" dia.) 1.250" OAL

| Powder | Sug. Starting Grains | Velocity F.P.S. | Pressure C.U.P. | Max. Grains | Velocity F.P.S. | Pressure C.U.P. |
|---|---|---|---|---|---|---|
| SR-4756 | 5.5 | 1042 | 18,700 | 7.6 | 1290 | 29,400 |
| Unique | 5.0 | 1030 | 18,000 | 6.3 | 1264 | 30,500 |
| Herco | 5.2 | 1014 | 19,300 | 6.4 | 1234 | 30,500 |
| Blue Dot | 7.0 | 1016 | 16,400 | 8.7 | 1283 | 29,800 |
| 630 | 8.3 | 1012 | 15,700 | *11.1 | 1307 | 29,300 |
| HS-7 | 7.7 | 991 | 17,000 | 10.0 | 1337 | 30,600 |
| ++AL-7 | 7.3 | 1044 | 18,700 | 9.8 | 1342 | 31,000 |

### 125 gr. Speer SP
#4005 (.355" dia.) 1.280" OAL

| Powder | Sug. Starting Grains | Velocity F.P.S. | Pressure C.U.P. | Max. Grains | Velocity F.P.S. | Pressure C.U.P. |
|---|---|---|---|---|---|---|
| Unique | 5.5 | 1028 | 20,800 | 7.0 | 1134 | 32,100 |
| Herco | 5.8 | 995 | 19,000 | 7.0 | 1137 | 30,900 |
| Blue Dot | 8.0 | 1090 | 20,300 | 10.1 | 1289 | 31,500 |
| SR-7625 | *4.6 | 970 | 20,100 | 6.5 | 1095 | 31,800 |
| **630 | 9.5 | 1073 | 20,800 | 11.5 | 1271 | 31,500 |
| HS-7 | 8.6 | 1070 | 20,600 | 11.0 | 1240 | 31,800 |
| **AL-7 | 7.7 | 1065 | 20,900 | 10.1 | 1297 | 30,100 |
| SR-4756 | 5.4 | 1039 | 20,600 | ‡7.7 | 1248 | 30,600 |

### 130 gr. Remington MC
#B22848 (.355" dia.) 1.280" OAL

| Powder | Sug. Starting Grains | Velocity F.P.S. | Pressure C.U.P. | Max. Grains | Velocity F.P.S. | Pressure C.U.P. |
|---|---|---|---|---|---|---|
| SR-4756 | 5.5 | 1048 | 21,600 | 7.3 | 1231 | 29,400 |
| Unique | 5.0 | 982 | 18,300 | 6.4 | 1207 | 29,700 |
| Herco | 5.0 | 1000 | 20,200 | 6.3 | 1180 | 29,700 |
| Blue Dot | 7.4 | 1074 | 19,300 | 9.1 | 1257 | 28,400 |
| 630 | 8.8 | 1060 | 19,900 | ‡*11.0 | 1283 | 30,300 |
| HS-7 | 7.8 | 987 | 17,200 | 9.8 | 1264 | 29,900 |
| AL-7 | 7.3 | 1068 | 20,800 | 9.0 | 1268 | 30,100 |

### 92 gr. Cast Bullet
#358242 (#2 Alloy) 1.175" OAL

| Powder | Sug. Starting Grains | Velocity F.P.S. | Pressure C.U.P. | Max. Grains | Velocity F.P.S. | Pressure C.U.P. |
|---|---|---|---|---|---|---|
| Bullseye | 4.2 | 1022 | 19,500 | 5.6 | 1261 | 30,900 |
| Unique | 5.1 | 1061 | 20,400 | 6.3 | 1267 | 30,500 |
| PB | 4.3 | 1009 | 18,900 | 5.7 | 1222 | 30,500 |
| SR-4756 | 5.9 | 1089 | 19,700 | 7.1 | 1263 | 30,300 |
| 231 | 4.7 | 1039 | 20,300 | *6.3 | 1282 | 31,600 |
| HP38 | 4.1 | 1026 | 20,600 | 5.8 | 1246 | 30,500 |
| HS-5 | 6.3 | 1046 | 19,700 | 8.0 | 1308 | 31,700 |
| AL-120 | 4.3 | 1030 | 20,600 | 5.8 | 1245 | 31,300 |

### 95 gr. Cast Bullet
#356404 (#2 Alloy) 1.130" OAL

| Powder | Sug. Starting Grains | Velocity F.P.S. | Pressure C.U.P. | Max. Grains | Velocity F.P.S. | Pressure C.U.P. |
|---|---|---|---|---|---|---|
| Bullseye | 4.5 | 1170 | 19,800 | 6.0 | 1425 | 31,000 |
| Unique | 5.5 | 1205 | 19,800 | 7.2 | 1460 | 31,000 |
| 700X | 4.2 | 1147 | 20,200 | 6.0 | 1399 | 30,100 |
| PB | 4.9 | 1195 | 20,400 | 6.7 | 1430 | 31,500 |
| SR-7625 | 5.4 | 1197 | 20,000 | 6.7 | 1429 | 31,100 |
| 231 | 5.2 | 1181 | 20,400 | 6.4 | 1433 | 30,100 |
| HP38 | *4.5 | 1159 | 20,500 | 6.2 | 1403 | 29,700 |
| AL-120 | 4.7 | 1160 | 20,000 | 6.3 | 1395 | 30,200 |

**Note:** Loads shown in shaded panels are maximum.
* Designates potentially most accurate load.
\+ Designates a compressed powder charge.
‡ Designates factory velocity duplication load.
++ Winchester 1½-108 Small Pistol Primer.
** CCI 550 Small Pistol Magnum Primer.

# .38 SUPER AUTO
## (Continued)

### 121 gr. Cast Bullet
#356402 (#2 Alloy) 1.230" OAL

| Powder | Sug. Starting Grains | Velocity F.P.S. | Pressure C.U.P. | Max. Grains | Velocity F.P.S. | Pressure C.U.P. |
|---|---|---|---|---|---|---|
| Bullseye | 4.0 | 1005 | 18,600 | 5.3 | 1235 | 30,000 |
| Unique | 5.1 | 1075 | 20,400 | 6.0 | 1255 | 31,000 |
| 700X | 4.0 | 1025 | 20,600 | 5.5 | 1252 | 29,900 |
| SR-7625 | 5.1 | 1099 | 20,600 | 6.6 | 1274 | 29,900 |
| 231 | 5.0 | 1092 | 20,700 | *6.5 | 1307 | 30,800 |
| HP38 | 4.4 | 1063 | 20,900 | 5.9 | 1267 | 29,700 |
| HS-5 | 6.8 | 1089 | 19,200 | 8.4 | 1341 | 30,700 |
| AL-120 | 4.5 | 1040 | 19,000 | 6.0 | 1235 | 29,800 |

### 121 gr. Cast Bullet
#358242 (#2 Alloy) 1.160" OAL

| Powder | Sug. Starting Grains | Velocity F.P.S. | Pressure C.U.P. | Max. Grains | Velocity F.P.S. | Pressure C.U.P. |
|---|---|---|---|---|---|---|
| Bullseye | 3.9 | 970 | 17,600 | 5.4 | 1214 | 29,200 |
| Unique | 4.9 | 1031 | 20,000 | 6.4 | 1281 | 28,800 |
| 700X | 3.7 | 979 | 19,800 | 5.3 | 1206 | 29,200 |
| SR-7625 | 4.2 | 978 | 18,100 | 5.8 | 1230 | 29,300 |
| 231 | 4.3 | 988 | 17,700 | 5.8 | 1233 | 29,600 |
| HP38 | 3.9 | 975 | 17,800 | 5.6 | 1230 | 29,400 |
| HS-5 | 6.1 | 1030 | 18,200 | 7.6 | 1269 | 30,600 |
| AL-120 | 3.8 | 952 | 17,200 | 5.4 | 1207 | 29,000 |

### 133 gr. Cast Bullet
#358480 (#2 Alloy) 1.115" OAL

| Powder | Sug. Starting Grains | Velocity F.P.S. | Pressure C.U.P. | Max. Grains | Velocity F.P.S. | Pressure C.U.P. |
|---|---|---|---|---|---|---|
| SR-4756 | 5.0 | 975 | 16,900 | 7.0 | 1237 | 29,400 |
| Bullseye | 4.0 | 1000 | 19,800 | 5.3 | 1180 | 31,500 |
| Unique | 5.1 | 1045 | 21,000 | 6.0 | 1200 | 30,000 |
| Blue Dot | 6.8 | 1042 | 18,500 | 8.9 | 1346 | 30,100 |
| 630 | 8.4 | 1054 | 18,100 | *10.6 | 1268 | 30,000 |
| HS-6 | 5.9 | 1026 | 18,000 | 7.8 | 1303 | 30,200 |
| AL-7 | 6.7 | 1041 | 18,700 | 8.6 | 1278 | 29,800 |

### 158 gr. Cast Bullet
#358311 (#2 Alloy) 1.245" OAL

| Powder | Sug. Starting Grains | Velocity F.P.S. | Pressure C.U.P. | Max. Grains | Velocity F.P.S. | Pressure C.U.P. |
|---|---|---|---|---|---|---|
| SR-4756 | 4.9 | 927 | 18,900 | 5.8 | 1085 | 29,800 |
| Unique | 4.4 | 958 | 20,100 | 5.1 | 1058 | 28,100 |
| Herco | 4.5 | 920 | 21,000 | 5.4 | 1058 | 29,000 |
| Blue Dot | 6.5 | 987 | 20,900 | 7.6 | 1133 | 28,400 |
| 630 | 7.8 | 985 | 20,900 | *9.3 | 1110 | 28,600 |
| HS-7 | 6.8 | 956 | 20,200 | 8.3 | 1100 | 28,200 |

**Note:** Loads shown in shaded panels are maximum.
* Designates potentially most accurate load.

# .380 AUTO
## (9mm CORTO & 9mm SHORT)

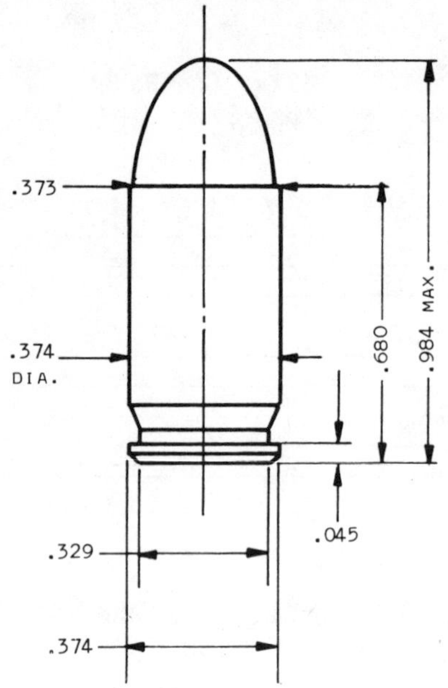


### TEST COMPONENTS

Cases ................ Federal; Super Vel (cast bullets only)
Trim-to Length ................................................ .677''
Primers ....................... Federal 100 (jacketed bullets); Winchester 1½-108 (cast bullets only)
Primer Size ....................................... Small Pistol
Lyman Shell Holder ................................... No. 26
Jacketed Bullets Used ............ Sierra HC #8100, 90 gr.
Speer HP #3983, 100 gr.
Hornady HP, 115 gr.
Cast Bullets Used ....................... (size to .355'' dia.)
#358242, 92 gr.
#358242, 121 gr.

### TEST SPECIFICATIONS
### (Velocity and Pressure)

Firearm Used ............................ Universal Receiver
Barrel Length ............................................... 3¾''
Twist ...................................................... 1-16''
Groove Diameter ........................................ .355''

### COMMENTS

While a wide variation in groove diameter may be encountered (as large as .362'') in pistols chambered for this cartridge, the use of bullets larger than .355'' is usually not possible. This is due to chamber and case dimensions. The larger diameter bullet will bulge the case to the point where the cartridge will not chamber. If the groove diameter of your pistol runs on the large size, accuracy will be poor and there is little you can do about it.

Heavier cast bullets of .355'' diameter will sometimes cause a slight case bulge with some lots of brass. This is due to the inside taper of the case. It may be ignored as long as the cartridge will chamber and function properly.

**90 gr. Sierra HC**
#8100 (.355'' dia.) .925'' OAL

| Powder | Sug. Starting Grains | Velocity F.P.S. | Pressure C.U.P. | Max. Grains | Velocity F.P.S. | Pressure C.U.P. |
|---|---|---|---|---|---|---|
| 700X | 1.8 | 704 | 9,100 | 2.5 | 959 | 15,900 |
| SR-7625 | 2.2 | 583 | 8,100 | 3.1 | 941 | 15,200 |
| Bullseye | 2.0 | 713 | 9,100 | *2.9 | 970 | 15,500 |
| Red Dot | 1.9 | 691 | 8,400 | 2.8 | 983 | 15,700 |
| Unique | 2.5 | 709 | 8,700 | 3.4 | 967 | 15,500 |
| 231 | 2.1 | 681 | 8,100 | 3.2 | 992 | 16,000 |
| HS-5 | 3.1 | 682 | 8,000 | 4.7 | 1006 | 16,000 |
| AL-5 | 2.8 | 646 | 7,200 | 4.4 | 981 | 15,700 |

**100 gr. Speer HP**
#3983 (.355'' dia.) .955'' OAL

| Powder | Sug. Starting Grains | Velocity F.P.S. | Pressure C.U.P. | Max. Grains | Velocity F.P.S. | Pressure C.U.P. |
|---|---|---|---|---|---|---|
| 700X | 1.9 | 704 | 9,100 | *2.5 | 889 | 14,700 |
| SR-7625 | 2.4 | 694 | 8,000 | 3.3 | 960 | 14,900 |
| Bullseye | 2.0 | 695 | 7,700 | 3.1 | 988 | 15,400 |
| Red Dot | 1.8 | 651 | 6,500 | 3.1 | 1026 | 15,900 |
| Unique | 2.6 | 725 | 8,600 | 3.6 | 1013 | 15,200 |
| 231 | 2.2 | 693 | 6,600 | 3.5 | 1024 | 15,600 |
| HS-5 | 3.2 | 693 | 6,900 | 5.0 | 1058 | 16,000 |
| AL-5 | 3.0 | 660 | 7,000 | 4.7 | 1015 | 15,300 |

**Note:** Loads shown in shaded panels are maximum.
* Designates potentially most accurate load.

# .380 AUTO
## (Continued)

### 115 gr. Hornady HP
.355" dia./.956" OAL

| Powder | Sug. Starting Grains | Velocity F.P.S. | Pressure C.U.P. | Max. Grains | Velocity F.P.S. | Pressure C.U.P. |
|---|---|---|---|---|---|---|
| 700X | 1.6 | 658 | 9,100 | 2.5 | 921 | 15,900 |
| SR-7625 | 2.2 | 667 | 8,700 | 3.1 | 928 | 15,800 |
| Bullseye | 2.0 | 711 | 9,900 | 2.9 | 937 | 15,100 |
| Red Dot | 1.8 | 670 | 8,700 | 2.8 | 945 | 15,500 |
| Unique | 2.4 | 718 | 9,500 | 3.4 | 973 | 15,900 |
| 231 | 2.0 | 664 | 8,300 | 3.1 | 934 | 15,300 |
| HS-6 | 3.1 | 677 | 8,700 | 4.5 | 945 | 15,200 |
| AL-5 | 2.8 | 657 | 8,000 | *4.3 | 939 | 15,800 |

### 92 gr. Cast Bullet
#358242 (#2 Alloy) .980" OAL

| Powder | Sug. Starting Grains | Velocity F.P.S. | Pressure C.U.P. | Max. Grains | Velocity F.P.S. | Pressure C.U.P. |
|---|---|---|---|---|---|---|
| 700X | 1.9 | 695 | 11,100 | 2.5 | 900 | 15,800 |
| PB | 2.2 | 645 | 10,300 | 2.8 | 860 | 15,200 |
| **SR-4756 | 2.8 | 653 | 6,900 | 4.1 | 970 | 15,600 |
| Bullseye | 2.4 | 750 | 11,800 | 3.0 | 915 | 16,000 |
| Red Dot | 2.0 | 690 | 11,000 | 2.6 | 880 | 15,200 |
| Green Dot | 2.1 | 655 | 10,000 | 3.0 | 910 | 16,000 |
| Unique | 3.1 | 755 | 11,800 | 3.7 | 920 | 16,000 |
| **Blue Dot | 4.0 | 667 | 7,000 | 5.5 | 961 | 14,600 |
| **231 | 2.3 | 701 | 7,900 | 3.5 | 1000 | 15,600 |
| **HS-5 | 3.1 | 658 | 6,400 | *5.0 | 1023 | 15,900 |
| **AL-5 | 3.0 | 644 | 6,600 | 5.0 | 1029 | 16,000 |

### 121 gr. Cast Bullet
#358242 (#2 Alloy) .980" OAL

| Powder | Sug. Starting Grains | Velocity F.P.S. | Pressure C.U.P. | Max. Grains | Velocity F.P.S. | Pressure C.U.P. |
|---|---|---|---|---|---|---|
| 700X | 1.4 | 610 | 11,300 | 1.9 | 750 | 15,800 |
| PB | 1.7 | 605 | 10,800 | 2.2 | 750 | 15,800 |
| **SR-4756 | 2.3 | 662 | 7,800 | 3.5 | 927 | 15,800 |
| Bullseye | 1.6 | 600 | 10,500 | 2.1 | 750 | 15,200 |
| Red Dot | 1.6 | 615 | 11,000 | 2.1 | 770 | 15,800 |
| Green Dot | 1.7 | 620 | 10,800 | 2.3 | 760 | 15,400 |
| Unique | 2.2 | 630 | 11,300 | 2.8 | 780 | 15,600 |
| **Blue Dot | 3.7 | 732 | 9,000 | 5.3 | 948 | 15,900 |
| **231 | 2.1 | 718 | 9,200 | 3.2 | 946 | 15,800 |
| **HS-5 | 2.6 | 619 | 6,300 | *4.2 | 934 | 15,000 |
| **AL-5 | 2.7 | 657 | 7,400 | 4.3 | 934 | 15,500 |

**Note:** Loads shown in shaded panels are maximum.
* Designates potentially most accurate load.
** Designates use of Federal cases.

# .38 SPECIAL

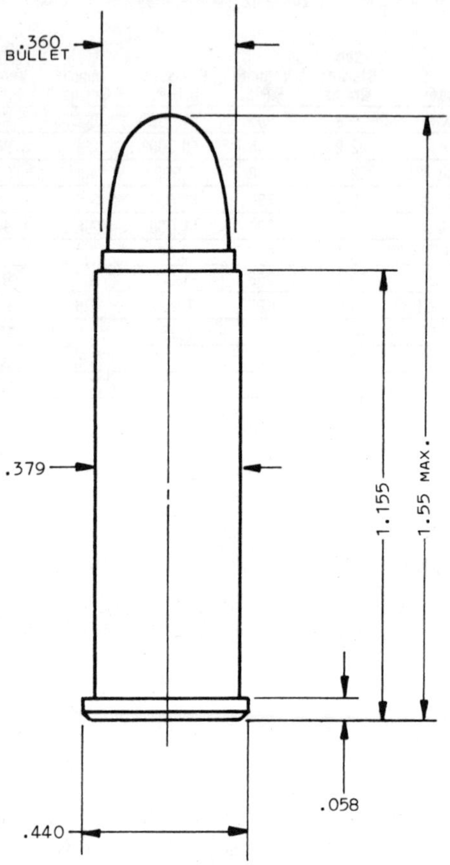


## TEST COMPONENTS

Cases .................................................... Federal
Trim-to Length .......................................... 1.149"
Primers ................................................ CCI 500
Primer Size ........................................ Small Pistol
Lyman Shell Holder ...................................... No. 1
Jacketed Bullets Used .. Remington SJHP #B22944, 95 gr.
Winchester JHP #B3573HP, 110 gr.
Hornady HP, 125 gr.
Speer JHP #4203, 140 gr.
Speer HJHP #4205, 146 gr.
Hornady HP, 158 gr.
Speer HJSP #9223, 160 gr.
Cast Bullets Used ...................... (size to .357" dia.)
*Gas Check
#358242, 92 gr.
#358242, 121 gr.
#356402, 121 gr.
#358495, 141 gr.
#358212, 146 gr.
#35863, 148 gr.
#35891, 148 gr.
*#358156, 158 gr.
#358311, 158 gr.
#357446, 162 gr.
#358429, 168 gr.

## TEST SPECIFICATIONS (Velocity and Pressure)

Firearm Used ......................... Universal Receiver and Special Vented Barrel
Barrel Length ............................................... 4"
Twist .................................................. 1-18¾"
Groove Diameter ......................................... .357"

## COMMENTS

Handguns of various makes and models will occasionally vary in groove diameter. It is wise to slug your barrel before selecting a sizing die. Size cast bullets to as near actual groove diameter as possible. Do not use a charge weight below the starting loads shown to prevent jackets from lodging in the barrel. If a crimp is desired on these half-jacketed bullets, it should be placed at the junction of the lead and the jacket.

Loads designated with the symbol (+P) are higher pressure loads to give greater velocities. These loads should be used only in arms recommended for (+P) type cartridges by the gun manufacturer. Repeated use of these high pressure loads in handguns with aluminum frames or aluminum cylinders or with lightweight steel frames or cylinders, is not recommended.

For mid-range loads, we suggest bullet No. 358495 loaded at starting load velocities. Bullet No. 358311 duplicates the factory 158 grain service bullet very closely while Bullet No. 358429 is an excellent "Keith type" hunting bullet.

# .38 SPECIAL
(Continued)

### 95 gr. Remington SJHP
#B22944 (.357" dia.) 1.470" OAL

| Powder | Sug. Starting Grains | Velocity F.P.S. | Pressure C.U.P. | Max. Grains | Velocity F.P.S. | Pressure C.U.P. |
|---|---|---|---|---|---|---|
| 700X | 3.8 | 737 | 10,100 | 4.8 | 995 | 16,800 |
| SR-7625 | 4.5 | 614 | 6,800 | 6.6 | 1048 | 16,500 |
| SR-4756 | 5.5 | 693 | 9,400 | 7.3 | 1036 | 16,800 |
| Bullseye | 3.2 | 613 | 7,900 | 5.4 | 1042 | 16,000 |
| | | | | *(+P)5.8 | 1095 | 17,300 |
| Unique | 4.5 | 671 | 9,100 | 6.6 | 1046 | 17,000 |
| | | | | (+P)7.1 | 1119 | 17,400 |
| 231 | 4.5 | 711 | 8,100 | 6.0 | 1016 | 15,300 |
| | | | | (+P)6.2 | 1060 | 18,400 |
| 630 | 8.0 | 714 | 8,100 | 11.0 | 1074 | 16,700 |
| | | | | (+P)12.3 | 1137 | 17,300 |
| HS-6 | 6.1 | 660 | 7,900 | 9.0 | 1072 | 16,500 |

### 110 gr. Winchester JHP
#B3573HP (.357" dia.) 1.478" OAL

| Powder | Sug. Starting Grains | Velocity F.P.S. | Pressure C.U.P. | Max. Grains | Velocity F.P.S. | Pressure C.U.P. |
|---|---|---|---|---|---|---|
| 700X | 3.7 | 596 | 7,300 | 5.1 | 955 | 16,800 |
| SR-4756 | 5.8 | 675 | 9,100 | 7.3 | 980 | 16,400 |
| | | | | ‡(+P)7.5 | 1027 | 17,500 |
| Bullseye | 3.4 | 528 | 6,300 | 5.4 | 946 | 16,600 |
| | | | | *(+P)5.6 | 981 | 18,000 |
| Unique | 4.7 | 580 | 7,600 | 6.7 | 967 | 16,600 |
| | | | | (+P)7.0 | 1026 | 18,200 |
| Blue Dot | 8.0 | 709 | 8,800 | 9.7 | 986 | 15,700 |
| 231 | 4.5 | 653 | 10,400 | 6.0 | 947 | 16,800 |
| | | | | (+P)6.3 | 985 | 18,400 |
| 630 | 8.4 | 681 | 9,200 | 11.4 | 1014 | 16,800 |
| | | | | (+P)11.8 | 1064 | 18,500 |
| HS-6 | 6.5 | 593 | 7,700 | 9.0 | 978 | 16,200 |
| | | | | (+P)9.5 | 1043 | 18,000 |
| AL-7 | 7.4 | 751 | 9,400 | 9.2 | 963 | 15,100 |

### 125 gr. Hornady HP
.357" dia./1.470" OAL

| Powder | Sug. Starting Grains | Velocity F.P.S. | Pressure C.U.P. | Max. Grains | Velocity F.P.S. | Pressure C.U.P. |
|---|---|---|---|---|---|---|
| 700X | 2.8 | 473 | 8,100 | *4.1 | 814 | 16,500 |
| | | | | (+P)4.5 | 877 | 17,300 |
| SR-4756 | 4.8 | 535 | 7,400 | 6.3 | 874 | 16,200 |
| Bullseye | 3.2 | 568 | 10,200 | (+P)5.0 | 920 | 18,300 |
| Unique | 4.0 | 498 | 8,000 | 6.0 | 895 | 16,700 |
| Blue Dot | 7.4 | 703 | 11,400 | 8.7 | 893 | 16,500 |
| 231 | 3.9 | 589 | 10,600 | 5.1 | 820 | 15,600 |
| | | | | (+P)5.4 | 882 | 18,300 |
| 630 | 6.8 | 558 | 8,800 | 9.6 | 886 | 16,300 |
| | | | | ‡(+P)10.1 | 951 | 18,200 |
| HS-6 | 5.6 | 600 | 9,400 | 7.4 | 862 | 16,100 |
| AL-8 | 6.0 | 464 | 6,800 | 8.7 | 878 | 16,300 |

### 140 gr. Speer JHP
#4203 (.357" dia.) 1.470" OAL

| Powder | Sug. Starting Grains | Velocity F.P.S. | Pressure C.U.P. | Max. Grains | Velocity F.P.S. | Pressure C.U.P. |
|---|---|---|---|---|---|---|
| SR-7625 | 4.2 | 505 | 10,800 | 5.1 | 715 | 16,100 |
| Bullseye | 3.5 | 533 | 11,900 | 4.3 | 714 | 16,500 |
| Unique | 4.3 | 537 | 11,200 | 5.3 | 709 | 15,900 |
| Blue Dot | 7.6 | 662 | 13,000 | 8.4 | 824 | 16,800 |
| 231 | 3.9 | 496 | 12,100 | *4.7 | 684 | 15,800 |
| 630 | 6.8 | 478 | 10,600 | 9.0 | 781 | 16,800 |
| HS-6 | 5.6 | 526 | 11,200 | 7.2 | 777 | 16,700 |
| AL-8 | 7.0 | 521 | 10,700 | 8.6 | 799 | 16,500 |

### 146 gr. Speer HJHP
#4205 (.357" dia.) 1.408" OAL

| Powder | Sug. Starting Grains | Velocity F.P.S. | Pressure C.U.P. | Max. Grains | Velocity F.P.S. | Pressure C.U.P. |
|---|---|---|---|---|---|---|
| SR-7625 | 4.7 | 694 | 15,100 | 4.9 | 740 | 16,800 |
| SR-4756 | 5.0 | 633 | 13,400 | 5.6 | 778 | 16,600 |
| Unique | 5.1 | 745 | 16,800 | (+P)5.4 | 795 | 17,600 |
| Herco | 5.0 | 711 | 15,100 | 5.2 | 744 | 15,900 |
| Blue Dot | 6.7 | 595 | 12,500 | 7.4 | 772 | 16,600 |
| 2400 | 9.0 | 681 | 13,400 | 10.0 | 911 | 16,300 |
| HS-6 | 6.2 | 691 | 14,800 | 6.7 | 782 | 16,000 |
| AL-8 | 7.7 | 784 | 16,300 | 8.0 | 802 | 16,500 |
| 630 | 9.2 | 847 | 17,500 | *(+P)9.3 | 866 | 18,700 |

### 158 gr. Hornady HP
.357" dia./1.480" OAL

| Powder | Sug. Starting Grains | Velocity F.P.S. | Pressure C.U.P. | Max. Grains | Velocity F.P.S. | Pressure C.U.P. |
|---|---|---|---|---|---|---|
| SR-4756 | 4.9 | 611 | 13,800 | 5.4 | 705 | 16,800 |
| SR-4759 | 9.0 | 685 | 14,800 | 9.5 | 756 | 16,700 |
| | | | | (+P)9.7 | 789 | 18,000 |
| Unique | 4.7 | 645 | 14,400 | 5.0 | 710 | 16,600 |
| | | | | *(+P)5.2 | 761 | 18,200 |
| Herco | 4.7 | 619 | 14,000 | 5.2 | 708 | 16,800 |
| Blue Dot | 6.0 | 546 | 12,200 | 7.1 | 721 | 16,300 |
| 2400 | 8.4 | 645 | 13,500 | 9.4 | 745 | 16,400 |
| 630 | 7.5 | 626 | 14,100 | ‡*8.5 | 754 | 16,600 |
| HS-6 | 5.5 | 572 | 12,400 | 6.5 | 726 | 16,600 |
| AL-8 | 7.0 | 595 | 12,500 | 7.5 | 676 | 15,500 |

**Note:** Loads shown in shaded panels are maximum.
* Designates potentially most accurate load.
\+ Designates a compressed powder charge.
‡ Designates factory velocity duplication load.
(+P) Designates higher pressure loads. See COMMENTS.

# .38 SPECIAL
## (Continued)

### 160 gr. Speer HJSP
#9223 (.357" dia.) 1.430" OAL

| Powder | Sug. Starting Grains | Velocity F.P.S. | Pressure C.U.P. | Max. Grains | Velocity F.P.S. | Pressure C.U.P. |
|---|---|---|---|---|---|---|
| SR-4759 | 9.1 | 742 | 14,900 | *9.6 | 792 | 16,600 |
| IMR-4227 | 9.6 | 747 | 15,300 | 10.0 | 793 | 16,700 |
| Unique | 5.2 | 769 | 16,600 | (+P)5.3 | 808 | 17,800 |
| Blue Dot | 6.4 | 621 | 11,900 | 7.2 | 786 | 16,500 |
| 2400 | 9.0 | 710 | 14,700 | 9.6 | 798 | 16,900 |
| 630 | 8.5 | 764 | 16,200 | 8.7 | 780 | 16,300 |
| HS-6 | 5.8 | 653 | 13,900 | 6.4 | 742 | 16,500 |
| AL-8 | 6.3 | 532 | 11,700 | 7.5 | 745 | 15,900 |

### 92 gr. Cast Bullet
#358242 (Linotype) 1.456" OAL

| Powder | Sug. Starting Grains | Velocity F.P.S. | Pressure C.U.P. | Max. Grains | Velocity F.P.S. | Pressure C.U.P. |
|---|---|---|---|---|---|---|
| 700X | 3.0 | 675 | 5,700 | *5.0 | 1107 | 15,700 |
| | | | | (+P)5.3 | 1168 | 18,300 |
| Bullseye | 3.3 | 765 | 7,600 | 5.3 | 1110 | 15,700 |
| | | | | (+P)5.6 | 1168 | 17,600 |
| Red Dot | 3.8 | 830 | 8,800 | 5.2 | 1108 | 16,300 |
| | | | | (+P)5.5 | 1154 | 17,700 |
| 231 | 3.5 | 689 | 5,900 | 5.7 | 1104 | 16,300 |
| | | | | (+P)6.0 | 1163 | 17,800 |
| HP38 | 3.6 | 836 | 9,100 | 5.0 | 1098 | 16,500 |
| HS-5 | 6.5 | 892 | 9,400 | 8.4 | 1168 | 16,800 |
| AL-5 | 6.0 | 911 | 10,400 | 7.9 | 1145 | 16,300 |

### 121 gr. Cast Bullet
#358242 (Linotype) 1.450" OAL

| Powder | Sug. Starting Grains | Velocity F.P.S. | Pressure C.U.P. | Max. Grains | Velocity F.P.S. | Pressure C.U.P. |
|---|---|---|---|---|---|---|
| 700X | 2.5 | 647 | 7,600 | 3.9 | 940 | 16,200 |
| | | | | (+P)4.2 | 987 | 17,900 |
| SR-7625 | 4.0 | 788 | 10,600 | 5.0 | 988 | 16,200 |
| Bullseye | 2.8 | 690 | 8,600 | *4.6 | 1001 | 16,400 |
| | | | | (+P)4.9 | 1045 | 18,400 |
| Red Dot | 3.3 | 749 | 10,100 | 4.6 | 974 | 16,200 |
| | | | | (+P)5.0 | 1047 | 18,400 |
| Unique | 4.0 | 725 | 9,000 | 5.7 | 1012 | 16,200 |
| 231 | 3.4 | 693 | 8,800 | 5.1 | 986 | 16,600 |
| | | | | 5.4 | 1036 | 17,900 |
| HP38 | 3.0 | 725 | 9,600 | 4.3 | 961 | 16,700 |
| | | | | (+P)4.8 | 1045 | 18,300 |
| HS-5 | 5.0 | 727 | 9,200 | 7.1 | 998 | 16,000 |
| AL-5 | 4.7 | 732 | 9,600 | 6.4 | 975 | 16,000 |

### 121 gr. Cast Bullet
#356402 (Linotype) 1.550" OAL

| Powder | Sug. Starting Grains | Velocity F.P.S. | Pressure C.U.P. | Max. Grains | Velocity F.P.S. | Pressure C.U.P. |
|---|---|---|---|---|---|---|
| 700X | 2.7 | 649 | 7,500 | 4.2 | 967 | 16,700 |
| | | | | (+P)4.5 | 1012 | 18,300 |
| SR-7625 | 4.0 | 725 | 8,500 | 5.3 | 1003 | 16,400 |
| Bullseye | 3.8 | 847 | 11,400 | *5.0 | 1048 | 16,500 |
| | | | | (+P)5.3 | 1068 | 18,200 |
| Unique | 3.9 | 774 | 9,700 | 5.0 | 994 | 15,700 |
| | | | | (+P)5.8 | 1083 | 17,900 |
| 231 | 3.5 | 765 | 9,500 | 4.8 | 1008 | 16,100 |
| | | | | (+P)5.2 | 1054 | 18,000 |
| HP38 | 3.5 | 794 | 10,900 | 4.5 | 1013 | 16,100 |
| | | | | (+P)4.9 | 1059 | 17,900 |
| HS-5 | 4.8 | 697 | 8,100 | 7.3 | 940 | 16,600 |
| AL-5 | 4.8 | 666 | 9,000 | 7.0 | 1012 | 15,800 |

### 141 gr. Cast Bullet
#358495 (Linotype) 1.310" OAL

| Powder | Sug. Starting Grains | Velocity F.P.S. | Pressure C.U.P. | Max. Grains | Velocity F.P.S. | Pressure C.U.P. |
|---|---|---|---|---|---|---|
| 700X | 2.2 | 636 | 8,300 | 4.0 | 994 | 16,600 |
| SR-7625 | 3.1 | 683 | 9,100 | 4.3 | 905 | 15,600 |
| Bullseye | 3.0 | 778 | 11,000 | 4.1 | 973 | 16,800 |
| Red Dot | 2.8 | 706 | 9,500 | 4.2 | 961 | 16,300 |
| Unique | 3.8 | 757 | 10,300 | 5.3 | 1013 | 16,700 |
| 231 | 3.0 | 667 | 8,500 | *4.9 | 1001 | 16,800 |
| HP38 | 2.7 | 683 | 9,000 | 4.5 | 1001 | 16,800 |
| HS-5 | 4.0 | 643 | 7,800 | 6.4 | 990 | 16,300 |
| AL-5 | 4.0 | 660 | 8,400 | 6.3 | 991 | 16,300 |

### 146 gr. Cast Bullet
#358212 (Linotype) 1.515" OAL

| Powder | Sug. Starting Grains | Velocity F.P.S. | Pressure C.U.P. | Max. Grains | Velocity F.P.S. | Pressure C.U.P. |
|---|---|---|---|---|---|---|
| 700X | 2.8 | 674 | 9,400 | *4.0 | 881 | 15,900 |
| SR-7625 | 3.5 | 594 | 6,900 | 5.0 | 899 | 16,100 |
| Bullseye | 2.7 | 612 | 8,000 | 4.3 | 906 | 15,800 |
| Red Dot | 2.8 | 650 | 8,800 | 4.3 | 917 | 16,400 |
| Unique | 3.7 | 652 | 9,000 | 5.4 | 928 | 15,900 |
| 231 | 3.2 | 617 | 7,900 | 5.0 | 930 | 16,700 |
| HP38 | 3.0 | 654 | 9,300 | 4.4 | 900 | 16,500 |
| HS-5 | 4.8 | 650 | 8,100 | 7.0 | 935 | 16,000 |
| AL-5 | 4.8 | 656 | 8,400 | 6.6 | 902 | 15,500 |

**Note:** Loads shown in shaded panels are maximum.
* Designates potentially most accurate load.
\+ Designates a compressed powder charge.
‡ Designates factory velocity duplication load.
(+P) Designates higher pressure loads. See COMMENTS.

# .38 SPECIAL
## (Continued)

### 148 gr. Cast Bullet
#35863 (Linotype) 1.152" OAL

| Powder | Sug. Starting Grains | Velocity F.P.S. | Pressure C.U.P. | Max. Grains | Velocity F.P.S. | Pressure C.U.P. |
|---|---|---|---|---|---|---|
| 700X | 2.0 | 584 | 6,900 | 3.5 | 961 | 16,300 |
| SR-7625 | 2.8 | 674 | 8,700 | 4.0 | 928 | 15,800 |
| Bullseye | 2.4 | 665 | 8,400 | 3.8 | 950 | 16,100 |
| Red Dot | 2.5 | 678 | 8,700 | 3.7 | 945 | 15,900 |
| Unique | 3.1 | 670 | 8,100 | 4.6 | 994 | 16,800 |
| 231 | 2.5 | 600 | 6,600 | 4.1 | 957 | 16,000 |
| HP38 | 2.7 | 750 | 10,400 | 3.9 | 972 | 16,400 |
| HS-5 | 4.0 | 778 | 10,900 | *5.8 | 981 | 16,400 |
| AL-5 | 3.3 | 588 | 6,700 | 5.4 | 988 | 15,700 |

### 148 gr. Cast Bullet
#35891 (Linotype) 1.317" OAL

| Powder | Sug. Starting Grains | Velocity F.P.S. | Pressure C.U.P. | Max. Grains | Velocity F.P.S. | Pressure C.U.P. |
|---|---|---|---|---|---|---|
| 700X | 2.0 | 512 | 6,000 | 3.6 | 889 | 15,900 |
| SR-7625 | 2.8 | 517 | 5,800 | 4.7 | 958 | 16,500 |
| Bullseye | 2.8 | 568 | 8,900 | 4.4 | 953 | 16,200 |
| Red Dot | 2.7 | 524 | 7,900 | 4.1 | 939 | 16,800 |
| Unique | 3.4 | 660 | 8,500 | 5.0 | 948 | 16,100 |
| 231 | 2.8 | 626 | 7,900 | *4.6 | 914 | 15,600 |
| HP38 | 2.0 | 635 | 8,700 | 4.1 | 911 | 16,700 |
| HS-5 | 4.2 | 657 | 8,800 | 6.4 | 965 | 16,400 |
| AL-5 | 4.3 | 647 | 8,300 | 6.7 | 986 | 16,800 |

### 158 gr. Cast Bullet
#358156 (Linotype) 1.460" OAL

| Powder | Sug. Starting Grains | Velocity F.P.S. | Pressure C.U.P. | Max. Grains | Velocity F.P.S. | Pressure C.U.P. |
|---|---|---|---|---|---|---|
| PB | 3.0 | 584 | 8,700 | 4.2 | 825 | 16,200 |
| SR-4756 | 4.2 | 630 | 9,500 | 5.5 | 898 | 16,700 |
| Bullseye | 2.8 | 638 | 10,000 | *4.0 | 858 | 16,000 |
| | | | | ‡(+P)4.4 | 915 | 18,100 |
| Unique | 3.4 | 601 | 8,900 | 5.1 | 895 | 16,100 |
| | | | | (+P)5.4 | 954 | 18,000 |
| Blue Dot | 6.0 | 686 | 10,400 | 7.2 | 867 | 15,700 |
| 231 | 3.5 | 680 | 11,100 | 4.7 | 885 | 16,800 |
| | | | | (+P)5.0 | 934 | 18,000 |
| 630 | 6.3 | 659 | 10,000 | 8.5 | 896 | 16,500 |
| HS-6 | 4.4 | 564 | 8,000 | 6.7 | 894 | 16,600 |
| AL-7 | 5.0 | 649 | 10,400 | 6.7 | 889 | 16,600 |
| HP38 | 2.8 | 642 | 10,200 | 3.9 | 854 | 16,300 |
| | | | | (+P)4.1 | 882 | 17,800 |

### 158 gr. Cast Bullet
#358311 (Linotype) 1.550" OAL

| Powder | Sug. Starting Grains | Velocity F.P.S. | Pressure C.U.P. | Max. Grains | Velocity F.P.S. | Pressure C.U.P. |
|---|---|---|---|---|---|---|
| PB | 3.0 | 604 | 7,900 | 4.5 | 871 | 16,100 |
| SR-4756 | 4.2 | 620 | 8,700 | 5.5 | 874 | 16,000 |
| Bullseye | 2.8 | 656 | 10,000 | 4.2 | 880 | 16,200 |
| | | | | ‡(+P)4.5 | 931 | 18,200 |
| Unique | 3.4 | 611 | 8,300 | 5.1 | 896 | 16,400 |
| | | | | (+P)5.4 | 946 | 18,000 |
| Blue Dot | 5.9 | 673 | 9,100 | 7.8 | 919 | 16,400 |
| 231 | 3.5 | 675 | 9,600 | *4.9 | 906 | 16,800 |
| | | | | (+P)5.2 | 956 | 18,500 |
| 630 | 6.3 | 663 | 9,000 | 8.9 | 895 | 16,000 |
| HS-6 | 4.5 | 610 | 8,200 | 6.7 | 901 | 16,200 |
| AL-7 | 5.8 | 716 | 10,000 | 7.0 | 897 | 15,700 |
| HP38 | 2.6 | 621 | 8,800 | 4.3 | 902 | 16,900 |
| | | | | (+P)4.5 | 944 | 18,400 |

### 162 gr. Cast Bullet
#357446 (Linotype) 1.457" OAL

| Powder | Sug. Starting Grains | Velocity F.P.S. | Pressure C.U.P. | Max. Grains | Velocity F.P.S. | Pressure C.U.P. |
|---|---|---|---|---|---|---|
| PB | 3.4 | 646 | 9,500 | 4.6 | 877 | 16,300 |
| SR-4756 | 4.6 | 694 | 10,400 | 5.8 | 921 | 16,800 |
| Bullseye | 3.0 | 680 | 10,100 | 4.2 | 889 | 16,900 |
| Unique | 3.5 | 639 | 8,900 | 5.2 | 914 | 16,200 |
| Blue Dot | 5.8 | 675 | 9,500 | 7.7 | 948 | 16,500 |
| 231 | 3.4 | 702 | 10,400 | 4.6 | 896 | 16,000 |
| 630 | 6.3 | 683 | 10,100 | *8.6 | 920 | 15,900 |
| HS-6 | 5.1 | 690 | 9,600 | 6.7 | 922 | 16,100 |
| AL-7 | 5.2 | 671 | 9,700 | 6.9 | 909 | 16,200 |

### 168 gr. Cast Bullet
#358429 (Linotype) 1.537" OAL

| Powder | Sug. Starting Grains | Velocity F.P.S. | Pressure C.U.P. | Max. Grains | Velocity F.P.S. | Pressure C.U.P. |
|---|---|---|---|---|---|---|
| SR-7625 | 3.4 | 621 | 9,700 | 4.6 | 821 | 16,300 |
| SR-4756 | 4.2 | 645 | 10,000 | 5.4 | 836 | 16,400 |
| Bullseye | 2.6 | 597 | 9,300 | 4.2 | 850 | 16,800 |
| Unique | 3.8 | 673 | 11,000 | 5.0 | 860 | 16,700 |
| Blue Dot | 5.5 | 647 | 10,000 | 7.3 | 869 | 16,500 |
| 231 | 3.0 | 599 | 9,300 | 4.6 | 855 | 16,900 |
| 630 | 6.1 | 660 | 9,900 | *8.4 | 869 | 16,500 |
| HS-6 | 5.0 | 677 | 11,000 | 6.4 | 856 | 16,800 |
| AL-7 | 5.4 | 717 | 11,900 | 6.9 | 879 | 16,700 |

**Note:** Loads shown in shaded panels are maximum.
* Designates potentially most accurate load.
\+ Designates a compressed powder charge.
‡ Designates factory velocity duplication load.
(+P) Designates higher pressure loads. See COMMENTS.

# .357 MAGNUM

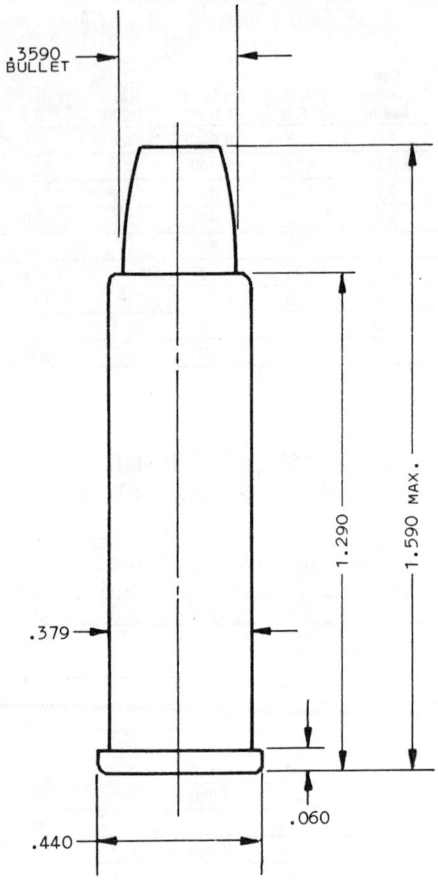


## TEST COMPONENTS

Cases .................................................... Federal
Trim-to Length ........................................ 1.285"
Primers ................................................ CCI 550
Primer Size ........................... Small Pistol Magnum
Lyman Shell Holder ..................................... No. 1
Jacketed Bullets Used .. Winchester JHP #B3573, 110 gr.
Hornady HP, 125 gr.
Speer JHP #4203, 140 gr.
Speer HJHP #4205, 146 gr.
Sierra HC #8330, 150 gr.
Hornady HP, 158 gr.
Speer HJSP #4223, 160 gr.
Cast Bullets Used ....................... (size to .357" dia.)
*Gas Check
#358242, 92 gr.
#358345, 115 gr.
#358242, 121 gr.
#356402, 121 gr.
#358480, 133 gr.
#358495, 141 gr.
#358477, 150 gr.
*#358156, 155 gr.
#358311, 158 gr.
#358429, 168 gr.

## TEST SPECIFICATIONS (Velocity and Pressure)

Firearm Used ........................ Universal Receiver and Special Vented Barrel
Barrel Length ................................................ 4"
Twist .................................................. 1-18¾"
Groove Diameter ...................................... .356"

## COMMENTS

Do not use .38 Special cases for .357 Magnum loads. Use of the smaller case will increase pressure.

Depending upon their manufacture, some variations in groove diameter exist in these handguns. Slug your bore before reloading and size cast bullets to exact groove diameter.

In order to hold the maximum overall length below 1.590", it is sometimes necessary to crimp cast bullets on the forward edge of the first driving band. Velocities for half-jacketed bullets must be kept above 750 feet per second to prevent jackets from lodging in the barrel.

For those desiring a mid-range load, we suggest bullet No. 358495 at starting load velocities. Bullet No. 358156 is extremely popular and an excellent choice for the heavier loads. Another great favorite is the "Keith type" bullet No. 358429. Refer to Opening Remarks when using Winchester 296 powder.

# .357 MAGNUM
## (Continued)

### 110 gr. Winchester JHP
#B3573 (.357" dia.) 1.590" OAL

| Powder | Sug. Starting Grains | Velocity F.P.S. | Pressure C.U.P. | Max. Grains | Velocity F.P.S. | Pressure C.U.P. |
|---|---|---|---|---|---|---|
| IMR-4227 | 14.8 | 1100 | 20,700 | *19.5+ | 1383 | 30,100 |
| Unique | 7.4 | 1052 | 17,300 | 10.0 | 1392 | 36,400 |
| Blue Dot | 10.0 | 930 | 12,400 | 14.6 | 1545 | 39,400 |
| 2400 | 13.5 | 1045 | 17,400 | 20.3+ | 1561 | 39,100 |
| 630 | 13.0 | 1148 | 19,200 | 18.0 | 1567 | 40,600 |
| AL-8 | 11.5 | 1075 | 17,400 | 14.5 | 1500 | 41,700 |

### 125 gr. Hornady HP
.357" dia./1.590" OAL

| Powder | Sug. Starting Grains | Velocity F.P.S. | Pressure C.U.P. | Max. Grains | Velocity F.P.S. | Pressure C.U.P. |
|---|---|---|---|---|---|---|
| SR-4759 | 13.0 | 963 | 19,000 | 17.0+ | 1214 | 29,300 |
| IMR-4227 | 15.6 | 1232 | 31,200 | 19.5+ | 1427 | 37,400 |
| Unique | 7.0 | 990 | 18,900 | 9.7 | 1359 | 41,600 |
| Blue Dot | 10.4 | 1113 | 20,000 | 13.3 | 1402 | 36,700 |
| 2400 | 13.0 | 1159 | 24,900 | ‡17.7 | 1478 | 40,600 |
| 630 | 12.5 | 1149 | 22,700 | 15.6 | 1415 | 40,200 |
| H110 | 18.9 | 1428 | 27,500 | 21.5 | 1645 | 41,300 |
| AL-8 | 10.7 | 1088 | 23,500 | 13.1 | 1386 | 41,500 |
| HS-7 | 10.9 | 1098 | 21,400 | 14.4 | 1403 | 39,200 |

### 140 gr. Speer JHP
#4203 (.357" dia.) 1.590" OAL

| Powder | Sug. Starting Grains | Velocity F.P.S. | Pressure C.U.P. | Max. Grains | Velocity F.P.S. | Pressure C.U.P. |
|---|---|---|---|---|---|---|
| IMR-4227 | 15.0 | 1106 | 28,800 | 17.8 | 1372 | 41,400 |
| Unique | 7.0 | 949 | 20,800 | 8.9 | 1233 | 40,000 |
| Blue Dot | 9.2 | 917 | 17,900 | 11.9 | 1272 | 35,200 |
| 2400 | 12.5 | 1047 | 26,500 | 16.5 | 1343 | 40,300 |
| 630 | 11.9 | 1058 | 24,800 | 14.9 | 1318 | 40,500 |
| **296 | — | — | — | 19.2 | 1469 | 41,300 |
| H110 | 15.2 | 1189 | 25,800 | 19.5 | 1427 | 37,000 |

### 146 gr. Speer HJHP
#4205 (.357" dia.) 1.590" OAL

| Powder | Sug. Starting Grains | Velocity F.P.S. | Pressure C.U.P. | Max. Grains | Velocity F.P.S. | Pressure C.U.P. |
|---|---|---|---|---|---|---|
| IMR-4227 | 11.5 | 914 | 21,800 | 16.2 | 1253 | 40,600 |
| Unique | 6.3 | 928 | 25,900 | 8.1 | 1200 | 41,100 |
| Blue Dot | 9.0 | 943 | 21,300 | 11.5 | 1273 | 39,700 |
| 2400 | 11.0 | 895 | 20,800 | 15.1 | 1242 | 39,100 |
| 630 | 10.2 | 926 | 20,800 | *13.7 | 1220 | 38,600 |
| **296 | — | — | — | 16.9 | 1389 | 41,800 |
| H110 | 13.0 | 1064 | 22,300 | 17.0 | 1380 | 41,900 |
| AL-8 | 8.8 | 854 | 20,200 | 12.0 | 1208 | 39,600 |

### 150 gr. Sierra HC
#8330 (.357" dia.) 1.590" OAL

| Powder | Sug. Starting Grains | Velocity F.P.S. | Pressure C.U.P. | Max. Grains | Velocity F.P.S. | Pressure C.U.P. |
|---|---|---|---|---|---|---|
| IMR-4227 | 12.5 | 896 | 20,700 | 16.8 | 1251 | 39,200 |
| Unique | 6.5 | 878 | 20,400 | 8.7 | 1204 | 41,000 |
| 2400 | 12.0 | 958 | 22,800 | 15.6 | 1311 | 40,700 |
| 630 | 10.5 | 985 | 21,200 | 14.0 | 1269 | 40,300 |
| **296 | — | — | — | 17.4 | 1366 | 40,300 |
| H110 | 13.9 | 1103 | 22,800 | 17.9 | 1359 | 38,400 |
| AL-8 | 8.8 | 855 | 17,500 | 11.6 | 1175 | 40,500 |

### 158 gr. Hornady HP
.357" dia./1.590" OAL

| Powder | Sug. Starting Grains | Velocity F.P.S. | Pressure C.U.P. | Max. Grains | Velocity F.P.S. | Pressure C.U.P. |
|---|---|---|---|---|---|---|
| IMR-4227 | 12.2 | 944 | 21,500 | ‡16.1 | 1249 | 41,400 |
| Unique | 6.4 | 887 | 18,900 | 8.3 | 1185 | 38,300 |
| Blue Dot | 9.6 | 995 | 20,100 | 11.9 | 1319 | 40,900 |
| 2400 | 11.3 | 938 | 20,700 | 14.9 | 1279 | 41,800 |
| 630 | 10.6 | 974 | 21,200 | 14.1 | 1277 | 39,900 |
| **296 | — | — | — | 17.1 | 1373 | 40,800 |
| H110 | 13.7 | 1087 | 22,300 | 17.7 | 1387 | 42,000 |
| AL-8 | 9.1 | 869 | 19,800 | 11.2 | 1215 | 41,800 |

**Note:** Loads shown in shaded panels are maximum.
* Designates potentially most accurate load.
+ Designates a compressed powder charge.
‡ Designates factory velocity duplication load.
** Reduced loads not recommended by Winchester.

# .357 MAGNUM
(Continued)

### 160 gr. Speer HJSP
#4223 (.357" dia.) 1.590" OAL

| Powder | Sug. Starting Grains | Velocity F.P.S. | Pressure C.U.P. | Max. Grains | Velocity F.P.S. | Pressure C.U.P. |
|---|---|---|---|---|---|---|
| IMR-4227 | 11.5 | 897 | 20,600 | 15.4 | 1242 | 40,900 |
| Unique | 6.1 | 858 | 19,400 | *8.2 | 1178 | 40,900 |
| Blue Dot | 9.2 | 996 | 20,100 | 11.2 | 1260 | 40,900 |
| 2400 | 11.4 | 870 | 23,600 | 14.8 | 1273 | 40,300 |
| 630 | 10.0 | 919 | 19,800 | 13.5 | 1266 | 40,400 |
| **296 | — | — | — | 16.4 | 1379 | 41,300 |
| H110 | 12.8 | 1029 | 20,700 | 16.8 | 1398 | 42,000 |
| AL-8 | 9.0 | 885 | 21,400 | 10.9 | 1175 | 41,300 |

### 92 gr. Cast Bullet
#358242 (Linotype) 1.585" OAL

| Powder | Sug. Starting Grains | Velocity F.P.S. | Pressure C.U.P. | Max. Grains | Velocity F.P.S. | Pressure C.U.P. |
|---|---|---|---|---|---|---|
| 700X | 6.0 | 1206 | 19,700 | *8.3 | 1581 | 41,500 |
| SR-7625 | 7.5 | 1262 | 20,700 | 9.8 | 1613 | 40,900 |
| Bullseye | 7.5 | 1365 | 25,900 | 9.5 | 1630 | 39,900 |
| 231 | 8.1 | 1329 | 25,600 | 10.1 | 1629 | 41,200 |
| 630 | 14.9 | 1385 | 23,900 | 18.1 | 1717 | 40,700 |
| HS-5 | 10.8 | 1404 | 25,700 | 13.6 | 1671 | 38,500 |
| AL-5 | 9.2 | 1283 | 21,900 | 12.2 | 1645 | 39,400 |

### 115 gr. Cast Bullet
#358345 (Linotype) 1.465" OAL

| Powder | Sug. Starting Grains | Velocity F.P.S. | Pressure C.U.P. | Max. Grains | Velocity F.P.S. | Pressure C.U.P. |
|---|---|---|---|---|---|---|
| SR-7625 | 5.4 | 1036 | 17,600 | *8.3 | 1446 | 41,900 |
| SR-4756 | 6.8 | 1137 | 19,600 | 12.0 | 1626 | 41,900 |
| Unique | 7.5 | 1233 | 24,300 | 9.2 | 1489 | 41,400 |
| Blue Dot | 9.4 | 1134 | 20,400 | 12.8 | 1586 | 41,200 |
| 231 | 6.4 | 1166 | 23,900 | 8.4 | 1442 | 40,600 |
| 630 | 11.9 | 1226 | 23,100 | 15.5 | 1564 | 42,000 |
| HS-5 | 8.0 | 1086 | 17,900 | 11.7 | 1548 | 41,700 |
| AL-5 | 7.5 | 1070 | 18,500 | 10.5 | 1494 | 40,700 |

### 121 gr. Cast Bullet
#358242 (Linotype) 1.585" OAL

| Powder | Sug. Starting Grains | Velocity F.P.S. | Pressure C.U.P. | Max. Grains | Velocity F.P.S. | Pressure C.U.P. |
|---|---|---|---|---|---|---|
| SR-7625 | 5.5 | 974 | 17,500 | 8.0 | 1335 | 41,100 |
| Unique | 6.9 | 1088 | 21,100 | 9.1 | 1409 | 42,000 |
| 2400 | 13.0 | 1178 | 25,700 | 16.4 | 1494 | 41,700 |
| 630 | *11.4 | 1102 | 20,300 | 14.9 | 1439 | 39,500 |
| HS-7 | 10.0 | 1053 | 18,400 | 13.2 | 1436 | 40,300 |
| AL-8 | 9.0 | 925 | 16,300 | 11.9 | 1383 | 40,500 |

### 121 gr. Cast Bullet
#356402 (Linotype) 1.590" OAL

| Powder | Sug. Starting Grains | Velocity F.P.S. | Pressure C.U.P. | Max. Grains | Velocity F.P.S. | Pressure C.U.P. |
|---|---|---|---|---|---|---|
| SR-7625 | 7.0 | 1205 | 23,000 | *8.8 | 1429 | 41,600 |
| Unique | 7.2 | 1134 | 15,800 | 9.4 | 1482 | 41,300 |
| Blue Dot | 10.8 | 1156 | 17,000 | 13.0 | 1564 | 41,400 |
| 2400 | 13.5 | 1140 | 17,300 | 19.5+ | 1565 | 37,200 |
| 630 | 12.4 | 1230 | 20,800 | 15.5 | 1520 | 40,200 |
| AL-8 | 10.3 | 1087 | 16,000 | 13.0 | 1481 | 40,300 |

### 133 gr. Cast Bullet
#358480 (Linotype) 1.508" OAL

| Powder | Sug. Starting Grains | Velocity F.P.S. | Pressure C.U.P. | Max. Grains | Velocity F.P.S. | Pressure C.U.P. |
|---|---|---|---|---|---|---|
| IMR-4227 | 12.0 | 1045 | 21,700 | 17.4 | 1421 | 40,200 |
| Unique | 6.2 | 1006 | 18,300 | 8.6 | 1356 | 40,300 |
| Blue Dot | 10.0 | 1162 | 18,800 | 11.6 | 1426 | 41,000 |
| 2400 | 11.8 | 1058 | 15,500 | 16.8 | 1501 | 40,200 |
| 630 | 11.2 | 1130 | 17,300 | *14.8 | 1466 | 40,900 |
| AL-8 | 9.5 | 1096 | 19,100 | 12.7 | 1461 | 41,500 |

**Note:** Loads shown in shaded panels are maximum.
* Designates potentially most accurate load.
+ Designates a compressed powder charge.
** Reduced loads not recommended by Winchester.

# .357 MAGNUM
(Continued)

## 141 gr. Cast Bullet
#358495 (Linotype) 1.435'' OAL

| Powder | Sug. Starting Grains | Velocity F.P.S. | Pressure C.U.P. | Max. Grains | Velocity F.P.S. | Pressure C.U.P. |
|---|---|---|---|---|---|---|
| IMR-4227 | 12.3 | 1084 | 22,200 | *16.3+ | 1365 | 39,600 |
| Unique | 5.8 | 960 | 14,600 | 7.9 | 1289 | 40,300 |
| Blue Dot | 9.6 | 1134 | 21,800 | 11.2 | 1382 | 40,600 |
| 2400 | 10.5 | 1015 | 18,200 | 14.6 | 1376 | 39,700 |
| 630 | 9.9 | 1032 | 16,600 | 13.5 | 1369 | 40,100 |
| HS-7 | 9.8 | 1130 | 23,200 | 11.8 | 1356 | 40,700 |
| AL-8 | 8.8 | 1012 | 19,800 | 11.6 | 1334 | 41,000 |

## 150 gr. Cast Bullet
#358477 (Linotype) 1.510'' OAL

| Powder | Sug. Starting Grains | Velocity F.P.S. | Pressure C.U.P. | Max. Grains | Velocity F.P.S. | Pressure C.U.P. |
|---|---|---|---|---|---|---|
| IMR-4227 | 11.8 | 1044 | 21,700 | 16.0+ | 1333 | 40,200 |
| Herco | 6.2 | 972 | 17,700 | 7.8 | 1225 | 41,900 |
| Blue Dot | 8.2 | 950 | 13,000 | 10.8 | 1356 | 41,200 |
| 2400 | 11.0 | 998 | 17,900 | 15.0 | 1362 | 41,400 |
| 630 | 10.0 | 1029 | 18,400 | *13.0 | 1311 | 39,900 |
| **296 | — | — | — | 17.8 | 1452 | 41,900 |
| H110 | 12.4 | 1075 | 17,300 | 17.7 | 1459 | 41,200 |
| AL-8 | 8.8 | 1046 | 23,000 | 10.4 | 1263 | 40,900 |

## 155 gr. Cast Bullet
#358156 (Linotype) 1.590'' OAL

| Powder | Sug. Starting Grains | Velocity F.P.S. | Pressure C.U.P. | Max. Grains | Velocity F.P.S. | Pressure C.U.P. |
|---|---|---|---|---|---|---|
| IMR-4227 | 11.4 | 973 | 24,100 | *15.2 | 1254 | 41,300 |
| Herco | 5.0 | 785 | 16,300 | 7.5 | 1151 | 41,000 |
| Blue Dot | 8.5 | 975 | 21,500 | 10.5 | 1277 | 40,800 |
| 2400 | 10.6 | 999 | 24,900 | 14.0 | 1299 | 41,900 |
| **296 | — | — | — | 15.6 | 1334 | 39,700 |
| 630 | 9.6 | 953 | 20,900 | 13.2 | 1286 | 42,000 |
| H110 | 11.6 | 1037 | 21,800 | 15.7 | 1363 | 40,300 |
| AL-8 | 8.2 | 919 | 22,100 | 10.7 | 1235 | 41,500 |

## 158 gr. Cast Bullet
#358311 (Linotype) 1.590''OAL

| Powder | Sug. Starting Grains | Velocity F.P.S. | Pressure C.U.P. | Max. Grains | Velocity F.P.S. | Pressure C.U.P. |
|---|---|---|---|---|---|---|
| SR-4759 | 10.0 | 874 | 12,000 | 14.0+ | 1189 | 30,100 |
| IMR-4227 | 11.8 | 977 | 19,600 | *17.0+ | 1345 | 40,600 |
| Herco | 6.3 | 963 | 18,700 | 7.9 | 1203 | 41,000 |
| Blue Dot | 8.2 | 888 | 11,600 | 10.9 | 1316 | 39,200 |
| 2400 | 11.4 | 1024 | 20,200 | 15.5 | 1344 | 39,700 |
| **296 | — | — | — | 18.0 | 1461 | 41,400 |
| 630 | 10.3 | 1016 | 18,000 | 13.6 | 1316 | 40,400 |
| H110 | 13.0 | 1115 | 19,100 | 18.3 | 1460 | 40,100 |
| AL-8 | 8.9 | 973 | 18,700 | ‡10.8 | 1257 | 39,400 |

## 168 gr. Cast Bullet
#358429 (Linotype) 1.553'' OAL

| Powder | Sug. Starting Grains | Velocity F.P.S. | Pressure C.U.P. | Max. Grains | Velocity F.P.S. | Pressure C.U.P. |
|---|---|---|---|---|---|---|
| SR-4759 | 9.8 | 942 | 18,400 | 13.5+ | 1206 | 40,600 |
| IMR-4227 | 9.8 | 835 | 14,100 | 14.5+ | 1233 | 40,800 |
| Herco | 5.6 | 885 | 17,900 | 7.1 | 1104 | 39,000 |
| Blue Dot | 8.3 | 970 | 18,400 | 10.0 | 1233 | 39,200 |
| 2400 | 9.7 | 879 | 15,900 | 13.5 | 1242 | 41,100 |
| **296 | — | — | — | 15.2 | 1307 | 39,900 |
| 630 | 9.7 | 1012 | 21,400 | 12.5 | 1247 | 41,100 |
| H110 | *11.8 | 1037 | 18,900 | 15.7 | 1318 | 39,800 |
| AL-8 | 8.3 | 882 | 19,500 | 11.2 | 1208 | 41,000 |

**Notes:** Loads shown in shaded panels are maximum.
* Designates potentially most accurate load.
+ Designates a compressed powder charge.
‡ Designates factory velocity duplication load.
** Reduced loads not recommended by Winchester.

# .38/40 WINCHESTER

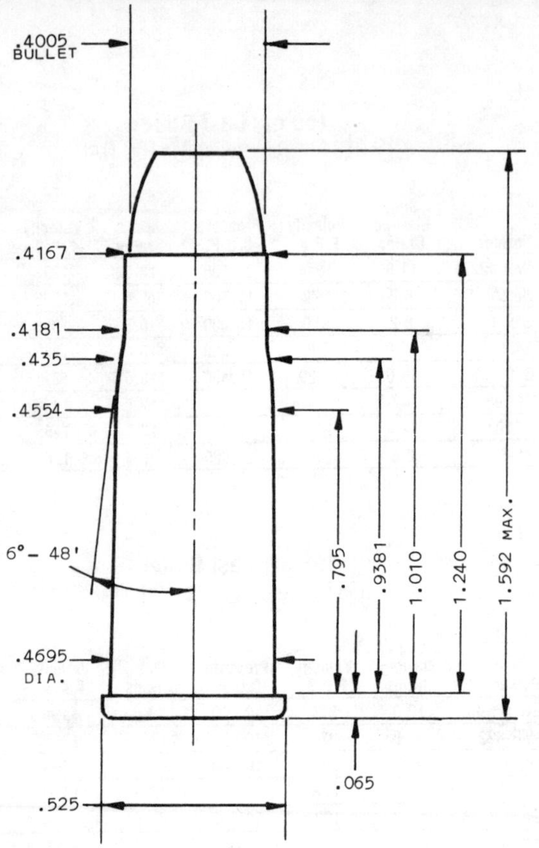


## TEST COMPONENTS

Cases ............ Winchester
Trim-to Length ............ 1.300''
Primers ............ Winchester 7-111
Primer Size ............ Large Pistol
Lyman Shell Holder ............ No. 14B
Jacketed Bullets Used .. Remington SP #B22880, 180 gr.
Cast Bullets Used ............ (size to .401'' dia.)
#40143, 172 gr.
(Can also use #40188, 170 gr.)

## TEST SPECIFICATIONS (Velocity Only)

Firearm Used ............ Colt S.A.A.
Barrel Length ............ 7½''
Twist ............ 1-36''
Groove Diameter ............ .400''

## COMMENTS

These loads are not to be used in handguns which were designed for black powder. A great many revolvers which are chambered for this cartridge have groove and bore dimensions which are so far oversized that it is impossible to obtain accuracy. Because chamber dimensions limit the diameter of the bullet, nothing can be done to make these oversize barrels shoot well.

**180 gr. Remington SP**
#B22880 (.400'' dia.) 1.592'' OAL

| Powder | Sug. Starting Grains | Velocity F.P.S. | Pressure C.U.P. | Max. Grains | Velocity F.P.S. | Pressure C.U.P. |
|---|---|---|---|---|---|---|
| Unique | *8.0 | 715 | — | ‡9.5 | 1000 | — |

**172 gr. Cast Bullet**
#40143 (#2 Alloy) 1.592'' OAL

| Powder | Sug. Starting Grains | Velocity F.P.S. | Pressure C.U.P. | Max. Grains | Velocity F.P.S. | Pressure C.U.P. |
|---|---|---|---|---|---|---|
| Bullseye | 4.0 | 740 | — | 5.9 | 965 | — |
| Unique | 7.0 | 830 | — | 10.0 | 1105 | — |

**Note:** Loads shown in shaded panels are maximum.
* Designates potentially most accurate load.
‡ Designates factory velocity duplication load.

# .41 MAGNUM

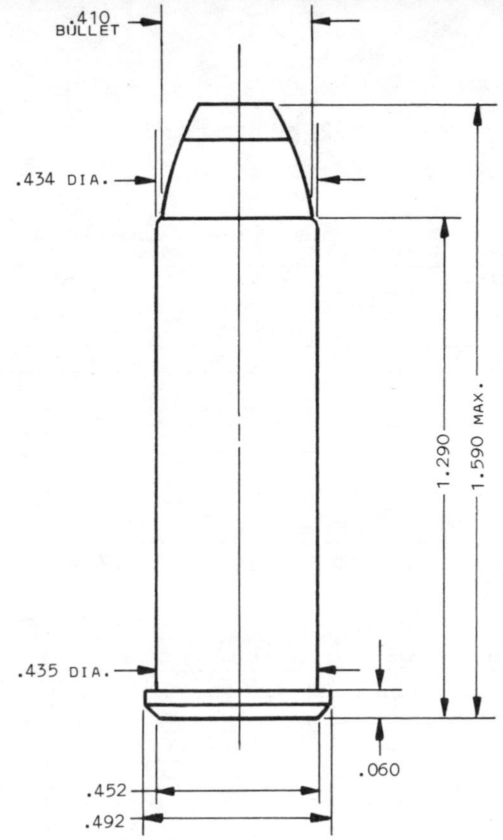


## TEST COMPONENTS

Cases ........ Winchester
Trim-to Length ........ 1.285''
Primers ........ Remington 2½
Primer Size ........ Large Pistol
Lyman Shell Holder ........ No. 30
Jacketed Bullets Used ........ Sierra HC #8500, 170 gr.
Speer HJHP #4405, 200 gr.
Hornady HP, 210 gr.
Speer HJSP #4417, 220 gr.
Cast Bullets Used ........ (size to .410'' dia.)
*Gas Check #41028, 212 gr.
*#410610, 215 gr.
#410459, 220 gr.
#410426, 240 gr.

## TEST SPECIFICATIONS (Velocity and Pressure)

Firearm Used ........ Universal Receiver and Special Vented Barrel
Barrel Length ........ 4''
Twist ........ 1-18¾''
Groove Diameter ........ .409''

## COMMENTS

Cast bullet No.'s 410459 and 410426 list overall lengths that exceed the maximum of 1.590'' and are suitable only for the Thompson/Center Contender or similar firearms that will accept cartridges of these dimensions. The data listed for these bullets take this extra length into consideration.

Overall length with bullet No. 410459 - 1.700''.

Overall length with bullet No. 410426 - 1.707''.

Bullet No. 410426 is extremely popular in this caliber.

**170 gr. Sierra HC**
#8500 (.410'' dia.) 1.580'' OAL

| Powder | Sug. Starting Grains | Velocity F.P.S. | Pressure C.U.P. | Max. Grains | Velocity F.P.S. | Pressure C.U.P. |
|---|---|---|---|---|---|---|
| SR-4756 | 10.6 | 1052 | 13,300 | 13.0 | 1360 | 36,700 |
| IMR-4227 | 20.5 | 1131 | 20,000 | 23.5+ | 1340 | 33,000 |
| 2400 | 17.5 | 1040 | 15,000 | 22.0+ | 1311 | 28,600 |
| Unique | 9.8 | 1056 | 13,700 | 12.4 | 1378 | 36,100 |
| 630 | 16.2 | 1216 | 22,400 | *19.0 | 1414 | 35,300 |
| H110 | 22.5 | 1217 | 15,200 | 26.3+ | 1507 | 33,700 |
| AL-8 | 15.7 | 1156 | 22,900 | 20.0+ | 1407 | 35,600 |

**200 gr. Speer HJHP**
#4405 (.410'' dia.) 1.590'' OAL

| Powder | Sug. Starting Grains | Velocity F.P.S. | Pressure C.U.P. | Max. Grains | Velocity F.P.S. | Pressure C.U.P. |
|---|---|---|---|---|---|---|
| IMR-4227 | 18.0 | 1043 | 21,200 | 21.0+ | 1254 | 35,800 |
| Blue Dot | 12.5 | 1018 | 15,900 | 15.4 | 1302 | 35,200 |
| 2400 | 15.3 | 966 | 15,300 | 20.0 | 1277 | 36,100 |
| Unique | 8.2 | 984 | 14,100 | 10.7 | 1218 | 35,400 |
| 630 | 13.9 | 1043 | 18,400 | *17.1 | 1272 | 35,600 |
| H110 | 19.6 | 1164 | 20,800 | 23.4 | 1421 | 37,000 |
| AL-8 | 12.5 | 960 | 16,600 | 17.0+ | 1260 | 35,400 |

**Note:** Loads shown in shaded panels are maximum.
* Designates potentially most accurate load.

\+ Designates a compressed powder charge.

# .41 MAGNUM
(Continued)

## 210 gr. Hornady HP
.410" dia./1.585" OAL

| Powder | Sug. Starting Grains | Velocity F.P.S. | Pressure C.U.P. | Max. Grains | Velocity F.P.S. | Pressure C.U.P. |
|---|---|---|---|---|---|---|
| IMR-4227 | 17.8 | 1016 | 19,100 | *21.0+ | 1232 | 36,500 |
| Blue Dot | 12.6 | 1029 | 16,800 | 15.8 | 1304 | 37,000 |
| 2400 | 15.3 | 966 | 14,600 | 20.0 | 1233 | 36,100 |
| Unique | 8.6 | 1011 | 15,600 | 11.0 | 1196 | 34,700 |
| 630 | 14.3 | 1019 | 19,700 | 17.4 | 1251 | 35,700 |
| H110 | 19.0 | 1105 | 19,800 | ‡22.9 | 1349 | 35,300 |

## 220 gr. Speer HJSP
#4417 (.410" dia.) 1.590" OAL

| Powder | Sug. Starting Grains | Velocity F.P.S. | Pressure C.U.P. | Max. Grains | Velocity F.P.S. | Pressure C.U.P. |
|---|---|---|---|---|---|---|
| IMR-4227 | 15.3 | 900 | 15,500 | *19.7 | 1190 | 36,900 |
| Blue Dot | 11.0 | 996 | 14,900 | 14.2 | 1219 | 34,900 |
| 2400 | 12.9 | 876 | 17,700 | 17.4 | 1172 | 36,700 |
| Unique | 7.5 | 810 | 13,100 | 10.0 | 1139 | 34,800 |
| 630 | 10.6 | 913 | 18,100 | 15.3 | 1172 | 36,900 |
| H110 | 15.7 | 1031 | 20,000 | 19.4 | 1261 | 36,000 |
| AL-8 | 11.7 | 880 | 19,900 | 15.2 | 1197 | 37,000 |

## 212 gr. Cast Bullet
#41028 (Linotype) 1.586" OAL

| Powder | Sug. Starting Grains | Velocity F.P.S. | Pressure C.U.P. | Max. Grains | Velocity F.P.S. | Pressure C.U.P. |
|---|---|---|---|---|---|---|
| SR-4756 | 8.8 | 1030 | 21,700 | 10.9 | 1221 | 36,300 |
| IMR-4227 | 16.2 | 1006 | 21,300 | 20.5 | 1247 | 35,900 |
| Unique | 8.2 | 998 | 20,300 | 10.7 | 1217 | 36,300 |
| Blue Dot | 12.2 | 1041 | 20,900 | 15.4 | 1333 | 36,500 |
| 2400 | 15.4 | 1030 | 21,700 | 20.0 | 1292 | 36,100 |
| 630 | *11.9 | 1023 | 19,500 | 16.5 | 1259 | 35,700 |
| HS-7 | 12.0 | 1031 | 21,500 | 15.1 | 1265 | 36,200 |
| AL-8 | 13.5 | 1050 | 25,300 | 16.7 | 1264 | 35,500 |

## 215 gr. Cast Bullet
#410610 (Linotype) 1.575" OAL

| Powder | Sug. Starting Grains | Velocity F.P.S. | Pressure C.U.P. | Max. Grains | Velocity F.P.S. | Pressure C.U.P. |
|---|---|---|---|---|---|---|
| SR-4756 | 8.0 | 909 | 17,100 | 10.8 | 1213 | 36,000 |
| IMR-4227 | 14.9 | 956 | 20,100 | 19.2 | 1221 | 35,200 |
| Unique | 7.8 | 979 | 19,800 | 10.4 | 1214 | 36,100 |
| Blue Dot | 11.4 | 1003 | 20,200 | 14.6 | 1170 | 35,800 |
| 2400 | 14.2 | 835 | 19,600 | 18.8 | 1272 | 37,000 |
| 630 | 11.5 | 923 | 18,200 | 16.0 | 1251 | 35,800 |
| HS-7 | 11.5 | 1007 | 20,900 | *14.5 | 1249 | 36,800 |
| AL-8 | 11.9 | 974 | 22,200 | 15.0+ | 1206 | 35,000 |

## 220 gr. Cast Bullet
#410459 (Linotype) 1.700" OAL

| Powder | Sug. Starting Grains | Velocity F.P.S. | Pressure C.U.P. | Max. Grains | Velocity F.P.S. | Pressure C.U.P. |
|---|---|---|---|---|---|---|
| SR-4756 | 9.8 | 1061 | 20,900 | *11.6 | 1246 | 33,800 |
| IMR-4227 | 17.3 | 1048 | 22,100 | 21.5+ | 1280 | 35,500 |
| Unique | 8.1 | 970 | 18,000 | 11.2 | 1261 | 36,200 |
| Blue Dot | 13.0 | 1100 | 21,700 | 16.0 | 1359 | 36,000 |
| 2400 | 15.7 | 1022 | 21,100 | 20.1 | 1330 | 35,400 |
| 630 | 13.0 | 1012 | 19,000 | 17.5 | 1306 | 36,100 |
| HS-7 | 13.0 | 1061 | 21,400 | 15.9 | 1313 | 36,900 |
| AL-8 | 13.7 | 1057 | 23,100 | 16.8 | 1291 | 36,000 |

## 240 gr. Cast Bullet
#410426 (Linotype) 1.707" OAL

| Powder | Sug. Starting Grains | Velocity F.P.S. | Pressure C.U.P. | Max. Grains | Velocity F.P.S. | Pressure C.U.P. |
|---|---|---|---|---|---|---|
| SR-4756 | 8.7 | 1000 | 23,900 | 10.5 | 1159 | 36,800 |
| IMR-4227 | 15.0 | 932 | 20,500 | 19.1 | 1177 | 35,800 |
| Unique | 6.9 | 879 | 18,200 | 10.0 | 1147 | 36,900 |
| Blue Dot | 11.5 | 991 | 19,400 | 14.5 | 1249 | 36,300 |
| 2400 | 14.0 | 915 | 18,200 | 18.5 | 1214 | 36,100 |
| 630 | 11.1 | 900 | 18,300 | *15.6 | 1195 | 36,000 |
| HS-7 | 11.5 | 1007 | 21,400 | 14.5 | 1211 | 36,100 |
| AL-8 | 12.2 | 936 | 23,000 | 15.5 | 1193 | 36,900 |

**Note:** Loads shown in shaded panels are maximum.
* Designates potentially most accurate load.
\+ Designates a compressed powder charge.
‡ Designates factory velocity duplication load.

# .44 SMITH & WESSON SPECIAL

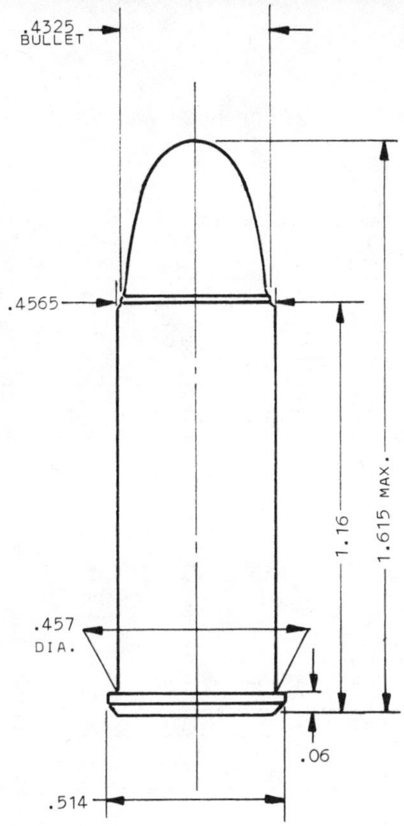


## TEST COMPONENTS

Cases ........ Winchester
Trim-to Length ........ 1.152''
Primers ........ CCI 300
Primer Size ........ Large Pistol
Lyman Shell Holder ........ No. 7
Jacketed Bullets Used .......... Sierra HC #8600, 180 gr.
Hornady HP, 200 gr.
Speer HJHP #4435, 225 gr.
Remington SP #B22906, 240 gr.
Cast Bullets Used ........ (size to .429'' dia.)
*Gas Check
#42798, 205 gr.
*#429215, 215 gr.
#429360, 232 gr.
#429421, 245 gr.
*#429244, 250 gr.

## TEST SPECIFICATIONS (Velocity and Pressure)

Firearm Used ........ Universal Receiver and Special Vented Barrel
Barrel Length ........ 4''
Twist ........ 1-20''
Groove Diameter ........ .429''

## COMMENTS

Considerable variations in groove diameters exist with handguns chambered for this cartridge. We recommend that you slug your barrel before reloading and size bullets to as near groove diameter as possible.

When loading half-jacketed bullets, do not use charge weights below suggested starting loads to avoid the danger of lodging bullets in the barrel.

Those wishing a ''Keith type'' bullet should choose bullet No. 429421.

**180 gr. Sierra HC**
#8600 (.4295'' dia.) 1.470'' OAL

| Powder | Sug. Starting Grains | Velocity F.P.S. | Pressure C.U.P. | Max. Grains | Velocity F.P.S. | Pressure C.U.P. |
|---|---|---|---|---|---|---|
| SR-4759 | 14.6 | 705 | 10,100 | 16.3+ | 869 | 12,800 |
| IMR-4227 | 15.1 | 685 | 9,400 | 16.8 | 862 | 13,100 |
| Red Dot | 6.4 | 793 | 11,600 | 7.1 | 850 | 13,700 |
| 2400 | *14.0 | 674 | 8,100 | 15.8 | 851 | 13,300 |
| 630 | 12.0 | 745 | 10,100 | 13.6 | 877 | 13,200 |
| 231 | 6.5 | 689 | 9,300 | 7.5 | 832 | 13,600 |
| HP38 | 5.6 | 733 | 10,700 | 6.4 | 858 | 14,000 |

**200 gr. Hornady HP**
.430'' dia./1.593'' OAL

| Powder | Sug. Starting Grains | Velocity F.P.S. | Pressure C.U.P. | Max. Grains | Velocity F.P.S. | Pressure C.U.P. |
|---|---|---|---|---|---|---|
| SR-4759 | 13.9 | 665 | 11,400 | 15.5 | 828 | 13,000 |
| IMR-4227 | 14.3 | 657 | 10,200 | 15.8 | 882 | 13,900 |
| Red Dot | 5.5 | 660 | 9,000 | 6.3 | 776 | 13,800 |
| 2400 | 12.0 | 599 | 8,000 | 13.5 | 838 | 13,700 |
| 630 | 11.7 | 700 | 10,300 | 13.0 | 808 | 13,400 |
| 231 | 6.4 | 707 | 11,100 | 7.1 | 797 | 13,800 |
| HP38 | 5.2 | 599 | 10,400 | 5.8 | 710 | 13,800 |

**Note:** Loads shown in shaded panels are maximum.
* Designates potentially most accurate load.

# .44 SMITH & WESSON SPECIAL
## (Continued)

### 225 gr. Speer HJHP
#4435 (.429" dia.) 1.615" OAL

| Powder | Sug. Starting Grains | Velocity F.P.S. | Pressure C.U.P. | Max. Grains | Velocity F.P.S. | Pressure C.U.P. |
|---|---|---|---|---|---|---|
| SR-4759 | 12.3 | 597 | 9,200 | 14.0 | 763 | 14,000 |
| IMR-4227 | *12.3 | 640 | 10,600 | 14.0 | 749 | 13,400 |
| Red Dot | 4.9 | 590 | 10,900 | 5.7 | 767 | 13,900 |
| 2400 | 11.6 | 653 | 10,600 | 13.3 | 790 | 13,800 |
| 630 | 9.5 | 632 | 10,400 | 11.0 | 763 | 13,600 |
| 231 | 5.0 | 585 | 11,300 | 5.8 | 729 | 13,900 |
| HP38 | 4.6 | 627 | 11,800 | 5.3 | 757 | 14,000 |

### 240 gr. Remington SP
#B22906 (.430" dia.) 1.475" OAL

| Powder | Sug. Starting Grains | Velocity F.P.S. | Pressure C.U.P. | Max. Grains | Velocity F.P.S. | Pressure C.U.P. |
|---|---|---|---|---|---|---|
| SR-4759 | 12.8 | 607 | 10,400 | 14.2+ | 769 | 13,300 |
| IMR-4227 | 13.3 | 616 | 11,100 | 14.8 | 711 | 13,300 |
| Red Dot | 4.8 | 571 | 10,500 | 5.6 | 713 | 13,600 |
| 2400 | 12.1 | 604 | 10,300 | 13.5 | 723 | 13,000 |
| 630 | 10.0 | 621 | 11,100 | 11.2 | 725 | 13,900 |
| 231 | 5.3 | 581 | 11,000 | 6.1 | 616 | 13,400 |
| HP38 | 4.5 | 581 | 11,000 | 5.2 | 703 | 13,800 |

### 205 gr. Cast Bullet
#42798 (Linotype) 1.537" OAL

| Powder | Sug. Starting Grains | Velocity F.P.S. | Pressure C.U.P. | Max. Grains | Velocity F.P.S. | Pressure C.U.P. |
|---|---|---|---|---|---|---|
| SR-7625 | 6.8 | 761 | 9,700 | 7.7 | 838 | 12,600 |
| SR-4759 | 13.2 | 651 | 9,200 | 15.2+ | 798 | 12,100 |
| Unique | 7.0 | 736 | 10,500 | *8.2 | 869 | 14,000 |
| Blue Dot | 10.5 | 747 | 10,000 | 12.0 | 855 | 13,300 |
| 2400 | 13.3 | 734 | 10,800 | 15.2 | 870 | 13,500 |
| 630 | 10.8 | 790 | 12,200 | 12.2 | 876 | 13,800 |
| HS-7 | 10.0 | 732 | 10,100 | 11.6 | 874 | 13,600 |
| AL-8 | 11.0 | 767 | 11,400 | 12.5 | 871 | 13,900 |

### 215 gr. Cast Bullet
#429215 (Linotype) 1.500" OAL

| Powder | Sug. Starting Grains | Velocity F.P.S. | Pressure C.U.P. | Max. Grains | Velocity F.P.S. | Pressure C.U.P. |
|---|---|---|---|---|---|---|
| SR-7625 | 6.2 | 732 | 10,900 | 6.8 | 801 | 13,500 |
| Unique | 6.3 | 729 | 9,500 | 7.7 | 884 | 13,600 |
| Blue Dot | 9.5 | 689 | 8,800 | 11.0 | 885 | 13,400 |
| 2400 | 13.0 | 792 | 11,300 | 14.5 | 905 | 14,000 |
| 630 | 7.5 | 534 | 5,500 | 12.1 | 895 | 13,900 |
| HS-7 | 9.0 | 659 | 8,100 | 11.0 | 841 | 12,600 |
| AL-8 | 9.4 | 628 | 8,600 | 10.9 | 828 | 13,700 |

### 232 gr. Cast Bullet
#429360 (Linotype) 1.525" OAL

| Powder | Sug. Starting Grains | Velocity F.P.S. | Pressure C.U.P. | Max. Grains | Velocity F.P.S. | Pressure C.U.P. |
|---|---|---|---|---|---|---|
| SR-4756 | 6.7 | 666 | 9,800 | 8.0 | 774 | 13,400 |
| SR-4759 | 12.5 | 647 | 10,300 | 14.3 | 803 | 13,500 |
| Unique | 6.0 | 604 | 10,400 | 7.1 | 781 | 13,500 |
| Blue Dot | 9.0 | 646 | 10,300 | 10.0 | 751 | 13,100 |
| 2400 | 12.0 | 732 | 10,500 | 13.8 | 848 | 13,800 |
| 630 | 9.2 | 716 | 10,400 | 11.0 | 838 | 13,800 |
| HS-7 | 9.5 | 729 | 10,500 | 11.0 | 853 | 13,900 |
| AL-8 | 9.5 | 699 | 9,900 | *11.0 | 825 | 13,300 |

### 245 gr. Cast Bullet
#429421 (Linotype) 1.571" OAL

| Powder | Sug. Starting Grains | Velocity F.P.S. | Pressure C.U.P. | Max. Grains | Velocity F.P.S. | Pressure C.U.P. |
|---|---|---|---|---|---|---|
| SR-4756 | 6.8 | 661 | 9,000 | 8.1 | 780 | 13,800 |
| SR-4759 | 11.9 | 652 | 10,500 | 13.4 | 789 | 14,000 |
| Unique | 6.0 | 649 | 10,300 | ‡6.9 | 767 | 13,300 |
| Blue Dot | 9.0 | 710 | 10,800 | 10.0 | 807 | 13,200 |
| 2400 | 11.4 | 704 | 10,400 | 13.2 | 797 | 13,800 |
| 630 | 9.6 | 716 | 10,300 | *10.8 | 819 | 13,700 |
| HS-7 | 9.1 | 728 | 11,300 | 10.4 | 828 | 14,000 |
| AL-8 | 8.9 | 609 | 10,000 | 10.3 | 773 | 12,700 |

**Note:** Loads shown in shaded panels are maximum.
* Designates potentially most accurate load.
\+ Designates a compressed powder charge.
‡ Designates factory velocity duplication load.

# .44 SMITH & WESSON SPECIAL
(Continued)

**250 gr. Cast Bullet**
#429244 (Linotype) 1.550" OAL

| Powder | Sug. Starting Grains | Velocity F.P.S. | Pressure C.U.P. | Max. Grains | Velocity F.P.S. | Pressure C.U.P. |
|---|---|---|---|---|---|---|
| SR-4756 | 6.2 | 584 | 8,400 | 7.5 | 765 | 14,000 |
| Unique | 5.7 | 635 | 10,200 | 6.6 | 747 | 13,600 |
| 2400 | 10.5 | 602 | 8,500 | 12.3 | 742 | 12,700 |
| 630 | 9.3 | 667 | 10,400 | 10.5 | 769 | 13,200 |
| HS-7 | 8.5 | 606 | 8,700 | *10.0 | 743 | 13,400 |
| AL-8 | 8.5 | 607 | 9,800 | 9.9 | 730 | 13,100 |
| SR-4759 | 11.6 | 658 | 11,600 | 13.1 | 752 | 13,200 |

**Note:** Loads shown in shaded panels are maximum.
* Designates potentially most accurate load.

# .44 REMINGTON MAGNUM

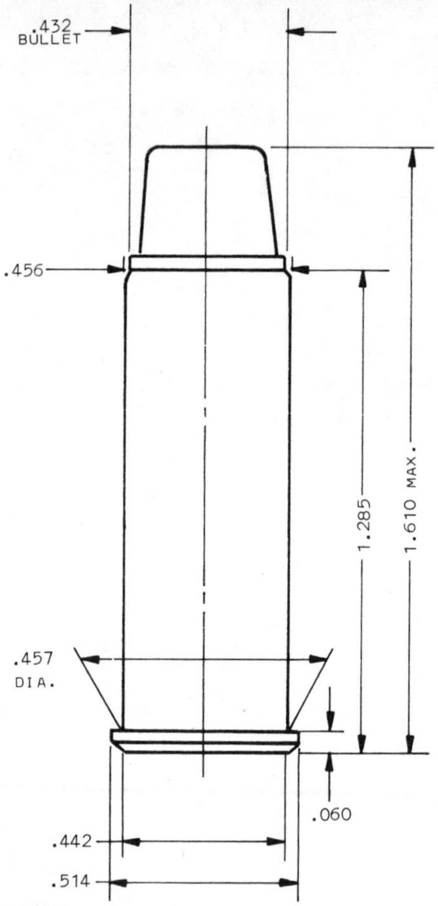


## TEST COMPONENTS

| | |
|---|---|
| Cases | Remington |
| Trim-to Length | 1.280'' |
| Primers | CCI 300 |
| Primer Size | Large Pistol |
| Lyman Shell Holder | No. 7 |
| Jacketed Bullets Used | Sierra HC #8600, 180 gr. |
| | Hornady HP, 200 gr. |
| | Speer HJHP #4435, 225 gr. |
| | Sierra HC #8610, 240 gr. |
| | Speer JHP #4453, 240 gr. |
| | Hornady FP, 265 gr. |
| Cast Bullets Used | (size to .429'' dia.) |
| *Gas Check | #429348, 180 gr. |
| | *#429303, 200 gr. |
| | *#429215, 215 gr. |
| | #429360, 232 gr. |
| | #429421, 245 gr. |
| | *#429244, 245 gr. |

## TEST SPECIFICATIONS
## (Velocity and Pressure)

| | |
|---|---|
| Firearm Used | Universal Receiver and Special Vented Barrel |
| Barrel Length | 4'' |
| Twist | 1-20'' |
| Groove Diameter | .429'' |

## COMMENTS

To prevent bullet movement when the gun is under recoil, the cartridge requires a hard crimp on all of the heavier loads.

When loading half-jacketed bullets, do not use charge weights below suggested starting loads to avoid the danger of lodging bullets in the barrel.

Cast bullet No.'s 429303, 429215, 429360, 429244, 429421 list overall lengths that exceed the maximum of 1.610'' and are suitable only for the Thonpson/Center Contender or similar firearms that will accept cartridges of these dimensions. The data listed for these bullets take this extra length into consideration.

Overall length with bullet No. 429303 - 1.692''.
Overall length with bullet No. 429215 - 1.645''.
Overall length with bullet No. 429360 - 1.660''.
Overall length with bullet No. 429244 - 1.680''.
Overall length with bullet No. 429421 - 1.710''.

Bullet No. 429348, loaded to starting load velocities, is recommended for mid-range use, while bullet No. 429215 is suggested as the best all around choice.

Refer to Opening Remarks when using Winchester 296 powder.

# .44 REMINGTON MAGNUM
(Continued)

### 180 gr. Sierra HC
#8600 (.4295" dia.) 1.610" OAL

| Powder | Sug. Starting Grains | Velocity F.P.S. | Pressure C.U.P. | Max. Grains | Velocity F.P.S. | Pressure C.U.P. |
|---|---|---|---|---|---|---|
| SR-4756 | 11.4 | 970 | 17,400 | 13.9 | 1262 | 34,400 |
| IMR-4227 | 24.2 | 1136 | 27,600 | 27.0 | 1317 | 34,800 |
| Unique | 12.0 | 1061 | 25,100 | 13.6 | 1301 | 35,700 |
| Blue Dot | 17.8 | 1126 | 25,800 | 20.2 | 1404 | 36,400 |
| 2400 | 21.3 | 1103 | 24,800 | 25.5 | 1370 | 36,400 |
| 231 | 10.2 | 1077 | 22,100 | 13.0 | 1267 | 36,300 |
| 630 | *18.1 | 1067 | 21,900 | 21.8 | 1327 | 35,900 |
| HS-7 | 17.8 | 1111 | 24,200 | 20.6 | 1341 | 36,300 |
| H110 | 28.7 | 1277 | 23,300 | 32.3 | 1510 | 36,500 |

### 200 gr. Hornady HP
.430" dia./1.610" OAL

| Powder | Sug. Starting Grains | Velocity F.P.S. | Pressure C.U.P. | Max. Grains | Velocity F.P.S. | Pressure C.U.P. |
|---|---|---|---|---|---|---|
| SR-4756 | 11.0 | 894 | 18,500 | 13.6 | 1164 | 36,300 |
| IMR-4227 | 22.8 | 1000 | 22,800 | 26.4 | 1261 | 36,600 |
| Unique | *10.5 | 905 | 21,100 | 13.2 | 1148 | 35,100 |
| Blue Dot | 15.7 | 942 | 19,600 | 18.7 | 1245 | 35,100 |
| 2400 | 19.5 | 972 | 23,100 | 23.6 | 1227 | 35,000 |
| 630 | 16.4 | 899 | 19,000 | 20.5 | 1209 | 35,300 |
| **296 | — | — | — | 28.3 | 1337 | 34,600 |
| HS-7 | 16.0 | 939 | 19,600 | 19.4 | 1209 | 34,700 |
| H110 | 24.5 | 1059 | 21,300 | 29.9 | 1378 | 35,800 |
| AL-8 | 16.0 | 931 | 22,200 | 20.5 | 1194 | 34,200 |

### 225 gr. Speer HJHP
#4435 (.429" dia.) 1.610" OAL

| Powder | Sug. Starting Grains | Velocity F.P.S. | Pressure C.U.P. | Max. Grains | Velocity F.P.S. | Pressure C.U.P. |
|---|---|---|---|---|---|---|
| SR-4756 | 10.2 | 865 | 20,800 | 12.6 | 1123 | 36,500 |
| IMR-4227 | 20.4 | 1010 | 24,800 | 24.0 | 1205 | 35,800 |
| Unique | 10.0 | 922 | 23,300 | 12.1 | 1127 | 35,500 |
| Blue Dot | 14.5 | 993 | 24,100 | 17.4 | 1210 | 35,700 |
| 2400 | 17.7 | 935 | 22,300 | 22.0 | 1199 | 36,000 |
| 630 | 15.7 | 936 | 23,300 | *19.8 | 1193 | 35,300 |
| **296 | — | — | — | 24.9 | 1261 | 36,700 |
| HS-7 | 14.8 | 916 | 21,400 | 18.2 | 1179 | 35,600 |
| H110 | 21.3 | 1028 | 23,300 | 26.0 | 1276 | 36,500 |
| AL-8 | 14.4 | 931 | 24,200 | 18.6 | 1190 | 36,000 |

### 240 gr. Sierra HC
#8610 (.4295" dia.) 1.610" OAL

| Powder | Sug. Starting Grains | Velocity F.P.S. | Pressure C.U.P. | Max. Grains | Velocity F.P.S. | Pressure C.U.P. |
|---|---|---|---|---|---|---|
| IMR-4227 | 19.3 | 921 | 22,600 | 23.1 | 1177 | 36,800 |
| Unique | 10.0 | 886 | 23,800 | 12.1 | 1101 | 35,600 |
| Blue Dot | 13.8 | 916 | 23,500 | 16.8 | 1141 | 35,800 |
| 2400 | 17.3 | 878 | 22,100 | 21.5 | 1160 | 36,800 |
| 630 | 15.3 | 895 | 24,100 | *19.3 | 1152 | 36,200 |
| **296 | — | — | — | ‡22.8 | 1178 | 36,200 |
| HS-7 | 14.2 | 878 | 22,000 | 17.6 | 1132 | 36,700 |
| H110 | 19.3 | 931 | 22,100 | 23.9 | 1191 | 35,100 |
| AL-8 | 14.0 | 879 | 22,000 | 18.2 | 1159 | 36,600 |
| SR-4756 | 9.9 | 854 | 23,700 | 12.3 | 1072 | 35,800 |

### 240 gr. Speer JHP
#4453 (.429" dia.) 1.610" OAL

| Powder | Sug. Starting Grains | Velocity F.P.S. | Pressure C.U.P. | Max. Grains | Velocity F.P.S. | Pressure C.U.P. |
|---|---|---|---|---|---|---|
| IMR-4227 | 19.3 | 884 | 23,100 | 23.1 | 1109 | 36,000 |
| Unique | *10.0 | 847 | 24,200 | 12.2 | 1042 | 37,000 |
| Blue Dot | 13.0 | 803 | 19,600 | 16.2 | 1068 | 35,300 |
| 2400 | 17.4 | 840 | 21,600 | 21.6 | 1119 | 35,500 |
| 630 | 15.3 | 833 | 23,700 | 19.3 | 1102 | 36,200 |
| **296 | — | — | — | ‡24.0 | 1175 | 36,100 |
| HS-7 | 14.0 | 929 | 21,700 | 17.6 | 1107 | 37,000 |
| H110 | 19.6 | 937 | 20,800 | 24.5 | 1177 | 36,000 |
| AL-8 | 14.2 | 854 | 22,100 | 18.4 | 1106 | 36,300 |

### 265 gr. Hornady FP
.430" dia./1.610" OAL

| Powder | Sug. Starting Grains | Velocity F.P.S. | Pressure C.U.P. | Max. Grains | Velocity F.P.S. | Pressure C.U.P. |
|---|---|---|---|---|---|---|
| IMR-4227 | 18.5 | 870 | 23,800 | 22.0 | 1088 | 34,800 |
| Blue Dot | 13.0 | 830 | 21,600 | 16.0 | 1086 | 35,100 |
| 2400 | 16.2 | 794 | 21,100 | 20.2 | 1096 | 35,200 |
| 630 | 15.0 | 870 | 24,300 | *19.0 | 1121 | 37,000 |
| **296 | — | — | — | 22.0 | 1128 | 36,000 |
| H110 | 17.9 | 871 | 19,700 | 22.8 | 1113 | 34,800 |
| AL-8 | 13.4 | 884 | 21,700 | 17.5 | 1090 | 36,400 |

**Note:** Loads shown in shaded panels are maximum.
* Designates potentially most accurate load.
‡ Designates factory velocity duplication load.
** Reduced loads not recommended by Winchester.

# .44 REMINGTON MAGNUM
## (Continued)

### 180 gr. Cast Bullet
#429348 (Linotype) 1.420" OAL

| Powder | Sug. Starting Grains | Velocity F.P.S. | Pressure C.U.P. | Max. Grains | Velocity F.P.S. | Pressure C.U.P. |
|---|---|---|---|---|---|---|
| SR-4756 | 11.9 | 1096 | 23,300 | 15.8 | 1359 | 36,900 |
| IMR-4227 | 22.7 | 1065 | 21,700 | 27.0+ | 1394 | 35,900 |
| Unique | 11.0 | 1079 | 22,500 | 13.5 | 1343 | 36,500 |
| Blue Dot | 16.1 | 1110 | 22,600 | 20.4 | 1408 | 35,800 |
| 2400 | 20.3 | 1190 | 22,000 | 25.3+ | 1410 | 36,100 |
| 231 | 9.0 | 1002 | 22,000 | 12.6 | 1315 | 36,700 |
| 630 | 17.6 | 1144 | 24,500 | *21.8 | 1390 | 36,000 |
| HS-6 | 14.2 | 1082 | 23,200 | 17.3 | 1352 | 36,400 |
| AL-7 | 15.7 | 1075 | 22,900 | 19.0 | 1361 | 36,500 |
| Herco | 11.7 | 1049 | 22,800 | 14.6 | 1344 | 36,400 |

### 200 gr. Cast Bullet
#429303 (Linotype) 1.692" OAL

| Powder | Sug. Starting Grains | Velocity F.P.S. | Pressure C.U.P. | Max. Grains | Velocity F.P.S. | Pressure C.U.P. |
|---|---|---|---|---|---|---|
| SR-4756 | 12.3 | 1043 | 22,600 | 15.5 | 1289 | 36,700 |
| IMR-4227 | 22.0 | 1038 | 22,100 | 24.5+ | 1201 | 29,600 |
| Unique | 11.0 | 1004 | 22,100 | 13.2 | 1265 | 36,700 |
| Herco | 11.4 | 1035 | 24,300 | 13.8 | 1245 | 36,700 |
| Blue Dot | 15.8 | 1062 | 22,800 | 19.8 | 1363 | 36,800 |
| 2400 | 19.0 | 991 | 20,400 | 24.0+ | 1314 | 34,600 |
| 231 | 9.1 | 913 | 19,600 | *12.6 | 1231 | 36,900 |
| 630 | 17.5 | 1049 | 23,300 | 21.5 | 1317 | 36,500 |
| HS-6 | 14.2 | 1020 | 22,000 | 17.3 | 1295 | 36,900 |
| AL-7 | 16.0 | 999 | 20,400 | 19.0 | 1318 | 36,300 |

### 215 gr. Cast Bullet
#429215 (Linotype) 1.645" OAL

| Powder | Sug. Starting Grains | Velocity F.P.S. | Pressure C.U.P. | Max. Grains | Velocity F.P.S. | Pressure C.U.P. |
|---|---|---|---|---|---|---|
| SR-4756 | 12.2 | 1036 | 24,900 | 15.2 | 1228 | 36,100 |
| IMR-4227 | 21.8 | 1017 | 22,100 | 25.5+ | 1227 | 32,800 |
| Unique | 10.0 | 930 | 21,000 | 13.2 | 1221 | 36,000 |
| Herco | 11.4 | 984 | 22,800 | 13.5 | 1205 | 36,000 |
| Blue Dot | 15.7 | 1032 | 23,500 | 19.5 | 1314 | 37,000 |
| 2400 | 20.0 | 986 | 21,800 | 25.0 | 1310 | 37,000 |
| 231 | *9.3 | 914 | 21,000 | 12.8 | 1191 | 37,000 |
| 630 | 17.8 | 1037 | 24,100 | 21.5 | 1269 | 36,500 |
| HS-6 | 13.5 | 985 | 21,700 | 16.5 | 1238 | 36,000 |
| AL-7 | 14.5 | 997 | 20,600 | 18.7 | 1280 | 34,900 |

### 232 gr. Cast Bullet
#429360 (Linotype) 1.655" OAL

| Powder | Sug. Starting Grains | Velocity F.P.S. | Pressure C.U.P. | Max. Grains | Velocity F.P.S. | Pressure C.U.P. |
|---|---|---|---|---|---|---|
| SR-4756 | 9.9 | 887 | 16,000 | 13.3 | 1145 | 36,100 |
| IMR-4227 | 20.2 | 961 | 21,900 | 24.5 | 1236 | 36,100 |
| Unique | 9.5 | 896 | 16,100 | 12.7 | 1159 | 36,200 |
| Herco | 10.7 | 937 | 23,300 | 13.5 | 1163 | 36,500 |
| Blue Dot | 14.4 | 952 | 20,400 | 18.2 | 1240 | 36,300 |
| 2400 | 18.5 | 998 | 22,700 | 23.0 | 1256 | 36,700 |
| 630 | 15.6 | 947 | 21,900 | *19.6 | 1201 | 36,100 |
| HS-6 | 12.5 | 921 | 20,800 | 16.2 | 1177 | 36,600 |
| AL-8 | 15.0 | 924 | 21,400 | 20.0+ | 1229 | 36,700 |

### 245 gr. Cast Bullet
#429421 (Linotype) 1.710" OAL

| Powder | Sug. Starting Grains | Velocity F.P.S. | Pressure C.U.P. | Max. Grains | Velocity F.P.S. | Pressure C.U.P. |
|---|---|---|---|---|---|---|
| SR-4756 | 11.4 | 927 | 20,500 | 13.1 | 1125 | 36,500 |
| IMR-4227 | 20.2 | 938 | 21,300 | 24.0+ | 1196 | 34,600 |
| Unique | 9.8 | 912 | 20,800 | 13.0 | 1147 | 36,800 |
| Herco | 10.9 | 918 | 22,900 | 13.1 | 1090 | 34,600 |
| Blue Dot | 15.0 | 941 | 20,000 | 18.4 | 1196 | 35,300 |
| 2400 | 19.4 | 974 | 23,500 | 23.4 | 1232 | 36,000 |
| 630 | 17.1 | 1009 | 25,300 | *20.7 | 1190 | 36,700 |
| HS-7 | 15.2 | 931 | 20,900 | 18.2 | 1171 | 35,300 |
| AL-8 | 14.6 | 856 | 18,700 | 19.6 | 1189 | 36,700 |

### 245 gr. Cast Bullet
#429244 (Linotype) 1.680" OAL

| Powder | Sug. Starting Grains | Velocity F.P.S. | Pressure C.U.P. | Max. Grains | Velocity F.P.S. | Pressure C.U.P. |
|---|---|---|---|---|---|---|
| SR-4756 | 10.5 | 892 | 21,500 | 12.2 | 1119 | 35,300 |
| IMR-4227 | 20.2 | 903 | 20,600 | 24.0+ | 1135 | 33,700 |
| Unique | 8.9 | 805 | 17,500 | *12.1 | 1100 | 36,400 |
| Herco | 11.0 | 927 | 24,500 | 12.4 | 1085 | 35,400 |
| Blue Dot | 14.0 | 911 | 20,300 | 17.4 | 1160 | 34,800 |
| 2400 | 18.2 | 915 | 21,600 | 22.2 | 1165 | 35,300 |
| 630 | 16.2 | 942 | 23,700 | 19.3 | 1146 | 35,000 |
| HS-7 | 14.5 | 921 | 22,600 | 17.5 | 1160 | 37,000 |
| AL-8 | 14.4 | 901 | 23,200 | 19.5 | 1113 | 34,000 |

**Note:** Loads shown in shaded panels are maximum.
* Designates potentially most accurate load.
\+ Designates a compressed powder charge.

# .44/40 WINCHESTER

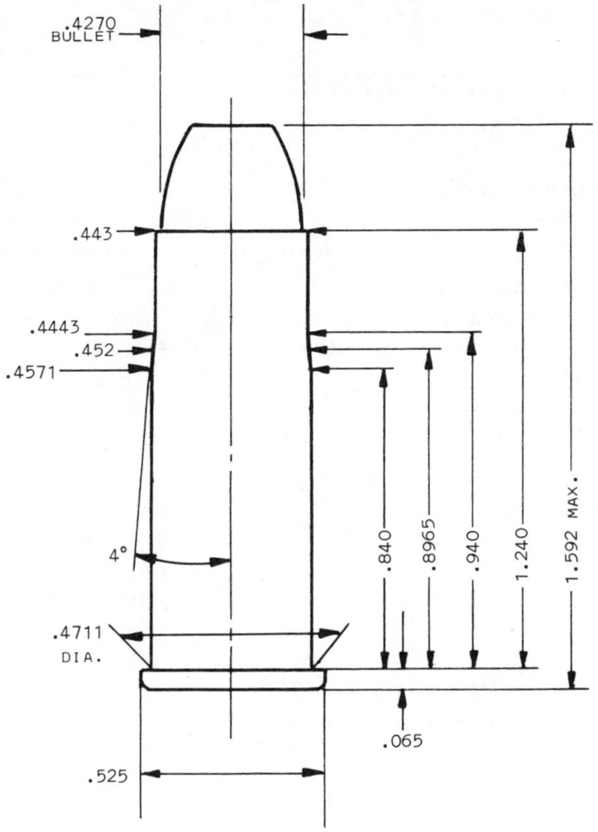


## TEST COMPONENTS

Cases ........ Winchester
Trim-to Length ........ 1.300''
Primers ........ Winchester 7-111
Primer Size ........ Large Pistol
Lyman Shell Holder ........ No. 14B
Cast Bullets Used ........ (size to .427'' dia.) #42798, 205 gr.

## TEST SPECIFICATIONS (Velocity Only)

Firearm Used ........ Colt S.A.A.
Barrel Length ........ 7½''
Twist ........ 1-20''
Groove Diameter ........ .426''

## COMMENTS

These loads are not to be used in handguns which were designed for black powder. A wide variation in groove diameters exists with handguns chambered for this cartridge. Slug your barrel before reloading and size cast bullets as near to actual groove diameter as possible. Jacketed bullets that are within .001'' of the groove diameter may be used successfully.

**205 gr. Cast Bullet**
#42798 (#2 Alloy) 1.592'' OAL

| Powder | Sug. Starting Grains | Velocity F.P.S. | Pressure C.U.P. | Max. Grains | Velocity F.P.S. | Pressure C.U.P. |
|---|---|---|---|---|---|---|
| Bullseye | 4.0 | 695 | — | 6.6 | 945 | — |
| Unique | 6.0 | 750 | — | 10.9 | 1095 | — |

**Note:** Loads shown in shaded panels are maximum.

# .45 A.C.P.

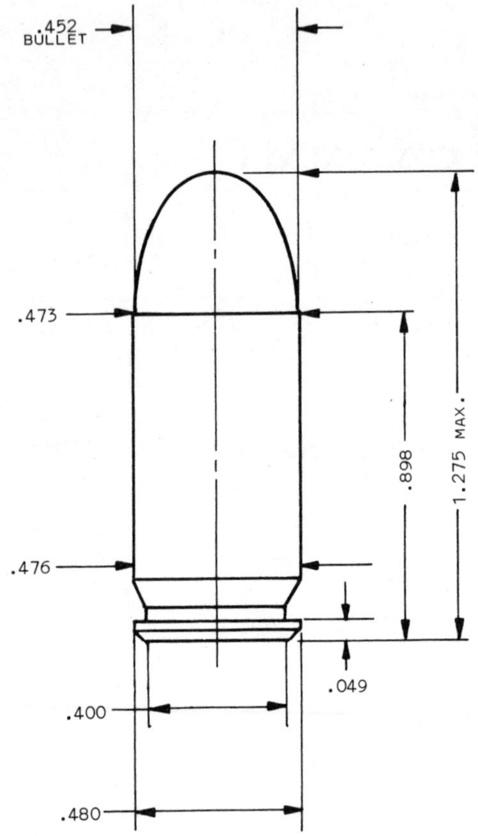


### TEST COMPONENTS

Cases ........................................ Federal; Remington (cast bullets only)
Trim-to Length ........................................ .895''
Primers ........................................ CCI 300; Remington 2½ (cast bullets only)
Primer Size ........................................ Large Pistol
Lyman Shell Holder ........................................ No. 2
Jacketed Bullets Used . Remington MC #B22890, 185 gr.
Hornady HP, 185 gr.
Speer HP #4477, 200 gr.
Speer HP #4479, 225 gr.
Remington MC #B22892, 230 gr.
Cast Bullets Used ........................ (size to .450'' dia.)
#452389, 185 gr.
#452488, 195 gr.
#452460, 200 gr.
#452374, 225 gr.

### TEST SPECIFICATIONS (Velocity and Pressure)

Firearm Used ........................................ Universal Receiver
Barrel Length ........................................ 5''
Twist ........................................ 1-16''
Groove Diameter ........................................ .450''

### COMMENTS

Because this cartridge headspaces on the case mouth, the trim-to length must be adhered to closely and the case should not be crimped. A short case, or a crimp, can cause headspace problems.

While groove dimensions for these handguns will range from .450'' - .453'', the reloader may have difficulty if he uses a bullet over .451'' diameter. Case and chamber dimensions usually provide for a bullet of .451'', but larger bullets may bulge cases to the point where they fail to chamber.

Most pistols (unless they are altered) will not feed wadcutter-type bullets such as No. 452389. Bullet No. 452460 feeds well in most pistols and, when loaded to starting load velocities, is recommended for mid-range shooting.

# .45 A.C.P.
## (Continued)

### 185 gr. Remington MC
#B22890 (.451" dia.) 1.255" OAL

| Powder | Sug. Starting Grains | Velocity F.P.S. | Pressure C.U.P. | Max. Grains | Velocity F.P.S. | Pressure C.U.P. |
|---|---|---|---|---|---|---|
| 700X | 3.4 | 675 | 7,900 | 5.0 | 967 | 17,800 |
| SR-7625 | 4.4 | 706 | 9,100 | 6.5 | 972 | 16,800 |
| SR-4756 | 5.7 | 666 | 7,900 | 7.7 | 1002 | 17,000 |
| Bullseye | 3.5 | 678 | 8,000 | *5.6 | 975 | 16,600 |
| Unique | 4.7 | 661 | 7,700 | 7.5 | 1055 | 18,000 |
| Blue Dot | 7.3 | 666 | 8,000 | 10.2 | 1015 | 17,800 |
| 231 | 4.1 | 680 | 9,100 | 6.1 | 981 | 16,600 |
| HS-6 | 6.6 | 689 | 8,000 | 9.1 | 1038 | 18,000 |
| AL-5 | 5.7 | 682 | 8,600 | 8.6 | 1000 | 17,700 |

### 185 gr. Hornady HP
.451" dia./1.175" OAL

| Powder | Sug. Starting Grains | Velocity F.P.S. | Pressure C.U.P. | Max. Grains | Velocity F.P.S. | Pressure C.U.P. |
|---|---|---|---|---|---|---|
| 700X | 3.5 | 612 | 7,500 | 5.5 | 933 | 17,800 |
| SR-7625 | 4.9 | 633 | 7,100 | 7.0 | 960 | 17,900 |
| Bullseye | 3.5 | 610 | 7,300 | 6.0 | 976 | 18,000 |
| Unique | 4.8 | 630 | 7,800 | 7.8 | 985 | 17,800 |
| 231 | 4.4 | 640 | 7,500 | ‡6.1 | 937 | 15,700 |
| PB | 4.6 | 660 | 9,100 | 6.9 | 968 | 17,700 |
| Red Dot | 3.5 | 607 | 7,200 | 6.1 | 942 | 17,600 |
| AL-120 | 4.5 | 635 | 7,800 | 6.3 | 958 | 17,800 |
| HP38 | 3.5 | 599 | 7,500 | *5.8 | 960 | 17,900 |

### 200 gr. Speer HP
#4477 (.451" dia.) 1.178" OAL

| Powder | Sug. Starting Grains | Velocity F.P.S. | Pressure C.U.P. | Max. Grains | Velocity F.P.S. | Pressure C.U.P. |
|---|---|---|---|---|---|---|
| 700X | 3.3 | 597 | 7,400 | 4.9 | 871 | 17,300 |
| SR-7625 | 4.5 | 601 | 7,300 | 6.7 | 935 | 17,400 |
| SR-4756 | 5.4 | 588 | 7,100 | 7.4 | 961 | 18,000 |
| Bullseye | 3.4 | 604 | 7,100 | *5.7 | 942 | 17,800 |
| Unique | 4.0 | 604 | 7,300 | 6.5 | 927 | 17,700 |
| Blue Dot | 7.2 | 620 | 7,100 | 10.2 | 956 | 17,500 |
| 231 | 4.0 | 636 | 6,800 | 6.5 | 970 | 17,800 |
| HS-6 | 5.7 | 566 | 6,500 | 9.0 | 951 | 17,600 |
| AL-5 | 5.8 | 636 | 7,000 | 8.6 | 949 | 17,500 |

### 225 gr. Speer HP
#4479 (.451" dia.) 1.243" OAL

| Powder | Sug. Starting Grains | Velocity F.P.S. | Pressure C.U.P. | Max. Grains | Velocity F.P.S. | Pressure C.U.P. |
|---|---|---|---|---|---|---|
| 700X | 3.1 | 549 | 7,500 | 4.8 | 796 | 17,200 |
| SR-7625 | 4.4 | 587 | 7,400 | 6.3 | 844 | 17,400 |
| SR-4756 | 4.8 | 552 | 6,700 | 7.3 | 882 | 17,300 |
| Bullseye | 3.1 | 554 | 7,300 | 5.3 | 840 | 17,400 |
| Unique | 4.1 | 548 | 6,800 | 6.7 | 883 | 17,500 |
| Blue Dot | 7.2 | 602 | 6,600 | 10.7 | 925 | 17,700 |
| 630 | 7.4 | 590 | 7,400 | *11.2 | 871 | 17,400 |
| HS-7 | 6.7 | 564 | 6,800 | 10.3 | 884 | 17,600 |
| AL-7 | 7.3 | 618 | 8,000 | 9.8 | 900 | 17,000 |

### 230 gr. Remington MC
#B22892 (.450" dia.) 1.275" OAL

| Powder | Sug. Starting Grains | Velocity F.P.S. | Pressure C.U.P. | Max. Grains | Velocity F.P.S. | Pressure C.U.P. |
|---|---|---|---|---|---|---|
| 700X | 3.6 | 667 | 11,200 | *4.8 | 867 | 17,900 |
| SR-7625 | 4.9 | 678 | 10,800 | ‡5.9 | 838 | 15,800 |
| SR-4756 | 6.0 | 695 | 11,200 | 7.0 | 866 | 17,400 |
| Bullseye | 3.8 | 662 | 10,400 | 5.3 | 878 | 17,100 |
| Unique | 5.0 | 677 | 10,000 | 6.8 | 916 | 16,900 |
| Blue Dot | 7.5 | 662 | 8,700 | 10.2 | 923 | 16,900 |
| 630 | 8.0 | 651 | 9,200 | 11.4 | 919 | 16,900 |
| HS-7 | 8.1 | 671 | 10,000 | 10.7 | 914 | 17,800 |
| AL-7 | 6.8 | 673 | 9,800 | 9.0 | 905 | 17,200 |

### 185 gr. Cast Bullet
#452389 (#2 Alloy) 1.255" OAL

| Powder | Sug. Starting Grains | Velocity F.P.S. | Pressure C.U.P. | Max. Grains | Velocity F.P.S. | Pressure C.U.P. |
|---|---|---|---|---|---|---|
| 700X | 4.0 | 760 | 8,800 | 5.5 | 1010 | 17,500 |
| PB | 4.6 | 740 | 8,300 | 5.5 | 920 | 13,500 |
| SR-7625 | 5.5 | 825 | 9,800 | 6.5 | 1030 | 17,400 |
| SR-4756 | 5.3 | 663 | 8,900 | 7.9 | 988 | 17,600 |
| Bullseye | 3.5 | 660 | 6,000 | 5.0 | 890 | 12,300 |
| Red Dot | 4.0 | 735 | 8,100 | 5.5 | 970 | 14,800 |
| Green Dot | 4.5 | 745 | 8,300 | 6.0 | 965 | 14,800 |
| Unique | 5.0 | 684 | 6,900 | 7.5 | 995 | 15,500 |
| Blue Dot | 7.0 | 636 | 7,600 | 9.2+ | 844 | 13,200 |
| 231 | 3.9 | 662 | 7,900 | 6.0 | 958 | 17,500 |
| HS-6 | 5.6 | 613 | 7,500 | 9.1 | 1034 | 17,100 |
| AL-5 | 5.1 | 621 | 7,200 | *8.3 | 974 | 17,300 |

**Note:** Loads shown in shaded panels are maximum.
* Designates potentially most accurate load.
\+ Designates a compressed powder charge.
‡ Designates factory velocity duplication load.

# .45 A.C.P.
## (Continued)

### 195 gr. Cast Bullet
#452488 (#2 Alloy) 1.275'' OAL

| Powder | Sug. Starting Grains | Velocity F.P.S. | Pressure C.U.P. | Max. Grains | Velocity F.P.S. | Pressure C.U.P. |
|---|---|---|---|---|---|---|
| 700X | 4.0 | 735 | 9,600 | 5.2 | 950 | 16,700 |
| PB | 4.6 | 765 | 10,100 | 5.5 | 910 | 14,800 |
| SR-7625 | 5.0 | 780 | 10,300 | 6.2 | 945 | 16,200 |
| SR-4756 | 5.3 | 639 | 8,600 | 7.8 | 977 | 17,500 |
| Bullseye | 3.5 | 680 | 7,700 | 5.0 | 895 | 13,900 |
| Red Dot | 4.0 | 730 | 9,000 | 5.5 | 950 | 16,200 |
| Green Dot | 4.5 | 740 | 9,400 | 6.0 | 955 | 15,800 |
| Unique | 5.0 | 695 | 7,900 | 7.5 | 1010 | 16,800 |
| Blue Dot | 7.0 | 646 | 8,200 | 10.7 | 1033 | 17,900 |
| 231 | 3.9 | 642 | 8,500 | 5.7 | 962 | 16,900 |
| HS-6 | 6.1 | 652 | 8,300 | 9.0 | 1027 | 17,900 |
| AL-5 | 5.8 | 722 | 8,600 | *8.5 | 1014 | 17,800 |

### 200 gr. Cast Bullet
#452460 (#2 Alloy) 1.161'' OAL

| Powder | Sug. Starting Grains | Velocity F.P.S. | Pressure C.U.P. | Max. Grains | Velocity F.P.S. | Pressure C.U.P. |
|---|---|---|---|---|---|---|
| 700X | 4.0 | 745 | 10,300 | 5.2 | 940 | 17,300 |
| PB | 4.6 | 725 | 9,000 | 5.5 | 880 | 14,200 |
| SR-7625 | 5.0 | 735 | 9,000 | 6.2 | 945 | 17,000 |
| **SR-4756 | 5.3 | 704 | 9,100 | 7.3 | 993 | 17,000 |
| Bullseye | 3.5 | 645 | 6,900 | 5.0 | 860 | 13,400 |
| Red Dot | 4.0 | 695 | 8,400 | 5.3 | 895 | 14,700 |
| Green Dot | 4.5 | 715 | 9,000 | 5.8 | 895 | 14,400 |
| Unique | 5.0 | 670 | 7,700 | 7.5 | 980 | 16,600 |
| **Blue Dot | 7.1 | 701 | 8,900 | 10.6 | 1012 | 17,200 |
| **231 | 4.0 | 694 | 9,200 | 6.0 | 987 | 18,000 |
| **HS-6 | 6.0 | 690 | 8,600 | 8.7 | 1016 | 17,700 |
| **AL-5 | 5.8 | 729 | 9,300 | *8.4 | 1008 | 17,400 |

### 225 gr. Cast Bullet
#452374 (#2 Alloy) 1.272'' OAL

| Powder | Sug. Starting Grains | Velocity F.P.S. | Pressure C.U.P. | Max. Grains | Velocity F.P.S. | Pressure C.U.P. |
|---|---|---|---|---|---|---|
| 700X | 4.0 | 695 | 10,500 | 5.0 | 855 | 17,300 |
| PB | 4.8 | 705 | 10,300 | 5.5 | 810 | 14,200 |
| SR-7625 | 5.0 | 675 | 9,000 | 6.0 | 850 | 15,200 |
| **SR-4756 | 5.5 | 662 | 8,800 | 7.5 | 955 | 18,000 |
| Bullseye | 4.0 | 680 | 10,100 | 5.0 | 815 | 14,400 |
| Red Dot | 4.3 | 705 | 11,000 | 5.3 | 835 | 15,300 |
| Green Dot | 4.8 | 725 | 11,400 | 5.8 | 845 | 15,100 |
| Unique | 5.5 | 695 | 10,100 | 7.3 | 905 | 16,500 |
| **Blue Dot | 7.5 | 660 | 8,600 | 10.7 | 964 | 17,300 |
| **231 | 4.0 | 661 | 9,200 | *5.8 | 902 | 17,500 |
| **HS-6 | 6.2 | 664 | 8,400 | 8.6 | 921 | 16,600 |
| **AL-5 | 5.5 | 657 | 8,200 | 8.7 | 943 | 16,800 |

**Note:** Loads shown in shaded panels are maximum.
* Designates potentially most accurate load.
** Designates use of Federal cases.

# .45 AUTO RIM

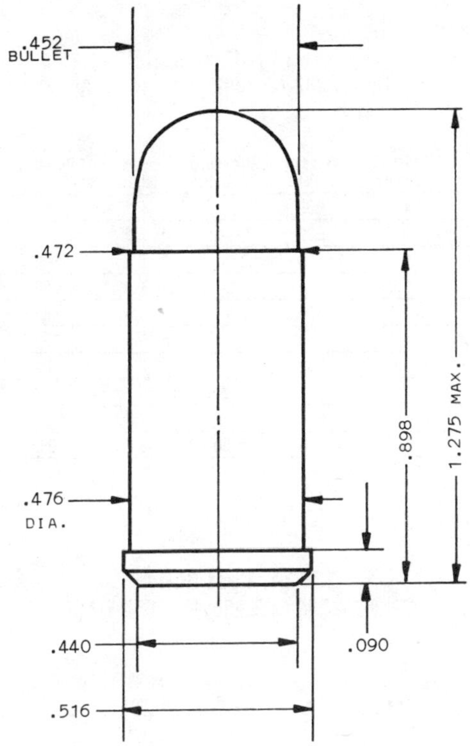


## TEST COMPONENTS

Cases ........ Remington
Trim-to Length ........ .894''
Primers ........ Remington 2½
Primer Size ........ Large Pistol
Lyman Shell Holder ........ No. 14A
Jacketed Bullets Used ........ Hornady HP, 185 gr.
Speer HP #4477, 200 gr.
Speer HP #4479, 225 gr.
Cast Bullets Used ........ (size to .451'' dia.)
#452389, 185 gr.
#452460, 200 gr.
#452374, 225 gr.
#452423, 238 gr.

## TEST SPECIFICATIONS (Velocity and Pressure)

Firearm Used ........ Special Test Barrel;
Smith & Wesson Model 25 (cast bullets only)
Barrel Length ........ 4½'' and 6½''
Twist ........ 1-16''
Groove Diameter ........ .451''

## COMMENTS

Heavier loads for this cartridge should be lightly crimped to prevent bullet movement when the pistol is under recoil.

**185 gr. Hornady HP**
.451'' dia./1.236'' OAL

| Powder | Sug. Starting Grains | Velocity F.P.S. | Pressure C.U.P. | Max. Grains | Velocity F.P.S. | Pressure C.U.P. |
|---|---|---|---|---|---|---|
| 700X | 4.5 | 713 | 8,200 | 5.2 | 838 | 11,900 |
| SR-7625 | 6.0 | 709 | 7,500 | 7.2 | 849 | 13,700 |
| Bullseye | 4.6 | 722 | 8,600 | 5.8 | 915 | 14,800 |
| Red Dot | 4.8 | 711 | 8,600 | *6.4 | 933 | 14,700 |
| Unique | 5.0 | 585 | 5,100 | 7.6 | 905 | 13,200 |
| 231 | 4.4 | 618 | 6,400 | 6.1 | 889 | 13,700 |
| HS-5 | 7.5 | 754 | 9,000 | 9.0 | 913 | 13,100 |
| AL-5 | 8.0 | 788 | 10,000 | 9.1 | 890 | 13,400 |

**200 gr. Speer HP**
#4477 (.451'' dia.) 1.180'' OAL

| Powder | Sug. Starting Grains | Velocity F.P.S. | Pressure C.U.P. | Max. Grains | Velocity F.P.S. | Pressure C.U.P. |
|---|---|---|---|---|---|---|
| 700X | 3.4 | 491 | 5,000 | 4.9 | 799 | 13,400 |
| SR-7625 | 4.8 | 608 | 7,200 | 6.0 | 809 | 12,400 |
| Bullseye | 4.0 | 614 | 7,800 | *5.4 | 838 | 13,700 |
| Red Dot | 4.0 | 615 | 7,700 | 5.4 | 852 | 14,600 |
| Unique | 5.8 | 719 | 10,100 | 7.1 | 882 | 14,500 |
| 231 | 4.0 | 570 | 7,800 | 5.7 | 852 | 14,800 |
| HS-6 | 6.1 | 558 | 6,400 | 8.7 | 846 | 13,600 |
| AL-5 | 7.0 | 727 | 9,900 | 8.5 | 884 | 14,900 |

**Note:** Loads shown in shaded panels are maximum.
* Designates potentially most accurate load.

# .45 AUTO RIM
(Continued)

## 225 gr. Speer HP
#4479 (.451" dia.) 1.255" OAL

| Powder | Sug. Starting Grains | Velocity F.P.S. | Pressure C.U.P. | Max. Grains | Velocity F.P.S. | Pressure C.U.P. |
|---|---|---|---|---|---|---|
| SR-7625 | 4.4 | 518 | 6,100 | 6.2 | 812 | 14,900 |
| Unique | 5.0 | 609 | 8,200 | 7.0 | 834 | 14,500 |
| Blue Dot | 8.5 | 718 | 10,200 | 10.0 | 864 | 14,500 |
| 231 | 3.7 | 496 | 6,500 | 5.2 | 773 | 13,800 |
| 630 | 9.0 | 690 | 9,800 | 10.8 | 823 | 13,700 |
| HS-6 | 6.7 | 634 | 8,800 | 8.3 | 812 | 13,600 |
| AL-7 | 7.8 | 692 | 10,300 | *9.0 | 809 | 13,500 |

## 185 gr. Cast Bullet
#452389 (#2 Alloy) 1.118" OAL

| Powder | Sug. Starting Grains | Velocity F.P.S. | Pressure C.U.P. | Max. Grains | Velocity F.P.S. | Pressure C.U.P. |
|---|---|---|---|---|---|---|
| 700X | 4.0 | 655 | — | 5.3 | 905 | — |
| PB | 4.5 | 620 | — | 5.8 | 855 | — |
| SR-7625 | 5.0 | 615 | — | 6.3 | 810 | — |
| Bullseye | 3.0 | 505 | — | 4.8 | 780 | — |
| Red Dot | 4.0 | 650 | — | 5.3 | 885 | — |
| Green Dot | 4.5 | 650 | — | 5.8 | 855 | — |
| Unique | 5.0 | 600 | — | 7.0 | 870 | — |
| **231 | 4.0 | 735 | 7,200 | 5.7 | 972 | 13,400 |
| **HS-6 | 5.5 | 674 | 5,300 | *8.8 | 1030 | 14,800 |
| **AL-5 | 6.5 | 816 | 8,500 | 8.7 | 1042 | 14,900 |

## 200 gr. Cast Bullet
#452460 (#2 Alloy) 1.161" OAL

| Powder | Sug. Starting Grains | Velocity F.P.S. | Pressure C.U.P. | Max. Grains | Velocity F.P.S. | Pressure C.U.P. |
|---|---|---|---|---|---|---|
| 700X | 4.0 | 675 | — | 5.0 | 875 | — |
| PB | 4.5 | 650 | — | 5.5 | 840 | — |
| SR-7625 | 5.0 | 660 | — | 6.1 | 850 | — |
| Bullseye | 3.0 | 510 | — | 4.7 | 775 | — |
| Red Dot | 4.0 | 645 | — | 5.2 | 845 | — |
| Green Dot | 4.5 | 660 | — | 5.6 | 840 | — |
| Unique | 5.0 | 610 | — | 6.8 | 845 | — |
| **231 | 4.0 | 694 | 7,200 | 5.5 | 936 | 14,100 |
| **HS-5 | 6.5 | 765 | 8,400 | *8.2 | 934 | 12,500 |
| **AL-5 | 5.7 | 697 | 6,800 | 8.4 | 977 | 14,400 |

## 225 gr. Cast Bullet
#452374 (#2 Alloy) 1.265" OAL

| Powder | Sug. Starting Grains | Velocity F.P.S. | Pressure C.U.P. | Max. Grains | Velocity F.P.S. | Pressure C.U.P. |
|---|---|---|---|---|---|---|
| 700X | 4.0 | 625 | — | 4.8 | 770 | — |
| PB | 4.5 | 565 | — | 5.4 | 735 | — |
| SR-7625 | 5.0 | 560 | — | 5.9 | 750 | — |
| Bullseye | 3.0 | 475 | — | 4.5 | 690 | — |
| Red Dot | 4.0 | 625 | — | 5.0 | 760 | — |
| Green Dot | 4.5 | 615 | — | 5.4 | 750 | — |
| Unique | 5.0 | 550 | — | 6.6 | 785 | — |
| **231 | 4.0 | 660 | 6,900 | *5.5 | 873 | 13,800 |
| **HS-6 | 7.3 | 799 | 9,400 | 8.6 | 922 | 13,800 |
| **AL-7 | 8.0 | 787 | 9,100 | 9.7 | 939 | 14,700 |

## 238 gr. Cast Bullet
#452423 (#2 Alloy) 1.200" OAL

| Powder | Sug. Starting Grains | Velocity F.P.S. | Pressure C.U.P. | Max. Grains | Velocity F.P.S. | Pressure C.U.P. |
|---|---|---|---|---|---|---|
| 700X | 3.7 | 620 | — | 4.5 | 755 | — |
| PB | 3.9 | 535 | — | 4.8 | 690 | — |
| SR-7625 | 4.1 | 505 | — | 4.9 | 655 | — |
| Bullseye | 3.0 | 500 | — | 4.5 | 720 | — |
| Red Dot | 3.5 | 590 | — | 4.3 | 715 | — |
| Green Dot | 4.0 | 605 | — | 4.8 | 715 | — |
| Unique | 4.8 | 585 | — | 6.0 | 750 | — |
| **Blue Dot | 6.5 | 654 | 8,100 | 8.5 | 855 | 13,300 |
| **630 | 7.5 | 694 | 9,100 | 10.0 | 894 | 14,900 |
| **HS-7 | 8.0 | 757 | 10,900 | *9.3 | 864 | 14,200 |
| **AL-8 | 7.0 | 628 | 8,300 | 9.6 | 814 | 13,800 |

**Note:** Loads shown in shaded panels are maximum.
* Designates potentially most accurate load.
** Designates use of a special test barrel.

# .45 COLT

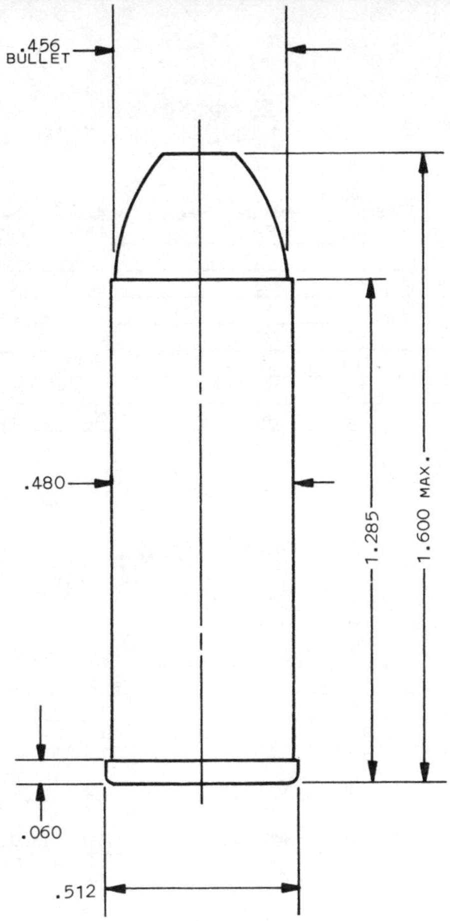


## TEST COMPONENTS

Cases ................................ Remington; Winchester
Trim-to Length ........................................ 1.280"
Primers .................................... Winchester 7-111; Remington 2½
Primer Size ....................................... Large Pistol
Lyman Shell Holder .................................... No. 11
Jacketed Bullets Used .......... Sierra HC #8800, 185 gr.
Speer HP #4477, 200 gr.
Speer HP #4479, 225 gr.
Sierra HC #8820, 240 gr.
Hornady HP, 250 gr.
Cast Bullets Used ........................ (size to .452" dia.)
#45468, 175 gr.
#454190, 250 gr.
#454424, 255 gr.

## TEST SPECIFICATIONS (Velocity Only)

Firearm Used ............................... Ruger Blackhawk
Barrel Length ............................................. 7½"
Twist ..................................................... 1-16"
Groove Diameter ....................................... .451"

## COMMENTS

Pre-World War II model revolvers normally have a groove diameter of .454", while post-war models usually run .451". It is wise to slug your barrel and size cast bullets to as near the exact diameter as possible.

Some sizing dies may not reduce cases enough to hold .451"-.452" diameter bullets tightly. An expander plug, not over .450", should be used when loading .451" - .452" bullets. Case life is very limited when using .45 Colt maximum loads, designed specifically for the Ruger and Thompson/Center Contender firearms. Many cases fired in the Lyman laboratory showed signs of longitudinal splits after only a few reloads.

In order to adhere to the maximum overall length listed, the seating depth of some cast bullets must be watched closely. Bullet No. 454190 should be seated to crimp on the ogive. Bullet No. 454424 should be seated to crimp on the forward edge of the first driving band.

If a crimp is desired on half-jacketed bullets, it should be placed at the junction of the jacket and lead nose.

Bullet No. 454424 loaded with Unique powder is popular for this cartridge.

# .45 COLT
## (Continued)

### 185 gr. Sierra HC
#8800 (.4515" dia.) 1.520" OAL

| Powder | Sug. Starting Grains | Velocity F.P.S. | Pressure C.U.P. | Max. Grains | Velocity F.P.S. | Pressure C.U.P. |
|---|---|---|---|---|---|---|
| Unique | 7.5 | 615 | — | 10.0 | 934 | — |
| Blue Dot | 12.0 | 692 | — | 13.5 | 826 | — |
| 700X | 6.0 | 658 | — | 7.0 | 884 | — |
| IMR-4227 | 16.5 | 711 | — | 20.0 | 875 | — |
| 630 | *13.0 | 700 | — | 15.0 | 935 | — |
| HS-7 | 13.0 | 700 | — | 14.5 | 908 | — |
| AL-7 | 10.5 | 691 | — | 13.0 | 950 | — |

### 200 gr. Speer HP
#4477 (.451" dia.) 1.557" OAL

| Powder | Sug. Starting Grains | Velocity F.P.S. | Pressure C.U.P. | Max. Grains | Velocity F.P.S. | Pressure C.U.P. |
|---|---|---|---|---|---|---|
| SR-4756 | 9.5 | 679 | — | 11.8 | 950 | — |
| Unique | 8.0 | 683 | — | 11.2 | 994 | — |
| Blue Dot | 11.6 | 560 | — | 12.8 | 835 | — |
| HS-6 | 10.3 | 740 | — | 13.3 | 993 | — |
| AL-7 | 11.4 | 639 | — | 13.0 | 964 | — |
| SR-7625 | 7.0 | 581 | — | 9.5 | 894 | — |
| 630 | 14.0 | 867 | — | 16.0 | 963 | — |

### 225 gr. Speer HP
#4479 (.451" dia.) 1.557" OAL

| Powder | Sug. Starting Grains | Velocity F.P.S. | Pressure C.U.P. | Max. Grains | Velocity F.P.S. | Pressure C.U.P. |
|---|---|---|---|---|---|---|
| SR-7625 | 7.0 | 482 | — | 9.3 | 810 | — |
| IMR-4227 | 16.5 | 581 | — | 21.0 | 914 | — |
| Unique | 8.3 | 696 | — | 10.8 | 926 | — |
| Blue Dot | 12.4 | 624 | — | 15.4 | 947 | — |
| 630 | 13.5 | 720 | — | 15.0 | 845 | — |
| HS-6 | 11.6 | 674 | — | 13.6 | 914 | — |
| AL-7 | 11.0 | 603 | — | 13.0 | 829 | — |

### 240 gr. Sierra HC
#8820 (.4515" dia.) 1.575" OAL

| Powder | Sug. Starting Grains | Velocity F.P.S. | Pressure C.U.P. | Max. Grains | Velocity F.P.S. | Pressure C.U.P. |
|---|---|---|---|---|---|---|
| Unique | 7.5 | 633 | — | 9.5 | 851 | — |
| 2400 | 15.0 | 671 | — | 18.5 | 899 | — |
| SR-4759 | 15.0 | 619 | — | 18.5 | 819 | — |
| IMR-4227 | 16.0 | 610 | — | 20.5 | 904 | — |
| 630 | *12.0 | 649 | — | 15.5 | 901 | — |
| HS-7 | 12.0 | 687 | — | 14.0 | 865 | — |
| AL-8 | 11.0 | 610 | — | 15.5 | 922 | — |

### 250 gr. Hornady HP
.452" dia./1.590" OAL

| Powder | Sug. Starting Grains | Velocity F.P.S. | Pressure C.U.P. | Max. Grains | Velocity F.P.S. | Pressure C.U.P. |
|---|---|---|---|---|---|---|
| Unique | *7.5 | 625 | — | 9.3 | 801 | — |
| 2400 | 15.2 | 696 | — | 17.5 | 814 | — |
| SR-4759 | 16.0 | 682 | — | 18.5 | 807 | — |
| IMR-4227 | 17.0 | 646 | — | 20.3 | 862 | — |
| 630 | 11.5 | 613 | — | 15.0 | 851 | — |
| HS-7 | 11.7 | 654 | — | 14.3 | 837 | — |
| AL-8 | 12.4 | 709 | — | 15.0 | 844 | — |

### 175 gr. Cast Bullet
#45468 (#2 Alloy) 1.560" OAL

| Powder | Sug. Starting Grains | Velocity F.P.S. | Pressure C.U.P. | Max. Grains | Velocity F.P.S. | Pressure C.U.P. |
|---|---|---|---|---|---|---|
| ++Red Dot | 4.0 | 475 | — | 7.0 | 885 | — |
| ++Green Dot | 4.5 | 490 | — | 7.5 | 935 | — |
| ++Unique | 6.0 | 595 | — | 10.0 | 1035 | — |
| ++700X | 4.0 | 435 | — | 7.0 | 910 | — |
| ++PB | 4.5 | 440 | — | 7.5 | 845 | — |
| ++SR-7625 | 6.0 | 510 | — | 9.5 | 925 | — |
| 231 | 7.0 | 770 | — | 9.0 | 1041 | — |
| HS-5 | 9.5 | 818 | — | 11.0 | 878 | — |
| AL-5 | 9.0 | 732 | — | 11.0 | 959 | — |

**Note:** Loads shown in shaded panels are maximum.
* Designates potentially most accurate load.
++Designates use of Winchester cases and primers

# .45 COLT
## (Continued)

**250 gr. Cast Bullet**
#454190 (#2 Alloy) 1.600" OAL

| Powder | Sug. Starting Grains | Velocity F.P.S. | Pressure C.U.P. | Max. Grains | Velocity F.P.S. | Pressure C.U.P. |
|---|---|---|---|---|---|---|
| ++Red Dot | 4.5 | 535 | — | 6.5 | 835 | — |
| ++Green Dot | 4.5 | 505 | — | 7.0 | 835 | — |
| ++Unique | 6.0 | 595 | — | ‡9.0 | 875 | — |
| ++700X | 4.5 | 535 | — | 6.0 | 800 | — |
| ++PB | 5.0 | 545 | — | 7.5 | 820 | — |
| ++SR-7625 | 6.0 | 555 | — | 8.5 | 890 | — |
| 231 | *6.5 | 762 | — | 7.4 | 931 | — |
| HS-5 | 8.0 | 696 | — | 9.8 | 859 | — |
| AL-5 | 8.0 | 693 | — | 9.6 | 869 | — |

**255 gr. Cast Bullet**
#454424 (#2 Alloy) 1.575" OAL

| Powder | Sug. Starting Grains | Velocity F.P.S. | Pressure C.U.P. | Max. Grains | Velocity F.P.S. | Pressure C.U.P. |
|---|---|---|---|---|---|---|
| ++Red Dot | 4.5 | 550 | — | 6.0 | 780 | — |
| ++Green Dot | 4.5 | 500 | — | 6.5 | 765 | — |
| ++Unique | 6.0 | 590 | — | 8.5 | 845 | — |
| ++700X | 4.5 | 535 | — | 6.0 | 785 | — |
| ++PB | 5.0 | 530 | — | 7.0 | 750 | — |
| ++SR-7625 | 6.0 | 555 | — | 8.0 | 835 | — |
| 630 | *9.3 | 650 | — | 11.5 | 768 | — |
| HS-6 | 7.8 | 630 | — | 9.5 | 761 | — |
| AL-7 | 7.9 | 647 | — | 10.0 | 756 | — |

**Note:** Loads shown in shaded panels are maximum.
* Designates potentially most accurate load.
‡ Designates factory velocity duplication load.
++Designates use of Winchester cases and primers.

# .45 COLT
## (Ruger and T/C Contender Only)

### 185 gr. Sierra HC
#8800 (.4515" dia.) 1.520" OAL

| Powder | Sug. Starting Grains | Velocity F.P.S. | Pressure C.U.P. | Max. Grains | Velocity F.P.S. | Pressure C.U.P. |
|---|---|---|---|---|---|---|
| Unique | 10.0 | 934 | — | 13.0 | 1221 | — |
| Blue Dot | 13.5 | 826 | — | 17.5 | 1188 | — |
| 700X | 7.0 | 884 | — | *10.0 | 1193 | — |
| IMR-4227 | 20.0 | 875 | — | 26.5 | 1188 | — |
| 630 | 15.0 | 935 | — | 20.0 | 1212 | — |
| HS-7 | 14.5 | 908 | — | 18.5 | 1226 | — |
| AL-7 | 13.0 | 950 | — | 16.5 | 1241 | — |

### 200 gr. Speer HP
#4477 (.451" dia.) 1.557" OAL

| Powder | Sug. Starting Grains | Velocity F.P.S. | Pressure C.U.P. | Max. Grains | Velocity F.P.S. | Pressure C.U.P. |
|---|---|---|---|---|---|---|
| SR-4756 | 11.8 | 950 | — | 12.5 | 1054 | — |
| Unique | 11.2 | 994 | — | 12.2 | 1153 | — |
| Blue Dot | 12.8 | 835 | — | 15.8 | 1110 | — |
| HS-6 | 13.3 | 993 | — | 16.0 | 1241 | — |
| AL-7 | 13.0 | 964 | — | *15.0 | 1134 | — |
| SR-7625 | 9.5 | 894 | — | 11.3 | 1141 | — |
| 630 | 16.0 | 963 | — | 19.1 | 1189 | — |

### 225 gr. Speer HP
#4479 (.451" dia.) 1.557" OAL

| Powder | Sug. Starting Grains | Velocity F.P.S. | Pressure C.U.P. | Max. Grains | Velocity F.P.S. | Pressure C.U.P. |
|---|---|---|---|---|---|---|
| SR-7625 | 9.3 | 810 | — | 11.0 | 1017 | — |
| IMR-4227 | 21.0 | 914 | — | 25.0 | 1039 | — |
| Unique | 10.8 | 926 | — | 11.5 | 1057 | — |
| Blue Dot | 15.4 | 947 | — | 16.8 | 1063 | — |
| 630 | 15.0 | 845 | — | 18.0 | 1022 | — |
| HS-6 | 13.6 | 914 | — | *14.2 | 1017 | — |
| AL-7 | 13.0 | 829 | — | 15.7 | 1117 | — |

### 240 gr. Sierra HC
#8820 (.4515" dia.) 1.575" OAL

| Powder | Sug. Starting Grains | Velocity F.P.S. | Pressure C.U.P. | Max. Grains | Velocity F.P.S. | Pressure C.U.P. |
|---|---|---|---|---|---|---|
| Unique | 9.5 | 851 | — | 12.5 | 1087 | — |
| 2400 | 18.5 | 899 | — | 23.0 | 1121 | — |
| SR-4759 | 18.5 | 819 | — | 24.0+ | 1139 | — |
| IMR-4227 | 20.5 | 904 | — | 25.0 | 1153 | — |
| 630 | 15.5 | 901 | — | *19.5 | 1168 | — |
| HS-7 | 14.0 | 865 | — | 19.5 | 1226 | — |
| AL-8 | 15.5 | 922 | — | 20.0 | 1141 | — |

### 250 gr. Hornady HP
.452" dia./1.590" OAL

| Powder | Sug. Starting Grains | Velocity F.P.S. | Pressure C.U.P. | Max. Grains | Velocity F.P.S. | Pressure C.U.P. |
|---|---|---|---|---|---|---|
| Unique | 9.3 | 801 | — | *12.2 | 1053 | — |
| 2400 | 17.5 | 814 | — | 22.0 | 1087 | — |
| SR-4759 | 18.5 | 807 | — | 24.0+ | 1102 | — |
| IMR-4227 | 20.3 | 862 | — | 24.0+ | 1032 | — |
| 630 | 15.0 | 851 | — | 19.0 | 1091 | — |
| HS-7 | 14.3 | 837 | — | 19.0 | 1145 | — |
| AL-8 | 15.0 | 844 | — | 19.5 | 1099 | — |

### 175 gr. Cast Bullet
#45468 (#2 Alloy) 1.560" OAL

| Powder | Sug. Starting Grains | Velocity F.P.S. | Pressure C.U.P. | Max. Grains | Velocity F.P.S. | Pressure C.U.P. |
|---|---|---|---|---|---|---|
| Red Dot | 7.0 | 885 | — | 9.0 | 1100 | — |
| Green Dot | 7.5 | 935 | — | 9.5 | 1098 | — |
| Unique | 10.0 | 1035 | — | 11.5 | 1192 | — |
| 700X | 7.0 | 910 | — | 9.0 | 1159 | — |
| PB | 7.5 | 845 | — | 10.4 | 1140 | — |
| SR-7625 | 9.5 | 925 | — | 10.5 | 1117 | — |
| 231 | 9.0 | 1041 | — | 10.0 | 1114 | — |
| HS-5 | 11.0 | 878 | — | 12.5 | 1081 | — |
| AL-5 | 11.0 | 959 | — | *13.0 | 1107 | — |

**Note:** Loads shown in shaded panels are maximum.
* Designates potentially most accurate load.
\+ Designates a compressed powder charge.

# .45 COLT
## (Ruger and T/C Contender Only—Continued)

**250 gr. Cast Bullet**
#454190 (#2 Alloy) 1.600" OAL

| Powder | Sug. Starting Grains | Velocity F.P.S. | Pressure C.U.P. | Max. Grains | Velocity F.P.S. | Pressure C.U.P. |
|---|---|---|---|---|---|---|
| ++Red Dot | 6.5 | 835 | — | 9.4 | 1097 | — |
| ++Unique | ‡9.0 | 875 | — | 11.5 | 1185 | — |
| ++700X | 6.0 | 800 | — | 9.0 | 1122 | — |
| ++PB | 7.5 | 820 | — | 10.0 | 1127 | — |
| ++SR-7625 | 8.5 | 890 | — | 10.5 | 1150 | — |
| 231 | 7.4 | 931 | — | 10.0 | 1088 | — |
| HS-5 | 9.8 | 859 | — | 12.0 | 1017 | — |
| AL-5 | 9.6 | 869 | — | *12.0 | 1003 | — |

**255 gr. Cast Bullet**
#454424 (#2 Alloy) 1.575" OAL

| Powder | Sug. Starting Grains | Velocity F.P.S. | Pressure C.U.P. | Max. Grains | Velocity F.P.S. | Pressure C.U.P. |
|---|---|---|---|---|---|---|
| ++Red Dot | 6.0 | 780 | — | 8.3 | 1018 | — |
| ++Unique | 8.5 | 845 | — | 10.2 | 1081 | — |
| ++700X | 6.0 | 785 | — | 8.0 | 1039 | — |
| ++PB | 7.0 | 750 | — | 8.5 | 986 | — |
| ++SR-7625 | 8.0 | 835 | — | 9.7 | 1060 | — |
| 630 | 11.5 | 768 | — | *15.5 | 1024 | — |
| HS-6 | 9.5 | 761 | — | 14.0 | 1145 | — |
| AL-7 | 10.0 | 756 | — | 14.0 | 1064 | — |

**Note:** Loads shown in shaded panels are maximum.
* Designates potentially most accurate load.
‡ Designates factory velocity duplication load.
++ Designates use of Winchester cases and primers.

# Reloader's Notes

# 11

# Cap and Ball Data Introduction

# Cap & Ball Revolver Data Introduction

**Off-the-shelf Lyman revolvers were used as the basis for these two pressure-sensing devices. The 1860 Army .44 is clamped in the "V-block" rest while the 1851 "Squareback" .36 sits atop the cast iron rest; the .36's piston is removed and the lapped hole into the chamber is clearly visible.**

**Introduction**

This section of revolver data presents for the first time a report of chamber pressures recorded in conjunction with velocity tests in two specially modified cap and ball revolvers. Velocity/Charge data is presented for .36, .44, and .45 revolvers with the .36 and .44 data being fired from the specially made P&V revolvers.

To investigate chamber pressures encountered during the firing of the two most popular calibers—.36 and .44—the Lyman technical staff designed and produced two unique handguns. Replicas of the 1851 Colt "Squareback" .36 and the Colt 1860 New Army .44 were chosen for their open-top frame configuration and a machined steel block was attached atop each cylinder with silver solder.

A hole was carefully drilled down through the block into the revolver chamber. The hole and a crusher piston of suitable dimensions were hand-lapped to a perfect fit. Next, an anvil was machined and fitted to the block. Finally, a "V-block" barrel vise was constructed and bolted to the oak test bed.

**Test Components**

Powder—Pyrodex, Curtis & Harvey and Gearhart-Owen powders were fired for almost all projectiles. Furthermore, both 3Fg and 4Fg were fired within each black powder brand for each projectile. In all cases, each and every charge was weighed before use and there was no variance toleration . . . a twenty grain charge weighed exactly twenty grains.

Caps—As a continuation of our standardization efforts within the major test sequences, we used only Remington caps in the appropriate size throughout the revolver work.

Lubricant—Here we deviated from our otherwise pervasive use of Crisco and used the new Lyman Revolver Lubricant. It performed very well and no abnormalities (jamming, multiple discharges, etc.) were experienced.

Chamber Filler—Plain old cornmeal was used with all below-maximum charges for both the round ball and conicals. This was an attempt to ensure uniformity by minimizing the distance the bullet had to travel before engaging the rifling. This technique of using filler has proven its merit to reloaders of brass centerfire ammunition. Cornmeal was specifically chosen because of its compressibility. Other cereals, like Cream of Wheat, require accurate portions since there is little compression possible.

Loading—The .45 Ruger was loaded and the bullets rammed with the integral rammer assembly. This gun was fired by hand without using the "V-block" rest. The .36 and .44 P&V revolvers, on the other hand, were dismantled into their main structural groups (grip and frame, cylinder, and barrel) and reloaded—a single shot only—on a free-standing loading tool. The rammer assembly had been removed from each pistol permitting a snugger fit in the "V-block". After each shot, the barrel wedge was tapped out and the barrel left in the "V-block" while the special cylinder was reloaded for the next shot.

**Test Procedures**

Revolver (Velocity Only)—Before each sequence the cylinder and bore were thoroughly cleaned. Then six rounds of the particular charge/projectile were loaded into the cylinder using normal loading procedures. Next, the first shot was fired but listed as a "warmer" shot and its readout was not included in the velocity average of the next five shots. Finally, the cylinder and bore were thoroughly cleaned in preparation for the next sequence.

**Barrel-to-cylinder gap was set by feeler gauge at .005″ while reassembling the gun for each shot. Every effort was made to duplicate conditions encountered by the plinker and target shooter. The camming action of the Colt barrel wedge allowed this clearance adjustment. Padding between barrel and clamp minimizes marring of the revolver barrel and gives the "V-clamp" better purchase.**

**Each gun was carefully bore-sighted to shoot out of the open firing port and through the two chronograph screen houses; the farthest twenty feet away. Here you see the .44 cocked but uncapped, with the brass loading spacer substituted for the lead crusher. The small knurled Allen bolt on the right of the anvil keeps the cylinder block and anvil in alignment.**

Revolver (Pressure and Velocity)—A total of six shots were fired with the first being a "warmer" which was not included in the final computations. Each sequence began with a clean chamber and bore. The entire shot string was fired without cleaning, as were the Velocity-Only guns, to approximate actual field use.

Before each shot was triggered (by a remote cable system) a .225″x.500″ lead crusher was positioned atop the piston and the anvil tightened to hold it snugly in place. After firing, the now-distorted crusher was removed, measured by micrometer and discarded. Later this measurement would be translated via a "Tarage Table" into Lead Units of Pressure (LUP). Remember that neither LUP nor CUP can accurately be called "p.s.i.". Conclusion: We had no difficulty with any phase of the revolver sequence. Each piece of equipment performed as it should have and there were no signs of strain, bulging or other pressure problems. Be sure to examine your revolver and follow only your manufacturer's loading recommendations.

**Remember: Use only black powder and be sure to wear shooting glasses.**

# Cleaning Cap & Ball Revolvers

No discussion about blackpowder shooting can be complete without mentioning the cleaning and preservation of these guns. Blackpowder shooting is messy, to say the least. After an afternoon enjoyment, both you and your gun are covered with an evil-smelling sooty residue which clings to everything it touches. It may not harm you, the shooter, but it can cause serious damage to your gun if left untended any length of time.

This blackpowder fouling is hygroscopic—it absorbs moisture from the air abnormally fast. As many shooters are aware, this moisture will cause the metal of your gun to rust in only a few hours with serious damage (pitting, etc.) following a short time later. The best thing to do with a gun that'll behave like that is to clean it! To aid you in finding the quickest and easiest way to preserve your particular piece, the Lyman staff offer the following suggestions:

These caplock wheelguns are fun to shoot but a bit tougher to clean. Fouling seems to collect in a new cranny or crevice each time a shot is fired. Superficial cleaning may get you past the first trip or two to the range but, sooner or later, you're going to have to dig into that handgun and scrub its internal parts. When that happens, here's how you go about it.

**Disassembly of the 1851 and 1860 pattern replicas:**

1. Using a brass drift, drive the barrel wedge from right to left until stopped by the wedge retaining screw.
2. Draw the hammer to half-cock, release the loading lever latch and rotate the cylinder slightly so a chamber wall is directly in front of the bullet plunger.
3. Operate the loading lever as if you were seating a ball. The plunger will be stopped by the cylinder wall and continuing pressure on the loading lever will cause the Barrel Assembly to separate from the two frame alignment pins and slide off the cylinder pin.
4. Slide the cylinder off the cylinder pin.

**NOTE:** The pistol is now in its major components which will allow satisfactory cleaning: Detailed disassembly follows.

Barrel and Lever Group: Remove the loading lever screw and draw the Lever Assembly from the Barrel Assembly. Removal of the plunger screw will separate the plunger from the loading lever proper. The 1860 Army has no loading lever screw. Removal of the plunger screw will seaprate the assembly from the barrel as well as the plunger from the loading lever proper. The latch assembly at the end of the loading lever usually is not dismantled.

Cylinder and Nipples Group: Removal of each nipple will completely break down this group.

Grip and Frame Group: Remove the Grip Assembly first. Undo the three backstrap screws—two rear and one lower. Now pull the one piece grips and backstrap away and separate. Next loosen the mainspring screw and lay aside the spring and screw. Unscrew the front and rear trigger guard screws and separate the trigger guard from the frame. The Grip Assembly is dismantled. Now for the Frame Assembly.

Undo the trigger and bolt spring screw and remove both screw and spring. Undo and remove the bolt screw. Remove the bolt from the frame. Undo and remove the trigger screw. Remove the trigger from the frame. Finally, undo and remove the hammer screw. Withdraw the hammer and hand downwards from the frame. The hand may be separated from the hammer but the hammer roller is seldom—if ever—removed.

Disassembly of the solid-frame Remington replicas is not as complicated as the Colt. Dropping the loading lever exposes

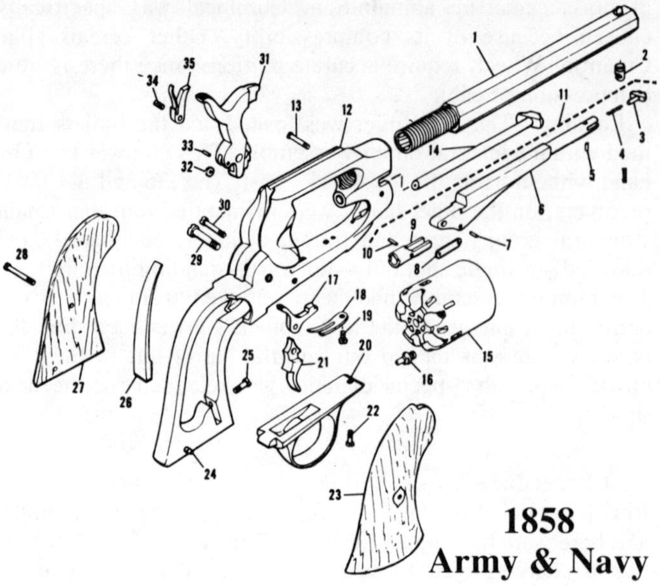


**1858**
**Army & Navy**

| Part No. | Description | Part No. | Description |
|---|---|---|---|
| 1 | Barrel (.36 Cal.) | 16 | Nipple |
| 1a | Barrel (.44 Cal.) | 17 | Cylinder Stop |
| 2 | Loading Lever Retainer | 18 | Trigger & Cylinder Stop Spring |
| 3 | Loading Lever Catch | | |
| 4 | Catch Spring | 19 | Spring Retainer Screw |
| 5 | Catch Pin | 20 | Trigger Guard |
| 6 | Loading Lever (.36 Cal.) | 21 | Trigger |
| 6a | Loading Lever (.44 Cal.) | 22 | Trigger Guard Screw |
| 7 | Link Pin | 23 | Right Grip |
| 8 | Link | 24 | Grip Pin |
| 9 | Bullet Rammer (.36 Cal.) | 25 | Main Spring Screw |
| | | 26 | Main Spring |
| 9a | Bullet Rammer (.44 Cal.) | 27 | Left Grip |
| | | 28 | Grip Screw |
| 10 | Rear Rammer Pin | 29 | Hammer Screw |
| 11 | Loading Lever Assembly (.36 Cal.) | 30 | Trigger & Cylinder Stop Screw |
| 11a | Loading Lever Assembly (.44 Cal.) | 31 | Hammer |
| | | 32 | Roller |
| 12 | Frame | 33 | Roller Pin |
| 13 | Loading Lever Screw | 34 | Hand Screw |
| 14 | Cylinder Pin | 35 | Hand |
| 15 | Cylinder (.36 Cal.) | | |
| 15a | Cylinder (.44 Cal.) | | |

the cylinder pin which needs only a pull to free the cylinder—after the hammer is at half-cock. This is far enough for most cleaning projects.

The Grip Assembly, with the exception of the grip panels proper, is integral with the frame. The Trigger Guard is released by removing one screw.

The bolt and trigger are hung from the same screw.

Removal of the hammer and hand follows removal of the mainspring. Push the hammer down through the frame as far as possible.

Look just below the bolt/trigger screw hole on the frame's left. You'll see a small screw holds the hand to the hammer.

Remove the small screw and lift the hammer out the top. Pull the hand out the bottom.

This completes disassembly of your revolver. Now—scrub it, flush it and dry it. Then oil and reassemble all the pieces by reversing the foregoing disassembly instructions.

Temporary protection can be secured by spraying the bore, chambers and other surfaces with a light mist of moisture displacing lubricant. Be sure to store the gun in the warmest driest place possible until it gets that good all-over cleaning it needs.

# 12

# Cap and Ball Data

# .36 CALIBER/7½" BBL.

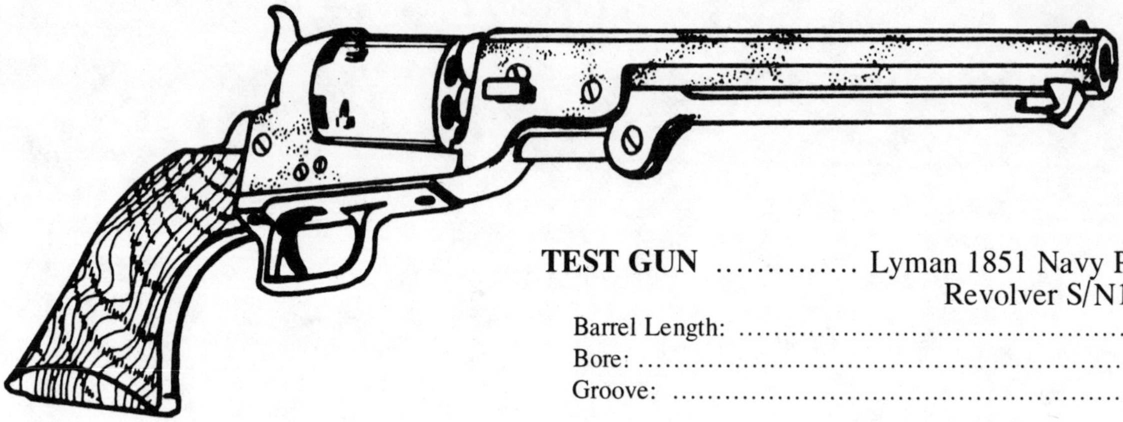

**TEST GUN** .............. Lyman 1851 Navy P & V Revolver S/N18916

Barrel Length: .............................................. 7½"
Bore: ........................................................ .369"
Groove: ...................................................... .378"

**PROJECTILE:** ........... R.B.
Diameter: .............. .375"
Weight: ............... 81 Grs.
Lube: .................. Lyman
Patch: .................... None

**PROPELLANT:**
Brand: .................... G-O
Granulation: ............ FFFg

**IGNITION:** ................. Cap
Brand: ................... Rem.
Size: ................... No. 11

| Charge (Grs.) | Velocity (FPS) | Pressure (LUP) | Muzzle Energy (Ft./lbs.) |
|---|---|---|---|
| 14.0 | 752 | 6,900 | 102 |
| 17.0 | 879 | 7,200 | 139 |
| 20.0 | 967 | 8,300 | 168 |
| 23.0 | 1017 | 8,560 | 186 |
| 26.0 | 1079 | 8,600 | 209 |
| *29.0 | 1097 | 9,440 | 216 |

*Maximum Powder Charge — All data from special pressure revolver.

---

**PROJECTILE:** ........... R.B.
Diameter: .............. .375"
Weight: ............... 81 Grs.
Lube: .................. Lyman
Patch: .................... None

**PROPELLANT:**
Brand: .................... G-O
Granulation: .......... FFFFg

**IGNITION:** ................. Cap
Brand: ................... Rem.
Size: ................... No. 11

| Charge (Grs.) | Velocity (FPS) | Pressure (LUP) | Muzzle Energy (Ft./lbs.) |
|---|---|---|---|
| 14.0 | 884 | 6,900 | 140 |
| 17.0 | 924 | 7,240 | 153 |
| 20.0 | 1006 | 8,460 | 182 |
| 23.0 | 1027 | 8,640 | 189 |
| 26.0 | 1033 | 8,820 | 192 |
| *27.5 | 1090 | 9,660 | 213 |

*Maximum Powder Charge — All data from special pressure revolver.

# .36 CALIBER/7½'' BBL.
(Continued)

**PROJECTILE:** .......... R.B.
Diameter: ............. .375''
Weight: .............. 81 Grs.
Lube: ................. Lyman
Patch: ................... None

**PROPELLANT:**
Brand: .................. C&H
Granulation: ........... FFFg

**IGNITION:** ................ Cap
Brand: .................. Rem.
Size: .................. No. 11

| Charge (Grs.) | Velocity (FPS) | Pressure (LUP) | Muzzle Energy (Ft./lbs.) |
|---|---|---|---|
| 14.0 | 675 | 4,700 | 82 |
| 17.0 | 765 | 5,300 | 105 |
| 20.0 | 841 | 5,900 | 127 |
| 23.0 | 869 | 6,100 | 136 |
| *26.0 | 899 | 6,340 | 145 |

*Maximum Powder Charge — Any increase will distort the projectile due to the high increase in seating pressure required.

---

**PROJECTILE:** .......... R.B.
Diameter: ............. .375''
Weight: .............. 81 Grs.
Lube: ................. Lyman
Patch: ................... None

**PROPELLANT:**
Brand: .................. C&H
Granulation: ......... FFFFg

**IGNITION:** ................ Cap
Brand: .................. Rem.
Size: .................. No. 11

| Charge (Grs.) | Velocity (FPS) | Pressure (LUP) | Muzzle Energy (Ft./lbs.) |
|---|---|---|---|
| 14.0 | 873 | 7,200 | 137 |
| 17.0 | 937 | 7,300 | 158 |
| 20.0 | 977 | 7,700 | 171 |
| 23.0 | 1039 | 8,100 | 194 |
| 26.0 | 1037 | 8,600 | 193 |
| *27.5 | 1002 | 8,800 | 180 |

*Maximum Powder Charge — All data from special pressure revolver.

---

**PROJECTILE:** .......... R.B.
Diameter .............. .380''
Weight ............... 82 Grs.
Lube .................. Lyman

**PROPELLANT:**
Brand ............... Pyrodex
Granulation ................ P

**IGNITION:** ................ Cap
Brand ................... Rem.
Size ................... No. 11

| Charge (Grs.) | Velocity (FPS) | Pressure (LUP) | Muzzle Energy (Ft./lbs.) |
|---|---|---|---|
| 10.0 | 461 | 4,260 | 39 |
| 13.0 | 584 | 4,640 | 63 |
| 15.0 | 650 | 4,660 | 77 |
| 18.0 | 738 | 4,720 | 99 |
| *21.0 | 823 | 5,680 | 123 |

*Maximum Powder Charge — All data from special pressure revolver.

**PROJECTILE:** ....... #37583
Diameter: .............. .375''
Weight: ............ 150 Grs.
Lube: ................. Lyman
Patch: ................... None

**PROPELLANT:**
Brand: ................... G-O
Granulation: ........... FFFg

**IGNITION:** ................ Cap
Brand: .................. Rem.
Size: .................. No. 11

| Charge (Grs.) | Velocity (FPS) | Pressure (LUP) | Muzzle Energy (Ft./lbs.) |
|---|---|---|---|
| 9.0 | 480 | 5,700 | 77 |
| 12.0 | 560 | 7,200 | 104 |
| *15.0 | 668 | 8,500 | 148 |

*Maximum Powder Charge — All data from special pressure revolver.

---

**PROJECTILE:** ....... #37583
Diameter: .............. .375''
Weight: ............ 150 Grs.
Lube: ................. Lyman
Patch: ................... None

**PROPELLANT:**
Brand: ................... G-O
Granulation: .......... FFFFg

**IGNITION:** ................ Cap
Brand: .................. Rem.
Size: .................. No. 11

| Charge (Grs.) | Velocity (FPS) | Pressure (LUP) | Muzzle Energy (Ft./lbs.) |
|---|---|---|---|
| 9.0 | 559 | 7,300 | 104 |
| 12.0 | 682 | 8,740 | 155 |
| *15.0 | 787 | 11,200 | 206 |

*Maximum Powder Charge — All data from special pressure revolver.

---

**PROJECTILE:** ....... #37583
Diameter: .............. .375''
Weight: ............ 150 Grs.
Lube: ................. Lyman
Patch: ................... None

**PROPELLANT:**
Brand: ................... C&H
Granulation: ........... FFFg

**IGNITION:** ................ Cap
Brand: .................. Rem.
Size: .................. No. 11

| Charge (Grs.) | Velocity (FPS) | Pressure (LUP) | Muzzle Energy (Ft./lbs.) |
|---|---|---|---|
| 9.0 | 439 | 5,000 | 64 |
| 12.0 | 529 | 5,900 | 93 |
| *15.0 | 601 | 6,400 | 120 |

*Maximum Powder Charge — All data from special pressure revolver.

# .36 CALIBER/7½" BBL.
(Continued)

**PROJECTILE:** ....... #37583
Diameter: .............. .375"
Weight: ............ 150 Grs.
Lube: ................. Lyman
Patch: .................... None

**PROPELLANT:**
Brand: .................. C&H
Granulation: ......... FFFFg

**IGNITION:** ................. Cap
Brand: .................. Rem.
Size: .................. No. 11

| Charge (Grs.) | Velocity (FPS) | Pressure (LUP) | Muzzle Energy (Ft./lbs.) |
|---|---|---|---|
| 9.0 | 544 | 6,800 | 98 |
| 12.0 | 653 | 9,200 | 142 |
| *15.0 | 758 | 10,600 | 191 |

*Maximum Powder Charge — All data from special pressure revolver.

# .44 CALIBER/8'' BBL.

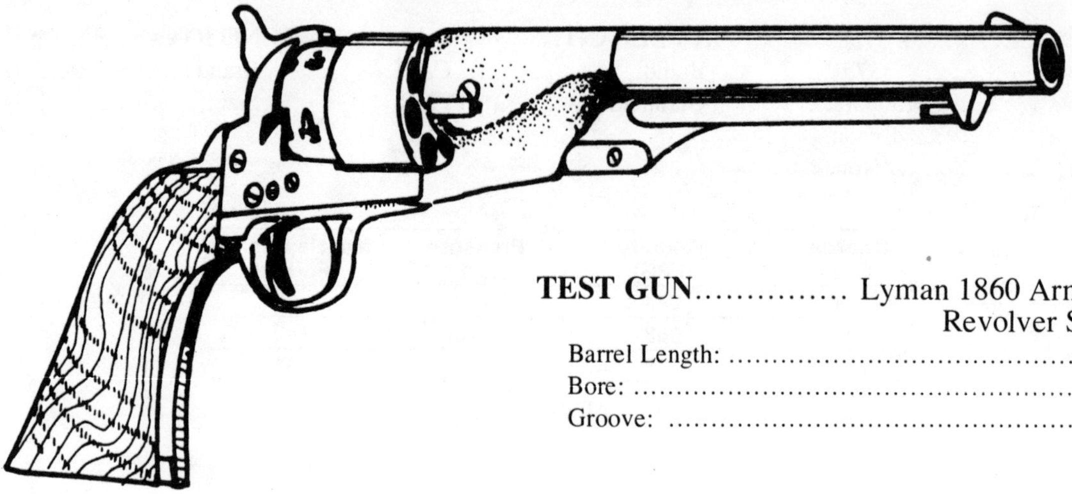

**TEST GUN**............... Lyman 1860 Army P & V Revolver S/N47676

Barrel Length: ................................................ 8''
Bore: ........................................................ .440''
Groove: .................................................... .448''

**PROJECTILE:** ........... R.B.
Diameter: .............. .451''
Weight: ............ 138 Grs.
Lube: ................. Lyman
Patch: .................... None

**PROPELLANT:**
Brand: .................... G-O
Granulation: ............ FFFg

**IGNITION:** ................. Cap
Brand: ................... Rem.
Size: ................... No. 11

| Charge (Grs.) | Velocity (FPS) | Pressure (LUP) | Muzzle Energy (Ft./lbs.) |
|---|---|---|---|
| 19.0 | 706 | 5,540 | 153 |
| 22.0 | 752 | 5,540 | 173 |
| 25.0 | 805 | 5,780 | 198 |
| 28.0 | 885 | 6,380 | 240 |
| 31.0 | 933 | 7,060 | 266 |
| 33.0 | 979 | 7,340 | 293 |
| *37.0 | 1032 | 7,940 | 326 |

*Maximum Powder Charge — All data from special pressure revolver.

---

**PROJECTILE:** ............ R.B.
Diameter: .............. .451''
Weight: ............. 138 Grs.
Lube: .................. Lyman
Patch: .................... None

**PROPELLANT:**
Brand: ................... C&H
Granulation: ............ FFFg

**IGNITION:**.................. Cap
Brand: ................... Rem.
Size: ................... No. 11

| Charge (Grs.) | Velocity (FPS) | Pressure (LUP) | Muzzle Energy (Ft./lbs.) |
|---|---|---|---|
| 19.0 | 660 | 4,740 | 133 |
| 22.0 | 694 | 4,620 | 147 |
| 25.0 | 734 | 4,600 | 165 |
| 28.0 | 765 | 4,820 | 179 |
| 31.0 | 819 | 5,220 | 205 |
| 34.0 | 826 | 5,380 | 209 |
| *37.0 | 852 | 5,360 | 222 |

*Maximum Powder Charge — All data from special pressure revolver.

# .44 CALIBER/8" BBL.
(Continued)

**PROJECTILE:** ........... R.B.
Diameter: .............. .451"
Weight: ............ 138 Grs.
Lube: .................. Lyman
Patch: .................... None

**PROPELLANT:**
Brand: .................... G-O
Granulation: .......... FFFFg

**IGNITION:** ................. Cap
Brand: .................... Rem.
Size: ................... No. 12

| Charge (Grs.) | Velocity (FPS) | Pressure (LUP) | Muzzle Energy (Ft./lbs.) |
|---|---|---|---|
| 19.0 | 772 | 5,240 | 182 |
| 22.0 | 797 | 5,420 | 194 |
| 25.0 | 868 | 6,080 | 231 |
| 28.0 | 881 | 6,320 | 238 |
| 31.0 | 859 | 6,500 | 226 |
| 34.0 | 953 | 7,240 | 278 |
| *37.0 | 960 | 7,420 | 282 |

*Maximum Powder Charge — All data from special pressure revolver.

**PROJECTILE:** ........... R.B.
Diameter: .............. .451"
Weight: ............ 138 Grs.
Lube: .................. Lyman
Patch: .................... None

**PROPELLANT:**
Brand: ................... C&H
Granulation: .......... FFFFg

**IGNITION:** ................. Cap
Brand: .................... Rem.
Size: ................... No. 11

| Charge (Grs.) | Velocity (FPS) | Pressure (LUP) | Muzzle Energy (Ft./lbs.) |
|---|---|---|---|
| 19.0 | 776 | 5,720 | 184 |
| 22.0 | 771 | 5,720 | 182 |
| 25.0 | 817 | 5,960 | 204 |
| 28.0 | 839 | 6,280 | 215 |
| 31.0 | 875 | 6,460 | 234 |
| 34.0 | 897 | 6,680 | 246 |
| *37.0 | 943 | 6,820 | 272 |

*Maximum Powder Charge — All data from special pressure revolver.

**PROJECTILE:** ........... R.B.
Diameter: .............. .451"
Weight: ............ 138 Grs.
Lube .................. Lyman

**PROPELLANT:**
Brand ................ Pyrodex
Granulation ................. P

**IGNITION:** ................. Cap
Brand: .................... Rem.
Size: ................... No. 11

| Charge (Grs.) | Velocity (FPS) | Pressure (LUP) | Muzzle Energy (Ft./lbs.) |
|---|---|---|---|
| 15.0 | 606 | 4,320 | 113 |
| 19.0 | 690 | 4,420 | 146 |
| 22.0 | 718 | 4,500 | 158 |
| 25.0 | 774 | 4,520 | 184 |
| *28.0 | 810 | 4,800 | 201 |

*Maximum Powder Charge — All data from special pressure revolver.

# .44 CALIBER/8" BBL.
(Continued)

**PROJECTILE:** ...... #450229
Diameter: .............. .450"
Weight: ............ 155 Grs.
Lube: ................. Lyman
Patch: .................... None

**PROPELLANT:**
Brand: .................... G-O
Granulation: ........... FFFg

**IGNITION:** ................ Cap
Brand: .................. Rem.
Size: .................. No. 11

| Charge (Grs.) | Velocity (FPS) | Pressure (LUP) | Muzzle Energy (Ft./lbs.) |
|---|---|---|---|
| 19.0 | 705 | 6,820 | 171 |
| 22.0 | 768 | 7,520 | 203 |
| 25.0 | 882 | 7,880 | 267 |
| *28.0 | 861 | 8,820 | 255 |

*Maximum Powder Charge — All data from special pressure revolver.

---

**PROJECTILE:** ...... #450229
Diameter: .............. .450"
Weight: ............ 155 Grs.
Lube: ................. Lyman
Patch: .................... None

**PROPELLANT:**
Brand: .................. C&H
Granulation: ........... FFFg

**IGNITION:** ................ Cap
Brand: .................. Rem.
Size: .................. No. 11

| Charge (Grs.) | Velocity (FPS) | Pressure (LUP) | Muzzle Energy (Ft./lbs.) |
|---|---|---|---|
| 19.0 | 632 | 5,020 | 137 |
| 22.0 | 681 | 5,460 | 159 |
| 25.0 | 741 | 6,280 | 189 |
| *28.0 | 785 | 6,080 | 212 |

*Maximum Powder Charge — All data from special pressure revolver.

---

**PROJECTILE:** ...... #450229
Diameter: .............. .450"
Weight: ............ 155 Grs.
Lube: ................. Lyman
Patch: .................... None

**PROPELLANT:**
Brand: .................... G-O
Granulation: .......... FFFFg

**IGNITION:** ................ Cap
Brand: .................. Rem.
Size: .................. No. 11

| Charge (Grs.) | Velocity (FPS) | Pressure (LUP) | Muzzle Energy (Ft./lbs.) |
|---|---|---|---|
| 19.0 | 759 | 6,060 | 198 |
| 22.0 | 814 | 7,400 | 228 |
| 25.0 | 835 | 7,660 | 240 |
| *28.0 | 885 | 8,480 | 269 |

*Maximum Powder Charge — All data from special pressure revolver.

# .44 CALIBER/8" BBL.
(Continued)

**PROJECTILE:** ...... #450229
Diameter: .............. .450"
Weight: ............ 155 Grs.
Lube: ................. Lyman
Patch: ................... None

**PROPELLANT:**
Brand: .................. C&H
Granulation: ......... FFFFg

**IGNITION:** ................ Cap
Brand: .................. Rem.
Size: .................. No. 11

| Charge (Grs.) | Velocity (FPS) | Pressure (LUP) | Muzzle Energy (Ft./lbs.) |
|---|---|---|---|
| 19.0 | 756 | 6,240 | 196 |
| 22.0 | 788 | 6,420 | 213 |
| 25.0 | 841 | 6,800 | 243 |
| *28.0 | 862 | 7,220 | 255 |

*Maximum Powder Charge — All data from special pressure revolver.

---

**PROJECTILE:**....... #450229
Diameter: .............. .451"
Weight: ............ 155 Grs.
Lube: ................. Lyman

**PROPELLANT:**
Brand: ................ Pyrodex
Granulation: ................ P

**IGNITION:** ................ Cap
Brand: .................. Rem.
Size: .................. No. 11

| Charge (Grs.) | Velocity (FPS) | Pressure (LUP) | Muzzle Energy (Ft./lbs.) |
|---|---|---|---|
| 18.0 | 615 | 5,200 | 140 |
| 21.0 | 743 | 5,960 | 205 |
| 24.0 | 778 | 6,160 | 225 |
| *27.0 | 825 | 6,320 | 252 |

*Maximum Powder Charge — All data from special pressure revolver.

# .45 CALIBER/7½" BBL.

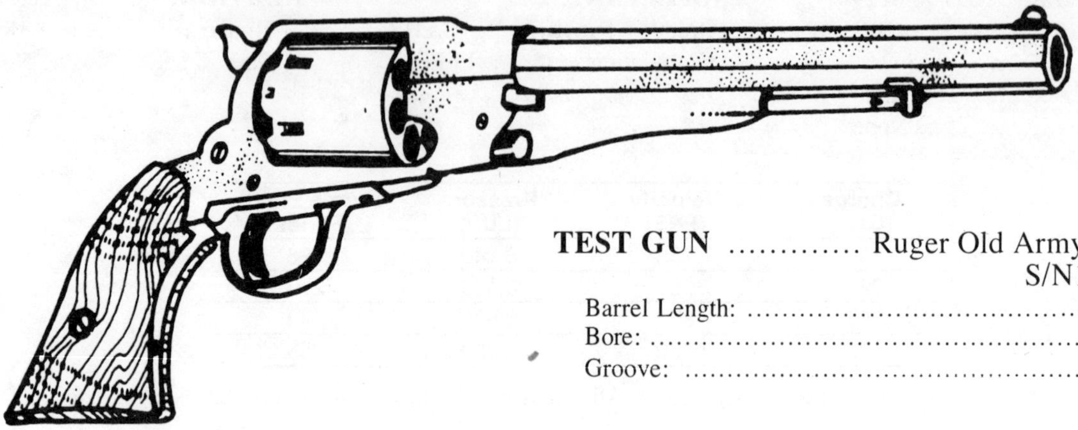

**TEST GUN** ............. Ruger Old Army Revolver S/N140-04089

Barrel Length: ............................................. 7½"
Bore: ........................................................ .442"
Groove: ................................................... .450"

**PROJECTILE:** ........... R.B.
Diameter: .............. .457"
Weight: ............ 143 Grs.
Lube: .................. Lyman
Patch: .................... None

**PROPELLANT:**
Brand: .................... G-O
Granulation: ............ FFFg

**IGNITION:** ................. Cap
Brand: ................... Rem.
Size: ................... No. 10

| Charge (Grs.) | Velocity (FPS) | Muzzle Energy (Ft./lbs.) |
|---|---|---|
| *30.0 | 858 | 233 |
| 40.0 | 917 | 263 |

*¾ of Maximum Charge Weight — Reduced volume taken up with corn meal filler material.

---

**PROJECTILE:** ........... R.B.
Diameter: .............. .457"
Weight: ............ 143 Grs.
Lube: .................. Lyman
Patch: .................... None

**PROPELLANT:**
Brand: .................. C&H
Granulation: ............ FFFg

**IGNITION:** ................. Cap
Brand: ................... Rem.
Size: ................... No. 10

| Charge (Grs.) | Velocity (FPS) | Muzzle Energy (Ft./lbs.) |
|---|---|---|
| *30.0 | 761 | 184 |
| 40.0 | 772 | 189 |

*¾ of Maximum Charge Weight — Reduced volume taken up with corn meal filler material.

# .45 CALIBER/7½" BBL.
## (Continued)

**PROJECTILE:** .......... R.B.
Diameter: .............. .457"
Weight: ............ 143 Grs.
Lube: ................ Lyman
Patch: .................. None

**PROPELLANT:**
Brand: .................. G-O
Granulation: .......... FFFFg

**IGNITION:** ................ Cap
Brand: .................. Rem.
Size: .................. No. 10

| Charge (Grs.) | Velocity (FPS) | Muzzle Energy (Ft./lbs.) |
|---|---|---|
| *31.0 | 964 | 295 |
| 41.0 | 1036 | 340 |

*¾ of Maximum Charge Weight — Reduced volume taken up with corn meal filler material.

---

**PROJECTILE:** .......... R.B.
Diameter: .............. .457"
Weight: ............ 143 Grs.
Lube: ................ Lyman
Patch: .................. None

**PROPELLANT:**
Brand: .................. C&H
Granulation: .......... FFFFg

**IGNITION:** ................ Cap
Brand: .................. Rem.
Size: .................. No. 10

| Charge (Grs.) | Velocity (FPS) | Muzzle Energy (Ft./lbs.) |
|---|---|---|
| *31.0 | 911 | 263 |
| 41.0 | 1021 | 331 |

*¾ of Maximum Charge Weight — Reduced volume taken up with corn meal filler material.

---

**PROJECTILE:**........... R.B.
Diameter: .............. .457"
Weight: ............ 143 Grs.
Lube: ................ Lyman
Patch: .................. None

**PROPELLANT:**
Brand: ............... Pyrodex
Granulation: ................ P

**IGNITION:** ................ Cap
Brand: .................. Rem.
Size: .................. No. 11

| Charge (Grs.) | Velocity (FPS) | Muzzle Energy (Ft./lbs.) |
|---|---|---|
| *26.0 | 831 | 219 |
| 34.0 | 869 | 240 |

*¾ of Maximum Charge Weight — Reduced volume taken up with corn meal filler material.

# .45 CALIBER/7½" BBL.
(Continued)

**PROJECTILE:** ....... #45468
Diameter: .............. .454"
Weight: ............ 185 Grs.
Lube: ................. Lyman
Patch: ................... None

**PROPELLANT:**
Brand: .................... G-O
Granulation: ........... FFFg

**IGNITION:** ................ Cap
Brand: .................. Rem.
Size: .................. No. 10

| Charge (Grs.) | Velocity (FPS) | Muzzle Energy (Ft./lbs.) |
|---|---|---|
| *25.5 | 717 | 210 |
| 34.0 | 813 | 271 |

*¾ of Maximum Charge Weight — Reduced volume taken up with corn meal filler material.

---

**PROJECTILE:** ....... #45468
Diameter: .............. .454"
Weight: ............ 185 Grs.
Lube: ................. Lyman
Patch: ................... None

**PROPELLANT:**
Brand: .................. C&H
Granulation: ........... FFFg

**IGNITION:** ................ Cap
Brand: .................. Rem.
Size: .................. No. 10

| Charge (Grs.) | Velocity (FPS) | Muzzle Energy (Ft./lbs.) |
|---|---|---|
| *25.5 | 659 | 178 |
| 34.0 | 701 | 201 |

*¾ of Maximum Charge Weight — Reduced volume taken up with corn meal filler material.

---

**PROJECTILE:** ....... #45468
Diameter: .............. .454"
Weight: ............ 185 Grs.
Lube: ................. Lyman
Patch: ................... None

**PROPELLANT:**
Brand: .................... G-O
Granulation: .......... FFFFg

**IGNITION:** ................ Cap
Brand: .................. Rem.
Size: .................. No. 10

| Charge (Grs.) | Velocity (FPS) | Muzzle Energy (Ft./lbs.) |
|---|---|---|
| *26.0 | 860 | 303 |
| 35.0 | 905 | 335 |

*¾ of Maximum Charge Weight — Reduced volume taken up with corn meal filler material.

# .45 CALIBER/7½" BBL.
(Continued)

**PROJECTILE:** ....... #45468
Diameter: .............. .454"
Weight: ............ 185 Grs.
Lube: ................. Lyman
Patch: ................... None

**PROPELLANT:**
Brand: .................. C&H
Granulation: .......... FFFFg

**IGNITION:** ................ Cap
Brand: .................. Rem.
Size: .................. No. 10

| Charge (Grs.) | Velocity (FPS) | Muzzle Energy (Ft./lbs.) |
|---|---|---|
| *26.0 | 778 | 248 |
| 35.0 | 898 | 330 |

*¾ of Maximum Charge Weight — Reduced volume taken up with corn meal filler material.

---

**PROJECTILE:** .........#45468
Diameter: .............. .454"
Weight: ............ 185 Grs.
Lube: ................. Lyman
Patch: ................... None

**PROPELLANT:**
Brand: ............... Pyrodex
Granulation: ................ P

**IGNITION:**................. Cap
Brand: .................. Rem.
Size: ................... No. 10

| Charge (Grs.) | Velocity (FPS) | Muzzle Energy (Ft./lbs.) |
|---|---|---|
| *24.0 | 741 | 213 |
| 32.0 | 765 | 227 |

*¾ of Maximum Charge Weight — Reduced volume taken up with corn meal filler material.

---

**PROJECTILE:** ........ #45467
Diameter: .............. .454"
Weight: ............. 190 Grs.
Lube: ................. Lyman
Patch: ................... None

**PROPELLANT:**
Brand: ................... G-O
Granulation: ........... FFFg

**IGNITION:**................. Cap
Brand: .................. Rem.
Size: .................. No. 10

| Charge (Grs.) | Velocity (FPS) | Muzzle Energy (Ft./lbs.) |
|---|---|---|
| *25.0 | 710 | 212 |
| 33.0 | 780 | 256 |

*¾ of Maximum Charge Weight — Reduced volume taken up with corn meal filler material.

# .45 CALIBER/7½" BBL.
(Continued)

**PROJECTILE:** ....... #45467
Diameter: .............. .454"
Weight: ............ 190 Grs.
Lube: ................. Lyman
Patch: ................... None

**PROPELLANT:**
Brand: .................. C&H
Granulation: ........... FFFg

**IGNITION:** ................ Cap
Brand: .................. Rem.
Size: .................. No. 10

| Charge (Grs.) | Velocity (FPS) | Muzzle Energy (Ft./lbs.) |
|---|---|---|
| *25.0 | 674 | 191 |
| 33.0 | 709 | 211 |

*¾ of Maximum Charge Weight — Reduced volume taken up with corn meal filler material.

---

**PROJECTILE:** ....... #45467
Diameter: .............. .454"
Weight: ............ 190 Grs.
Lube: ................. Lyman
Patch: ................... None

**PROPELLANT:**
Brand: .................. G-O
Granulation: .......... FFFFg

**IGNITION:** ................ Cap
Brand: .................. Rem.
Size: .................. No. 10

| Charge (Grs.) | Velocity (FPS) | Muzzle Energy (Ft./lbs.) |
|---|---|---|
| *25.5 | 841 | 297 |
| 34.0 | 863 | 313 |

*¾ of Maximum Charge Weight — Reduced volume taken up with corn meal filler material.

---

**PROJECTILE:** ....... #45467
Diameter: .............. .454"
Weight: ............ 190 Grs.
Lube: ................. Lyman
Patch: ................... None

**PROPELLANT:**
Brand: .................. C&H
Granulation: .......... FFFFg

**IGNITION:** ................ Cap
Brand: .................. Rem.
Size: .................. No. 10

| Charge (Grs.) | Velocity (FPS) | Muzzle Energy (Ft./lbs.) |
|---|---|---|
| *25.5 | 840 | 297 |
| 34.0 | 898 | 339 |

*¾ of Maximum Charge Weight — Reduced volume taken up with corn meal filler material.

# .45 CALIBER/7½" BBL.
(Continued)

**PROJECTILE:** ....... #45467
Diameter: .............. .454"
Weight: ............ 190 Grs.
Lube: ..................Lyman
Patch: .....................None

**PROPELLANT:**
Brand: ................ Pyrodex
Granulation: ................ P

**IGNITION:** ................ Cap
Brand: .................. Rem.
Size: .................... No. 10

| Charge (Grs.) | Velocity (FPS) | Muzzle Energy (Ft./lbs.) |
|---|---|---|
| *24.0 | 721 | 219 |
| 31.5 | 894 | 337 |

* ¾ of Maximum Charge Weight — Reduced volume taken up with corn meal filler material.

# Reloader's Notes

# 13

# Exterior Ballistics

# Handgun Ballistics Data Introduction

W. T. McDonald & T. C. Almgren

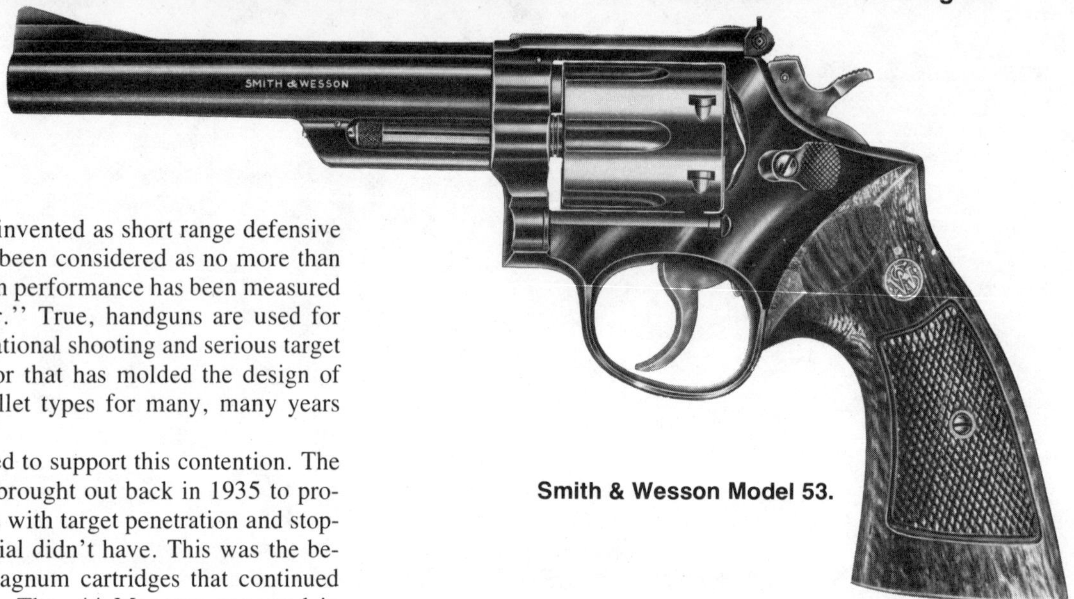


Smith & Wesson Model 53.

**Introduction:**

Handguns were originally invented as short range defensive weapons and they have long been considered as no more than that. For many years, handgun performance has been measured in terms of "stopping power." True, handguns are used for other purposes, such as recreational shooting and serious target shooting, but the major factor that has molded the design of new guns, cartridges and bullet types for many, many years has been stopping power.

Many examples can be cited to support this contention. The .357 Magnum handgun was brought out back in 1935 to provide law enforcement officers with target penetration and stopping power that the .38 Special didn't have. This was the beginning of a trend toward magnum cartridges that continued strongly after World War II. The .44 Magnum appeared in 1956, billed as "the world's most powerful handgun." Then the .41 Magnum was introduced in 1964 expressly to provide law enforcement officers with penetration and stopping power adequate for the modern day. New handgun designs with greatly increased firepower (and therefore more stopping power) also appeared. The .357 and .44 Auto Mags, the Smith and Wesson Model 59, and other 9mm semiautos that carry fifteen or sixteen cartridges are typical examples of these arms.

About the biggest thing that has happened in handgun bullet design in the last fifty years is the jacketed hollow point bullet. This type of bullet was expressly invented for expansion and hydrostatic shock in soft flesh; again, to improve stopping power. Hollow points had been used in lead bullets for many years to improve expansion, but the jacketed bullets could be fired at higher velocities and expansion results were much more spectacular.

Up until very recent years, no one was much interested in the exterior ballistic performance of handgun bullets. The external shapes of handgun bullets today are not much different than they were a century ago. If anything, bullet shapes have become worse from an exterior ballistics point of view. Since handguns have traditionally been considered short range defensive weapons, little attention was paid to how well a bullet would hold its velocity over a range of one hundred or two hundred yards.

A gradual change in this line of thinking began in the early 1960's, and it has been gathering momentum ever since. This change is a real, sustained interest in long range handgun shooting for targets and hunting. This interest created a market for some specialized handguns, handgun telescope sights, some specialized cartridges and even a few specialized bullet designs in very recent years.

We believe that this new trend in handgun thinking first made itself evident in 1961, when Smith and Wesson brought out the Model 53 revolver chambered for the .22 Remington Jet cartridge, a necked-down version of the .357 Magnum case. This revolver/cartridge development was a team effort by Remington and Smith and Wesson to offer the first small caliber, high performance handgun to the shooting public. At first, shooters greeted it enthusiastically, but then disappointments developed. In the first place, bore diameter was .223 inch, the same as .22 rimfire cartridges which the Model 53 could shoot interchangeably with cylinder chamber inserts. The selection of jacketed bullets was limited to the 40 and 45 grain .223 Hornet bullets. The wide variety of .224 diameter bullets would not deliver accurate, high velocity handloads in the Model 53. Also, velocity was a bit disappointing. Although the handgun and cartridge were designed for 40,000 psi chamber pressure, the 45 grain Hornet bullet could be driven out the six inch barrel only at about 1700 fps. But the real disappointment with the Model 53 was a functional problem. When the high pressure Jet cartridge fired, there was a tendency for it to set back against the recoil shield, preventing cylinder rotation. The cure for this problem was thorough degreasing of both cylinder and loaded cartridges, but this functional deficiency gave the gun a bad reputation and prevented it from reaching the popularity it should have achieved.

In 1962, Remington introduced a most unconventional handgun, the bolt action single-shot XP-100 firing the .221 Remington Fireball cartridge, a shortened version of the popular .222 Remington rifle cartridge. The XP-100 took .224 diameter bullets, so all the jacketed .22 caliber bullets were available for handloading. The Fireball cartridge would drive a 45 grain bullet at 2700 fps from the 10½-inch barrel, a full

Remington Model XP-100

1000 fps faster than the Jet in the Model 53. The XP-100 soon established a record for amazing accuracy. Some intrepid souls mounted telescope sights, worked over the trigger pull, and found they could shoot one-inch groups at 200 yards from a bench rest. The unconventional XP-100 was very successful just where the S & W Model 53 failed to be: with a growing fraternity of handgunners interested in shooting targets and varmints at very long handgun ranges.

In 1966, the Thompson/Center Contender pistol reached the market. This is a tip-up action, single-shot pistol with interchangeable barrels for a large selection of centerfire and rimfire cartridges, and it accommodates telescope sights. The T/C Contender has been very successful, especially since it offers large calibers and is easily convertible for wildcat cartridges. It has popularized several wildcat cartridges, including the .30 and .357 Herretts, the .357-44 B & D, and the .22 K-Hornet, and barrels are even available for the .25-35, .30-30, and .35 Remington rifle cartridges. The T/C Contender certainly is the most popular single-shot pistol for hunting, target, and handgun silhouette shooting in America today.

The interest in long range handgun shooting for targets and hunting has grown throughout the 1970's. Handgun hunting for small and medium game has become very popular, particularly in the western states. One of the most rapidly growing shooting sports today is handgun silhouette shooting. The metal silhouette targets are chickens at 50 meters, javelina at 100, turkeys at 150, and rams at 200 meters, and the rules require iron sights. Since a meter is slightly more than a yard, this competition challenges the best handgunners. The steel targets must be tumbled over (not just struck) in order to score, so most competitors use big bore handguns with heavy bullets. Silhouette shooting, though only about three years old in this country, has had an impact on manufacturers already. Sierra Bullets announced in its 1978 catalog a 170 grain full metal jacket .357 bullet specifically for handgun silhouette shooting. Thompson/Center has just announced its new Super Contender 14, initially available in six calibers, five of which should be excellent for the silhouette competition.

Today, then, stopping power is no longer the only measure of handgun performance. A steadily increasing number of shooters is interested in the exterior ballistic performance of loads shot at ranges up to 300 yards. Very little exterior ballistics information for handgun bullets exists in today's shooting literature. Ballistic coefficients of handgun bullets are not generally published by bullet manufacturers, so even if a shooter takes the trouble to chronograph a load, there is no way to compute bullet trajectory and wind deflection at various ranges.

Lyman has reversed that situation in this handbook. A very extensive set of firing tests was performed to determine the ballistic coefficients of the bullets and round balls included in this handbook, and the Ballistics Tables show velocity, energy, drop, bullet path height, time of flight and crosswind deflections for each projectile at several muzzle velocities to a range of 300 yards. These exterior ballistics data should answer many questions and settle many arguments about handgun performance at long ranges.

The last section of this article talks about the exterior ballistics of handguns. It describes what a ballistic coefficient is and why it is important; describes Lyman's firing tests and points out what a shooter needs to worry about under windy conditions and changes in altitude above sea level.

**Exterior Ballistics**

After a bullet exits the gun barrel, two physical forces control its flight: gravity and air drag. Air drag is much the stronger force, but gravity cannot be neglected when bullet trajectories are computed. To a first approximation, air drag causes the bullet to slow down; gravity causes the trajectory to bend down (drop). This approximation gives a useful mental picture of how the two forces act, but when the trajectory is not precisely level, there is an interaction between gravity and drag forces, and this must be accounted for in accurate ballistics computations.

Drag is a very complicated function of the weight, shape and velocity of a bullet and of the temperature and density of the air through which it travels. A great deal of scientific and engineering research has been devoted to air drag effects on projectiles—from the days of Sir Isaac Newton right up to the present time—most of it paid for by military and government agencies searching for better firepower. Strictly speaking, every bullet type has a unique drag characteristic. To measure it, though, requires a very extensive firing test program, complex instrumentation and many detailed calculations. Once the

Thompson/Center Contender, standard model

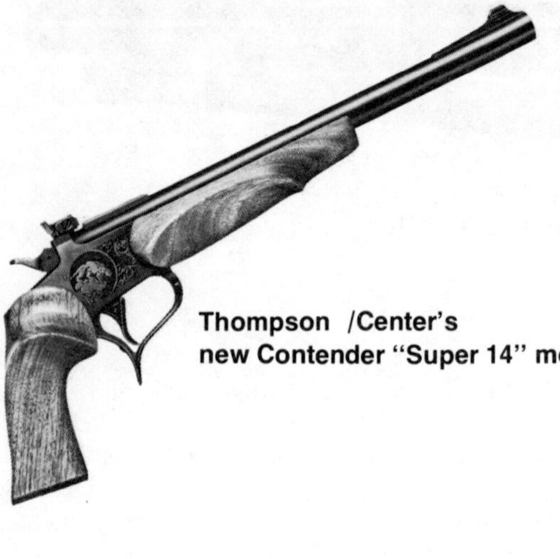

Thompson /Center's
new Contender "Super 14" model

drag characteristic (drag versus velocity) of a given bullet type is obtained, other detailed calculations are necessary to obtain trajectory data for the bullet. Nowadays we have high speed electronic computers to perform these calculations, but the amount of work and cost involved in measuring the drag characteristic for each and every bullet type is impractical. If these measurements had to be made, handloaders would have a very small selection of bullets from which to choose and very little information about their ballistic performance. Luckily, we have a way around these problems, inherited from ballisticians who did their work more than a century ago.

Firearms developed very rapidly in the 1800's. Rifled barrels and elongated bullets were widely adopted in military small arms in the early 1800's. Percussion ignition also appeared in the early 1800's, replacing flintlocks, and it in turn was replaced by metallic cartridges in the mid-1800's. Smokeless powder (smokeless propellant, properly speaking) appeared in 1885. These advances greatly increased the need for good exterior ballistics so that accurate long range fire could be obtained. Ballisticians worked very hard during this period to understand the nature of drag.

Theoretical drag investigations never did develop to the point where the drag on a particular bullet could be accurately predicted without using firing test data, and this is largely true even today. Also, ballisticians of the 1800's didn't have computers to perform tedious computations. About 1850 ballisticians developed the concept of a "standard bullet," a remarkable achievement which simplified both experimental determination of drag and calculations of bullet trajectories. By 1850 bullets had evolved to the familiar shape of a cylinder with a pointed nose. Theory indicated that a bullet of such a shape could be adopted as "standard," and precise drag measurements could be made for that standard bullet. Then, the drag force on any other bullet of the same shape would be related to the "standard drag" by a constant factor, which has come to be known as the "ballistic coefficient." The first advantage of this concept is that the ballistic coefficient can be measured much more easily than its entire drag characteristic. About 1880 an Italian ballistician named Siacci discovered the second major advantage. He found a way to use the same ballistic coefficient to compute the ballistics of an actual bullet from the ballistics of the standard bullet. The trajectory of the standard bullet had to be computed laboriously, using the calculus, but the trajectory of any actual bullet could be determined from the standard trajectory with simple algebra. Siacci's method shortcut a great amount of computation, produced reasonably accurate bullet trajectories before the age of computers and it has been used widely in ballistics for nearly a hundred years.

Between about 1865 and 1930 many firing tests were conducted almost worldwide to determine the drag characteristics of standard bullets adopted by different countries. Of particular note were tests made by Krupp in Germany in 1881 and by the Gavre Commission in France from 1873 to 1898. The Gavre Commission work was very comprehensive: featuring extensive firing tests; going up to a velocity of 6000 fps and concluding with a comprehensive survey of data available from tests in other countries. Unfortunately, the effects of varying atmospheric conditions during the many tests were not well understood at the time, and the Gavre drag function contained errors due to these effects. Just a few years after the Gavre Commission finished its work, ballisticians adopted standard atmospheric conditions for the expression of drag data. These standard conditions are in use today.

The Krupp test data turned out to be the basis of ballistics tables for small arms right up to the present time. The standard bullet used by Krupp was a flat base design, 3 calibers long with a 2 caliber ogive head. After the Krupp drag data were published, a Russian Army Colonel named Mayevski constructed a mathematical model for the standard drag deceleration for this bullet. Colonel James M. Ingalls, U.S. Army, used Mayevski's analytical model and computed and published his now-famous Ingalls Tables in 1917. The Ingalls Tables are the ballistics of a standard bullet of the Krupp shape. It turned out that this standard bullet has been a very fine model for use in computing the ballistics of most bullets for sporting use.

After World War II the Ballistics Research Laboratories of the Aberdeen Proving Ground undertook another long series of firing tests on military and sporting bullets and developed four standard drag models corresponding to four general types of bullets. These are lead nose hollow point bullets, boattail bullets, very sharp pointed full patch bullets and all others except those in the above categories. It was found that the drag model for the last category, which includes the vast majority of hunting and target bullets in use today, is almost identical to the model developed from the Krupp firing data. This model was adopted for the Lyman firing tests and ballistics computations for this handbook.

**Figure 1** shows this standard drag model over a velocity range appropriate for handgun bullets. The standard bullet has a ballistic coefficient of 1.00 by definition, and we'll describe how the ballistic coefficient relates standard drag to actual drag of other bullets a little later. Note that **Figure 1** shows *drag deceleration* as a function of velocity. Deceleration is the rate at which the bullet slows down. For example, if the standard bullet were traveling at 1500 fps velocity, **Figure 1** shows that air drag would be slowing it down at an instantaneous rate of about 300 fps for each second of flight. This deceleration applies only at the instant that the standard bullet has the 1500 fps velocity. When it slows down to 1400 fps (which would take a little more than a third of a second), the deceleration drops to about 260 fps per second of flight.

**Figure 1** shows that drag is relatively small when the bullet travels at velocities well below the speed of sound and very large when bullet velocity is above the speed of sound. When the bullet moves at a speed such that the air molecules which it pushes aside must move at speeds which exceed the speed of

**Figure 1 Drag Model for the Standard Bullet**

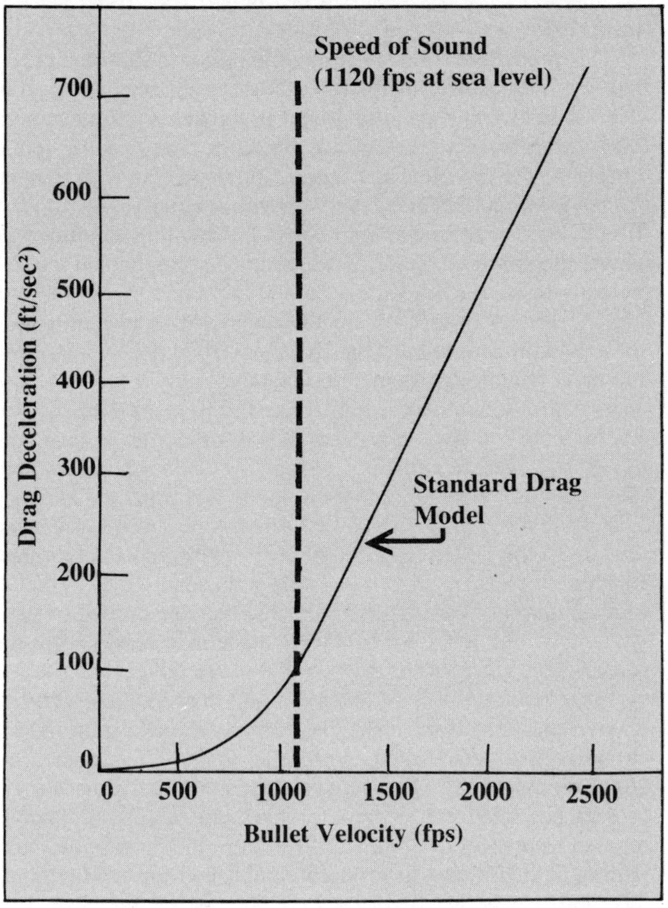


sound, a shock wave begins to form in the air. This begins to happen for the standard bullet at about 800 fps. At this velocity the bullet travels slower than the speed of sound by more than 300 fps, but the air molecules which must flow around the bullet point and out of its way must move at velocities more than 1120 fps. When the bullet velocity exceeds 1120 fps, a bowl-shaped shock wave surrounds the bullet.

It is the buildup of this shock wave that causes the rapid change in drag near the speed of sound. The bullet must push this shock wave as it travels and provide the energy to move the air along the entire bowl-shaped surface of the shock. The shock wave bowl grows larger in diameter as the bullet velocity increases and the curvature of the shock front increases; that is, the bowl deepens. All these effects add to the drag on the bullet.

There is also a region of turbulence (turbulent wake) extending a certain distance behind a bullet. As the bullet moves, it pushes the air aside, leaving a void immediately behind it. Air then flows back into this void, and very often this causes the formation of another bowl-shaped shock wave emanating from this wake and following behind the bullet. High speed photographs of bullets in flight show these shock fronts (see, for example, page 167 of the Lyman *Black Powder Handbook*).

The drag on an actual handgun bullet behaves very much like the drag characteristic in **Figure 1** for the standard bullet, except that the drag magnitude is much larger. The ballistic coefficient is the factor that relates the two. The standard drag is *divided* by the ballistic coefficient of an actual bullet to obtain its drag. For example, if a handgun bullet has a ballistic coefficient of 0.2, its drag deceleration at any velocity is five times ($1/0.2 = 5$) the standard drag deceleration at that same velocity given by the curve in **Figure 1.** Bullets with low ballistic coefficients suffer very high drag, while those with higher ballistic coefficients have lower drag.

The shape of the standard drag function in **Figure 1** is approximately correct for nearly all bullets, but it is not exactly correct, for several reasons. Almost any kind of surface irregularity on a bullet will cause a shock wave to form. The lubricating grooves on cast bullets frequently are causes of shock waves. Secondary shock waves add to the drag in certain velocity ranges, and cause the shape of the actual drag characteristic to be different from the standard drag model. Also, point shapes and tail shapes of actual bullets differ from the point and tail shapes of the standard bullet.

This situation immediately raises a question. If the drag function of the standard bullet does not fit the actual bullet, how can we use the standard bullet drag to calculate trajectories of the actual bullet? The answer is that we use three different values of ballistic coefficient for each handgun bullet for three ranges of bullet velocity. One value is used when the velocity is above 1300 fps, a second value is used when the velocity is between 800 and 1300 fps, and the third applies when the velocity drops below 800 fps.

Experiments have shown that this method will allow trajectories to be calculated with a high degree of accuracy. Using a different ballistic coefficient for each velocity range allows the drag function of the standard bullet to be matched to the drag characteristic of the actual bullet in a piecewise manner. To do this, the three values of ballistic coefficient for each bullet must be known in each velocity range; these were established in the Lyman firing tests for this handbook.

Ballistic coefficients can be determined accurately only by means of firing tests. There are methods of calculating the ballistic coefficient from bullet shape, but the authors' experience shows that these calculated values are often in error by a significant amount; particularly for bullets fired at velocities around the speed of sound. A firing test program to determine ballistic coefficients for a certain bullet type requires several test shots (typically five) at each of two or three muzzle velocity levels. For handgun bullets, one muzzle velocity level would be well above the speed of sound, around 1400 or 1500 fps. The second level would be at a muzzle velocity well below the speed of sound, around 800 fps. If the ballistic coefficients determined for these velocities differed by twenty percent or more, then a third set of firing test data might be shot at a velocity around 1100 fps. Such a test program depends to some extent on the weight, caliber and muzzle velocity range for typical cartridges using the bullet under test. For example, a 250-grain bullet for the .45 ACP or .45 Colt cartridges wouldn't be tested at 1500 fps, because those cartridges are not loaded to the velocity with a bullet that heavy. Similarly, a 50-grain bullet for the .221 Remington Fireball would be tested at about 2500 fps.

Lyman conducted a firing test program for all the bullets in this handbook. For each test shot fired, muzzle velocity and time of flight over a measured 50-yard range were recorded. Photoelectric screens and high-precision electronic counters were used to measure these quantities for each bullet fired. The ballistic coefficient of each fired bullet was calculated from these measurements. This calculation is difficult, and a digital computer was used. The basic idea is that the trajectory of the standard bullet is calculated using the muzzle velocity of the actual bullet and then compared to the actual bullet trajectory. The ballistic coefficient of the test bullet is derived from this comparison. Normally, five or more test shots are used at each velocity level. An average of the five individual ballistic coef-

**Table 1. Handgun Ballistic Coefficients from Lyman Firing Tests**

| Caliber | Bullet | Weight (gr) | Ballistic Coefficient below 800 fps | 800 to 1300 fps | above 1300 fps |
|---|---|---|---|---|---|
| .22 | Speer Spire Point | 40 | .124 | .132 | .140 |
| | Remington H.P. 22702 | 45 | .178 | .189 | .140 |
| | Remington H.P. 22710 | 50 | .220 | .220 | .225 |
| | Lyman #225438 | 43 | .160 | .150 | .120 |
| | Lyman #225415 | 46 | .200 | .208 | .162 |
| .25 | Hornady Soft Point | 60 | .125 | .125 | .095 |
| | Sierra Hollow Point | 75 | .210 | .220 | .240 |
| | Hornady Spire Point | 87 | .220 | .220 | .220 |
| | Lyman #257420 | 65 | .150 | .200 | .160 |
| | Lyman #257312 | 85 | .210 | .200 | .185 |
| .38 | Winchester JHP | 110 | .080 | .130 | .111 |
| (.357) | Hornady JHP | 125 | .090 | .100 | .101 |
| | Speer JHP | 140 | .200 | .215 | .192 |
| | Speer SWCHP | 146 | .170 | .173 | .172 |
| | Sierra JHC | 150 | .125 | .153 | .180 |
| | Hornady JHP | 158 | .120 | .158 | .248 |
| | Speer SWCSP | 160 | .200 | .212 | .213 |
| | Lyman #358242 | 93 | .100 | .103 | .105 |
| | Lyman #358345 | 105 | .100 | .118 | .121 |
| | Lyman #358242 | 121 | .200 | .210 | .154 |
| | Lyman #356402 | 116 | .100 | .110 | .146 |
| | Lyman #358480 | 123 | .100 | .125 | .145 |
| | Lyman #358495 | 138 | .055 | .066 | .096 |
| | Lyman #358477 | 146 | .130 | .150 | .183 |
| | Lyman #358156 | 147 | .100 | .160 | .213 |
| | Lyman #358311 | 150 | .220 | .220 | .228 |
| | Lyman #358429 | 162 | .240 | .265 | .286 |
| .38 | Lyman .375'' Round Ball | 81 | .037 | .044 | .053 |
| (.375) | Lyman #37583 | 150 | .095 | .095 | — |
| .41 | Sierra JHC | 170 | .120 | .122 | .123 |
| | Speer SWCHP | 200 | .140 | .146 | .149 |
| | Hornady JHP | 210 | .160 | .169 | .168 |
| | Speer SWCSP | 220 | .160 | .200 | .170 |
| | Lyman #41028 | 206 | .150 | .170 | .160 |
| | Lyman #410610 | 204 | .150 | .170 | .158 |
| | Lyman #410459 | 203 | .160 | .195 | .150 |
| | Lyman #410426 | 234 | .235 | .235 | .235 |
| .44 | Sierra JHC | 180 | .130 | .133 | .133 |
| | Hornady JHP | 200 | .085 | .135 | .185 |
| | Speer SWCHP | 225 | .090 | .120 | .140 |
| | Sierra JHC | 240 | .165 | .172 | .183 |
| | Speer JHP | 240 | .135 | .210 | .210 |
| | Hornady Flat Nose | 265 | .265 | .265 | .267 |
| | Lyman #429348 | 175 | .040 | .045 | .065 |
| | Lyman #429303 | 189 | .090 | .095 | .105 |
| | Lyman #429215 | 203 | .185 | .215 | .188 |
| | Lyman #429360 | 225 | .170 | .195 | .169 |
| | Lyman #429421 | 235 | .200 | .250 | .209 |
| | Lyman #429244 | 237 | .200 | .280 | .201 |
| .45 | Sierra JHC | 185 | .130 | .120 | — |
| (.452) | Speer JHP | 200 | .100 | .165 | — |
| | Speer JHP | 225 | .105 | .175 | — |
| | Sierra JHC | 240 | .165 | .168 | — |
| | Hornady HP | 250 | .125 | .237 | — |
| | Lyman .451'' Round Ball | 138 | .045 | .052 | .063 |
| | Lyman #450229 | 155 | .050 | .050 | — |
| | Lyman #45468 | 161 | .130 | .105 | — |
| | Lyman #45467 | 190 | .112 | .112 | — |
| | Lyman #454190 | 238 | .279 | .269 | — |
| | Lyman #454424 | 245 | .160 | .210 | — |
| .45 (.457) | Lyman .457'' Round Ball | 145 | .045 | .053 | — |

ficients is used to characterize that bullet type at that velocity level.

This procedure sounds very straightforward, but in a practical sense measuring ballistic coefficients of bullets traveling near or below the speed of sound is fraught with difficulties. Very often when a test shot was fired, the photoelectric screen closest to the muzzle was triggered prematurely, giving an erroneous measurement of both muzzle velocity and time of flight. This situation did not happen for muzzle velocities well above the speed of sound, but it certainly complicated the low velocity tests. Baffles were arranged to protect the first photoelectric screen from both sound waves and smoke, but even these measures were not very effective. Often only a small timing error occurred, causing, for example, only a few fps error in muzzle velocity. Everything looked all right until the ballistic coefficients had been calculated. Then, ridiculous answers occurred because the ballistic coefficient calculation is very sensitive to small errors in muzzle velocity and time of flight.

Despite these difficulties, the firing test program was carried out for all the bullets, and the ballistic coefficients determined for these bullets are shown in **Table 1.** In some cases in the table, estimations were made for the intermediate or lower velocity ranges, but measurements were made in at least two of the three velocity ranges for every bullet in the table.

These firing tests were extensive and they required very intensive analysis efforts also. The purpose was to determine accurate ballistic coefficients so that accurate bullet trajectories could be calculated. We believe this is the first time such a project has been carried out for handgun bullets. Lyman is pleased to present the Ballistics Tables in this handbook, containing ballistics data never before published for handgun bullets, and which have a very high degree of accuracy based on these ballistic coefficient measurements for every bullet.

Before we leave the subject of exterior ballistics, there are two other effects on bullet trajectories that need to be mentioned. The first is the effect of altitude above sea level, and the second is the effect of winds, especially crosswinds.

Drag depends upon the density of the air and on the speed of sound; these in turn depend upon altitude, temperature and barometric pressure. At higher altitudes, air is thinner and colder than at lower altitudes. Since the air is thinner, the drag on the projectile is less. It also turns out that, since the air is colder, the speed of sound at higher altitudes is less than at lower altitudes. Consequently, the velocity at which shock waves form around the bullet is lower. The standard drag function shown back in **Figure 1** can be used to visualize these effects. The curve shown in the figure is for sea level standard conditions. When altitude increases, the curve in the figure shrinks in height and shifts toward the left.

Ballistics tables for small arms are nearly always computed for sea level altitude, and the Ballistics Tables in this handbook were calculated for these conditions. If you happen to live at a location several thousand feet above sea level, since air drag is lower, you can expect your handguns to shoot a little flatter than the trajectories in the Lyman Ballistics Tables. This effect is not terrifically large, as **Table 2** shows. The changes in bullet drop that occur between sea level and 5,000 and 10,000 feet altitudes are shown there for four typical handgun cartridges. At 100 yards range, drop changes less than an inch from sea level to 10,000 feet altitude; at 200 yards range, the change in drop is less than half a foot in all cases. These four cartridges pretty well span the range of conditions a handgun shooter might encounter (low ballistic coefficient, moderate velocity to high ballistic coefficient, high velocity). The data in the table give a pretty good idea of the changes that might be expected. If you shoot at extreme handgun ranges, it would be a good idea to check the zero on your gun if you change altitude by a few thousand feet.

Ballistics tables are also computed for "standard atmospheric conditions," which are particular reference values for temperature, barometric pressure and relative humidity. These have been defined for the U.S. by the Ballistic Research Laboratories of the Army's Aberdeen Proving Ground. On any given day, the actual atmospheric conditions may be different from standard, and these differences may affect bullet ballistics. Fortunately, these effects are small enough to be neglected for almost all handgun shooting. Also, it turns out that weather conditions are such that departures from standard air temperature and standard air pressure generally tend to have cancelling effects. On a warm balmy day, for example, the barometric pressure also tends to be high. The higher temperature tends to decrease drag slightly while the higher pressure tends to increase it; so the two effects tend to offset each other.

A bullet's reaction to a wind depends very much on both the speed of the wind and the direction in which it blows. A headwind or a tailwind causes a vertical deflection (a change in drop) of the bullet, while a crosswind causes a horizontal deflection. The effects of a headwind or tailwind can be neglected for almost all handgun shooting situations.

**Table 2. Effects of Altitude Above Sea Level on Bullet Trajectory**

| Bullet (Cartridge) | Muzzle Vel. (fps) | Altitude (ft) | Drop (inches) at Range (yards) | | | |
|---|---|---|---|---|---|---|
| | | | 50 | 100 | 150 | 200 |
| Lyman #225438 (.221 Fireball) | 2000 | Sea Level | 1.18 | 5.41 | 13.89 | 28.08 |
| | | 5,000 | 1.16 | 5.19 | 13.09 | 26.06 |
| | | 10,000 | 1.13 | 5.02 | 12.44 | 24.36 |
| Sierra .357 150 gr JHC (.38 Special) | 1000 | Sea level | 4.45 | 18.61 | 43.66 | 80.72 |
| | | 5,000 | 4.43 | 18.43 | 43.06 | 79.30 |
| | | 10,000 | 4.40 | 18.26 | 42.53 | 78.08 |
| Speer .429 240 gr JHP (.44 Magnum) | 1400 | Sea level | 2.31 | 9.98 | 23.96 | 45.22 |
| | | 5,000 | 2.28 | 9.78 | 23.34 | 43.82 |
| | | 10,000 | 2.25 | 9.60 | 22.79 | 42.55 |
| Hornady .452 250 gr H.P. (.45 ACP) | 850 | Sea level | 5.74 | 24.21 | 56.21 | 103.06 |
| | | 5,000 | 5.66 | 23.93 | 55.48 | 101.28 |
| | | 10,000 | 5.58 | 23.71 | 54.96 | 99.97 |

Crosswinds, though, are important for hunting as well as target shooting. The crosswind deflection data in the Ballistics Tables show how much crosswinds affect each bullet, and it is clear that both hunters and target shooters need to be aware of the importance of crosswinds.

**Table 3** has been prepared to show the effects of different winds on two bullets, a .38 caliber cast lead bullet and a .45 caliber jacketed bullet. The data shown in the table are typical examples of general behavior in all calibers. The first two entries for each bullet in **Table 3** show the trajectory deflections caused by a 20 mph headwind or tailwind only. Three points may be noted from these entries:

First, a headwind or tailwind acting alone (no crosswind) causes only a vertical deflection (change in drop).

Second, these vertical deflections are relatively small at reasonable ranges for handguns. For hunting situations and most target shooting situations, a vertical correction would be unnecessary unless the wind was very strong and the range very long.

Third, the table shows that a headwind always causes a slightly larger vertical deflection than a tailwind of equal speed. The reason for this has to do with the relatively complicated way that drag depends on the speed of the projectile relative to the air.

**Table 3** also shows the horizontal deflections caused by a 20 mph crosswind on the two example bullets. The effects of a 20 mph crosswind can be compared directly with the effects of a 20 mph headwind or tailwind, and it can be seen in each case that the horizontal bullet deflection caused by the crosswind is much larger than the vertical bullet deflection caused by a headwind or tailwind. Crosswinds have such strong effects that the Ballistics Tables include the deflections for each bullet at each muzzle velocity caused by crosswinds of 5, 10, 20, and 30 mph. These entries in the Ballistics Tables tell a shooter exactly what wind-bucking performance to expect from his handloads.

### The Ballistics Tables

The tables in the following pages contain ballistics data for handgun bullets. These data have been calculated in a large digital computer using highly accurate computational methods. The ballistic properties (ballistic coefficients) of each projectile have been determined by extensive firing tests and calculations, as explained earlier in the Handgun Ballistics article.

A brief explanation of the terms used in the tables is given below. Many shooters are already familiar with these terms. For those who are not, this explanation may help them both to use these tables and to improve their shooting skills.

**Table 3. Wind Effects on Two Handgun Bullets**

| Bullet | Muzzle Velocity (fps) | Headwind (HW) or Tailwind (TW) Velocity (mph) | Crosswind Velocity (mph) | Range (yards) | Crossrange Deflections Vertical (inches) | Horizontal (inches) |
|---|---|---|---|---|---|---|
| Lyman #358477 (.38 Special) | 1000 | 20 HW | 0 | 100 | − 0.26 | 0.00 |
| | | | | 200 | − 1.95 | 0.00 |
| | | 20 TW | 0 | 100 | + 0.23 | 0.00 |
| | | | | 200 | + 1.72 | 0.00 |
| | | 0 | 20 | 100 | 0.00 | 6.73 |
| | | | | 200 | 0.00 | 25.83 |
| Sierra .4515 185 gr JHC (.45 ACP) | 1000 | 20 HW | 0 | 100 | − 0.32 | 0.00 |
| | | | | 200 | − 2.33 | 0.00 |
| | | 20 TW | 0 | 100 | + 0.28 | 0.00 |
| | | | | 200 | + 2.01 | 0.00 |
| | | 0 | 20 | 100 | 0.00 | 8.31 |
| | | | | 200 | 0.00 | 31.83 |

***Note:* A positive (+) vertical deflection means that the bullet strikes higher on the target than it does with no wind, while a negative (−) deflection means that it strikes lower.**

## Terms Used in the Tables

*Trajectory:* The trajectory of a projectile is the actual path which that projectile follows after leaving the muzzle of a firearm. **Figure 1** illustrates a projectile trajectory. As soon as the projectile leaves the muzzle of the firearm, gravity acting upon it causes it to begin to fall away from its line of departure.

### FIGURE 1—PARTS OF A TRAJECTORY

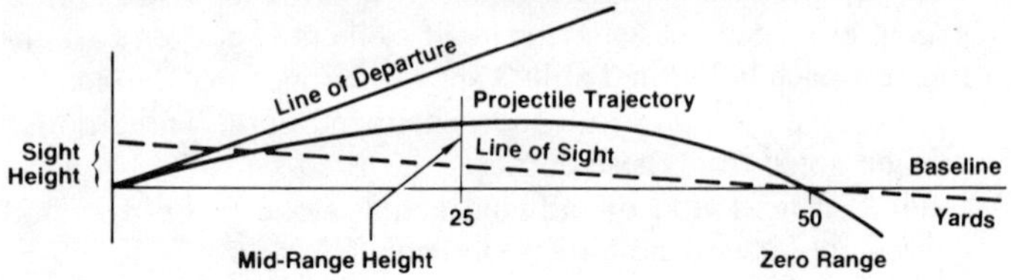


This causes the drop discussed later. The line of departure is an imaginary line extending along the bore in the direction which the bullet is traveling. Muzzle velocity, air drag and gravity are the major contributors to the trajectory shaping. If a crosswind blows, the projectile trajectory will also be curved in the direction of the wind, that is, it will curve into the paper or out from it in Figure 1.

In order to give some specific examples of the meaning of terms in the Ballistics Tables, let us suppose that we are shooting the Lyman #257312 handgun bullet at a muzzle velocity of 1600 fps. The numerical data from the Ballistics Tables for this bullet at this muzzle velocity will be used to illustrate the following discussion.

*Velocity:* Velocity is the speed of the projectile as it moves along its trajectory. Velocity is measured in feet per second (fps). The projectile moves slower and slower as it flies farther and farther from the gun muzzle, because air drag slows it down. The velocity of the projectile is listed in the table at each value of range from the muzzle. For example, for our Lyman #257312 bullet fired at 1600 fps muzzle velocity, the table shows that the velocity drops to 1514 fps at 25 yards from the muzzle, 1434 fps at 50 yards, 1358 fps at 75 yards, and so forth.

*Energy:* A moving projectile has kinetic energy, that is to say, energy of motion. Ordinarily, the term ''kinetic'' is dropped, and we speak of projectile energy. Energy is measured in foot-pounds (ft-lb), and it is usually regarded as an indication of the killing power of a projectile.

The energy of a projectile is determined by its mass and its velocity. The mass of a projectile is not its weight, but mass can be calculated from weight by the following formula:

$$\mathbf{m} = \frac{\mathbf{w}}{\mathbf{7000\ g}}$$

where: **w** = projectile weight in grains
and **g** = 32.174 ft./sec$^2$ (acceleration due to gravity).
(The factor 7000 converts weight from grains to pounds.)

The kinetic energy of the projectile can then be found from the formula:

$$\mathbf{K.E.} = \frac{\mathbf{mV^2}}{\mathbf{2}}$$

where: **m** = mass of the projectile from the formula above
and **V** = remaining velocity in fps.

Returning to the Ballistics Tables for our example, Lyman #25712 bullet at 1600 fps muzzle velocity, we see that the energy at the muzzle is 483 ft-lb. At 25 yards, it drops to 433 ft-lb., and at 50 yards to 388 ft-lb. At longer ranges energy continues to drop off, and this happens because bullet velocity grows progressively smaller.

### FIGURE 2—DROP

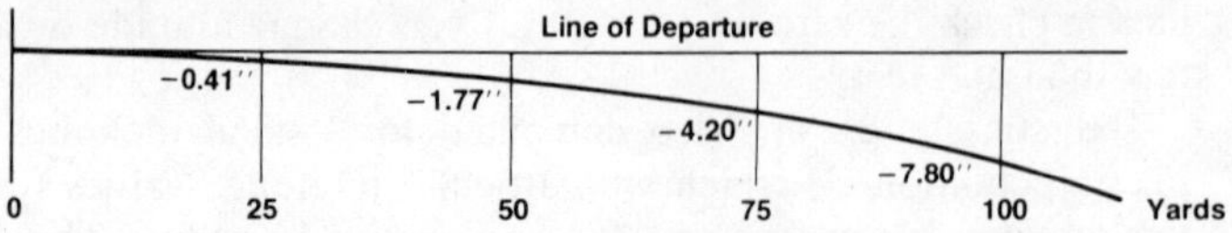


*Drop:* When a projectile leaves the muzzle it begins to drop, just as it would if we were to hold it up and then let it go. It drops with about the same speed when it is fired as it would if we dropped it. (There are some aerodynamic forces acting upon the projectile when it is fired which keep it from falling quite as fast, and these have been taken into account in the tables.)

If we sit on the side of a hill and fire our handgun such that the line of departure (bore line) is level, then **Figure 2** shows how far below the line of departure our example bullet will be at each specific range. The line of departure is an imaginary line extending through the bore of the gun and is the line on which the bullet leaves the muzzle. Drop is usually measured in inches, although at long ranges from the muzzle it can grow to be several feet. The minus signs on the drop values in the table signify that the trajectory passes below the line of departure.

### FIGURE 3—ZEROING IN

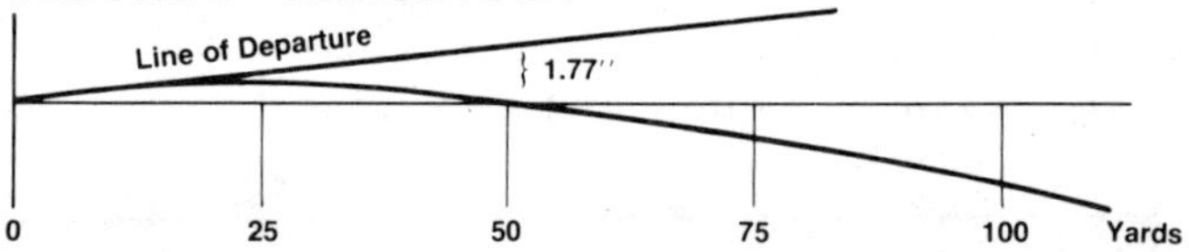


When we zero in a rifle or handgun, we adjust the sights to tilt the barrel up (elevate the line of departure). When we have just the right amount of tilt, the trajectory is rotated upward to the point where the projectile just crosses the level baseline at the desired zero range. Then, the line of departure is above the baseline by just the amount of drop at the zero range.

**Figure 3** shows this situation for our example bullet. When the barrel of the handgun is tilted upward so that the line of departure passes 1.77 inches above the baseline at 50 yards, then the Lyman #257312 bullet at 1600 fps muzzle velocity will be zeroed in at 50 yards.

*Mid-Range:* When a rifle or handgun is zeroed in at some particular range (as in Figure 3), the projectile passes above the baseline at ranges shorter than the zero range. It turns out that the projectile reaches its maximum height above the baseline at a range which is very near the mid-point between the muzzle and the zero range. We call the height of the trajectory at that mid-point the Mid-Range Height, or more frequently, the Mid-Range. The Mid-Range is very nearly the maximum height the projectile reaches in its flight between the muzzle and the zero range.

### FIGURE 4—MID-RANGE

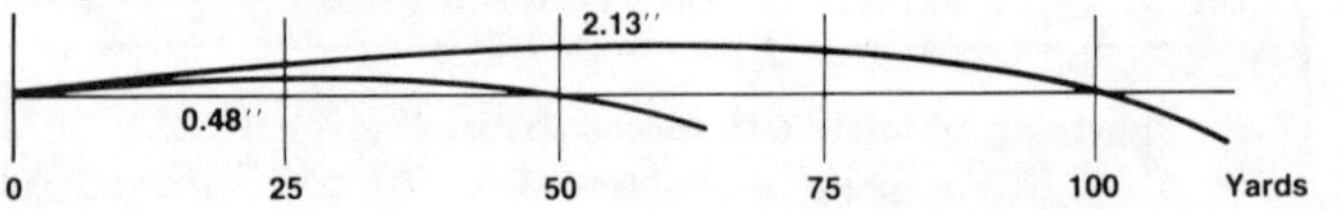


**Figure 4** illustrates two Mid-Range values for Lyman bullet #257312 at 1600 fps muzzle velocity. If the handgun is zeroed in at 50 yards, the Mid-Range is 0.48 inch (at 25 yards). If the handgun is sighted in at 100 yards, the Mid-Range is 2.13 inches (at 50 yards).

Note that in the Ballistics Tables a Mid-Range value is quoted at each range entry. In each case, this is the Mid-Range height if the gun is zeroed in at the range value at the top of the column. The Mid-Range height occurs halfway between the muzzle and that range.

Mid-Range height is important for a shooter to know. When he is selecting a zero range for his gun, Mid-Range tells him how high the bullet will rise between the muzzle and the range where he zeroes his handgun in.

*Bullet Path:* The sights on a rifle or handgun are mounted above the bore. The distance between the centerline of the bore and a line through the sights (line of sight) is called the sight height. On most handguns sight height turns out to be very close to ⅝-inch.

**Figure 5** shows the relationship of the line of sight to the baseline and the trajectory. The numerical values given in **Figure 5** are for our example Lyman #257312 bullet from an arbitrary handgun. When the gun is zeroed in, the line of sight intersects the baseline and the trajectory at the zero range. A shooter's real interest is where his bullet is relative to his line of sight, rather than the baseline. This is called the Bullet Path Height, usually shortened to Bullet Path. As shown in **Figure 5,** the bullet leaves our handgun 0.63 inch below the line of sight. So, we say that the Bullet Path is -0.63 inch at the muzzle. Since we have our handgun zeroed in at 50 yards, the bullet is above the line of sight by 0.16 inch at 25 yards, on at 50 yards, 1.22 inches low at 75 yards, and so forth.

*Time of Flight:* When a projectile leaves the muzzle, it requires a certain amount of time to reach a target. This is called the Time of Flight, and it depends both on range to the target and on muzzle velocity. The farther away the target is, the more the time required to reach it. The higher the muzzle velocity, the shorter the time of flight. The Ballistics Tables list the time of flight to reach each range for each muzzle velocity case. Our example bullet which leaves the muzzle at 1600 fps requires 0.099 second to reach 50 yards, 0.209 second to reach 100 yards, and 0.755 second to reach 300 yards.

**FIGURE 5—BULLET PATH**

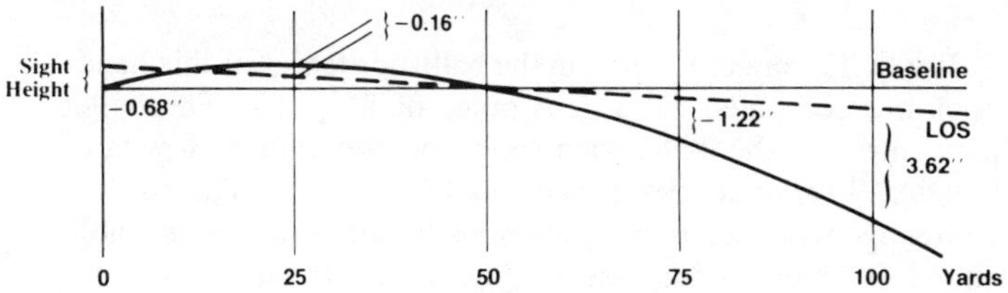


*Wind Deflection:* Projectiles are very susceptible to crosswinds, as explained in the Handgun Ballistics article in this manual. For that reason the deflections caused by crosswinds of 5, 10, 20, and 30 mph have been calculated for each projectile, and these crosswind deflections are listed in the Ballistics Tables for each muzzle velocity case.

A crosswind tends to drag the projectile along with it, so the deflections can be to the right or the left, depending on the direction of the crosswind. For our example bullet, a 5 mph crosswind will move the bullet sideways 1.94 inches at 100 yards, and 7.82 inches at 200 yards. If the crosswind is 30 mph, these deflections grow to 11.65 inches at 100 yards and 46.91 inches at 200 yards.

It is evident from the sizes of the figures in the Wind Deflection tables that strong crosswinds can cause very large misses, if windage corrections are not made. It is hoped that these data will help shooters select loads and improve their "guesstimation" abilities for windy day shooting.

**Ballistics staffers service two of the three screens which recorded simultaneous velocity and time-of-flight at 50 yards.**

**Plywood shield reduced, but didn't eliminate, microphonics problems. Ransom rest visible at four o'clock in screen aperture.**

# Changing the Zero Range

The bullet trajectory data in the ballistics tables of this handbook are computed for a zero range of 50 yards. The bullet path entries in the tables then show you how high or low your bullet will shoot at other ranges out to 300 yards. The 50 yard zero range was used in this handbook because it is a reasonable choice for most handgunners. Quite often, though, a shooter would like to use a different zero range, and two questions then arise. The first usually goes something like this:

> "I want to set my zero range at 125 yards, but I want to use a 50 yard target to do it. How high should I shoot at 50 yards to have my gun zeroed at 125 yards?"

The second question is then:

> "After I zero in at 125 yards, where does my gun shoot at other ranges?"

The answers to these questions are given by bullet path data calculated for the new zero range. However, it was not possible in this handbook to include bullet path data in the ballistics tables for all possible choices of zero range, because the tables would have been huge! There is a relatively easy method that any shooter can use to calculate bullet path for any cartridge and any load, for any choice of zero range, and for any sight height as well, which is important if he uses a telescope sight instead of iron sights. This method is explained below.

There is a simple and accurate formula for calculation of bullet path height, and it works for both handguns and rifles. To use the formula, you must know drop as a function of range for your cartridge and load, and this is always listed in ballistics tables. You also must know what your sight height is, and you must choose a zero range which is one of the entries in the ballistics table. The calculations can be done by hand, and they are especially easy if you have one of the new electronic calculators. The formula for bullet path height is the following:

$$y_b(R) = y(R) - h_s + \frac{R}{R_z}(h_s - y_z)$$

where $y_b(R)$ is bullet path height in *inches* at the range R (positive values for bullet above line of sight, and negative values for bullet below line of sight);

R is the range from the muzzle in *yards* at which bullet path height is to be calculated;

y (R) is the bullet drop (always negative) at the range R, measured in *inches*, and taken from the ballistics tables;

$h_s$ is the sight height above the bore centerline in *inches* (always positive);

$R_z$ is the zero range in *yards;*

$y_z$ is the drop in *inches* at the zero range R (always negative), and also taken from the drop figures in the ballistics tables.

This formula is good for ranges greater than $R_z$ as well as for ranges less than the zero range. The accuracy of the formula is very good for practical shooting purposes. The calculated bullet path will be correct within a small fraction of an inch for pistol bullets to ranges of 400 or 500 yards, and for rifle bullets to ranges exceeding 1000 yards.

The sketch below will help define the meaning of the terms in the equation above. It is necessary to manipulate negative numbers in the calculations, which might be a little unfamiliar if you're not a whiz at math. The following example calculation shows how these negative and positive numbers combine together. If you follow the method in the example, you can do these calculations even if you aren't a whiz at math!

### Example Calculation

Cartridge: .357 Magnum with Speer 140 grain Jacketed Hollow Cavity bullet
Muzzle velocity: 1400 fps (7½ inch barrel)
Zero range: 125 yards
Sight height: 1¼ inches (telescope sight)
Calculations: (1) bullet path height at 50 yards to use for sighting in
(2) bullet path heights at 75, 100, 150, and 200 yards
Ballistics table for reference: Lyman Pistol & Revolver Handbook

For $R_z$ = 125 yards, $y_z = -16.25$ inches from the ballistics table. The sight height $h_s$ = 1.25 inches. These numbers are the same for all the calculations below.

(1) At R = 50 yards, y(50) = −2.33 inches. Using the formula:

$$y_b(50) = -2.33 - 1.25 + \frac{50}{125}(1.25 - (-16.25))$$

$$= -3.58 + \frac{50}{125}(+17.50) = -3.58 + 7.00 = 3.42 \text{ inches}$$

Setting the telescope sight to shoot 3.4 inches high at 50 yards will zero the gun in at 125 yards.

(2) At R = 75 yards, y(75) = −5.46 inches

$$y_b(75) = -5.46 - 1.25 + \frac{75}{125}(17.50) = -6.71 + 10.50 = 3.79 \text{ inches}$$

At R = 100 yards, y(100) = −10.06 inches

$$y_b(100) = -10.06 - 1.25 + \frac{100}{125}(17.50) = -11.31 + 14.00 = 2.69 \text{ inches}$$

At R = 150 yards, y(150) = −24.15 inches

$$y_b(150) = -24.15 - 1.25 + \frac{150}{125}(17.50) = -25.40 + 21.00 = -4.40 \text{ inches (low)}$$

At R = 200 yards, y(200) = −45.53 inches

$$y_b(200) = -45.53 - 1.25 + \frac{200}{125}(17.50) = -46.78 + 28.00 = -18.78 \text{ inches (low)}$$

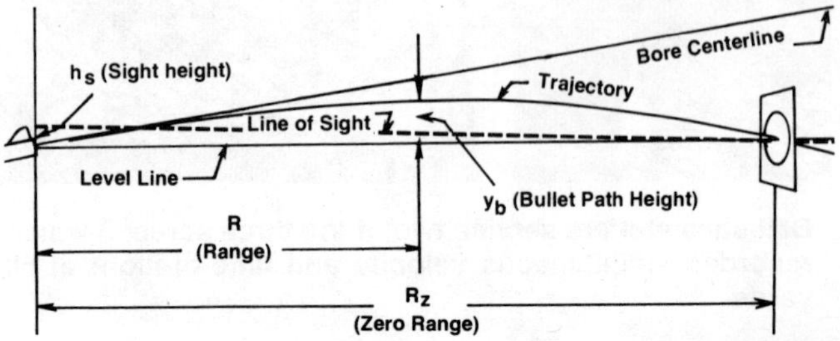

# 14

# Trajectory Tables

## Metallic
## Black Powder

## Bullet: Speer .224 Dia. 40 Gr. Spire Pt.
## Ballistic Coefficient: .140

| RANGE (YDS) | MUZZLE | 25 | 50 | 75 | 100 | 125 | 150 | 175 | 200 | 225 | 250 | 275 | 300 |
|---|---|---|---|---|---|---|---|---|---|---|---|---|---|
| VELOCITY (FPS) | 2800. | 2637. | 2480. | 2328. | 2182. | 2041. | 1905. | 1776. | 1652. | 1537. | 1429. | 1331. | 1241. |
| ENERGY (FT-LB) | 696. | 618. | 546. | 481. | 423. | 370. | 322. | 280. | 242. | 210. | 181. | 157. | 137. |
| DROP (IN) | .00 | -.13 | -.58 | -1.38 | -2.58 | -4.23 | -6.41 | -9.18 | -12.64 | -16.90 | -22.07 | -28.32 | -35.80 |
| MID-RANGE (IN) | .00 | .04 | .16 | .37 | .71 | 1.18 | 1.82 | 2.66 | 3.74 | 5.10 | 6.80 | 8.91 | 11.49 |
| BULLET PATH (IN) | -.63 | -.16 | .00 | -.19 | -.79 | -1.84 | -3.41 | -5.57 | -8.43 | -12.08 | -16.65 | -22.29 | -29.16 |
| TIME OF FLIGHT (SEC) | .000000 | .027601 | .056928 | .088139 | .121413 | .156956 | .194993 | .235775 | .279567 | .326642 | .377262 | .431653 | .490013 |
| WIND DEFLECTION (IN) | | | | | | | | | | | | | |
| 5 MPH | .00 | .07 | .30 | .68 | 1.26 | 2.03 | 3.02 | 4.25 | 5.74 | 7.53 | 9.63 | 12.06 | 14.84 |
| 10 MPH | .00 | .14 | .59 | 1.37 | 2.51 | 4.05 | 6.03 | 8.50 | 11.49 | 15.06 | 19.26 | 24.11 | 29.67 |
| 20 MPH | .00 | .29 | 1.18 | 2.74 | 5.02 | 8.11 | 12.07 | 16.99 | 22.98 | 30.12 | 38.51 | 48.23 | 59.34 |
| 30 MPH | .00 | .43 | 1.77 | 4.11 | 7.53 | 12.16 | 18.10 | 25.49 | 34.47 | 45.18 | 57.77 | 72.34 | 89.01 |
| VELOCITY (FPS) | 2700. | 2541. | 2387. | 2239. | 2095. | 1957. | 1825. | 1700. | 1581. | 1470. | 1368. | 1275. | 1191. |
| ENERGY (FT-LB) | 647. | 573. | 506. | 445. | 390. | 340. | 296. | 257. | 222. | 192. | 166. | 144. | 126. |
| DROP (IN) | .00 | -.15 | -.63 | -1.49 | -2.78 | -4.57 | -6.92 | -9.93 | -13.69 | -18.31 | -23.95 | -30.75 | -38.89 |
| MID-RANGE (IN) | .00 | .04 | .17 | .40 | .76 | 1.28 | 1.98 | 2.89 | 4.06 | 5.55 | 7.41 | 9.70 | 12.52 |
| BULLET PATH (IN) | -.63 | -.15 | .00 | -.23 | -.90 | -2.06 | -3.79 | -6.16 | -9.29 | -13.29 | -18.30 | -24.47 | -31.98 |
| TIME OF FLIGHT (SEC) | .000000 | .028636 | .059091 | .091537 | .126169 | .163205 | .202884 | .245469 | .291231 | .340438 | .393334 | .450112 | .511005 |
| WIND DEFLECTION (IN) | | | | | | | | | | | | | |
| 5 MPH | .00 | .08 | .31 | .72 | 1.33 | 2.14 | 3.19 | 4.49 | 6.07 | 7.96 | 10.17 | 12.72 | 15.64 |
| 10 MPH | .00 | .15 | .62 | 1.44 | 2.65 | 4.28 | 6.37 | 8.98 | 12.15 | 15.92 | 20.34 | 25.44 | 31.27 |
| 20 MPH | .00 | .30 | 1.24 | 2.89 | 5.30 | 8.56 | 12.75 | 17.96 | 24.29 | 31.83 | 40.68 | 50.88 | 62.54 |
| 30 MPH | .00 | .45 | 1.87 | 4.33 | 7.95 | 12.84 | 19.12 | 26.94 | 36.44 | 47.75 | 61.01 | 76.33 | 93.81 |
| VELOCITY (FPS) | 2600. | 2444. | 2294. | 2149. | 2009. | 1874. | 1746. | 1625. | 1511. | 1406. | 1310. | 1221. | 1146. |
| ENERGY (FT-LB) | 600. | 531. | 467. | 410. | 358. | 312. | 271. | 234. | 203. | 175. | 152. | 132. | 117. |
| DROP (IN) | .00 | -.16 | -.68 | -1.61 | -3.01 | -4.95 | -7.51 | -10.77 | -14.86 | -19.91 | -26.06 | -33.48 | -42.36 |
| MID-RANGE (IN) | .00 | .04 | .18 | .44 | .83 | 1.39 | 2.15 | 3.14 | 4.42 | 6.05 | 8.08 | 10.60 | 13.67 |
| BULLET PATH (IN) | -.63 | -.13 | .00 | -.28 | -1.03 | -2.31 | -4.22 | -6.83 | -10.27 | -14.67 | -20.16 | -26.93 | -35.15 |
| TIME OF FLIGHT (SEC) | .000000 | .029752 | .061428 | .095213 | .131317 | .169973 | .211434 | .255968 | .303849 | .355332 | .410629 | .469954 | .533411 |
| WIND DEFLECTION (IN) | | | | | | | | | | | | | |
| 5 MPH | .00 | .08 | .33 | .76 | 1.40 | 2.27 | 3.38 | 4.76 | 6.43 | 8.42 | 10.75 | 13.43 | 16.48 |
| 10 MPH | .00 | .16 | .66 | 1.53 | 2.80 | 4.53 | 6.75 | 9.51 | 12.86 | 16.85 | 21.50 | 26.87 | 32.96 |
| 20 MPH | .00 | .32 | 1.31 | 3.05 | 5.61 | 9.06 | 13.50 | 19.02 | 25.72 | 33.69 | 43.00 | 53.73 | 65.91 |
| 30 MPH | .00 | .48 | 1.97 | 4.58 | 8.41 | 13.59 | 20.25 | 28.54 | 38.59 | 50.54 | 64.50 | 80.60 | 98.87 |
| VELOCITY (FPS) | 2500. | 2348. | 2200. | 2059. | 1922. | 1792. | 1668. | 1551. | 1443. | 1343. | 1252. | 1172. | 1105. |
| ENERGY (FT-LB) | 555. | 489. | 430. | 376. | 328. | 285. | 247. | 214. | 185. | 160. | 139. | 122. | 108. |
| DROP (IN) | .00 | -.17 | -.73 | -1.74 | -3.27 | -5.38 | -8.16 | -11.73 | -16.20 | -21.72 | -28.44 | -36.56 | -46.26 |
| MID-RANGE (IN) | .00 | .05 | .20 | .47 | .90 | 1.51 | 2.34 | 3.43 | 4.83 | 6.62 | 8.84 | 11.60 | 14.96 |
| BULLET PATH (IN) | -.63 | -.12 | .00 | -.33 | -1.17 | -2.60 | -4.71 | -7.59 | -11.38 | -16.22 | -22.26 | -29.70 | -38.71 |
| TIME OF FLIGHT (SEC) | .000000 | .030959 | .063959 | .099198 | .136902 | .177317 | .220707 | .267346 | .317498 | .371394 | .429224 | .491193 | .557161 |
| WIND DEFLECTION (IN) | | | | | | | | | | | | | |
| 5 MPH | .00 | .08 | .35 | .81 | 1.49 | 2.40 | 3.58 | 5.05 | 6.82 | 8.92 | 11.37 | 14.18 | 17.35 |
| 10 MPH | .00 | .17 | .70 | 1.62 | 2.97 | 4.81 | 7.16 | 10.09 | 13.64 | 17.85 | 22.74 | 28.37 | 34.70 |
| 20 MPH | .00 | .34 | 1.39 | 3.24 | 5.95 | 9.62 | 14.33 | 20.19 | 27.28 | 35.69 | 45.49 | 56.74 | 69.40 |
| 30 MPH | .00 | .51 | 2.09 | 4.86 | 8.92 | 14.42 | 21.49 | 30.28 | 40.92 | 53.54 | 68.23 | 85.11 | 104.10 |

# Bullet: Speer .224 Dia. 40 Gr. Spire Pt.
## Ballistic Coefficient: .140

| RANGE (YDS) | MUZZLE | 25 | 50 | 75 | 100 | 125 | 150 | 175 | 200 | 225 | 250 | 275 | 300 |
|---|---|---|---|---|---|---|---|---|---|---|---|---|---|
| VELOCITY (FPS) | 2400. | 2251. | 2107. | 1969. | 1837. | 1710. | 1591. | 1479. | 1377. | 1283. | 1198. | 1127. | 1069. |
| ENERGY (FT-LB) | 512. | 450. | 394. | 344. | 300. | 260. | 225. | 194. | 168. | 146. | 127. | 113. | 102. |
| DROP (IN) | .00 | -.18 | -.80 | -1.90 | -3.56 | -5.86 | -8.91 | -12.81 | -17.71 | -23.76 | -31.14 | -40.03 | -50.62 |
| MID-RANGE (IN) | .00 | .05 | .21 | .52 | .98 | 1.65 | 2.56 | 3.75 | 5.30 | 7.26 | 9.71 | 12.73 | 16.40 |
| BULLET PATH (IN) | -.63 | -.10 | .00 | -.39 | -1.34 | -2.93 | -5.26 | -8.45 | -12.64 | -17.97 | -24.63 | -32.81 | -42.69 |
| TIME OF FLIGHT (SEC) | .000000 | .032268 | .066703 | .103522 | .142964 | .185288 | .230765 | .279665 | .332235 | .388669 | .449197 | .513806 | .582198 |
| WIND DEFLECTION (IN) | | | | | | | | | | | | | |
| 5 MPH | .00 | .09 | .37 | .86 | 1.58 | 2.56 | 3.81 | 5.36 | 7.24 | 9.45 | 12.03 | 14.96 | 18.23 |
| 10 MPH | .00 | .18 | .74 | 1.72 | 3.16 | 5.11 | 7.61 | 10.72 | 14.47 | 18.91 | 24.06 | 29.93 | 36.47 |
| 20 MPH | .00 | .36 | 1.48 | 3.44 | 6.32 | 10.22 | 15.23 | 21.44 | 28.95 | 37.81 | 48.12 | 59.86 | 72.93 |
| 30 MPH | .00 | .54 | 2.22 | 5.16 | 9.48 | 15.33 | 22.84 | 32.16 | 43.42 | 56.72 | 72.18 | 89.79 | 109.40 |
| VELOCITY (FPS) | 2300. | 2155. | 2014. | 1880. | 1751. | 1630. | 1515. | 1410. | 1314. | 1225. | 1149. | 1087. | 1037. |
| ENERGY (FT-LB) | 470. | 412. | 360. | 314. | 272. | 236. | 204. | 177. | 153. | 133. | 117. | 105. | 96. |
| DROP (IN) | .00 | -.20 | -.87 | -2.07 | -3.89 | -6.41 | -9.76 | -14.05 | -19.44 | -26.09 | -34.19 | -43.93 | -55.51 |
| MID-RANGE (IN) | .00 | .05 | .23 | .56 | 1.08 | 1.81 | 2.80 | 4.12 | 5.83 | 7.99 | 10.68 | 13.99 | 18.00 |
| BULLET PATH (IN) | -.63 | -.08 | .00 | -.45 | -1.52 | -3.30 | -5.89 | -9.43 | -14.07 | -19.97 | -27.32 | -36.31 | -47.14 |
| TIME OF FLIGHT (SEC) | .000000 | .033692 | .069693 | .108236 | .149572 | .193970 | .241704 | .293029 | .348161 | .407307 | .470588 | .537770 | .608461 |
| WIND DEFLECTION (IN) | | | | | | | | | | | | | |
| 5 MPH | .00 | .10 | .39 | .92 | 1.68 | 2.72 | 4.05 | 5.70 | 7.68 | 10.02 | 12.72 | 15.76 | 19.11 |
| 10 MPH | .00 | .19 | .79 | 1.83 | 3.37 | 5.44 | 8.11 | 11.40 | 15.36 | 20.03 | 25.43 | 31.52 | 38.22 |
| 20 MPH | .00 | .38 | 1.58 | 3.66 | 6.74 | 10.89 | 16.21 | 22.80 | 30.73 | 40.07 | 50.86 | 63.03 | 76.44 |
| 30 MPH | .00 | .57 | 2.36 | 5.50 | 10.10 | 16.33 | 24.32 | 34.20 | 46.09 | 60.10 | 76.30 | 94.55 | 114.66 |

## Bullet: Remington .224 Dia. 45 GR. H.P. (22702)
## Ballistic Coefficient: .140

| RANGE (YDS) | MUZZLE | 25 | 50 | 75 | 100 | 125 | 150 | 175 | 200 | 225 | 250 | 275 | 300 |
|---|---|---|---|---|---|---|---|---|---|---|---|---|---|
| VELOCITY (FPS) | 2700. | 2541. | 2387. | 2239. | 2095. | 1957. | 1825. | 1700. | 1581. | 1470. | 1368. | 1282. | 1221. |
| ENERGY (FT-LB) | 728. | 645. | 569. | 501. | 439. | 383. | 333. | 289. | 250. | 216. | 187. | 164. | 149. |
| DROP (IN) | .00 | -.15 | -.63 | -1.49 | -2.78 | -4.57 | -6.92 | -9.93 | -13.69 | -18.31 | -23.95 | -30.75 | -38.86 |
| MID-RANGE (IN) | .00 | .04 | .17 | .40 | .76 | 1.28 | 1.98 | 2.89 | 4.06 | 5.55 | 7.41 | 9.70 | 12.50 |
| BULLET PATH (IN) | -.63 | -.15 | .00 | -.23 | -.90 | -2.06 | -3.79 | -6.16 | -9.29 | -13.29 | -18.30 | -24.47 | -31.95 |
| TIME OF FLIGHT (SEC) | .000000 | .028636 | .059091 | .091537 | .126169 | .163205 | .202884 | .245469 | .291231 | .340438 | .393334 | .450071 | .510018 |
| WIND DEFLECTION (IN) | | | | | | | | | | | | | |
| 5 MPH | .00 | .08 | .31 | .72 | 1.33 | 2.14 | 3.19 | 4.49 | 6.07 | 7.96 | 10.17 | 12.72 | 15.55 |
| 10 MPH | .00 | .15 | .62 | 1.44 | 2.65 | 4.28 | 6.37 | 8.98 | 12.15 | 15.92 | 20.34 | 25.43 | 31.10 |
| 20 MPH | .00 | .30 | 1.24 | 2.89 | 5.30 | 8.56 | 12.75 | 17.96 | 24.29 | 31.83 | 40.68 | 50.87 | 62.19 |
| 30 MPH | .00 | .45 | 1.87 | 4.33 | 7.95 | 12.84 | 19.12 | 26.94 | 36.44 | 47.75 | 61.01 | 76.30 | 93.29 |
| VELOCITY (FPS) | 2600. | 2444. | 2294. | 2149. | 2009. | 1874. | 1746. | 1625. | 1511. | 1406. | 1310. | 1244. | 1187. |
| ENERGY (FT-LB) | 675. | 597. | 526. | 461. | 403. | 351. | 305. | 264. | 228. | 197. | 171. | 155. | 141. |
| DROP (IN) | .00 | -.16 | -.68 | -1.61 | -3.01 | -4.95 | -7.51 | -10.77 | -14.86 | -19.91 | -26.06 | -33.47 | -42.28 |
| MID-RANGE (IN) | .00 | .04 | .18 | .44 | .83 | 1.39 | 2.15 | 3.14 | 4.42 | 6.05 | 8.08 | 10.59 | 13.64 |
| BULLET PATH (IN) | -.63 | -.13 | .00 | -.28 | -1.03 | -2.31 | -4.22 | -6.83 | -10.27 | -14.67 | -20.16 | -26.91 | -35.08 |
| TIME OF FLIGHT (SEC) | .000000 | .029752 | .061428 | .095213 | .131317 | .169973 | .211434 | .255968 | .303849 | .355332 | .410629 | .469462 | .531216 |
| WIND DEFLECTION (IN) | | | | | | | | | | | | | |
| 5 MPH | .00 | .08 | .33 | .76 | 1.40 | 2.27 | 3.38 | 4.76 | 6.43 | 8.42 | 10.75 | 13.39 | 16.29 |
| 10 MPH | .00 | .16 | .66 | 1.53 | 2.80 | 4.53 | 6.75 | 9.51 | 12.86 | 16.85 | 21.50 | 26.78 | 32.57 |
| 20 MPH | .00 | .32 | 1.31 | 3.05 | 5.61 | 9.06 | 13.50 | 19.02 | 25.72 | 33.69 | 43.00 | 53.56 | 65.14 |
| 30 MPH | .00 | .48 | 1.97 | 4.58 | 8.41 | 13.59 | 20.25 | 28.54 | 38.59 | 50.54 | 64.50 | 80.34 | 97.71 |
| VELOCITY (FPS) | 2500. | 2348. | 2200. | 2059. | 1922. | 1792. | 1668. | 1551. | 1443. | 1343. | 1265. | 1206. | 1154. |
| ENERGY (FT-LB) | 624. | 551. | 484. | 423. | 369. | 321. | 278. | 240. | 208. | 180. | 160. | 145. | 133. |
| DROP (IN) | .00 | -.17 | -.73 | -1.74 | -3.27 | -5.38 | -8.16 | -11.73 | -16.20 | -21.72 | -28.44 | -36.52 | -46.09 |
| MID-RANGE (IN) | .00 | .05 | .20 | .47 | .90 | 1.51 | 2.34 | 3.43 | 4.83 | 6.62 | 8.84 | 11.58 | 14.88 |
| BULLET PATH (IN) | -.63 | -.12 | .00 | -.33 | -1.17 | -2.60 | -4.71 | -7.59 | -11.38 | -16.22 | -22.26 | -29.65 | -38.55 |
| TIME OF FLIGHT (SEC) | .000000 | .030959 | .063959 | .099198 | .136902 | .177317 | .220707 | .267346 | .317498 | .371394 | .429079 | .489826 | .553445 |
| WIND DEFLECTION (IN) | | | | | | | | | | | | | |
| 5 MPH | .00 | .08 | .35 | .81 | 1.49 | 2.40 | 3.58 | 5.05 | 6.82 | 8.92 | 11.36 | 14.06 | 17.02 |
| 10 MPH | .00 | .17 | .70 | 1.62 | 2.97 | 4.81 | 7.16 | 10.09 | 13.64 | 17.85 | 22.72 | 28.13 | 34.05 |
| 20 MPH | .00 | .34 | 1.39 | 3.24 | 5.95 | 9.62 | 14.33 | 20.19 | 27.28 | 35.69 | 45.44 | 56.26 | 68.09 |
| 30 MPH | .00 | .51 | 2.09 | 4.86 | 8.92 | 14.42 | 21.49 | 30.28 | 40.92 | 53.54 | 68.15 | 84.39 | 102.14 |
| VELOCITY (FPS) | 2400. | 2251. | 2107. | 1969. | 1837. | 1710. | 1591. | 1479. | 1377. | 1287. | 1226. | 1171. | 1123. |
| ENERGY (FT-LB) | 575. | 506. | 444. | 387. | 337. | 292. | 253. | 219. | 189. | 166. | 150. | 137. | 126. |
| DROP (IN) | .00 | -.18 | -.80 | -1.90 | -3.56 | -5.86 | -8.91 | -12.81 | -17.71 | -23.76 | -31.11 | -39.92 | -50.30 |
| MID-RANGE (IN) | .00 | .05 | .21 | .52 | .98 | 1.65 | 2.56 | 3.75 | 5.30 | 7.26 | 9.69 | 12.67 | 16.24 |
| BULLET PATH (IN) | -.63 | -.10 | .00 | -.39 | -1.34 | -2.93 | -5.26 | -8.45 | -12.64 | -17.97 | -24.61 | -32.70 | -42.37 |
| TIME OF FLIGHT (SEC) | .000000 | .032268 | .066703 | .103522 | .142964 | .185288 | .230765 | .279665 | .332235 | .388657 | .448381 | .511002 | .576421 |
| WIND DEFLECTION (IN) | | | | | | | | | | | | | |
| 5 MPH | .00 | .09 | .37 | .86 | 1.58 | 2.56 | 3.81 | 5.36 | 7.24 | 9.45 | 11.96 | 14.72 | 17.73 |
| 10 MPH | .00 | .18 | .74 | 1.72 | 3.16 | 5.11 | 7.61 | 10.72 | 14.47 | 18.90 | 23.91 | 29.44 | 35.45 |
| 20 MPH | .00 | .36 | 1.48 | 3.44 | 6.32 | 10.22 | 15.23 | 21.44 | 28.95 | 37.81 | 47.83 | 58.87 | 70.90 |
| 30 MPH | .00 | .54 | 2.22 | 5.16 | 9.48 | 15.33 | 22.84 | 32.16 | 43.42 | 56.71 | 71.74 | 88.31 | 106.35 |

# Bullet: Remington .224 Dia. 45 Gr. H.P. (22702)
## Ballistic Coefficient: .140

| RANGE (YDS) | MUZZLE | 25 | 50 | 75 | 100 | 125 | 150 | 175 | 200 | 225 | 250 | 275 | 300 |
|---|---|---|---|---|---|---|---|---|---|---|---|---|---|
| VELOCITY (FPS) | 2300. | 2155. | 2014. | 1880. | 1751. | 1630. | 1515. | 1410. | 1314. | 1246. | 1189. | 1139. | 1096. |
| ENERGY (FT-LB) | 528. | 464. | 405. | 353. | 306. | 265. | 229. | 199. | 172. | 155. | 141. | 130. | 120. |
| DROP (IN) | .00 | -.20 | -.87 | -2.07 | -3.89 | -6.41 | -9.76 | -14.05 | -19.44 | -26.07 | -34.12 | -43.69 | -54.94 |
| MID-RANGE (IN) | .00 | .05 | .23 | .56 | 1.08 | 1.81 | 2.80 | 4.12 | 5.83 | 7.98 | 10.64 | 13.87 | 17.72 |
| BULLET PATH (IN) | -.63 | -.08 | .00 | -.45 | -1.52 | -3.30 | -5.89 | -9.43 | -14.07 | -19.95 | -27.25 | -36.07 | -46.57 |
| TIME OF FLIGHT (SEC) | .000000 | .033692 | .069693 | .108236 | .149572 | .193970 | .241704 | .293029 | .348161 | .406877 | .468525 | .533012 | .600192 |
| WIND DEFLECTION (IN) | | | | | | | | | | | | | |
| 5 MPH | .00 | .10 | .39 | .92 | 1.68 | 2.72 | 4.05 | 5.70 | 7.68 | 9.98 | 12.53 | 15.34 | 18.38 |
| 10 MPH | .00 | .19 | .79 | 1.83 | 3.37 | 5.44 | 8.11 | 11.40 | 15.36 | 19.96 | 25.07 | 30.68 | 36.76 |
| 20 MPH | .00 | .38 | 1.58 | 3.66 | 6.74 | 10.89 | 16.21 | 22.80 | 30.73 | 39.92 | 50.14 | 61.36 | 73.53 |
| 30 MPH | .00 | .57 | 2.36 | 5.50 | 10.10 | 16.33 | 24.32 | 34.20 | 46.09 | 59.87 | 75.21 | 92.04 | 110.29 |
| VELOCITY (FPS) | 2200. | 2058. | 1922. | 1791. | 1667. | 1551. | 1442. | 1343. | 1265. | 1206. | 1153. | 1108. | 1069. |
| ENERGY (FT-LB) | 484. | 423. | 369. | 321. | 278. | 240. | 208. | 180. | 160. | 145. | 133. | 123. | 114. |
| DROP (IN) | .00 | -.22 | -.95 | -2.27 | -4.27 | -7.05 | -10.73 | -15.46 | -21.39 | -28.68 | -37.47 | -47.89 | -60.07 |
| MID-RANGE (IN) | .00 | .06 | .26 | .62 | 1.18 | 1.99 | 3.09 | 4.55 | 6.43 | 8.79 | 11.69 | 15.18 | 19.30 |
| BULLET PATH (IN) | -.63 | -.06 | .00 | -.53 | -1.73 | -3.72 | -6.61 | -10.55 | -15.69 | -22.19 | -30.18 | -39.81 | -51.20 |
| TIME OF FLIGHT (SEC) | .000000 | .035247 | .072959 | .113383 | .156784 | .203433 | .253597 | .307506 | .365203 | .425963 | .489593 | .555969 | .624903 |
| WIND DEFLECTION (IN) | | | | | | | | | | | | | |
| 5 MPH | .00 | .10 | .42 | .98 | 1.80 | 2.90 | 4.32 | 6.06 | 8.14 | 10.48 | 13.08 | 15.93 | 18.99 |
| 10 MPH | .00 | .20 | .84 | 1.96 | 3.59 | 5.80 | 8.63 | 12.12 | 16.28 | 20.97 | 26.17 | 31.85 | 37.98 |
| 20 MPH | .00 | .41 | 1.68 | 3.91 | 7.19 | 11.61 | 17.27 | 24.24 | 32.55 | 41.94 | 52.34 | 63.70 | 75.97 |
| 30 MPH | .00 | .61 | 2.52 | 5.87 | 10.78 | 17.41 | 25.90 | 36.36 | 48.83 | 62.91 | 78.51 | 95.55 | 113.95 |
| VELOCITY (FPS) | 2100. | 1962. | 1830. | 1704. | 1585. | 1474. | 1372. | 1284. | 1223. | 1169. | 1121. | 1081. | 1046. |
| ENERGY (FT-LB) | 441. | 385. | 334. | 290. | 251. | 217. | 188. | 165. | 149. | 136. | 126. | 117. | 109. |
| DROP (IN) | .00 | -.24 | -1.05 | -2.50 | -4.70 | -7.77 | -11.85 | -17.07 | -23.61 | -31.61 | -41.19 | -52.50 | -65.66 |
| MID-RANGE (IN) | .00 | .07 | .28 | .68 | 1.30 | 2.20 | 3.42 | 5.03 | 7.10 | 9.68 | 12.82 | 16.57 | 20.98 |
| BULLET PATH (IN) | -.63 | -.03 | .00 | -.61 | -1.98 | -4.20 | -7.44 | -11.83 | -17.52 | -24.68 | -33.42 | -43.89 | -56.21 |
| TIME OF FLIGHT (SEC) | .000000 | .035950 | .076536 | .119019 | .164668 | .213755 | .266522 | .323140 | .383012 | .445778 | .511335 | .579513 | .650105 |
| WIND DEFLECTION (IN) | | | | | | | | | | | | | |
| 5 MPH | .00 | .11 | .45 | 1.05 | 1.92 | 3.10 | 4.60 | 6.44 | 8.56 | 10.94 | 13.57 | 16.43 | 19.49 |
| 10 MPH | .00 | .22 | .90 | 2.09 | 3.84 | 6.19 | 9.19 | 12.87 | 17.12 | 21.89 | 27.14 | 32.85 | 38.99 |
| 20 MPH | .00 | .44 | 1.80 | 4.18 | 7.68 | 12.38 | 18.39 | 25.75 | 34.25 | 43.77 | 54.28 | 65.70 | 77.98 |
| 30 MPH | .00 | .65 | 2.70 | 6.27 | 11.52 | 18.58 | 27.58 | 38.62 | 51.37 | 65.66 | 81.41 | 98.55 | 116.97 |

# Bullet: Remington .224 Dia. 50 Gr. H.P. (22710)
## Ballistic Coefficient: .225

| RANGE (YDS) | MUZZLE | 25 | 50 | 75 | 100 | 125 | 150 | 175 | 200 | 225 | 250 | 275 | 300 |
|---|---|---|---|---|---|---|---|---|---|---|---|---|---|
| VELOCITY (FPS) | 2600. | 2502. | 2407. | 2313. | 2222. | 2133. | 2045. | 1960. | 1877. | 1797. | 1719. | 1643. | 1571. |
| ENERGY (FT-LB) | 750. | 695. | 643. | 594. | 548. | 505. | 464. | 427. | 391. | 358. | 328. | 300. | 274. |
| DROP (IN) | .00 | -.15 | -.64 | -1.51 | -2.79 | -4.50 | -6.69 | -9.39 | -12.66 | -16.55 | -21.10 | -26.39 | -32.48 |
| MID-RANGE (IN) | .00 | .04 | .17 | .41 | .75 | 1.22 | 1.83 | 2.60 | 3.55 | 4.69 | 6.05 | 7.66 | 9.55 |
| BULLET PATH (IN) | -.63 | -.14 | .00 | -.23 | -.87 | -1.94 | -3.50 | -5.57 | -8.20 | -11.45 | -15.36 | -20.02 | -25.47 |
| TIME OF FLIGHT (SEC) | .000000 | .029403 | .059963 | .091746 | .124825 | .159278 | .195188 | .232645 | .271743 | .312582 | .355267 | .399902 | .446593 |
| WIND DEFLECTION (IN) | | | | | | | | | | | | | |
| 5 MPH | .00 | .05 | .20 | .46 | .83 | 1.32 | 1.95 | 2.70 | 3.61 | 4.66 | 5.88 | 7.27 | 8.84 |
| 10 MPH | .00 | .10 | .40 | .92 | 1.66 | 2.65 | 3.89 | 5.41 | 7.21 | 9.32 | 11.76 | 14.54 | 17.68 |
| 20 MPH | .00 | .20 | .80 | 1.83 | 3.32 | 5.30 | 7.78 | 10.81 | 14.42 | 18.64 | 23.52 | 29.07 | 35.35 |
| 30 MPH | .00 | .29 | 1.20 | 2.75 | 4.98 | 7.94 | 11.67 | 16.22 | 21.63 | 27.97 | 35.27 | 43.61 | 53.03 |
| VELOCITY (FPS) | 2500. | 2405. | 2311. | 2220. | 2130. | 2043. | 1958. | 1875. | 1795. | 1717. | 1641. | 1569. | 1499. |
| ENERGY (FT-LB) | 694. | 642. | 593. | 547. | 504. | 463. | 426. | 390. | 358. | 327. | 299. | 273. | 250. |
| DROP (IN) | .00 | -.16 | -.70 | -1.64 | -3.02 | -4.88 | -7.26 | -10.20 | -13.76 | -17.99 | -22.96 | -28.73 | -35.37 |
| MID-RANGE (IN) | .00 | .05 | .19 | .44 | .81 | 1.32 | 1.99 | 2.83 | 3.86 | 5.11 | 6.60 | 8.36 | 10.43 |
| BULLET PATH (IN) | -.63 | -.13 | .00 | -.28 | -1.00 | -2.19 | -3.91 | -6.18 | -9.08 | -12.64 | -16.95 | -22.05 | -28.04 |
| TIME OF FLIGHT (SEC) | .000000 | .030590 | .062406 | .095519 | .130007 | .165956 | .203453 | .242594 | .283480 | .326212 | .370898 | .417643 | .466550 |
| WIND DEFLECTION (IN) | | | | | | | | | | | | | |
| 5 MPH | .00 | .05 | .21 | .49 | .88 | 1.40 | 2.06 | 2.87 | 3.83 | 4.95 | 6.24 | 7.71 | 9.38 |
| 10 MPH | .00 | .10 | .42 | .97 | 1.76 | 2.81 | 4.13 | 5.74 | 7.65 | 9.89 | 12.48 | 15.43 | 18.75 |
| 20 MPH | .00 | .21 | .85 | 1.94 | 3.52 | 5.62 | 8.26 | 11.47 | 15.30 | 19.79 | 24.96 | 30.85 | 37.51 |
| 30 MPH | .00 | .31 | 1.27 | 2.91 | 5.28 | 8.42 | 12.38 | 17.21 | 22.96 | 29.68 | 37.43 | 46.28 | 56.26 |
| VELOCITY (FPS) | 2400. | 2307. | 2215. | 2126. | 2039. | 1954. | 1871. | 1791. | 1713. | 1638. | 1565. | 1496. | 1430. |
| ENERGY (FT-LB) | 639. | 591. | 545. | 502. | 462. | 424. | 389. | 356. | 326. | 298. | 272. | 248. | 227. |
| DROP (IN) | .00 | -.18 | -.76 | -1.78 | -3.29 | -5.31 | -7.90 | -11.11 | -15.00 | -19.62 | -25.05 | -31.37 | -38.67 |
| MID-RANGE (IN) | .00 | .05 | .20 | .48 | .88 | 1.44 | 2.17 | 3.08 | 4.21 | 5.58 | 7.21 | 9.15 | 11.43 |
| BULLET PATH (IN) | -.63 | -.11 | .00 | -.33 | -1.14 | -2.47 | -4.37 | -6.88 | -10.07 | -14.00 | -18.74 | -24.36 | -30.97 |
| TIME OF FLIGHT (SEC) | .000000 | .031876 | .065053 | .099609 | .135629 | .173203 | .212425 | .253396 | .296219 | .341001 | .387847 | .436859 | .488135 |
| WIND DEFLECTION (IN) | | | | | | | | | | | | | |
| 5 MPH | .00 | .06 | .22 | .52 | .94 | 1.49 | 2.19 | 3.05 | 4.07 | 5.26 | 6.63 | 8.19 | 9.96 |
| 10 MPH | .00 | .11 | .45 | 1.03 | 1.87 | 2.98 | 4.39 | 6.10 | 8.13 | 10.52 | 13.26 | 16.39 | 19.91 |
| 20 MPH | .00 | .22 | .90 | 2.06 | 3.74 | 5.97 | 8.77 | 12.20 | 16.27 | 21.03 | 26.52 | 32.77 | 39.82 |
| 30 MPH | .00 | .33 | 1.35 | 3.09 | 5.61 | 8.95 | 13.16 | 18.29 | 24.40 | 31.55 | 39.78 | 49.16 | 59.74 |
| VELOCITY (FPS) | 2300. | 2209. | 2120. | 2033. | 1948. | 1865. | 1785. | 1707. | 1632. | 1560. | 1491. | 1426. | 1364. |
| ENERGY (FT-LB) | 587. | 542. | 499. | 459. | 421. | 386. | 354. | 324. | 296. | 270. | 247. | 226. | 206. |
| DROP (IN) | .00 | -.19 | -.83 | -1.95 | -3.59 | -5.80 | -8.64 | -12.15 | -16.41 | -21.48 | -27.44 | -34.39 | -42.41 |
| MID-RANGE (IN) | .00 | .05 | .22 | .52 | .97 | 1.58 | 2.37 | 3.38 | 4.62 | 6.12 | 7.92 | 10.06 | 12.57 |
| BULLET PATH (IN) | -.63 | -.09 | .00 | -.39 | -1.30 | -2.79 | -4.89 | -7.68 | -11.21 | -15.55 | -20.78 | -27.01 | -34.30 |
| TIME OF FLIGHT (SEC) | .000000 | .033275 | .067935 | .104065 | .141756 | .181102 | .222205 | .265167 | .310096 | .357096 | .406271 | .457715 | .511511 |
| WIND DEFLECTION (IN) | | | | | | | | | | | | | |
| 5 MPH | .00 | .06 | .24 | .55 | 1.00 | 1.59 | 2.34 | 3.25 | 4.33 | 5.60 | 7.06 | 8.71 | 10.58 |
| 10 MPH | .00 | .12 | .48 | 1.10 | 1.99 | 3.18 | 4.67 | 6.50 | 8.66 | 11.20 | 14.11 | 17.43 | 21.16 |
| 20 MPH | .00 | .23 | .96 | 2.20 | 3.98 | 6.36 | 9.35 | 12.99 | 17.33 | 22.39 | 28.22 | 34.85 | 42.31 |
| 30 MPH | .00 | .35 | 1.43 | 3.29 | 5.98 | 9.53 | 14.02 | 19.49 | 25.99 | 33.59 | 42.34 | 52.28 | 63.47 |

## Bullet: Remington .224 Dia. 50 Gr. H.P. (22710)
## Ballistic Coefficient: .225

| RANGE (YDS) | MUZZLE | 25 | 50 | 75 | 100 | 125 | 150 | 175 | 200 | 225 | 250 | 275 | 300 |
|---|---|---|---|---|---|---|---|---|---|---|---|---|---|
| VELOCITY (FPS) | 2200. | 2111. | 2024. | 1940. | 1857. | 1777. | 1700. | 1625. | 1553. | 1485. | 1419. | 1358. | 1300. |
| ENERGY (FT-LB) | 537. | 495. | 455. | 418. | 383. | 351. | 321. | 293. | 268. | 245. | 224. | 205. | 188. |
| DROP (IN) | .00 | -.21 | -.91 | -2.13 | -3.93 | -6.36 | -9.47 | -13.34 | -18.02 | -23.60 | -30.18 | -37.84 | -46.68 |
| MID-RANGE (IN) | .00 | .06 | .24 | .57 | 1.06 | 1.73 | 2.61 | 3.71 | 5.08 | 6.74 | 8.73 | 11.09 | 13.87 |
| BULLET PATH (IN) | -.63 | -.07 | .00 | -.46 | -1.49 | -3.15 | -5.50 | -8.59 | -12.50 | -17.32 | -23.13 | -30.02 | -38.09 |
| TIME OF FLIGHT (SEC) | .000000 | .034802 | .071082 | .108932 | .148447 | .189729 | .232882 | .278010 | .325221 | .374615 | .426288 | .480321 | .536778 |
| WIND DEFLECTION (IN) | | | | | | | | | | | | | |
| 5 MPH | .00 | .06 | .26 | .59 | 1.06 | 1.70 | 2.49 | 3.46 | 4.62 | 5.97 | 7.51 | 9.27 | 11.24 |
| 10 MPH | .00 | .13 | .51 | 1.17 | 2.13 | 3.39 | 4.99 | 6.93 | 9.24 | 11.93 | 15.03 | 18.54 | 22.47 |
| 20 MPH | .00 | .25 | 1.02 | 2.34 | 4.25 | 6.78 | 9.97 | 13.86 | 18.48 | 23.86 | 30.05 | 37.07 | 44.95 |
| 30 MPH | .00 | .38 | 1.53 | 3.52 | 6.38 | 10.18 | 14.96 | 20.79 | 27.72 | 35.80 | 45.08 | 55.61 | 67.42 |
| VELOCITY (FPS) | 2100. | 2013. | 1929. | 1847. | 1767. | 1690. | 1616. | 1544. | 1476. | 1411. | 1350. | 1293. | 1239. |
| ENERGY (FT-LB) | 490. | 450. | 413. | 379. | 347. | 317. | 290. | 265. | 242. | 221. | 202. | 186. | 171. |
| DROP (IN) | .00 | -.23 | -1.00 | -2.34 | -4.33 | -7.00 | -10.43 | -14.70 | -19.87 | -26.06 | -33.33 | -41.79 | -51.55 |
| MID-RANGE (IN) | .00 | .06 | .27 | .63 | 1.17 | 1.91 | 2.87 | 4.10 | 5.61 | 7.46 | 9.66 | 12.28 | 15.34 |
| BULLET PATH (IN) | -.63 | -.05 | .00 | -.53 | -1.70 | -3.57 | -6.19 | -9.64 | -14.00 | -19.37 | -25.83 | -33.48 | -42.42 |
| TIME OF FLIGHT (SEC) | .000000 | .036474 | .074529 | .114263 | .155776 | .199172 | .244559 | .292040 | .341718 | .393685 | .448022 | .504789 | .564045 |
| WIND DEFLECTION (IN) | | | | | | | | | | | | | |
| 5 MPH | .00 | .07 | .27 | .63 | 1.14 | 1.81 | 2.66 | 3.70 | 4.93 | 6.36 | 8.00 | 9.85 | 11.92 |
| 10 MPH | .00 | .13 | .55 | 1.25 | 2.27 | 3.63 | 5.33 | 7.40 | 9.86 | 12.72 | 15.99 | 19.70 | 23.84 |
| 20 MPH | .00 | .27 | 1.09 | 2.51 | 4.55 | 7.25 | 10.66 | 14.80 | 19.71 | 25.43 | 31.99 | 39.40 | 47.69 |
| 30 MPH | .00 | .40 | 1.64 | 3.76 | 6.82 | 10.88 | 15.98 | 22.20 | 29.57 | 38.15 | 47.98 | 59.10 | 71.53 |
| VELOCITY (FPS) | 2000. | 1916. | 1834. | 1755. | 1678. | 1604. | 1533. | 1466. | 1402. | 1341. | 1284. | 1231. | 1183. |
| ENERGY (FT-LB) | 444. | 407. | 373. | 342. | 313. | 286. | 261. | 238. | 218. | 200. | 183. | 168. | 155. |
| DROP (IN) | .00 | -.25 | -1.10 | -2.59 | -4.78 | -7.74 | -11.54 | -16.27 | -22.03 | -28.88 | -36.94 | -46.32 | -57.14 |
| MID-RANGE (IN) | .00 | .07 | .30 | .69 | 1.29 | 2.11 | 3.18 | 4.54 | 6.23 | 8.28 | 10.73 | 13.63 | 17.03 |
| BULLET PATH (IN) | -.63 | -.02 | .00 | -.62 | -1.95 | -4.05 | -6.98 | -10.85 | -15.74 | -21.73 | -28.92 | -37.44 | -47.39 |
| TIME OF FLIGHT (SEC) | .000000 | .038314 | .078322 | .120125 | .163829 | .209540 | .257362 | .307396 | .359733 | .414450 | .471605 | .531261 | .593413 |
| WIND DEFLECTION (IN) | | | | | | | | | | | | | |
| 5 MPH | .00 | .07 | .29 | .67 | 1.22 | 1.94 | 2.85 | 3.95 | 5.26 | 6.77 | 8.50 | 10.45 | 12.62 |
| 10 MPH | .00 | .14 | .58 | 1.34 | 2.43 | 3.88 | 5.70 | 7.90 | 10.51 | 13.54 | 17.00 | 20.90 | 25.24 |
| 20 MPH | .00 | .29 | 1.17 | 2.68 | 4.87 | 7.76 | 11.39 | 15.80 | 21.03 | 27.09 | 34.01 | 41.80 | 50.48 |
| 30 MPH | .00 | .43 | 1.75 | 4.03 | 7.30 | 11.64 | 17.09 | 23.71 | 31.54 | 40.63 | 51.01 | 62.71 | 75.72 |

# Bullet: Lyman #225438, 43 Gr.
## Ballistic Coefficient: .120

| RANGE (YDS) | MUZZLE | 25 | 50 | 75 | 100 | 125 | 150 | 175 | 200 | 225 | 250 | 275 | 300 |
|---|---|---|---|---|---|---|---|---|---|---|---|---|---|
| VELOCITY (FPS) | 2700. | 2515. | 2337. | 2166. | 2003. | 1847. | 1700. | 1562. | 1435. | 1321. | 1236. | 1167. | 1108. |
| ENERGY (FT-LB) | 696. | 604. | 521. | 448. | 383. | 326. | 276. | 233. | 197. | 167. | 146. | 130. | 117. |
| DROP (IN) | .00 | -.15 | -.64 | -1.53 | -2.88 | -4.77 | -7.30 | -10.57 | -14.75 | -19.98 | -26.45 | -34.35 | -43.84 |
| MID-RANGE (IN) | .00 | .04 | .17 | .42 | .80 | 1.35 | 2.12 | 3.15 | 4.50 | 6.24 | 8.46 | 11.23 | 14.62 |
| BULLET PATH (IN) | -.63 | -.14 | -.00 | -.25 | -.97 | -2.23 | -4.12 | -6.76 | -10.30 | -14.90 | -20.74 | -28.00 | -36.86 |
| TIME OF FLIGHT (SEC) | .000000 | .028784 | .059722 | .093057 | .129067 | .168066 | .210402 | .256445 | .306557 | .361052 | .419865 | .482366 | .548372 |
| WIND DEFLECTION (IN) | | | | | | | | | | | | | |
| 5 MPH | .00 | .09 | .37 | .86 | 1.58 | 2.57 | 3.85 | 5.46 | 7.42 | 9.77 | 12.50 | 15.56 | 18.92 |
| 10 MPH | .00 | .18 | .73 | 1.71 | 3.16 | 5.14 | 7.70 | 10.91 | 14.84 | 19.55 | 25.01 | 31.12 | 37.85 |
| 20 MPH | .00 | .35 | 1.47 | 3.42 | 6.32 | 10.27 | 15.39 | 21.82 | 29.69 | 39.09 | 50.01 | 62.24 | 75.69 |
| 30 MPH | .00 | .53 | 2.20 | 5.13 | 9.48 | 15.41 | 23.09 | 32.74 | 44.53 | 58.64 | 75.02 | 93.36 | 113.54 |
| VELOCITY (FPS) | 2600. | 2419. | 2245. | 2078. | 1918. | 1767. | 1625. | 1493. | 1373. | 1272. | 1198. | 1134. | 1082. |
| ENERGY (FT-LB) | 645. | 558. | 481. | 412. | 351. | 298. | 252. | 213. | 180. | 155. | 137. | 123. | 112. |
| DROP (IN) | .00 | -.16 | -.69 | -1.65 | -3.11 | -5.17 | -7.92 | -11.49 | -16.04 | -21.75 | -28.79 | -37.35 | -47.60 |
| MID-RANGE (IN) | .00 | .04 | .19 | .45 | .87 | 1.47 | 2.31 | 3.43 | 4.91 | 6.81 | 9.23 | 12.23 | 15.88 |
| BULLET PATH (IN) | -.63 | -.13 | .00 | -.30 | -1.10 | -2.50 | -4.59 | -7.50 | -11.39 | -16.44 | -22.82 | -30.72 | -40.31 |
| TIME OF FLIGHT (SEC) | .000000 | .029908 | .062096 | .096824 | .134391 | .175129 | .219401 | .267577 | .319999 | .376884 | .437669 | .502050 | .569800 |
| WIND DEFLECTION (IN) | | | | | | | | | | | | | |
| 5 MPH | .00 | .09 | .39 | .91 | 1.67 | 2.72 | 4.08 | 5.78 | 7.85 | 10.32 | 13.13 | 16.26 | 19.68 |
| 10 MPH | .00 | .19 | .78 | 1.81 | 3.35 | 5.44 | 8.15 | 11.56 | 15.70 | 20.64 | 26.26 | 32.51 | 39.36 |
| 20 MPH | .00 | .37 | 1.55 | 3.62 | 6.69 | 10.88 | 16.31 | 23.11 | 31.41 | 41.28 | 52.52 | 65.03 | 78.72 |
| 30 MPH | .00 | .56 | 2.33 | 5.43 | 10.04 | 16.31 | 24.46 | 34.67 | 47.11 | 61.92 | 78.78 | 97.54 | 118.09 |
| VELOCITY (FPS) | 2500. | 2323. | 2153. | 1990. | 1835. | 1688. | 1551. | 1425. | 1313. | 1230. | 1162. | 1104. | 1057. |
| ENERGY (FT-LB) | 597. | 515. | 442. | 378. | 321. | 272. | 230. | 194. | 164. | 144. | 129. | 116. | 107. |
| DROP (IN) | .00 | -.17 | -.75 | -1.79 | -3.38 | -5.62 | -8.62 | -12.52 | -17.50 | -23.73 | -31.40 | -40.68 | -51.74 |
| MID-RANGE (IN) | .00 | .05 | .20 | .49 | .94 | 1.60 | 2.52 | 3.75 | 5.37 | 7.45 | 10.08 | 13.33 | 17.25 |
| BULLET PATH (IN) | -.63 | -.11 | .00 | -.35 | -1.26 | -2.80 | -5.11 | -8.33 | -12.62 | -18.16 | -25.15 | -33.74 | -44.10 |
| TIME OF FLIGHT (SEC) | .000000 | .031125 | .064668 | .100909 | .140167 | .182792 | .229154 | .279613 | .334477 | .393592 | .456371 | .522637 | .592103 |
| WIND DEFLECTION (IN) | | | | | | | | | | | | | |
| 5 MPH | .00 | .10 | .41 | .96 | 1.77 | 2.89 | 4.33 | 6.13 | 8.31 | 10.88 | 13.76 | 16.95 | 20.43 |
| 10 MPH | .00 | .20 | .82 | 1.92 | 3.55 | 5.77 | 8.65 | 12.25 | 16.63 | 21.75 | 27.52 | 33.90 | 40.85 |
| 20 MPH | .00 | .40 | 1.64 | 3.84 | 7.10 | 11.54 | 17.30 | 24.50 | 33.26 | 43.50 | 55.04 | 67.81 | 81.70 |
| 30 MPH | .00 | .59 | 2.46 | 5.76 | 10.65 | 17.31 | 25.95 | 36.76 | 49.88 | 65.26 | 82.56 | 101.71 | 122.55 |
| VELOCITY (FPS) | 2400. | 2227. | 2061. | 1902. | 1752. | 1610. | 1479. | 1361. | 1264. | 1191. | 1128. | 1077. | 1034. |
| ENERGY (FT-LB) | 550. | 473. | 405. | 345. | 293. | 248. | 209. | 177. | 152. | 135. | 122. | 111. | 102. |
| DROP (IN) | .00 | -.19 | -.81 | -1.95 | -3.69 | -6.13 | -9.41 | -13.69 | -19.15 | -25.96 | -34.31 | -44.36 | -56.27 |
| MID-RANGE (IN) | .00 | .05 | .22 | .53 | 1.03 | 1.75 | 2.76 | 4.11 | 5.89 | 8.17 | 11.03 | 14.52 | 18.73 |
| BULLET PATH (IN) | -.63 | -.10 | .00 | -.41 | -1.43 | -3.15 | -5.71 | -9.27 | -14.00 | -20.10 | -27.72 | -37.05 | -48.24 |
| TIME OF FLIGHT (SEC) | .000000 | .032444 | .067457 | .105342 | .146437 | .191104 | .239713 | .292602 | .349941 | .411121 | .475878 | .543973 | .615087 |
| WIND DEFLECTION (IN) | | | | | | | | | | | | | |
| 5 MPH | .00 | .11 | .44 | 1.02 | 1.89 | 3.07 | 4.59 | 6.50 | 8.79 | 11.43 | 14.38 | 17.62 | 21.13 |
| 10 MPH | .00 | .21 | .87 | 2.04 | 3.77 | 6.13 | 9.19 | 13.00 | 17.59 | 22.86 | 28.75 | 35.24 | 42.26 |
| 20 MPH | .00 | .42 | 1.74 | 4.08 | 7.55 | 12.27 | 18.38 | 26.00 | 35.18 | 45.71 | 57.51 | 70.48 | 84.51 |
| 30 MPH | .00 | .63 | 2.62 | 6.12 | 11.32 | 18.40 | 27.57 | 38.99 | 52.77 | 68.57 | 86.26 | 105.72 | 126.77 |

## Bullet: Lyman #225438, 43 Gr.
### Ballistic Coefficient: .120

| RANGE (YDS) | MUZZLE | 25 | 50 | 75 | 100 | 125 | 150 | 175 | 200 | 225 | 250 | 275 | 300 |
|---|---|---|---|---|---|---|---|---|---|---|---|---|---|
| VELOCITY (FPS) | 2300. | 2131. | 1969. | 1815. | 1669. | 1534. | 1410. | 1299. | 1221. | 1154. | 1098. | 1052. | 1013. |
| ENERGY (FT-LB) | 505. | 433. | 370. | 314. | 266. | 225. | 190. | 161. | 142. | 127. | 115. | 106. | 98. |
| DROP (IN) | .00 | -.20 | -.89 | -2.13 | -4.03 | -6.71 | -10.32 | -15.03 | -21.01 | -28.45 | -37.53 | -48.40 | -61.22 |
| MID-RANGE (IN) | .00 | .06 | .24 | .58 | 1.13 | 1.92 | 3.03 | 4.52 | 6.47 | 8.96 | 12.05 | 15.81 | 20.29 |
| BULLET PATH (IN) | -.63 | -.08 | .00 | -.48 | -1.63 | -3.55 | -6.40 | -10.35 | -15.57 | -22.26 | -30.57 | -40.68 | -52.75 |
| TIME OF FLIGHT (SEC) | .000000 | .033879 | .070497 | .110175 | .153269 | .200147 | .251168 | .306626 | .366217 | .429448 | .496136 | .565982 | .638683 |
| WIND DEFLECTION (IN) | | | | | | | | | | | | | |
| 5 MPH | .00 | .11 | .46 | 1.09 | 2.01 | 3.27 | 4.89 | 6.90 | 9.27 | 11.97 | 14.96 | 18.24 | 21.77 |
| 10 MPH | .00 | .22 | .93 | 2.17 | 4.02 | 6.53 | 9.77 | 13.79 | 18.54 | 23.93 | 29.93 | 36.48 | 43.54 |
| 20 MPH | .00 | .45 | 1.86 | 4.35 | 8.04 | 13.06 | 19.54 | 27.58 | 37.08 | 47.86 | 59.86 | 72.96 | 87.08 |
| 30 MPH | .00 | .67 | 2.79 | 6.52 | 12.06 | 19.59 | 29.31 | 41.38 | 55.62 | 71.79 | 89.79 | 109.45 | 130.62 |
| VELOCITY (FPS) | 2200. | 2035. | 1878. | 1729. | 1589. | 1460. | 1343. | 1251. | 1180. | 1119. | 1069. | 1028. | 993. |
| ENERGY (FT-LB) | 462. | 395. | 337. | 285. | 241. | 203. | 172. | 149. | 133. | 120. | 109. | 101. | 94. |
| DROP (IN) | .00 | -.22 | -.97 | -2.34 | -4.43 | -7.38 | -11.36 | -16.54 | -23.11 | -31.24 | -41.09 | -52.85 | -66.65 |
| MID-RANGE (IN) | .00 | .06 | .26 | .64 | 1.24 | 2.12 | 3.34 | 4.99 | 7.13 | 9.83 | 13.17 | 17.19 | 21.97 |
| BULLET PATH (IN) | -.63 | -.05 | .00 | -.56 | -1.85 | -4.00 | -7.18 | -11.56 | -17.33 | -24.66 | -33.71 | -44.66 | -57.66 |
| TIME OF FLIGHT (SEC) | .000000 | .035447 | .073817 | .115452 | .160719 | .209985 | .263576 | .321579 | .383347 | .448662 | .517266 | .588840 | .663126 |
| WIND DEFLECTION (IN) | | | | | | | | | | | | | |
| 5 MPH | .00 | .12 | .50 | 1.16 | 2.14 | 3.48 | 5.19 | 7.30 | 9.73 | 12.48 | 15.52 | 18.82 | 22.36 |
| 10 MPH | .00 | .24 | .99 | 2.32 | 4.29 | 6.96 | 10.39 | 14.60 | 19.47 | 24.96 | 31.04 | 37.64 | 44.71 |
| 20 MPH | .00 | .48 | 1.98 | 4.64 | 8.57 | 13.91 | 20.78 | 29.20 | 38.94 | 49.93 | 62.08 | 75.27 | 89.42 |
| 30 MPH | .00 | .72 | 2.98 | 6.96 | 12.86 | 20.87 | 31.17 | 43.79 | 58.41 | 74.89 | 93.12 | 112.91 | 134.13 |
| VELOCITY (FPS) | 2100. | 1940. | 1787. | 1643. | 1510. | 1388. | 1283. | 1207. | 1142. | 1088. | 1044. | 1006. | 973. |
| ENERGY (FT-LB) | 421. | 359. | 305. | 258. | 218. | 184. | 157. | 139. | 125. | 113. | 104. | 97. | 90. |
| DROP (IN) | .00 | -.25 | -1.07 | -2.57 | -4.88 | -8.15 | -12.54 | -18.25 | -25.46 | -34.32 | -45.02 | -57.70 | -72.52 |
| MID-RANGE (IN) | .00 | .07 | .29 | .71 | 1.37 | 2.34 | 3.70 | 5.51 | 7.85 | 10.78 | 14.36 | 18.66 | 23.72 |
| BULLET PATH (IN) | -.63 | -.03 | .00 | -.65 | -2.11 | -4.53 | -8.07 | -12.94 | -19.29 | -27.31 | -37.15 | -48.98 | -62.96 |
| TIME OF FLIGHT (SEC) | .000000 | .037164 | .077453 | .121226 | .168853 | .220683 | .276983 | .337294 | .401220 | .468551 | .538977 | .612208 | .688032 |
| WIND DEFLECTION (IN) | | | | | | | | | | | | | |
| 5 MPH | .00 | .13 | .53 | 1.24 | 2.29 | 3.71 | 5.52 | 7.68 | 10.16 | 12.95 | 16.00 | 19.30 | 22.83 |
| 10 MPH | .00 | .26 | 1.06 | 2.48 | 4.58 | 7.41 | 11.03 | 15.36 | 20.33 | 25.89 | 32.00 | 38.61 | 45.67 |
| 20 MPH | .00 | .51 | 2.12 | 4.96 | 9.15 | 14.82 | 22.07 | 30.73 | 40.66 | 51.79 | 64.01 | 77.21 | 91.33 |
| 30 MPH | .00 | .77 | 3.18 | 7.44 | 13.73 | 22.23 | 33.10 | 46.09 | 60.99 | 77.68 | 96.01 | 115.82 | 137.00 |
| VELOCITY (FPS) | 2000. | 1844. | 1697. | 1560. | 1433. | 1319. | 1235. | 1166. | 1108. | 1060. | 1020. | 986. | 955. |
| ENERGY (FT-LB) | 382. | 325. | 275. | 232. | 196. | 166. | 146. | 130. | 117. | 107. | 99. | 93. | 87. |
| DROP (IN) | .00 | -.27 | -1.18 | -2.85 | -5.41 | -9.03 | -13.90 | -20.19 | -28.08 | -37.73 | -49.31 | -62.97 | -78.86 |
| MID-RANGE (IN) | .00 | .07 | .32 | .78 | 1.52 | 2.60 | 4.10 | 6.09 | 8.63 | 11.79 | 15.62 | 20.19 | 25.53 |
| BULLET PATH (IN) | -.63 | .00 | .00 | -.76 | -2.41 | -5.13 | -9.09 | -14.48 | -21.46 | -30.20 | -40.88 | -53.63 | -68.62 |
| TIME OF FLIGHT (SEC) | .000000 | .039053 | .081450 | .127560 | .177745 | .232317 | .291173 | .353706 | .419742 | .488999 | .561164 | .635998 | .713324 |
| WIND DEFLECTION (IN) | | | | | | | | | | | | | |
| 5 MPH | .00 | .14 | .57 | 1.33 | 2.44 | 3.94 | 5.82 | 8.03 | 10.54 | 13.33 | 16.38 | 19.67 | 23.17 |
| 10 MPH | .00 | .27 | 1.14 | 2.65 | 4.88 | 7.89 | 11.65 | 16.05 | 21.07 | 26.66 | 32.76 | 39.34 | 46.35 |
| 20 MPH | .00 | .55 | 2.27 | 5.30 | 9.77 | 15.78 | 23.29 | 32.10 | 42.15 | 53.33 | 65.53 | 78.67 | 92.69 |
| 30 MPH | .00 | .82 | 3.41 | 7.95 | 14.65 | 23.66 | 34.94 | 48.16 | 63.22 | 79.99 | 98.29 | 118.01 | 139.04 |

# Bullet: Lyman #225438, 43 Gr.
## Ballistic Coefficient: .120

| RANGE (YDS) | MUZZLE | 25 | 50 | 75 | 100 | 125 | 150 | 175 | 200 | 225 | 250 | 275 | 300 |
|---|---|---|---|---|---|---|---|---|---|---|---|---|---|
| VELOCITY (FPS) | 1900. | 1750. | 1608. | 1478. | 1359. | 1263. | 1190. | 1128. | 1076. | 1034. | 998. | 966. | 938. |
| ENERGY (FT-LB) | 345. | 292. | 247. | 208. | 176. | 152. | 135. | 121. | 111. | 102. | 95. | 89. | 84. |
| DROP (IN) | .00 | -.30 | -1.31 | -3.16 | -6.02 | -10.05 | -15.44 | -22.36 | -30.99 | -41.48 | -54.00 | -68.70 | -85.70 |
| MID-RANGE (IN) | .00 | .08 | .35 | .87 | 1.70 | 2.90 | 4.55 | 6.72 | 9.47 | 12.87 | 16.95 | 21.79 | 27.41 |
| BULLET PATH (IN) | -.63 | .04 | .00 | -.88 | -2.76 | -5.82 | -10.24 | -16.19 | -23.84 | -33.36 | -44.91 | -58.64 | -74.67 |
| TIME OF FLIGHT (SEC) | .000000 | .041140 | .085857 | .134521 | .187468 | .244852 | .306053 | .370832 | .438946 | .510077 | .583955 | .660385 | .739220 |
| WIND DEFLECTION (IN) | | | | | | | | | | | | | |
| 5 MPH | .00 | .15 | .61 | 1.42 | 2.60 | 4.18 | 6.09 | 8.32 | 10.84 | 13.62 | 16.65 | 19.90 | 23.37 |
| 10 MPH | .00 | .29 | 1.22 | 2.83 | 5.20 | 8.36 | 12.18 | 16.63 | 21.68 | 27.25 | 33.30 | 39.81 | 46.73 |
| 20 MPH | .00 | .59 | 2.43 | 5.67 | 10.41 | 16.71 | 24.36 | 33.27 | 43.35 | 54.49 | 66.60 | 79.61 | 93.47 |
| 30 MPH | .00 | .88 | 3.65 | 8.50 | 15.61 | 25.07 | 36.54 | 49.90 | 65.03 | 81.74 | 99.91 | 119.42 | 140.20 |
| VELOCITY (FPS) | 1800. | 1656. | 1521. | 1398. | 1291. | 1214. | 1148. | 1093. | 1048. | 1009. | 976. | 947. | 920. |
| ENERGY (FT-LB) | 309. | 262. | 221. | 187. | 159. | 141. | 126. | 114. | 105. | 97. | 91. | 86. | 81. |
| DROP (IN) | .00 | -.34 | -1.47 | -3.54 | -6.73 | -11.22 | -17.18 | -24.79 | -34.21 | -45.60 | -59.12 | -74.91 | -93.08 |
| MID-RANGE (IN) | .00 | .09 | .40 | .98 | 1.90 | 3.23 | 5.05 | 7.41 | 10.37 | 14.00 | 18.34 | 23.45 | 29.36 |
| BULLET PATH (IN) | -.63 | .08 | .00 | -1.03 | -3.17 | -6.60 | -11.52 | -18.08 | -26.45 | -36.80 | -49.27 | -64.01 | -81.13 |
| TIME OF FLIGHT (SEC) | .000000 | .043454 | .090729 | .142179 | .198086 | .258040 | .321621 | .388634 | .458773 | .531741 | .607320 | .685350 | .765712 |
| WIND DEFLECTION (IN) | | | | | | | | | | | | | |
| 5 MPH | .00 | .16 | .65 | 1.51 | 2.76 | 4.37 | 6.30 | 8.53 | 11.04 | 13.79 | 16.78 | 19.98 | 23.38 |
| 10 MPH | .00 | .31 | 1.30 | 3.02 | 5.53 | 8.75 | 12.61 | 17.07 | 22.08 | 27.59 | 33.56 | 39.95 | 46.77 |
| 20 MPH | .00 | .63 | 2.60 | 6.05 | 11.06 | 17.50 | 25.21 | 34.13 | 44.15 | 55.17 | 67.11 | 79.91 | 93.53 |
| 30 MPH | .00 | .94 | 3.90 | 9.07 | 16.59 | 26.24 | 37.82 | 51.20 | 66.23 | 82.76 | 100.67 | 119.86 | 140.30 |
| VELOCITY (FPS) | 1700. | 1562. | 1436. | 1321. | 1236. | 1167. | 1108. | 1061. | 1020. | 986. | 956. | 928. | 903. |
| ENERGY (FT-LB) | 276. | 233. | 197. | 167. | 146. | 130. | 117. | 107. | 99. | 93. | 87. | 82. | 78. |
| DROP (IN) | .00 | -.38 | -1.65 | -3.98 | -7.55 | -12.55 | -19.14 | -27.49 | -37.77 | -50.12 | -64.71 | -81.63 | -101.06 |
| MID-RANGE (IN) | .00 | .10 | .45 | 1.10 | 2.12 | 3.60 | 5.59 | 8.14 | 11.33 | 15.20 | 19.80 | 25.18 | 31.39 |
| BULLET PATH (IN) | -.63 | .13 | .00 | -1.19 | -3.62 | -7.48 | -12.92 | -20.13 | -29.27 | -40.48 | -53.93 | -69.71 | -88.00 |
| TIME OF FLIGHT (SEC) | .000000 | .046032 | .096133 | .150616 | .209416 | .271904 | .337897 | .407116 | .479247 | .554048 | .631344 | .711004 | .792933 |
| WIND DEFLECTION (IN) | | | | | | | | | | | | | |
| 5 MPH | .00 | .17 | .69 | 1.61 | 2.90 | 4.52 | 6.44 | 8.65 | 11.11 | 13.82 | 16.73 | 19.86 | 23.19 |
| 10 MPH | .00 | .34 | 1.39 | 3.21 | 5.80 | 9.03 | 12.88 | 17.30 | 22.23 | 27.63 | 33.47 | 39.72 | 46.38 |
| 20 MPH | .00 | .67 | 2.78 | 6.43 | 11.60 | 18.06 | 25.76 | 34.60 | 44.46 | 55.26 | 66.94 | 79.45 | 92.76 |
| 30 MPH | .00 | 1.01 | 4.17 | 9.64 | 17.39 | 27.09 | 38.65 | 51.90 | 66.69 | 82.89 | 100.41 | 119.17 | 139.14 |
| VELOCITY (FPS) | 1600. | 1470. | 1352. | 1258. | 1186. | 1124. | 1074. | 1031. | 996. | 964. | 936. | 910. | 887. |
| ENERGY (FT-LB) | 244. | 206. | 175. | 151. | 134. | 121. | 110. | 102. | 95. | 89. | 84. | 79. | 75. |
| DROP (IN) | .00 | -.43 | -1.87 | -4.50 | -8.50 | -14.05 | -21.31 | -30.45 | -41.62 | -54.98 | -70.65 | -88.77 | -109.50 |
| MID-RANGE (IN) | .00 | .12 | .51 | 1.23 | 2.38 | 4.00 | 6.15 | 8.90 | 12.31 | 16.42 | 21.28 | 26.93 | 33.44 |
| BULLET PATH (IN) | -.63 | .19 | .00 | -1.38 | -4.13 | -8.42 | -14.43 | -22.31 | -32.23 | -44.34 | -58.76 | -75.63 | -95.11 |
| TIME OF FLIGHT (SEC) | .000000 | .048920 | .102141 | .159785 | .221218 | .286216 | .354530 | .425842 | .499887 | .576471 | .655454 | .736731 | .820226 |
| WIND DEFLECTION (IN) | | | | | | | | | | | | | |
| 5 MPH | .00 | .18 | .74 | 1.69 | 2.97 | 4.56 | 6.45 | 8.60 | 10.99 | 13.60 | 16.43 | 19.46 | 22.68 |
| 10 MPH | .00 | .36 | 1.48 | 3.37 | 5.93 | 9.12 | 12.90 | 17.20 | 21.98 | 27.21 | 32.86 | 38.91 | 45.36 |
| 20 MPH | .00 | .72 | 2.95 | 6.74 | 11.87 | 18.25 | 25.79 | 34.40 | 43.96 | 54.42 | 65.72 | 77.83 | 90.72 |
| 30 MPH | .00 | 1.08 | 4.43 | 10.12 | 17.80 | 27.37 | 38.69 | 51.59 | 65.94 | 81.63 | 98.58 | 116.74 | 136.08 |

# Bullet: Lyman #225438, 43 Gr.
## Ballistic Coefficient: .120

| RANGE (YDS) | MUZZLE | 25 | 50 | 75 | 100 | 125 | 150 | 175 | 200 | 225 | 250 | 275 | 300 |
|---|---|---|---|---|---|---|---|---|---|---|---|---|---|
| VELOCITY (FPS) | 1500. | 1379. | 1277. | 1202. | 1138. | 1085. | 1041. | 1004. | 971. | 942. | 916. | 892. | 870. |
| ENERGY (FT-LB) | 215. | 182. | 156. | 138. | 124. | 112. | 103. | 96. | 90. | 85. | 80. | 76. | 72. |
| DROP (IN) | .00 | -.50 | -2.14 | -5.10 | -9.58 | -15.73 | -23.71 | -33.70 | -45.84 | -60.26 | -77.09 | -96.51 | -118.64 |
| MID-RANGE (IN) | .00 | .13 | .57 | 1.39 | 2.65 | 4.42 | 6.75 | 9.71 | 13.34 | 17.70 | 22.82 | 28.77 | 35.61 |
| BULLET PATH (IN) | -.63 | .26 | .00 | -1.59 | -4.68 | -9.44 | -16.05 | -24.65 | -35.41 | -48.45 | -63.89 | -81.92 | -102.67 |
| TIME OF FLIGHT (SEC) | .000000 | .052167 | .108803 | .169377 | .233555 | .301119 | .371753 | .445175 | .521177 | .599607 | .680353 | .763332 | .848487 |
| WIND DEFLECTION (IN) | | | | | | | | | | | | | |
| 5 MPH | .00 | .19 | .77 | 1.71 | 2.95 | 4.50 | 6.31 | 8.38 | 10.66 | 13.17 | 15.87 | 18.77 | 21.87 |
| 10 MPH | .00 | .38 | 1.55 | 3.41 | 5.91 | 9.00 | 12.63 | 16.75 | 21.33 | 26.33 | 31.74 | 37.55 | 43.73 |
| 20 MPH | .00 | .76 | 3.10 | 6.82 | 11.81 | 17.99 | 25.26 | 33.50 | 42.65 | 52.66 | 63.48 | 75.09 | 87.47 |
| 30 MPH | .00 | 1.14 | 4.65 | 10.23 | 17.72 | 26.99 | 37.89 | 50.25 | 63.98 | 78.99 | 95.23 | 112.64 | 131.20 |
| VELOCITY (FPS) | 1400. | 1291. | 1214. | 1148. | 1093. | 1048. | 1010. | 977. | 947. | 921. | 896. | 874. | 853. |
| ENERGY (FT-LB) | 187. | 159. | 141. | 126. | 114. | 105. | 97. | 91. | 86. | 81. | 77. | 73. | 69. |
| DROP (IN) | .00 | -.57 | -2.44 | -5.78 | -10.77 | -17.56 | -26.33 | -37.23 | -50.39 | -65.93 | -84.03 | -104.82 | -128.44 |
| MID-RANGE (IN) | .00 | .15 | .65 | 1.56 | 2.95 | 4.87 | 7.39 | 10.56 | 14.43 | 19.04 | 24.45 | 30.71 | 37.88 |
| BULLET PATH (IN) | -.63 | .33 | .00 | -1.81 | -5.26 | -10.53 | -17.76 | -27.12 | -38.75 | -52.76 | -69.32 | -88.57 | -110.66 |
| TIME OF FLIGHT (SEC) | .000000 | .055838 | .115765 | .179321 | .246310 | .316426 | .389374 | .464934 | .542945 | .623289 | .705878 | .790653 | .877570 |
| WIND DEFLECTION (IN) | | | | | | | | | | | | | |
| 5 MPH | .00 | .20 | .76 | 1.64 | 2.82 | 4.27 | 5.98 | 7.91 | 10.06 | 12.42 | 14.97 | 17.72 | 20.65 |
| 10 MPH | .00 | .40 | 1.52 | 3.27 | 5.64 | 8.55 | 11.96 | 15.83 | 20.13 | 24.84 | 29.95 | 35.44 | 41.31 |
| 20 MPH | .00 | .80 | 3.04 | 6.55 | 11.27 | 17.10 | 23.92 | 31.66 | 40.26 | 49.68 | 59.90 | 70.88 | 82.62 |
| 30 MPH | .00 | 1.20 | 4.55 | 9.82 | 16.91 | 25.64 | 35.88 | 47.49 | 60.39 | 74.52 | 89.85 | 106.32 | 123.93 |

# Bullet: Lyman #225415, 46 Gr.
## Ballistic Coefficient: .162

| RANGE (YDS) | MUZZLE | 25 | 50 | 75 | 100 | 125 | 150 | 175 | 200 | 225 | 250 | 275 | 300 |
|---|---|---|---|---|---|---|---|---|---|---|---|---|---|
| VELOCITY (FPS) | 2500. | 2368. | 2240. | 2116. | 1996. | 1880. | 1768. | 1662. | 1561. | 1466. | 1378. | 1297. | 1240. |
| ENERGY (FT-LB) | 638. | 573. | 512. | 457. | 407. | 361. | 319. | 282. | 249. | 219. | 194. | 172. | 157. |
| DROP (IN) | .00 | -.17 | -.72 | -1.71 | -3.18 | -5.19 | -7.82 | -11.14 | -15.25 | -20.25 | -26.27 | -33.43 | -41.87 |
| MID-RANGE (IN) | .00 | .05 | .19 | .46 | .87 | 1.44 | 2.20 | 3.19 | 4.45 | 6.01 | 7.94 | 10.29 | 13.11 |
| BULLET PATH (IN) | -.63 | -.12 | .00 | -.31 | -1.11 | -2.44 | -4.40 | -7.04 | -10.47 | -14.80 | -20.14 | -26.62 | -34.39 |
| TIME OF FLIGHT (SEC) | .000000 | .030826 | .063392 | .097844 | .134345 | .173070 | .214210 | .257964 | .304537 | .354128 | .406915 | .463040 | .522194 |
| WIND DEFLECTION (IN) | | | | | | | | | | | | | |
| 5 MPH | .00 | .07 | .30 | .69 | 1.26 | 2.03 | 3.01 | 4.22 | 5.68 | 7.40 | 9.41 | 11.71 | 14.27 |
| 10 MPH | .00 | .15 | .60 | 1.38 | 2.52 | 4.06 | 6.02 | 8.44 | 11.36 | 14.81 | 18.82 | 23.42 | 28.55 |
| 20 MPH | .00 | .29 | 1.19 | 2.76 | 5.05 | 8.12 | 12.04 | 16.88 | 22.72 | 29.61 | 37.63 | 46.83 | 57.09 |
| 30 MPH | .00 | .44 | 1.79 | 4.14 | 7.57 | 12.18 | 18.06 | 25.33 | 34.08 | 44.42 | 56.45 | 70.25 | 85.64 |
| VELOCITY (FPS) | 2400. | 2271. | 2146. | 2025. | 1908. | 1795. | 1688. | 1585. | 1489. | 1399. | 1316. | 1253. | 1200. |
| ENERGY (FT-LB) | 588. | 527. | 470. | 419. | 372. | 329. | 291. | 257. | 226. | 200. | 177. | 160. | 147. |
| DROP (IN) | .00 | -.18 | -.78 | -1.86 | -3.46 | -5.66 | -8.53 | -12.16 | -16.65 | -22.14 | -28.73 | -36.57 | -45.80 |
| MID-RANGE (IN) | .00 | .05 | .21 | .50 | .95 | 1.57 | 2.41 | 3.49 | 4.87 | 6.59 | 8.71 | 11.28 | 14.37 |
| BULLET PATH (IN) | -.63 | -.10 | .00 | -.37 | -1.26 | -2.75 | -4.91 | -7.84 | -11.62 | -16.40 | -22.29 | -29.42 | -37.95 |
| TIME OF FLIGHT (SEC) | .000000 | .032126 | .066100 | .102082 | .140244 | .180772 | .223865 | .269727 | .318559 | .370546 | .425841 | .484335 | .545511 |
| WIND DEFLECTION (IN) | | | | | | | | | | | | | |
| 5 MPH | .00 | .08 | .32 | .73 | 1.34 | 2.16 | 3.20 | 4.49 | 6.03 | 7.86 | 9.97 | 12.37 | 15.00 |
| 10 MPH | .00 | .15 | .63 | 1.47 | 2.68 | 4.32 | 6.40 | 8.97 | 12.07 | 15.72 | 19.95 | 24.74 | 30.01 |
| 20 MPH | .00 | .31 | 1.27 | 2.93 | 5.37 | 8.63 | 12.80 | 17.94 | 24.13 | 31.43 | 39.90 | 49.49 | 60.02 |
| 30 MPH | .00 | .46 | 1.90 | 4.40 | 8.05 | 12.95 | 19.20 | 26.92 | 36.20 | 47.15 | 59.84 | 74.23 | 90.03 |
| VELOCITY (FPS) | 2300. | 2174. | 2052. | 1934. | 1820. | 1712. | 1608. | 1510. | 1419. | 1334. | 1266. | 1212. | 1164. |
| ENERGY (FT-LB) | 540. | 483. | 430. | 382. | 338. | 299. | 264. | 233. | 206. | 182. | 164. | 150. | 138. |
| DROP (IN) | .00 | -.20 | -.86 | -2.03 | -3.78 | -6.19 | -9.33 | -13.31 | -18.26 | -24.29 | -31.53 | -40.12 | -50.20 |
| MID-RANGE (IN) | .00 | .05 | .23 | .55 | 1.03 | 1.72 | 2.64 | 3.83 | 5.35 | 7.25 | 9.58 | 12.40 | 15.77 |
| BULLET PATH (IN) | -.63 | -.09 | .00 | -.43 | -1.44 | -3.10 | -5.50 | -8.74 | -12.95 | -18.23 | -24.73 | -32.58 | -41.92 |
| TIME OF FLIGHT (SEC) | .000000 | .033541 | .069051 | .106701 | .146674 | .189166 | .234380 | .282519 | .333773 | .388305 | .446117 | .506670 | .569842 |
| WIND DEFLECTION (IN) | | | | | | | | | | | | | |
| 5 MPH | .00 | .08 | .34 | .78 | 1.43 | 2.30 | 3.41 | 4.77 | 6.42 | 8.34 | 10.56 | 13.02 | 15.71 |
| 10 MPH | .00 | .16 | .67 | 1.56 | 2.86 | 4.60 | 6.82 | 9.55 | 12.83 | 16.69 | 21.13 | 26.04 | 31.42 |
| 20 MPH | .00 | .33 | 1.35 | 3.12 | 5.72 | 9.19 | 13.63 | 19.10 | 25.66 | 33.38 | 42.25 | 52.09 | 62.85 |
| 30 MPH | .00 | .49 | 2.02 | 4.69 | 8.57 | 13.79 | 20.45 | 28.65 | 38.49 | 50.07 | 63.38 | 78.13 | 94.27 |
| VELOCITY (FPS) | 2200. | 2077. | 1958. | 1844. | 1734. | 1629. | 1530. | 1437. | 1351. | 1279. | 1223. | 1174. | 1130. |
| ENERGY (FT-LB) | 494. | 441. | 392. | 347. | 307. | 271. | 239. | 211. | 186. | 167. | 153. | 141. | 130. |
| DROP (IN) | .00 | -.22 | -.94 | -2.22 | -4.15 | -6.79 | -10.25 | -14.64 | -20.09 | -26.73 | -34.69 | -44.11 | -55.11 |
| MID-RANGE (IN) | .00 | .06 | .25 | .60 | 1.14 | 1.89 | 2.90 | 4.22 | 5.90 | 7.99 | 10.56 | 13.65 | 17.30 |
| BULLET PATH (IN) | -.63 | -.06 | .00 | -.50 | -1.64 | -3.50 | -6.18 | -9.79 | -14.45 | -20.30 | -27.48 | -36.12 | -46.33 |
| TIME OF FLIGHT (SEC) | .000000 | .035086 | .072274 | .111746 | .153695 | .198324 | .245836 | .296425 | .350261 | .407425 | .467416 | .530039 | .595211 |
| WIND DEFLECTION (IN) | | | | | | | | | | | | | |
| 5 MPH | .00 | .09 | .36 | .83 | 1.53 | 2.45 | 3.63 | 5.09 | 6.82 | 8.85 | 11.13 | 13.64 | 16.38 |
| 10 MPH | .00 | .18 | .72 | 1.67 | 3.05 | 4.90 | 7.27 | 10.17 | 13.65 | 17.71 | 22.27 | 27.29 | 32.76 |
| 20 MPH | .00 | .35 | 1.44 | 3.33 | 6.10 | 9.81 | 14.53 | 20.34 | 27.29 | 35.41 | 44.53 | 54.57 | 65.51 |
| 30 MPH | .00 | .53 | 2.16 | 5.00 | 9.15 | 14.71 | 21.80 | 30.51 | 40.94 | 53.12 | 66.80 | 81.86 | 98.27 |

## Bullet: Lyman #225415, 46 Gr.
**Ballistic Coefficient: .162**

| RANGE (YDS) | MUZZLE | 25 | 50 | 75 | 100 | 125 | 150 | 175 | 200 | 225 | 250 | 275 | 300 |
|---|---|---|---|---|---|---|---|---|---|---|---|---|---|
| VELOCITY (FPS) | 2100. | 1980. | 1865. | 1754. | 1649. | 1548. | 1454. | 1367. | 1289. | 1233. | 1182. | 1137. | 1098. |
| ENERGY (FT-LB) | 450. | 401. | 355. | 314. | 278. | 245. | 216. | 191. | 170. | 155. | 143. | 132. | 123. |
| DROP (IN) | .00 | -.24 | -1.03 | -2.45 | -4.56 | -7.48 | -11.31 | -16.17 | -22.19 | -29.51 | -38.26 | -48.57 | -60.55 |
| MID-RANGE (IN) | .00 | .07 | .28 | .66 | 1.25 | 2.08 | 3.21 | 4.67 | 6.53 | 8.84 | 11.65 | 15.01 | 18.97 |
| BULLET PATH (IN) | -.63 | -.04 | .00 | -.58 | -1.87 | -3.96 | -6.95 | -10.98 | -16.17 | -22.66 | -30.59 | -40.06 | -51.21 |
| TIME OF FLIGHT (SEC) | .000000 | .036778 | .075804 | .117270 | .161377 | .208329 | .258323 | .311535 | .368091 | .427587 | .489724 | .554429 | .621577 |
| WIND DEFLECTION (IN) | | | | | | | | | | | | | |
| 5 MPH | .00 | .09 | .39 | .89 | 1.63 | 2.62 | 3.88 | 5.42 | 7.25 | 9.34 | 11.67 | 14.22 | 16.98 |
| 10 MPH | .00 | .19 | .77 | 1.78 | 3.26 | 5.24 | 7.75 | 10.83 | 14.50 | 18.68 | 23.33 | 28.44 | 33.97 |
| 20 MPH | .00 | .37 | 1.54 | 3.56 | 6.52 | 10.47 | 15.50 | 21.66 | 29.00 | 37.37 | 46.67 | 56.87 | 67.94 |
| 30 MPH | .00 | .56 | 2.31 | 5.35 | 9.78 | 15.71 | 23.25 | 32.49 | 43.49 | 56.05 | 70.00 | 85.31 | 101.91 |
| VELOCITY (FPS) | 2000. | 1884. | 1772. | 1666. | 1565. | 1469. | 1381. | 1300. | 1243. | 1191. | 1145. | 1104. | 1069. |
| ENERGY (FT-LB) | 408. | 362. | 321. | 283. | 250. | 220. | 195. | 173. | 158. | 145. | 134. | 125. | 117. |
| DROP (IN) | .00 | -.26 | -1.14 | -2.70 | -5.05 | -8.28 | -12.53 | -17.92 | -24.58 | -32.66 | -42.26 | -53.52 | -66.56 |
| MID-RANGE (IN) | .00 | .07 | .31 | .73 | 1.39 | 2.31 | 3.56 | 5.19 | 7.24 | 9.78 | 12.85 | 16.49 | 20.75 |
| BULLET PATH (IN) | -.63 | -.01 | .00 | -.68 | -2.14 | -4.49 | -7.86 | -12.36 | -18.13 | -25.33 | -34.05 | -44.42 | -56.57 |
| TIME OF FLIGHT (SEC) | .000000 | .038639 | .079686 | .123340 | .169805 | .219281 | .271947 | .327946 | .386977 | .448655 | .512918 | .579650 | .648690 |
| WIND DEFLECTION (IN) | | | | | | | | | | | | | |
| 5 MPH | .00 | .10 | .41 | .95 | 1.74 | 2.80 | 4.13 | 5.76 | 7.65 | 9.78 | 12.14 | 14.71 | 17.48 |
| 10 MPH | .00 | .20 | .82 | 1.91 | 3.49 | 5.59 | 8.26 | 11.52 | 15.31 | 19.56 | 24.27 | 29.42 | 34.97 |
| 20 MPH | .00 | .40 | 1.65 | 3.82 | 6.97 | 11.19 | 16.53 | 23.04 | 30.62 | 39.13 | 48.55 | 58.84 | 69.94 |
| 30 MPH | .00 | .60 | 2.47 | 5.72 | 10.46 | 16.78 | 24.79 | 34.56 | 45.92 | 58.69 | 72.82 | 88.26 | 104.91 |
| VELOCITY (FPS) | 1900. | 1788. | 1680. | 1578. | 1482. | 1393. | 1311. | 1249. | 1197. | 1150. | 1109. | 1074. | 1042. |
| ENERGY (FT-LB) | 369. | 326. | 288. | 254. | 224. | 198. | 175. | 159. | 146. | 135. | 126. | 118. | 111. |
| DROP (IN) | .00 | -.29 | -1.26 | -3.00 | -5.61 | -9.22 | -13.95 | -19.93 | -27.31 | -36.21 | -46.75 | -59.04 | -73.22 |
| MID-RANGE (IN) | .00 | .08 | .34 | .81 | 1.54 | 2.58 | 3.97 | 5.77 | 8.05 | 10.82 | 14.15 | 18.08 | 22.66 |
| BULLET PATH (IN) | -.63 | .02 | .00 | -.79 | -2.45 | -5.12 | -8.90 | -13.93 | -20.37 | -28.32 | -37.91 | -49.26 | -62.49 |
| TIME OF FLIGHT (SEC) | .000000 | .040696 | .083971 | .130029 | .179069 | .231277 | .286801 | .345488 | .406843 | .470793 | .537228 | .605994 | .676918 |
| WIND DEFLECTION (IN) | | | | | | | | | | | | | |
| 5 MPH | .00 | .11 | .44 | 1.02 | 1.86 | 2.98 | 4.40 | 6.09 | 8.01 | 10.17 | 12.54 | 15.12 | 17.88 |
| 10 MPH | .00 | .22 | .88 | 2.04 | 3.73 | 5.97 | 8.79 | 12.17 | 16.03 | 20.33 | 25.08 | 30.23 | 35.77 |
| 20 MPH | .00 | .43 | 1.77 | 4.09 | 7.45 | 11.94 | 17.59 | 24.35 | 32.05 | 40.67 | 50.16 | 60.47 | 71.54 |
| 30 MPH | .00 | .65 | 2.65 | 6.13 | 11.18 | 17.90 | 26.38 | 36.52 | 48.08 | 61.00 | 75.24 | 90.70 | 107.31 |
| VELOCITY (FPS) | 1800. | 1692. | 1589. | 1493. | 1402. | 1320. | 1256. | 1203. | 1156. | 1114. | 1078. | 1046. | 1018. |
| ENERGY (FT-LB) | 331. | 292. | 258. | 228. | 201. | 178. | 161. | 148. | 136. | 127. | 119. | 112. | 106. |
| DROP (IN) | .00 | -.33 | -1.41 | -3.35 | -6.28 | -10.31 | -15.58 | -22.24 | -30.39 | -40.17 | -51.69 | -65.08 | -80.45 |
| MID-RANGE (IN) | .00 | .09 | .38 | .91 | 1.73 | 2.89 | 4.44 | 6.43 | 8.92 | 11.94 | 15.53 | 19.76 | 24.64 |
| BULLET PATH (IN) | -.63 | .06 | .00 | -.92 | -2.83 | -5.84 | -10.09 | -15.73 | -22.86 | -31.62 | -42.12 | -54.49 | -68.84 |
| TIME OF FLIGHT (SEC) | .000000 | .042979 | .088717 | .137417 | .189264 | .244414 | .302760 | .363788 | .427422 | .493558 | .562044 | .632710 | .705407 |
| WIND DEFLECTION (IN) | | | | | | | | | | | | | |
| 5 MPH | .00 | .12 | .47 | 1.09 | 1.99 | 3.18 | 4.64 | 6.35 | 8.28 | 10.43 | 12.79 | 15.35 | 18.08 |
| 10 MPH | .00 | .23 | .95 | 2.19 | 3.98 | 6.35 | 9.29 | 12.69 | 16.56 | 20.87 | 25.59 | 30.69 | 36.15 |
| 20 MPH | .00 | .46 | 1.89 | 4.37 | 7.95 | 12.70 | 18.57 | 25.39 | 33.12 | 41.73 | 51.17 | 61.38 | 72.30 |
| 30 MPH | .00 | .69 | 2.84 | 6.56 | 11.93 | 19.05 | 27.86 | 38.08 | 49.68 | 62.60 | 76.76 | 92.07 | 108.45 |

# Bullet: Lyman #225415, 46 Gr.
## Ballistic Coefficient: .162

| RANGE (YDS) | MUZZLE | 25 | 50 | 75 | 100 | 125 | 150 | 175 | 200 | 225 | 250 | 275 | 300 |
|---|---|---|---|---|---|---|---|---|---|---|---|---|---|
| VELOCITY (FPS) | 1700. | 1597. | 1500. | 1409. | 1325. | 1261. | 1207. | 1159. | 1117. | 1080. | 1048. | 1020. | 995. |
| ENERGY (FT-LB) | 295. | 260. | 230. | 203. | 179. | 162. | 149. | 137. | 127. | 119. | 112. | 106. | 101. |
| DROP (IN) | .00 | -.37 | -1.58 | -3.78 | -7.07 | -11.58 | -17.47 | -24.85 | -33.85 | -44.58 | -57.16 | -71.72 | -88.36 |
| MID-RANGE (IN) | .00 | .10 | .42 | 1.03 | 1.95 | 3.24 | 4.96 | 7.15 | 9.86 | 13.13 | 17.00 | 21.51 | 26.71 |
| BULLET PATH (IN) | -.63 | .11 | .00 | -1.09 | -3.27 | -6.68 | -11.46 | -17.74 | -25.62 | -35.25 | -46.73 | -60.18 | -75.71 |
| TIME OF FLIGHT (SEC) | .000000 | .045524 | .093995 | .145600 | .200498 | .258622 | .319446 | .382881 | .448827 | .517137 | .587639 | .660183 | .734651 |
| WIND DEFLECTION (IN) | | | | | | | | | | | | | |
| 5 MPH | .00 | .12 | .51 | 1.17 | 2.11 | 3.35 | 4.82 | 6.52 | 8.44 | 10.57 | 12.89 | 15.39 | 18.06 |
| 10 MPH | .00 | .25 | 1.01 | 2.33 | 4.23 | 6.69 | 9.63 | 13.03 | 16.88 | 21.13 | 25.78 | 30.78 | 36.12 |
| 20 MPH | .00 | .50 | 2.03 | 4.66 | 8.46 | 13.39 | 19.27 | 26.07 | 33.75 | 42.27 | 51.55 | 61.56 | 72.24 |
| 30 MPH | .00 | .74 | 3.04 | 6.99 | 12.69 | 20.08 | 28.90 | 39.10 | 50.63 | 63.40 | 77.33 | 92.34 | 108.37 |
| VELOCITY (FPS) | 1600. | 1503. | 1412. | 1328. | 1262. | 1208. | 1160. | 1118. | 1081. | 1049. | 1021. | 995. | 972. |
| ENERGY (FT-LB) | 261. | 231. | 204. | 180. | 163. | 149. | 137. | 128. | 119. | 112. | 106. | 101. | 97. |
| DROP (IN) | .00 | -.41 | -1.80 | -4.28 | -7.98 | -13.05 | -19.62 | -27.79 | -37.70 | -49.46 | -63.19 | -78.99 | -96.99 |
| MID-RANGE (IN) | .00 | .11 | .49 | 1.16 | 2.19 | 3.63 | 5.53 | 7.92 | 10.87 | 14.39 | 18.54 | 23.36 | 28.88 |
| BULLET PATH (IN) | -.63 | .17 | .00 | -1.26 | -3.75 | -7.61 | -12.95 | -19.91 | -28.60 | -39.14 | -51.66 | -66.25 | -83.03 |
| TIME OF FLIGHT (SEC) | .000000 | .048379 | .099886 | .154682 | .212736 | .273507 | .336891 | .402789 | .471054 | .541514 | .614018 | .688448 | .764708 |
| WIND DEFLECTION (IN) | | | | | | | | | | | | | |
| 5 MPH | .00 | .13 | .54 | 1.24 | 2.22 | 3.44 | 4.90 | 6.57 | 8.45 | 10.53 | 12.78 | 15.21 | 17.79 |
| 10 MPH | .00 | .26 | 1.08 | 2.47 | 4.44 | 6.89 | 9.79 | 13.14 | 16.91 | 21.06 | 25.57 | 30.42 | 35.59 |
| 20 MPH | .00 | .53 | 2.16 | 4.95 | 8.88 | 13.77 | 19.59 | 26.28 | 33.81 | 42.11 | 51.13 | 60.83 | 71.18 |
| 30 MPH | .00 | .79 | 3.24 | 7.42 | 13.32 | 20.66 | 29.38 | 39.42 | 50.72 | 63.17 | 76.70 | 91.25 | 106.77 |
| VELOCITY (FPS) | 1500. | 1409. | 1326. | 1261. | 1207. | 1159. | 1117. | 1080. | 1048. | 1020. | 995. | 972. | 951. |
| ENERGY (FT-LB) | 230. | 203. | 179. | 162. | 149. | 137. | 127. | 119. | 112. | 106. | 101. | 96. | 92. |
| DROP (IN) | .00 | -.49 | -2.06 | -4.87 | -9.05 | -14.72 | -22.01 | -31.03 | -41.90 | -54.75 | -69.68 | -86.79 | -106.20 |
| MID-RANGE (IN) | .00 | .13 | .55 | 1.31 | 2.46 | 4.05 | 6.13 | 8.73 | 11.90 | 15.68 | 20.12 | 25.24 | 31.09 |
| BULLET PATH (IN) | -.63 | .23 | .00 | -1.46 | -4.29 | -8.62 | -14.55 | -22.23 | -31.75 | -43.25 | -56.83 | -72.60 | -90.66 |
| TIME OF FLIGHT (SEC) | .000000 | .051596 | .106485 | .164600 | .225414 | .288840 | .354777 | .423078 | .493572 | .566109 | .640568 | .716857 | .794898 |
| WIND DEFLECTION (IN) | | | | | | | | | | | | | |
| 5 MPH | .00 | .14 | .57 | 1.28 | 2.24 | 3.42 | 4.82 | 6.43 | 8.23 | 10.22 | 12.37 | 14.68 | 17.15 |
| 10 MPH | .00 | .28 | 1.14 | 2.57 | 4.47 | 6.84 | 9.64 | 12.86 | 16.47 | 20.44 | 24.74 | 29.37 | 34.30 |
| 20 MPH | .00 | .56 | 2.28 | 5.14 | 8.95 | 13.67 | 19.28 | 25.72 | 32.94 | 40.87 | 49.48 | 58.73 | 68.60 |
| 30 MPH | .00 | .84 | 3.42 | 7.71 | 13.42 | 20.51 | 28.92 | 38.59 | 49.41 | 61.31 | 74.22 | 88.10 | 102.91 |
| VELOCITY (FPS) | 1400. | 1317. | 1254. | 1201. | 1154. | 1112. | 1076. | 1045. | 1017. | 992. | 969. | 948. | 929. |
| ENERGY (FT-LB) | 200. | 177. | 161. | 147. | 136. | 126. | 118. | 112. | 106. | 101. | 96. | 92. | 88. |
| DROP (IN) | .00 | -.56 | -2.36 | -5.55 | -10.25 | -16.57 | -24.64 | -34.58 | -46.51 | -60.53 | -76.75 | -95.28 | -116.16 |
| MID-RANGE (IN) | .00 | .15 | .62 | 1.48 | 2.76 | 4.51 | 6.77 | 9.59 | 13.01 | 17.07 | 21.81 | 27.26 | 33.44 |
| BULLET PATH (IN) | -.63 | .31 | .00 | -1.70 | -4.90 | -9.72 | -16.30 | -24.75 | -35.18 | -47.70 | -62.43 | -79.46 | -98.85 |
| TIME OF FLIGHT (SEC) | .000000 | .055246 | .113691 | .174817 | .238545 | .304769 | .373338 | .444080 | .516848 | .591528 | .668026 | .746269 | .826197 |
| WIND DEFLECTION (IN) | | | | | | | | | | | | | |
| 5 MPH | .00 | .15 | .58 | 1.24 | 2.13 | 3.25 | 4.57 | 6.08 | 7.77 | 9.63 | 11.64 | 13.81 | 16.13 |
| 10 MPH | .00 | .29 | 1.15 | 2.48 | 4.27 | 6.50 | 9.14 | 12.16 | 15.54 | 19.25 | 23.29 | 27.63 | 32.27 |
| 20 MPH | .00 | .59 | 2.30 | 4.96 | 8.54 | 12.99 | 18.27 | 24.32 | 31.07 | 38.50 | 46.57 | 55.26 | 64.54 |
| 30 MPH | .00 | .88 | 3.46 | 7.45 | 12.81 | 19.49 | 27.41 | 36.47 | 46.61 | 57.76 | 69.86 | 82.89 | 96.80 |

## Bullet: Hornady .257 Dia. 60 Gr. S.P.
## Ballistic Coefficient: .095

| RANGE (YDS) | MUZZLE | 25 | 50 | 75 | 100 | 125 | 150 | 175 | 200 | 225 | 250 | 275 | 300 |
|---|---|---|---|---|---|---|---|---|---|---|---|---|---|
| VELOCITY (FPS) | 2400. | 2182. | 1976. | 1783. | 1603. | 1440. | 1298. | 1206. | 1130. | 1069. | 1020. | 979. | 944. |
| ENERGY (FT-LB) | 767. | 634. | 520. | 423. | 342. | 276. | 224. | 194. | 170. | 152. | 139. | 128. | 119. |
| DROP (IN) | .00 | -.19 | -.84 | -2.05 | -3.94 | -6.67 | -10.46 | -15.53 | -22.10 | -30.36 | -40.51 | -52.75 | -67.25 |
| MID-RANGE (IN) | .00 | .05 | .23 | .57 | 1.13 | 1.97 | 3.19 | 4.87 | 7.11 | 9.99 | 13.58 | 17.95 | 23.16 |
| BULLET PATH (IN) | -.63 | -.09 | .00 | -.47 | -1.62 | -3.63 | -6.68 | -11.01 | -16.84 | -24.37 | -33.79 | -45.29 | -59.05 |
| TIME OF FLIGHT (SEC) | .000000 | .032772 | .068889 | .108855 | .153236 | .202618 | .257514 | .317503 | .381822 | .450150 | .522040 | .597134 | .675185 |
| WIND DEFLECTION (IN) | | | | | | | | | | | | | |
| 5 MPH | .00 | .13 | .56 | 1.33 | 2.48 | 4.08 | 6.16 | 8.69 | 11.60 | 14.86 | 18.44 | 22.30 | 26.42 |
| 10 MPH | .00 | .27 | 1.12 | 2.66 | 4.97 | 8.16 | 12.32 | 17.38 | 23.20 | 29.73 | 36.88 | 44.60 | 52.83 |
| 20 MPH | .00 | .54 | 2.25 | 5.32 | 9.94 | 16.32 | 24.64 | 34.76 | 46.40 | 59.45 | 73.76 | 89.19 | 105.67 |
| 30 MPH | .00 | .80 | 3.37 | 7.98 | 14.91 | 24.48 | 36.97 | 52.14 | 69.60 | 89.18 | 110.64 | 133.79 | 158.50 |
| VELOCITY (FPS) | 2300. | 2087. | 1887. | 1699. | 1527. | 1373. | 1255. | 1170. | 1100. | 1046. | 1001. | 963. | 929. |
| ENERGY (FT-LB) | 705. | 580. | 474. | 385. | 311. | 251. | 210. | 182. | 161. | 146. | 133. | 123. | 115. |
| DROP (IN) | .00 | -.21 | -.92 | -2.24 | -4.31 | -7.32 | -11.48 | -17.02 | -24.15 | -33.07 | -43.97 | -57.03 | -72.40 |
| MID-RANGE (IN) | .00 | .06 | .25 | .62 | 1.24 | 2.17 | 3.50 | 5.34 | 7.77 | 10.85 | 14.66 | 19.27 | 24.72 |
| BULLET PATH (IN) | -.63 | -.07 | .00 | -.55 | -1.84 | -4.08 | -7.47 | -12.23 | -18.59 | -26.73 | -36.85 | -49.14 | -63.74 |
| TIME OF FLIGHT (SEC) | .000000 | .034230 | .072023 | .113917 | .160492 | .212317 | .269677 | .331642 | .397820 | .467807 | .541178 | .617629 | .696954 |
| WIND DEFLECTION (IN) | | | | | | | | | | | | | |
| 5 MPH | .00 | .14 | .60 | 1.42 | 2.65 | 4.34 | 6.51 | 9.10 | 12.05 | 15.34 | 18.93 | 22.79 | 26.90 |
| 10 MPH | .00 | .29 | 1.20 | 2.83 | 5.29 | 8.67 | 13.03 | 18.20 | 24.10 | 30.68 | 37.86 | 45.57 | 53.79 |
| 20 MPH | .00 | .57 | 2.40 | 5.66 | 10.58 | 17.34 | 26.06 | 36.39 | 48.21 | 61.36 | 75.71 | 91.14 | 107.59 |
| 30 MPH | .00 | .86 | 3.59 | 8.50 | 15.87 | 26.02 | 39.09 | 54.59 | 72.31 | 92.05 | 113.57 | 136.72 | 161.38 |
| VELOCITY (FPS) | 2200. | 1993. | 1798. | 1617. | 1453. | 1309. | 1213. | 1135. | 1073. | 1024. | 982. | 947. | 915. |
| ENERGY (FT-LB) | 645. | 529. | 431. | 348. | 281. | 228. | 196. | 172. | 153. | 140. | 129. | 119. | 112. |
| DROP (IN) | .00 | -.23 | -1.01 | -2.46 | -4.74 | -8.06 | -12.63 | -18.69 | -26.42 | -36.03 | -47.71 | -61.64 | -77.95 |
| MID-RANGE (IN) | .00 | .06 | .27 | .69 | 1.36 | 2.39 | 3.86 | 5.86 | 8.47 | 11.76 | 15.80 | 20.65 | 26.34 |
| BULLET PATH (IN) | -.63 | -.04 | .00 | -.63 | -2.09 | -4.59 | -8.35 | -13.59 | -20.50 | -29.30 | -40.16 | -53.26 | -68.75 |
| TIME OF FLIGHT (SEC) | .000000 | .035821 | .075446 | .119438 | .168383 | .222807 | .282419 | .346382 | .414390 | .485995 | .560830 | .638639 | .719247 |
| WIND DEFLECTION (IN) | | | | | | | | | | | | | |
| 5 MPH | .00 | .15 | .64 | 1.51 | 2.82 | 4.61 | 6.85 | 9.48 | 12.47 | 15.77 | 19.35 | 23.20 | 27.29 |
| 10 MPH | .00 | .30 | 1.28 | 3.02 | 5.64 | 9.21 | 13.71 | 18.96 | 24.93 | 31.54 | 38.71 | 46.40 | 54.59 |
| 20 MPH | .00 | .61 | 2.56 | 6.04 | 11.27 | 18.43 | 27.41 | 37.93 | 49.87 | 63.07 | 77.41 | 92.80 | 109.17 |
| 30 MPH | .00 | .91 | 3.84 | 9.06 | 16.91 | 27.64 | 41.12 | 56.89 | 74.80 | 94.61 | 116.12 | 139.20 | 163.76 |
| VELOCITY (FPS) | 2100. | 1899. | 1710. | 1537. | 1382. | 1260. | 1174. | 1104. | 1048. | 1003. | 965. | 931. | 901. |
| ENERGY (FT-LB) | 587. | 480. | 390. | 315. | 254. | 211. | 184. | 162. | 146. | 134. | 124. | 115. | 108. |
| DROP (IN) | .00 | -.25 | -1.11 | -2.71 | -5.23 | -8.90 | -13.93 | -20.53 | -28.92 | -39.27 | -51.78 | -66.59 | -83.89 |
| MID-RANGE (IN) | .00 | .07 | .30 | .76 | 1.51 | 2.64 | 4.26 | 6.43 | 9.23 | 12.73 | 16.99 | 22.07 | 28.02 |
| BULLET PATH (IN) | -.63 | -.01 | .00 | -.73 | -2.38 | -5.18 | -9.34 | -15.08 | -22.59 | -32.07 | -43.71 | -57.66 | -74.09 |
| TIME OF FLIGHT (SEC) | .000000 | .037564 | .079193 | .125469 | .176963 | .234020 | .295748 | .361705 | .431496 | .504691 | .580981 | .660153 | .742060 |
| WIND DEFLECTION (IN) | | | | | | | | | | | | | |
| 5 MPH | .00 | .16 | .68 | 1.61 | 3.00 | 4.88 | 7.17 | 9.83 | 12.83 | 16.13 | 19.70 | 23.52 | 27.59 |
| 10 MPH | .00 | .33 | 1.37 | 3.23 | 6.00 | 9.76 | 14.34 | 19.66 | 25.66 | 32.25 | 39.40 | 47.04 | 55.17 |
| 20 MPH | .00 | .65 | 2.73 | 6.45 | 12.01 | 19.52 | 28.67 | 39.32 | 51.32 | 64.51 | 78.79 | 94.09 | 110.35 |
| 30 MPH | .00 | .98 | 4.10 | 9.68 | 18.01 | 29.28 | 43.01 | 58.98 | 76.97 | 96.76 | 118.19 | 141.13 | 165.52 |

## Bullet: Hornady .257 Dia. 60 Gr. S.P.
## Ballistic Coefficient: .095

| RANGE (YDS) | MUZZLE | 25 | 50 | 75 | 100 | 125 | 150 | 175 | 200 | 225 | 250 | 275 | 300 |
|---|---|---|---|---|---|---|---|---|---|---|---|---|---|
| VELOCITY (FPS) | 2000. | 1805. | 1623. | 1459. | 1314. | 1216. | 1138. | 1075. | 1025. | 984. | 948. | 916. | 888. |
| ENERGY (FT-LB) | 533. | 434. | 351. | 283. | 230. | 197. | 172. | 154. | 140. | 129. | 120. | 112. | 105. |
| DROP (IN) | .00 | -.28 | -1.23 | -3.00 | -5.80 | -9.85 | -15.37 | -22.57 | -31.64 | -42.76 | -56.13 | -71.88 | -90.20 |
| MID-RANGE (IN) | .00 | .07 | .33 | .84 | 1.67 | 2.93 | 4.69 | 7.03 | 10.02 | 13.73 | 18.22 | 23.52 | 29.73 |
| BULLET PATH (IN) | -.63 | .02 | .00 | -.84 | -2.72 | -5.84 | -10.43 | -16.70 | -24.84 | -35.04 | -47.48 | -62.30 | -79.69 |
| TIME OF FLIGHT (SEC) | .000000 | .039481 | .083309 | .132069 | .186291 | .245748 | .309567 | .377443 | .448933 | .523662 | .601372 | .681887 | .765083 |
| WIND DEFLECTION (IN) | | | | | | | | | | | | | |
| 5 MPH | .00 | .17 | .73 | 1.72 | 3.19 | 5.13 | 7.44 | 10.12 | 13.11 | 16.38 | 19.92 | 23.71 | 27.73 |
| 10 MPH | .00 | .35 | 1.46 | 3.44 | 6.39 | 10.25 | 14.88 | 20.23 | 26.21 | 32.76 | 39.84 | 47.41 | 55.45 |
| 20 MPH | .00 | .70 | 2.92 | 6.89 | 12.77 | 20.50 | 29.77 | 40.46 | 52.42 | 65.53 | 79.68 | 94.82 | 110.91 |
| 30 MPH | .00 | 1.05 | 4.39 | 10.33 | 19.16 | 30.76 | 44.65 | 60.69 | 78.64 | 98.29 | 119.52 | 142.24 | 166.36 |
| VELOCITY (FPS) | 1900. | 1712. | 1538. | 1383. | 1261. | 1175. | 1104. | 1049. | 1003. | 965. | 931. | 901. | 874. |
| ENERGY (FT-LB) | 481. | 390. | 315. | 255. | 212. | 184. | 162. | 146. | 134. | 124. | 116. | 108. | 102. |
| DROP (IN) | .00 | -.31 | -1.36 | -3.34 | -6.45 | -10.93 | -16.98 | -24.81 | -34.61 | -46.56 | -60.81 | -77.55 | -96.95 |
| MID-RANGE (IN) | .00 | .08 | .37 | .94 | 1.86 | 3.24 | 5.15 | 7.67 | 10.85 | 14.77 | 19.47 | 25.03 | 31.49 |
| BULLET PATH (IN) | -.63 | .06 | .00 | -.98 | -3.10 | -6.58 | -11.64 | -18.47 | -27.27 | -38.22 | -51.48 | -67.22 | -85.62 |
| TIME OF FLIGHT (SEC) | .000000 | .041599 | .087839 | .139294 | .196309 | .257996 | .323915 | .393673 | .466837 | .543099 | .622244 | .704125 | .788642 |
| WIND DEFLECTION (IN) | | | | | | | | | | | | | |
| 5 MPH | .00 | .19 | .78 | 1.84 | 3.38 | 5.34 | 7.66 | 10.33 | 13.29 | 16.53 | 20.02 | 23.75 | 27.72 |
| 10 MPH | .00 | .37 | 1.56 | 3.67 | 6.76 | 10.67 | 15.32 | 20.65 | 26.58 | 33.06 | 40.04 | 47.50 | 55.43 |
| 20 MPH | .00 | .75 | 3.13 | 7.35 | 13.52 | 21.34 | 30.65 | 41.31 | 53.17 | 66.12 | 80.08 | 95.01 | 110.87 |
| 30 MPH | .00 | 1.12 | 4.69 | 11.02 | 20.28 | 32.01 | 45.97 | 61.96 | 79.75 | 99.18 | 120.12 | 142.51 | 166.30 |
| VELOCITY (FPS) | 1800. | 1619. | 1455. | 1310. | 1214. | 1136. | 1074. | 1024. | 983. | 947. | 916. | 887. | 861. |
| ENERGY (FT-LB) | 432. | 349. | 282. | 229. | 196. | 172. | 154. | 140. | 129. | 119. | 112. | 105. | 99. |
| DROP (IN) | .00 | -.35 | -1.52 | -3.73 | -7.20 | -12.14 | -18.76 | -27.26 | -37.82 | -50.62 | -65.80 | -83.56 | -104.08 |
| MID-RANGE (IN) | .00 | .09 | .42 | 1.05 | 2.08 | 3.58 | 5.65 | 8.33 | 11.71 | 15.83 | 20.76 | 26.55 | 33.28 |
| BULLET PATH (IN) | -.63 | .10 | .00 | -1.13 | -3.52 | -7.39 | -12.93 | -20.35 | -29.83 | -41.55 | -55.66 | -72.35 | -91.78 |
| TIME OF FLIGHT (SEC) | .000000 | .043945 | .092837 | .147202 | .206756 | .270662 | .338617 | .410177 | .484970 | .562740 | .643310 | .726560 | .812409 |
| WIND DEFLECTION (IN) | | | | | | | | | | | | | |
| 5 MPH | .00 | .20 | .84 | 1.95 | 3.53 | 5.48 | 7.80 | 10.43 | 13.34 | 16.52 | 19.94 | 23.60 | 27.49 |
| 10 MPH | .00 | .40 | 1.67 | 3.91 | 7.06 | 10.97 | 15.60 | 20.86 | 26.69 | 33.04 | 39.89 | 47.21 | 54.98 |
| 20 MPH | .00 | .80 | 3.35 | 7.82 | 14.11 | 21.94 | 31.19 | 41.72 | 53.38 | 66.08 | 79.78 | 94.42 | 109.97 |
| 30 MPH | .00 | 1.20 | 5.02 | 11.72 | 21.17 | 32.91 | 46.79 | 62.57 | 80.06 | 99.13 | 119.67 | 141.62 | 164.95 |

## Bullet: Sierra .257 Dia. 75 Gr. H.P.
## Ballistic Coefficient: .240

| RANGE (YDS) | MUZZLE | 25 | 50 | 75 | 100 | 125 | 150 | 175 | 200 | 225 | 250 | 275 | 300 |
|---|---|---|---|---|---|---|---|---|---|---|---|---|---|
| VELOCITY (FPS) | 2200. | 2117. | 2035. | 1955. | 1878. | 1802. | 1729. | 1657. | 1589. | 1523. | 1460. | 1400. | 1343. |
| ENERGY (FT-LB) | 806. | 746. | 690. | 637. | 587. | 541. | 498. | 457. | 420. | 386. | 355. | 326. | 300. |
| DROP (IN) | .00 | -.21 | -.90 | -2.12 | -3.90 | -6.29 | -9.36 | -13.14 | -17.72 | -23.15 | -29.53 | -36.93 | -45.45 |
| MID-RANGE (IN) | .00 | .06 | .24 | .57 | 1.05 | 1.71 | 2.56 | 3.64 | 4.96 | 6.56 | 8.47 | 10.73 | 13.37 |
| BULLET PATH (IN) | -.63 | -.07 | .00 | -.45 | -1.47 | -3.10 | -5.40 | -8.42 | -12.23 | -16.89 | -22.51 | -29.15 | -36.89 |
| TIME OF FLIGHT (SEC) | .000000 | .034756 | .070894 | .108492 | .147634 | .188408 | .230905 | .275217 | .321439 | .369660 | .419969 | .472444 | .527153 |
| WIND DEFLECTION (IN) | | | | | | | | | | | | | |
| 5 MPH | .00 | .06 | .24 | .55 | .99 | 1.58 | 2.32 | 3.22 | 4.29 | 5.53 | 6.96 | 8.58 | 10.39 |
| 10 MPH | .00 | .12 | .48 | 1.09 | 1.98 | 3.16 | 4.64 | 6.44 | 8.57 | 11.06 | 13.91 | 17.15 | 20.78 |
| 20 MPH | .00 | .23 | .95 | 2.19 | 3.97 | 6.32 | 9.28 | 12.88 | 17.15 | 22.12 | 27.83 | 34.30 | 41.56 |
| 30 MPH | .00 | .35 | 1.43 | 3.28 | 5.95 | 9.48 | 13.92 | 19.31 | 25.72 | 33.18 | 41.74 | 51.45 | 62.34 |
| VELOCITY (FPS) | 2100. | 2019. | 1940. | 1862. | 1787. | 1714. | 1643. | 1575. | 1510. | 1447. | 1388. | 1332. | 1278. |
| ENERGY (FT-LB) | 734. | 679. | 626. | 577. | 532. | 489. | 450. | 413. | 380. | 349. | 321. | 295. | 272. |
| DROP (IN) | .00 | -.23 | -.99 | -2.33 | -4.29 | -6.93 | -10.31 | -14.48 | -19.53 | -25.55 | -32.60 | -40.79 | -50.19 |
| MID-RANGE (IN) | .00 | .06 | .27 | .62 | 1.15 | 1.88 | 2.83 | 4.01 | 5.48 | 7.25 | 9.37 | 11.87 | 14.79 |
| BULLET PATH (IN) | -.63 | -.05 | .00 | -.53 | -1.68 | -3.51 | -6.07 | -9.44 | -13.68 | -18.89 | -25.13 | -32.50 | -41.10 |
| TIME OF FLIGHT (SEC) | .000000 | .036426 | .074328 | .113792 | .154906 | .197762 | .242452 | .289069 | .337705 | .388445 | .441366 | .496532 | .553998 |
| WIND DEFLECTION (IN) | | | | | | | | | | | | | |
| 5 MPH | .00 | .06 | .26 | .59 | 1.06 | 1.69 | 2.48 | 3.44 | 4.58 | 5.90 | 7.41 | 9.12 | 11.04 |
| 10 MPH | .00 | .13 | .51 | 1.17 | 2.12 | 3.38 | 4.96 | 6.88 | 9.15 | 11.79 | 14.82 | 18.25 | 22.08 |
| 20 MPH | .00 | .25 | 1.02 | 2.34 | 4.24 | 6.76 | 9.91 | 13.75 | 18.30 | 23.59 | 29.65 | 36.49 | 44.15 |
| 30 MPH | .00 | .38 | 1.53 | 3.51 | 6.36 | 10.13 | 14.87 | 20.63 | 27.45 | 35.38 | 44.47 | 54.74 | 66.23 |
| VELOCITY (FPS) | 2000. | 1921. | 1844. | 1770. | 1697. | 1627. | 1560. | 1495. | 1433. | 1375. | 1319. | 1265. | 1214. |
| ENERGY (FT-LB) | 666. | 615. | 566. | 521. | 480. | 441. | 405. | 372. | 342. | 315. | 290. | 267. | 245. |
| DROP (IN) | .00 | -.25 | -1.09 | -2.57 | -4.74 | -7.66 | -11.40 | -16.03 | -21.64 | -28.32 | -36.14 | -45.21 | -55.64 |
| MID-RANGE (IN) | .00 | .07 | .29 | .69 | 1.28 | 2.08 | 3.13 | 4.45 | 6.08 | 8.06 | 10.41 | 13.18 | 16.42 |
| BULLET PATH (IN) | -.63 | -.02 | .00 | -.62 | -1.92 | -3.98 | -6.86 | -10.62 | -15.38 | -21.19 | -28.15 | -36.35 | -45.92 |
| TIME OF FLIGHT (SEC) | .000000 | .038262 | .078106 | .119622 | .162901 | .208036 | .255120 | .304243 | .355489 | .408933 | .464635 | .522671 | .583205 |
| WIND DEFLECTION (IN) | | | | | | | | | | | | | |
| 5 MPH | .00 | .07 | .27 | .63 | 1.14 | 1.81 | 2.65 | 3.67 | 4.88 | 6.29 | 7.89 | 9.70 | 11.72 |
| 10 MPH | .00 | .13 | .55 | 1.25 | 2.27 | 3.61 | 5.30 | 7.35 | 9.77 | 12.57 | 15.78 | 19.39 | 23.44 |
| 20 MPH | .00 | .27 | 1.09 | 2.51 | 4.54 | 7.23 | 10.60 | 14.69 | 19.53 | 25.14 | 31.55 | 38.78 | 46.89 |
| 30 MPH | .00 | .40 | 1.64 | 3.76 | 6.81 | 10.84 | 15.90 | 22.04 | 29.30 | 37.72 | 47.33 | 58.17 | 70.33 |
| VELOCITY (FPS) | 1900. | 1824. | 1750. | 1678. | 1608. | 1542. | 1478. | 1417. | 1359. | 1305. | 1251. | 1201. | 1156. |
| ENERGY (FT-LB) | 601. | 554. | 510. | 469. | 431. | 396. | 364. | 334. | 308. | 284. | 260. | 240. | 223. |
| DROP (IN) | .00 | -.28 | -1.21 | -2.85 | -5.26 | -8.51 | -12.67 | -17.84 | -24.09 | -31.52 | -40.21 | -50.31 | -61.92 |
| MID-RANGE (IN) | .00 | .08 | .33 | .76 | 1.42 | 2.31 | 3.48 | 4.96 | 6.78 | 8.98 | 11.60 | 14.69 | 18.29 |
| BULLET PATH (IN) | -.63 | .01 | .00 | -.72 | -2.21 | -4.53 | -7.77 | -12.02 | -17.35 | -23.85 | -31.62 | -40.80 | -51.48 |
| TIME OF FLIGHT (SEC) | .000000 | .040292 | .082280 | .126056 | .171714 | .219346 | .269042 | .320881 | .374936 | .431262 | .489976 | .551197 | .614879 |
| WIND DEFLECTION (IN) | | | | | | | | | | | | | |
| 5 MPH | .00 | .07 | .29 | .67 | 1.22 | 1.93 | 2.83 | 3.92 | 5.20 | 6.69 | 8.38 | 10.29 | 12.43 |
| 10 MPH | .00 | .14 | .59 | 1.34 | 2.43 | 3.87 | 5.67 | 7.84 | 10.41 | 13.38 | 16.76 | 20.59 | 24.85 |
| 20 MPH | .00 | .29 | 1.17 | 2.69 | 4.86 | 7.74 | 11.33 | 15.69 | 20.82 | 26.75 | 33.52 | 41.18 | 49.70 |
| 30 MPH | .00 | .43 | 1.76 | 4.03 | 7.30 | 11.60 | 17.00 | 23.53 | 31.23 | 40.13 | 50.29 | 61.77 | 74.55 |

# Bullet: Sierra .257 Dia. 75 Gr. H.P.
## Ballistic Coefficient: .240

| RANGE (YDS) | MUZZLE | 25 | 50 | 75 | 100 | 125 | 150 | 175 | 200 | 225 | 250 | 275 | 300 |
|---|---|---|---|---|---|---|---|---|---|---|---|---|---|
| VELOCITY (FPS) | 1800. | 1727. | 1656. | 1587. | 1521. | 1458. | 1398. | 1342. | 1288. | 1234. | 1186. | 1143. | 1105. |
| ENERGY (FT-LB) | 539. | 496. | 456. | 419. | 385. | 354. | 326. | 300. | 276. | 254. | 234. | 218. | 203. |
| DROP (IN) | .00 | -.31 | -1.35 | -3.19 | -5.88 | -9.51 | -14.18 | -19.95 | -26.93 | -35.22 | -44.95 | -56.22 | -69.15 |
| MID-RANGE (IN) | .00 | .09 | .36 | .85 | 1.58 | 2.59 | 3.90 | 5.56 | 7.59 | 10.04 | 12.96 | 16.40 | 20.40 |
| BULLET PATH (IN) | -.63 | .05 | .00 | -.84 | -2.54 | -5.18 | -8.86 | -13.64 | -19.62 | -26.92 | -35.66 | -45.94 | -57.87 |
| TIME OF FLIGHT (SEC) | .000000 | .042545 | .086907 | .133181 | .181458 | .231824 | .284359 | .339128 | .396187 | .455694 | .517701 | .582142 | .648913 |
| WIND DEFLECTION (IN) | | | | | | | | | | | | | |
| 5 MPH | .00 | .08 | .31 | .72 | 1.30 | 2.07 | 3.02 | 4.18 | 5.53 | 7.10 | 8.89 | 10.90 | 13.10 |
| 10 MPH | .00 | .15 | .63 | 1.44 | 2.60 | 4.13 | 6.05 | 8.35 | 11.06 | 14.20 | 17.78 | 21.79 | 26.21 |
| 20 MPH | .00 | .31 | 1.26 | 2.88 | 5.21 | 8.27 | 12.09 | 16.71 | 22.12 | 28.40 | 35.56 | 43.58 | 52.42 |
| 30 MPH | .00 | .46 | 1.89 | 4.32 | 7.81 | 12.40 | 18.14 | 25.06 | 33.19 | 42.61 | 53.35 | 65.37 | 78.63 |
| VELOCITY (FPS) | 1700. | 1630. | 1562. | 1497. | 1436. | 1377. | 1322. | 1268. | 1216. | 1170. | 1128. | 1092. | 1061. |
| ENERGY (FT-LB) | 481. | 442. | 406. | 373. | 343. | 316. | 291. | 268. | 246. | 228. | 212. | 199. | 187. |
| DROP (IN) | .00 | -.35 | -1.52 | -3.57 | -6.62 | -10.71 | -15.94 | -22.42 | -30.26 | -39.57 | -50.47 | -63.06 | -77.47 |
| MID-RANGE (IN) | .00 | .10 | .41 | .96 | 1.79 | 2.92 | 4.40 | 6.24 | 8.51 | 11.26 | 14.52 | 18.35 | 22.79 |
| BULLET PATH (IN) | -.63 | .09 | .00 | -.98 | -2.95 | -5.96 | -10.12 | -15.53 | -22.29 | -30.53 | -40.35 | -51.87 | -65.20 |
| TIME OF FLIGHT (SEC) | .000000 | .045060 | .092064 | .141105 | .192265 | .245620 | .301231 | .359164 | .419585 | .482487 | .547791 | .615374 | .685085 |
| WIND DEFLECTION (IN) | | | | | | | | | | | | | |
| 5 MPH | .00 | .08 | .34 | .77 | 1.39 | 2.20 | 3.21 | 4.43 | 5.86 | 7.52 | 9.38 | 11.45 | 13.70 |
| 10 MPH | .00 | .17 | .67 | 1.54 | 2.78 | 4.41 | 6.43 | 8.86 | 11.73 | 15.04 | 18.76 | 22.89 | 27.40 |
| 20 MPH | .00 | .33 | 1.35 | 3.08 | 5.56 | 8.81 | 12.86 | 17.72 | 23.46 | 30.07 | 37.53 | 45.79 | 54.80 |
| 30 MPH | .00 | .50 | 2.02 | 4.62 | 8.34 | 13.22 | 19.29 | 26.58 | 35.19 | 45.11 | 56.29 | 68.68 | 82.20 |

## Bullet: Hornady .257 Dia. 87 Gr. Spire Pt.
## Ballistic Coefficient: .220

| RANGE (YDS) | MUZZLE | 25 | 50 | 75 | 100 | 125 | 150 | 175 | 200 | 225 | 250 | 275 | 300 |
|---|---|---|---|---|---|---|---|---|---|---|---|---|---|
| VELOCITY (FPS) | 2100. | 2012. | 1925. | 1842. | 1760. | 1682. | 1606. | 1533. | 1464. | 1399. | 1337. | 1280. | 1227. |
| ENERGY (FT-LB) | 852. | 782. | 716. | 655. | 598. | 546. | 498. | 454. | 414. | 378. | 345. | 316. | 291. |
| DROP (IN) | .00 | -.23 | -1.00 | -2.35 | -4.34 | -7.03 | -10.48 | -14.77 | -20.00 | -26.24 | -33.60 | -42.16 | -52.06 |
| MID-RANGE (IN) | .00 | .06 | .27 | .63 | 1.17 | 1.92 | 2.89 | 4.13 | 5.66 | 7.53 | 9.77 | 12.43 | 15.55 |
| BULLET PATH (IN) | -.63 | -.05 | .00 | -.54 | -1.71 | -3.59 | -6.23 | -9.71 | -14.12 | -19.55 | -26.09 | -33.84 | -42.92 |
| TIME OF FLIGHT (SEC) | .000000 | .036492 | .074603 | .114435 | .156094 | .199689 | .245332 | .293132 | .343194 | .395613 | .450470 | .507822 | .567689 |
| WIND DEFLECTION (IN) | | | | | | | | | | | | | |
| 5 MPH | .00 | .07 | .28 | .64 | 1.16 | 1.86 | 2.73 | 3.80 | 5.06 | 6.53 | 8.21 | 10.12 | 12.24 |
| 10 MPH | .00 | .14 | .56 | 1.28 | 2.33 | 3.72 | 5.46 | 7.59 | 10.12 | 13.06 | 16.43 | 20.23 | 24.48 |
| 20 MPH | .00 | .27 | 1.12 | 2.57 | 4.66 | 7.43 | 10.93 | 15.18 | 20.23 | 26.11 | 32.85 | 40.47 | 48.97 |
| 30 MPH | .00 | .41 | 1.68 | 3.85 | 6.99 | 11.15 | 16.39 | 22.77 | 30.35 | 39.17 | 49.28 | 60.70 | 73.45 |
| VELOCITY (FPS) | 2000. | 1914. | 1831. | 1750. | 1671. | 1596. | 1524. | 1455. | 1390. | 1329. | 1272. | 1221. | 1174. |
| ENERGY (FT-LB) | 773. | 708. | 647. | 591. | 540. | 492. | 449. | 409. | 373. | 341. | 313. | 288. | 266. |
| DROP (IN) | .00 | -.26 | -1.10 | -2.59 | -4.79 | -7.77 | -11.59 | -16.36 | -22.17 | -29.09 | -37.24 | -46.73 | -57.69 |
| MID-RANGE (IN) | .00 | .07 | .30 | .70 | 1.30 | 2.12 | 3.20 | 4.58 | 6.29 | 8.36 | 10.85 | 13.80 | 17.25 |
| BULLET PATH (IN) | -.63 | -.02 | .00 | -.63 | -1.96 | -4.07 | -7.03 | -10.93 | -15.87 | -21.93 | -29.21 | -37.84 | -47.93 |
| TIME OF FLIGHT (SEC) | .000000 | .038333 | .078400 | .120309 | .164169 | .210091 | .258185 | .308553 | .361291 | .416475 | .474160 | .534359 | .597044 |
| WIND DEFLECTION (IN) | | | | | | | | | | | | | |
| 5 MPH | .00 | .07 | .30 | .69 | 1.25 | 1.99 | 2.92 | 4.05 | 5.39 | 6.95 | 8.73 | 10.72 | 12.94 |
| 10 MPH | .00 | .15 | .60 | 1.37 | 2.49 | 3.98 | 5.84 | 8.11 | 10.79 | 13.90 | 17.45 | 21.45 | 25.88 |
| 20 MPH | .00 | .29 | 1.20 | 2.75 | 4.99 | 7.95 | 11.68 | 16.21 | 21.57 | 27.80 | 34.90 | 42.89 | 51.76 |
| 30 MPH | .00 | .44 | 1.80 | 4.12 | 7.48 | 11.93 | 17.52 | 24.32 | 32.36 | 41.70 | 52.36 | 64.34 | 77.64 |
| VELOCITY (FPS) | 1900. | 1817. | 1736. | 1659. | 1584. | 1512. | 1444. | 1380. | 1319. | 1263. | 1212. | 1166. | 1125. |
| ENERGY (FT-LB) | 697. | 638. | 582. | 531. | 485. | 442. | 403. | 368. | 336. | 308. | 284. | 263. | 245. |
| DROP (IN) | .00 | -.28 | -1.22 | -2.88 | -5.32 | -8.63 | -12.90 | -18.21 | -24.67 | -32.37 | -41.43 | -51.98 | -64.12 |
| MID-RANGE (IN) | .00 | .08 | .33 | .77 | 1.44 | 2.36 | 3.57 | 5.11 | 7.01 | 9.32 | 12.08 | 15.35 | 19.16 |
| BULLET PATH (IN) | -.63 | .01 | .00 | -.73 | -2.25 | -4.63 | -7.97 | -12.36 | -17.89 | -24.66 | -32.80 | -42.42 | -53.63 |
| TIME OF FLIGHT (SEC) | .000000 | .040368 | .082595 | .126791 | .173068 | .221535 | .272293 | .325434 | .381034 | .439138 | .499756 | .562850 | .628337 |
| WIND DEFLECTION (IN) | | | | | | | | | | | | | |
| 5 MPH | .00 | .08 | .32 | .74 | 1.34 | 2.13 | 3.12 | 4.32 | 5.74 | 7.38 | 9.24 | 11.32 | 13.61 |
| 10 MPH | .00 | .16 | .64 | 1.47 | 2.67 | 4.25 | 6.24 | 8.64 | 11.48 | 14.76 | 18.48 | 22.64 | 27.22 |
| 20 MPH | .00 | .31 | 1.28 | 2.95 | 5.34 | 8.51 | 12.48 | 17.29 | 22.97 | 29.52 | 36.97 | 45.28 | 54.44 |
| 30 MPH | .00 | .47 | 1.93 | 4.42 | 8.01 | 12.76 | 18.72 | 25.93 | 34.45 | 44.29 | 55.45 | 67.92 | 81.66 |
| VELOCITY (FPS) | 1800. | 1720. | 1643. | 1569. | 1498. | 1430. | 1367. | 1307. | 1252. | 1202. | 1157. | 1118. | 1083. |
| ENERGY (FT-LB) | 626. | 571. | 521. | 475. | 433. | 395. | 361. | 330. | 303. | 279. | 259. | 241. | 226. |
| DROP (IN) | .00 | -.32 | -1.36 | -3.21 | -5.94 | -9.66 | -14.43 | -20.37 | -27.57 | -36.17 | -46.27 | -57.99 | -71.44 |
| MID-RANGE (IN) | .00 | .09 | .37 | .86 | 1.61 | 2.65 | 4.00 | 5.72 | 7.84 | 10.41 | 13.48 | 17.09 | 21.29 |
| BULLET PATH (IN) | -.63 | .05 | .00 | -.85 | -2.59 | -5.30 | -9.08 | -14.02 | -20.23 | -27.83 | -36.93 | -47.65 | -60.11 |
| TIME OF FLIGHT (SEC) | .000000 | .042626 | .087245 | .133967 | .182900 | .234144 | .287789 | .343905 | .402530 | .463665 | .527263 | .593231 | .661439 |
| WIND DEFLECTION (IN) | | | | | | | | | | | | | |
| 5 MPH | .00 | .08 | .34 | .79 | 1.43 | 2.27 | 3.33 | 4.60 | 6.09 | 7.80 | 9.73 | 11.87 | 14.21 |
| 10 MPH | .00 | .17 | .69 | 1.58 | 2.86 | 4.54 | 6.65 | 9.19 | 12.18 | 15.61 | 19.46 | 23.74 | 28.41 |
| 20 MPH | .00 | .34 | 1.38 | 3.16 | 5.71 | 9.09 | 13.30 | 18.39 | 24.36 | 31.21 | 38.93 | 47.48 | 56.83 |
| 30 MPH | .00 | .51 | 2.07 | 4.73 | 8.57 | 13.63 | 19.95 | 27.58 | 36.54 | 46.82 | 58.39 | 71.23 | 85.24 |

## Bullet: Hornady .257 Dia. 87 Gr. Spire Pt.
## Ballistic Coefficient: .220

| RANGE (YDS) | MUZZLE | 25 | 50 | 75 | 100 | 125 | 150 | 175 | 200 | 225 | 250 | 275 | 300 |
|---|---|---|---|---|---|---|---|---|---|---|---|---|---|
| VELOCITY (FPS) | 1700. | 1624. | 1550. | 1480. | 1414. | 1351. | 1293. | 1239. | 1190. | 1147. | 1108. | 1075. | 1045. |
| ENERGY (FT-LB) | 558. | 509. | 464. | 423. | 386. | 353. | 323. | 297. | 274. | 254. | 237. | 223. | 211. |
| DROP (IN) | .00 | -.35 | -1.53 | -3.61 | -6.70 | -10.87 | -16.23 | -22.88 | -30.96 | -40.57 | -51.83 | -64.86 | -79.75 |
| MID-RANGE (IN) | .00 | .10 | .41 | .97 | 1.82 | 2.98 | 4.50 | 6.42 | 8.79 | 11.65 | 15.05 | 19.04 | 23.65 |
| BULLET PATH (IN) | -.63 | .10 | .00 | -1.00 | -3.01 | -6.10 | -10.37 | -15.95 | -22.95 | -31.48 | -41.66 | -53.61 | -67.42 |
| TIME OF FLIGHT (SEC) | .000000 | .045147 | .092425 | .141941 | .193792 | .248064 | .304820 | .364090 | .425862 | .490077 | .556632 | .625387 | .696188 |
| WIND DEFLECTION (IN) | | | | | | | | | | | | | |
| 5 MPH | .00 | .09 | .37 | .84 | 1.52 | 2.42 | 3.53 | 4.86 | 6.42 | 8.19 | 10.16 | 12.33 | 14.68 |
| 10 MPH | .00 | .18 | .74 | 1.69 | 3.05 | 4.84 | 7.06 | 9.73 | 12.83 | 16.37 | 20.32 | 24.66 | 29.35 |
| 20 MPH | .00 | .36 | 1.47 | 3.37 | 6.10 | 9.67 | 14.12 | 19.45 | 25.67 | 32.74 | 40.64 | 49.31 | 58.71 |
| 30 MPH | .00 | .54 | 2.21 | 5.06 | 9.15 | 14.51 | 21.18 | 29.18 | 38.50 | 49.11 | 60.96 | 73.97 | 88.06 |
| VELOCITY (FPS) | 1600. | 1528. | 1459. | 1394. | 1332. | 1275. | 1223. | 1176. | 1134. | 1097. | 1065. | 1036. | 1011. |
| ENERGY (FT-LB) | 494. | 451. | 411. | 375. | 343. | 314. | 289. | 267. | 248. | 232. | 219. | 207. | 197. |
| DROP (IN) | .00 | -.40 | -1.74 | -4.11 | -7.59 | -12.29 | -18.34 | -25.84 | -34.91 | -45.66 | -58.22 | -72.68 | -89.16 |
| MID-RANGE (IN) | .00 | .11 | .47 | 1.11 | 2.06 | 3.36 | 5.06 | 7.22 | 9.87 | 13.05 | 16.81 | 21.20 | 26.24 |
| BULLET PATH (IN) | -.63 | .15 | .00 | -1.18 | -3.48 | -7.00 | -11.86 | -18.18 | -26.07 | -35.64 | -47.00 | -60.28 | -75.58 |
| TIME OF FLIGHT (SEC) | .000000 | .047976 | .098222 | .150832 | .205886 | .263437 | .323503 | .386057 | .451027 | .518296 | .587715 | .659135 | .732436 |
| WIND DEFLECTION (IN) | | | | | | | | | | | | | |
| 5 MPH | .00 | .10 | .39 | .90 | 1.62 | 2.56 | 3.72 | 5.10 | 6.69 | 8.49 | 10.47 | 12.63 | 14.95 |
| 10 MPH | .00 | .19 | .79 | 1.80 | 3.24 | 5.11 | 7.44 | 10.20 | 13.38 | 16.97 | 20.94 | 25.26 | 29.91 |
| 20 MPH | .00 | .39 | 1.57 | 3.59 | 6.47 | 10.23 | 14.87 | 20.39 | 26.76 | 33.94 | 41.88 | 50.52 | 59.82 |
| 30 MPH | .00 | .58 | 2.36 | 5.39 | 9.71 | 15.34 | 22.31 | 30.59 | 40.14 | 50.91 | 62.81 | 75.77 | 89.73 |

# Bullet: Lyman #257420, 65 Gr.
## Ballistic Coefficient: .160

| RANGE (YDS) | MUZZLE | 25 | 50 | 75 | 100 | 125 | 150 | 175 | 200 | 225 | 250 | 275 | 300 |
|---|---|---|---|---|---|---|---|---|---|---|---|---|---|
| VELOCITY (FPS) | 2000. | 1883. | 1770. | 1662. | 1560. | 1464. | 1375. | 1294. | 1235. | 1183. | 1136. | 1095. | 1060. |
| ENERGY (FT-LB) | 577. | 511. | 452. | 399. | 351. | 309. | 273. | 242. | 220. | 202. | 186. | 173. | 162. |
| DROP (IN) | .00 | -.26 | -1.14 | -2.71 | -5.06 | -8.31 | -12.58 | -18.00 | -24.70 | -32.84 | -42.53 | -53.89 | -67.07 |
| MID-RANGE (IN) | .00 | .07 | .31 | .73 | 1.39 | 2.32 | 3.58 | 5.22 | 7.29 | 9.86 | 12.95 | 16.64 | 20.95 |
| BULLET PATH (IN) | -.63 | -.01 | .00 | -.68 | -2.15 | -4.51 | -7.90 | -12.43 | -18.25 | -25.50 | -34.31 | -44.79 | -57.08 |
| TIME OF FLIGHT (SEC) | .000000 | .038654 | .079748 | .123487 | .170080 | .219731 | .272622 | .328894 | .388220 | .450294 | .515038 | .582317 | .651952 |
| WIND DEFLECTION (IN) | | | | | | | | | | | | | |
| 5 MPH | .00 | .10 | .42 | .97 | 1.77 | 2.84 | 4.19 | 5.84 | 7.76 | 9.93 | 12.32 | 14.94 | 17.77 |
| 10 MPH | .00 | .20 | .84 | 1.93 | 3.53 | 5.67 | 8.38 | 11.69 | 15.53 | 19.85 | 24.65 | 29.89 | 35.54 |
| 20 MPH | .00 | .41 | 1.67 | 3.87 | 7.07 | 11.35 | 16.76 | 23.37 | 31.05 | 39.70 | 49.29 | 59.78 | 71.09 |
| 30 MPH | .00 | .61 | 2.51 | 5.80 | 10.60 | 17.02 | 25.14 | 35.06 | 46.58 | 59.55 | 73.94 | 89.66 | 106.63 |
| VELOCITY (FPS) | 1900. | 1786. | 1678. | 1575. | 1478. | 1388. | 1305. | 1244. | 1190. | 1142. | 1101. | 1065. | 1034. |
| ENERGY (FT-LB) | 521. | 461. | 406. | 358. | 315. | 278. | 246. | 223. | 204. | 188. | 175. | 164. | 154. |
| DROP (IN) | .00 | -.29 | -1.27 | -3.01 | -5.62 | -9.25 | -14.00 | -20.01 | -27.44 | -36.40 | -47.02 | -59.43 | -73.74 |
| MID-RANGE (IN) | .00 | .08 | .34 | .81 | 1.55 | 2.59 | 3.99 | 5.81 | 8.10 | 10.90 | 14.26 | 18.24 | 22.87 |
| BULLET PATH (IN) | -.63 | .02 | .00 | -.79 | -2.46 | -5.14 | -8.94 | -14.01 | -20.49 | -28.50 | -38.17 | -49.63 | -63.00 |
| TIME OF FLIGHT (SEC) | .000000 | .040712 | .084037 | .130185 | .179361 | .231752 | .287508 | .346412 | .408072 | .472418 | .539323 | .608613 | .680106 |
| WIND DEFLECTION (IN) | | | | | | | | | | | | | |
| 5 MPH | .00 | .11 | .45 | 1.04 | 1.89 | 3.03 | 4.46 | 6.17 | 8.12 | 10.31 | 12.72 | 15.35 | 18.17 |
| 10 MPH | .00 | .22 | .90 | 2.07 | 3.78 | 6.05 | 8.92 | 12.34 | 16.24 | 20.62 | 25.45 | 30.69 | 36.33 |
| 20 MPH | .00 | .44 | 1.79 | 4.14 | 7.56 | 12.10 | 17.83 | 24.67 | 32.48 | 41.24 | 50.89 | 61.39 | 72.66 |
| 30 MPH | .00 | .65 | 2.69 | 6.21 | 11.33 | 18.15 | 26.75 | 37.01 | 48.73 | 61.86 | 76.34 | 92.08 | 108.99 |
| VELOCITY (FPS) | 1800. | 1691. | 1587. | 1489. | 1398. | 1315. | 1251. | 1196. | 1148. | 1106. | 1069. | 1038. | 1009. |
| ENERGY (FT-LB) | 468. | 413. | 363. | 320. | 282. | 249. | 226. | 207. | 190. | 176. | 165. | 155. | 147. |
| DROP (IN) | .00 | -.33 | -1.41 | -3.36 | -6.29 | -10.34 | -15.64 | -22.33 | -30.54 | -40.39 | -52.02 | -65.54 | -81.07 |
| MID-RANGE (IN) | .00 | .09 | .38 | .91 | 1.74 | 2.90 | 4.46 | 6.47 | 8.97 | 12.02 | 15.67 | 19.94 | 24.90 |
| BULLET PATH (IN) | -.63 | .06 | .00 | -.92 | -2.84 | -5.87 | -10.14 | -15.81 | -23.00 | -31.83 | -42.44 | -54.94 | -69.45 |
| TIME OF FLIGHT (SEC) | .000000 | .042995 | .088788 | .137583 | .189572 | .244912 | .303473 | .364806 | .428838 | .495446 | .564460 | .635699 | .709012 |
| WIND DEFLECTION (IN) | | | | | | | | | | | | | |
| 5 MPH | .00 | .12 | .48 | 1.11 | 2.02 | 3.22 | 4.71 | 6.44 | 8.40 | 10.60 | 13.01 | 15.61 | 18.39 |
| 10 MPH | .00 | .23 | .96 | 2.21 | 4.03 | 6.44 | 9.41 | 12.87 | 16.81 | 21.20 | 26.01 | 31.22 | 36.79 |
| 20 MPH | .00 | .47 | 1.92 | 4.43 | 8.06 | 12.88 | 18.82 | 25.75 | 33.62 | 42.40 | 52.02 | 62.43 | 73.57 |
| 30 MPH | .00 | .70 | 2.88 | 6.64 | 12.09 | 19.31 | 28.23 | 38.62 | 50.43 | 63.60 | 78.04 | 93.65 | 110.36 |
| VELOCITY (FPS) | 1700. | 1596. | 1497. | 1406. | 1321. | 1255. | 1200. | 1151. | 1109. | 1072. | 1040. | 1011. | 986. |
| ENERGY (FT-LB) | 417. | 367. | 324. | 285. | 252. | 227. | 208. | 191. | 177. | 166. | 156. | 148. | 140. |
| DROP (IN) | .00 | -.37 | -1.58 | -3.78 | -7.08 | -11.62 | -17.53 | -24.96 | -34.02 | -44.85 | -57.56 | -72.27 | -89.10 |
| MID-RANGE (IN) | .00 | .10 | .43 | 1.03 | 1.96 | 3.25 | 4.98 | 7.19 | 9.93 | 13.24 | 17.16 | 21.74 | 27.02 |
| BULLET PATH (IN) | -.63 | .11 | .00 | -1.09 | -3.28 | -6.71 | -11.52 | -17.84 | -25.80 | -35.51 | -47.12 | -60.72 | -76.45 |
| TIME OF FLIGHT (SEC) | .000000 | .045542 | .094070 | .145776 | .200821 | .259150 | .320278 | .384110 | .450531 | .519371 | .590448 | .663610 | .738736 |
| WIND DEFLECTION (IN) | | | | | | | | | | | | | |
| 5 MPH | .00 | .13 | .51 | 1.18 | 2.14 | 3.39 | 4.89 | 6.63 | 8.59 | 10.76 | 13.14 | 15.69 | 18.42 |
| 10 MPH | .00 | .25 | 1.03 | 2.36 | 4.29 | 6.79 | 9.78 | 13.25 | 17.18 | 21.53 | 26.27 | 31.38 | 36.84 |
| 20 MPH | .00 | .50 | 2.05 | 4.72 | 8.57 | 13.57 | 19.56 | 26.50 | 34.35 | 43.05 | 52.54 | 62.77 | 73.68 |
| 30 MPH | .00 | .75 | 3.08 | 7.09 | 12.86 | 20.36 | 29.34 | 39.75 | 51.53 | 64.58 | 78.82 | 94.15 | 110.52 |

## Bullet: Lyman #257420, 65 Gr.
**Ballistic Coefficient: .160**

| RANGE (YDS) | MUZZLE | 25 | 50 | 75 | 100 | 125 | 150 | 175 | 200 | 225 | 250 | 275 | 300 |
|---|---|---|---|---|---|---|---|---|---|---|---|---|---|
| VELOCITY (FPS) | 1600. | 1501. | 1409. | 1325. | 1258. | 1203. | 1154. | 1111. | 1074. | 1041. | 1013. | 987. | 964. |
| ENERGY (FT-LB) | 369. | 325. | 287. | 253. | 228. | 209. | 192. | 178. | 166. | 156. | 148. | 141. | 134. |
| DROP (IN) | .00 | -.41 | -1.80 | -4.29 | -8.00 | -13.09 | -19.68 | -27.91 | -37.88 | -49.74 | -63.59 | -79.56 | -97.75 |
| MID-RANGE (IN) | .00 | .11 | .49 | 1.16 | 2.20 | 3.65 | 5.56 | 7.97 | 10.94 | 14.50 | 18.70 | 23.59 | 29.19 |
| BULLET PATH (IN) | -.63 | .17 | .00 | -1.27 | -3.76 | -7.64 | -13.01 | -20.02 | -28.78 | -39.42 | -52.05 | -66.80 | -83.77 |
| TIME OF FLIGHT (SEC) | .000000 | .048398 | .099966 | .154867 | .213050 | .274033 | .337724 | .404011 | .472728 | .543690 | .616744 | .691768 | .768667 |
| WIND DEFLECTION (IN) | | | | | | | | | | | | | |
| 5 MPH | .00 | .13 | .55 | 1.25 | 2.25 | 3.49 | 4.97 | 6.68 | 8.60 | 10.72 | 13.02 | 15.50 | 18.14 |
| 10 MPH | .00 | .27 | 1.09 | 2.51 | 4.50 | 6.98 | 9.94 | 13.36 | 17.20 | 21.44 | 26.05 | 31.00 | 36.29 |
| 20 MPH | .00 | .54 | 2.19 | 5.01 | 8.99 | 13.96 | 19.88 | 26.71 | 34.40 | 42.88 | 52.09 | 62.00 | 72.57 |
| 30 MPH | .00 | .80 | 3.28 | 7.52 | 13.49 | 20.94 | 29.82 | 40.07 | 51.60 | 64.32 | 78.14 | 93.00 | 108.86 |

## Bullet: Lyman #257312, 85 Gr.
**Ballistic Coefficient: .185**

| RANGE (YDS) | MUZZLE | 25 | 50 | 75 | 100 | 125 | 150 | 175 | 200 | 225 | 250 | 275 | 300 |
|---|---|---|---|---|---|---|---|---|---|---|---|---|---|
| VELOCITY (FPS) | 1500. | 1420. | 1346. | 1279. | 1221. | 1170. | 1125. | 1086. | 1052. | 1022. | 996. | 972. | 950. |
| ENERGY (FT-LB) | 425. | 381. | 342. | 309. | 281. | 258. | 239. | 222. | 209. | 197. | 187. | 178. | 170. |
| DROP (IN) | .00 | -.48 | -2.04 | -4.79 | -8.88 | -14.42 | -21.55 | -30.39 | -41.07 | -53.71 | -68.42 | -85.31 | -104.50 |
| MID-RANGE (IN) | .00 | .13 | .54 | 1.28 | 2.40 | 3.95 | 5.98 | 8.54 | 11.66 | 15.40 | 19.79 | 24.88 | 30.70 |
| BULLET PATH (IN) | -.63 | .22 | .00 | -1.42 | -4.17 | -8.38 | -14.18 | -21.69 | -31.03 | -42.33 | -55.71 | -71.27 | -89.13 |
| TIME OF FLIGHT (SEC) | .000000 | .051395 | .105658 | .162871 | .222909 | .285681 | .351094 | .418999 | .489210 | .561562 | .635922 | .712187 | .790274 |
| WIND DEFLECTION (IN) | | | | | | | | | | | | | |
| 5 MPH | .00 | .12 | .50 | 1.13 | 2.02 | 3.14 | 4.50 | 6.07 | 7.85 | 9.82 | 11.96 | 14.27 | 16.74 |
| 10 MPH | .00 | .25 | 1.00 | 2.27 | 4.03 | 6.28 | 8.99 | 12.14 | 15.70 | 19.63 | 23.92 | 28.54 | 33.49 |
| 20 MPH | .00 | .49 | 1.99 | 4.53 | 8.06 | 12.56 | 17.99 | 24.29 | 31.40 | 39.27 | 47.84 | 57.09 | 66.98 |
| 30 MPH | .00 | .74 | 2.99 | 6.80 | 12.10 | 18.84 | 26.98 | 36.43 | 47.10 | 58.90 | 71.77 | 85.63 | 100.46 |
| VELOCITY (FPS) | 1400. | 1327. | 1263. | 1207. | 1157. | 1114. | 1076. | 1044. | 1015. | 989. | 966. | 944. | 925. |
| ENERGY (FT-LB) | 370. | 332. | 301. | 275. | 253. | 234. | 219. | 206. | 194. | 185. | 176. | 168. | 161. |
| DROP (IN) | .00 | -.55 | -2.34 | -5.49 | -10.13 | -16.39 | -24.39 | -34.27 | -46.13 | -60.10 | -76.28 | -94.77 | -115.65 |
| MID-RANGE (IN) | .00 | .15 | .61 | 1.46 | 2.73 | 4.46 | 6.71 | 9.52 | 12.94 | 17.00 | 21.75 | 27.22 | 33.43 |
| BULLET PATH (IN) | -.63 | .30 | .00 | -1.67 | -4.83 | -9.60 | -16.13 | -24.52 | -34.90 | -47.38 | -62.08 | -79.09 | -98.49 |
| TIME OF FLIGHT (SEC) | .000000 | .055036 | .113012 | .173775 | .237254 | .303338 | .371868 | .442656 | .515547 | .590417 | .667168 | .745722 | .826018 |
| WIND DEFLECTION (IN) | | | | | | | | | | | | | |
| 5 MPH | .00 | .13 | .52 | 1.15 | 2.02 | 3.12 | 4.44 | 5.95 | 7.65 | 9.53 | 11.57 | 13.77 | 16.12 |
| 10 MPH | .00 | .26 | 1.03 | 2.30 | 4.04 | 6.24 | 8.88 | 11.91 | 15.31 | 19.06 | 23.14 | 27.53 | 32.24 |
| 20 MPH | .00 | .52 | 2.07 | 4.60 | 8.08 | 12.49 | 17.75 | 23.82 | 30.62 | 38.11 | 46.27 | 55.07 | 64.47 |
| 30 MPH | .00 | .77 | 3.10 | 6.90 | 12.13 | 18.73 | 26.63 | 35.72 | 45.92 | 57.17 | 69.41 | 82.60 | 96.71 |

# Bullet: Lyman #257312, 85 Gr.
## Ballistic Coefficient: .185

| RANGE (YDS) | MUZZLE | 25 | 50 | 75 | 100 | 125 | 150 | 175 | 200 | 225 | 250 | 275 | 300 |
|---|---|---|---|---|---|---|---|---|---|---|---|---|---|
| VELOCITY (FPS) | 1900. | 1801. | 1707. | 1616. | 1529. | 1448. | 1371. | 1301. | 1241. | 1187. | 1140. | 1099. | 1063. |
| ENERGY (FT-LB) | 681. | 612. | 550. | 493. | 441. | 395. | 355. | 319. | 291. | 266. | 245. | 228. | 213. |
| DROP (IN) | .00 | -.29 | -1.25 | -2.95 | -5.48 | -8.94 | -13.45 | -19.11 | -26.05 | -34.41 | -44.31 | -55.88 | -69.24 |
| MID-RANGE (IN) | .00 | .08 | .33 | .79 | 1.49 | 2.47 | 3.78 | 5.45 | 7.55 | 10.12 | 13.21 | 16.88 | 21.17 |
| BULLET PATH (IN) | -.63 | .02 | .00 | -.76 | -2.36 | -4.88 | -8.45 | -13.17 | -19.18 | -26.60 | -35.56 | -46.19 | -58.61 |
| TIME OF FLIGHT (SEC) | .000000 | .040541 | .083317 | .128486 | .176203 | .226616 | .279857 | .336029 | .395092 | .456908 | .521405 | .588451 | .657872 |
| WIND DEFLECTION (IN) | | | | | | | | | | | | | |
| 5 MPH | .00 | .09 | .38 | .89 | 1.61 | 2.57 | 3.79 | 5.25 | 6.98 | 8.94 | 11.15 | 13.57 | 16.21 |
| 10 MPH | .00 | .19 | .77 | 1.77 | 3.22 | 5.15 | 7.57 | 10.51 | 13.96 | 17.89 | 22.29 | 27.15 | 32.42 |
| 20 MPH | .00 | .38 | 1.54 | 3.54 | 6.44 | 10.30 | 15.14 | 21.02 | 27.91 | 35.78 | 44.59 | 54.29 | 64.83 |
| 30 MPH | .00 | .56 | 2.31 | 5.31 | 9.67 | 15.44 | 22.71 | 31.53 | 41.87 | 53.67 | 66.88 | 81.44 | 97.25 |
| VELOCITY (FPS) | 1800. | 1705. | 1614. | 1528. | 1446. | 1370. | 1300. | 1240. | 1187. | 1139. | 1098. | 1063. | 1032. |
| ENERGY (FT-LB) | 611. | 549. | 492. | 441. | 395. | 354. | 319. | 290. | 266. | 245. | 228. | 213. | 201. |
| DROP (IN) | .00 | -.32 | -1.39 | -3.29 | -6.13 | -10.01 | -15.04 | -21.35 | -29.09 | -38.37 | -49.31 | -62.05 | -76.71 |
| MID-RANGE (IN) | .00 | .09 | .37 | .89 | 1.67 | 2.77 | 4.23 | 6.10 | 8.42 | 11.26 | 14.65 | 18.66 | 23.32 |
| BULLET PATH (IN) | -.63 | .06 | .00 | -.89 | -2.71 | -5.58 | -9.61 | -14.91 | -21.64 | -29.90 | -39.84 | -51.57 | -65.21 |
| TIME OF FLIGHT (SEC) | .000000 | .042812 | .088018 | .135774 | .186229 | .239514 | .295731 | .354822 | .416666 | .481189 | .548260 | .617703 | .689339 |
| WIND DEFLECTION (IN) | | | | | | | | | | | | | |
| 5 MPH | .00 | .10 | .41 | .95 | 1.72 | 2.74 | 4.02 | 5.56 | 7.33 | 9.34 | 11.58 | 14.02 | 16.66 |
| 10 MPH | .00 | .20 | .82 | 1.90 | 3.44 | 5.49 | 8.05 | 11.12 | 14.67 | 18.69 | 23.16 | 28.05 | 33.32 |
| 20 MPH | .00 | .40 | 1.65 | 3.79 | 6.89 | 10.98 | 16.10 | 22.23 | 29.33 | 37.38 | 46.32 | 56.10 | 66.65 |
| 30 MPH | .00 | .60 | 2.47 | 5.69 | 10.33 | 16.46 | 24.15 | 33.35 | 44.00 | 56.07 | 69.48 | 84.15 | 99.97 |
| VELOCITY (FPS) | 1700. | 1609. | 1523. | 1442. | 1366. | 1296. | 1237. | 1184. | 1137. | 1096. | 1061. | 1030. | 1003. |
| ENERGY (FT-LB) | 545. | 489. | 438. | 392. | 352. | 317. | 289. | 264. | 244. | 227. | 212. | 200. | 190. |
| DROP (IN) | .00 | -.36 | -1.56 | -3.70 | -6.90 | -11.25 | -16.89 | -23.96 | -32.58 | -42.87 | -54.97 | -68.99 | -85.05 |
| MID-RANGE (IN) | .00 | .10 | .42 | 1.00 | 1.89 | 3.12 | 4.74 | 6.82 | 9.39 | 12.52 | 16.23 | 20.59 | 25.63 |
| BULLET PATH (IN) | -.63 | .10 | .00 | -1.05 | -3.14 | -6.40 | -10.95 | -16.92 | -24.44 | -33.64 | -44.64 | -57.56 | -72.53 |
| TIME OF FLIGHT (SEC) | .000000 | .045345 | .093250 | .143861 | .197308 | .253692 | .312942 | .374942 | .439615 | .506828 | .576401 | .648158 | .721959 |
| WIND DEFLECTION (IN) | | | | | | | | | | | | | |
| 5 MPH | .00 | .11 | .44 | 1.01 | 1.83 | 2.91 | 4.24 | 5.82 | 7.63 | 9.66 | 11.90 | 14.33 | 16.94 |
| 10 MPH | .00 | 22 | .88 | 2.03 | 3.67 | 5.83 | 8.49 | 11.64 | 15.25 | 19.32 | 23.80 | 28.66 | 33.89 |
| 20 MPH | .00 | .43 | 1.77 | 4.05 | 7.33 | 11.65 | 16.98 | 23.27 | 30.51 | 38.64 | 47.60 | 57.33 | 67.78 |
| 30 MPH | .00 | .65 | 2.65 | 6.08 | 11.00 | 17.48 | 25.47 | 34.91 | 45.76 | 57.96 | 71.40 | 85.99 | 101.66 |
| VELOCITY (FPS) | 1600. | 1514. | 1434. | 1358. | 1289. | 1231. | 1178. | 1132. | 1092. | 1057. | 1027. | 1000. | 976. |
| ENERGY (FT-LB) | 483. | 433. | 388. | 348. | 314. | 286. | 262. | 242. | 225. | 211. | 199. | 189. | 180. |
| DROP (IN) | .00 | -.41 | -1.77 | -4.20 | -7.80 | -12.71 | -19.05 | -26.96 | -36.56 | -47.97 | -61.32 | -76.72 | -94.29 |
| MID-RANGE (IN) | .00 | .11 | .48 | 1.14 | 2.13 | 3.50 | 5.33 | 7.63 | 10.48 | 13.90 | 17.95 | 22.67 | 28.09 |
| BULLET PATH (IN) | -.63 | .16 | .00 | -1.22 | -3.62 | -7.32 | -12.47 | -19.17 | -27.57 | -37.78 | -49.93 | -64.13 | -80.50 |
| TIME OF FLIGHT (SEC) | .000000 | .048138 | .099098 | .152856 | .209558 | .269113 | .331412 | .396372 | .463853 | .533674 | .605662 | .679679 | .755618 |
| WIND DEFLECTION (IN) | | | | | | | | | | | | | |
| 5 MPH | .00 | .12 | .47 | 1.08 | 1.94 | 3.06 | 4.41 | 6.01 | 7.82 | 9.84 | 12.05 | 14.44 | 16.99 |
| 10 MPH | .00 | .23 | .94 | 2.15 | 3.88 | 6.11 | 8.83 | 12.01 | 15.64 | 19.68 | 24.10 | 28.87 | 33.99 |
| 20 MPH | .00 | .46 | 1.88 | 4.31 | 7.76 | 12.23 | 17.66 | 24.02 | 31.28 | 39.35 | 48.19 | 57.75 | 67.98 |
| 30 MPH | .00 | .69 | 2.82 | 6.46 | 11.65 | 18.34 | 26.49 | 36.03 | 46.91 | 59.03 | 72.29 | 86.62 | 101.97 |

## Bullet: Winchester .357 Dia. 110 Gr. JHP
## Ballistic Coefficient: .111

| RANGE (YDS) | MUZZLE | 25 | 50 | 75 | 100 | 125 | 150 | 175 | 200 | 225 | 250 | 275 | 300 |
|---|---|---|---|---|---|---|---|---|---|---|---|---|---|
| VELOCITY (FPS) | 1700. | 1552. | 1416. | 1296. | 1207. | 1133. | 1074. | 1026. | 986. | 951. | 920. | 892. | 867. |
| ENERGY (FT-LB) | 706. | 588. | 490. | 410. | 356. | 314. | 282. | 257. | 237. | 221. | 207. | 194. | 183. |
| DROP (IN) | .00 | -.38 | -1.67 | -4.05 | -7.71 | -12.87 | -19.71 | -28.43 | -39.21 | -52.22 | -67.59 | -85.51 | -106.16 |
| MID-RANGE (IN) | .00 | .10 | .45 | 1.12 | 2.18 | 3.72 | 5.81 | 8.51 | 11.89 | 16.02 | 20.93 | 26.69 | 33.37 |
| BULLET PATH (IN) | -.63 | .14 | .00 | -1.23 | -3.74 | -7.75 | -13.44 | -21.01 | -30.63 | -42.49 | -56.71 | -73.49 | -92.98 |
| TIME OF FLIGHT (SEC) | .000000 | .046193 | .096813 | .152215 | .212217 | .276386 | .344434 | .415944 | .490572 | .568078 | .648292 | .731097 | .816413 |
| WIND DEFLECTION (IN) | | | | | | | | | | | | | |
| 5 MPH | .00 | .18 | .75 | 1.75 | 3.15 | 4.91 | 7.02 | 9.43 | 12.11 | 15.05 | 18.23 | 21.63 | 25.26 |
| 10 MPH | .00 | .37 | 1.51 | 3.50 | 6.29 | 9.82 | 14.03 | 18.85 | 24.22 | 30.10 | 36.45 | 43.26 | 50.51 |
| 20 MPH | .00 | .73 | 3.02 | 6.99 | 12.58 | 19.64 | 28.06 | 37.71 | 48.45 | 60.20 | 72.90 | 86.52 | 101.02 |
| 30 MPH | .00 | 1.10 | 4.53 | 10.49 | 18.87 | 29.46 | 42.10 | 56.56 | 72.67 | 90.30 | 109.36 | 129.78 | 151.54 |
| VELOCITY (FPS) | 1600. | 1460. | 1334. | 1236. | 1157. | 1092. | 1041. | 999. | 962. | 930. | 901. | 875. | 851. |
| ENERGY (FT-LB) | 625. | 521. | 435. | 373. | 327. | 291. | 265. | 243. | 226. | 211. | 198. | 187. | 177. |
| DROP (IN) | .00 | -.43 | -1.89 | -4.57 | -8.67 | -14.40 | -21.94 | -31.47 | -43.18 | -57.21 | -73.72 | -92.90 | -114.90 |
| MID-RANGE (IN) | .00 | .12 | .51 | 1.26 | 2.44 | 4.13 | 6.40 | 9.30 | 12.91 | 17.28 | 22.46 | 28.52 | 35.52 |
| BULLET PATH (IN) | -.63 | .20 | .00 | -1.41 | -4.26 | -8.72 | -14.99 | -23.27 | -33.71 | -46.48 | -61.73 | -79.64 | -100.39 |
| TIME OF FLIGHT (SEC) | .000000 | .049090 | .102856 | .161384 | .224167 | .290948 | .361332 | .434938 | .511493 | .590806 | .672746 | .757219 | .844168 |
| WIND DEFLECTION (IN) | | | | | | | | | | | | | |
| 5 MPH | .00 | .19 | .80 | 1.83 | 3.23 | 4.98 | 7.05 | 9.40 | 12.01 | 14.87 | 17.95 | 21.26 | 24.79 |
| 10 MPH | .00 | .39 | 1.60 | 3.65 | 6.45 | 9.96 | 14.09 | 18.80 | 24.02 | 29.73 | 35.90 | 42.52 | 49.57 |
| 20 MPH | .00 | .78 | 3.21 | 7.31 | 12.91 | 19.91 | 28.19 | 37.60 | 48.05 | 59.46 | 71.81 | 85.04 | 99.15 |
| 30 MPH | .00 | 1.17 | 4.81 | 10.96 | 19.36 | 29.87 | 42.28 | 56.40 | 72.07 | 89.20 | 107.71 | 127.56 | 148.72 |
| VELOCITY (FPS) | 1500. | 1370. | 1262. | 1178. | 1110. | 1055. | 1010. | 972. | 939. | 910. | 883. | 858. | 835. |
| ENERGY (FT-LB) | 549. | 458. | 389. | 339. | 301. | 272. | 249. | 231. | 215. | 202. | 190. | 180. | 170. |
| DROP (IN) | .00 | -.50 | -2.16 | -5.18 | -9.77 | -16.12 | -24.41 | -34.82 | -47.51 | -62.64 | -80.38 | -100.90 | -124.36 |
| MID-RANGE (IN) | .00 | .13 | .58 | 1.42 | 2.73 | 4.57 | 7.02 | 10.14 | 13.99 | 18.61 | 24.07 | 30.44 | 37.77 |
| BULLET PATH (IN) | -.63 | .26 | .00 | -1.63 | -4.82 | -9.77 | -16.67 | -25.69 | -36.99 | -50.72 | -67.07 | -86.19 | -108.26 |
| TIME OF FLIGHT (SEC) | .000000 | .052347 | .109523 | .171086 | .236733 | .306109 | .378805 | .454520 | .533043 | .614226 | .697965 | .784194 | .872873 |
| WIND DEFLECTION (IN) | | | | | | | | | | | | | |
| 5 MPH | .00 | .21 | .84 | 1.86 | 3.23 | 4.94 | 6.93 | 9.20 | 11.71 | 14.45 | 17.42 | 20.61 | 24.01 |
| 10 MPH | .00 | .41 | 1.68 | 3.71 | 6.46 | 9.88 | 13.87 | 18.40 | 23.42 | 28.90 | 34.84 | 41.22 | 48.03 |
| 20 MPH | .00 | .83 | 3.35 | 7.42 | 12.93 | 19.75 | 27.74 | 36.79 | 46.83 | 57.81 | 69.68 | 82.44 | 96.05 |
| 30 MPH | .00 | 1.24 | 5.03 | 11.13 | 19.39 | 29.63 | 41.61 | 55.19 | 70.25 | 86.71 | 104.53 | 123.65 | 144.08 |
| VELOCITY (FPS) | 1400. | 1284. | 1197. | 1125. | 1067. | 1020. | 981. | 947. | 916. | 889. | 864. | 840. | 818. |
| ENERGY (FT-LB) | 479. | 403. | 350. | 309. | 278. | 254. | 235. | 219. | 205. | 193. | 182. | 172. | 163. |
| DROP (IN) | .00 | -.57 | -2.46 | -5.87 | -10.99 | -18.00 | -27.10 | -38.45 | -52.18 | -68.48 | -87.53 | -109.47 | -134.49 |
| MID-RANGE (IN) | .00 | .15 | .66 | 1.59 | 3.03 | 5.04 | 7.68 | 11.02 | 15.10 | 20.00 | 25.76 | 32.45 | 40.14 |
| BULLET PATH (IN) | -.63 | .34 | .00 | -1.86 | -5.43 | -10.90 | -18.45 | -28.25 | -40.43 | -55.19 | -72.68 | -93.08 | -116.55 |
| TIME OF FLIGHT (SEC) | .000000 | .056019 | .116553 | .181229 | .249735 | .321652 | .396653 | .474506 | .555050 | .638172 | .723797 | .811880 | .902397 |
| WIND DEFLECTION (IN) | | | | | | | | | | | | | |
| 5 MPH | .00 | .22 | .83 | 1.81 | 3.12 | 4.73 | 6.62 | 8.76 | 11.13 | 13.73 | 16.55 | 19.59 | 22.84 |
| 10 MPH | .00 | .43 | 1.66 | 3.61 | 6.24 | 9.47 | 13.24 | 17.51 | 22.26 | 27.46 | 33.10 | 39.18 | 45.68 |
| 20 MPH | .00 | .86 | 3.31 | 7.22 | 12.48 | 18.94 | 26.48 | 35.03 | 44.52 | 54.92 | 66.21 | 78.35 | 91.36 |
| 30 MPH | .00 | 1.29 | 4.97 | 10.83 | 18.72 | 28.40 | 39.72 | 52.54 | 66.78 | 82.38 | 99.31 | 117.53 | 137.04 |

# Bullet: Winchester .357 Dia. 110 Gr. JHP
## Ballistic Coefficient: .111

| RANGE (YDS) | MUZZLE | 25 | 50 | 75 | 100 | 125 | 150 | 175 | 200 | 225 | 250 | 275 | 300 |
|---|---|---|---|---|---|---|---|---|---|---|---|---|---|
| VELOCITY (FPS) | 1300. | 1211. | 1136. | 1076. | 1028. | 987. | 952. | 921. | 893. | 868. | 844. | 822. | 800. |
| ENERGY (FT-LB) | 413. | 358. | 315. | 283. | 258. | 238. | 221. | 207. | 195. | 184. | 174. | 165. | 156. |
| DROP (IN) | .00 | -.66 | -2.80 | -6.63 | -12.32 | -20.06 | -30.02 | -42.34 | -57.21 | -74.79 | -95.24 | -118.73 | -145.44 |
| MID-RANGE (IN) | .00 | .17 | .74 | 1.78 | 3.35 | 5.54 | 8.39 | 11.95 | 16.29 | 21.47 | 27.56 | 34.62 | 42.70 |
| BULLET PATH (IN) | -.63 | .43 | .00 | -2.10 | -6.08 | -12.10 | -20.35 | -30.96 | -44.10 | -59.97 | -78.70 | -100.48 | -125.47 |
| TIME OF FLIGHT (SEC) | .000000 | .059820 | .123815 | .191704 | .263073 | .337572 | .414957 | .495056 | .577749 | .662956 | .750626 | .840732 | .933270 |
| WIND DEFLECTION (IN) | | | | | | | | | | | | | |
| 5 MPH | .00 | .19 | .74 | 1.64 | 2.84 | 4.32 | 6.05 | 8.03 | 10.23 | 12.65 | 15.29 | 18.14 | 21.20 |
| 10 MPH | .00 | .37 | 1.48 | 3.28 | 5.69 | 8.64 | 12.11 | 16.05 | 20.45 | 25.30 | 30.57 | 36.28 | 42.41 |
| 20 MPH | .00 | .75 | 2.97 | 6.56 | 11.37 | 17.29 | 24.22 | 32.11 | 40.91 | 50.59 | 61.14 | 72.55 | 84.82 |
| 30 MPH | .00 | 1.12 | 4.45 | 9.83 | 17.06 | 25.93 | 36.33 | 48.16 | 61.36 | 75.89 | 91.71 | 108.83 | 127.23 |
| VELOCITY (FPS) | 1200. | 1127. | 1069. | 1022. | 982. | 948. | 917. | 890. | 864. | 841. | 819. | 796. | 764. |
| ENERGY (FT-LB) | 352. | 310. | 279. | 255. | 236. | 219. | 206. | 193. | 183. | 173. | 164. | 155. | 143. |
| DROP (IN) | .00 | -.77 | -3.25 | -7.63 | -14.07 | -22.77 | -33.83 | -47.47 | -63.84 | -83.11 | -105.45 | -131.01 | -159.98 |
| MID-RANGE (IN) | .00 | .20 | .85 | 2.02 | 3.78 | 6.19 | 9.29 | 13.16 | 17.85 | 23.43 | 29.96 | 37.52 | 46.15 |
| BULLET PATH (IN) | -.63 | .54 | .00 | -2.43 | -6.94 | -13.69 | -22.81 | -34.51 | -48.94 | -66.27 | -86.66 | -110.28 | -137.31 |
| TIME OF FLIGHT (SEC) | .000000 | .064540 | .132923 | .204731 | .279631 | .357390 | .437844 | .520879 | .606419 | .694416 | .784849 | .877718 | .973923 |
| WIND DEFLECTION (IN) | | | | | | | | | | | | | |
| 5 MPH | .00 | .18 | .70 | 1.52 | 2.61 | 3.95 | 5.53 | 7.34 | 9.36 | 11.61 | 14.07 | 16.74 | 19.71 |
| 10 MPH | .00 | .36 | 1.39 | 3.03 | 5.22 | 7.90 | 11.06 | 14.67 | 18.73 | 23.22 | 28.13 | 33.48 | 39.41 |
| 20 MPH | .00 | .72 | 2.79 | 6.07 | 10.43 | 15.80 | 22.12 | 29.35 | 37.46 | 46.43 | 56.27 | 66.96 | 78.82 |
| 30 MPH | .00 | 1.08 | 4.18 | 9.10 | 15.65 | 23.70 | 33.18 | 44.02 | 56.19 | 69.65 | 84.40 | 100.44 | 118.23 |
| VELOCITY (FPS) | 1100. | 1047. | 1004. | 967. | 934. | 905. | 878. | 854. | 831. | 809. | 782. | 750. | 721. |
| ENERGY (FT-LB) | 295. | 268. | 246. | 228. | 213. | 200. | 188. | 178. | 169. | 160. | 149. | 138. | 127. |
| DROP (IN) | .00 | -.91 | -3.79 | -8.83 | -16.17 | -25.96 | -38.41 | -53.65 | -71.87 | -93.22 | -117.87 | -146.05 | -178.06 |
| MID-RANGE (IN) | .00 | 24 | .99 | 2.32 | 4.29 | 6.95 | 10.37 | 14.62 | 19.77 | 25.87 | 32.97 | 41.18 | 50.63 |
| BULLET PATH (IN) | -.63 | 67 | .00 | -2.82 | -7.95 | -15.53 | -25.76 | -38.80 | -54.80 | -73.94 | -96.38 | -122.34 | -152.15 |
| TIME OF FLIGHT (SEC) | .000000 | .069938 | .143139 | .219320 | .298280 | .379880 | .464023 | .550646 | .639714 | .731213 | .825371 | .923358 | 1.025395 |
| WIND DEFLECTION (IN) | | | | | | | | | | | | | |
| 5 MPH | .00 | .15 | .60 | 1.30 | 2.25 | 3.43 | 4.83 | 6.46 | 8.29 | 10.35 | 12.63 | 15.26 | 18.23 |
| 10 MPH | .00 | .31 | 1.19 | 2.60 | 4.50 | 6.86 | 9.67 | 12.91 | 16.59 | 20.69 | 25.27 | 30.51 | 36.47 |
| 20 MPH | .00 | .62 | 2.39 | 5.20 | 8.99 | 13.72 | 19.34 | 25.83 | 33.18 | 41.39 | 50.53 | 61.02 | 72.94 |
| 30 MPH | .00 | .93 | 3.58 | 7.80 | 13.49 | 20.58 | 29.00 | 38.74 | 49.77 | 62.08 | 75.80 | 91.53 | 109.41 |
| VELOCITY (FPS) | 1000. | 963. | 931. | 902. | 876. | 852. | 829. | 807. | 780. | 749. | 719. | 692. | 665. |
| ENERGY (FT-LB) | 244. | 227. | 212. | 199. | 187. | 177. | 168. | 159. | 149. | 137. | 126. | 117. | 108. |
| DROP (IN) | .00 | -1.08 | -4.48 | -10.36 | -18.90 | -30.25 | -44.60 | -62.09 | -82.90 | -107.25 | -135.45 | -167.84 | -204.76 |
| MID-RANGE (IN) | .00 | .28 | 1.16 | 2.69 | 4.97 | 8.03 | 11.94 | 16.76 | 22.55 | 29.41 | 37.47 | 46.88 | 57.78 |
| BULLET PATH (IN) | -.63 | .84 | .00 | -3.32 | -9.30 | -18.10 | -29.89 | -44.83 | -63.08 | -84.87 | -110.52 | -140.35 | -174.72 |
| TIME OF FLIGHT (SEC) | .000000 | .076441 | .155646 | .237480 | .321850 | .408697 | .497985 | .589703 | .684080 | .782287 | .884548 | .990957 | 1.101631 |
| WIND DEFLECTION (IN) | | | | | | | | | | | | | |
| 5 MPH | .00 | .13 | .50 | 1.10 | 1.92 | 2.97 | 4.22 | 5.69 | 7.40 | 9.44 | 11.84 | 14.60 | 17.74 |
| 10 MPH | .00 | .25 | .99 | 2.20 | 3.85 | 5.93 | 8.45 | 11.39 | 14.80 | 18.88 | 23.68 | 29.21 | 35.49 |
| 20 MPH | .00 | .51 | 1.99 | 4.39 | 7.69 | 11.86 | 16.89 | 22.78 | 29.60 | 37.77 | 47.36 | 58.42 | 70.97 |
| 30 MPH | .00 | .76 | 2.98 | 6.59 | 11.54 | 17.79 | 25.34 | 34.16 | 44.39 | 56.65 | 71.04 | 87.63 | 106.46 |

## Bullet: Hornady .357 Dia. 125 Gr. JHP
## Ballistic Coefficient: .101

| RANGE (YDS) | MUZZLE | 25 | 50 | 75 | 100 | 125 | 150 | 175 | 200 | 225 | 250 | 275 | 300 |
|---|---|---|---|---|---|---|---|---|---|---|---|---|---|
| VELOCITY (FPS) | 1600. | 1447. | 1311. | 1196. | 1106. | 1038. | 984. | 940. | 902. | 868. | 838. | 809. | 780. |
| ENERGY (FT-LB) | 710. | 581. | 477. | 397. | 339. | 299. | 269. | 245. | 226. | 209. | 195. | 182. | 169. |
| DROP (IN) | .00 | -.44 | -1.92 | -4.67 | -8.94 | -14.98 | -23.02 | -33.30 | -46.01 | -61.38 | -79.61 | -100.94 | -125.52 |
| MID-RANGE (IN) | .00 | .12 | .52 | 1.30 | 2.55 | 4.37 | 6.84 | 10.05 | 14.07 | 18.97 | 24.83 | 31.74 | 39.74 |
| BULLET PATH (IN) | -.63 | .21 | .00 | -1.47 | -4.46 | -9.23 | -16.00 | -25.00 | -36.43 | -50.52 | -67.48 | -87.53 | -110.84 |
| TIME OF FLIGHT (SEC) | .000000 | .049317 | .103805 | .163751 | .229063 | .299181 | .373464 | .451482 | .532966 | .617749 | .705735 | .796886 | .891278 |
| WIND DEFLECTION (IN) | | | | | | | | | | | | | |
| 5 MPH | .00 | .21 | .88 | 2.04 | 3.66 | 5.70 | 8.11 | 10.86 | 13.90 | 17.24 | 20.85 | 24.75 | 28.93 |
| 10 MPH | .00 | .43 | 1.77 | 4.07 | 7.32 | 11.41 | 16.23 | 21.71 | 27.80 | 34.47 | 41.71 | 49.50 | 57.86 |
| 20 MPH | .00 | .86 | 3.54 | 8.14 | 14.63 | 22.81 | 32.46 | 43.42 | 55.60 | 68.95 | 83.42 | 99.00 | 115.73 |
| 30 MPH | .00 | 1.29 | 5.31 | 12.21 | 21.95 | 34.22 | 48.69 | 65.13 | 83.41 | 103.42 | 125.13 | 148.51 | 173.59 |
| VELOCITY (FPS) | 1500. | 1358. | 1235. | 1135. | 1060. | 1002. | 955. | 915. | 880. | 848. | 819. | 791. | 762. |
| ENERGY (FT-LB) | 624. | 512. | 423. | 358. | 312. | 279. | 253. | 232. | 215. | 200. | 186. | 174. | 161. |
| DROP (IN) | .00 | -.51 | -2.19 | -5.30 | -10.10 | -16.81 | -25.69 | -36.92 | -50.72 | -67.32 | -86.94 | -109.76 | -135.99 |
| MID-RANGE (IN) | .00 | .13 | .59 | 1.47 | 2.86 | 4.86 | 7.54 | 10.99 | 15.27 | 20.46 | 26.65 | 33.91 | 42.30 |
| BULLET PATH (IN) | -.63 | .27 | .00 | -1.70 | -5.09 | -10.40 | -17.86 | -27.68 | -40.07 | -55.27 | -73.47 | -94.88 | -119.70 |
| TIME OF FLIGHT (SEC) | .000000 | .052585 | .110555 | .173984 | .242458 | .315305 | .392019 | .472282 | .555894 | .642737 | .732753 | .825946 | .922588 |
| WIND DEFLECTION (IN) | | | | | | | | | | | | | |
| 5 MPH | .00 | .23 | .93 | 2.11 | 3.74 | 5.75 | 8.10 | 10.76 | 13.72 | 16.96 | 20.48 | 24.28 | 28.39 |
| 10 MPH | .00 | .46 | 1.86 | 4.22 | 7.47 | 11.49 | 16.20 | 21.52 | 27.44 | 33.92 | 40.96 | 48.57 | 56.78 |
| 20 MPH | .00 | .91 | 3.72 | 8.44 | 14.95 | 22.99 | 32.39 | 43.04 | 54.87 | 67.84 | 81.93 | 97.13 | 113.55 |
| 30 MPH | .00 | 1.37 | 5.57 | 12.66 | 22.42 | 34.48 | 48.59 | 64.57 | 82.31 | 101.77 | 122.89 | 145.70 | 170.33 |
| VELOCITY (FPS) | 1400. | 1271. | 1164. | 1081. | 1019. | 969. | 927. | 890. | 858. | 828. | 800. | 771. | 744. |
| ENERGY (FT-LB) | 544. | 448. | 376. | 324. | 288. | 260. | 238. | 220. | 204. | 190. | 178. | 165. | 153. |
| DROP (IN) | .00 | -.58 | -2.51 | -6.04 | -11.42 | -18.89 | -28.65 | -40.92 | -55.92 | -73.86 | -94.96 | -119.37 | -147.43 |
| MID-RANGE (IN) | .00 | .15 | .67 | 1.66 | 3.20 | 5.39 | 8.29 | 11.98 | 16.54 | 22.05 | 28.59 | 36.21 | 45.06 |
| BULLET PATH (IN) | -.63 | .36 | .00 | -1.96 | -5.77 | -11.67 | -19.87 | -30.57 | -43.99 | -60.36 | -79.90 | -102.74 | -129.23 |
| TIME OF FLIGHT (SEC) | .000000 | .056273 | .118029 | .185007 | .256562 | .332117 | .411305 | .493893 | .579741 | .668774 | .760969 | .856514 | .955620 |
| WIND DEFLECTION (IN) | | | | | | | | | | | | | |
| 5 MPH | .00 | .24 | .96 | 2.14 | 3.72 | 5.65 | 7.91 | 10.46 | 13.30 | 16.42 | 19.82 | 23.52 | 27.52 |
| 10 MPH | .00 | .48 | 1.92 | 4.28 | 7.44 | 11.31 | 15.82 | 20.93 | 26.61 | 32.85 | 39.64 | 47.03 | 55.05 |
| 20 MPH | .00 | .95 | 3.83 | 8.55 | 14.88 | 22.62 | 31.64 | 41.85 | 53.21 | 65.69 | 79.29 | 94.06 | 110.09 |
| 30 MPH | .00 | 1.43 | 5.75 | 12.83 | 22.32 | 33.93 | 47.45 | 62.78 | 79.82 | 98.54 | 118.93 | 141.10 | 165.14 |
| VELOCITY (FPS) | 1300. | 1187. | 1098. | 1032. | 980. | 936. | 899. | 865. | 835. | 806. | 778. | 750. | 724. |
| ENERGY (FT-LB) | 469. | 391. | 335. | 296. | 266. | 243. | 224. | 208. | 193. | 181. | 168. | 156. | 145. |
| DROP (IN) | .00 | -.67 | -2.89 | -6.90 | -12.94 | -21.23 | -31.97 | -45.38 | -61.69 | -81.11 | -103.80 | -130.06 | -160.16 |
| MID-RANGE (IN) | .00 | .18 | .77 | 1.88 | 3.58 | 5.96 | 9.09 | 13.04 | 17.91 | 23.76 | 30.67 | 38.75 | 48.11 |
| BULLET PATH (IN) | -.63 | .46 | .00 | -2.25 | -6.53 | -13.06 | -22.04 | -33.70 | -48.24 | -65.90 | -86.84 | -111.33 | -139.67 |
| TIME OF FLIGHT (SEC) | .000000 | .060453 | .126236 | .196762 | .271403 | .349749 | .431542 | .516622 | .604899 | .696339 | .791051 | .889304 | .991158 |
| WIND DEFLECTION (IN) | | | | | | | | | | | | | |
| 5 MPH | .00 | .24 | .95 | 2.08 | 3.58 | 5.39 | 7.51 | 9.92 | 12.62 | 15.59 | 18.84 | 22.41 | 26.30 |
| 10 MPH | .00 | .49 | 1.91 | 4.17 | 7.15 | 10.79 | 15.03 | 19.85 | 25.23 | 31.17 | 37.69 | 44.83 | 52.60 |
| 20 MPH | .00 | .97 | 3.82 | 8.34 | 14.30 | 21.57 | 30.06 | 39.70 | 50.46 | 62.34 | 75.37 | 89.65 | 105.20 |
| 30 MPH | .00 | 1.46 | 5.73 | 12.51 | 21.45 | 32.36 | 45.09 | 59.55 | 75.69 | 93.51 | 113.06 | 134.48 | 157.79 |

## Bullet: Hornady .357 Dia. 125 Gr. JHP
## Ballistic Coefficient: .101

| RANGE (YDS) | MUZZLE | 25 | 50 | 75 | 100 | 125 | 150 | 175 | 200 | 225 | 250 | 275 | 300 |
|---|---|---|---|---|---|---|---|---|---|---|---|---|---|
| VELOCITY (FPS) | 1200. | 1108. | 1040. | 986. | 941. | 903. | 869. | 839. | 810. | 782. | 754. | 727. | 702. |
| ENERGY (FT-LB) | 400. | 341. | 300. | 270. | 246. | 226. | 210. | 195. | 182. | 170. | 158. | 147. | 137. |
| DROP (IN) | .00 | -.79 | -3.34 | -7.89 | -14.66 | -23.86 | -35.71 | -50.42 | -68.21 | -89.27 | -113.83 | -142.20 | -174.66 |
| MID-RANGE (IN) | .00 | .20 | .88 | 2.12 | 4.00 | 6.58 | 9.96 | 14.22 | 19.44 | 25.69 | 33.06 | 41.66 | 51.62 |
| BULLET PATH (IN) | -.63 | .57 | .00 | -2.57 | -7.36 | -14.57 | -24.44 | -37.16 | -52.97 | -72.05 | -94.63 | -121.01 | -151.49 |
| TIME OF FLIGHT (SEC) | .000000 | .065131 | .135093 | .209238 | .287130 | .368494 | .453162 | .541035 | .632072 | .726325 | .824085 | .925436 | 1.030448 |
| WIND DEFLECTION (IN) | | | | | | | | | | | | | |
| 5 MPH | .00 | .23 | .89 | 1.91 | 3.27 | 4.93 | 6.88 | 9.11 | 11.62 | 14.42 | 17.52 | 20.94 | 24.68 |
| 10 MPH | .00 | .46 | 1.78 | 3.83 | 6.53 | 9.86 | 13.76 | 18.22 | 23.24 | 28.83 | 35.04 | 41.88 | 49.36 |
| 20 MPH | .00 | .93 | 3.55 | 7.65 | 13.07 | 19.71 | 27.51 | 36.44 | 46.49 | 57.67 | 70.08 | 83.75 | 98.72 |
| 30 MPH | .00 | 1.39 | 5.33 | 11.48 | 19.60 | 29.57 | 41.27 | 54.67 | 69.73 | 86.50 | 105.12 | 125.63 | 148.08 |
| VELOCITY (FPS) | 1100. | 1033. | 981. | 937. | 899. | 866. | 835. | 807. | 779. | 751. | 725. | 700. | 676. |
| ENERGY (FT-LB) | 336. | 296. | 267. | 244. | 224. | 208. | 194. | 181. | 168. | 156. | 146. | 136. | 127. |
| DROP (IN) | .00 | -.92 | -3.87 | -9.07 | -16.71 | -27.02 | -40.23 | -56.53 | -76.11 | -99.24 | -126.20 | -157.29 | -192.81 |
| MID-RANGE (IN) | .00 | .24 | 1.01 | 2.40 | 4.48 | 7.33 | 11.05 | 15.70 | 21.35 | 28.10 | 36.08 | 45.40 | 56.18 |
| BULLET PATH (IN) | -.63 | .70 | .00 | -2.94 | -8.33 | -16.39 | -27.35 | -41.40 | -58.73 | -79.60 | -104.31 | -133.15 | -166.41 |
| TIME OF FLIGHT (SEC) | .000000 | .070439 | .145003 | .223279 | .305005 | .390020 | .478233 | .569609 | .664226 | .762371 | .864114 | .969526 | 1.078695 |
| WIND DEFLECTION (IN) | | | | | | | | | | | | | |
| 5 MPH | .00 | .20 | .76 | 1.65 | 2.84 | 4.32 | 6.08 | 8.13 | 10.45 | 13.09 | 16.04 | 19.32 | 22.93 |
| 10 MPH | .00 | .40 | 1.52 | 3.30 | 5.68 | 8.64 | 12.17 | 16.25 | 20.90 | 26.18 | 32.08 | 38.64 | 45.85 |
| 20 MPH | .00 | .79 | 3.04 | 6.59 | 11.36 | 17.29 | 24.34 | 32.50 | 41.81 | 52.35 | 64.17 | 77.27 | 91.70 |
| 30 MPH | .00 | 1.19 | 4.56 | 9.89 | 17.04 | 25.93 | 36.51 | 48.75 | 62.71 | 78.53 | 96.25 | 115.91 | 137.55 |
| VELOCITY (FPS) | 1000. | 953. | 914. | 879. | 847. | 818. | 790. | 761. | 735. | 709. | 685. | 662. | 639. |
| ENERGY (FT-LB) | 278. | 252. | 232. | 214. | 199. | 186. | 173. | 161. | 150. | 140. | 130. | 121. | 113. |
| DROP (IN) | .00 | -1.10 | -4.56 | -10.60 | -19.45 | -31.32 | -46.42 | -64.92 | -87.15 | -113.40 | -143.94 | -179.11 | -219.23 |
| MID-RANGE (IN) | .00 | .28 | 1.18 | 2.78 | 5.17 | 8.42 | 12.61 | 17.80 | 24.12 | 31.70 | 40.65 | 51.09 | 63.20 |
| BULLET PATH (IN) | -.63 | .86 | .00 | -3.45 | -9.71 | -18.99 | -31.49 | -47.40 | -67.04 | -90.69 | -118.64 | -151.22 | -188.74 |
| TIME OF FLIGHT (SEC) | .000000 | .076861 | .157260 | .241001 | .327968 | .418106 | .511418 | .608170 | .708498 | .812464 | .920151 | 1.031658 | 1.147099 |
| WIND DEFLECTION (IN) | | | | | | | | | | | | | |
| 5 MPH | .00 | .16 | .64 | 1.41 | 2.46 | 3.79 | 5.40 | 7.32 | 9.55 | 12.10 | 14.97 | 18.19 | 21.74 |
| 10 MPH | .00 | .33 | 1.28 | 2.82 | 4.92 | 7.59 | 10.81 | 14.64 | 19.10 | 24.19 | 29.95 | 36.37 | 43.49 |
| 20 MPH | .00 | .66 | 2.56 | 5.63 | 9.84 | 15.17 | 21.62 | 29.28 | 38.19 | 48.39 | 59.89 | 72.74 | 86.98 |
| 30 MPH | .00 | .98 | 3.83 | 8.45 | 14.77 | 22.76 | 32.43 | 43.91 | 57.29 | 72.58 | 89.84 | 109.12 | 130.47 |
| VELOCITY (FPS) | 900. | 866. | 836. | 807. | 779. | 751. | 725. | 700. | 676. | 653. | 631. | 610. | 589. |
| ENERGY (FT-LB) | 225. | 208. | 194. | 181. | 168. | 157. | 146. | 136. | 127. | 118. | 110. | 103. | 96. |
| DROP (IN) | .00 | -1.31 | -5.50 | -12.79 | -23.36 | -37.47 | -55.41 | -77.46 | -103.93 | -135.14 | -171.44 | -213.18 | -260.75 |
| MID-RANGE (IN) | .00 | .35 | 1.44 | 3.37 | 6.18 | 9.99 | 14.92 | 21.07 | 28.61 | 37.62 | 48.25 | 60.65 | 74.97 |
| BULLET PATH (IN) | -.63 | 1.13 | -.00 | -4.23 | -11.74 | -22.78 | -37.66 | -56.65 | -80.06 | -108.21 | -141.44 | -180.12 | -224.63 |
| TIME OF FLIGHT (SEC) | .000000 | .084958 | .173115 | .264435 | .358994 | .457079 | .558760 | .664107 | .773209 | .886171 | 1.003114 | 1.124171 | 1.249492 |
| WIND DEFLECTION (IN) | | | | | | | | | | | | | |
| 5 MPH | .00 | .14 | .57 | 1.27 | 2.26 | 3.56 | 5.17 | 7.11 | 9.38 | 11.98 | 14.94 | 18.26 | 21.96 |
| 10 MPH | .00 | .29 | 1.13 | 2.54 | 4.52 | 7.11 | 10.34 | 14.22 | 18.75 | 23.97 | 29.88 | 36.52 | 43.91 |
| 20 MPH | .00 | .57 | 2.27 | 5.08 | 9.03 | 14.23 | 20.68 | 28.43 | 37.50 | 47.93 | 59.76 | 73.04 | 87.82 |
| 30 MPH | .00 | .86 | 3.40 | 7.62 | 13.55 | 21.34 | 31.03 | 42.65 | 56.25 | 71.90 | 89.64 | 109.56 | 131.73 |

# Bullet: Speer .357 Dia. 140 Gr. JHP
## Ballistic Coefficient: .192

| RANGE (YDS) | MUZZLE | 25 | 50 | 75 | 100 | 125 | 150 | 175 | 200 | 225 | 250 | 275 | 300 |
|---|---|---|---|---|---|---|---|---|---|---|---|---|---|
| VELOCITY (FPS) | 1500. | 1423. | 1351. | 1286. | 1231. | 1182. | 1139. | 1100. | 1067. | 1038. | 1011. | 988. | 966. |
| ENERGY (FT-LB) | 699. | 629. | 567. | 514. | 471. | 435. | 403. | 376. | 354. | 335. | 318. | 303. | 290. |
| DROP (IN) | .00 | -.48 | -2.03 | -4.77 | -8.82 | -14.31 | -21.35 | -30.07 | -40.57 | -52.97 | -67.39 | -83.92 | -102.67 |
| MID-RANGE (IN) | .00 | .13 | .54 | 1.27 | 2.38 | 3.91 | 5.91 | 8.41 | 11.46 | 15.10 | 19.38 | 24.33 | 29.98 |
| BULLET PATH (IN) | -.63 | .22 | .00 | -1.41 | -4.13 | -8.29 | -14.00 | -21.38 | -30.55 | -41.63 | -54.71 | -69.91 | -87.34 |
| TIME OF FLIGHT (SEC) | .000000 | .051343 | .105447 | .162391 | .222011 | .284184 | .348842 | .415869 | .485112 | .556412 | .629642 | .704700 | .781503 |
| WIND DEFLECTION (IN) | | | | | | | | | | | | | |
| 5 MPH | .00 | .12 | .48 | 1.09 | 1.94 | 3.01 | 4.30 | 5.80 | 7.49 | 9.36 | 11.41 | 13.61 | 15.97 |
| 10 MPH | .00 | .24 | .96 | 2.18 | 3.87 | 6.02 | 8.60 | 11.59 | 14.98 | 18.73 | 22.82 | 27.23 | 31.94 |
| 20 MPH | .00 | .47 | 1.92 | 4.36 | 7.75 | 12.03 | 17.19 | 23.19 | 29.96 | 37.46 | 45.63 | 54.45 | 63.89 |
| 30 MPH | .00 | .71 | 2.88 | 6.54 | 11.62 | 18.05 | 25.79 | 34.78 | 44.94 | 56.19 | 68.45 | 81.68 | 95.83 |
| VELOCITY (FPS) | 1400. | 1330. | 1268. | 1216. | 1168. | 1126. | 1090. | 1058. | 1029. | 1004. | 981. | 960. | 940. |
| ENERGY (FT-LB) | 609. | 550. | 500. | 459. | 424. | 394. | 369. | 348. | 329. | 313. | 299. | 286. | 275. |
| DROP (IN) | .00 | -.55 | -2.33 | -5.46 | -10.06 | -16.25 | -24.15 | -33.87 | -45.53 | -59.23 | -75.09 | -93.19 | -113.62 |
| MID-RANGE (IN) | .00 | .15 | .61 | 1.45 | 2.70 | 4.41 | 6.62 | 9.37 | 12.70 | 16.67 | 21.29 | 26.62 | 32.66 |
| BULLET PATH (IN) | -.63 | .30 | .00 | -1.65 | -4.78 | -9.49 | -15.91 | -24.15 | -34.33 | -46.55 | -60.93 | -77.55 | -96.51 |
| TIME OF FLIGHT (SEC) | .000000 | .054982 | .112785 | .173196 | .236144 | .301545 | .369271 | .439162 | .511068 | .584870 | .660473 | .737798 | .816783 |
| WIND DEFLECTION (IN) | | | | | | | | | | | | | |
| 5 MPH | .00 | .12 | .50 | 1.10 | 1.92 | 2.96 | 4.21 | 5.65 | 7.26 | 9.04 | 10.98 | 13.07 | 15.31 |
| 10 MPH | .00 | .25 | .99 | 2.20 | 3.85 | 5.93 | 8.42 | 11.29 | 14.52 | 18.08 | 21.96 | 26.14 | 30.61 |
| 20 MPH | .00 | .50 | 1.99 | 4.39 | 7.69 | 11.86 | 16.84 | 22.59 | 29.04 | 36.16 | 43.91 | 52.28 | 61.22 |
| 30 MPH | .00 | .75 | 2.98 | 6.59 | 11.54 | 17.79 | 25.26 | 33.88 | 43.56 | 54.24 | 65.87 | 78.41 | 91.83 |
| VELOCITY (FPS) | 1300. | 1244. | 1194. | 1149. | 1109. | 1075. | 1045. | 1018. | 993. | 971. | 951. | 932. | 914. |
| ENERGY (FT-LB) | 525. | 481. | 443. | 410. | 383. | 359. | 339. | 322. | 307. | 293. | 281. | 270. | 260. |
| DROP (IN) | .00 | -.63 | -2.68 | -6.25 | -11.46 | -18.42 | -27.27 | -38.09 | -51.01 | -66.13 | -83.54 | -103.30 | -125.54 |
| MID-RANGE (IN) | .00 | .17 | .71 | 1.65 | 3.05 | 4.95 | 7.39 | 10.41 | 14.05 | 18.35 | 23.35 | 29.05 | 35.51 |
| BULLET PATH (IN) | -.63 | .39 | .00 | -1.91 | -5.47 | -10.78 | -17.97 | -27.14 | -38.40 | -51.87 | -67.62 | -85.73 | -106.32 |
| TIME OF FLIGHT (SEC) | .000000 | .058980 | .120524 | .184574 | .251025 | .319731 | .390532 | .463292 | .537903 | .614277 | .692344 | .772048 | .853342 |
| WIND DEFLECTION (IN) | | | | | | | | | | | | | |
| 5 MPH | .00 | .11 | .45 | 1.01 | 1.78 | 2.75 | 3.91 | 5.23 | 6.72 | 8.36 | 10.16 | 12.09 | 14.17 |
| 10 MPH | .00 | .23 | .90 | 2.02 | 3.56 | 5.50 | 7.81 | 10.46 | 13.44 | 16.73 | 20.31 | 24.19 | 28.34 |
| 20 MPH | .00 | .45 | 1.81 | 4.05 | 7.13 | 11.01 | 15.62 | 20.92 | 26.88 | 33.46 | 40.63 | 48.38 | 56.68 |
| 30 MPH | .00 | .68 | 2.71 | 6.07 | 10.69 | 16.51 | 23.43 | 31.39 | 40.32 | 50.18 | 60.94 | 72.56 | 85.03 |
| VELOCITY (FPS) | 1200. | 1154. | 1114. | 1079. | 1048. | 1021. | 996. | 974. | 953. | 934. | 916. | 899. | 883. |
| ENERGY (FT-LB) | 448. | 414. | 386. | 362. | 341. | 324. | 308. | 295. | 282. | 271. | 261. | 251. | 243. |
| DROP (IN) | .00 | -.75 | -3.14 | -7.26 | -13.25 | -21.21 | -31.24 | -43.46 | -57.97 | -74.81 | -94.12 | -116.01 | -140.58 |
| MID-RANGE (IN) | .00 | .20 | .81 | 1.90 | 3.49 | 5.63 | 8.36 | 11.72 | 15.74 | 20.43 | 25.85 | 32.05 | 39.05 |
| BULLET PATH (IN) | -.63 | .50 | .00 | -2.24 | -6.35 | -12.42 | -20.58 | -30.91 | -43.54 | -58.50 | -75.92 | -95.93 | -118.62 |
| TIME OF FLIGHT (SEC) | .000000 | .063745 | .129907 | .198344 | .268894 | .341419 | .415805 | .491965 | .569826 | .649329 | .730427 | .813083 | .897267 |
| WIND DEFLECTION (IN) | | | | | | | | | | | | | |
| 5 MPH | .00 | .11 | .43 | .95 | 1.66 | 2.54 | 3.59 | 4.79 | 6.14 | 7.64 | 9.28 | 11.05 | 12.96 |
| 10 MPH | .00 | .22 | .86 | 1.91 | 3.33 | 5.09 | 7.18 | 9.59 | 12.29 | 15.28 | 18.56 | 22.10 | 25.92 |
| 20 MPH | .00 | .44 | 1.73 | 3.82 | 6.65 | 10.18 | 14.36 | 19.17 | 24.58 | 30.56 | 37.11 | 44.21 | 51.84 |
| 30 MPH | .00 | .66 | 2.59 | 5.73 | 9.98 | 15.27 | 21.55 | 28.76 | 36.87 | 45.85 | 55.67 | 66.31 | 77.76 |

## Bullet: Speer .357 Dia. 140 Gr. JHP
## Ballistic Coefficient: .192

| RANGE (YDS) | MUZZLE | 25 | 50 | 75 | 100 | 125 | 150 | 175 | 200 | 225 | 250 | 275 | 300 |
|---|---|---|---|---|---|---|---|---|---|---|---|---|---|
| VELOCITY (FPS) | 1100. | 1067. | 1037. | 1011. | 987. | 966. | 946. | 927. | 910. | 893. | 877. | 862. | 848. |
| ENERGY (FT-LB) | 376. | 354. | 334. | 318. | 303. | 290. | 278. | 267. | 257. | 248. | 239. | 231. | 224. |
| DROP (IN) | .00 | -.89 | -3.68 | -8.48 | -15.40 | -24.55 | -36.00 | -49.83 | -66.18 | -85.14 | -106.83 | -131.32 | -158.72 |
| MID-RANGE (IN) | .00 | .23 | .95 | 2.20 | 4.02 | 6.45 | 9.52 | 13.24 | 17.69 | 22.88 | 28.87 | 35.67 | 43.36 |
| BULLET PATH (IN) | -.63 | .64 | .00 | -2.65 | -7.42 | -14.41 | -23.71 | -35.38 | -49.58 | -66.39 | -85.92 | -108.26 | -133.51 |
| TIME OF FLIGHT (SEC) | .000000 | .069260 | .140577 | .213822 | .288893 | .365708 | .444199 | .524313 | .606007 | .689245 | .774003 | .860259 | .947999 |
| WIND DEFLECTION (IN) | | | | | | | | | | | | | |
| 5 MPH | .00 | .09 | .37 | .82 | 1.42 | 2.18 | 3.09 | 4.14 | 5.33 | 6.65 | 8.11 | 9.70 | 11.42 |
| 10 MPH | .00 | .19 | .74 | 1.63 | 2.85 | 4.36 | 6.18 | 8.28 | 10.66 | 13.31 | 16.22 | 19.41 | 22.85 |
| 20 MPH | .00 | .38 | 1.48 | 3.27 | 5.69 | 8.73 | 12.36 | 16.56 | 21.31 | 26.61 | 32.45 | 38.81 | 45.70 |
| 30 MPH | .00 | .57 | 2.22 | 4.90 | 8.54 | 13.09 | 18.54 | 24.84 | 31.97 | 39.92 | 48.67 | 58.22 | 68.54 |
| VELOCITY (FPS) | 1000. | 977. | 956. | 937. | 919. | 902. | 886. | 870. | 856. | 842. | 828. | 815. | 802. |
| ENERGY (FT-LB) | 311. | 297. | 284. | 273. | 262. | 253. | 244. | 235. | 228. | 220. | 213. | 206. | 200. |
| DROP (IN) | .00 | -1.06 | -4.38 | -10.05 | -18.16 | -28.83 | -42.16 | -58.24 | -77.19 | -99.08 | -124.04 | -152.15 | -183.53 |
| MID-RANGE (IN) | .00 | .28 | 1.13 | 2.59 | 4.70 | 7.49 | 11.03 | 15.33 | 20.44 | 26.38 | 33.19 | 40.92 | 49.61 |
| BULLET PATH (IN) | -.63 | .82 | .00 | -3.16 | -8.76 | -16.93 | -27.75 | -41.33 | -57.77 | -77.16 | -99.61 | -125.22 | -154.09 |
| TIME OF FLIGHT (SEC) | .000000 | .075879 | .153469 | .232709 | .313551 | .395954 | .479890 | .565333 | .652267 | .740679 | .830564 | .921918 | 1.014743 |
| WIND DEFLECTION (IN) | | | | | | | | | | | | | |
| 5 MPH | .00 | .08 | .31 | .68 | 1.19 | 1.84 | 2.63 | 3.55 | 4.60 | 5.78 | 7.09 | 8.53 | 10.10 |
| 10 MPH | .00 | .15 | .61 | 1.36 | 2.38 | 3.69 | 5.26 | 7.10 | 9.20 | 11.56 | 14.18 | 17.06 | 20.19 |
| 20 MPH | .00 | .31 | 1.22 | 2.71 | 4.77 | 7.38 | 10.52 | 14.20 | 18.40 | 23.12 | 28.36 | 34.12 | 40.39 |
| 30 MPH | .00 | .46 | 1.83 | 4.07 | 7.15 | 11.06 | 15.78 | 21.30 | 27.60 | 34.68 | 42.54 | 51.17 | 60.58 |
| VELOCITY (FPS) | 900. | 884. | 869. | 854. | 840. | 826. | 813. | 801. | 788. | 775. | 762. | 750. | 738. |
| ENERGY (FT-LB) | 252. | 243. | 235. | 227. | 219. | 212. | 206. | 199. | 193. | 187. | 181. | 175. | 169. |
| DROP (IN) | .00 | -1.22 | -5.20 | -12.05 | -21.87 | -34.75 | -50.80 | -70.13 | -92.75 | -118.76 | -148.37 | -181.71 | -218.89 |
| MID-RANGE (IN) | .00 | .34 | 1.38 | 3.17 | 5.73 | 9.11 | 13.35 | 18.48 | 24.50 | 31.46 | 39.43 | 48.48 | 58.64 |
| BULLET PATH (IN) | -.63 | 1.07 | -.00 | -3.93 | -10.83 | -20.80 | -33.94 | -50.35 | -70.05 | -93.15 | -119.84 | -150.26 | -184.52 |
| TIME OF FLIGHT (SEC) | .000000 | .084093 | .169692 | .256778 | .345343 | .435377 | .526880 | .619854 | .714335 | .810404 | .908073 | 1.007350 | 1.108249 |
| WIND DEFLECTION (IN) | | | | | | | | | | | | | |
| 5 MPH | .00 | .07 | .27 | .60 | 1.06 | 1.65 | 2.37 | 3.21 | 4.19 | 5.32 | 6.58 | 7.98 | 9.53 |
| 10 MPH | .00 | .13 | .53 | 1.19 | 2.11 | 3.29 | 4.73 | 6.43 | 8.39 | 10.63 | 13.15 | 15.96 | 19.05 |
| 20 MPH | .00 | .27 | 1.06 | 2.39 | 4.23 | 6.59 | 9.46 | 12.86 | 16.78 | 21.26 | 26.31 | 31.92 | 38.10 |
| 30 MPH | .00 | .40 | 1.60 | 3.58 | 6.34 | 9.88 | 14.19 | 19.28 | 25.17 | 31.89 | 39.46 | 47.88 | 57.16 |
| VELOCITY (FPS) | 800. | 787. | 774. | 761. | 749. | 737. | 725. | 714. | 703. | 692. | 682. | 671. | 661. |
| ENERGY (FT-LB) | 199. | 192. | 186. | 180. | 174. | 169. | 164. | 158. | 154. | 149. | 144. | 140. | 136. |
| DROP (IN) | .00 | -1.41 | -6.21 | -14.62 | -26.75 | -42.73 | -62.70 | -86.77 | -115.10 | -147.80 | -185.03 | -226.92 | -273.63 |
| MID-RANGE (IN) | .00 | .38 | 1.70 | 3.94 | 7.17 | 11.41 | 16.73 | 23.18 | 30.80 | 39.65 | 49.78 | 61.25 | 74.12 |
| BULLET PATH (IN) | -.63 | 1.38 | -.00 | -4.99 | -13.70 | -26.26 | -42.81 | -63.46 | -88.36 | -117.65 | -151.46 | -189.93 | -233.22 |
| TIME OF FLIGHT (SEC) | .000000 | .094544 | .190679 | .288414 | .387756 | .488717 | .591312 | .695555 | .801464 | .909057 | 1.018355 | 1.129380 | 1.242154 |
| WIND DEFLECTION (IN) | | | | | | | | | | | | | |
| 5 MPH | .00 | .07 | .28 | .63 | 1.12 | 1.76 | 2.54 | 3.46 | 4.53 | 5.75 | 7.12 | 8.64 | 10.31 |
| 10 MPH | .00 | .14 | .56 | 1.26 | 2.24 | 3.51 | 5.07 | 6.92 | 9.06 | 11.49 | 14.23 | 17.27 | 20.62 |
| 20 MPH | .00 | .28 | 1.12 | 2.52 | 4.49 | 7.03 | 10.14 | 13.84 | 18.12 | 22.99 | 28.46 | 34.54 | 41.24 |
| 30 MPH | .00 | .42 | 1.68 | 3.78 | 6.73 | 10.54 | 15.21 | 20.75 | 27.17 | 34.48 | 42.69 | 51.81 | 61.86 |

## Bullet: Speer .357 Dia. 146 Gr. HJHP(SWC)
## Ballistic Coefficient: .172

| RANGE (YDS) | MUZZLE | 25 | 50 | 75 | 100 | 125 | 150 | 175 | 200 | 225 | 250 | 275 | 300 |
|---|---|---|---|---|---|---|---|---|---|---|---|---|---|
| VELOCITY (FPS) | 1500. | 1414. | 1335. | 1263. | 1199. | 1143. | 1096. | 1055. | 1021. | 991. | 964. | 939. | 917. |
| ENERGY (FT-LB) | 729. | 648. | 578. | 517. | 466. | 424. | 389. | 361. | 338. | 318. | 301. | 286. | 272. |
| DROP (IN) | .00 | -.48 | -2.05 | -4.84 | -8.99 | -14.66 | -21.98 | -31.11 | -42.18 | -55.33 | -70.68 | -88.36 | -108.45 |
| MID-RANGE (IN) | .00 | .13 | .54 | 1.29 | 2.44 | 4.04 | 6.15 | 8.82 | 12.10 | 16.04 | 20.69 | 26.08 | 32.25 |
| BULLET PATH (IN) | -.63 | .23 | .00 | -1.44 | -4.26 | -8.58 | -14.56 | -22.35 | -32.08 | -43.89 | -57.90 | -74.24 | -92.99 |
| TIME OF FLIGHT (SEC) | .000000 | .051502 | .106098 | .163881 | .224859 | .288958 | .356017 | .425801 | .498080 | .572675 | .649452 | .728307 | .809161 |
| WIND DEFLECTION (IN) | | | | | | | | | | | | | |
| 5 MPH | .00 | .13 | .54 | 1.22 | 2.19 | 3.43 | 4.93 | 6.67 | 8.63 | 10.80 | 13.15 | 15.69 | 18.41 |
| 10 MPH | .00 | .26 | 1.07 | 2.44 | 4.38 | 6.86 | 9.86 | 13.34 | 17.26 | 21.59 | 26.30 | 31.38 | 36.81 |
| 20 MPH | .00 | .53 | 2.15 | 4.89 | 8.75 | 13.71 | 19.72 | 26.68 | 34.52 | 43.18 | 52.61 | 62.76 | 73.62 |
| 30 MPH | .00 | .79 | 3.22 | 7.33 | 13.13 | 20.57 | 29.58 | 40.02 | 51.79 | 64.77 | 78.91 | 94.15 | 110.44 |
| VELOCITY (FPS) | 1400. | 1322. | 1251. | 1189. | 1134. | 1088. | 1049. | 1015. | 986. | 959. | 935. | 913. | 892. |
| ENERGY (FT-LB) | 635. | 566. | 507. | 458. | 417. | 384. | 357. | 334. | 315. | 298. | 283. | 270. | 258. |
| DROP (IN) | .00 | -.56 | -2.35 | -5.54 | -10.27 | -16.68 | -24.93 | -35.14 | -47.45 | -61.99 | -78.85 | -98.17 | -120.09 |
| MID-RANGE (IN) | .00 | .15 | .62 | 1.48 | 2.78 | 4.58 | 6.92 | 9.86 | 13.45 | 17.73 | 22.74 | 28.52 | 35.12 |
| BULLET PATH (IN) | -.63 | .31 | .00 | -1.70 | -4.94 | -9.86 | -16.61 | -25.33 | -36.15 | -49.20 | -64.57 | -82.40 | -102.83 |
| TIME OF FLIGHT (SEC) | .000000 | .055148 | .113492 | .175024 | .239656 | .307210 | .377446 | .450141 | .525127 | .602274 | .681483 | .762678 | .845803 |
| WIND DEFLECTION (IN) | | | | | | | | | | | | | |
| 5 MPH | .00 | .14 | .56 | 1.26 | 2.23 | 3.46 | 4.93 | 6.61 | 8.50 | 10.57 | 12.83 | 15.26 | 17.86 |
| 10 MPH | .00 | .28 | 1.12 | 2.52 | 4.47 | 6.93 | 9.86 | 13.22 | 16.99 | 21.14 | 25.66 | 30.52 | 35.72 |
| 20 MPH | .00 | .56 | 2.23 | 5.04 | 8.93 | 13.85 | 19.72 | 26.45 | 33.99 | 42.29 | 51.31 | 61.03 | 71.44 |
| 30 MPH | .00 | .83 | 3.35 | 7.56 | 13.40 | 20.78 | 29.58 | 39.67 | 50.98 | 63.43 | 76.97 | 91.55 | 107.16 |
| VELOCITY (FPS) | 1300. | 1232. | 1172. | 1120. | 1076. | 1039. | 1006. | 978. | 952. | 928. | 907. | 887. | 867. |
| ENERGY (FT-LB) | 548. | 492. | 445. | 406. | 375. | 350. | 328. | 310. | 294. | 279. | 266. | 255. | 244. |
| DROP (IN) | .00 | -.65 | -2.73 | -6.39 | -11.78 | -19.04 | -28.31 | -39.72 | -53.40 | -69.42 | -87.95 | -109.11 | -133.03 |
| MID-RANGE (IN) | .00 | .17 | .72 | 1.70 | 3.16 | 5.17 | 7.76 | 11.00 | 14.92 | 19.54 | 24.93 | 31.14 | 38.20 |
| BULLET PATH (IN) | -.63 | .40 | .00 | -1.98 | -5.69 | -11.28 | -18.87 | -28.59 | -40.59 | -54.94 | -71.78 | -91.27 | -113.51 |
| TIME OF FLIGHT (SEC) | .000000 | .059293 | .121756 | .187276 | .255650 | .326634 | .400023 | .475660 | .553426 | .633229 | .715000 | .798685 | .884247 |
| WIND DEFLECTION (IN) | | | | | | | | | | | | | |
| 5 MPH | .00 | .14 | .56 | 1.25 | 2.19 | 3.36 | 4.74 | 6.32 | 8.09 | 10.03 | 12.15 | 14.44 | 16.89 |
| 10 MPH | .00 | .28 | 1.12 | 2.50 | 4.38 | 6.72 | 9.48 | 12.64 | 16.17 | 20.06 | 24.30 | 28.88 | 33.78 |
| 20 MPH | .00 | .56 | 2.24 | 5.00 | 8.76 | 13.44 | 18.96 | 25.28 | 32.34 | 40.13 | 48.60 | 57.75 | 67.56 |
| 30 MPH | .00 | .84 | 3.36 | 7.50 | 13.14 | 20.16 | 28.44 | 37.92 | 48.52 | 60.19 | 72.90 | 86.63 | 101.34 |
| VELOCITY (FPS) | 1200. | 1144. | 1096. | 1056. | 1022. | 991. | 964. | 940. | 917. | 896. | 877. | 858. | 841. |
| ENERGY (FT-LB) | 467. | 424. | 390. | 362. | 338. | 319. | 301. | 286. | 273. | 260. | 249. | 239. | 229. |
| DROP (IN) | .00 | -.76 | -3.18 | -7.40 | -13.56 | -21.80 | -32.24 | -45.00 | -60.17 | -77.92 | -98.37 | -121.63 | -147.84 |
| MID-RANGE (IN) | .00 | .20 | .83 | 1.94 | 3.60 | 5.84 | 8.72 | 12.27 | 16.52 | 21.55 | 27.38 | 34.08 | 41.68 |
| BULLET PATH (IN) | -.63 | .51 | .00 | -2.31 | -6.57 | -12.90 | -21.44 | -32.29 | -45.56 | -61.40 | -79.94 | -101.30 | -125.60 |
| TIME OF FLIGHT (SEC) | .000000 | .064042 | .131048 | .200782 | .273015 | .347566 | .424302 | .503116 | .583929 | .666681 | .751326 | .837832 | .926179 |
| WIND DEFLECTION (IN) | | | | | | | | | | | | | |
| 5 MPH | .00 | .14 | .53 | 1.17 | 2.03 | 3.09 | 4.34 | 5.77 | 7.39 | 9.17 | 11.12 | 13.23 | 15.50 |
| 10 MPH | .00 | .27 | 1.06 | 2.34 | 4.05 | 6.17 | 8.68 | 11.55 | 14.77 | 18.34 | 22.23 | 26.46 | 31.01 |
| 20 MPH | .00 | .54 | 2.13 | 4.68 | 8.10 | 12.34 | 17.35 | 23.10 | 29.54 | 36.67 | 44.47 | 52.92 | 62.01 |
| 30 MPH | .00 | .81 | 3.19 | 7.01 | 12.15 | 18.52 | 26.03 | 34.65 | 44.31 | 55.01 | 66.70 | 79.38 | 93.02 |

## Bullet: Speer .357 Dia. 146 Gr. HJHP(SWC)
### Ballistic Coefficient: .172

| RANGE (YDS) | MUZZLE | 25 | 50 | 75 | 100 | 125 | 150 | 175 | 200 | 225 | 250 | 275 | 300 |
|---|---|---|---|---|---|---|---|---|---|---|---|---|---|
| VELOCITY (FPS) | 1100. | 1059. | 1024. | 994. | 966. | 942. | 919. | 898. | 878. | 860. | 842. | 825. | 809. |
| ENERGY (FT-LB) | 392. | 364. | 340. | 320. | 303. | 287. | 274. | 261. | 250. | 240. | 230. | 221. | 212. |
| DROP (IN) | .00 | -.90 | -3.72 | -8.62 | -15.70 | -25.09 | -36.89 | -51.25 | -68.30 | -88.16 | -110.94 | -136.78 | -165.80 |
| MID-RANGE (IN) | .00 | .23 | .97 | 2.25 | 4.13 | 6.64 | 9.83 | 13.75 | 18.45 | 23.97 | 30.38 | 37.71 | 46.01 |
| BULLET PATH (IN) | -.63 | .65 | .00 | -2.71 | -7.62 | -14.84 | -24.46 | -36.64 | -51.52 | -69.19 | -89.80 | -113.46 | -140.31 |
| TIME OF FLIGHT (SEC) | .000000 | .069514 | .141544 | .215906 | .292461 | .371103 | .451750 | .534339 | .618824 | .705172 | .793362 | .883382 | .975228 |
| WIND DEFLECTION (IN) | | | | | | | | | | | | | |
| 5 MPH | .00 | .12 | .46 | 1.00 | 1.74 | 2.66 | 3.75 | 5.02 | 6.46 | 8.06 | 9.82 | 11.74 | 13.82 |
| 10 MPH | .00 | .23 | .91 | 2.00 | 3.47 | 5.31 | 7.51 | 10.04 | 12.91 | 16.11 | 19.63 | 23.48 | 27.64 |
| 20 MPH | .00 | .47 | 1.82 | 4.00 | 6.95 | 10.63 | 15.02 | 20.09 | 25.83 | 32.22 | 39.26 | 46.95 | 55.28 |
| 30 MPH | .00 | .70 | 2.74 | 6.00 | 10.42 | 15.94 | 22.52 | 30.13 | 38.74 | 48.33 | 58.90 | 70.43 | 82.92 |
| VELOCITY (FPS) | 1000. | 972. | 947. | 924. | 902. | 882. | 863. | 846. | 829. | 812. | 797. | 781. | 766. |
| ENERGY (FT-LB) | 324. | 306. | 291. | 277. | 264. | 252. | 242. | 232. | 223. | 214. | 206. | 198. | 190. |
| DROP (IN) | .00 | -1.07 | -4.43 | -10.16 | -18.42 | -29.35 | -43.06 | -59.67 | -79.30 | -102.09 | -128.16 | -157.51 | -190.40 |
| MID-RANGE (IN) | .00 | .28 | 1.15 | 2.62 | 4.78 | 7.69 | 11.37 | 15.87 | 21.23 | 27.50 | 34.73 | 42.91 | 52.14 |
| BULLET PATH (IN) | -.63 | .83 | .00 | -3.20 | -8.94 | -17.34 | -28.52 | -42.60 | -59.70 | -79.96 | -103.51 | -130.33 | -160.69 |
| TIME OF FLIGHT (SEC) | .000000 | .076089 | .154285 | .234501 | .316671 | .400746 | .486690 | .574478 | .664098 | .755543 | .848819 | .943952 | 1.040964 |
| WIND DEFLECTION (IN) | | | | | | | | | | | | | |
| 5 MPH | .00 | .10 | .38 | .84 | 1.47 | 2.27 | 3.23 | 4.35 | 5.64 | 7.09 | 8.70 | 10.47 | 12.40 |
| 10 MPH | .00 | .19 | .75 | 1.67 | 2.93 | 4.53 | 6.46 | 8.71 | 11.28 | 14.18 | 17.39 | 20.94 | 24.81 |
| 20 MPH | .00 | .38 | 1.51 | 3.34 | 5.87 | 9.06 | 12.91 | 17.42 | 22.56 | 28.35 | 34.78 | 41.87 | 49.62 |
| 30 MPH | .00 | .57 | 2.26 | 5.02 | 8.80 | 13.59 | 19.37 | 26.12 | 33.84 | 42.53 | 52.18 | 62.81 | 74.43 |
| VELOCITY (FPS) | 900. | 880. | 861. | 844. | 827. | 810. | 795. | 779. | 764. | 750. | 736. | 722. | 709. |
| ENERGY (FT-LB) | 263. | 251. | 241. | 231. | 222. | 213. | 205. | 197. | 189. | 182. | 176. | 169. | 163. |
| DROP (IN) | .00 | -1.24 | -5.29 | -12.25 | -22.25 | -35.43 | -51.87 | -71.61 | -94.92 | -121.93 | -152.78 | -187.63 | -226.63 |
| MID-RANGE (IN) | .00 | .34 | 1.40 | 3.22 | 5.84 | 9.32 | 13.69 | 18.95 | 25.21 | 32.53 | 40.96 | 50.57 | 61.45 |
| BULLET PATH (IN) | -.63 | 1.08 | -.00 | -4.00 | -11.05 | -21.26 | -34.75 | -51.54 | -71.88 | -95.93 | -123.83 | -155.72 | -191.76 |
| TIME OF FLIGHT (SEC) | .000000 | .084276 | .170418 | .258402 | .348216 | .439855 | .533324 | .628654 | .725862 | .824960 | .925963 | 1.028891 | 1.133766 |
| WIND DEFLECTION (IN) | | | | | | | | | | | | | |
| 5 MPH | .00 | .08 | .33 | .74 | 1.31 | 2.04 | 2.93 | 3.99 | 5.21 | 6.60 | 8.15 | 9.88 | 11.77 |
| 10 MPH | .00 | .17 | .66 | 1.48 | 2.62 | 4.08 | 5.87 | 7.98 | 10.42 | 13.19 | 16.30 | 19.75 | 23.54 |
| 20 MPH | .00 | .33 | 1.32 | 2.96 | 5.24 | 8.16 | 11.73 | 15.95 | 20.84 | 26.39 | 32.61 | 39.50 | 47.09 |
| 30 MPH | .00 | .50 | 1.98 | 4.44 | 7.86 | 12.24 | 17.60 | 23.93 | 31.26 | 39.58 | 48.91 | 59.25 | 70.63 |
| VELOCITY (FPS) | 800. | 784. | 769. | 755. | 741. | 727. | 713. | 700. | 688. | 675. | 663. | 651. | 640. |
| ENERGY (FT-LB) | 207. | 199. | 192. | 185. | 178. | 171. | 165. | 159. | 153. | 148. | 142. | 137. | 133. |
| DROP (IN) | .00 | -1.45 | -6.38 | -14.96 | -27.33 | -43.65 | -64.06 | -88.71 | -117.78 | -151.43 | -189.83 | -233.16 | -281.61 |
| MID-RANGE (IN) | .00 | .41 | 1.74 | 4.01 | 7.29 | 11.62 | 17.07 | 23.69 | 31.56 | 40.72 | 51.26 | 63.25 | 76.74 |
| BULLET PATH (IN) | -.63 | 1.42 | -.00 | -5.07 | -13.94 | -26.75 | -43.66 | -64.81 | -90.37 | -120.51 | -155.40 | -195.23 | -240.17 |
| TIME OF FLIGHT (SEC) | .000000 | .094684 | .191242 | .289685 | .390026 | .492283 | .596477 | .702632 | .810774 | .920931 | 1.033134 | 1.147416 | 1.263814 |
| WIND DEFLECTION (IN) | | | | | | | | | | | | | |
| 5 MPH | .00 | .08 | .33 | .74 | 1.32 | 2.07 | 2.99 | 4.08 | 5.35 | 6.79 | 8.42 | 10.22 | 12.22 |
| 10 MPH | .00 | .16 | .66 | 1.48 | 2.64 | 4.14 | 5.98 | 8.16 | 10.70 | 13.58 | 16.83 | 20.45 | 24.43 |
| 20 MPH | .00 | .33 | 1.32 | 2.97 | 5.29 | 8.28 | 11.96 | 16.33 | 21.39 | 27.17 | 33.66 | 40.89 | 48.86 |
| 30 MPH | .00 | .49 | 1.98 | 4.45 | 7.93 | 12.43 | 17.94 | 24.49 | 32.09 | 40.75 | 50.49 | 61.34 | 73.29 |

# Bullet: Sierra .357 Dia. 150 Gr. JHC
## Ballistic Coefficient: .180

| RANGE (YDS) | MUZZLE | 25 | 50 | 75 | 100 | 125 | 150 | 175 | 200 | 225 | 250 | 275 | 300 |
|---|---|---|---|---|---|---|---|---|---|---|---|---|---|
| VELOCITY (FPS) | 1500. | 1418. | 1342. | 1268. | 1195. | 1133. | 1082. | 1039. | 1003. | 971. | 943. | 917. | 894. |
| ENERGY (FT-LB) | 749. | 670. | 600. | 535. | 476. | 428. | 390. | 360. | 335. | 314. | 296. | 280. | 266. |
| DROP (IN) | .00 | -.48 | -2.04 | -4.81 | -8.94 | -14.58 | -21.92 | -31.10 | -42.29 | -55.63 | -71.26 | -89.29 | -109.90 |
| MID-RANGE (IN) | .00 | .13 | .54 | 1.29 | 2.43 | 4.03 | 6.15 | 8.86 | 12.21 | 16.25 | 21.05 | 26.62 | 33.03 |
| BULLET PATH (IN) | -.63 | .23 | .00 | -1.43 | -4.22 | -8.53 | -14.53 | -22.38 | -32.23 | -44.24 | -58.53 | -75.22 | -94.49 |
| TIME OF FLIGHT (SEC) | .000000 | .051434 | .105820 | .163281 | .224237 | .288713 | .356488 | .427256 | .500748 | .576766 | .655166 | .735838 | .818702 |
| WIND DEFLECTION (IN) | | | | | | | | | | | | | |
| 5 MPH | .00 | .13 | .51 | 1.17 | 2.13 | 3.41 | 4.97 | 6.80 | 8.87 | 11.16 | 13.65 | 16.35 | 19.25 |
| 10 MPH | .00 | .25 | 1.02 | 2.34 | 4.27 | 6.81 | 9.94 | 13.60 | 17.73 | 22.31 | 27.31 | 32.71 | 38.49 |
| 20 MPH | .00 | .50 | 2.05 | 4.67 | 8.53 | 13.63 | 19.88 | 27.19 | 35.46 | 44.62 | 54.62 | 65.41 | 76.98 |
| 30 MPH | .00 | .76 | 3.07 | 7.01 | 12.80 | 20.44 | 29.83 | 40.79 | 53.19 | 66.93 | 81.93 | 98.12 | 115.47 |
| VELOCITY (FPS) | 1400. | 1325. | 1250. | 1180. | 1120. | 1071. | 1030. | 995. | 964. | 937. | 912. | 888. | 867. |
| ENERGY (FT-LB) | 653. | 585. | 520. | 463. | 418. | 382. | 353. | 330. | 310. | 292. | 277. | 263. | 250. |
| DROP (IN) | .00 | -.55 | -2.35 | -5.54 | -10.29 | -16.76 | -25.12 | -35.53 | -48.12 | -63.02 | -80.37 | -100.32 | -123.01 |
| MID-RANGE (IN) | .00 | .15 | .62 | 1.48 | 2.80 | 4.62 | 7.02 | 10.06 | 13.77 | 18.21 | 23.42 | 29.46 | 36.38 |
| BULLET PATH (IN) | -.63 | .30 | .00 | -1.70 | -4.97 | -9.95 | -16.83 | -25.74 | -36.85 | -50.26 | -66.12 | -84.59 | -105.79 |
| TIME OF FLIGHT (SEC) | .000000 | .055077 | .113317 | .175120 | .240401 | .308912 | .380345 | .454450 | .531043 | .609988 | .691184 | .774557 | .860056 |
| WIND DEFLECTION (IN) | | | | | | | | | | | | | |
| 5 MPH | .00 | .13 | .54 | 1.27 | 2.30 | 3.61 | 5.18 | 6.99 | 9.02 | 11.25 | 13.68 | 16.30 | 19.11 |
| 10 MPH | .00 | .27 | 1.09 | 2.54 | 4.60 | 7.23 | 10.37 | 13.98 | 18.04 | 22.50 | 27.36 | 32.61 | 38.23 |
| 20 MPH | .00 | .53 | 2.17 | 5.07 | 9.19 | 14.45 | 20.74 | 27.97 | 36.07 | 45.00 | 54.73 | 65.22 | 76.45 |
| 30 MPH | .00 | .80 | 3.26 | 7.61 | 13.79 | 21.68 | 31.11 | 41.95 | 54.11 | 67.50 | 82.09 | 97.82 | 114.68 |
| VELOCITY (FPS) | 1300. | 1223. | 1157. | 1102. | 1056. | 1017. | 984. | 954. | 927. | 903. | 880. | 859. | 840. |
| ENERGY (FT-LB) | 563. | 498. | 446. | 404. | 371. | 344. | 322. | 303. | 286. | 272. | 258. | 246. | 235. |
| DROP (IN) | .00 | -.65 | -2.76 | -6.48 | -11.99 | -19.45 | -29.00 | -40.78 | -54.92 | -71.56 | -90.86 | -112.94 | -137.96 |
| MID-RANGE (IN) | .00 | .17 | .73 | 1.73 | 3.24 | 5.32 | 8.01 | 11.39 | 15.46 | 20.31 | 25.98 | 32.52 | 39.99 |
| BULLET PATH (IN) | -.63 | .41 | .00 | -2.03 | -5.85 | -11.61 | -19.46 | -29.55 | -41.99 | -56.94 | -74.54 | -94.94 | -118.26 |
| TIME OF FLIGHT (SEC) | .000000 | .059501 | .122574 | .189050 | .258645 | .331062 | .406079 | .483530 | .563293 | .645277 | .729418 | .815670 | .904004 |
| WIND DEFLECTION (IN) | | | | | | | | | | | | | |
| 5 MPH | .00 | .16 | .63 | 1.41 | 2.45 | 3.75 | 5.27 | 7.01 | 8.95 | 11.09 | 13.42 | 15.93 | 18.63 |
| 10 MPH | .00 | .32 | 1.27 | 2.81 | 4.91 | 7.50 | 10.55 | 14.02 | 17.91 | 22.18 | 26.84 | 31.87 | 37.26 |
| 20 MPH | .00 | .64 | 2.53 | 5.62 | 9.81 | 15.00 | 21.09 | 28.05 | 35.82 | 44.37 | 53.68 | 63.73 | 74.52 |
| 30 MPH | .00 | .96 | 3.80 | 8.43 | 14.72 | 22.49 | 31.64 | 42.07 | 53.73 | 66.55 | 80.52 | 95.60 | 111.78 |
| VELOCITY (FPS) | 1200. | 1137. | 1085. | 1042. | 1005. | 973. | 945. | 919. | 895. | 873. | 853. | 833. | 815. |
| ENERGY (FT-LB) | 480. | 431. | 392. | 362. | 337. | 315. | 297. | 281. | 267. | 254. | 242. | 231. | 221. |
| DROP (IN) | .00 | -.77 | -3.21 | -7.49 | -13.77 | -22.19 | -32.89 | -45.98 | -61.63 | -79.99 | -101.17 | -125.35 | -152.64 |
| MID-RANGE (IN) | .00 | .20 | .84 | 1.98 | 3.67 | 5.98 | 8.95 | 12.62 | 17.05 | 22.29 | 28.40 | 35.43 | 43.43 |
| BULLET PATH (IN) | -.63 | .52 | .00 | -2.36 | -6.72 | -13.22 | -22.00 | -33.17 | -46.90 | -63.33 | -82.60 | -104.86 | -130.23 |
| TIME OF FLIGHT (SEC) | .000000 | .064240 | .131797 | .202368 | .275679 | .351528 | .429766 | .510282 | .592995 | .677846 | .764798 | .853825 | .944919 |
| WIND DEFLECTION (IN) | | | | | | | | | | | | | |
| 5 MPH | .00 | .15 | .60 | 1.31 | 2.26 | 3.43 | 4.82 | 6.40 | 8.18 | 10.15 | 12.30 | 14.64 | 17.15 |
| 10 MPH | .00 | .31 | 1.20 | 2.62 | 4.52 | 6.87 | 9.64 | 12.81 | 16.37 | 20.30 | 24.60 | 29.27 | 34.31 |
| 20 MPH | .00 | .61 | 2.39 | 5.23 | 9.04 | 13.74 | 19.28 | 25.62 | 32.73 | 40.60 | 49.21 | 58.55 | 68.61 |
| 30 MPH | .00 | .92 | 3.59 | 7.85 | 13.56 | 20.61 | 28.92 | 38.43 | 49.10 | 60.90 | 73.81 | 87.82 | 102.92 |

## Bullet: Sierra .357 Dia. 150 Gr. JHC
### Ballistic Coefficient: .180

| RANGE (YDS) | MUZZLE | 25 | 50 | 75 | 100 | 125 | 150 | 175 | 200 | 225 | 250 | 275 | 300 |
|---|---|---|---|---|---|---|---|---|---|---|---|---|---|
| VELOCITY (FPS) | 1100. | 1054. | 1016. | 983. | 953. | 927. | 902. | 880. | 859. | 839. | 820. | 802. | 782. |
| ENERGY (FT-LB) | 403. | 370. | 344. | 322. | 303. | 286. | 271. | 258. | 246. | 234. | 224. | 214. | 203. |
| DROP (IN) | .00 | -.90 | -3.75 | -8.70 | -15.89 | -25.44 | -37.50 | -52.21 | -69.72 | -90.16 | -113.69 | -140.44 | -170.49 |
| MID-RANGE (IN) | .00 | .23 | .97 | 2.28 | 4.19 | 6.76 | 10.05 | 14.10 | 18.97 | 24.72 | 31.41 | 39.08 | 47.75 |
| BULLET PATH (IN) | -.63 | .66 | .00 | -2.76 | -7.76 | -15.11 | -24.98 | -37.50 | -52.82 | -71.07 | -92.41 | -116.97 | -144.83 |
| TIME OF FLIGHT (SEC) | .000000 | .069683 | .142180 | .217271 | .294792 | .374620 | .456668 | .540869 | .627180 | .715573 | .806033 | .898559 | .993315 |
| WIND DEFLECTION (IN) | | | | | | | | | | | | | |
| 5 MPH | .00 | .13 | .51 | 1.12 | 1.94 | 2.97 | 4.19 | 5.60 | 7.19 | 8.97 | 10.93 | 13.07 | 15.41 |
| 10 MPH | .00 | .26 | 1.02 | 2.24 | 3.88 | 5.93 | 8.37 | 11.19 | 14.38 | 17.94 | 21.86 | 26.15 | 30.82 |
| 20 MPH | .00 | .53 | 2.05 | 4.48 | 7.77 | 11.87 | 16.75 | 22.39 | 28.77 | 35.88 | 43.72 | 52.29 | 61.65 |
| 30 MPH | .00 | .79 | 3.07 | 6.72 | 11.65 | 17.80 | 25.12 | 33.58 | 43.15 | 53.82 | 65.59 | 78.44 | 92.47 |
| VELOCITY (FPS) | 1000. | 969. | 941. | 915. | 892. | 870. | 849. | 830. | 812. | 793. | 772. | 752. | 733. |
| ENERGY (FT-LB) | 333. | 312. | 295. | 279. | 265. | 252. | 240. | 229. | 219. | 209. | 198. | 188. | 179. |
| DROP (IN) | .00 | -1.08 | -4.45 | -10.24 | -18.62 | -29.71 | -43.67 | -60.62 | -80.73 | -104.09 | -130.84 | -161.22 | -195.42 |
| MID-RANGE (IN) | .00 | .28 | 1.15 | 2.65 | 4.86 | 7.83 | 11.59 | 16.21 | 21.75 | 28.23 | 35.71 | 44.29 | 54.04 |
| BULLET PATH (IN) | -.63 | .83 | .00 | -3.25 | -9.09 | -17.65 | -29.06 | -43.48 | -61.04 | -81.87 | -106.08 | -133.92 | -165.58 |
| TIME OF FLIGHT (SEC) | .000000 | .076228 | .154826 | .235686 | .318732 | .403909 | .491181 | .580525 | .671934 | .765430 | .861349 | .959830 | 1.060901 |
| WIND DEFLECTION (IN) | | | | | | | | | | | | | |
| 5 MPH | .00 | .11 | .42 | .94 | 1.65 | 2.54 | 3.62 | 4.89 | 6.33 | 7.96 | 9.80 | 11.87 | 14.16 |
| 10 MPH | .00 | .22 | .85 | 1.88 | 3.30 | 5.09 | 7.25 | 9.77 | 12.66 | 15.92 | 19.60 | 23.73 | 28.32 |
| 20 MPH | .00 | .43 | 1.70 | 3.76 | 6.59 | 10.18 | 14.50 | 19.54 | 25.32 | 31.83 | 39.19 | 47.46 | 56.64 |
| 30 MPH | .00 | .65 | 2.55 | 5.64 | 9.89 | 15.26 | 21.74 | 29.32 | 37.98 | 47.75 | 58.79 | 71.19 | 84.96 |
| VELOCITY (FPS) | 900. | 878. | 857. | 837. | 818. | 800. | 780. | 760. | 740. | 722. | 704. | 686. | 670. |
| ENERGY (FT-LB) | 270. | 256. | 244. | 233. | 223. | 213. | 202. | 192. | 182. | 173. | 165. | 157. | 149. |
| DROP (IN) | .00 | -1.26 | -5.32 | -12.34 | -22.45 | -35.80 | -52.46 | -72.66 | -96.60 | -124.49 | -156.53 | -192.95 | -233.96 |
| MID-RANGE (IN) | .00 | .34 | 1.40 | 3.24 | 5.90 | 9.45 | 13.89 | 19.33 | 25.85 | 33.54 | 42.47 | 52.78 | 64.52 |
| BULLET PATH (IN) | -.63 | 1.09 | -.00 | -4.04 | -11.17 | -21.55 | -35.24 | -52.46 | -73.42 | -98.34 | -127.40 | -160.84 | -198.88 |
| TIME OF FLIGHT (SEC) | .000000 | .084399 | .170903 | .259486 | .350135 | .442848 | .537792 | .635281 | .735347 | .838021 | .943343 | 1.051360 | 1.162123 |
| WIND DEFLECTION (IN) | | | | | | | | | | | | | |
| 5 MPH | .00 | .09 | .37 | .83 | 1.48 | 2.30 | 3.33 | 4.57 | 6.04 | 7.75 | 9.68 | 11.85 | 14.27 |
| 10 MPH | .00 | .19 | .75 | 1.67 | 2.96 | 4.61 | 6.65 | 9.14 | 12.09 | 15.49 | 19.36 | 23.71 | 28.53 |
| 20 MPH | .00 | .38 | 1.49 | 3.34 | 5.91 | 9.22 | 13.30 | 18.29 | 24.18 | 30.98 | 38.72 | 47.41 | 57.07 |
| 30 MPH | .00 | .56 | 2.24 | 5.01 | 8.87 | 13.82 | 19.95 | 27.43 | 36.26 | 46.48 | 58.09 | 71.12 | 85.60 |
| VELOCITY (FPS) | 800. | 779. | 759. | 739. | 721. | 703. | 685. | 669. | 652. | 636. | 621. | 606. | 591. |
| ENERGY (FT-LB) | 213. | 202. | 192. | 182. | 173. | 164. | 156. | 149. | 142. | 135. | 128. | 122. | 116. |
| DROP (IN) | .00 | -1.51 | -6.58 | -15.40 | -28.19 | -45.14 | -66.47 | -92.37 | -123.11 | -158.94 | -200.13 | -246.94 | -299.67 |
| MID-RANGE (IN) | .00 | .43 | 1.78 | 4.11 | 7.51 | 12.06 | 17.83 | 24.90 | 33.37 | 43.34 | 54.93 | 68.23 | 83.36 |
| BULLET PATH (IN) | -.63 | 1.46 | -.00 | -5.22 | -14.40 | -27.75 | -45.47 | -67.77 | -94.90 | -127.13 | -164.71 | -207.92 | -257.05 |
| TIME OF FLIGHT (SEC) | .000000 | .095021 | .192595 | .292746 | .395506 | .500913 | .609014 | .719861 | .833514 | .950035 | 1.069495 | 1.191967 | 1.317530 |
| WIND DEFLECTION (IN) | | | | | | | | | | | | | |
| 5 MPH | .00 | .11 | .45 | 1.01 | 1.80 | 2.83 | 4.09 | 5.60 | 7.35 | 9.35 | 11.62 | 14.14 | 16.94 |
| 10 MPH | .00 | .22 | .90 | 2.02 | 3.61 | 5.66 | 8.19 | 11.20 | 14.70 | 18.71 | 23.23 | 28.29 | 33.89 |
| 20 MPH | .00 | .45 | 1.79 | 4.05 | 7.22 | 11.32 | 16.37 | 22.39 | 29.40 | 37.41 | 46.46 | 56.57 | 67.77 |
| 30 MPH | .00 | .67 | 2.69 | 6.07 | 10.83 | 16.98 | 24.56 | 33.59 | 44.10 | 56.12 | 69.69 | 84.86 | 101.66 |

# Bullet: Hornady .357 Dia. 158 Gr. JHP
## Ballistic Coefficient: .248

| RANGE (YDS) | MUZZLE | 25 | 50 | 75 | 100 | 125 | 150 | 175 | 200 | 225 | 250 | 275 | 300 |
|---|---|---|---|---|---|---|---|---|---|---|---|---|---|
| VELOCITY (FPS) | 1500. | 1440. | 1383. | 1329. | 1268. | 1197. | 1137. | 1086. | 1044. | 1008. | 977. | 949. | 923. |
| ENERGY (FT-LB) | 789. | 727. | 671. | 619. | 564. | 503. | 453. | 414. | 382. | 357. | 335. | 316. | 299. |
| DROP (IN) | .00 | -.47 | -1.98 | -4.63 | -8.52 | -13.77 | -20.53 | -28.97 | -39.25 | -51.52 | -65.91 | -82.58 | -101.61 |
| MID-RANGE (IN) | .00 | .13 | .53 | 1.23 | 2.27 | 3.72 | 5.63 | 8.07 | 11.11 | 14.80 | 19.19 | 24.34 | 30.27 |
| BULLET PATH (IN) | -.63 | .21 | .00 | -1.34 | -3.92 | -7.86 | -13.32 | -20.46 | -29.43 | -40.38 | -53.47 | -68.83 | -86.55 |
| TIME OF FLIGHT (SEC) | .000000 | .051037 | .104192 | .159526 | .217177 | .278093 | .342423 | .409963 | .480428 | .553559 | .629162 | .707094 | .787246 |
| WIND DEFLECTION (IN) | | | | | | | | | | | | | |
| 5 MPH | .00 | .09 | .37 | .84 | 1.51 | 2.47 | 3.73 | 5.28 | 7.08 | 9.11 | 11.37 | 13.82 | 16.48 |
| 10 MPH | .00 | .18 | .74 | 1.68 | 3.02 | 4.94 | 7.47 | 10.55 | 14.16 | 18.23 | 22.73 | 27.65 | 32.96 |
| 20 MPH | .00 | .36 | 1.48 | 3.35 | 6.05 | 9.89 | 14.93 | 21.11 | 28.31 | 36.45 | 45.47 | 55.30 | 65.91 |
| 30 MPH | .00 | .55 | 2.21 | 5.03 | 9.07 | 14.83 | 22.40 | 31.66 | 42.47 | 54.68 | 68.20 | 82.95 | 98.87 |
| VELOCITY (FPS) | 1400. | 1345. | 1291. | 1218. | 1154. | 1101. | 1056. | 1019. | 986. | 957. | 931. | 907. | 885. |
| ENERGY (FT-LB) | 688. | 635. | 585. | 520. | 467. | 425. | 391. | 364. | 341. | 321. | 304. | 289. | 275. |
| DROP (IN) | .00 | -.54 | -2.28 | -5.34 | -9.87 | -16.03 | -23.98 | -33.86 | -45.84 | -60.04 | -76.57 | -95.59 | -117.25 |
| MID-RANGE (IN) | .00 | .14 | .60 | 1.42 | 2.66 | 4.38 | 6.65 | 9.52 | 13.04 | 17.28 | 22.25 | 28.03 | 34.65 |
| BULLET PATH (IN) | -.63 | .28 | .00 | -1.61 | -4.68 | -9.39 | -15.88 | -24.31 | -34.83 | -47.57 | -62.65 | -80.22 | -100.42 |
| TIME OF FLIGHT (SEC) | .000000 | .054662 | .111536 | .171383 | .234681 | .301263 | .370861 | .443196 | .518056 | .595285 | .674764 | .756405 | .840144 |
| WIND DEFLECTION (IN) | | | | | | | | | | | | | |
| 5 MPH | .00 | .10 | .39 | .94 | 1.79 | 2.94 | 4.35 | 6.00 | 7.87 | 9.96 | 12.24 | 14.71 | 17.36 |
| 10 MPH | .00 | .19 | .77 | 1.88 | 3.59 | 5.88 | 8.70 | 12.00 | 15.75 | 19.91 | 24.47 | 29.41 | 34.72 |
| 20 MPH | .00 | .38 | 1.55 | 3.76 | 7.18 | 11.76 | 17.40 | 24.00 | 31.50 | 39.83 | 48.95 | 58.83 | 69.44 |
| 30 MPH | .00 | .58 | 2.32 | 5.63 | 10.77 | 17.64 | 26.10 | 36.01 | 47.25 | 59.74 | 73.42 | 88.24 | 104.17 |
| VELOCITY (FPS) | 1300. | 1226. | 1161. | 1106. | 1061. | 1023. | 990. | 960. | 934. | 910. | 887. | 867. | 847. |
| ENERGY (FT-LB) | 593. | 527. | 473. | 429. | 395. | 367. | 344. | 323. | 306. | 290. | 276. | 263. | 252. |
| DROP (IN) | .00 | -.65 | -2.75 | -6.46 | -11.94 | -19.34 | -28.81 | -40.49 | -54.50 | -70.98 | -90.06 | -111.90 | -136.62 |
| MID-RANGE (IN) | .00 | .17 | .73 | 1.72 | 3.22 | 5.28 | 7.95 | 11.28 | 15.31 | 20.10 | 25.69 | 32.14 | 39.50 |
| BULLET PATH (IN) | -.63 | .41 | .00 | -2.02 | -5.81 | -11.52 | -19.30 | -29.29 | -41.61 | -56.39 | -73.79 | -93.93 | -116.96 |
| TIME OF FLIGHT (SEC) | .000000 | .059444 | .122351 | .188567 | .257833 | .329865 | .404443 | .481405 | .560629 | .642023 | .725521 | .811076 | .898655 |
| WIND DEFLECTION (IN) | | | | | | | | | | | | | |
| 5 MPH | .00 | .15 | .61 | 1.36 | 2.38 | 3.64 | 5.13 | 6.83 | 8.72 | 10.81 | 13.08 | 15.53 | 18.16 |
| 10 MPH | .00 | .31 | 1.23 | 2.73 | 4.76 | 7.29 | 10.26 | 13.65 | 17.44 | 21.61 | 26.15 | 31.06 | 36.32 |
| 20 MPH | .00 | .62 | 2.45 | 5.45 | 9.53 | 14.57 | 20.52 | 27.30 | 34.88 | 43.22 | 52.31 | 62.11 | 72.63 |
| 30 MPH | .00 | .92 | 3.68 | 8.18 | 14.29 | 21.86 | 30.78 | 40.95 | 52.32 | 64.83 | 78.46 | 93.17 | 108.95 |
| VELOCITY (FPS) | 1200. | 1139. | 1088. | 1046. | 1010. | 978. | 950. | 924. | 901. | 879. | 859. | 840. | 822. |
| ENERGY (FT-LB) | 505. | 455. | 415. | 384. | 358. | 336. | 317. | 300. | 285. | 271. | 259. | 247. | 237. |
| DROP (IN) | .00 | -.77 | -3.20 | -7.47 | -13.71 | -22.08 | -32.72 | -45.71 | -61.23 | -79.42 | -100.41 | -124.33 | -151.33 |
| MID-RANGE (IN) | .00 | .20 | .83 | 1.97 | 3.65 | 5.94 | 8.89 | 12.52 | 16.90 | 22.09 | 28.12 | 35.06 | 42.95 |
| BULLET PATH (IN) | -.63 | .52 | .00 | -2.35 | -6.68 | -13.13 | -21.85 | -32.92 | -46.53 | -62.80 | -81.88 | -103.88 | -128.97 |
| TIME OF FLIGHT (SEC) | .000000 | .064186 | .131593 | .201938 | .274957 | .350455 | .428286 | .508341 | .590539 | .674820 | .761144 | .849484 | .939826 |
| WIND DEFLECTION (IN) | | | | | | | | | | | | | |
| 5 MPH | .00 | .15 | .58 | 1.27 | 2.20 | 3.34 | 4.69 | 6.23 | 7.97 | 9.88 | 11.98 | 14.25 | 16.70 |
| 10 MPH | .00 | .30 | 1.16 | 2.54 | 4.39 | 6.68 | 9.38 | 12.47 | 15.93 | 19.77 | 23.96 | 28.51 | 33.41 |
| 20 MPH | .00 | .59 | 2.32 | 5.08 | 8.78 | 13.36 | 18.76 | 24.94 | 31.87 | 39.54 | 47.92 | 57.02 | 66.82 |
| 30 MPH | .00 | .89 | 3.48 | 7.62 | 13.18 | 20.04 | 28.14 | 37.40 | 47.80 | 59.31 | 71.88 | 85.53 | 100.23 |

## Bullet: Hornady .357 Dia. 158 Gr. JHP
## Ballistic Coefficient: .248

| RANGE (YDS) | MUZZLE | 25 | 50 | 75 | 100 | 125 | 150 | 175 | 200 | 225 | 250 | 275 | 300 |
|---|---|---|---|---|---|---|---|---|---|---|---|---|---|
| VELOCITY (FPS) | 1100. | 1056. | 1018. | 986. | 957. | 931. | 907. | 885. | 864. | 844. | 826. | 808. | 789. |
| ENERGY (FT-LB) | 424. | 391. | 364. | 341. | 321. | 304. | 288. | 275. | 262. | 250. | 239. | 229. | 218. |
| DROP (IN) | .00 | -.90 | -3.74 | -8.68 | -15.84 | -25.35 | -37.33 | -51.95 | -69.33 | -89.62 | -112.94 | -139.44 | -169.24 |
| MID-RANGE (IN) | .00 | .23 | .97 | 2.27 | 4.18 | 6.73 | 9.99 | 14.00 | 18.82 | 24.51 | 31.13 | 38.70 | 47.28 |
| BULLET PATH (IN) | -.63 | .66 | .00 | -2.75 | -7.72 | -15.04 | -24.84 | -37.27 | -52.46 | -70.56 | -91.70 | -116.01 | -143.62 |
| TIME OF FLIGHT (SEC) | .000000 | .069637 | .142007 | .216900 | .294158 | .373664 | .455331 | .539094 | .624908 | .712743 | .802582 | .894422 | .988328 |
| WIND DEFLECTION (IN) | | | | | | | | | | | | | |
| 5 MPH | .00 | .13 | .50 | 1.09 | 1.89 | 2.88 | 4.07 | 5.44 | 6.99 | 8.72 | 10.63 | 12.71 | 14.97 |
| 10 MPH | .00 | .26 | .99 | 2.17 | 3.77 | 5.76 | 8.14 | 10.88 | 13.98 | 17.44 | 21.25 | 25.42 | 29.95 |
| 20 MPH | .00 | .51 | 1.99 | 4.35 | 7.54 | 11.53 | 16.28 | 21.76 | 27.97 | 34.89 | 42.51 | 50.84 | 59.89 |
| 30 MPH | .00 | .77 | 2.98 | 6.52 | 11.32 | 17.29 | 24.41 | 32.64 | 41.95 | 52.33 | 63.76 | 76.25 | 89.84 |
| VELOCITY (FPS) | 1000. | 970. | 942. | 917. | 895. | 873. | 853. | 834. | 816. | 798. | 777. | 756. | 736. |
| ENERGY (FT-LB) | 351. | 330. | 311. | 295. | 281. | 267. | 255. | 244. | 234. | 224. | 212. | 200. | 190. |
| DROP (IN) | .00 | -1.07 | -4.44 | -10.22 | -18.56 | -29.61 | -43.49 | -60.36 | -80.33 | -103.56 | -130.12 | -160.27 | -194.20 |
| MID-RANGE (IN) | .00 | .28 | 1.15 | 2.64 | 4.84 | 7.79 | 11.53 | 16.12 | 21.61 | 28.04 | 35.45 | 43.95 | 53.61 |
| BULLET PATH (IN) | -.63 | .83 | -.00 | -3.24 | -9.05 | -17.56 | -28.92 | -43.25 | -60.69 | -81.38 | -105.41 | -133.02 | -164.42 |
| TIME OF FLIGHT (SEC) | .000000 | .076190 | .154678 | .235363 | .318170 | .403046 | .489955 | .578875 | .669794 | .762716 | .858011 | .955970 | 1.056621 |
| WIND DEFLECTION (IN) | | | | | | | | | | | | | |
| 5 MPH | .00 | .10 | .41 | .91 | 1.60 | 2.47 | 3.52 | 4.74 | 6.14 | 7.72 | 9.50 | 11.53 | 13.78 |
| 10 MPH | .00 | .21 | .82 | 1.82 | 3.20 | 4.94 | 7.03 | 9.48 | 12.28 | 15.44 | 19.01 | 23.05 | 27.57 |
| 20 MPH | .00 | .42 | 1.65 | 3.65 | 6.40 | 9.87 | 14.06 | 18.96 | 24.57 | 30.88 | 38.02 | 46.10 | 55.13 |
| 30 MPH | .00 | .63 | 2.47 | 5.47 | 9.59 | 14.81 | 21.10 | 28.45 | 36.85 | 46.31 | 57.03 | 69.15 | 82.70 |
| VELOCITY (FPS) | 900. | 878. | 858. | 839. | 820. | 803. | 782. | 761. | 741. | 721. | 703. | 684. | 667. |
| ENERGY (FT-LB) | 284. | 271. | 258. | 247. | 236. | 226. | 214. | 203. | 192. | 182. | 173. | 164. | 156. |
| DROP (IN) | .00 | -1.25 | -5.31 | -12.31 | -22.39 | -35.69 | -52.30 | -72.42 | -96.28 | -124.08 | -156.05 | -192.40 | -233.38 |
| MID-RANGE (IN) | .00 | .34 | 1.40 | 3.23 | 5.89 | 9.41 | 13.84 | 19.25 | 25.75 | 33.41 | 42.33 | 52.63 | 64.39 |
| BULLET PATH (IN) | -.63 | 1.09 | -.00 | -4.03 | -11.15 | -21.48 | -35.11 | -52.27 | -73.16 | -97.99 | -126.99 | -160.37 | -198.38 |
| TIME OF FLIGHT (SEC) | .000000 | .084365 | .170770 | .259190 | .349610 | .442029 | .536665 | .633948 | .733914 | .836596 | .942038 | 1.050289 | 1.161408 |
| WIND DEFLECTION (IN) | | | | | | | | | | | | | |
| 5 MPH | .00 | .09 | .36 | .81 | 1.43 | 2.23 | 3.23 | 4.45 | 5.92 | 7.62 | 9.57 | 11.76 | 14.20 |
| 10 MPH | .00 | .18 | .72 | 1.62 | 2.86 | 4.46 | 6.45 | 8.91 | 11.84 | 15.24 | 19.13 | 23.52 | 28.41 |
| 20 MPH | .00 | .36 | 1.44 | 3.23 | 5.73 | 8.93 | 12.91 | 17.82 | 23.67 | 30.48 | 38.26 | 47.04 | 56.82 |
| 30 MPH | .00 | .54 | 2.17 | 4.85 | 8.59 | 13.39 | 19.36 | 26.72 | 35.51 | 45.72 | 57.40 | 70.55 | 85.22 |
| VELOCITY (FPS) | 800. | 778. | 757. | 737. | 718. | 699. | 681. | 664. | 647. | 630. | 614. | 599. | 584. |
| ENERGY (FT-LB) | 224. | 212. | 201. | 191. | 181. | 171. | 163. | 155. | 147. | 139. | 132. | 126. | 120. |
| DROP (IN) | .00 | -1.52 | -6.60 | -15.45 | -28.29 | -45.33 | -66.81 | -92.91 | -123.94 | -160.14 | -201.80 | -249.20 | -302.66 |
| MID-RANGE (IN) | .00 | .43 | 1.78 | 4.13 | 7.55 | 12.12 | 17.95 | 25.09 | 33.67 | 43.79 | 55.56 | 69.10 | 84.52 |
| BULLET PATH (IN) | -.63 | 1.46 | -.00 | -5.24 | -14.46 | -27.89 | -45.74 | -68.24 | -95.64 | -128.23 | -166.27 | -210.06 | -259.91 |
| TIME OF FLIGHT (SEC) | .000000 | .095074 | .192809 | .293230 | .396373 | .502281 | .611007 | .722608 | .837150 | .954704 | 1.075346 | 1.199159 | 1.326232 |
| WIND DEFLECTION (IN) | | | | | | | | | | | | | |
| 5 MPH | .00 | .12 | .47 | 1.05 | 1.88 | 2.95 | 4.27 | 5.84 | 7.67 | 9.76 | 12.13 | 14.78 | 17.71 |
| 10 MPH | .00 | .23 | .93 | 2.11 | 3.76 | 5.90 | 8.54 | 11.68 | 15.34 | 19.53 | 24.26 | 29.55 | 35.42 |
| 20 MPH | .00 | .47 | 1.87 | 4.22 | 7.52 | 11.80 | 17.07 | 23.36 | 30.68 | 39.06 | 48.52 | 59.10 | 70.83 |
| 30 MPH | .00 | .70 | 2.80 | 6.33 | 11.28 | 17.70 | 25.61 | 35.04 | 46.02 | 58.58 | 72.78 | 88.66 | 106.25 |

# Bullet: Speer .357 Dia. 160 Gr. HJSP(SWC)
## Ballistic Coefficient: .213

| RANGE (YDS) | MUZZLE | 25 | 50 | 75 | 100 | 125 | 150 | 175 | 200 | 225 | 250 | 275 | 300 |
|---|---|---|---|---|---|---|---|---|---|---|---|---|---|
| VELOCITY (FPS) | 1500. | 1430. | 1365. | 1304. | 1247. | 1196. | 1150. | 1110. | 1075. | 1044. | 1017. | 992. | 970. |
| ENERGY (FT-LB) | 799. | 727. | 662. | 604. | 552. | 508. | 470. | 437. | 410. | 387. | 367. | 350. | 334. |
| DROP (IN) | .00 | -.47 | -2.01 | -4.71 | -8.69 | -14.07 | -20.98 | -29.52 | -39.82 | -51.99 | -66.16 | -82.41 | -100.87 |
| MID-RANGE (IN) | .00 | .13 | .53 | 1.25 | 2.33 | 3.83 | 5.78 | 8.22 | 11.22 | 14.80 | 19.00 | 23.88 | 29.46 |
| BULLET PATH (IN) | -.63 | .22 | .00 | -1.38 | -4.04 | -8.10 | -13.68 | -20.90 | -29.88 | -40.74 | -53.58 | -68.52 | -85.66 |
| TIME OF FLIGHT (SEC) | .000000 | .051209 | .104898 | .161139 | .219979 | .281420 | .345406 | .411829 | .480541 | .551376 | .624195 | .698887 | .775365 |
| WIND DEFLECTION (IN) | | | | | | | | | | | | | |
| 5 MPH | .00 | .11 | .43 | .98 | 1.76 | 2.76 | 4.00 | 5.44 | 7.09 | 8.92 | 10.93 | 13.10 | 15.43 |
| 10 MPH | .00 | .21 | .86 | 1.96 | 3.52 | 5.53 | 7.99 | 10.88 | 14.18 | 17.84 | 21.86 | 26.20 | 30.86 |
| 20 MPH | .00 | .43 | 1.72 | 3.92 | 7.03 | 11.06 | 15.98 | 21.76 | 28.35 | 35.68 | 43.72 | 52.41 | 61.73 |
| 30 MPH | .00 | .64 | 2.59 | 5.88 | 10.55 | 16.59 | 23.97 | 32.65 | 42.53 | 53.53 | 65.57 | 78.61 | 92.59 |
| VELOCITY (FPS) | 1400. | 1336. | 1277. | 1223. | 1174. | 1131. | 1093. | 1060. | 1031. | 1005. | 982. | 960. | 940. |
| ENERGY (FT-LB) | 696. | 634. | 579. | 531. | 490. | 454. | 424. | 399. | 378. | 359. | 342. | 328. | 314. |
| DROP (IN) | .00 | -.55 | -2.31 | -5.41 | -9.96 | -16.09 | -23.91 | -33.54 | -45.10 | -58.70 | -74.44 | -92.43 | -112.74 |
| MID-RANGE (IN) | .00 | .14 | .61 | 1.43 | 2.67 | 4.36 | 6.55 | 9.28 | 12.59 | 16.53 | 21.13 | 26.44 | 32.46 |
| BULLET PATH (IN) | -.63 | .29 | .00 | -1.63 | -4.71 | -9.37 | -15.72 | -23.88 | -33.97 | -46.10 | -60.37 | -76.89 | -95.74 |
| TIME OF FLIGHT (SEC) | .000000 | .054842 | .112262 | .172288 | .234897 | .300006 | .367491 | .437187 | .508939 | .582621 | .658134 | .735398 | .814347 |
| WIND DEFLECTION (IN) | | | | | | | | | | | | | |
| 5 MPH | .00 | .11 | .45 | 1.02 | 1.81 | 2.83 | 4.05 | 5.47 | 7.07 | 8.84 | 10.77 | 12.86 | 15.09 |
| 10 MPH | .00 | .22 | .90 | 2.04 | 3.63 | 5.66 | 8.11 | 10.94 | 14.14 | 17.68 | 21.55 | 25.72 | 30.18 |
| 20 MPH | .00 | .45 | 1.80 | 4.07 | 7.26 | 11.32 | 16.21 | 21.89 | 28.29 | 35.37 | 43.09 | 51.43 | 60.36 |
| 30 MPH | .00 | .67 | 2.70 | 6.11 | 10.88 | 16.97 | 24.32 | 32.83 | 42.43 | 53.05 | 64.64 | 77.15 | 90.55 |
| VELOCITY (FPS) | 1300. | 1244. | 1193. | 1147. | 1107. | 1073. | 1042. | 1015. | 991. | 969. | 948. | 929. | 911. |
| ENERGY (FT-LB) | 600. | 549. | 505. | 468. | 436. | 409. | 386. | 366. | 349. | 333. | 319. | 307. | 295. |
| DROP (IN) | .00 | -.63 | -2.68 | -6.26 | -11.47 | -18.46 | -27.33 | -38.19 | -51.15 | -66.33 | -83.81 | -103.64 | -125.98 |
| MID-RANGE (IN) | .00 | .17 | .71 | 1.65 | 3.05 | 4.96 | 7.41 | 10.44 | 14.10 | 18.42 | 23.44 | 29.17 | 35.66 |
| BULLET PATH (IN) | -.63 | .39 | .00 | -1.92 | -5.48 | -10.81 | -18.02 | -27.23 | -38.53 | -52.05 | -67.88 | -86.05 | -106.74 |
| TIME OF FLIGHT (SEC) | .000000 | .058999 | .120596 | .184733 | .251299 | .320143 | .391101 | .464035 | .538837 | .615418 | .693708 | .773651 | .855200 |
| WIND DEFLECTION (IN) | | | | | | | | | | | | | |
| 5 MPH | .00 | .11 | .46 | 1.03 | 1.81 | 2.79 | 3.96 | 5.30 | 6.80 | 8.46 | 10.28 | 12.24 | 14.33 |
| 10 MPH | .00 | .23 | .92 | 2.05 | 3.61 | 5.58 | 7.91 | 10.59 | 13.60 | 16.93 | 20.55 | 24.47 | 28.67 |
| 20 MPH | .00 | .46 | 1.83 | 4.10 | 7.23 | 11.15 | 15.82 | 21.19 | 27.21 | 33.86 | 41.11 | 48.94 | 57.34 |
| 30 MPH | .00 | .69 | 2.75 | 6.15 | 10.84 | 16.73 | 23.73 | 31.78 | 40.81 | 50.79 | 61.66 | 73.41 | 86.01 |
| VELOCITY (FPS) | 1200. | 1154. | 1113. | 1077. | 1047. | 1019. | 994. | 972. | 951. | 932. | 914. | 897. | 881. |
| ENERGY (FT-LB) | 512. | 473. | 440. | 412. | 389. | 369. | 351. | 335. | 321. | 308. | 297. | 286. | 276. |
| DROP (IN) | .00 | -.75 | -3.14 | -7.27 | -13.27 | -21.24 | -31.30 | -43.56 | -58.10 | -75.00 | -94.37 | -116.35 | -141.01 |
| MID-RANGE (IN) | .00 | .20 | .81 | 1.90 | 3.50 | 5.64 | 8.38 | 11.75 | 15.78 | 20.50 | 25.94 | 32.17 | 39.20 |
| BULLET PATH (IN) | -.63 | .50 | .00 | -2.25 | -6.36 | -12.45 | -20.63 | -31.00 | -43.66 | -58.67 | -76.16 | -96.25 | -119.03 |
| TIME OF FLIGHT (SEC) | .000000 | .063763 | .129974 | .198489 | .269139 | .341785 | .416313 | .492631 | .570669 | .650366 | .731675 | .814560 | .898990 |
| WIND DEFLECTION (IN) | | | | | | | | | | | | | |
| 5 MPH | .00 | .11 | .44 | .97 | 1.68 | 2.58 | 3.64 | 4.85 | 6.22 | 7.73 | 9.39 | 11.18 | 13.11 |
| 10 MPH | .00 | .22 | .88 | 1.93 | 3.37 | 5.15 | 7.27 | 9.70 | 12.44 | 15.46 | 18.77 | 22.36 | 26.22 |
| 20 MPH | .00 | .44 | 1.75 | 3.87 | 6.74 | 10.31 | 14.54 | 19.41 | 24.88 | 30.93 | 37.55 | 44.73 | 52.44 |
| 30 MPH | .00 | .67 | 2.63 | 5.80 | 10.11 | 15.46 | 21.81 | 29.11 | 37.31 | 46.39 | 56.32 | 67.09 | 78.67 |

## Bullet: Speer .357 Dia. 160 Gr. HJSP(SWC)
### Ballistic Coefficient: .213

| RANGE (YDS) | MUZZLE | 25 | 50 | 75 | 100 | 125 | 150 | 175 | 200 | 225 | 250 | 275 | 300 |
|---|---|---|---|---|---|---|---|---|---|---|---|---|---|
| VELOCITY (FPS) | 1100. | 1066. | 1037. | 1010. | 986. | 964. | 944. | 925. | 908. | 891. | 875. | 860. | 846. |
| ENERGY (FT-LB) | 430. | 404. | 382. | 362. | 345. | 330. | 317. | 304. | 293. | 282. | 272. | 263. | 254. |
| DROP (IN) | .00 | -.89 | -3.68 | -8.49 | -15.42 | -24.58 | -36.06 | -49.92 | -66.31 | -85.32 | -107.07 | -131.64 | -159.14 |
| MID-RANGE (IN) | .00 | .23 | .95 | 2.21 | 4.03 | 6.46 | 9.54 | 13.28 | 17.73 | 22.95 | 28.95 | 35.79 | 43.51 |
| BULLET PATH (IN) | -.63 | .64 | .00 | -2.65 | -7.43 | -14.44 | -23.76 | -35.46 | -49.70 | -66.56 | -86.15 | -108.57 | -133.91 |
| TIME OF FLIGHT (SEC) | .000000 | .069275 | .140634 | .213945 | .289104 | .366027 | .444646 | .524907 | .606766 | .690189 | .775149 | .861626 | .949606 |
| WIND DEFLECTION (IN) | | | | | | | | | | | | | |
| 5 MPH | .00 | .10 | .38 | .83 | 1.44 | 2.21 | 3.13 | 4.19 | 5.40 | 6.74 | 8.21 | 9.82 | 11.57 |
| 10 MPH | .00 | .19 | .75 | 1.65 | 2.88 | 4.42 | 6.26 | 8.38 | 10.79 | 13.47 | 16.43 | 19.65 | 23.13 |
| 20 MPH | .00 | .38 | 1.50 | 3.31 | 5.76 | 8.84 | 12.52 | 16.77 | 21.58 | 26.95 | 32.85 | 39.29 | 46.26 |
| 30 MPH | .00 | .58 | 2.25 | 4.96 | 8.65 | 13.26 | 18.77 | 25.15 | 32.37 | 40.42 | 49.28 | 58.94 | 69.39 |
| VELOCITY (FPS) | 1000. | 977. | 956. | 936. | 918. | 901. | 884. | 869. | 854. | 840. | 826. | 813. | 800. |
| ENERGY (FT-LB) | 355. | 339. | 324. | 311. | 299. | 288. | 278. | 268. | 259. | 251. | 242. | 235. | 227. |
| DROP (IN) | .00 | -1.06 | -4.38 | -10.06 | -18.17 | -28.86 | -42.21 | -58.33 | -77.31 | -99.27 | -124.29 | -152.48 | -183.96 |
| MID-RANGE (IN) | .00 | .28 | 1.14 | 2.60 | 4.70 | 7.50 | 11.05 | 15.37 | 20.48 | 26.44 | 33.28 | 41.04 | 49.76 |
| BULLET PATH (IN) | -.63 | .82 | .00 | -3.16 | -8.78 | -16.95 | -27.80 | -41.41 | -57.89 | -77.33 | -99.85 | -125.53 | -154.50 |
| TIME OF FLIGHT (SEC) | .000000 | .075891 | .153517 | .232814 | .313734 | .396236 | .480289 | .565870 | .652962 | .741551 | .831634 | .923206 | 1.016271 |
| WIND DEFLECTION (IN) | | | | | | | | | | | | | |
| 5 MPH | .00 | .08 | .31 | .69 | 1.21 | 1.87 | 2.67 | 3.60 | 4.66 | 5.86 | 7.18 | 8.64 | 10.23 |
| 10 MPH | .00 | .16 | .62 | 1.38 | 2.42 | 3.74 | 5.33 | 7.19 | 9.32 | 11.71 | 14.37 | 17.28 | 20.46 |
| 20 MPH | .00 | .31 | 1.24 | 2.75 | 4.83 | 7.48 | 10.66 | 14.39 | 18.64 | 23.43 | 28.73 | 34.57 | 40.93 |
| 30 MPH | .00 | .47 | 1.86 | 4.13 | 7.25 | 11.21 | 15.99 | 21.58 | 27.96 | 35.14 | 43.10 | 51.85 | 61.39 |
| VELOCITY (FPS) | 900. | 884. | 868. | 853. | 839. | 825. | 812. | 799. | 786. | 773. | 761. | 749. | 737. |
| ENERGY (FT-LB) | 288. | 277. | 268. | 259. | 250. | 242. | 234. | 227. | 219. | 212. | 206. | 199. | 193. |
| DROP (IN) | .00 | -1.22 | -5.21 | -12.07 | -21.90 | -34.80 | -50.88 | -70.24 | -92.81 | -118.87 | -148.55 | -181.96 | -219.24 |
| MID-RANGE (IN) | .00 | .34 | 1.38 | 3.17 | 5.74 | 9.13 | 13.37 | 18.51 | 24.51 | 31.48 | 39.48 | 48.55 | 58.74 |
| BULLET PATH (IN) | -.63 | 1.07 | -.00 | -3.94 | -10.85 | -20.83 | -33.99 | -50.43 | -70.08 | -93.22 | -119.98 | -150.48 | -184.83 |
| TIME OF FLIGHT (SEC) | .000000 | .084104 | .169734 | .256873 | .345510 | .435638 | .527256 | .620365 | .715017 | .811262 | .909106 | 1.008562 | 1.109639 |
| WIND DEFLECTION (IN) | | | | | | | | | | | | | |
| 5 MPH | .00 | .07 | .27 | .60 | 1.07 | 1.67 | 2.40 | 3.26 | 4.25 | 5.39 | 6.67 | 8.09 | 9.65 |
| 10 MPH | .00 | .14 | .54 | 1.21 | 2.14 | 3.34 | 4.80 | 6.52 | 8.51 | 10.78 | 13.34 | 16.17 | 19.30 |
| 20 MPH | .00 | .27 | 1.08 | 2.42 | 4.29 | 6.68 | 9.59 | 13.04 | 17.02 | 21.56 | 26.67 | 32.35 | 38.59 |
| 30 MPH | .00 | .41 | 1.62 | 3.63 | 6.43 | 10.02 | 14.39 | 19.55 | 25.53 | 32.35 | 40.01 | 48.52 | 57.89 |
| VELOCITY (FPS) | 800. | 787. | 774. | 761. | 749. | 737. | 725. | 714. | 703. | 692. | 682. | 671. | 661. |
| ENERGY (FT-LB) | 227. | 220. | 213. | 206. | 199. | 193. | 187. | 181. | 176. | 170. | 165. | 160. | 155. |
| DROP (IN) | .00 | -1.41 | -6.21 | -14.62 | -26.75 | -42.73 | -62.70 | -86.77 | -115.10 | -147.80 | -185.03 | -226.92 | -273.63 |
| MID-RANGE (IN) | .00 | .38 | 1.70 | 3.94 | 7.17 | 11.41 | 16.73 | 23.18 | 30.80 | 39.65 | 49.78 | 61.25 | 74.12 |
| BULLET PATH (IN) | -.63 | 1.38 | -.00 | -4.99 | -13.70 | -26.26 | -42.81 | -63.46 | -88.36 | -117.65 | -151.46 | -189.93 | -233.22 |
| TIME OF FLIGHT (SEC) | .000000 | .094544 | .190679 | .288414 | .387756 | .488717 | .591312 | .695555 | .801464 | .909057 | 1.018355 | 1.129380 | 1.242154 |
| WIND DEFLECTION (IN) | | | | | | | | | | | | | |
| 5 MPH | .00 | .07 | .28 | .63 | 1.12 | 1.76 | 2.54 | 3.46 | 4.53 | 5.75 | 7.12 | 8.64 | 10.31 |
| 10 MPH | .00 | .14 | .56 | 1.26 | 2.24 | 3.51 | 5.07 | 6.92 | 9.06 | 11.49 | 14.23 | 17.27 | 20.62 |
| 20 MPH | .00 | .28 | 1.12 | 2.52 | 4.49 | 7.03 | 10.14 | 13.84 | 18.12 | 22.99 | 28.46 | 34.54 | 41.24 |
| 30 MPH | .00 | .42 | 1.68 | 3.78 | 6.73 | 10.54 | 15.21 | 20.75 | 27.17 | 34.48 | 42.69 | 51.81 | 61.86 |

## Bullet: Lyman #358242 93 Gr.
### Ballistic Coefficient: .105

| RANGE (YDS) | MUZZLE | 25 | 50 | 75 | 100 | 125 | 150 | 175 | 200 | 225 | 250 | 275 | 300 |
|---|---|---|---|---|---|---|---|---|---|---|---|---|---|
| VELOCITY (FPS) | 1700. | 1543. | 1402. | 1277. | 1171. | 1089. | 1026. | 976. | 935. | 898. | 866. | 836. | 808. |
| ENERGY (FT-LB) | 597. | 492. | 406. | 336. | 283. | 245. | 217. | 197. | 180. | 167. | 155. | 144. | 135. |
| DROP (IN) | .00 | -.38 | -1.69 | -4.10 | -7.84 | -13.18 | -20.33 | -29.55 | -41.02 | -54.95 | -71.57 | -91.08 | -113.69 |
| MID-RANGE (IN) | .00 | .10 | .46 | 1.14 | 2.24 | 3.85 | 6.07 | 8.98 | 12.67 | 17.18 | 22.61 | 29.03 | 36.51 |
| BULLET PATH (IN) | -.63 | .14 | .00 | -1.25 | -3.84 | -8.02 | -14.02 | -22.07 | -32.39 | -45.16 | -60.62 | -78.97 | -100.42 |
| TIME OF FLIGHT (SEC) | .000000 | .046316 | .097335 | .153444 | .214864 | .281392 | .352434 | .427422 | .505983 | .587881 | .672968 | .761162 | .852426 |
| WIND DEFLECTION (IN) | | | | | | | | | | | | | |
| 5 MPH | .00 | .19 | .80 | 1.86 | 3.38 | 5.35 | 7.72 | 10.44 | 13.47 | 16.79 | 20.40 | 24.28 | 28.43 |
| 10 MPH | .00 | .39 | 1.60 | 3.71 | 6.76 | 10.70 | 15.44 | 20.87 | 26.94 | 33.58 | 40.80 | 48.55 | 56.85 |
| 20 MPH | .00 | .77 | 3.20 | 7.42 | 13.51 | 21.40 | 30.88 | 41.75 | 53.87 | 67.17 | 81.59 | 97.11 | 113.70 |
| 30 MPH | .00 | 1.16 | 4.80 | 11.14 | 20.27 | 32.10 | 46.32 | 62.62 | 80.81 | 100.75 | 122.39 | 145.66 | 170.55 |
| VELOCITY (FPS) | 1600. | 1452. | 1321. | 1208. | 1117. | 1048. | 994. | 949. | 911. | 877. | 847. | 819. | 792. |
| ENERGY (FT-LB) | 529. | 436. | 360. | 301. | 257. | 227. | 204. | 186. | 171. | 159. | 148. | 138. | 129. |
| DROP (IN) | .00 | -.44 | -1.91 | -4.63 | -8.84 | -14.79 | -22.71 | -32.82 | -45.31 | -60.41 | -78.31 | -99.24 | -123.38 |
| MID-RANGE (IN) | .00 | .12 | .52 | 1.28 | 2.51 | 4.30 | 6.73 | 9.88 | 13.81 | 18.62 | 24.37 | 31.13 | 38.98 |
| BULLET PATH (IN) | -.63 | .20 | .00 | -1.45 | -4.39 | -9.07 | -15.72 | -24.57 | -35.78 | -49.61 | -66.24 | -85.90 | -108.77 |
| TIME OF FLIGHT (SEC) | .000000 | .049221 | .103403 | .162840 | .227521 | .296961 | .370532 | .447796 | .528473 | .612385 | .699428 | .789550 | .882746 |
| WIND DEFLECTION (IN) | | | | | | | | | | | | | |
| 5 MPH | .00 | .21 | .85 | 1.95 | 3.52 | 5.51 | 7.86 | 10.53 | 13.51 | 16.76 | 20.30 | 24.11 | 28.18 |
| 10 MPH | .00 | .41 | 1.70 | 3.91 | 7.04 | 11.02 | 15.71 | 21.06 | 27.01 | 33.53 | 40.60 | 48.21 | 56.36 |
| 20 MPH | .00 | .83 | 3.40 | 7.82 | 14.09 | 22.03 | 31.43 | 42.12 | 54.02 | 67.06 | 81.20 | 96.42 | 112.73 |
| 30 MPH | .00 | 1.24 | 5.10 | 11.73 | 21.13 | 33.05 | 47.14 | 63.19 | 81.03 | 100.59 | 121.80 | 144.63 | 169.09 |
| VELOCITY (FPS) | 1500. | 1363. | 1243. | 1144. | 1069. | 1010. | 963. | 923. | 888. | 857. | 828. | 801. | 775. |
| ENERGY (FT-LB) | 465. | 384. | 319. | 270. | 236. | 211. | 192. | 176. | 163. | 152. | 141. | 132. | 124. |
| DROP (IN) | .00 | -.50 | -2.18 | -5.26 | -10.00 | -16.63 | -25.39 | -36.46 | -50.05 | -66.39 | -85.69 | -108.14 | -133.89 |
| MID-RANGE (IN) | .00 | .13 | .59 | 1.45 | 2.82 | 4.79 | 7.43 | 10.82 | 15.03 | 20.13 | 26.21 | 33.34 | 41.56 |
| BULLET PATH (IN) | -.63 | .27 | .00 | -1.68 | -5.02 | -10.25 | -17.60 | -27.26 | -39.46 | -54.39 | -72.28 | -93.33 | -117.68 |
| TIME OF FLIGHT (SEC) | .000000 | .052485 | .110145 | .173109 | .241034 | .313284 | .389352 | .468913 | .551760 | .637766 | .726862 | .819027 | .914309 |
| WIND DEFLECTION (IN) | | | | | | | | | | | | | |
| 5 MPH | .00 | .22 | .89 | 2.03 | 3.61 | 5.57 | 7.86 | 10.46 | 13.35 | 16.52 | 19.96 | 23.67 | 27.66 |
| 10 MPH | .00 | .44 | 1.79 | 4.07 | 7.22 | 11.14 | 15.73 | 20.93 | 26.71 | 33.05 | 39.93 | 47.35 | 55.32 |
| 20 MPH | .00 | .87 | 3.57 | 8.13 | 14.44 | 22.28 | 31.45 | 41.86 | 53.42 | 66.09 | 79.86 | 94.70 | 110.64 |
| 30 MPH | .00 | 1.31 | 5.36 | 12.20 | 21.67 | 33.41 | 47.18 | 62.79 | 80.13 | 99.14 | 119.78 | 142.05 | 165.96 |
| VELOCITY (FPS) | 1400. | 1275. | 1170. | 1088. | 1026. | 976. | 934. | 898. | 865. | 836. | 808. | 782. | 757. |
| ENERGY (FT-LB) | 405. | 336. | 283. | 244. | 217. | 197. | 180. | 166. | 155. | 144. | 135. | 126. | 118. |
| DROP (IN) | .00 | -.58 | -2.50 | -6.00 | -11.33 | -18.72 | -28.38 | -40.49 | -55.30 | -72.99 | -93.79 | -117.85 | -145.40 |
| MID-RANGE (IN) | .00 | .15 | .67 | 1.65 | 3.17 | 5.33 | 8.19 | 11.83 | 16.32 | 21.74 | 28.17 | 35.67 | 44.32 |
| BULLET PATH (IN) | -.63 | .36 | .00 | -1.94 | -5.71 | -11.54 | -19.63 | -30.18 | -43.42 | -59.55 | -78.79 | -101.28 | -127.27 |
| TIME OF FLIGHT (SEC) | .000000 | .056170 | .117650 | .184234 | .255324 | .330354 | .408954 | .490888 | .576010 | .664237 | .755535 | .849927 | .947504 |
| WIND DEFLECTION (IN) | | | | | | | | | | | | | |
| 5 MPH | .00 | .23 | .92 | 2.07 | 3.61 | 5.50 | 7.70 | 10.20 | 12.97 | 16.02 | 19.34 | 22.94 | 26.81 |
| 10 MPH | .00 | .46 | 1.85 | 4.14 | 7.22 | 11.00 | 15.40 | 20.40 | 25.95 | 32.05 | 38.69 | 45.87 | 53.62 |
| 20 MPH | .00 | .91 | 3.70 | 8.28 | 14.45 | 22.00 | 30.81 | 40.79 | 51.90 | 64.10 | 77.38 | 91.75 | 107.24 |
| 30 MPH | .00 | 1.37 | 5.55 | 12.42 | 21.67 | 33.00 | 46.21 | 61.19 | 77.85 | 96.15 | 116.07 | 137.62 | 160.85 |

## Bullet: Lyman #358242 93 Gr.
**Ballistic Coefficient: .105**

| RANGE (YDS) | MUZZLE | 25 | 50 | 75 | 100 | 125 | 150 | 175 | 200 | 225 | 250 | 275 | 300 |
|---|---|---|---|---|---|---|---|---|---|---|---|---|---|
| VELOCITY (FPS) | 1300. | 1190. | 1103. | 1037. | 985. | 942. | 905. | 872. | 842. | 814. | 787. | 762. | 738. |
| ENERGY (FT-LB) | 349. | 292. | 251. | 222. | 200. | 183. | 169. | 157. | 146. | 137. | 128. | 120. | 112. |
| DROP (IN) | .00 | -.67 | -2.88 | -6.86 | -12.86 | -21.09 | -31.72 | -45.00 | -61.13 | -80.32 | -102.75 | -128.60 | -158.17 |
| MID-RANGE (IN) | .00 | .18 | .77 | 1.86 | 3.55 | 5.91 | 9.00 | 12.90 | 17.70 | 23.47 | 30.29 | 38.21 | 47.36 |
| BULLET PATH (IN) | -.63 | .45 | .00 | -2.23 | -6.47 | -12.94 | -21.83 | -33.35 | -47.73 | -65.16 | -85.83 | -109.93 | -137.75 |
| TIME OF FLIGHT (SEC) | .000000 | .060373 | .125935 | .196140 | .270385 | .348264 | .429518 | .513985 | .601569 | .692227 | .785964 | .882870 | .982994 |
| WIND DEFLECTION (IN) | | | | | | | | | | | | | |
| 5 MPH | .00 | .24 | .93 | 2.03 | 3.49 | 5.26 | 7.34 | 9.69 | 12.32 | 15.22 | 18.40 | 21.85 | 25.58 |
| 10 MPH | .00 | .47 | 1.86 | 4.06 | 6.97 | 10.53 | 14.67 | 19.38 | 24.65 | 30.45 | 36.79 | 43.69 | 51.16 |
| 20 MPH | .00 | .94 | 3.71 | 8.12 | 13.94 | 21.05 | 29.34 | 38.77 | 49.29 | 60.89 | 73.58 | 87.39 | 102.32 |
| 30 MPH | .00 | 1.42 | 5.57 | 12.18 | 20.92 | 31.58 | 44.02 | 58.15 | 73.94 | 91.34 | 110.37 | 131.08 | 153.48 |
| VELOCITY (FPS) | 1200. | 1111. | 1043. | 990. | 946. | 909. | 875. | 845. | 816. | 790. | 764. | 740. | 717. |
| ENERGY (FT-LB) | 297. | 255. | 225. | 202. | 185. | 170. | 158. | 147. | 138. | 129. | 121. | 113. | 106. |
| DROP (IN) | .00 | -.79 | -3.33 | -7.86 | -14.59 | -23.72 | -35.47 | -50.05 | -67.66 | -88.50 | -112.74 | -140.67 | -172.56 |
| MID-RANGE (IN) | .00 | .20 | .88 | 2.11 | 3.97 | 6.53 | 9.88 | 14.09 | 19.24 | 25.41 | 32.65 | 41.08 | 50.80 |
| BULLET PATH (IN) | -.63 | .56 | .00 | -2.55 | -7.31 | -14.46 | -24.24 | -36.83 | -52.47 | -71.33 | -93.59 | -119.54 | -149.45 |
| TIME OF FLIGHT (SEC) | .000000 | .065057 | .134824 | .208683 | .286208 | .367130 | .451277 | .538548 | .628895 | .722317 | .818897 | .918691 | 1.021748 |
| WIND DEFLECTION (IN) | | | | | | | | | | | | | |
| 5 MPH | .00 | .23 | .86 | 1.86 | 3.19 | 4.81 | 6.71 | 8.89 | 11.34 | 14.06 | 17.06 | 20.34 | 23.91 |
| 10 MPH | .00 | .45 | 1.73 | 3.73 | 6.37 | 9.61 | 13.42 | 17.78 | 22.69 | 28.13 | 34.13 | 40.69 | 47.83 |
| 20 MPH | .00 | .90 | 3.46 | 7.46 | 12.75 | 19.23 | 26.85 | 35.57 | 45.37 | 56.26 | 68.25 | 81.38 | 95.66 |
| 30 MPH | .00 | 1.35 | 5.19 | 11.18 | 19.12 | 28.84 | 40.27 | 53.35 | 68.06 | 84.38 | 102.38 | 122.07 | 143.48 |
| VELOCITY (FPS) | 1100. | 1035. | 984. | 941. | 904. | 871. | 840. | 813. | 786. | 761. | 737. | 714. | 691. |
| ENERGY (FT-LB) | 250. | 221. | 200. | 183. | 169. | 156. | 146. | 136. | 128. | 119. | 112. | 105. | 99. |
| DROP (IN) | .00 | -.92 | -3.86 | -9.04 | -16.64 | -26.89 | -40.00 | -56.17 | -75.58 | -98.43 | -125.02 | -155.60 | -190.43 |
| MID-RANGE (IN) | .00 | .24 | 1.01 | 2.39 | 4.46 | 7.28 | 10.96 | 15.56 | 21.15 | 27.80 | 35.62 | 44.72 | 55.22 |
| BULLET PATH (IN) | -.63 | .70 | .00 | -2.93 | -8.28 | -16.29 | -27.15 | -41.07 | -58.23 | -78.84 | -103.18 | -131.51 | -164.10 |
| TIME OF FLIGHT (SEC) | .000000 | .070376 | .144772 | .222788 | .304171 | .388762 | .476467 | .567245 | .661107 | .758144 | .858401 | .961929 | 1.068792 |
| WIND DEFLECTION (IN) | | | | | | | | | | | | | |
| 5 MPH | .00 | .19 | .74 | 1.61 | 2.77 | 4.21 | 5.93 | 7.92 | 10.18 | 12.72 | 15.54 | 18.65 | 22.05 |
| 10 MPH | .00 | .39 | 1.48 | 3.21 | 5.53 | 8.42 | 11.86 | 15.84 | 20.35 | 25.43 | 31.08 | 37.30 | 44.11 |
| 20 MPH | .00 | .77 | 2.96 | 6.42 | 11.07 | 16.84 | 23.72 | 31.67 | 40.71 | 50.87 | 62.16 | 74.60 | 88.21 |
| 30 MPH | .00 | 1.16 | 4.44 | 9.63 | 16.60 | 25.27 | 35.57 | 47.51 | 61.06 | 76.30 | 93.24 | 111.90 | 132.32 |
| VELOCITY (FPS) | 1000. | 955. | 916. | 881. | 851. | 822. | 795. | 769. | 745. | 721. | 699. | 677. | 657. |
| ENERGY (FT-LB) | 206. | 188. | 173. | 160. | 149. | 139. | 131. | 122. | 114. | 107. | 101. | 95. | 89. |
| DROP (IN) | .00 | -1.10 | -4.55 | -10.57 | -19.38 | -31.19 | -46.20 | -64.53 | -86.52 | -112.41 | -142.46 | -176.94 | -216.15 |
| MID-RANGE (IN) | .00 | .28 | 1.18 | 2.77 | 5.15 | 8.37 | 12.53 | 17.65 | 23.88 | 31.30 | 40.04 | 50.19 | 61.88 |
| BULLET PATH (IN) | -.63 | .86 | .00 | -3.44 | -9.66 | -18.88 | -31.30 | -47.05 | -66.45 | -89.74 | -117.21 | -149.10 | -185.72 |
| TIME OF FLIGHT (SEC) | .000000 | .076808 | .157058 | .240560 | .327203 | .416927 | .509720 | .605655 | .704797 | .807189 | .912892 | 1.021980 | 1.134537 |
| WIND DEFLECTION (IN) | | | | | | | | | | | | | |
| 5 MPH | .00 | .16 | .62 | 1.37 | 2.39 | 3.69 | 5.26 | 7.10 | 9.22 | 11.63 | 14.33 | 17.33 | 20.64 |
| 10 MPH | .00 | .32 | 1.24 | 2.74 | 4.79 | 7.38 | 10.51 | 14.20 | 18.44 | 23.27 | 28.67 | 34.67 | 41.28 |
| 20 MPH | .00 | .64 | 2.48 | 5.48 | 9.58 | 14.76 | 21.02 | 28.39 | 36.89 | 46.53 | 57.34 | 69.34 | 82.56 |
| 30 MPH | .00 | .95 | 3.73 | 8.22 | 14.36 | 22.14 | 31.53 | 42.59 | 55.33 | 69.80 | 86.01 | 104.01 | 123.84 |

## Bullet: Lyman #358345 105 Gr.
### Ballistic Coefficient: .121

| RANGE (YDS) | MUZZLE | 25 | 50 | 75 | 100 | 125 | 150 | 175 | 200 | 225 | 250 | 275 | 300 |
|---|---|---|---|---|---|---|---|---|---|---|---|---|---|
| VELOCITY (FPS) | 1600. | 1471. | 1354. | 1250. | 1161. | 1090. | 1034. | 989. | 951. | 917. | 887. | 859. | 834. |
| ENERGY (FT-LB) | 597. | 504. | 427. | 364. | 314. | 277. | 249. | 228. | 211. | 196. | 183. | 172. | 162. |
| DROP (IN) | .00 | -.43 | -1.87 | -4.50 | -8.52 | -14.15 | -21.60 | -31.07 | -42.76 | -56.82 | -73.45 | -92.83 | -115.14 |
| MID-RANGE (IN) | .00 | .12 | .51 | 1.23 | 2.39 | 4.05 | 6.30 | 9.22 | 12.86 | 17.29 | 22.57 | 28.78 | 35.97 |
| BULLET PATH (IN) | -.63 | .19 | .00 | -1.37 | -4.15 | -8.52 | -14.73 | -22.95 | -33.39 | -46.19 | -61.57 | -79.70 | -100.76 |
| TIME OF FLIGHT (SEC) | .000000 | .048902 | .102069 | .159747 | .222066 | .288807 | .359500 | .433694 | .511079 | .591448 | .674661 | .760625 | .849291 |
| WIND DEFLECTION (IN) | | | | | | | | | | | | | |
| 5 MPH | .00 | .18 | .73 | 1.68 | 3.04 | 4.79 | 6.89 | 9.29 | 11.97 | 14.92 | 18.12 | 21.56 | 25.24 |
| 10 MPH | .00 | .36 | 1.46 | 3.37 | 6.08 | 9.58 | 13.77 | 18.58 | 23.95 | 29.84 | 36.24 | 43.12 | 50.48 |
| 20 MPH | .00 | .71 | 2.93 | 6.73 | 12.17 | 19.16 | 27.54 | 37.16 | 47.90 | 59.69 | 72.48 | 86.24 | 100.95 |
| 30 MPH | .00 | 1.07 | 4.39 | 10.10 | 18.25 | 28.74 | 41.32 | 55.74 | 71.85 | 89.53 | 108.72 | 129.36 | 151.43 |
| VELOCITY (FPS) | 1500. | 1380. | 1273. | 1180. | 1105. | 1046. | 999. | 959. | 924. | 893. | 865. | 839. | 815. |
| ENERGY (FT-LB) | 524. | 444. | 378. | 325. | 285. | 255. | 233. | 214. | 199. | 186. | 175. | 164. | 155. |
| DROP (IN) | .00 | -.50 | -2.13 | -5.12 | -9.66 | -15.97 | -24.27 | -34.73 | -47.53 | -62.84 | -80.87 | -101.79 | -125.79 |
| MID-RANGE (IN) | .00 | .13 | .57 | 1.40 | 2.69 | 4.54 | 7.02 | 10.18 | 14.10 | 18.84 | 24.46 | 31.04 | 38.63 |
| BULLET PATH (IN) | -.63 | .26 | .00 | -1.60 | -4.76 | -9.69 | -16.60 | -25.68 | -37.10 | -51.03 | -67.68 | -87.22 | -109.83 |
| TIME OF FLIGHT (SEC) | .000000 | .052149 | .108760 | .170010 | .235763 | .305585 | .378999 | .455666 | .535358 | .617921 | .703253 | .791295 | .882019 |
| WIND DEFLECTION (IN) | | | | | | | | | | | | | |
| 5 MPH | .00 | .19 | .77 | 1.76 | 3.15 | 4.89 | 6.95 | 9.30 | 11.91 | 14.78 | 17.89 | 21.23 | 24.82 |
| 10 MPH | .00 | .38 | 1.54 | 3.52 | 6.29 | 9.78 | 13.90 | 18.60 | 23.82 | 29.55 | 35.77 | 42.47 | 49.64 |
| 20 MPH | .00 | .76 | 3.08 | 7.04 | 12.59 | 19.57 | 27.81 | 37.19 | 47.65 | 59.11 | 71.55 | 84.94 | 99.27 |
| 30 MPH | .00 | 1.13 | 4.63 | 10.57 | 18.88 | 29.35 | 41.71 | 55.79 | 71.47 | 88.66 | 107.32 | 127.40 | 148.91 |
| VELOCITY (FPS) | 1400. | 1291. | 1195. | 1117. | 1056. | 1007. | 966. | 930. | 899. | 870. | 844. | 819. | 795. |
| ENERGY (FT-LB) | 457. | 388. | 333. | 291. | 260. | 236. | 217. | 202. | 188. | 176. | 166. | 156. | 147. |
| DROP (IN) | .00 | -.57 | -2.45 | -5.85 | -10.99 | -18.06 | -27.28 | -38.81 | -52.82 | -69.50 | -89.06 | -111.65 | -137.47 |
| MID-RANGE (IN) | .00 | .15 | .65 | 1.59 | 3.04 | 5.09 | 7.79 | 11.22 | 15.42 | 20.48 | 26.46 | 33.43 | 41.45 |
| BULLET PATH (IN) | -.63 | .34 | .00 | -1.86 | -5.46 | -11.00 | -18.67 | -28.66 | -41.13 | -56.28 | -74.29 | -95.35 | -119.62 |
| TIME OF FLIGHT (SEC) | .000000 | .055821 | .116262 | .181255 | .250406 | .323222 | .399342 | .478520 | .560590 | .645445 | .733017 | .823274 | .916217 |
| WIND DEFLECTION (IN) | | | | | | | | | | | | | |
| 5 MPH | .00 | .20 | .80 | 1.81 | 3.18 | 4.87 | 6.86 | 9.11 | 11.62 | 14.37 | 17.36 | 20.59 | 24.06 |
| 10 MPH | .00 | .40 | 1.60 | 3.62 | 6.36 | 9.74 | 13.71 | 18.22 | 23.24 | 28.74 | 34.73 | 41.18 | 48.11 |
| 20 MPH | .00 | .79 | 3.21 | 7.23 | 12.71 | 19.49 | 27.43 | 36.44 | 46.47 | 57.48 | 69.45 | 82.36 | 96.22 |
| 30 MPH | .00 | 1.19 | 4.81 | 10.85 | 19.07 | 29.23 | 41.14 | 54.66 | 69.71 | 86.22 | 104.18 | 123.55 | 144.33 |
| VELOCITY (FPS) | 1300. | 1203. | 1123. | 1060. | 1010. | 969. | 933. | 901. | 872. | 846. | 821. | 798. | 772. |
| ENERGY (FT-LB) | 394. | 337. | 294. | 262. | 238. | 219. | 203. | 189. | 177. | 167. | 157. | 148. | 139. |
| DROP (IN) | .00 | -.66 | -2.83 | -6.72 | -12.53 | -20.46 | -30.69 | -43.38 | -58.74 | -76.94 | -98.18 | -122.62 | -150.39 |
| MID-RANGE (IN) | .00 | .18 | .75 | 1.81 | 3.43 | 5.68 | 8.63 | 12.32 | 16.84 | 22.25 | 28.63 | 36.03 | 44.51 |
| BULLET PATH (IN) | -.63 | .44 | .00 | -2.15 | -6.23 | -12.43 | -20.93 | -31.89 | -45.52 | -61.99 | -81.49 | -104.20 | -130.24 |
| TIME OF FLIGHT (SEC) | .000000 | .060035 | .124644 | .193454 | .265968 | .341811 | .420730 | .502553 | .587168 | .674505 | .764528 | .857229 | .952881 |
| WIND DEFLECTION (IN) | | | | | | | | | | | | | |
| 5 MPH | .00 | .21 | .81 | 1.79 | 3.10 | 4.69 | 6.56 | 8.69 | 11.06 | 13.66 | 16.51 | 19.59 | 22.93 |
| 10 MPH | .00 | .41 | 1.63 | 3.59 | 6.19 | 9.39 | 13.13 | 17.37 | 22.11 | 27.33 | 33.02 | 39.18 | 45.86 |
| 20 MPH | .00 | .82 | 3.26 | 7.17 | 12.39 | 18.78 | 26.25 | 34.74 | 44.22 | 54.66 | 66.04 | 78.36 | 91.72 |
| 30 MPH | .00 | 1.24 | 4.89 | 10.76 | 18.58 | 28.17 | 39.38 | 52.12 | 66.33 | 81.98 | 99.06 | 117.54 | 137.58 |

# Bullet: Lyman #358345 105 Gr.
## Ballistic Coefficient: .121

| RANGE (YDS) | MUZZLE | 25 | 50 | 75 | 100 | 125 | 150 | 175 | 200 | 225 | 250 | 275 | 300 |
|---|---|---|---|---|---|---|---|---|---|---|---|---|---|
| VELOCITY (FPS) | 1200. | 1121. | 1059. | 1009. | 968. | 932. | 900. | 872. | 845. | 820. | 797. | 771. | 746. |
| ENERGY (FT-LB) | 336. | 293. | 261. | 237. | 218. | 202. | 189. | 177. | 166. | 157. | 148. | 139. | 130. |
| DROP (IN) | .00 | -.78 | -3.28 | -7.72 | -14.28 | -23.14 | -34.47 | -48.48 | -65.33 | -85.22 | -108.32 | -134.76 | -164.84 |
| MID-RANGE (IN) | .00 | .20 | .86 | 2.06 | 3.86 | 6.32 | 9.52 | 13.52 | 18.39 | 24.19 | 31.02 | 38.90 | 47.94 |
| BULLET PATH (IN) | -.63 | .55 | .00 | -2.48 | -7.08 | -13.99 | -23.37 | -35.41 | -50.31 | -68.24 | -89.39 | -113.88 | -141.99 |
| TIME OF FLIGHT (SEC) | .000000 | .064742 | .133670 | .206288 | .282226 | .361234 | .443142 | .527838 | .615255 | .705357 | .798137 | .893868 | .992804 |
| WIND DEFLECTION (IN) | | | | | | | | | | | | | |
| 5 MPH | .00 | .20 | .76 | 1.65 | 2.84 | 4.29 | 6.00 | 7.95 | 10.14 | 12.57 | 15.24 | 18.16 | 21.37 |
| 10 MPH | .00 | .39 | 1.53 | 3.31 | 5.67 | 8.58 | 11.99 | 15.90 | 20.28 | 25.14 | 30.47 | 36.32 | 42.73 |
| 20 MPH | .00 | .79 | 3.05 | 6.61 | 11.34 | 17.15 | 23.99 | 31.80 | 40.57 | 50.29 | 60.94 | 72.64 | 85.47 |
| 30 MPH | .00 | 1.18 | 4.58 | 9.92 | 17.02 | 25.73 | 35.98 | 47.70 | 60.85 | 75.43 | 91.42 | 108.96 | 128.20 |
| VELOCITY (FPS) | 1100. | 1042. | 996. | 956. | 922. | 891. | 863. | 837. | 813. | 788. | 763. | 739. | 716. |
| ENERGY (FT-LB) | 282. | 253. | 231. | 213. | 198. | 185. | 174. | 163. | 154. | 145. | 136. | 127. | 119. |
| DROP (IN) | .00 | -.91 | -3.82 | -8.91 | -16.35 | -26.33 | -39.03 | -54.63 | -73.33 | -95.26 | -120.63 | -149.72 | -182.77 |
| MID-RANGE (IN) | .00 | .24 | 1.00 | 2.35 | 4.35 | 7.08 | 10.60 | 14.99 | 20.31 | 26.62 | 33.99 | 42.53 | 52.36 |
| BULLET PATH (IN) | -.63 | .68 | .00 | -2.87 | -8.08 | -15.83 | -26.30 | -39.68 | -56.15 | -75.86 | -99.01 | -125.86 | -156.69 |
| TIME OF FLIGHT (SEC) | .000000 | .070109 | .143778 | .220680 | .300592 | .383365 | .468900 | .557141 | .648063 | .741714 | .838474 | .938451 | 1.041694 |
| WIND DEFLECTION (IN) | | | | | | | | | | | | | |
| 5 MPH | .00 | .17 | .65 | 1.42 | 2.45 | 3.74 | 5.26 | 7.03 | 9.03 | 11.27 | 13.79 | 16.58 | 19.67 |
| 10 MPH | .00 | .34 | 1.30 | 2.84 | 4.90 | 7.47 | 10.53 | 14.06 | 18.06 | 22.54 | 27.57 | 33.17 | 39.34 |
| 20 MPH | .00 | .68 | 2.61 | 5.68 | 9.81 | 14.94 | 21.05 | 28.11 | 36.12 | 45.08 | 55.14 | 66.33 | 78.68 |
| 30 MPH | .00 | 1.02 | 3.91 | 8.52 | 14.71 | 22.42 | 31.58 | 42.17 | 54.18 | 67.62 | 82.71 | 99.50 | 118.01 |
| VELOCITY (FPS) | 1000. | 960. | 925. | 894. | 866. | 840. | 815. | 791. | 765. | 741. | 718. | 696. | 674. |
| ENERGY (FT-LB) | 233. | 215. | 200. | 186. | 175. | 164. | 155. | 146. | 137. | 128. | 120. | 113. | 106. |
| DROP (IN) | .00 | -1.09 | -4.51 | -10.45 | -19.09 | -30.62 | -45.22 | -63.06 | -84.29 | -109.21 | -138.07 | -171.13 | -208.67 |
| MID-RANGE (IN) | .00 | .28 | 1.17 | 2.72 | 5.03 | 8.16 | 12.16 | 17.11 | 23.06 | 30.13 | 38.42 | 48.04 | 59.11 |
| BULLET PATH (IN) | -.63 | .85 | .00 | -3.37 | -9.44 | -18.40 | -30.43 | -45.70 | -64.36 | -86.71 | -113.00 | -143.49 | -178.46 |
| TIME OF FLIGHT (SEC) | .000000 | .076584 | .156197 | .238682 | .323938 | .411904 | .502552 | .595901 | .692315 | .791941 | .894826 | 1.001033 | 1.110637 |
| WIND DEFLECTION (IN) | | | | | | | | | | | | | |
| 5 MPH | .00 | .14 | .55 | 1.20 | 2.11 | 3.25 | 4.62 | 6.24 | 8.12 | 10.29 | 12.74 | 15.49 | 18.54 |
| 10 MPH | .00 | .28 | 1.09 | 2.41 | 4.21 | 6.50 | 9.25 | 12.48 | 16.25 | 20.58 | 25.49 | 30.98 | 37.07 |
| 20 MPH | .00 | .56 | 2.18 | 4.82 | 8.43 | 12.99 | 18.50 | 24.96 | 32.49 | 41.16 | 50.98 | 61.96 | 74.14 |
| 30 MPH | .00 | .84 | 3.27 | 7.22 | 12.64 | 19.49 | 27.75 | 37.44 | 48.74 | 61.74 | 76.47 | 92.95 | 111.22 |
| VELOCITY (FPS) | 900. | 871. | 845. | 820. | 796. | 770. | 746. | 722. | 700. | 678. | 658. | 638. | 618. |
| ENERGY (FT-LB) | 189. | 177. | 166. | 157. | 148. | 138. | 130. | 122. | 114. | 107. | 101. | 95. | 89. |
| DROP (IN) | .00 | -1.29 | -5.42 | -12.60 | -22.99 | -36.72 | -54.08 | -75.33 | -100.73 | -130.55 | -165.07 | -204.61 | -249.47 |
| MID-RANGE (IN) | .00 | .34 | 1.43 | 3.31 | 6.07 | 9.74 | 14.44 | 20.28 | 27.37 | 35.85 | 45.81 | 57.37 | 70.65 |
| BULLET PATH (IN) | -.63 | 1.11 | -.00 | -4.15 | -11.52 | -22.22 | -36.55 | -54.77 | -77.14 | -103.93 | -135.43 | -171.94 | -213.78 |
| TIME OF FLIGHT (SEC) | .000000 | .084712 | .172143 | .262258 | .355052 | .450819 | .549788 | .652005 | .757529 | .866433 | .978800 | 1.094728 | 1.214322 |
| WIND DEFLECTION (IN) | | | | | | | | | | | | | |
| 5 MPH | .00 | .12 | .48 | 1.08 | 1.91 | 3.01 | 4.38 | 6.04 | 8.00 | 10.25 | 12.80 | 15.67 | 18.86 |
| 10 MPH | .00 | .24 | .96 | 2.16 | 3.82 | 6.01 | 8.76 | 12.09 | 15.99 | 20.49 | 25.60 | 31.34 | 37.72 |
| 20 MPH | .00 | .49 | 1.93 | 4.31 | 7.65 | 12.02 | 17.53 | 24.17 | 31.98 | 40.98 | 51.20 | 62.68 | 75.44 |
| 30 MPH | .00 | .73 | 2.89 | 6.47 | 11.47 | 18.03 | 26.29 | 36.26 | 47.98 | 61.48 | 76.81 | 94.02 | 113.16 |

# Bullet: Lyman #358242 121 Gr.
## Ballistic Coefficient: .154

| RANGE (YDS) | MUZZLE | 25 | 50 | 75 | 100 | 125 | 150 | 175 | 200 | 225 | 250 | 275 | 300 |
|---|---|---|---|---|---|---|---|---|---|---|---|---|---|
| VELOCITY (FPS) | 1600. | 1498. | 1403. | 1316. | 1254. | 1201. | 1154. | 1113. | 1077. | 1046. | 1018. | 994. | 971. |
| ENERGY (FT-LB) | 688. | 603. | 528. | 465. | 422. | 388. | 358. | 333. | 312. | 294. | 279. | 265. | 253. |
| DROP (IN) | .00 | -.42 | -1.81 | -4.31 | -8.06 | -13.19 | -19.83 | -28.10 | -38.11 | -49.99 | -63.85 | -79.80 | -97.94 |
| MID-RANGE (IN) | .00 | .11 | .49 | 1.17 | 2.21 | 3.68 | 5.60 | 8.03 | 11.00 | 14.55 | 18.73 | 23.58 | 29.14 |
| BULLET PATH (IN) | -.63 | .18 | .00 | -1.28 | -3.80 | -7.71 | -13.13 | -20.17 | -28.96 | -39.62 | -52.26 | -66.99 | -83.91 |
| TIME OF FLIGHT (SEC) | .000000 | .048458 | .100218 | .155451 | .213947 | .275089 | .338809 | .405003 | .473522 | .544198 | .616886 | .691472 | .767864 |
| WIND DEFLECTION (IN) | | | | | | | | | | | | | |
| 5 MPH | .00 | .14 | .57 | 1.30 | 2.33 | 3.58 | 5.07 | 6.77 | 8.67 | 10.76 | 13.04 | 15.47 | 18.07 |
| 10 MPH | .00 | .28 | 1.14 | 2.61 | 4.65 | 7.17 | 10.13 | 13.53 | 17.34 | 21.53 | 26.07 | 30.95 | 36.14 |
| 20 MPH | .00 | .56 | 2.28 | 5.22 | 9.31 | 14.33 | 20.26 | 27.06 | 34.68 | 43.06 | 52.14 | 61.90 | 72.29 |
| 30 MPH | .00 | .84 | 3.42 | 7.83 | 13.96 | 21.50 | 30.39 | 40.59 | 52.02 | 64.59 | 78.22 | 92.85 | 108.43 |
| VELOCITY (FPS) | 1500. | 1405. | 1317. | 1254. | 1202. | 1155. | 1113. | 1078. | 1046. | 1019. | 994. | 971. | 950. |
| ENERGY (FT-LB) | 604. | 530. | 466. | 422. | 388. | 358. | 333. | 312. | 294. | 279. | 265. | 253. | 243. |
| DROP (IN) | .00 | -.49 | -2.08 | -4.90 | -9.12 | -14.85 | -22.19 | -31.29 | -42.25 | -55.19 | -70.21 | -87.44 | -106.96 |
| MID-RANGE (IN) | .00 | .13 | .55 | 1.32 | 2.49 | 4.10 | 6.19 | 8.81 | 12.00 | 15.80 | 20.26 | 25.41 | 31.28 |
| BULLET PATH (IN) | -.63 | .24 | .00 | -1.48 | -4.34 | -8.71 | -14.71 | -22.45 | -32.06 | -43.64 | -57.32 | -73.19 | -91.35 |
| TIME OF FLIGHT (SEC) | .000000 | .051681 | .106831 | .165292 | .226412 | .290111 | .356285 | .424785 | .495443 | .568115 | .642685 | .719061 | .797171 |
| WIND DEFLECTION (IN) | | | | | | | | | | | | | |
| 5 MPH | .00 | .15 | .60 | 1.35 | 2.32 | 3.53 | 4.95 | 6.58 | 8.40 | 10.39 | 12.56 | 14.88 | 17.35 |
| 10 MPH | .00 | .30 | 1.20 | 2.69 | 4.65 | 7.06 | 9.91 | 13.16 | 16.80 | 20.79 | 25.11 | 29.75 | 34.70 |
| 20 MPH | .00 | .59 | 2.40 | 5.38 | 9.30 | 14.12 | 19.81 | 26.32 | 33.60 | 41.58 | 50.23 | 59.51 | 69.40 |
| 30 MPH | .00 | .89 | 3.61 | 8.07 | 13.95 | 21.18 | 29.72 | 39.49 | 50.39 | 62.36 | 75.34 | 89.26 | 104.11 |
| VELOCITY (FPS) | 1400. | 1313. | 1251. | 1199. | 1153. | 1112. | 1076. | 1045. | 1017. | 993. | 970. | 949. | 930. |
| ENERGY (FT-LB) | 527. | 463. | 421. | 386. | 357. | 332. | 311. | 293. | 278. | 265. | 253. | 242. | 232. |
| DROP (IN) | .00 | -.56 | -2.37 | -5.57 | -10.29 | -16.63 | -24.73 | -34.70 | -46.65 | -60.69 | -76.94 | -95.49 | -116.38 |
| MID-RANGE (IN) | .00 | .15 | .62 | 1.49 | 2.77 | 4.53 | 6.79 | 9.62 | 13.04 | 17.10 | 21.84 | 27.29 | 33.46 |
| BULLET PATH (IN) | -.63 | .31 | .00 | -1.70 | -4.92 | -9.76 | -16.36 | -24.83 | -35.28 | -47.83 | -62.58 | -79.63 | -99.02 |
| TIME OF FLIGHT (SEC) | .000000 | .055334 | .113934 | .175179 | .238998 | .305286 | .373893 | .444649 | .517413 | .592070 | .668530 | .746720 | .826580 |
| WIND DEFLECTION (IN) | | | | | | | | | | | | | |
| 5 MPH | .00 | .16 | .60 | 1.27 | 2.17 | 3.29 | 4.62 | 6.13 | 7.82 | 9.67 | 11.69 | 13.85 | 16.17 |
| 10 MPH | .00 | .31 | 1.20 | 2.55 | 4.35 | 6.59 | 9.23 | 12.26 | 15.64 | 19.35 | 23.38 | 27.71 | 32.34 |
| 20 MPH | .00 | .62 | 2.39 | 5.09 | 8.70 | 13.17 | 18.47 | 24.52 | 31.27 | 38.69 | 46.75 | 55.42 | 64.67 |
| 30 MPH | .00 | .93 | 3.59 | 7.64 | 13.05 | 19.76 | 27.70 | 36.77 | 46.91 | 58.04 | 70.13 | 83.13 | 97.01 |
| VELOCITY (FPS) | 1300. | 1243. | 1192. | 1146. | 1106. | 1071. | 1041. | 1013. | 989. | 967. | 946. | 927. | 909. |
| ENERGY (FT-LB) | 454. | 415. | 382. | 353. | 329. | 308. | 291. | 276. | 263. | 251. | 241. | 231. | 222. |
| DROP (IN) | .00 | -.63 | -2.68 | -6.26 | -11.49 | -18.48 | -27.37 | -38.26 | -51.25 | -66.46 | -83.98 | -103.87 | -126.28 |
| MID-RANGE (IN) | .00 | .17 | .71 | 1.65 | 3.06 | 4.97 | 7.42 | 10.47 | 14.14 | 18.47 | 23.51 | 29.25 | 35.77 |
| BULLET PATH (IN) | -.63 | .39 | .00 | -1.92 | -5.49 | -10.83 | -18.06 | -27.29 | -38.62 | -52.18 | -68.04 | -86.27 | -107.02 |
| TIME OF FLIGHT (SEC) | 000000 | .059011 | .120645 | .184841 | .251486 | .320424 | .391488 | .464541 | .539472 | .616194 | .694636 | .774740 | .856463 |
| WIND DEFLECTION (IN) | | | | | | | | | | | | | |
| 5 MPH | .00 | .12 | .46 | 1.04 | 1.82 | 2.81 | 3.99 | 5.34 | 6.86 | 8.53 | 10.36 | 12.33 | 14.45 |
| 10 MPH | .00 | .23 | .93 | 2.07 | 3.65 | 5.63 | 7.98 | 10.68 | 13.72 | 17.07 | 20.72 | 24.66 | 28.89 |
| 20 MPH | .00 | .46 | 1.85 | 4.14 | 7.29 | 11.25 | 15.96 | 21.36 | 27.43 | 34.13 | 41.43 | 49.32 | 57.78 |
| 30 MPH | .00 | .70 | 2.78 | 6.21 | 10.94 | 16.88 | 23.94 | 32.05 | 41.15 | 51.20 | 62.15 | 73.99 | 86.67 |

## Bullet: Lyman #358242 121 Gr.
### Ballistic Coefficient: .154

| RANGE (YDS) | MUZZLE | 25 | 50 | 75 | 100 | 125 | 150 | 175 | 200 | 225 | 250 | 275 | 300 |
|---|---|---|---|---|---|---|---|---|---|---|---|---|---|
| VELOCITY (FPS) | 1200. | 1153. | 1112. | 1077. | 1045. | 1018. | 993. | 970. | 949. | 930. | 912. | 895. | 879. |
| ENERGY (FT-LB) | 387. | 357. | 332. | 311. | 294. | 278. | 265. | 253. | 242. | 232. | 223. | 215. | 208. |
| DROP (IN) | .00 | -.76 | -3.14 | -7.28 | -13.28 | -21.27 | -31.34 | -43.62 | -58.19 | -75.12 | -94.54 | -116.56 | -141.30 |
| MID-RANGE (IN) | .00 | .20 | .81 | 1.90 | 3.50 | 5.65 | 8.40 | 11.77 | 15.82 | 20.54 | 26.00 | 32.25 | 39.31 |
| BULLET PATH (IN) | -.63 | .50 | .00 | -2.25 | -6.37 | -12.47 | -20.66 | -31.05 | -43.74 | -58.78 | -76.32 | -96.46 | -119.31 |
| TIME OF FLIGHT (SEC) | .000000 | .063775 | .130020 | .198587 | .269306 | .342035 | .416658 | .493085 | .571243 | .651072 | .732525 | .815565 | .900164 |
| WIND DEFLECTION (IN) | | | | | | | | | | | | | |
| 5 MPH | .00 | .11 | .44 | .98 | 1.70 | 2.60 | 3.67 | 4.89 | 6.27 | 7.79 | 9.46 | 11.27 | 13.21 |
| 10 MPH | .00 | .22 | .88 | 1.95 | 3.40 | 5.20 | 7.33 | 9.78 | 12.54 | 15.59 | 18.92 | 22.54 | 26.43 |
| 20 MPH | .00 | .45 | 1.77 | 3.90 | 6.80 | 10.40 | 14.66 | 19.57 | 25.08 | 31.18 | 37.85 | 45.08 | 52.86 |
| 30 MPH | .00 | .67 | 2.65 | 5.85 | 10.19 | 15.59 | 22.00 | 29.35 | 37.62 | 46.77 | 56.77 | 67.62 | 79.29 |
| VELOCITY (FPS) | 1100. | 1066. | 1036. | 1009. | 985. | 963. | 943. | 924. | 906. | 890. | 874. | 859. | 844. |
| ENERGY (FT-LB) | 325. | 305. | 288. | 274. | 261. | 249. | 239. | 229. | 221. | 213. | 205. | 198. | 191. |
| DROP (IN) | .00 | -.89 | -3.68 | -8.49 | -15.43 | -24.60 | -36.09 | -49.98 | -66.39 | -85.44 | -107.23 | -131.86 | -159.43 |
| MID-RANGE (IN) | .00 | .23 | .95 | 2.21 | 4.03 | 6.47 | 9.55 | 13.30 | 17.76 | 22.99 | 29.01 | 35.87 | 43.62 |
| BULLET PATH (IN) | -.63 | .64 | .00 | -2.66 | -7.44 | -14.45 | -23.79 | -35.51 | -49.77 | -66.67 | -86.30 | -108.77 | -134.18 |
| TIME OF FLIGHT (SEC) | .000000 | .069286 | .140673 | .214029 | .289248 | .366245 | .444951 | .525312 | .607284 | .690832 | .775929 | .862557 | .950702 |
| WIND DEFLECTION (IN) | | | | | | | | | | | | | |
| 5 MPH | .00 | .10 | .38 | .83 | 1.45 | 2.23 | 3.16 | 4.23 | 5.44 | 6.79 | 8.28 | 9.91 | 11.66 |
| 10 MPH | .00 | .19 | .76 | 1.67 | 2.91 | 4.46 | 6.31 | 8.45 | 10.88 | 13.59 | 16.56 | 19.81 | 23.32 |
| 20 MPH | .00 | .39 | 1.52 | 3.34 | 5.82 | 8.92 | 12.62 | 16.91 | 21.76 | 27.17 | 33.13 | 39.62 | 46.65 |
| 30 MPH | .00 | .58 | 2.28 | 5.01 | 8.72 | 13.38 | 18.93 | 25.36 | 32.65 | 40.76 | 49.69 | 59.43 | 69.97 |
| VELOCITY (FPS) | 1000. | 977. | 955. | 936. | 917. | 900. | 883. | 868. | 853. | 839. | 825. | 811. | 799. |
| ENERGY (FT-LB) | 269. | 256. | 245. | 235. | 226. | 218. | 210. | 202. | 195. | 189. | 183. | 177. | 171. |
| DROP (IN) | .00 | -1.06 | -4.39 | -10.05 | -18.18 | -28.87 | -42.24 | -58.38 | -77.39 | -99.38 | -124.45 | -152.70 | -184.24 |
| MID-RANGE (IN) | .00 | .28 | 1.14 | 2.60 | 4.70 | 7.51 | 11.07 | 15.39 | 20.52 | 26.49 | 33.35 | 41.13 | 49.88 |
| BULLET PATH (IN) | -.63 | .82 | .00 | -3.16 | -8.78 | -16.96 | -27.83 | -41.46 | -57.96 | -77.44 | -100.00 | -125.74 | -154.77 |
| TIME OF FLIGHT (SEC) | .000000 | .075900 | .153549 | .232886 | .313859 | .396428 | .480562 | .566237 | .653435 | .742146 | .832363 | .924084 | 1.017313 |
| WIND DEFLECTION (IN) | | | | | | | | | | | | | |
| 5 MPH | .00 | .08 | .31 | .69 | 1.22 | 1.89 | 2.69 | 3.63 | 4.70 | 5.91 | 7.25 | 8.72 | 10.32 |
| 10 MPH | .00 | .16 | .62 | 1.39 | 2.44 | 3.77 | 5.38 | 7.26 | 9.40 | 11.82 | 14.50 | 17.44 | 20.65 |
| 20 MPH | .00 | .32 | 1.25 | 2.78 | 4.88 | 7.54 | 10.76 | 14.52 | 18.81 | 23.64 | 28.99 | 34.88 | 41.29 |
| 30 MPH | .00 | .48 | 1.87 | 4.16 | 7.32 | 11.31 | 16.14 | 21.77 | 28.21 | 35.45 | 43.49 | 52.32 | 61.94 |
| VELOCITY (FPS) | 900. | 884. | 868. | 853. | 839. | 825. | 811. | 798. | 785. | 772. | 760. | 748. | 736. |
| ENERGY (FT-LB) | 218. | 210. | 202. | 195. | 189. | 183. | 177. | 171. | 166. | 160. | 155. | 150. | 146. |
| DROP (IN) | .00 | -1.22 | -5.21 | -12.08 | -21.92 | -34.83 | -50.93 | -70.31 | -92.91 | -119.01 | -148.73 | -182.19 | -219.52 |
| MID-RANGE (IN) | .00 | .34 | 1.39 | 3.17 | 5.74 | 9.13 | 13.38 | 18.54 | 24.54 | 31.52 | 39.53 | 48.62 | 58.83 |
| BULLET PATH (IN) | -.63 | 1.07 | -.00 | -3.94 | -10.86 | -20.85 | -34.02 | -50.48 | -70.16 | -93.34 | -120.13 | -150.67 | -185.08 |
| TIME OF FLIGHT (SEC) | .000000 | .084111 | .169763 | .256938 | .345625 | .435817 | .527512 | .620714 | .715467 | .811812 | .909759 | 1.009316 | 1.110496 |
| WIND DEFLECTION (IN) | | | | | | | | | | | | | |
| 5 MPH | .00 | .07 | .27 | .61 | 1.08 | 1.69 | 2.42 | 3.29 | 4.29 | 5.44 | 6.73 | 8.15 | 9.72 |
| 10 MPH | .00 | .14 | .54 | 1.22 | 2.16 | 3.37 | 4.84 | 6.58 | 8.59 | 10.88 | 13.45 | 16.31 | 19.45 |
| 20 MPH | .00 | .27 | 1.09 | 2.44 | 4.33 | 6.74 | 9.68 | 13.16 | 17.18 | 21.76 | 26.90 | 32.61 | 38.89 |
| 30 MPH | .00 | .41 | 1.63 | 3.66 | 6.49 | 10.11 | 14.53 | 19.74 | 25.77 | 32.64 | 40.35 | 48.92 | 58.34 |

# Bullet: Lyman #356402 116 Gr.
## Ballistic Coefficient: .146

| RANGE (YDS) | MUZZLE | 25 | 50 | 75 | 100 | 125 | 150 | 175 | 200 | 225 | 250 | 275 | 300 |
|---|---|---|---|---|---|---|---|---|---|---|---|---|---|
| VELOCITY (FPS) | 1600. | 1492. | 1393. | 1302. | 1199. | 1115. | 1050. | 999. | 957. | 920. | 887. | 858. | 830. |
| ENERGY (FT-LB) | 659. | 574. | 499. | 437. | 370. | 320. | 284. | 257. | 236. | 218. | 203. | 189. | 178. |
| DROP (IN) | .00 | -.42 | -1.83 | -4.35 | -8.17 | -13.50 | -20.57 | -29.61 | -40.81 | -54.36 | -70.46 | -89.30 | -111.10 |
| MID-RANGE (IN) | .00 | .11 | .49 | 1.18 | 2.26 | 3.81 | 5.94 | 8.72 | 12.24 | 16.55 | 21.73 | 27.85 | 34.97 |
| BULLET PATH (IN) | -.63 | .18 | .00 | -1.30 | -3.89 | -7.99 | -13.84 | -21.64 | -31.62 | -43.94 | -58.81 | -76.43 | -96.99 |
| TIME OF FLIGHT (SEC) | .000000 | .048547 | .100587 | .156306 | .216347 | .281279 | .350658 | .423928 | .500690 | .580685 | .663741 | .749752 | .838659 |
| WIND DEFLECTION (IN) | | | | | | | | | | | | | |
| 5 MPH | .00 | .15 | .60 | 1.38 | 2.54 | 4.13 | 6.11 | 8.43 | 11.06 | 13.98 | 17.16 | 20.60 | 24.30 |
| 10 MPH | .00 | .29 | 1.20 | 2.76 | 5.08 | 8.26 | 12.22 | 16.86 | 22.12 | 27.95 | 34.32 | 41.21 | 48.60 |
| 20 MPH | .00 | .59 | 2.41 | 5.52 | 10.15 | 16.51 | 24.43 | 33.72 | 44.24 | 55.90 | 68.64 | 82.41 | 97.21 |
| 30 MPH | .00 | .88 | 3.61 | 8.28 | 15.23 | 24.77 | 36.65 | 50.58 | 66.36 | 83.85 | 102.96 | 123.62 | 145.81 |
| VELOCITY (FPS) | 1500. | 1400. | 1308. | 1206. | 1120. | 1054. | 1002. | 959. | 922. | 889. | 860. | 832. | 807. |
| ENERGY (FT-LB) | 579. | 505. | 441. | 375. | 323. | 286. | 259. | 237. | 219. | 204. | 190. | 178. | 168. |
| DROP (IN) | .00 | -.49 | -2.09 | -4.97 | -9.34 | -15.44 | -23.49 | -33.69 | -46.22 | -61.29 | -79.09 | -99.82 | -123.69 |
| MID-RANGE (IN) | .00 | .13 | .55 | 1.34 | 2.58 | 4.37 | 6.78 | 9.89 | 13.77 | 18.48 | 24.11 | 30.71 | 38.36 |
| BULLET PATH (IN) | -.63 | .24 | .00 | -1.52 | -4.53 | -9.27 | -15.96 | -24.80 | -35.98 | -49.68 | -66.13 | -85.51 | -108.01 |
| TIME OF FLIGHT (SEC) | .000000 | .051775 | .107217 | .166910 | .231508 | .300591 | .373600 | .450125 | .529898 | .612741 | .698545 | .787247 | .878823 |
| WIND DEFLECTION (IN) | | | | | | | | | | | | | |
| 5 MPH | .00 | .16 | .64 | 1.49 | 2.77 | 4.45 | 6.48 | 8.81 | 11.43 | 14.32 | 17.47 | 20.88 | 24.54 |
| 10 MPH | .00 | .31 | 1.27 | 2.98 | 5.55 | 8.90 | 12.95 | 17.62 | 22.86 | 28.64 | 34.94 | 41.76 | 49.07 |
| 20 MPH | .00 | .62 | 2.54 | 5.95 | 11.09 | 17.81 | 25.91 | 35.24 | 45.72 | 57.28 | 69.89 | 83.51 | 98.15 |
| 30 MPH | .00 | .94 | 3.81 | 8.93 | 16.64 | 26.71 | 38.86 | 52.87 | 68.59 | 85.93 | 104.83 | 125.27 | 147.22 |
| VELOCITY (FPS) | 1400. | 1309. | 1206. | 1121. | 1055. | 1003. | 959. | 922. | 889. | 860. | 832. | 807. | 781. |
| ENERGY (FT-LB) | 505. | 441. | 375. | 323. | 286. | 259. | 237. | 219. | 204. | 190. | 178. | 168. | 157. |
| DROP (IN) | .00 | -.56 | -2.41 | -5.74 | -10.80 | -17.81 | -26.98 | -38.47 | -52.50 | -69.26 | -88.96 | -111.78 | -137.87 |
| MID-RANGE (IN) | .00 | .15 | .64 | 1.56 | 3.00 | 5.03 | 7.75 | 11.19 | 15.44 | 20.58 | 26.67 | 33.78 | 41.96 |
| BULLET PATH (IN) | -.63 | .32 | .00 | -1.82 | -5.36 | -10.86 | -18.50 | -28.48 | -40.99 | -56.24 | -74.41 | -95.72 | -120.29 |
| TIME OF FLIGHT (SEC) | .000000 | .055432 | .115115 | .179703 | .248778 | .321779 | .398297 | .478062 | .560899 | .646696 | .735391 | .826960 | .921490 |
| WIND DEFLECTION (IN) | | | | | | | | | | | | | |
| 5 MPH | .00 | .16 | .70 | 1.67 | 3.04 | 4.75 | 6.76 | 9.07 | 11.64 | 14.48 | 17.57 | 20.92 | 24.52 |
| 10 MPH | .00 | .33 | 1.40 | 3.34 | 6.07 | 9.49 | 13.53 | 18.14 | 23.29 | 28.96 | 35.14 | 41.83 | 49.04 |
| 20 MPH | .00 | .65 | 2.81 | 6.68 | 12.14 | 18.98 | 27.06 | 36.28 | 46.58 | 57.92 | 70.29 | 83.66 | 98.08 |
| 30 MPH | .00 | .98 | 4.21 | 10.03 | 18.21 | 28.47 | 40.59 | 54.42 | 69.87 | 86.88 | 105.43 | 125.49 | 147.12 |
| VELOCITY (FPS) | 1300. | 1196. | 1113. | 1048. | 998. | 955. | 919. | 886. | 857. | 830. | 804. | 778. | 753. |
| ENERGY (FT-LB) | 435. | 369. | 319. | 283. | 256. | 235. | 217. | 202. | 189. | 177. | 166. | 156. | 146. |
| DROP (IN) | .00 | -.67 | -2.86 | -6.79 | -12.70 | -20.78 | -31.21 | -44.19 | -59.94 | -78.63 | -100.47 | -125.60 | -154.28 |
| MID-RANGE (IN) | .00 | .18 | .76 | 1.84 | 3.49 | 5.80 | 8.81 | 12.61 | 17.27 | 22.86 | 29.46 | 37.11 | 45.93 |
| BULLET PATH (IN) | -.63 | .44 | .00 | -2.19 | -6.35 | -12.69 | -21.38 | -32.62 | -46.62 | -63.57 | -83.67 | -107.06 | -133.99 |
| TIME OF FLIGHT (SEC) | .000000 | .060204 | .125291 | .194807 | .268196 | .345067 | .425164 | .508317 | .594422 | .683421 | .775293 | .870126 | .968146 |
| WIND DEFLECTION (IN) | | | | | | | | | | | | | |
| 5 MPH | .00 | .22 | .87 | 1.91 | 3.29 | 4.98 | 6.95 | 9.19 | 11.69 | 14.45 | 17.46 | 20.72 | 24.27 |
| 10 MPH | .00 | .44 | 1.74 | 3.82 | 6.59 | 9.96 | 13.91 | 18.39 | 23.39 | 28.90 | 34.91 | 41.45 | 48.55 |
| 20 MPH | .00 | .88 | 3.49 | 7.65 | 13.17 | 19.93 | 27.81 | 36.77 | 46.78 | 57.79 | 69.83 | 82.90 | 97.09 |
| 30 MPH | .00 | 1.33 | 5.23 | 11.47 | 19.76 | 29.89 | 41.72 | 55.16 | 70.16 | 86.69 | 104.74 | 124.35 | 145.64 |

## Bullet: Lyman #356402 116 Gr.
### Ballistic Coefficient: .146

| RANGE (YDS) | MUZZLE | 25 | 50 | 75 | 100 | 125 | 150 | 175 | 200 | 225 | 250 | 275 | 300 |
|---|---|---|---|---|---|---|---|---|---|---|---|---|---|
| VELOCITY (FPS) | 1200. | 1116. | 1051. | 999. | 957. | 920. | 887. | 858. | 831. | 805. | 779. | 754. | 730. |
| ENERGY (FT-LB) | 371. | 321. | 284. | 257. | 236. | 218. | 203. | 190. | 178. | 167. | 156. | 146. | 137. |
| DROP (IN) | .00 | -.78 | -3.30 | -7.79 | -14.44 | -23.43 | -34.97 | -49.26 | -66.50 | -86.87 | -110.52 | -137.71 | -168.70 |
| MID-RANGE (IN) | .00 | .20 | .87 | 2.08 | 3.91 | 6.43 | 9.70 | 13.80 | 18.81 | 24.81 | 31.83 | 39.98 | 49.37 |
| BULLET PATH (IN) | -.63 | .55 | .00 | -2.51 | -7.20 | -14.22 | -23.80 | -36.12 | -51.39 | -69.79 | -91.48 | -116.70 | -145.72 |
| TIME OF FLIGHT (SEC) | .000000 | .C64900 | .134250 | .207493 | .284231 | .364202 | .447236 | .533224 | .622107 | .713863 | .808580 | .906481 | 1.007615 |
| WIND DEFLECTION (IN) | | | | | | | | | | | | | |
| 5 MPH | .00 | .21 | .81 | 1.76 | 3.01 | 4.55 | 6.36 | 8.42 | 10.75 | 13.32 | 16.16 | 19.27 | 22.67 |
| 10 MPH | .00 | .42 | 1.63 | 3.52 | 6.02 | 9.10 | 12.71 | 16.85 | 21.49 | 26.64 | 32.31 | 38.54 | 45.34 |
| 20 MPH | .00 | .84 | 3.26 | 7.04 | 12.05 | 18.20 | 25.43 | 33.69 | 42.98 | 53.28 | 64.62 | 77.08 | 90.68 |
| 30 MPH | .00 | 1.27 | 4.88 | 10.56 | 18.07 | 27.30 | 38.14 | 50.54 | 64.47 | 79.92 | 96.93 | 115.62 | 136.02 |
| VELOCITY (FPS) | 1100. | 1039. | 990. | 948. | 913. | 881. | 852. | 825. | 800. | 773. | 749. | 725. | 703. |
| ENERGY (FT-LB) | 312. | 278. | 252. | 232. | 214. | 200. | 187. | 175. | 165. | 154. | 144. | 135. | 127. |
| DROP (IN) | .00 | -.92 | -3.84 | -8.98 | -16.50 | -26.61 | -39.51 | -55.40 | -74.47 | -96.84 | -122.83 | -152.67 | -186.63 |
| MID-RANGE (IN) | .00 | .24 | 1.00 | 2.37 | 4.40 | 7.18 | 10.78 | 15.27 | 20.74 | 27.21 | 34.80 | 43.63 | 53.80 |
| BULLET PATH (IN) | -.63 | .69 | .00 | -2.90 | -8.18 | -16.06 | -26.72 | -40.37 | -57.21 | -77.35 | -101.09 | -128.70 | -160.42 |
| TIME OF FLIGHT (SEC) | .000000 | .070243 | .144277 | .221738 | .302389 | .386074 | .472696 | .562206 | .654588 | .750000 | .848613 | .950469 | 1.055626 |
| WIND DEFLECTION (IN) | | | | | | | | | | | | | |
| 5 MPH | .00 | .18 | .70 | 1.51 | 2.61 | 3.97 | 5.60 | 7.47 | 9.60 | 12.00 | 14.68 | 17.64 | 20.90 |
| 10 MPH | .00 | .36 | 1.39 | 3.03 | 5.22 | 7.95 | 11.19 | 14.95 | 19.21 | 24.00 | 29.36 | 35.28 | 41.79 |
| 20 MPH | .00 | .73 | 2.79 | 6.05 | 10.44 | 15.90 | 22.39 | 29.90 | 38.41 | 48.00 | 58.71 | 70.57 | 83.58 |
| 30 MPH | .00 | 1.09 | 4.18 | 9.08 | 15.66 | 23.85 | 33.58 | 44.84 | 57.62 | 72.00 | 88.07 | 105.85 | 125.37 |
| VELOCITY (FPS) | 1000. | 957. | 920. | 888. | 858. | 831. | 805. | 779. | 754. | 730. | 708. | 686. | 665. |
| ENERGY (FT-LB) | 258. | 236. | 218. | 203. | 190. | 178. | 167. | 156. | 147. | 137. | 129. | 121. | 114. |
| DROP (IN) | .00 | -1.09 | -4.53 | -10.51 | -19.24 | -30.90 | -45.70 | -63.78 | -85.39 | -110.80 | -140.27 | -174.07 | -212.47 |
| MID-RANGE (IN) | .00 | .28 | 1.17 | 2.75 | 5.09 | 8.26 | 12.34 | 17.37 | 23.46 | 30.71 | 39.24 | 49.14 | 60.53 |
| BULLET PATH (IN) | -.63 | .85 | .00 | -3.40 | -9.55 | -18.64 | -30.87 | -46.36 | -65.40 | -88.23 | -115.12 | -146.34 | -182.16 |
| TIME OF FLIGHT (SEC) | .000000 | .076696 | .156628 | .239623 | .325573 | .414418 | .506135 | .600813 | .698673 | .799765 | .904142 | 1.011873 | 1.123039 |
| WIND DEFLECTION (IN) | | | | | | | | | | | | | |
| 5 MPH | .00 | .15 | .58 | 1.29 | 2.25 | 3.47 | 4.94 | 6.67 | 8.68 | 10.98 | 13.56 | 16.44 | 19.63 |
| 10 MPH | .00 | .30 | 1.17 | 2.57 | 4.50 | 6.94 | 9.88 | 13.34 | 17.37 | 21.96 | 27.13 | 32.89 | 39.25 |
| 20 MPH | .00 | .60 | 2.33 | 5.15 | 9.00 | 13.88 | 19.76 | 26.69 | 34.73 | 43.92 | 54.26 | 65.78 | 78.51 |
| 30 MPH | .00 | .90 | 3.50 | 7.72 | 13.50 | 20.81 | 29.64 | 40.03 | 52.10 | 65.88 | 81.39 | 98.67 | 117.76 |
| VELOCITY (FPS) | 900. | 869. | 841. | 815. | 789. | 764. | 739. | 716. | 694. | 673. | 652. | 632. | 613. |
| ENERGY (FT-LB) | 209. | 195. | 182. | 171. | 160. | 150. | 141. | 132. | 124. | 117. | 110. | 103. | 97. |
| DROP (IN) | .00 | -1.30 | -5.45 | -12.67 | -23.13 | -36.99 | -54.55 | -76.07 | -101.81 | -132.05 | -167.07 | -207.18 | -252.72 |
| MID-RANGE (IN) | .00 | .35 | 1.43 | 3.33 | 6.11 | 9.83 | 14.61 | 20.54 | 27.78 | 36.41 | 46.54 | 58.30 | 71.81 |
| BULLET PATH (IN) | -.63 | 1.12 | -.00 | -4.18 | -11.59 | -22.41 | -36.93 | -55.41 | -78.10 | -105.30 | -137.28 | -174.35 | -216.85 |
| TIME OF FLIGHT (SEC) | .000000 | .084812 | .172536 | .263138 | .356633 | .453254 | .553087 | .656183 | .762603 | .872424 | .985734 | 1.102631 | 1.223224 |
| WIND DEFLECTION (IN) | | | | | | | | | | | | | |
| 5 MPH | .00 | .13 | .52 | 1.16 | 2.05 | 3.22 | 4.67 | 6.41 | 8.44 | 10.77 | 13.41 | 16.36 | 19.64 |
| 10 MPH | .00 | .26 | 1.03 | 2.31 | 4.10 | 6.44 | 9.34 | 12.82 | 16.88 | 21.55 | 26.82 | 32.73 | 39.29 |
| 20 MPH | .00 | .52 | 2.07 | 4.62 | 8.20 | 12.88 | 18.69 | 25.64 | 33.77 | 43.09 | 53.64 | 65.46 | 78.57 |
| 30 MPH | .00 | .78 | 3.10 | 6.94 | 12.30 | 19.32 | 28.03 | 38.46 | 50.65 | 64.64 | 80.47 | 98.19 | 117.86 |

# Bullet: Lyman #358480 123 Gr.
## Ballistic Coefficient: .145

| RANGE (YDS) | MUZZLE | 25 | 50 | 75 | 100 | 125 | 150 | 175 | 200 | 225 | 250 | 275 | 300 |
|---|---|---|---|---|---|---|---|---|---|---|---|---|---|
| VELOCITY (FPS) | 1600. | 1492. | 1391. | 1300. | 1209. | 1132. | 1070. | 1021. | 980. | 945. | 914. | 885. | 859. |
| ENERGY (FT-LB) | 699. | 608. | 529. | 462. | 399. | 350. | 313. | 285. | 262. | 244. | 228. | 214. | 202. |
| DROP (IN) | .00 | -.42 | -1.83 | -4.36 | -8.18 | -13.49 | -20.49 | -29.38 | -40.35 | -53.56 | -69.18 | -87.38 | -108.35 |
| MID-RANGE (IN) | .00 | .11 | .49 | 1.18 | 2.26 | 3.80 | 5.89 | 8.60 | 12.00 | 16.15 | 21.10 | 26.93 | 33.69 |
| BULLET PATH (IN) | -.63 | .18 | .00 | -1.30 | -3.89 | -7.97 | -13.75 | -21.41 | -31.15 | -43.14 | -57.52 | -74.50 | -94.24 |
| TIME OF FLIGHT (SEC) | .000000 | .048559 | .100636 | .156420 | .216277 | .280468 | .348681 | .420468 | .495469 | .573432 | .654187 | .737614 | .823636 |
| WIND DEFLECTION (IN) | | | | | | | | | | | | | |
| 5 MPH | .00 | .15 | .61 | 1.39 | 2.53 | 4.06 | 5.93 | 8.13 | 10.60 | 13.34 | 16.32 | 19.54 | 22.98 |
| 10 MPH | .00 | .30 | 1.21 | 2.78 | 5.06 | 8.11 | 11.87 | 16.25 | 21.20 | 26.67 | 32.64 | 39.07 | 45.96 |
| 20 MPH | .00 | .59 | 2.42 | 5.56 | 10.13 | 16.22 | 23.74 | 32.50 | 42.40 | 53.35 | 65.27 | 78.14 | 91.92 |
| 30 MPH | .00 | .89 | 3.64 | 8.34 | 15.19 | 24.34 | 35.60 | 48.76 | 63.61 | 80.02 | 97.91 | 117.21 | 137.88 |
| VELOCITY (FPS) | 1500. | 1399. | 1307. | 1215. | 1137. | 1075. | 1025. | 983. | 947. | 916. | 887. | 861. | 837. |
| ENERGY (FT-LB) | 614. | 535. | 467. | 403. | 353. | 315. | 287. | 264. | 245. | 229. | 215. | 203. | 191. |
| DROP (IN) | .00 | -.49 | -2.09 | -4.97 | -9.32 | -15.34 | -23.25 | -33.21 | -45.41 | -60.00 | -77.16 | -97.07 | -119.90 |
| MID-RANGE (IN) | .00 | .13 | .56 | 1.34 | 2.57 | 4.32 | 6.66 | 9.66 | 13.39 | 17.89 | 23.23 | 29.49 | 36.70 |
| BULLET PATH (IN) | -.63 | .24 | .00 | -1.52 | -4.51 | -9.18 | -15.72 | -24.32 | -35.17 | -48.39 | -64.19 | -82.74 | -104.21 |
| TIME OF FLIGHT (SEC) | .000000 | .051787 | .107268 | .166786 | .230648 | .298565 | .370090 | .444851 | .522592 | .603134 | .686356 | .772179 | .860553 |
| WIND DEFLECTION (IN) | | | | | | | | | | | | | |
| 5 MPH | .00 | .16 | .64 | 1.48 | 2.70 | 4.27 | 6.17 | 8.35 | 10.79 | 13.48 | 16.40 | 19.55 | 22.93 |
| 10 MPH | .00 | .31 | 1.28 | 2.95 | 5.39 | 8.55 | 12.34 | 16.69 | 21.58 | 26.95 | 32.80 | 39.10 | 45.86 |
| 20 MPH | .00 | .63 | 2.56 | 5.91 | 10.79 | 17.09 | 24.67 | 33.39 | 43.15 | 53.90 | 65.60 | 78.21 | 91.71 |
| 30 MPH | .00 | .94 | 3.84 | 8.86 | 16.18 | 25.64 | 37.01 | 50.08 | 64.73 | 80.85 | 98.40 | 117.31 | 137.57 |
| VELOCITY (FPS) | 1400. | 1308. | 1216. | 1138. | 1075. | 1025. | 984. | 948. | 916. | 888. | 861. | 837. | 814. |
| ENERGY (FT-LB) | 535. | 467. | 404. | 353. | 316. | 287. | 264. | 245. | 229. | 215. | 203. | 191. | 181. |
| DROP (IN) | .00 | -.56 | -2.40 | -5.71 | -10.70 | -17.56 | -26.47 | -37.63 | -51.17 | -67.28 | -86.14 | -107.92 | -132.79 |
| MID-RANGE (IN) | .00 | .15 | .64 | 1.55 | 2.95 | 4.92 | 7.52 | 10.83 | 14.89 | 19.76 | 25.51 | 32.21 | 39.92 |
| BULLET PATH (IN) | -.63 | .32 | .00 | -1.80 | -5.26 | -10.61 | -18.01 | -27.65 | -39.67 | -54.27 | -71.61 | -91.88 | -115.23 |
| TIME OF FLIGHT (SEC) | .000000 | .055445 | .114926 | .178754 | .246639 | .318136 | .392871 | .470587 | .551107 | .634307 | .720107 | .808460 | .899342 |
| WIND DEFLECTION (IN) | | | | | | | | | | | | | |
| 5 MPH | .00 | .16 | .68 | 1.59 | 2.85 | 4.42 | 6.29 | 8.41 | 10.78 | 13.39 | 16.23 | 19.29 | 22.57 |
| 10 MPH | .00 | .33 | 1.37 | 3.18 | 5.69 | 8.85 | 12.57 | 16.82 | 21.57 | 26.78 | 32.45 | 38.57 | 45.14 |
| 20 MPH | .00 | .66 | 2.74 | 6.35 | 11.39 | 17.70 | 25.15 | 33.65 | 43.13 | 53.56 | 64.91 | 77.15 | 90.28 |
| 30 MPH | .00 | .99 | 4.11 | 9.53 | 17.08 | 26.55 | 37.72 | 50.47 | 64.70 | 80.34 | 97.36 | 115.72 | 135.42 |
| VELOCITY (FPS) | 1300. | 1208. | 1131. | 1070. | 1021. | 980. | 945. | 913. | 885. | 859. | 835. | 812. | 788. |
| ENERGY (FT-LB) | 461. | 398. | 349. | 312. | 285. | 262. | 244. | 228. | 214. | 202. | 190. | 180. | 170. |
| DROP (IN) | .00 | -.66 | -2.82 | -6.66 | -12.40 | -20.22 | -30.29 | -42.75 | -57.81 | -75.64 | -96.40 | -120.26 | -147.39 |
| MID-RANGE (IN) | .00 | .17 | .75 | 1.79 | 3.38 | 5.59 | 8.48 | 12.09 | 16.51 | 21.78 | 27.98 | 35.17 | 43.41 |
| BULLET PATH (IN) | -.63 | .43 | .00 | -2.12 | -6.14 | -12.23 | -20.58 | -31.33 | -44.66 | -60.76 | -79.80 | -101.94 | -127.35 |
| TIME OF FLIGHT (SEC) | .000000 | .059905 | .124142 | .192397 | .264221 | .339254 | .417248 | .498031 | .581485 | .667533 | .756130 | .847256 | .940969 |
| WIND DEFLECTION (IN) | | | | | | | | | | | | | |
| 5 MPH | .00 | .19 | .77 | 1.70 | 2.94 | 4.47 | 6.26 | 8.29 | 10.56 | 13.05 | 15.77 | 18.71 | 21.88 |
| 10 MPH | .00 | .39 | 1.54 | 3.40 | 5.89 | 8.94 | 12.51 | 16.58 | 21.11 | 26.10 | 31.54 | 37.42 | 43.76 |
| 20 MPH | .00 | .78 | 3.08 | 6.80 | 11.77 | 17.88 | 25.03 | 33.15 | 42.22 | 52.20 | 63.08 | 74.85 | 87.53 |
| 30 MPH | .00 | 1.17 | 4.62 | 10.20 | 17.66 | 26.82 | 37.54 | 49.73 | 63.33 | 78.30 | 94.62 | 112.27 | 131.29 |

## Bullet: Lyman #358480 123 Gr.
### Ballistic Coefficient: .145

| RANGE (YDS) | MUZZLE | 25 | 50 | 75 | 100 | 125 | 150 | 175 | 200 | 225 | 250 | 275 | 300 |
|---|---|---|---|---|---|---|---|---|---|---|---|---|---|
| VELOCITY (FPS) | 1200. | 1125. | 1065. | 1017. | 976. | 942. | 911. | 883. | 857. | 833. | 810. | 786. | 761. |
| ENERGY (FT-LB) | 393. | 345. | 310. | 282. | 260. | 242. | 226. | 213. | 200. | 189. | 179. | 169. | 158. |
| DROP (IN) | .00 | -.78 | -3.26 | -7.66 | -14.15 | -22.91 | -34.09 | -47.87 | -64.43 | -83.94 | -106.58 | -132.49 | -161.86 |
| MID-RANGE (IN) | .00 | .20 | .86 | 2.04 | 3.81 | 6.24 | 9.38 | 13.30 | 18.06 | 23.73 | 30.38 | 38.06 | 46.84 |
| BULLET PATH (IN) | -.63 | .54 | .00 | -2.45 | -7.00 | -13.81 | -23.03 | -34.87 | -49.48 | -67.04 | -87.73 | -111.70 | -139.12 |
| TIME OF FLIGHT (SEC) | .000000 | .064620 | .133218 | .205347 | .280659 | .358913 | .439942 | .523634 | .609915 | .698741 | .790097 | .884039 | .981057 |
| WIND DEFLECTION (IN) | | | | | | | | | | | | | |
| 5 MPH | .00 | .19 | .72 | 1.57 | 2.70 | 4.08 | 5.71 | 7.58 | 9.67 | 11.99 | 14.53 | 17.30 | 20.33 |
| 10 MPH | .00 | .37 | 1.45 | 3.14 | 5.40 | 8.17 | 11.43 | 15.16 | 19.35 | 23.98 | 29.06 | 34.59 | 40.67 |
| 20 MPH | .00 | .75 | 2.89 | 6.28 | 10.79 | 16.34 | 22.86 | 30.32 | 38.69 | 47.96 | 58.11 | 69.18 | 81.33 |
| 30 MPH | .00 | 1.12 | 4.34 | 9.42 | 16.19 | 24.51 | 34.29 | 45.48 | 58.04 | 71.94 | 87.17 | 103.77 | 122.00 |
| VELOCITY (FPS) | 1100. | 1045. | 1000. | 962. | 929. | 899. | 872. | 847. | 824. | 802. | 776. | 751. | 728. |
| ENERGY (FT-LB) | 330. | 298. | 273. | 253. | 236. | 221. | 208. | 196. | 185. | 175. | 164. | 154. | 145. |
| DROP (IN) | .00 | -.91 | -3.81 | -8.86 | -16.24 | -26.11 | -38.65 | -54.04 | -72.44 | -94.04 | -118.95 | -147.44 | -179.75 |
| MID-RANGE (IN) | .00 | .24 | .99 | 2.33 | 4.31 | 7.00 | 10.46 | 14.76 | 19.98 | 26.17 | 33.37 | 41.68 | 51.23 |
| BULLET PATH (IN) | -.63 | .68 | .00 | -2.84 | -8.00 | -15.65 | -25.97 | -39.14 | -55.33 | -74.70 | -97.40 | -123.67 | -153.77 |
| TIME OF FLIGHT (SEC) | .000000 | .070005 | .143392 | .219857 | .299194 | .381257 | .465950 | .553211 | .643009 | .735334 | .830429 | .928714 | 1.030239 |
| WIND DEFLECTION (IN) | | | | | | | | | | | | | |
| 5 MPH | .00 | .16 | .62 | 1.35 | 2.33 | 3.55 | 5.00 | 6.68 | 8.58 | 10.71 | 13.08 | 15.73 | 18.66 |
| 10 MPH | .00 | .32 | 1.24 | 2.69 | 4.66 | 7.10 | 10.01 | 13.37 | 17.17 | 21.42 | 26.16 | 31.45 | 37.32 |
| 20 MPH | .00 | .64 | 2.47 | 5.39 | 9.32 | 14.20 | 20.01 | 26.73 | 34.34 | 42.84 | 52.31 | 62.91 | 74.64 |
| 30 MPH | .00 | .96 | 3.71 | 8.08 | 13.97 | 21.30 | 30.02 | 40.10 | 51.51 | 64.26 | 78.47 | 94.36 | 111.97 |
| VELOCITY (FPS) | 1000. | 962. | 929. | 899. | 872. | 847. | 824. | 801. | 776. | 751. | 727. | 705. | 683. |
| ENERGY (FT-LB) | 273. | 253. | 236. | 221. | 208. | 196. | 185. | 175. | 164. | 154. | 144. | 136. | 127. |
| DROP (IN) | .00 | -1.09 | -4.49 | -10.39 | -18.97 | -30.40 | -44.84 | -62.48 | -83.44 | -107.96 | -136.32 | -168.78 | -205.60 |
| MID-RANGE (IN) | .00 | .28 | 1.16 | 2.70 | 4.99 | 8.08 | 12.03 | 16.90 | 22.75 | 29.66 | 37.76 | 47.16 | 57.95 |
| BULLET PATH (IN) | -.63 | .85 | .00 | -3.34 | -9.35 | -18.22 | -30.10 | -45.18 | -63.57 | -85.54 | -111.34 | -141.23 | -175.49 |
| TIME OF FLIGHT (SEC) | .000000 | .076497 | .155863 | .237954 | .322674 | .409961 | .499784 | .592134 | .687253 | .785563 | .887111 | .991953 | 1.100161 |
| WIND DEFLECTION (IN) | | | | | | | | | | | | | |
| 5 MPH | .00 | .13 | .52 | 1.14 | 2.00 | 3.08 | 4.38 | 5.91 | 7.68 | 9.73 | 12.07 | 14.69 | 17.61 |
| 10 MPH | .00 | .26 | 1.03 | 2.28 | 3.99 | 6.15 | 8.76 | 11.82 | 15.36 | 19.46 | 24.13 | 29.38 | 35.23 |
| 20 MPH | .00 | .53 | 2.06 | 4.56 | 7.98 | 12.31 | 17.52 | 23.63 | 30.71 | 38.92 | 48.26 | 58.77 | 70.46 |
| 30 MPH | .00 | .79 | 3.10 | 6.84 | 11.97 | 18.46 | 26.29 | 35.45 | 46.07 | 58.38 | 72.39 | 88.15 | 105.68 |
| VELOCITY (FPS) | 900. | 873. | 848. | 824. | 802. | 776. | 751. | 728. | 705. | 683. | 662. | 642. | 623. |
| ENERGY (FT-LB) | 221. | 208. | 196. | 185. | 176. | 165. | 154. | 145. | 136. | 128. | 120. | 113. | 106. |
| DROP (IN) | .00 | -1.28 | -5.41 | -12.55 | -22.87 | -36.51 | -53.72 | -74.74 | -99.86 | -129.33 | -163.44 | -202.49 | -246.80 |
| MID-RANGE (IN) | .00 | .34 | 1.42 | 3.29 | 6.03 | 9.67 | 14.31 | 20.07 | 27.06 | 35.39 | 45.21 | 56.59 | 69.68 |
| BULLET PATH (IN) | -.63 | 1.11 | -.00 | -4.12 | -11.43 | -22.05 | -36.24 | -54.25 | -76.34 | -102.80 | -133.89 | -169.92 | -211.21 |
| TIME OF FLIGHT (SEC) | .000000 | .084636 | .171840 | .261580 | .353847 | .448882 | .547106 | .648565 | .753316 | .861429 | .972985 | 1.088079 | 1.206813 |
| WIND DEFLECTION (IN) | | | | | | | | | | | | | |
| 5 MPH | .00 | .11 | .46 | 1.02 | 1.81 | 2.83 | 4.15 | 5.74 | 7.63 | 9.81 | 12.29 | 15.08 | 18.20 |
| 10 MPH | .00 | .23 | .91 | 2.04 | 3.61 | 5.67 | 8.29 | 11.48 | 15.25 | 19.61 | 24.58 | 30.17 | 36.40 |
| 20 MPH | .00 | .46 | 1.82 | 4.08 | 7.22 | 11.34 | 16.58 | 22.96 | 30.50 | 39.22 | 49.16 | 60.34 | 72.80 |
| 30 MPH | .00 | .69 | 2.73 | 6.11 | 10.83 | 17.01 | 24.87 | 34.44 | 45.75 | 58.83 | 73.74 | 90.51 | 109.20 |

## Bullet: Lyman #358495 135 Gr.
**Ballistic Coefficient: .096**

| RANGE (YDS) | MUZZLE | 25 | 50 | 75 | 100 | 125 | 150 | 175 | 200 | 225 | 250 | 275 | 300 |
|---|---|---|---|---|---|---|---|---|---|---|---|---|---|
| VELOCITY (FPS) | 1500. | 1351. | 1194. | 1067. | 982. | 918. | 865. | 820. | 776. | 731. | 690. | 652. | 617. |
| ENERGY (FT-LB) | 689. | 559. | 437. | 348. | 295. | 258. | 229. | 206. | 184. | 164. | 146. | 130. | 116. |
| DROP (IN) | .00 | -.51 | -2.21 | -5.45 | -10.58 | -17.95 | -27.89 | -40.71 | -56.73 | -76.34 | -99.98 | -128.17 | -161.45 |
| MID-RANGE (IN) | .00 | .13 | .60 | 1.53 | 3.08 | 5.36 | 8.50 | 12.60 | 17.79 | 24.20 | 32.04 | 41.51 | 52.84 |
| BULLET PATH (IN) | -.63 | .28 | .00 | -1.81 | -5.52 | -11.47 | -19.98 | -31.38 | -45.99 | -64.17 | -86.39 | -113.16 | -145.02 |
| TIME OF FLIGHT (SEC) | .000000 | .052723 | .111588 | .178291 | .251746 | .330858 | .415074 | .504144 | .598110 | .697752 | .803367 | .915193 | 1.033541 |
| WIND DEFLECTION (IN) | | | | | | | | | | | | | |
| 5 MPH | .00 | .24 | 1.02 | 2.49 | 4.55 | 7.12 | 10.13 | 13.56 | 17.43 | 21.80 | 26.70 | 32.14 | 38.15 |
| 10 MPH | .00 | .48 | 2.04 | 4.98 | 9.11 | 14.23 | 20.25 | 27.13 | 34.87 | 43.60 | 53.39 | 64.27 | 76.30 |
| 20 MPH | .00 | .96 | 4.08 | 9.96 | 18.21 | 28.46 | 40.51 | 54.26 | 69.73 | 87.21 | 106.79 | 128.55 | 152.61 |
| 30 MPH | .00 | 1.44 | 6.12 | 14.94 | 27.32 | 42.69 | 60.76 | 81.39 | 104.60 | 130.81 | 160.18 | 192.82 | 228.91 |
| VELOCITY (FPS) | 1400. | 1251. | 1104. | 1008. | 938. | 882. | 835. | 792. | 746. | 704. | 665. | 628. | 594. |
| ENERGY (FT-LB) | 600. | 479. | 374. | 311. | 270. | 239. | 214. | 192. | 170. | 152. | 135. | 121. | 108. |
| DROP (IN) | .00 | -.58 | -2.57 | -6.33 | -12.22 | -20.55 | -31.67 | -45.88 | -63.52 | -85.04 | -110.94 | -141.74 | -178.02 |
| MID-RANGE (IN) | .00 | .15 | .70 | 1.79 | 3.54 | 6.07 | 9.50 | 13.95 | 19.54 | 26.46 | 34.92 | 45.13 | 57.35 |
| BULLET PATH (IN) | -.63 | .39 | .00 | -2.16 | -6.45 | -13.18 | -22.69 | -35.31 | -51.34 | -71.27 | -95.57 | -124.76 | -159.45 |
| TIME OF FLIGHT (SEC) | .000000 | .056503 | .120545 | .191838 | .269088 | .351589 | .439004 | .531228 | .628897 | .732475 | .842176 | .958286 | 1.081157 |
| WIND DEFLECTION (IN) | | | | | | | | | | | | | |
| 5 MPH | .00 | .26 | 1.18 | 2.74 | 4.82 | 7.37 | 10.35 | 13.75 | 17.63 | 22.03 | 26.97 | 32.47 | 38.57 |
| 10 MPH | .00 | .52 | 2.36 | 5.48 | 9.65 | 14.74 | 20.69 | 27.50 | 35.26 | 44.06 | 53.94 | 64.94 | 77.14 |
| 20 MPH | .00 | 1.03 | 4.72 | 10.96 | 19.29 | 29.47 | 41.39 | 54.99 | 70.51 | 88.12 | 107.87 | 129.89 | 154.28 |
| 30 MPH | .00 | 1.55 | 7.08 | 16.43 | 28.94 | 44.21 | 62.08 | 82.49 | 105.77 | 132.18 | 161.81 | 194.83 | 231.42 |
| VELOCITY (FPS) | 1300. | 1138. | 1030. | 955. | 897. | 847. | 804. | 758. | 715. | 675. | 638. | 603. | 570. |
| ENERGY (FT-LB) | 518. | 397. | 325. | 280. | 246. | 220. | 198. | 176. | 157. | 140. | 125. | 112. | 100. |
| DROP (IN) | .00 | -.70 | -3.07 | -7.48 | -14.24 | -23.70 | -36.17 | -51.96 | -71.51 | -95.30 | -123.82 | -157.67 | -197.48 |
| MID-RANGE (IN) | .00 | .19 | .84 | 2.09 | 4.05 | 6.85 | 10.61 | 15.44 | 21.51 | 29.03 | 38.21 | 49.30 | 62.57 |
| BULLET PATH (IN) | -.63 | .52 | .00 | -2.56 | -7.47 | -15.08 | -25.70 | -39.64 | -57.34 | -79.27 | -105.95 | -137.95 | -175.90 |
| TIME OF FLIGHT (SEC) | .000000 | .061845 | .131317 | .207040 | .288159 | .374258 | .465169 | .561230 | .663146 | .771116 | .885409 | 1.006359 | 1.134357 |
| WIND DEFLECTION (IN) | | | | | | | | | | | | | |
| 5 MPH | .00 | .37 | 1.40 | 2.99 | 5.05 | 7.55 | 10.47 | 13.85 | 17.74 | 22.17 | 27.15 | 32.71 | 38.90 |
| 10 MPH | .00 | .73 | 2.80 | 5.98 | 10.10 | 15.10 | 20.95 | 27.70 | 35.48 | 44.33 | 54.29 | 65.43 | 77.80 |
| 20 MPH | .00 | 1.46 | 5.61 | 11.95 | 20.20 | 30.20 | 41.89 | 55.40 | 70.97 | 88.66 | 108.59 | 130.85 | 155.60 |
| 30 MPH | .00 | 2.19 | 8.41 | 17.93 | 30.30 | 45.30 | 62.84 | 83.10 | 106.45 | 133.00 | 162.88 | 196.28 | 233.40 |

# Bullet: Lyman #358495 135 Gr.
## Ballistic Coefficient: .096

| RANGE (YDS) | MUZZLE | 25 | 50 | 75 | 100 | 125 | 150 | 175 | 200 | 225 | 250 | 275 | 300 |
|---|---|---|---|---|---|---|---|---|---|---|---|---|---|
| VELOCITY (FPS) | 1100. | 1005. | 936. | 881. | 834. | 790. | 744. | 702. | 664. | 627. | 593. | 560. | 529. |
| ENERGY (FT-LB) | 371. | 309. | 268. | 238. | 213. | 191. | 170. | 151. | 135. | 120. | 108. | 96. | 86. |
| DROP (IN) | .00 | -.95 | -4.04 | -9.58 | -17.91 | -29.36 | -44.25 | -63.04 | -86.21 | -114.31 | -147.92 | -187.70 | -234.40 |
| MID-RANGE (IN) | .00 | .25 | 1.07 | 2.59 | 4.92 | 8.20 | 12.54 | 18.14 | 25.19 | 33.93 | 44.60 | 57.51 | 72.96 |
| BULLET PATH (IN) | -.63 | .76 | .00 | -3.21 | -9.21 | -18.33 | -30.88 | -47.34 | -68.18 | -93.95 | -125.22 | -162.67 | -207.04 |
| TIME OF FLIGHT (SEC) | .000000 | .071522 | .148966 | .231644 | .319228 | .411627 | .509498 | .613285 | .723202 | .839539 | .962649 | 1.092940 | 1.230874 |
| WIND DEFLECTION (IN) | | | | | | | | | | | | | |
| 5 MPH | .00 | .29 | 1.11 | 2.38 | 4.09 | 6.22 | 8.84 | 11.97 | 15.64 | 19.88 | 24.71 | 30.18 | 36.32 |
| 10 MPH | .00 | .59 | 2.22 | 4.77 | 8.18 | 12.45 | 17.67 | 23.94 | 31.28 | 39.76 | 49.43 | 60.36 | 72.63 |
| 20 MPH | .00 | 1.18 | 4.44 | 9.54 | 16.37 | 24.89 | 35.34 | 47.88 | 62.57 | 79.52 | 98.85 | 120.71 | 145.27 |
| 30 MPH | .00 | 1.76 | 6.65 | 14.31 | 24.55 | 37.34 | 53.01 | 71.81 | 93.85 | 119.28 | 148.28 | 181.07 | 217.90 |
| VELOCITY (FPS) | 1000. | 932. | 878. | 831. | 787. | 741. | 700. | 661. | 625. | 591. | 558. | 527. | 498. |
| ENERGY (FT-LB) | 306. | 266. | 236. | 211. | 190. | 168. | 150. | 134. | 120. | 107. | 95. | 85. | 76. |
| DROP (IN) | .00 | -1.12 | -4.72 | -11.12 | -20.66 | -33.66 | -50.61 | -71.97 | -98.28 | -130.14 | -168.23 | -213.28 | -266.15 |
| MID-RANGE (IN) | .00 | .29 | 1.24 | 2.97 | 5.61 | 9.28 | 14.18 | 20.50 | 28.48 | 38.38 | 50.45 | 65.03 | 82.47 |
| BULLET PATH (IN) | -.63 | .92 | -.00 | -3.73 | -10.59 | -20.93 | -35.20 | -53.88 | -77.52 | -106.71 | -142.13 | -184.51 | -234.70 |
| TIME OF FLIGHT (SEC) | .000000 | .077782 | .160769 | .248650 | .341360 | .439606 | .543778 | .654097 | .770856 | .894411 | 1.025176 | 1.163613 | 1.310238 |
| WIND DEFLECTION (IN) | | | | | | | | | | | | | |
| 5 MPH | .00 | .24 | .95 | 2.08 | 3.64 | 5.69 | 8.25 | 11.36 | 15.04 | 19.31 | 24.22 | 29.80 | 36.10 |
| 10 MPH | .00 | .49 | 1.90 | 4.16 | 7.28 | 11.37 | 16.50 | 22.72 | 30.07 | 38.62 | 48.43 | 59.60 | 72.20 |
| 20 MPH | .00 | .98 | 3.79 | 8.32 | 14.56 | 22.74 | 33.01 | 45.44 | 60.14 | 77.23 | 96.86 | 119.19 | 144.40 |
| 30 MPH | .00 | 1.47 | 5.69 | 12.49 | 21.84 | 34.11 | 49.51 | 68.16 | 90.21 | 115.85 | 145.29 | 178.79 | 216.61 |
| VELOCITY (FPS) | 900. | 350. | 807. | 761. | 718. | 678. | 641. | 606. | 572. | 541. | 511. | 482. | 455. |
| ENERGY (FT-LB) | 248. | 221. | 199. | 177. | 158. | 141. | 126. | 112. | 100. | 90. | 80. | 71. | 63. |
| DROP (IN) | .00 | -1.34 | -5.69 | -13.33 | -24.70 | -40.27 | -60.56 | -86.11 | -117.57 | -155.67 | -201.18 | -255.03 | -318.24 |
| MID-RANGE (IN) | .00 | .35 | 1.50 | 3.54 | 6.66 | 11.06 | 16.95 | 24.54 | 34.09 | 45.91 | 60.32 | 77.73 | 98.56 |
| BULLET PATH (IN) | -.63 | 1.18 | -.00 | -4.48 | -12.70 | -25.11 | -42.24 | -64.63 | -92.94 | -127.88 | -170.23 | -220.93 | -280.97 |
| TIME OF FLIGHT (SEC) | .000000 | .085784 | .176386 | .272072 | .373588 | .481142 | .594997 | .715482 | .842985 | .977946 | 1.120859 | 1.272265 | 1.432756 |
| WIND DEFLECTION (IN) | | | | | | | | | | | | | |
| 5 MPH | .00 | .22 | .86 | 1.94 | 3.54 | 5.67 | 8.36 | 11.63 | 15.52 | 20.06 | 25.30 | 31.29 | 38.08 |
| 10 MPH | .00 | .43 | 1.71 | 3.88 | 7.08 | 11.35 | 16.72 | 23.26 | 31.03 | 40.12 | 50.60 | 62.59 | 76.17 |
| 20 MPH | .00 | .86 | 3.42 | 7.77 | 14.17 | 22.70 | 33.44 | 46.52 | 62.06 | 80.24 | 101.21 | 125.17 | 152.33 |
| 30 MPH | .00 | 1.29 | 5.13 | 11.65 | 21.25 | 34.04 | 50.16 | 69.77 | 93.10 | 120.36 | 151.81 | 187.76 | 228.50 |
| VELOCITY (FPS) | 800. | 753. | 711. | 671. | 634. | 600. | 567. | 535. | 506. | 477. | 451. | 425. | 401. |
| ENERGY (FT-LB) | 196. | 174. | 155. | 138. | 123. | 110. | 98. | 88. | 78. | 70. | 62. | 55. | 49. |
| DROP (IN) | .00 | -1.66 | -7.12 | -16.87 | -31.43 | -51.37 | -77.34 | -110.05 | -150.33 | -199.10 | -257.42 | -326.47 | -407.59 |
| MID-RANGE (IN) | .00 | .45 | 1.90 | 4.55 | 8.59 | 14.25 | 21.80 | 31.51 | 43.73 | 58.86 | 77.34 | 99.67 | 126.46 |
| BULLET PATH (IN) | -.63 | 1.59 | -.00 | -5.88 | -16.56 | -32.62 | -54.71 | -83.54 | -119.94 | -164.84 | -219.28 | -284.45 | -361.70 |
| TIME OF FLIGHT (SEC) | .000000 | .096647 | .199174 | .307778 | .422733 | .544379 | .673113 | .809383 | .953691 | 1.106585 | 1.268665 | 1.440585 | 1.623060 |
| WIND DEFLECTION (IN) | | | | | | | | | | | | | |
| 5 MPH | .00 | .25 | 1.03 | 2.33 | 4.20 | 6.66 | 9.73 | 13.48 | 17.92 | 23.13 | 29.14 | 36.02 | 43.83 |
| 10 MPH | .00 | .51 | 2.05 | 4.67 | 8.40 | 13.31 | 19.47 | 26.95 | 35.85 | 46.26 | 58.28 | 72.04 | 87.66 |
| 20 MPH | .00 | 1.02 | 4.11 | 9.34 | 16.80 | 26.62 | 38.94 | 53.90 | 71.70 | 92.52 | 116.57 | 144.09 | 175.32 |
| 30 MPH | .00 | 1.53 | 6.16 | 14.01 | 25.20 | 39.93 | 58.40 | 80.85 | 107.55 | 138.78 | 174.85 | 216.13 | 262.98 |

# Bullet: Lyman #358477 146 Gr.
## Ballistic Coeffecient: .183

| RANGE (YDS) | MUZZLE | 25 | 50 | 75 | 100 | 125 | 150 | 175 | 200 | 225 | 250 | 275 | 300 |
|---|---|---|---|---|---|---|---|---|---|---|---|---|---|
| VELOCITY (FPS) | 1500. | 1419. | 1344. | 1270. | 1196. | 1133. | 1081. | 1037. | 1001. | 969. | 940. | 914. | 890. |
| ENERGY (FT-LB) | 729. | 653. | 586. | 523. | 464. | 416. | 378. | 349. | 325. | 304. | 286. | 271. | 257. |
| DROP (IN) | .00 | -.48 | -2.04 | -4.80 | -8.92 | -14.55 | -21.87 | -31.05 | -42.24 | -55.59 | -71.24 | -89.31 | -109.98 |
| MID-RANGE (IN) | .00 | .13 | .54 | 1.26 | 2.42 | 4.01 | 6.14 | 8.85 | 12.20 | 16.26 | 21.07 | 26.67 | 33.11 |
| BULLET PATH (IN) | -.63 | .22 | .00 | -1.43 | -4.21 | -8.51 | -14.49 | -22.34 | -32.19 | -44.21 | -58.53 | -75.26 | -94.59 |
| TIME OF FLIGHT (SEC) | .000000 | .051411 | .105722 | .163064 | .223948 | .288422 | .356258 | .427138 | .500785 | .576998 | .655628 | .736565 | .819731 |
| WIND DEFLECTION (IN) | | | | | | | | | | | | | |
| 5 MPH | .00 | .12 | .50 | 1.15 | 2.11 | 3.38 | 4.95 | 6.79 | 8.87 | 11.18 | 13.70 | 16.42 | 19.34 |
| 10 MPH | .00 | .25 | 1.01 | 2.30 | 4.21 | 6.76 | 9.90 | 13.58 | 17.74 | 22.35 | 27.39 | 32.84 | 38.67 |
| 20 MPH | .00 | .50 | 2.01 | 4.60 | 8.43 | 13.52 | 19.80 | 27.15 | 35.48 | 44.70 | 54.78 | 65.67 | 77.35 |
| 30 MPH | .00 | .74 | 3.02 | 6.90 | 12.64 | 20.29 | 29.70 | 40.73 | 53.21 | 67.05 | 82.17 | 98.51 | 116.02 |
| VELOCITY (FPS) | 1400. | 1326. | 1251. | 1180. | 1119. | 1069. | 1028. | 993. | 961. | 934. | 908. | 885. | 863. |
| ENERGY (FT-LB) | 635. | 570. | 508. | 451. | 406. | 371. | 343. | 319. | 300. | 282. | 267. | 254. | 241. |
| DROP (IN) | .00 | -.55 | -2.34 | -5.53 | -10.27 | -16.74 | -25.11 | -35.52 | -48.14 | -63.07 | -80.47 | -100.49 | -123.28 |
| MID-RANGE (IN) | .00 | .15 | .62 | 1.48 | 2.79 | 4.62 | 7.03 | 10.07 | 13.80 | 18.25 | 23.49 | 29.57 | 36.53 |
| BULLET PATH (IN) | -.63 | .30 | .00 | -1.70 | -4.96 | -9.94 | -16.82 | -25.75 | -36.88 | -50.33 | -66.24 | -84.78 | -106.08 |
| TIME OF FLIGHT (SEC) | .000000 | .055053 | .113209 | .174969 | .240277 | .308874 | .380441 | .454721 | .531526 | .610719 | .692197 | .775888 | .861740 |
| WIND DEFLECTION (IN) | | | | | | | | | | | | | |
| 5 MPH | .00 | .13 | .53 | 1.25 | 2.29 | 3.61 | 5.19 | 7.02 | 9.06 | 11.31 | 13.77 | 16.42 | 19.26 |
| 10 MPH | .00 | .26 | 1.07 | 2.51 | 4.57 | 7.22 | 10.39 | 14.03 | 18.12 | 22.63 | 27.54 | 32.84 | 38.52 |
| 20 MPH | .00 | .52 | 2.14 | 5.02 | 9.15 | 14.44 | 20.77 | 28.06 | 36.24 | 45.26 | 55.08 | 65.68 | 77.05 |
| 30 MPH | .00 | .78 | 3.20 | 7.53 | 13.72 | 21.66 | 31.16 | 42.09 | 54.36 | 67.89 | 82.62 | 98.53 | 115.57 |
| VELOCITY (FPS) | 1300. | 1222. | 1155. | 1099. | 1052. | 1013. | 980. | 950. | 923. | 899. | 876. | 855. | 835. |
| ENERGY (FT-LB) | 548. | 484. | 432. | 391. | 359. | 333. | 311. | 293. | 276. | 262. | 249. | 237. | 226. |
| DROP (IN) | .00 | -.65 | -2.76 | -6.50 | -12.03 | -19.52 | -29.11 | -40.97 | -55.18 | -71.93 | -91.36 | -113.60 | -138.81 |
| MID-RANGE (IN) | .00 | .17 | .73 | 1.73 | 3.25 | 5.34 | 8.06 | 11.45 | 15.56 | 20.44 | 26.16 | 32.76 | 40.29 |
| BULLET PATH (IN) | -.63 | .41 | .00 | -2.04 | -5.87 | -11.66 | -19.56 | -29.72 | -42.23 | -57.28 | -75.02 | -95.56 | -119.08 |
| TIME OF FLIGHT (SEC) | .000000 | .059537 | .122715 | .189354 | .259155 | .331814 | .407105 | .484863 | .564964 | .647318 | .731863 | .818553 | .907364 |
| WIND DEFLECTION (IN) | | | | | | | | | | | | | |
| 5 MPH | .00 | .16 | .65 | 1.43 | 2.50 | 3.82 | 5.36 | 7.13 | 9.10 | 11.27 | 13.63 | 16.19 | 18.92 |
| 10 MPH | .00 | .32 | 1.29 | 2.86 | 5.00 | 7.63 | 10.73 | 14.26 | 18.20 | 22.54 | 27.27 | 32.37 | 37.85 |
| 20 MPH | .00 | .65 | 2.58 | 5.73 | 9.99 | 15.26 | 21.45 | 28.52 | 36.41 | 45.09 | 54.54 | 64.75 | 75.70 |
| 30 MPH | .00 | .97 | 3.87 | 8.59 | 14.99 | 22.89 | 32.18 | 42.78 | 54.61 | 67.63 | 81.81 | 97.12 | 113.55 |
| VELOCITY (FPS) | 1200. | 1136. | 1083. | 1040. | 1003. | 970. | 942. | 916. | 892. | 869. | 849. | 829. | 810. |
| ENERGY (FT-LB) | 467. | 418. | 380. | 350. | 326. | 305. | 287. | 272. | 258. | 245. | 233. | 223. | 213. |
| DROP (IN) | .00 | -.77 | -3.22 | -7.51 | -13.80 | -22.25 | -32.99 | -46.15 | -61.88 | -80.34 | -101.66 | -125.99 | -153.47 |
| MID-RANGE (IN) | .00 | .20 | .84 | 1.98 | 3.69 | 6.01 | 8.99 | 12.68 | 17.14 | 22.42 | 28.58 | 35.66 | 43.74 |
| BULLET PATH (IN) | -.63 | .52 | .00 | -2.37 | -6.74 | -13.27 | -22.09 | -33.32 | -47.13 | -63.66 | -83.06 | -105.47 | -131.03 |
| TIME OF FLIGHT (SEC) | .000000 | .064274 | .131926 | .202639 | .276133 | .352203 | .430696 | .511501 | .594537 | .679748 | .767095 | .856557 | .948125 |
| WIND DEFLECTION (IN) | | | | | | | | | | | | | |
| 5 MPH | .00 | .16 | .61 | 1.33 | 2.30 | 3.49 | 4.90 | 6.51 | 8.32 | 10.32 | 12.50 | 14.88 | 17.44 |
| 10 MPH | .00 | .31 | 1.22 | 2.66 | 4.60 | 6.99 | 9.80 | 13.02 | 16.64 | 20.64 | 25.01 | 29.75 | 34.87 |
| 20 MPH | .00 | .62 | 2.44 | 5.33 | 9.20 | 13.98 | 19.60 | 26.05 | 33.28 | 41.27 | 50.02 | 59.51 | 69.74 |
| 30 MPH | .00 | .94 | 3.66 | 7.99 | 13.80 | 20.96 | 29.41 | 39.07 | 49.92 | 61.91 | 75.03 | 89.26 | 104.61 |

## Bullet: Lyman #358477 146 Gr.
### Ballistic Coefficient: .183

| RANGE (YDS) | MUZZLE | 25 | 50 | 75 | 100 | 125 | 150 | 175 | 200 | 225 | 250 | 275 | 300 |
|---|---|---|---|---|---|---|---|---|---|---|---|---|---|
| VELOCITY (FPS) | 1100. | 1054. | 1014. | 981. | 951. | 924. | 899. | 877. | 856. | 835. | 816. | 798. | 778. |
| ENERGY (FT-LB) | 392. | 360. | 334. | 312. | 293. | 277. | 262. | 249. | 237. | 226. | 216. | 206. | 196. |
| DROP (IN) | .00 | -.90 | -3.76 | -8.72 | -15.93 | -25.50 | -37.60 | -52.38 | -69.96 | -90.51 | -114.16 | -141.07 | -171.28 |
| MID-RANGE (IN) | .00 | .23 | .98 | 2.28 | 4.21 | 6.78 | 10.08 | 14.16 | 19.05 | 24.85 | 31.58 | 39.31 | 48.04 |
| BULLET PATH (IN) | -.63 | .66 | .00 | -2.77 | -7.78 | -15.16 | -25.07 | -37.65 | -53.04 | -71.40 | -92.86 | -117.57 | -145.58 |
| TIME OF FLIGHT (SEC) | .000000 | .069712 | .142290 | .217505 | .295191 | .375222 | .457509 | .541987 | .628611 | .717356 | .808208 | .901171 | .996451 |
| WIND DEFLECTION (IN) | | | | | | | | | | | | | |
| 5 MPH | .00 | .13 | .52 | 1.14 | 1.98 | 3.02 | 4.26 | 5.69 | 7.32 | 9.13 | 11.12 | 13.30 | 15.69 |
| 10 MPH | .00 | .27 | 1.04 | 2.28 | 3.95 | 6.04 | 8.52 | 11.39 | 14.64 | 18.25 | 22.24 | 26.61 | 31.38 |
| 20 MPH | .00 | .54 | 2.09 | 4.56 | 7.91 | 12.08 | 17.04 | 22.78 | 29.27 | 36.51 | 44.49 | 53.21 | 62.75 |
| 30 MPH | .00 | .81 | 3.13 | 6.84 | 11.86 | 18.12 | 25.56 | 34.17 | 43.91 | 54.76 | 66.73 | 79.82 | 94.13 |
| VELOCITY (FPS) | 1000. | 968. | 939. | 914. | 890. | 868. | 847. | 827. | 809. | 789. | 770. | 751. | 732. |
| ENERGY (FT-LB) | 324. | 304. | 286. | 271. | 257. | 244. | 233. | 222. | 212. | 202. | 192. | 183. | 174. |
| DROP (IN) | .00 | -1.08 | -4.45 | -10.26 | -18.65 | -29.78 | -43.77 | -60.79 | -80.97 | -104.43 | -131.30 | -161.81 | -196.16 |
| MID-RANGE (IN) | .00 | .28 | 1.15 | 2.66 | 4.87 | 7.85 | 11.63 | 16.27 | 21.84 | 28.35 | 35.87 | 44.50 | 54.31 |
| BULLET PATH (IN) | -.63 | .83 | .00 | -3.26 | -9.12 | -17.70 | -29.16 | -43.63 | -61.28 | -82.19 | -106.52 | -134.49 | -166.29 |
| TIME OF FLIGHT (SEC) | .000000 | .076252 | .154919 | .235890 | .319087 | .404454 | .491954 | .581567 | .673285 | .767151 | .863419 | .962154 | 1.063381 |
| WIND DEFLECTION (IN) | | | | | | | | | | | | | |
| 5 MPH | .00 | .11 | .43 | .96 | 1.68 | 2.59 | 3.69 | 4.98 | 6.45 | 8.11 | 9.98 | 12.07 | 14.38 |
| 10 MPH | .00 | .22 | .87 | 1.92 | 3.36 | 5.18 | 7.38 | 9.96 | 12.90 | 16.22 | 19.96 | 24.14 | 28.76 |
| 20 MPH | .00 | .44 | 1.73 | 3.83 | 6.72 | 10.37 | 14.77 | 19.91 | 25.80 | 32.44 | 39.92 | 48.28 | 57.51 |
| 30 MPH | .00 | .66 | 2.60 | 5.75 | 10.08 | 15.55 | 22.15 | 29.87 | 38.69 | 48.66 | 59.89 | 72.42 | 86.27 |
| VELOCITY (FPS) | 900. | 877. | 856. | 836. | 817. | 798. | 778. | 759. | 740. | 722. | 705. | 688. | 672. |
| ENERGY (FT-LB) | 263. | 249. | 237. | 226. | 216. | 207. | 196. | 187. | 178. | 169. | 161. | 154. | 146. |
| DROP (IN) | .00 | -1.26 | -5.33 | -12.36 | -22.48 | -35.86 | -52.56 | -72.80 | -96.79 | -124.72 | -156.80 | -193.23 | -234.24 |
| MID-RANGE (IN) | .00 | .34 | 1.40 | 3.24 | 5.91 | 9.47 | 13.92 | 19.38 | 25.91 | 33.60 | 42.54 | 52.84 | 64.56 |
| BULLET PATH (IN) | -.63 | 1.09 | -.00 | -4.04 | -11.19 | -21.59 | -35.30 | -52.56 | -73.57 | -98.52 | -127.62 | -161.07 | -199.09 |
| TIME OF FLIGHT (SEC) | .000000 | .084420 | .170987 | .259673 | .350466 | .443366 | .538545 | .636181 | .736295 | .838917 | .944084 | 1.051838 | 1.162226 |
| WIND DEFLECTION (IN) | | | | | | | | | | | | | |
| 5 MPH | .00 | .10 | .38 | .85 | 1.51 | 2.35 | 3.39 | 4.65 | 6.13 | 7.82 | 9.75 | 11.90 | 14.28 |
| 10 MPH | .00 | .19 | .76 | 1.70 | 3.02 | 4.70 | 6.78 | 9.30 | 12.25 | 15.65 | 19.49 | 23.79 | 28.55 |
| 20 MPH | .00 | .38 | 1.52 | 3.41 | 6.03 | 9.40 | 13.57 | 18.60 | 24.51 | 31.30 | 38.98 | 47.58 | 57.10 |
| 30 MPH | .00 | .57 | 2.28 | 5.11 | 9.05 | 14.10 | 20.35 | 27.90 | 36.76 | 46.95 | 58.48 | 71.37 | 85.66 |
| VELOCITY (FPS) | 800. | 780. | 760. | 742. | 724. | 706. | 689. | 673. | 657. | 642. | 627. | 612. | 598. |
| ENERGY (FT-LB) | 207. | 197. | 187. | 178. | 170. | 162. | 154. | 147. | 140. | 134. | 127. | 121. | 116. |
| DROP (IN) | .00 | -1.50 | -6.56 | -15.35 | -28.08 | -44.95 | -66.16 | -91.85 | -122.33 | -157.83 | -198.58 | -244.86 | -296.92 |
| MID-RANGE (IN) | .00 | .43 | 1.77 | 4.10 | 7.49 | 12.00 | 17.73 | 24.71 | 33.08 | 42.93 | 54.34 | 67.43 | 82.31 |
| BULLET PATH (IN) | -.63 | 1.46 | -.00 | -5.20 | -14.34 | -27.61 | -45.23 | -67.33 | -94.22 | -126.12 | -163.28 | -205.96 | -254.44 |
| TIME OF FLIGHT (SEC) | .000000 | .094972 | .192399 | .292300 | .394706 | .499652 | .607179 | .717335 | .830172 | .945748 | 1.064127 | 1.185373 | 1.309560 |
| WIND DEFLECTION (IN) | | | | | | | | | | | | | |
| 5 MPH | .00 | .11 | .43 | .97 | 1.73 | 2.72 | 3.93 | 5.38 | 7.06 | 8.98 | 11.14 | 13.56 | 16.24 |
| 10 MPH | .00 | .22 | .86 | 1.94 | 3.47 | 5.44 | 7.86 | 10.75 | 14.11 | 17.95 | 22.29 | 27.13 | 32.48 |
| 20 MPH | .00 | .43 | 1.72 | 3.89 | 6.94 | 10.88 | 15.73 | 21.50 | 28.22 | 35.90 | 44.57 | 54.25 | 64.97 |
| 30 MPH | .00 | .65 | 2.59 | 5.83 | 10.40 | 16.32 | 23.59 | 32.25 | 42.33 | 53.86 | 66.86 | 81.38 | 97.45 |

# Bullet: Lyman #358477 146 Gr.
## Ballistic Coefficient: .183

| RANGE (YDS) | MUZZLE | 25 | 50 | 75 | 100 | 125 | 150 | 175 | 200 | 225 | 250 | 275 | 300 |
|---|---|---|---|---|---|---|---|---|---|---|---|---|---|
| VELOCITY (FPS) | 700. | 683. | 667. | 651. | 636. | 621. | 606. | 592. | 578. | 565. | 552. | 539. | 526. |
| ENERGY (FT-LB) | 159. | 151. | 144. | 138. | 131. | 125. | 119. | 114. | 108. | 103. | 99. | 94. | 90. |
| DROP (IN) | .00 | -1.94 | -8.51 | -19.95 | -36.49 | -58.38 | -85.89 | -119.21 | -158.71 | -204.69 | -257.48 | -317.40 | -384.82 |
| MID-RANGE (IN) | .00 | .57 | 2.32 | 5.34 | 9.73 | 15.59 | 23.00 | 32.05 | 42.87 | 55.60 | 70.36 | 87.28 | 106.52 |
| BULLET PATH (IN) | -.63 | 2.00 | -.00 | -6.87 | -18.84 | -36.16 | -59.11 | -87.86 | -122.79 | -164.20 | -212.41 | -267.77 | -330.62 |
| TIME OF FLIGHT (SEC) | .000000 | .108449 | .219543 | .333338 | .449891 | .569266 | .691532 | .816762 | .945030 | 1.076419 | 1.211013 | 1.348900 | 1.490174 |
| WIND DEFLECTION (IN) | | | | | | | | | | | | | |
| 5 MPH | .00 | .11 | .46 | 1.05 | 1.88 | 2.95 | 4.28 | 5.88 | 7.73 | 9.87 | 12.28 | 14.99 | 17.99 |
| 10 MPH | .00 | .23 | .93 | 2.10 | 3.75 | 5.91 | 8.57 | 11.75 | 15.47 | 19.74 | 24.57 | 29.98 | 35.98 |
| 20 MPH | .00 | .46 | 1.85 | 4.19 | 7.50 | 11.81 | 17.13 | 23.50 | 30.94 | 39.47 | 49.13 | 59.96 | 71.97 |
| 30 MPH | .00 | .69 | 2.78 | 6.29 | 11.26 | 17.72 | 25.70 | 35.25 | 46.40 | 59.21 | 73.70 | 89.93 | 107.95 |
| VELOCITY (FPS) | 600. | 586. | 572. | 558. | 545. | 532. | 520. | 507. | 495. | 484. | 472. | 461. | 450. |
| ENERGY (FT-LB) | 117. | 111. | 106. | 101. | 96. | 92. | 88. | 83. | 80. | 76. | 72. | 69. | 66. |
| DROP (IN) | .00 | -2.63 | -11.57 | -27.14 | -49.65 | -79.44 | -116.90 | -162.27 | -216.09 | -278.76 | -350.73 | -432.48 | -524.51 |
| MID-RANGE (IN) | .00 | .77 | 3.15 | 7.27 | 13.25 | 21.21 | 31.31 | 43.64 | 58.40 | 75.77 | 95.92 | 119.05 | 145.36 |
| BULLET PATH (IN) | -.63 | 2.84 | .00 | -9.46 | -25.87 | -49.56 | -80.91 | -120.19 | -167.90 | -224.47 | -290.34 | -365.99 | -451.91 |
| TIME OF FLIGHT (SEC) | .000000 | .126509 | .256089 | .388821 | .524791 | .664087 | .806804 | .953038 | 1.102891 | 1.256468 | 1.413879 | 1.575238 | 1.740664 |
| WIND DEFLECTION (IN) | | | | | | | | | | | | | |
| 5 MPH | .00 | .13 | .54 | 1.22 | 2.18 | 3.44 | 5.00 | 6.87 | 9.05 | 11.57 | 14.42 | 17.62 | 21.18 |
| 10 MPH | .00 | .27 | 1.07 | 2.43 | 4.36 | 6.88 | 10.00 | 13.73 | 18.11 | 23.14 | 28.84 | 35.24 | 42.36 |
| 20 MPH | .00 | .53 | 2.14 | 4.86 | 8.73 | 13.76 | 19.99 | 27.47 | 36.22 | 46.28 | 57.69 | 70.48 | 84.71 |
| 30 MPH | .00 | .80 | 3.21 | 7.30 | 13.09 | 20.64 | 29.99 | 41.20 | 54.33 | 69.42 | 86.53 | 105.73 | 127.07 |

# Bullet: Lyman #358156 147 Gr.
## Ballistic Coefficient: .213

| RANGE (YDS) | MUZZLE | 25 | 50 | 75 | 100 | 125 | 150 | 175 | 200 | 225 | 250 | 275 | 300 |
|---|---|---|---|---|---|---|---|---|---|---|---|---|---|
| VELOCITY (FPS) | 1400. | 1336. | 1271. | 1201. | 1141. | 1090. | 1048. | 1012. | 981. | 953. | 927. | 904. | 882. |
| ENERGY (FT-LB) | 640. | 583. | 527. | 471. | 425. | 388. | 358. | 334. | 314. | 296. | 281. | 267. | 254. |
| DROP (IN) | .00 | -.55 | -2.31 | -5.43 | -10.05 | -16.34 | -24.45 | -34.54 | -46.73 | -61.18 | -77.99 | -97.30 | -119.26 |
| MID-RANGE (IN) | .00 | .14 | .61 | 1.45 | 2.72 | 4.48 | 6.80 | 9.72 | 13.31 | 17.61 | 22.65 | 28.49 | 35.18 |
| BULLET PATH (IN) | -.63 | .29 | .00 | -1.65 | -4.80 | -9.62 | -16.26 | -24.87 | -35.60 | -48.58 | -63.91 | -81.76 | -102.25 |
| TIME OF FLIGHT (SEC) | .000000 | .054842 | .112306 | .173026 | .237125 | .304413 | .374615 | .447472 | .522788 | .600419 | .680256 | .762216 | .846240 |
| WIND DEFLECTION (IN) | | | | | | | | | | | | | |
| 5 MPH | .00 | .11 | .45 | 1.08 | 2.01 | 3.22 | 4.68 | 6.38 | 8.29 | 10.41 | 12.72 | 15.22 | 17.90 |
| 10 MPH | .00 | .22 | .91 | 2.17 | 4.02 | 6.43 | 9.36 | 12.76 | 16.58 | 20.82 | 25.44 | 30.44 | 35.80 |
| 20 MPH | .00 | .45 | 1.82 | 4.33 | 8.04 | 12.87 | 18.72 | 25.51 | 33.16 | 41.63 | 50.88 | 60.87 | 71.59 |
| 30 MPH | .00 | .67 | 2.73 | 6.50 | 12.06 | 19.30 | 28.08 | 38.27 | 49.75 | 62.45 | 76.32 | 91.31 | 107.39 |
| VELOCITY (FPS) | 1300. | 1227. | 1163. | 1108. | 1063. | 1025. | 992. | 963. | 936. | 912. | 890. | 869. | 850. |
| ENERGY (FT-LB) | 552. | 491. | 441. | 401. | 369. | 343. | 321. | 302. | 286. | 272. | 259. | 247. | 236. |
| DROP (IN) | .00 | -.65 | -2.75 | -6.45 | -11.91 | -19.30 | -28.74 | -40.38 | -54.35 | -70.75 | -89.76 | -111.50 | -136.10 |
| MID-RANGE (IN) | .00 | .17 | .72 | 1.72 | 3.21 | 5.26 | 7.92 | 11.24 | 15.26 | 20.02 | 25.58 | 32.00 | 39.31 |
| BULLET PATH (IN) | -.63 | .41 | .00 | -2.01 | -5.79 | -11.48 | -19.24 | -29.19 | -41.47 | -56.19 | -73.50 | -93.55 | -116.47 |
| TIME OF FLIGHT (SEC) | .000000 | .059422 | .122266 | .188382 | .257521 | .329404 | .403813 | .480586 | .559602 | .640770 | .724021 | .809308 | .896598 |
| WIND DEFLECTION (IN) | | | | | | | | | | | | | |
| 5 MPH | .00 | .15 | .61 | 1.35 | 2.35 | 3.60 | 5.07 | 6.75 | 8.63 | 10.70 | 12.94 | 15.37 | 17.98 |
| 10 MPH | .00 | .30 | 1.21 | 2.69 | 4.71 | 7.21 | 10.15 | 13.51 | 17.26 | 21.39 | 25.89 | 30.75 | 35.96 |
| 20 MPH | .00 | .61 | 2.42 | 5.39 | 9.42 | 14.41 | 20.30 | 27.01 | 34.52 | 42.78 | 51.78 | 61.49 | 71.91 |
| 30 MPH | .00 | .91 | 3.63 | 8.08 | 14.12 | 21.62 | 30.44 | 40.52 | 51.78 | 64.17 | 77.67 | 92.24 | 107.87 |
| VELOCITY (FPS) | 1200. | 1140. | 1089. | 1047. | 1011. | 980. | 952. | 927. | 903. | 882. | 862. | 842. | 824. |
| ENERGY (FT-LB) | 470. | 424. | 387. | 358. | 334. | 313. | 296. | 280. | 266. | 254. | 242. | 232. | 222. |
| DROP (IN) | .00 | -.77 | -3.20 | -7.46 | -13.69 | -22.04 | -32.65 | -45.61 | -61.08 | -79.20 | -100.11 | -123.94 | -150.83 |
| MID-RANGE (IN) | .00 | .20 | .83 | 1.96 | 3.65 | 5.93 | 8.87 | 12.48 | 16.85 | 22.01 | 28.02 | 34.92 | 42.77 |
| BULLET PATH (IN) | -.63 | .52 | .00 | -2.34 | -6.66 | -13.10 | -21.79 | -32.83 | -46.39 | -62.60 | -81.60 | -103.51 | -128.48 |
| TIME OF FLIGHT (SEC) | .000000 | .064165 | .131515 | .201772 | .274679 | .350042 | .427717 | .507595 | .589595 | .673657 | .759740 | .847816 | .937872 |
| WIND DEFLECTION (IN) | | | | | | | | | | | | | |
| 5 MPH | .00 | .15 | .57 | 1.26 | 2.17 | 3.30 | 4.64 | 6.17 | 7.88 | 9.78 | 11.86 | 14.11 | 16.53 |
| 10 MPH | .00 | .29 | 1.15 | 2.51 | 4.34 | 6.61 | 9.28 | 12.34 | 15.77 | 19.56 | 23.71 | 28.22 | 33.07 |
| 20 MPH | .00 | .59 | 2.29 | 5.02 | 8.69 | 13.21 | 18.56 | 24.67 | 31.54 | 39.13 | 47.43 | 56.43 | 66.13 |
| 30 MPH | .00 | .88 | 3.44 | 7.54 | 13.03 | 19.82 | 27.83 | 37.01 | 47.31 | 58.69 | 71.14 | 84.65 | 99.20 |
| VELOCITY (FPS) | 1100. | 1056. | 1019. | 987. | 958. | 932. | 908. | 887. | 866. | 847. | 828. | 811. | 791. |
| ENERGY (FT-LB) | 395. | 364. | 339. | 318. | 300. | 284. | 269. | 256. | 245. | 234. | 224. | 214. | 204. |
| DROP (IN) | .00 | -.90 | -3.74 | -8.67 | -15.82 | -25.31 | -37.27 | -51.85 | -69.18 | -89.41 | -112.65 | -139.06 | -168.75 |
| MID-RANGE (IN) | .00 | .23 | .97 | 2.27 | 4.17 | 6.72 | 9.97 | 13.97 | 18.77 | 24.43 | 31.02 | 38.56 | 47.11 |
| BULLET PATH (IN) | -.63 | .66 | .00 | -2.74 | -7.71 | -15.01 | -24.78 | -37.18 | -52.32 | -70.36 | -91.42 | -115.64 | -143.15 |
| TIME OF FLIGHT (SEC) | .000000 | .069619 | .141941 | .216757 | .293915 | .373297 | .454818 | .538413 | .624036 | .711657 | .801259 | .892836 | .986428 |
| WIND DEFLECTION (IN) | | | | | | | | | | | | | |
| 5 MPH | .00 | .13 | .49 | 1.07 | 1.86 | 2.85 | 4.02 | 5.38 | 6.92 | 8.63 | 10.51 | 12.57 | 14.81 |
| 10 MPH | .00 | .25 | .98 | 2.15 | 3.73 | 5.70 | 8.05 | 10.76 | 13.83 | 17.25 | 21.02 | 25.14 | 29.61 |
| 20 MPH | .00 | .51 | 1.96 | 4.30 | 7.46 | 11.40 | 16.10 | 21.52 | 27.66 | 34.50 | 42.04 | 50.28 | 59.22 |
| 30 MPH | .00 | .76 | 2.94 | 6.45 | 11.19 | 17.10 | 24.14 | 32.28 | 41.4 | 51.75 | 63.06 | 75.42 | 88.83 |

# Bullet: Lyman #358156 147 Gr.
## Ballistic Coefficient: .213

| RANGE (YDS) | MUZZLE | 25 | 50 | 75 | 100 | 125 | 150 | 175 | 200 | 225 | 250 | 275 | 300 |
|---|---|---|---|---|---|---|---|---|---|---|---|---|---|
| VELOCITY (FPS) | 1000. | 970. | 943. | 918. | 896. | 875. | 855. | 836. | 818. | 801. | 776. | 751. | 728. |
| ENERGY (FT-LB) | 326. | 307. | 290. | 275. | 262. | 250. | 238. | 228. | 218. | 209. | 197. | 184. | 173. |
| DROP (IN) | .00 | -1.07 | -4.44 | -10.21 | -18.55 | -29.57 | -43.44 | -60.26 | -80.19 | -103.36 | -129.88 | -159.98 | -193.90 |
| MID-RANGE (IN) | .00 | .28 | 1.15 | 2.64 | 4.83 | 7.77 | 11.50 | 16.08 | 21.55 | 27.96 | 35.37 | 43.85 | 53.51 |
| BULLET PATH (IN) | -.63 | .83 | .00 | -3.23 | -9.03 | -17.52 | -28.85 | -43.14 | -60.53 | -81.16 | -105.15 | -132.71 | -164.10 |
| TIME OF FLIGHT (SEC) | .000000 | .076176 | .154621 | .235239 | .317955 | .402716 | .489486 | .578242 | .668975 | .761682 | .856765 | .955025 | 1.056524 |
| WIND DEFLECTION (IN) | | | | | | | | | | | | | |
| 5 MPH | .00 | .10 | .41 | .90 | 1.58 | 2.44 | 3.47 | 4.69 | 6.07 | 7.63 | 9.40 | 11.44 | 13.77 |
| 10 MPH | .00 | .21 | .81 | 1.80 | 3.16 | 4.88 | 6.95 | 9.37 | 12.14 | 15.26 | 18.79 | 22.88 | 27.55 |
| 20 MPH | .00 | .41 | 1.63 | 3.60 | 6.32 | 9.76 | 13.90 | 18.74 | 24.28 | 30.51 | 37.58 | 45.77 | 55.10 |
| 30 MPH | .00 | .62 | 2.44 | 5.41 | 9.48 | 14.63 | 20.85 | 28.11 | 36.42 | 45.77 | 56.37 | 68.65 | 82.64 |
| VELOCITY (FPS) | 900. | 879. | 858. | 839. | 821. | 804. | 781. | 756. | 732. | 709. | 687. | 666. | 646. |
| ENERGY (FT-LB) | 264. | 252. | 241. | 230. | 220. | 211. | 199. | 187. | 175. | 164. | 154. | 145. | 136. |
| DROP (IN) | .00 | -1.25 | -5.30 | -12.30 | -22.37 | -35.66 | -52.27 | -72.40 | -96.31 | -124.25 | -156.50 | -193.34 | -235.06 |
| MID-RANGE (IN) | .00 | .34 | 1.40 | 3.23 | 5.88 | 9.40 | 13.83 | 19.26 | 25.78 | 33.52 | 42.60 | 53.13 | 65.26 |
| BULLET PATH (IN) | -.63 | 1.09 | -.00 | -4.03 | -11.13 | -21.45 | -35.10 | -52.26 | -73.20 | -98.18 | -127.46 | -161.33 | -200.08 |
| TIME OF FLIGHT (SEC) | .000000 | .084353 | .170720 | .259076 | .349409 | .441715 | .536233 | .633856 | .734707 | .838838 | .946318 | 1.057225 | 1.171652 |
| WIND DEFLECTION (IN) | | | | | | | | | | | | | |
| 5 MPH | .00 | .09 | .36 | .80 | 1.41 | 2.20 | 3.19 | 4.45 | 5.99 | 7.82 | 9.94 | 12.37 | 15.11 |
| 10 MPH | .00 | .18 | .71 | 1.60 | 2.83 | 4.41 | 6.38 | 8.89 | 11.98 | 15.64 | 19.89 | 24.74 | 30.21 |
| 20 MPH | .00 | .36 | 1.43 | 3.19 | 5.66 | 8.82 | 12.75 | 17.78 | 23.95 | 31.27 | 39.77 | 49.48 | 60.42 |
| 30 MPH | .00 | .54 | 2.14 | 4.79 | 8.49 | 13.23 | 19.13 | 26.68 | 35.93 | 46.91 | 59.66 | 74.21 | 90.63 |
| VELOCITY (FPS) | 800. | 774. | 749. | 725. | 703. | 681. | 660. | 640. | 621. | 602. | 583. | 566. | 549. |
| ENERGY (FT-LB) | 209. | 195. | 183. | 172. | 161. | 151. | 142. | 134. | 126. | 118. | 111. | 104. | 98. |
| DROP (IN) | .00 | -1.55 | -6.69 | -15.67 | -28.77 | -46.25 | -68.40 | -95.52 | -127.93 | -165.96 | -209.99 | -260.38 | -317.57 |
| MID-RANGE (IN) | .00 | .43 | 1.79 | 4.18 | 7.70 | 12.45 | 18.53 | 26.07 | 35.19 | 46.03 | 58.74 | 73.47 | 90.38 |
| BULLET PATH (IN) | -.63 | 1.48 | -.00 | -5.33 | -14.77 | -28.59 | -47.09 | -70.55 | -99.30 | -133.67 | -174.04 | -220.78 | -274.31 |
| TIME OF FLIGHT (SEC) | .000000 | .095340 | .193878 | .295655 | .400730 | .509172 | .621064 | .736501 | .855588 | .978439 | 1.105177 | 1.235936 | 1.370857 |
| WIND DEFLECTION (IN) | | | | | | | | | | | | | |
| 5 MPH | .00 | .14 | .56 | 1.27 | 2.26 | 3.56 | 5.15 | 7.06 | 9.29 | 11.85 | 14.76 | 18.01 | 21.64 |
| 10 MPH | .00 | .28 | 1.12 | 2.54 | 4.53 | 7.11 | 10.31 | 14.12 | 18.58 | 23.71 | 29.51 | 36.02 | 43.27 |
| 20 MPH | .00 | .56 | 2.24 | 5.07 | 9.06 | 14.23 | 20.61 | 28.25 | 37.17 | 47.41 | 59.02 | 72.05 | 86.54 |
| 30 MPH | .00 | .84 | 3.37 | 7.61 | 13.59 | 21.34 | 30.92 | 42.37 | 55.75 | 71.12 | 88.53 | 108.07 | 129.81 |
| VELOCITY (FPS) | 700. | 678. | 658. | 637. | 618. | 599. | 581. | 563. | 546. | 529. | 513. | 497. | 482. |
| ENERGY (FT-LB) | 160. | 150. | 141. | 133. | 125. | 117. | 110. | 104. | 97. | 91. | 86. | 81. | 76. |
| DROP (IN) | .00 | -1.98 | -8.65 | -20.34 | -37.35 | -60.02 | -88.74 | -123.87 | -165.83 | -215.08 | -272.07 | -337.33 | -411.38 |
| MID-RANGE (IN) | .00 | .57 | 2.35 | 5.46 | 10.02 | 16.16 | 24.03 | 33.78 | 45.57 | 59.58 | 76.01 | 95.06 | 116.95 |
| BULLET PATH (IN) | -.63 | 2.04 | .00 | -7.04 | -19.41 | -37.45 | -61.52 | -92.00 | -129.33 | -173.93 | -226.29 | -286.90 | -356.31 |
| TIME OF FLIGHT (SEC) | .000000 | .108844 | .221148 | .337007 | .456527 | .579823 | .707019 | .838248 | .973653 | 1.113383 | 1.257599 | 1.406467 | 1.560164 |
| WIND DEFLECTION (IN) | | | | | | | | | | | | | |
| 5 MPH | .00 | .15 | .60 | 1.37 | 2.46 | 3.88 | 5.65 | 7.77 | 10.25 | 13.12 | 16.38 | 20.05 | 24.15 |
| 10 MPH | .00 | .30 | 1.21 | 2.74 | 4.92 | 7.76 | 11.29 | 15.53 | 20.51 | 26.24 | 32.77 | 40.11 | 48.30 |
| 20 MPH | .00 | .60 | 2.42 | 5.48 | 9.84 | 15.53 | 22.58 | 31.06 | 41.01 | 52.48 | 65.53 | 80.22 | 96.61 |
| 30 MPH | .00 | .90 | 3.62 | 8.23 | 14.76 | 23.29 | 33.88 | 46.59 | 61.52 | 78.72 | 98.30 | 120.33 | 144.91 |

# Bullet: Lyman #358311 150 Gr.
## Ballistic Coefficient: .228

| RANGE (YDS) | MUZZLE | 25 | 50 | 75 | 100 | 125 | 150 | 175 | 200 | 225 | 250 | 275 | 300 |
|---|---|---|---|---|---|---|---|---|---|---|---|---|---|
| VELOCITY (FPS) | 1500. | 1435. | 1373. | 1315. | 1260. | 1209. | 1164. | 1123. | 1088. | 1057. | 1029. | 1004. | 982. |
| ENERGY (FT-LB) | 749. | 686. | 628. | 576. | 529. | 487. | 451. | 420. | 394. | 372. | 353. | 336. | 321. |
| DROP (IN) | .00 | -.47 | -2.00 | -4.68 | -8.61 | -13.92 | -20.71 | -29.11 | -39.22 | -51.16 | -65.04 | -80.97 | -99.05 |
| MID-RANGE (IN) | .00 | .13 | .53 | 1.24 | 2.30 | 3.77 | 5.68 | 8.07 | 11.00 | 14.50 | 18.60 | 23.37 | 28.82 |
| BULLET PATH (IN) | -.63 | .21 | .00 | -1.36 | -3.98 | -7.97 | -13.45 | -20.53 | -29.33 | -39.96 | -52.52 | -67.14 | -83.90 |
| TIME OF FLIGHT (SEC) | .000000 | .051129 | .104568 | .160386 | .218645 | .279413 | .342653 | .408280 | .476167 | .546159 | .618117 | .691926 | .767496 |
| WIND DEFLECTION (IN) | | | | | | | | | | | | | |
| 5 MPH | .00 | .10 | .40 | .91 | 1.64 | 2.59 | 3.75 | 5.13 | 6.70 | 8.46 | 10.39 | 12.49 | 14.74 |
| 10 MPH | .00 | .20 | .80 | 1.83 | 3.28 | 5.18 | 7.51 | 10.26 | 13.41 | 16.92 | 20.79 | 24.98 | 29.48 |
| 20 MPH | .00 | .40 | 1.61 | 3.66 | 6.56 | 10.35 | 15.01 | 20.51 | 26.81 | 33.85 | 41.58 | 49.96 | 58.96 |
| 30 MPH | .00 | .60 | 2.41 | 5.48 | 9.84 | 15.53 | 22.52 | 30.77 | 40.22 | 50.77 | 62.37 | 74.94 | 88.44 |
| VELOCITY (FPS) | 1400. | 1340. | 1284. | 1231. | 1183. | 1141. | 1103. | 1070. | 1041. | 1015. | 991. | 970. | 950. |
| ENERGY (FT-LB) | 653. | 598. | 549. | 505. | 466. | 433. | 405. | 381. | 361. | 343. | 327. | 313. | 301. |
| DROP (IN) | .00 | -.55 | -2.30 | -5.37 | -9.88 | -15.94 | -23.66 | -33.17 | -44.56 | -57.96 | -73.46 | -91.17 | -111.18 |
| MID-RANGE (IN) | .00 | .14 | .60 | 1.42 | 2.64 | 4.31 | 6.46 | 9.15 | 12.40 | 16.27 | 20.80 | 26.00 | 31.93 |
| BULLET PATH (IN) | -.63 | .29 | .00 | -1.61 | -4.66 | -9.25 | -15.51 | -23.55 | -33.49 | -45.42 | -59.46 | -75.70 | -94.25 |
| TIME OF FLIGHT (SEC) | .000000 | .054758 | .111923 | .171576 | .233725 | .298303 | .365203 | .434278 | .505377 | .578375 | .653171 | .729684 | .807847 |
| WIND DEFLECTION (IN) | | | | | | | | | | | | | |
| 5 MPH | .00 | .10 | .42 | .96 | 1.71 | 2.68 | 3.85 | 5.22 | 6.76 | 8.47 | 10.34 | 12.36 | 14.52 |
| 10 MPH | .00 | .21 | .84 | 1.91 | 3.42 | 5.36 | 7.70 | 10.43 | 13.52 | 16.94 | 20.67 | 24.71 | 29.04 |
| 20 MPH | .00 | .42 | 1.68 | 3.82 | 6.84 | 10.72 | 15.41 | 20.87 | 27.04 | 33.87 | 41.34 | 49.42 | 58.08 |
| 30 MPH | .00 | .63 | 2.52 | 5.73 | 10.26 | 16.08 | 23.11 | 31.30 | 40.55 | 50.81 | 62.02 | 74.13 | 87.11 |
| VELOCITY (FPS) | 1300. | 1246. | 1196. | 1152. | 1113. | 1079. | 1049. | 1022. | 998. | 976. | 955. | 936. | 919. |
| ENERGY (FT-LB) | 563. | 517. | 477. | 442. | 412. | 387. | 366. | 348. | 331. | 317. | 304. | 292. | 281. |
| DROP (IN) | .00 | -.63 | -2.67 | -6.23 | -11.42 | -18.37 | -27.17 | -37.94 | -50.79 | -65.81 | -83.11 | -102.75 | -124.83 |
| MID-RANGE (IN) | .00 | .17 | .71 | 1.65 | 3.04 | 4.93 | 7.35 | 10.35 | 13.97 | 18.24 | 23.19 | 28.85 | 35.25 |
| BULLET PATH (IN) | -.63 | .39 | .00 | -1.91 | -5.45 | -10.73 | -17.88 | -27.00 | -38.20 | -51.57 | -67.22 | -85.20 | -105.64 |
| TIME OF FLIGHT (SEC) | .000000 | .058951 | .120408 | .184318 | .250583 | .319067 | .389614 | .462093 | .536395 | .612434 | .690142 | .769460 | .850342 |
| WIND DEFLECTION (IN) | | | | | | | | | | | | | |
| 5 MPH | .00 | .11 | .44 | .99 | 1.74 | 2.69 | 3.82 | 5.13 | 6.59 | 8.20 | 9.96 | 11.87 | 13.91 |
| 10 MPH | .00 | .22 | .88 | 1.98 | 3.49 | 5.39 | 7.65 | 10.25 | 13.17 | 16.40 | 19.93 | 23.73 | 27.81 |
| 20 MPH | .00 | .44 | 1.77 | 3.96 | 6.97 | 10.77 | 15.30 | 20.50 | 26.35 | 32.81 | 39.85 | 47.47 | 55.63 |
| 30 MPH | .00 | .66 | 2.65 | 5.94 | 10.46 | 16.16 | 22.95 | 30.75 | 39.52 | 49.21 | 59.78 | 71.20 | 83.44 |
| VELOCITY (FPS) | 1200. | 1155. | 1116. | 1081. | 1051. | 1024. | 999. | 977. | 957. | 938. | 920. | 903. | 888. |
| ENERGY (FT-LB) | 480. | 444. | 415. | 389. | 368. | 349. | 333. | 318. | 305. | 293. | 282. | 272. | 262. |
| DROP (IN) | .00 | -.75 | -3.13 | -7.25 | -13.22 | -21.15 | -31.15 | -43.32 | -57.75 | -74.52 | -93.72 | -115.48 | -139.90 |
| MID-RANGE (IN) | .00 | .20 | .81 | 1.89 | 3.48 | 5.61 | 8.33 | 11.66 | 15.66 | 20.33 | 25.71 | 31.86 | 38.80 |
| BULLET PATH (IN) | -.63 | .50 | .00 | -2.24 | -6.33 | -12.38 | -20.50 | -30.78 | -43.34 | -58.22 | -75.55 | -95.43 | -117.97 |
| TIME OF FLIGHT (SEC) | .000000 | .063717 | .129799 | .198112 | .268500 | .340828 | .414987 | .490890 | .568465 | .647655 | .728413 | .810700 | .894488 |
| WIND DEFLECTION (IN) | | | | | | | | | | | | | |
| 5 MPH | .00 | .11 | .42 | .93 | 1.63 | 2.49 | 3.52 | 4.70 | 6.02 | 7.49 | 9.10 | 10.84 | 12.71 |
| 10 MPH | .00 | .21 | .84 | 1.87 | 3.26 | 4.99 | 7.04 | 9.40 | 12.05 | 14.99 | 18.20 | 21.68 | 25.43 |
| 20 MPH | .00 | .43 | 1.69 | 3.74 | 6.51 | 9.97 | 14.08 | 18.79 | 24.10 | 29.97 | 36.40 | 43.37 | 50.86 |
| 30 MPH | .00 | .64 | 2.53 | 5.60 | 9.77 | 14.96 | 21.11 | 28.19 | 36.15 | 44.96 | 54.60 | 65.05 | 76.29 |

# Bullet: Lyman #358311 150 Gr.
## Ballistic Coefficient: .228

| RANGE (YDS) | MUZZLE | 25 | 50 | 75 | 100 | 125 | 150 | 175 | 200 | 225 | 250 | 275 | 300 |
|---|---|---|---|---|---|---|---|---|---|---|---|---|---|
| VELOCITY (FPS) | 1100. | 1067. | 1039. | 1013. | 990. | 968. | 948. | 930. | 913. | 896. | 881. | 866. | 852. |
| ENERGY (FT-LB) | 403. | 379. | 359. | 342. | 326. | 312. | 300. | 288. | 277. | 268. | 258. | 250. | 242. |
| DROP (IN) | .00 | -.89 | -3.67 | -8.47 | -15.37 | -24.49 | -35.93 | -49.70 | -65.99 | -84.87 | -106.44 | -130.81 | -158.06 |
| MID-RANGE (IN) | .00 | .23 | .95 | 2.20 | 4.01 | 6.43 | 9.49 | 13.20 | 17.62 | 22.78 | 28.73 | 35.49 | 43.11 |
| BULLET PATH (IN) | -.63 | .64 | .00 | -2.64 | -7.40 | -14.37 | -23.65 | -35.27 | -49.40 | -66.14 | -85.56 | -107.77 | -132.87 |
| TIME OF FLIGHT (SEC) | .000000 | .069237 | .140486 | .213624 | .288553 | .365193 | .443478 | .523356 | .604783 | .687725 | .772157 | .858057 | .945411 |
| WIND DEFLECTION (IN) | | | | | | | | | | | | | |
| 5 MPH | .00 | .09 | .36 | .80 | 1.39 | 2.14 | 3.03 | 4.06 | 5.22 | 6.52 | 7.95 | 9.51 | 11.20 |
| 10 MPH | .00 | .19 | .73 | 1.60 | 2.79 | 4.27 | 6.05 | 8.11 | 10.44 | 13.04 | 15.90 | 19.02 | 22.39 |
| 20 MPH | .00 | .37 | 1.45 | 3.20 | 5.57 | 8.55 | 12.10 | 16.22 | 20.88 | 26.08 | 31.80 | 38.04 | 44.78 |
| 30 MPH | .00 | .56 | 2.18 | 4.79 | 8.36 | 12.82 | 18.16 | 24.33 | 31.33 | 39.12 | 47.70 | 57.05 | 67.18 |
| VELOCITY (FPS) | 1000. | 978. | 957. | 938. | 920. | 904. | 888. | 873. | 858. | 844. | 831. | 818. | 806. |
| ENERGY (FT-LB) | 333. | 318. | 305. | 293. | 282. | 272. | 262. | 254. | 245. | 237. | 230. | 223. | 216. |
| DROP (IN) | .00 | -1.06 | -4.38 | -10.03 | -18.13 | -28.77 | -42.07 | -58.10 | -76.98 | -98.79 | -123.63 | -151.62 | -182.85 |
| MID-RANGE (IN) | .00 | .28 | 1.13 | 2.59 | 4.69 | 7.47 | 11.00 | 15.28 | 20.36 | 26.27 | 33.04 | 40.73 | 49.36 |
| BULLET PATH (IN) | -.63 | .82 | .00 | -3.16 | -8.74 | -16.89 | -27.68 | -41.21 | -57.58 | -76.89 | -99.23 | -124.72 | -153.44 |
| TIME OF FLIGHT (SEC) | .000000 | .075859 | .153392 | .232540 | .313256 | .395501 | .479247 | .564469 | .651149 | .739277 | .828845 | .919848 | 1.012289 |
| WIND DEFLECTION (IN) | | | | | | | | | | | | | |
| 5 MPH | .00 | .08 | .30 | .66 | 1.17 | 1.80 | 2.57 | 3.47 | 4.50 | 5.66 | 6.94 | 8.35 | 9.88 |
| 10 MPH | .00 | .15 | .60 | 1.33 | 2.33 | 3.61 | 5.15 | 6.95 | 9.00 | 11.31 | 13.88 | 16.69 | 19.76 |
| 20 MPH | .00 | .30 | 1.19 | 2.65 | 4.67 | 7.22 | 10.29 | 13.89 | 18.00 | 22.63 | 27.75 | 33.39 | 39.53 |
| 30 MPH | .00 | .45 | 1.79 | 3.98 | 7.00 | 10.82 | 15.44 | 20.84 | 27.01 | 33.94 | 41.63 | 50.08 | 59.29 |
| VELOCITY (FPS) | 900. | 884. | 869. | 855. | 841. | 828. | 815. | 803. | 790. | 779. | 767. | 756. | 745. |
| ENERGY (FT-LB) | 270. | 260. | 252. | 243. | 236. | 228. | 221. | 214. | 208. | 202. | 196. | 190. | 185. |
| DROP (IN) | .00 | -1.21 | -5.19 | -12.03 | -21.82 | -34.68 | -50.69 | -69.97 | -92.50 | -118.40 | -147.86 | -181.01 | -217.95 |
| MID-RANGE (IN) | .00 | .34 | 1.38 | 3.16 | 5.72 | 9.09 | 13.32 | 18.44 | 24.43 | 31.34 | 39.26 | 48.22 | 58.28 |
| BULLET PATH (IN) | -.63 | 1.07 | -.00 | -3.93 | -10.81 | -20.76 | -33.87 | -50.24 | -69.85 | -92.84 | -119.40 | -149.64 | -183.67 |
| TIME OF FLIGHT (SEC) | .000000 | .084076 | .169623 | .256626 | .345073 | .434957 | .526277 | .619033 | .713231 | .808875 | .905971 | 1.004526 | 1.104549 |
| WIND DEFLECTION (IN) | | | | | | | | | | | | | |
| 5 MPH | .00 | .07 | .26 | .58 | 1.03 | 1.61 | 2.31 | 3.14 | 4.10 | 5.18 | 6.39 | 7.73 | 9.20 |
| 10 MPH | .00 | .13 | .52 | 1.17 | 2.07 | 3.22 | 4.62 | 6.28 | 8.20 | 10.36 | 12.78 | 15.46 | 18.40 |
| 20 MPH | .00 | .26 | 1.04 | 2.33 | 4.13 | 6.44 | 9.25 | 12.57 | 16.39 | 20.72 | 25.57 | 30.93 | 36.80 |
| 30 MPH | .00 | .39 | 1.56 | 3.50 | 6.20 | 9.66 | 13.87 | 18.85 | 24.59 | 31.09 | 38.35 | 46.39 | 55.20 |
| VELOCITY (FPS) | 800. | 788. | 776. | 765. | 753. | 742. | 732. | 721. | 711. | 701. | 691. | 682. | 672. |
| ENERGY (FT-LB) | 213. | 207. | 201. | 195. | 189. | 184. | 178. | 173. | 168. | 164. | 159. | 155. | 150. |
| DROP (IN) | .00 | -1.39 | -6.15 | -14.51 | -26.57 | -42.45 | -62.25 | -86.09 | -114.09 | -146.37 | -183.06 | -224.29 | -270.19 |
| MID-RANGE (IN) | .00 | .38 | 1.69 | 3.93 | 7.13 | 11.35 | 16.61 | 22.97 | 30.47 | 39.16 | 49.08 | 60.30 | 72.85 |
| BULLET PATH (IN) | -.63 | 1.37 | -.00 | -4.97 | -13.63 | -26.12 | -42.53 | -62.97 | -87.58 | -116.47 | -149.77 | -187.61 | -230.11 |
| TIME OF FLIGHT (SEC) | .000000 | .094472 | .190390 | .287759 | .386589 | .486886 | .588663 | .691930 | .796702 | .902993 | 1.010818 | 1.120195 | 1.231142 |
| WIND DEFLECTION (IN) | | | | | | | | | | | | | |
| 5 MPH | .00 | .06 | .25 | .57 | 1.02 | 1.60 | 2.30 | 3.14 | 4.11 | 5.21 | 6.45 | 7.83 | 9.34 |
| 10 MPH | .00 | .13 | .51 | 1.15 | 2.04 | 3.19 | 4.60 | 6.28 | 8.22 | 10.43 | 12.90 | 15.65 | 18.68 |
| 20 MPH | .00 | .25 | 1.02 | 2.29 | 4.08 | 6.38 | 9.21 | 12.56 | 16.44 | 20.85 | 25.81 | 31.31 | 37.36 |
| 30 MPH | .00 | .38 | 1.53 | 3.44 | 6.12 | 9.58 | 13.81 | 18.84 | 24.66 | 31.28 | 38.71 | 46.96 | 56.04 |

## Bullet: Lyman #358311 150 Gr.
### Ballistic Coefficient: .228

| RANGE (YDS) | MUZZLE | 25 | 50 | 75 | 100 | 125 | 150 | 175 | 200 | 225 | 250 | 275 | 300 |
|---|---|---|---|---|---|---|---|---|---|---|---|---|---|
| VELOCITY (FPS) | 700. | 690. | 680. | 671. | 661. | 652. | 643. | 634. | 625. | 617. | 608. | 600. | 592. |
| ENERGY (FT-LB) | 163. | 159. | 154. | 150. | 146. | 142. | 138. | 134. | 130. | 127. | 123. | 120. | 117. |
| DROP (IN) | .00 | -1.67 | -7.89 | -18.78 | -34.49 | -55.14 | -80.88 | -111.86 | -148.21 | -190.12 | -237.72 | -291.19 | -350.69 |
| MID-RANGE (IN) | .00 | .56 | 2.27 | 5.19 | 9.35 | 14.83 | 21.66 | 29.90 | 39.62 | 50.87 | 63.72 | 78.23 | 94.46 |
| BULLET PATH (IN) | -.63 | 1.96 | -.00 | -6.63 | -18.08 | -34.47 | -55.96 | -82.67 | -114.77 | -152.41 | -195.75 | -244.96 | -300.20 |
| TIME OF FLIGHT (SEC) | .000000 | .107912 | .217376 | .328409 | .441029 | .555257 | .671113 | .788621 | .907802 | 1.028681 | 1.151283 | 1.275634 | 1.401759 |
| WIND DEFLECTION (IN) | | | | | | | | | | | | | |
| 5 MPH | .00 | .07 | .27 | .61 | 1.10 | 1.72 | 2.49 | 3.40 | 4.46 | 5.67 | 7.03 | 8.54 | 10.21 |
| 10 MPH | .00 | .14 | .54 | 1.23 | 2.19 | 3.44 | 4.97 | 6.80 | 8.92 | 11.33 | 14.05 | 17.08 | 20.42 |
| 20 MPH | .00 | .27 | 1.09 | 2.46 | 4.38 | 6.88 | 9.95 | 13.59 | 17.83 | 22.67 | 28.11 | 34.17 | 40.85 |
| 30 MPH | .00 | .41 | 1.63 | 3.69 | 6.58 | 10.32 | 14.92 | 20.39 | 26.75 | 34.00 | 42.16 | 51.25 | 61.27 |
| VELOCITY (FPS) | 600. | 592. | 583. | 575. | 567. | 559. | 552. | 544. | 536. | 529. | 522. | 515. | 508. |
| ENERGY (FT-LB) | 120. | 117. | 113. | 110. | 107. | 104. | 101. | 99. | 96. | 93. | 91. | 88. | 86. |
| DROP (IN) | .00 | -2.28 | -10.74 | -25.56 | -46.93 | -75.03 | -110.06 | -152.20 | -201.68 | -258.71 | -323.50 | -396.29 | -477.31 |
| MID-RANGE (IN) | .00 | .76 | 3.09 | 7.06 | 12.73 | 20.17 | 29.47 | 40.69 | 53.91 | 69.23 | 86.72 | 106.48 | 128.60 |
| BULLET PATH (IN) | -.63 | 2.78 | .00 | -9.14 | -24.82 | -47.24 | -76.59 | -113.05 | -156.84 | -208.19 | -267.30 | -334.40 | -409.74 |
| TIME OF FLIGHT (SEC) | .000000 | .125889 | .253573 | .383080 | .514439 | .647679 | .782830 | .919923 | 1.058992 | 1.200068 | 1.343185 | 1.488380 | 1.635686 |
| WIND DEFLECTION (IN) | | | | | | | | | | | | | |
| 5 MPH | .00 | .08 | .31 | .71 | 1.27 | 2.00 | 2.89 | 3.95 | 5.19 | 6.61 | 8.20 | 9.98 | 11.94 |
| 10 MPH | .00 | .16 | .63 | 1.42 | 2.54 | 3.99 | 5.78 | 7.91 | 10.38 | 13.21 | 16.40 | 19.95 | 23.88 |
| 20 MPH | .00 | .31 | 1.26 | 2.84 | 5.08 | 7.98 | 11.56 | 15.81 | 20.76 | 26.42 | 32.80 | 39.91 | 47.76 |
| 30 MPH | .00 | .47 | 1.89 | 4.27 | 7.62 | 11.97 | 17.33 | 23.72 | 31.15 | 39.64 | 49.20 | 59.86 | 71.64 |

# Bullet: Lyman #358429 162 Gr.
## Ballistic Coefficient: .286

| RANGE (YDS) | MUZZLE | 25 | 50 | 75 | 100 | 125 | 150 | 175 | 200 | 225 | 250 | 275 | 300 |
|---|---|---|---|---|---|---|---|---|---|---|---|---|---|
| VELOCITY (FPS) | 1400. | 1352. | 1307. | 1262. | 1219. | 1180. | 1144. | 1112. | 1084. | 1058. | 1035. | 1013. | 994. |
| ENERGY (FT-LB) | 705. | 658. | 614. | 573. | 534. | 501. | 471. | 445. | 422. | 402. | 385. | 369. | 355. |
| DROP (IN) | .00 | -.54 | -2.26 | -5.25 | -9.61 | -15.44 | -22.83 | -31.88 | -42.67 | -55.31 | -69.88 | -86.48 | -105.19 |
| MID-RANGE (IN) | .00 | .14 | .59 | 1.38 | 2.55 | 4.13 | 6.17 | 8.69 | 11.72 | 15.31 | 19.50 | 24.30 | 29.76 |
| BULLET PATH (IN) | -.63 | .28 | .00 | -1.55 | -4.47 | -8.86 | -14.80 | -22.40 | -31.75 | -42.95 | -56.08 | -71.24 | -88.50 |
| TIME OF FLIGHT (SEC) | .000000 | .054516 | .110943 | .169346 | .229832 | .292381 | .356943 | .423442 | .491782 | .561854 | .633561 | .706824 | .781576 |
| WIND DEFLECTION (IN) | | | | | | | | | | | | | |
| 5 MPH | .00 | .08 | .33 | .76 | 1.37 | 2.16 | 3.13 | 4.26 | 5.56 | 7.01 | 8.61 | 10.34 | 12.21 |
| 10 MPH | .00 | .17 | .67 | 1.52 | 2.74 | 4.32 | 6.25 | 8.53 | 11.13 | 14.03 | 17.22 | 20.69 | 24.41 |
| 20 MPH | .00 | .33 | 1.34 | 3.04 | 5.47 | 8.63 | 12.50 | 17.05 | 22.25 | 28.06 | 34.44 | 41.37 | 48.83 |
| 30 MPH | .00 | .50 | 2.01 | 4.56 | 8.21 | 12.95 | 18.75 | 25.58 | 33.38 | 42.09 | 51.66 | 62.06 | 73.24 |
| VELOCITY (FPS) | 1300. | 1255. | 1212. | 1174. | 1139. | 1107. | 1079. | 1054. | 1031. | 1010. | 991. | 973. | 956. |
| ENERGY (FT-LB) | 608. | 566. | 529. | 496. | 466. | 441. | 419. | 399. | 382. | 367. | 353. | 340. | 329. |
| DROP (IN) | .00 | -.62 | -2.64 | -6.13 | -11.20 | -17.93 | -26.43 | -36.80 | -49.11 | -63.46 | -79.93 | -98.61 | -119.58 |
| MID-RANGE (IN) | .00 | .17 | .70 | 1.61 | 2.96 | 4.77 | 7.09 | 9.94 | 13.36 | 17.38 | 22.03 | 27.35 | 33.36 |
| BULLET PATH (IN) | -.63 | .38 | .00 | -1.86 | -5.29 | -10.39 | -17.26 | -25.99 | -36.67 | -49.38 | -64.22 | -81.27 | -100.60 |
| TIME OF FLIGHT (SEC) | .000000 | .058737 | .119560 | .182439 | .247320 | .314124 | .382751 | .453093 | .525056 | .598563 | .673549 | .749958 | .827745 |
| WIND DEFLECTION (IN) | | | | | | | | | | | | | |
| 5 MPH | .00 | .09 | .37 | .82 | 1.46 | 2.26 | 3.22 | 4.33 | 5.59 | 6.98 | 8.50 | 10.15 | 11.92 |
| 10 MPH | .00 | .18 | .73 | 1.65 | 2.91 | 4.52 | 6.44 | 8.67 | 11.18 | 13.96 | 17.01 | 20.30 | 23.84 |
| 20 MPH | .00 | .37 | 1.47 | 3.30 | 5.83 | 9.03 | 12.88 | 17.33 | 22.36 | 27.93 | 34.01 | 40.60 | 47.67 |
| 30 MPH | .00 | .55 | 2.20 | 4.94 | 8.74 | 13.55 | 19.32 | 26.00 | 33.54 | 41.89 | 51.02 | 60.90 | 71.51 |
| VELOCITY (FPS) | 1200. | 1163. | 1129. | 1098. | 1071. | 1047. | 1024. | 1004. | 985. | 968. | 951. | 936. | 921. |
| ENERGY (FT-LB) | 518. | 486. | 458. | 434. | 413. | 394. | 377. | 362. | 349. | 337. | 325. | 315. | 305. |
| DROP (IN) | .00 | -.75 | -3.10 | -7.15 | -12.99 | -20.73 | -30.44 | -42.22 | -56.15 | -72.31 | -90.79 | -111.60 | -134.87 |
| MID-RANGE (IN) | .00 | .19 | .80 | 1.86 | 3.40 | 5.46 | 8.07 | 11.27 | 15.08 | 19.53 | 24.66 | 30.47 | 36.99 |
| BULLET PATH (IN) | -.63 | .49 | .00 | -2.19 | -6.17 | -12.04 | -19.89 | -29.81 | -41.87 | -56.17 | -72.78 | -91.73 | -113.14 |
| TIME OF FLIGHT (SEC) | .000000 | .063513 | .129005 | .196391 | .265566 | .336425 | .408878 | .482853 | .558288 | .635132 | .713339 | .792875 | .873708 |
| WIND DEFLECTION (IN) | | | | | | | | | | | | | |
| 5 MPH | .00 | .09 | .35 | .78 | 1.37 | 2.11 | 2.98 | 3.99 | 5.13 | 6.39 | 7.77 | 9.27 | 10.89 |
| 10 MPH | .00 | .18 | .70 | 1.56 | 2.74 | 4.21 | 5.96 | 7.98 | 10.26 | 12.78 | 15.55 | 18.55 | 21.77 |
| 20 MPH | .00 | .36 | 1.41 | 3.13 | 5.48 | 8.42 | 11.93 | 15.96 | 20.52 | 25.57 | 31.10 | 37.09 | 43.55 |
| 30 MPH | .00 | .53 | 2.11 | 4.69 | 8.22 | 12.63 | 17.89 | 23.95 | 30.78 | 38.35 | 46.64 | 55.64 | 65.32 |
| VELOCITY (FPS) | 1100. | 1073. | 1048. | 1026. | 1005. | 986. | 969. | 952. | 936. | 922. | 908. | 894. | 882. |
| ENERGY (FT-LB) | 435. | 414. | 395. | 378. | 363. | 350. | 337. | 326. | 315. | 306. | 296. | 288. | 280. |
| DROP (IN) | .00 | -.88 | -3.64 | -8.37 | -15.16 | -24.10 | -35.26 | -48.73 | -64.54 | -82.80 | -103.61 | -127.04 | -153.19 |
| MID-RANGE (IN) | .00 | .23 | .94 | 2.17 | 3.94 | 6.30 | 9.26 | 12.86 | 17.11 | 22.05 | 27.71 | 34.13 | 41.33 |
| BULLET PATH (IN) | -.63 | .63 | .00 | -2.60 | -7.26 | -14.06 | -23.09 | -34.43 | -48.10 | -64.23 | -82.90 | -104.20 | -128.22 |
| TIME OF FLIGHT (SEC) | .000000 | .069061 | .139812 | .212163 | .286040 | .361381 | .438132 | .516250 | .595697 | .676443 | .758462 | .841731 | .926235 |
| WIND DEFLECTION (IN) | | | | | | | | | | | | | |
| 5 MPH | .00 | .08 | .30 | .67 | 1.17 | 1.80 | 2.56 | 3.43 | 4.42 | 5.53 | 6.74 | 8.07 | 9.51 |
| 10 MPH | .00 | .15 | .61 | 1.34 | 2.34 | 3.60 | 5.11 | 6.86 | 8.84 | 11.05 | 13.49 | 16.14 | 19.02 |
| 20 MPH | .00 | .31 | 1.21 | 2.68 | 4.69 | 7.21 | 10.22 | 13.72 | 17.69 | 22.11 | 26.98 | 32.29 | 38.03 |
| 30 MPH | .00 | .46 | 1.82 | 4.02 | 7.03 | 10.81 | 15.33 | 20.58 | 26.53 | 33.16 | 40.47 | 48.43 | 57.05 |

# Bullet: Lyman #358429 162 Gr.
## Ballistic Coefficient: .286

| RANGE (YDS) | MUZZLE | 25 | 50 | 75 | 100 | 125 | 150 | 175 | 200 | 225 | 250 | 275 | 300 |
|---|---|---|---|---|---|---|---|---|---|---|---|---|---|
| VELOCITY (FPS) | 1000. | 981. | 964. | 948. | 932. | 918. | 904. | 891. | 878. | 866. | 854. | 843. | 832. |
| ENERGY (FT-LB) | 360. | 346. | 334. | 323. | 313. | 303. | 294. | 285. | 277. | 270. | 262. | 255. | 249. |
| DROP (IN) | .00 | -1.04 | -4.34 | -9.96 | -17.92 | -28.37 | -41.39 | -57.06 | -75.46 | -96.67 | -120.77 | -147.84 | -177.96 |
| MID-RANGE (IN) | .00 | .28 | 1.13 | 2.58 | 4.62 | 7.33 | 10.73 | 14.89 | 19.81 | 25.51 | 32.02 | 39.37 | 47.59 |
| BULLET PATH (IN) | -.63 | .81 | .00 | -3.14 | -8.62 | -16.58 | -27.12 | -40.31 | -56.22 | -74.95 | -96.56 | -121.15 | -148.78 |
| TIME OF FLIGHT (SEC) | .000000 | .075715 | .152828 | .231298 | .311087 | .392168 | .474514 | .558105 | .642926 | .728963 | .816206 | .904648 | .994285 |
| WIND DEFLECTION (IN) | | | | | | | | | | | | | |
| 5 MPH | .00 | .06 | .25 | .55 | .98 | 1.51 | 2.16 | 2.91 | 3.78 | 4.75 | 5.83 | 7.01 | 8.30 |
| 10 MPH | .00 | .13 | .50 | 1.11 | 1.95 | 3.02 | 4.31 | 5.83 | 7.55 | 9.50 | 11.65 | 14.02 | 16.59 |
| 20 MPH | .00 | .25 | 1.00 | 2.22 | 3.90 | 6.04 | 8.63 | 11.65 | 15.11 | 18.99 | 23.30 | 28.04 | 33.19 |
| 30 MPH | .00 | .38 | 1.49 | 3.33 | 5.85 | 9.06 | 12.94 | 17.48 | 22.66 | 28.49 | 34.96 | 42.05 | 49.78 |
| VELOCITY (FPS) | 900. | 887. | 874. | 862. | 851. | 839. | 828. | 818. | 807. | 797. | 786. | 775. | 765. |
| ENERGY (FT-LB) | 291. | 283. | 275. | 267. | 260. | 253. | 247. | 240. | 234. | 228. | 222. | 216. | 210. |
| DROP (IN) | .00 | -1.18 | -5.12 | -11.89 | -21.58 | -34.26 | -50.02 | -68.92 | -91.06 | -116.53 | -145.23 | -177.44 | -213.25 |
| MID-RANGE (IN) | .00 | .34 | 1.38 | 3.14 | 5.67 | 8.98 | 13.12 | 18.10 | 23.95 | 30.73 | 38.36 | 46.97 | 56.60 |
| BULLET PATH (IN) | -.63 | 1.06 | -.00 | -3.90 | -10.71 | -20.52 | -33.40 | -49.43 | -68.70 | -91.29 | -117.12 | -146.45 | -179.39 |
| TIME OF FLIGHT (SEC) | .000000 | .033950 | .169125 | .255512 | .343101 | .431888 | .521867 | .613039 | .705402 | .798961 | .893789 | .989945 | 1.087434 |
| WIND DEFLECTION (IN) | | | | | | | | | | | | | |
| 5 MPH | .00 | .05 | .22 | .49 | .86 | 1.34 | 1.92 | 2.61 | 3.41 | 4.31 | 5.32 | 6.45 | 7.69 |
| 10 MPH | .00 | .11 | .43 | .97 | 1.72 | 2.68 | 3.85 | 5.23 | 6.82 | 8.62 | 10.64 | 12.90 | 15.39 |
| 20 MPH | .00 | .22 | .87 | 1.94 | 3.44 | 5.36 | 7.70 | 10.46 | 13.63 | 17.23 | 21.28 | 25.79 | 30.78 |
| 30 MPH | .00 | .33 | 1.30 | 2.91 | 5.16 | 8.04 | 11.55 | 15.68 | 20.45 | 25.85 | 31.92 | 38.69 | 46.17 |
| VELOCITY (FPS) | 800. | 789. | 778. | 767. | 757. | 747. | 737. | 727. | 718. | 709. | 700. | 691. | 682. |
| ENERGY (FT-LB) | 230. | 224. | 218. | 212. | 206. | 201. | 195. | 190. | 185. | 181. | 176. | 172. | 167. |
| DROP (IN) | .00 | -1.39 | -6.14 | -14.48 | -26.49 | -42.27 | -61.94 | -85.57 | -113.29 | -145.22 | -181.46 | -222.13 | -267.35 |
| MID-RANGE (IN) | .00 | .38 | 1.69 | 3.91 | 7.10 | 11.28 | 16.49 | 22.77 | 30.16 | 38.71 | 48.46 | 59.45 | 71.74 |
| BULLET PATH (IN) | -.63 | 1.37 | -.00 | -4.95 | -13.57 | -25.97 | -42.25 | -62.49 | -86.83 | -115.37 | -148.22 | -185.51 | -227.34 |
| TIME OF FLIGHT (SEC) | .000000 | .094411 | .190148 | .287215 | .385618 | .485364 | .586461 | .688920 | .792750 | .897963 | 1.004573 | 1.112592 | 1.222035 |
| WIND DEFLECTION (IN) | | | | | | | | | | | | | |
| 5 MPH | .00 | .06 | .23 | .52 | .93 | 1.46 | 2.11 | 2.87 | 3.76 | 4.77 | 5.90 | 7.16 | 8.54 |
| 10 MPH | .00 | .12 | .47 | 1.05 | 1.87 | 2.92 | 4.22 | 5.75 | 7.52 | 9.54 | 11.80 | 14.32 | 17.08 |
| 20 MPH | .00 | .23 | .93 | 2.10 | 3.74 | 5.85 | 8.43 | 11.50 | 15.05 | 19.08 | 23.61 | 28.63 | 34.16 |
| 30 MPH | .00 | .35 | 1.40 | 3.15 | 5.61 | 8.77 | 12.65 | 17.25 | 22.57 | 28.62 | 35.41 | 42.95 | 51.23 |
| VELOCITY (FPS) | 700. | 691. | 682. | 673. | 665. | 656. | 648. | 639. | 631. | 623. | 616. | 608. | 600. |
| ENERGY (FT-LB) | 176. | 172. | 167. | 163. | 159. | 155. | 151. | 147. | 143. | 140. | 136. | 133. | 130. |
| DROP (IN) | .00 | -1.67 | -7.88 | -18.74 | -34.38 | -54.92 | -80.50 | -111.21 | -147.22 | -188.66 | -235.69 | -288.45 | -347.09 |
| MID-RANGE (IN) | .00 | .56 | 2.27 | 5.17 | 9.31 | 14.74 | 21.51 | 29.65 | 39.23 | 50.30 | 62.92 | 77.15 | 93.04 |
| BULLET PATH (IN) | -.63 | 1.95 | -.00 | -6.61 | -17.99 | -34.28 | -55.61 | -82.06 | -113.81 | -151.01 | -193.78 | -242.29 | -296.67 |
| TIME OF FLIGHT (SEC) | .000000 | .107848 | .217117 | .327821 | .439976 | .553598 | .668706 | .785317 | .903449 | 1.023124 | 1.144361 | 1.267181 | 1.391607 |
| WIND DEFLECTION (IN) | | | | | | | | | | | | | |
| 5 MPH | .00 | .06 | .25 | .56 | 1.00 | 1.57 | 2.27 | 3.11 | 4.07 | 5.18 | 6.42 | 7.80 | 9.32 |
| 10 MPH | .00 | .12 | .50 | 1.13 | 2.01 | 3.15 | 4.55 | 6.22 | 8.15 | 10.36 | 12.84 | 15.60 | 18.64 |
| 20 MPH | .00 | .25 | 1.00 | 2.25 | 4.01 | 6.30 | 9.10 | 12.43 | 16.30 | 20.71 | 25.67 | 31.19 | 37.27 |
| 30 MPH | .00 | .37 | 1.49 | 3.38 | 6.02 | 9.44 | 13.65 | 18.65 | 24.45 | 31.07 | 38.51 | 46.79 | 55.91 |

# Bullet: Sierra .410 Dia. 170 Gr. JHC
## Ballistic Coefficient: .123

| RANGE (YDS) | MUZZLE | 25 | 50 | 75 | 100 | 125 | 150 | 175 | 200 | 225 | 250 | 275 | 300 |
|---|---|---|---|---|---|---|---|---|---|---|---|---|---|
| VELOCITY (FPS) | 1600. | 1473. | 1358. | 1255. | 1168. | 1098. | 1042. | 997. | 959. | 925. | 895. | 868. | 842. |
| ENERGY (FT-LB) | 966. | 819. | 696. | 595. | 515. | 455. | 410. | 375. | 347. | 323. | 302. | 284. | 268. |
| DROP (IN) | .00 | -.43 | -1.87 | -4.48 | -8.48 | -14.07 | -21.46 | -30.83 | -42.39 | -56.28 | -72.69 | -91.81 | -113.80 |
| MID-RANGE (IN) | .00 | .12 | .50 | 1.23 | 2.37 | 4.02 | 6.25 | 9.12 | 12.71 | 17.08 | 22.28 | 28.38 | 35.44 |
| BULLET PATH (IN) | -.63 | .19 | .00 | -1.37 | -4.12 | -8.46 | -14.59 | -22.72 | -33.03 | -45.68 | -60.84 | -78.70 | -99.44 |
| TIME OF FLIGHT (SEC) | .000000 | .048868 | .101927 | .159412 | .221403 | .287703 | .357885 | .431510 | .508269 | .587950 | .670411 | .755554 | .843322 |
| WIND DEFLECTION (IN) | | | | | | | | | | | | | |
| 5 MPH | .00 | .18 | .72 | 1.65 | 2.98 | 4.69 | 6.74 | 9.10 | 11.73 | 14.61 | 17.75 | 21.11 | 24.71 |
| 10 MPH | .00 | .35 | 1.44 | 3.31 | 5.97 | 9.39 | 13.49 | 18.20 | 23.46 | 29.23 | 35.49 | 42.23 | 49.42 |
| 20 MPH | .00 | .70 | 2.88 | 6.61 | 11.93 | 18.77 | 26.98 | 36.39 | 46.91 | 58.46 | 70.98 | 84.46 | 98.85 |
| 30 MPH | .00 | 1.05 | 4.32 | 9.92 | 17.90 | 28.16 | 40.46 | 54.59 | 70.37 | 87.69 | 106.48 | 126.68 | 148.27 |
| VELOCITY (FPS) | 1500. | 1382. | 1277. | 1186. | 1112. | 1053. | 1006. | 967. | 932. | 901. | 873. | 848. | 824. |
| ENERGY (FT-LB) | 849. | 721. | 615. | 531. | 467. | 419. | 382. | 353. | 328. | 307. | 288. | 271. | 256. |
| DROP (IN) | .00 | -.50 | -2.13 | -5.10 | -9.62 | -15.88 | -24.10 | -34.45 | -47.11 | -62.25 | -80.04 | -100.68 | -124.33 |
| MID-RANGE (IN) | .00 | .13 | .57 | 1.39 | 2.68 | 4.51 | 6.95 | 10.08 | 13.94 | 18.61 | 24.14 | 30.61 | 38.07 |
| BULLET PATH (IN) | -.63 | .25 | .00 | -1.59 | -4.73 | -9.61 | -16.45 | -25.42 | -36.70 | -50.46 | -66.87 | -86.13 | -108.40 |
| TIME OF FLIGHT (SEC) | .000000 | .052113 | .108611 | .169616 | .234998 | .304368 | .377265 | .453355 | .532407 | .614265 | .698824 | .786017 | .875809 |
| WIND DEFLECTION (IN) | | | | | | | | | | | | | |
| 5 MPH | .00 | .19 | .76 | 1.73 | 3.08 | 4.78 | 6.80 | 9.10 | 11.65 | 14.46 | 17.50 | 20.77 | 24.27 |
| 10 MPH | .00 | .37 | 1.52 | 3.45 | 6.16 | 9.57 | 13.60 | 18.19 | 23.30 | 28.91 | 34.99 | 41.54 | 48.54 |
| 20 MPH | .00 | .74 | 3.03 | 6.90 | 12.32 | 19.14 | 27.20 | 36.38 | 46.61 | 57.82 | 69.99 | 83.08 | 97.08 |
| 30 MPH | .00 | 1.12 | 4.55 | 10.36 | 18.48 | 28.71 | 40.80 | 54.57 | 69.91 | 86.73 | 104.98 | 124.62 | 145.63 |
| VELOCITY (FPS) | 1400. | 1293. | 1199. | 1123. | 1062. | 1013. | 972. | 937. | 906. | 878. | 852. | 827. | 805. |
| ENERGY (FT-LB) | 740. | 631. | 543. | 476. | 426. | 387. | 357. | 332. | 310. | 291. | 274. | 258. | 244. |
| DROP (IN) | .00 | -.57 | -2.44 | -5.83 | -10.93 | -17.95 | -27.09 | -38.50 | -52.35 | -68.84 | -88.14 | -110.43 | -135.88 |
| MID-RANGE (IN) | .00 | .15 | .65 | 1.58 | 3.02 | 5.04 | 7.72 | 11.10 | 15.25 | 20.23 | 26.12 | 32.97 | 40.85 |
| BULLET PATH (IN) | -.63 | .34 | .00 | -1.85 | -5.41 | -10.90 | -18.50 | -28.37 | -40.69 | -55.64 | -73.41 | -94.15 | -118.07 |
| TIME OF FLIGHT (SEC) | .000000 | .055784 | .116070 | .180774 | .249539 | .321898 | .397495 | .476086 | .557503 | .641635 | .728410 | .817787 | .909755 |
| WIND DEFLECTION (IN) | | | | | | | | | | | | | |
| 5 MPH | .00 | .19 | .79 | 1.77 | 3.10 | 4.76 | 6.69 | 8.90 | 11.35 | 14.04 | 16.96 | 20.11 | 23.49 |
| 10 MPH | .00 | .39 | 1.57 | 3.53 | 6.20 | 9.51 | 13.39 | 17.79 | 22.69 | 28.07 | 33.91 | 40.22 | 46.97 |
| 20 MPH | .00 | .78 | 3.14 | 7.06 | 12.41 | 19.02 | 26.78 | 35.58 | 45.38 | 56.14 | 67.83 | 80.43 | 93.95 |
| 30 MPH | .00 | 1.17 | 4.71 | 10.59 | 18.61 | 28.53 | 40.16 | 53.37 | 68.08 | 84.21 | 101.74 | 120.65 | 140.92 |
| VELOCITY (FPS) | 1300. | 1206. | 1128. | 1066. | 1016. | 975. | 940. | 908. | 880. | 854. | 829. | 806. | 784. |
| ENERGY (FT-LB) | 638. | 549. | 480. | 429. | 390. | 359. | 333. | 311. | 292. | 275. | 259. | 245. | 232. |
| DROP (IN) | .00 | -.66 | -2.82 | -6.69 | -12.45 | -20.32 | -30.45 | -43.02 | -58.20 | -76.18 | -97.14 | -121.24 | -148.61 |
| MID-RANGE (IN) | .00 | .18 | .75 | 1.80 | 3.40 | 5.63 | 8.54 | 12.19 | 16.64 | 21.98 | 28.25 | 35.52 | 43.85 |
| BULLET PATH (IN) | -.63 | .43 | .00 | -2.14 | -6.18 | -12.32 | -20.73 | -31.56 | -45.02 | -61.27 | -80.50 | -102.88 | -128.52 |
| TIME OF FLIGHT (SEC) | .000000 | .059959 | .124350 | .192836 | .264948 | .340319 | .418699 | .499915 | .583852 | .670435 | .759624 | .851403 | .945790 |
| WIND DEFLECTION (IN) | | | | | | | | | | | | | |
| 5 MPH | .00 | .20 | .79 | 1.74 | 3.01 | 4.56 | 6.38 | 8.45 | 10.76 | 13.31 | 16.08 | 19.08 | 22.31 |
| 10 MPH | .00 | .40 | 1.58 | 3.48 | 6.02 | 9.13 | 12.77 | 16.91 | 21.53 | 26.61 | 32.16 | 38.15 | 44.61 |
| 20 MPH | .00 | .80 | 3.16 | 6.96 | 12.03 | 18.25 | 25.54 | 33.82 | 43.05 | 53.22 | 64.31 | 76.31 | 89.23 |
| 30 MPH | .00 | 1.20 | 4.73 | 10.43 | 18.05 | 27.38 | 38.30 | 50.72 | 64.58 | 79.84 | 96.47 | 114.46 | 133.84 |

## Bullet: Sierra .410 Dia. 170 Gr. JHC
### Ballistic Coefficient: .123

| RANGE (YDS) | MUZZLE | 25 | 50 | 75 | 100 | 125 | 150 | 175 | 200 | 225 | 250 | 275 | 300 |
|---|---|---|---|---|---|---|---|---|---|---|---|---|---|
| VELOCITY (FPS) | 1200. | 1123. | 1062. | 1014. | 973. | 938. | 906. | 878. | 852. | 828. | 805. | 783. | 762. |
| ENERGY (FT-LB) | 543. | 476. | 426. | 388. | 357. | 332. | 310. | 291. | 274. | 258. | 244. | 231. | 219. |
| DROP (IN) | .00 | -.78 | -3.27 | -7.69 | -14.21 | -23.01 | -34.25 | -48.12 | -64.80 | -84.47 | -107.30 | -133.40 | -163.01 |
| MID-RANGE (IN) | .00 | .20 | .86 | 2.05 | 3.83 | 6.28 | 9.44 | 13.39 | 18.20 | 23.92 | 30.64 | 38.39 | 47.26 |
| BULLET PATH (IN) | -.63 | .54 | .00 | -2.46 | -7.03 | -13.88 | -23.17 | -35.09 | -49.82 | -67.54 | -88.42 | -112.57 | -140.23 |
| TIME OF FLIGHT (SEC) | .000000 | .064670 | .133406 | .205739 | .281311 | .359879 | .441274 | .525383 | .612136 | .701491 | .793437 | .887992 | .985204 |
| WIND DEFLECTION (IN) | | | | | | | | | | | | | |
| 5 MPH | .00 | .19 | .74 | 1.60 | 2.76 | 4.17 | 5.83 | 7.73 | 9.87 | 12.23 | 14.82 | 17.64 | 20.70 |
| 10 MPH | .00 | .38 | 1.48 | 3.21 | 5.51 | 8.34 | 11.66 | 15.47 | 19.74 | 24.46 | 29.64 | 35.29 | 41.40 |
| 20 MPH | .00 | .76 | 2.96 | 6.42 | 11.02 | 16.68 | 23.33 | 30.93 | 39.47 | 48.92 | 59.29 | 70.57 | 82.79 |
| 30 MPH | .00 | 1.15 | 4.44 | 9.63 | 16.53 | 25.02 | 34.99 | 46.40 | 59.21 | 73.39 | 88.93 | 105.86 | 124.19 |
| VELOCITY (FPS) | 1100. | 1044. | 998. | 960. | 926. | 896. | 869. | 843. | 819. | 797. | 775. | 754. | 734. |
| ENERGY (FT-LB) | 457. | 411. | 376. | 348. | 324. | 303. | 285. | 268. | 253. | 240. | 227. | 215. | 204. |
| DROP (IN) | .00 | -.91 | -3.81 | -8.88 | -16.29 | -26.20 | -38.81 | -54.29 | -72.81 | -94.56 | -119.61 | -148.27 | -180.73 |
| MID-RANGE (IN) | .00 | .24 | .99 | 2.34 | 4.33 | 7.03 | 10.52 | 14.86 | 20.12 | 26.36 | 33.61 | 41.98 | 51.56 |
| BULLET PATH (IN) | -.63 | .68 | .00 | -2.85 | -8.03 | -15.72 | -26.11 | -39.37 | -55.67 | -75.20 | -98.03 | -124.47 | -154.71 |
| TIME OF FLIGHT (SEC) | .000000 | .070048 | .143552 | .220199 | .299775 | .382133 | .467176 | .554844 | .645107 | .737962 | .833448 | .931600 | 1.032445 |
| WIND DEFLECTION (IN) | | | | | | | | | | | | | |
| 5 MPH | .00 | .16 | .63 | 1.38 | 2.38 | 3.63 | 5.11 | 6.83 | 8.77 | 10.94 | 13.34 | 15.98 | 18.86 |
| 10 MPH | .00 | .33 | 1.27 | 2.76 | 4.76 | 7.26 | 10.22 | 13.65 | 17.54 | 21.88 | 26.69 | 31.96 | 37.71 |
| 20 MPH | .00 | .66 | 2.53 | 5.51 | 9.52 | 14.51 | 20.45 | 27.31 | 35.08 | 43.76 | 53.37 | 63.92 | 75.42 |
| 30 MPH | .00 | .99 | 3.80 | 8.27 | 14.28 | 21.77 | 30.67 | 40.96 | 52.62 | 65.64 | 80.06 | 95.88 | 113.13 |

# Bullet: Speer .410 Dia. 200 Gr. HJHP (SWC)
## Ballistic Coefficient: .149

| RANGE (YDS) | MUZZLE | 25 | 50 | 75 | 100 | 125 | 150 | 175 | 200 | 225 | 250 | 275 | 300 |
|---|---|---|---|---|---|---|---|---|---|---|---|---|---|
| VELOCITY (FPS) | 1500. | 1402. | 1312. | 1231. | 1160. | 1102. | 1054. | 1014. | 979. | 949. | 921. | 896. | 873. |
| ENERGY (FT-LB) | 999. | 872. | 764. | 672. | 598. | 539. | 493. | 456. | 426. | 400. | 377. | 357. | 339. |
| DROP (IN) | .00 | -.49 | -2.08 | -4.94 | -9.24 | -15.14 | -22.83 | -32.47 | -44.22 | -58.23 | -74.60 | -93.52 | -115.14 |
| MID-RANGE (IN) | .00 | .13 | .55 | 1.33 | 2.54 | 4.23 | 6.48 | 9.34 | 12.87 | 17.14 | 22.16 | 28.01 | 34.73 |
| BULLET PATH (IN) | -.63 | .24 | .00 | -1.50 | -4.44 | -8.99 | -15.32 | -23.61 | -34.00 | -46.65 | -61.66 | -79.23 | -99.49 |
| TIME OF FLIGHT (SEC) | .000000 | .051738 | .107067 | .166119 | .228922 | .295302 | .364953 | .437551 | .512855 | .590691 | .670930 | .753479 | .838273 |
| WIND DEFLECTION (IN) | | | | | | | | | | | | | |
| 5 MPH | .00 | .15 | .62 | 1.42 | 2.55 | 3.99 | 5.72 | 7.70 | 9.93 | 12.38 | 15.04 | 17.91 | 20.97 |
| 10 MPH | .00 | .31 | 1.24 | 2.84 | 5.09 | 7.97 | 11.43 | 15.41 | 19.86 | 24.76 | 30.08 | 35.81 | 41.94 |
| 20 MPH | .00 | .61 | 2.49 | 5.67 | 10.18 | 15.95 | 22.86 | 30.82 | 39.73 | 49.52 | 60.17 | 71.62 | 83.87 |
| 30 MPH | .00 | .92 | 3.73 | 8.51 | 15.27 | 23.92 | 34.30 | 46.23 | 59.59 | 74.28 | 90.25 | 107.44 | 125.81 |
| VELOCITY (FPS) | 1400. | 1310. | 1229. | 1159. | 1101. | 1053. | 1013. | 979. | 948. | 921. | 896. | 873. | 851. |
| ENERGY (FT-LB) | 870. | 762. | 671. | 597. | 538. | 492. | 456. | 425. | 399. | 377. | 356. | 338. | 322. |
| DROP (IN) | .00 | -.56 | -2.39 | -5.66 | -10.54 | -17.21 | -25.84 | -36.57 | -49.56 | -64.93 | -82.84 | -103.46 | -126.91 |
| MID-RANGE (IN) | .00 | .15 | .63 | 1.52 | 2.88 | 4.77 | 7.26 | 10.40 | 14.24 | 18.82 | 24.21 | 30.46 | 37.62 |
| BULLET PATH (IN) | -.63 | .32 | .00 | -1.76 | -5.13 | -10.29 | -17.41 | -26.63 | -38.12 | -51.97 | -68.38 | -87.48 | -109.42 |
| TIME OF FLIGHT (SEC) | .000000 | .055394 | .114516 | .177388 | .243832 | .313540 | .386190 | .461543 | .539424 | .619706 | .702297 | .787131 | .874166 |
| WIND DEFLECTION (IN) | | | | | | | | | | | | | |
| 5 MPH | .00 | .16 | .65 | 1.47 | 2.60 | 4.02 | 5.70 | 7.62 | 9.76 | 12.11 | 14.66 | 17.41 | 20.36 |
| 10 MPH | .00 | .32 | 1.30 | 2.93 | 5.20 | 8.04 | 11.40 | 15.23 | 19.51 | 24.21 | 29.32 | 34.82 | 40.71 |
| 20 MPH | .00 | .64 | 2.60 | 5.87 | 10.40 | 16.08 | 22.80 | 30.46 | 39.02 | 48.42 | 58.64 | 69.64 | 81.42 |
| 30 MPH | .00 | .96 | 3.89 | 8.80 | 15.60 | 24.12 | 34.19 | 45.69 | 58.53 | 72.63 | 87.96 | 104.46 | 122.13 |
| VELOCITY (FPS) | 1300. | 1220. | 1151. | 1094. | 1048. | 1009. | 975. | 945. | 918. | 893. | 870. | 849. | 829. |
| ENERGY (FT-LB) | 750. | 661. | 589. | 532. | 487. | 452. | 422. | 396. | 374. | 354. | 336. | 320. | 305. |
| DROP (IN) | .00 | -.65 | -2.77 | -6.52 | -12.08 | -19.62 | -29.28 | -41.21 | -55.54 | -72.44 | -92.05 | -114.52 | -140.00 |
| MID-RANGE (IN) | .00 | .17 | .73 | 1.74 | 3.27 | 5.38 | 8.12 | 11.54 | 15.69 | 20.63 | 26.41 | 33.09 | 40.72 |
| BULLET PATH (IN) | -.63 | .42 | .00 | -2.05 | -5.91 | -11.74 | -19.70 | -29.94 | -42.57 | -57.76 | -75.67 | -96.44 | -120.22 |
| TIME OF FLIGHT (SEC) | .000000 | .059588 | .122912 | .189777 | .259863 | .332857 | .408529 | .486711 | .567281 | .650151 | .735257 | .822559 | .912034 |
| WIND DEFLECTION (IN) | | | | | | | | | | | | | |
| 5 MPH | .00 | .17 | .66 | 1.47 | 2.56 | 3.91 | 5.49 | 7.29 | 9.31 | 11.52 | 13.93 | 16.54 | 19.34 |
| 10 MPH | .00 | .33 | 1.32 | 2.94 | 5.12 | 7.81 | 10.98 | 14.58 | 18.61 | 23.04 | 27.87 | 33.08 | 38.67 |
| 20 MPH | .00 | .67 | 2.65 | 5.88 | 10.24 | 15.63 | 21.96 | 29.17 | 37.22 | 46.08 | 55.73 | 66.16 | 77.34 |
| 30 MPH | .00 | 1.00 | 3.97 | 8.82 | 15.36 | 23.44 | 32.93 | 43.75 | 55.83 | 69.13 | 83.60 | 99.23 | 116.02 |
| VELOCITY (FPS) | 1200. | 1135. | 1081. | 1036. | 999. | 966. | 937. | 911. | 887. | 864. | 843. | 823. | 804. |
| ENERGY (FT-LB) | 639. | 572. | 519. | 477. | 443. | 414. | 390. | 368. | 349. | 332. | 316. | 301. | 287. |
| DROP (IN) | .00 | -.77 | -3.22 | -7.53 | -13.85 | -22.35 | -33.14 | -46.38 | -62.23 | -80.84 | -102.34 | -126.88 | -154.63 |
| MID-RANGE (IN) | .00 | .20 | .84 | 1.99 | 3.70 | 6.04 | 9.04 | 12.76 | 17.27 | 22.60 | 28.82 | 35.99 | 44.17 |
| BULLET PATH (IN) | -.63 | .53 | .00 | -2.38 | -6.78 | -13.34 | -22.22 | -33.53 | -47.45 | -64.13 | -83.71 | -106.33 | -132.15 |
| TIME OF FLIGHT (SEC) | .000000 | .064322 | .132105 | .203016 | .276765 | .353140 | .431987 | .513194 | .596682 | .682392 | .770291 | .860358 | .952590 |
| WIND DEFLECTION (IN) | | | | | | | | | | | | | |
| 5 MPH | .00 | .16 | .63 | 1.37 | 2.36 | 3.58 | 5.01 | 6.66 | 8.51 | 10.55 | 12.79 | 15.21 | 17.83 |
| 10 MPH | .00 | .32 | 1.25 | 2.73 | 4.71 | 7.15 | 10.03 | 13.32 | 17.02 | 21.10 | 25.57 | 30.42 | 35.66 |
| 20 MPH | .00 | .64 | 2.50 | 5.46 | 9.42 | 14.31 | 20.06 | 26.64 | 34.03 | 42.20 | 51.14 | 60.85 | 71.31 |
| 30 MPH | .00 | .96 | 3.75 | 8.19 | 14.13 | 21.46 | 30.09 | 39.97 | 51.05 | 63.30 | 76.71 | 91.27 | 106.97 |

# Bullet: Speer .410 Dia. 200 Gr. HJHP (SWC)
## Ballistic Coefficient: .149

| RANGE (YDS) | MUZZLE | 25 | 50 | 75 | 100 | 125 | 150 | 175 | 200 | 225 | 250 | 275 | 300 |
|---|---|---|---|---|---|---|---|---|---|---|---|---|---|
| VELOCITY (FPS) | 1100. | 1052. | 1013. | 978. | 948. | 921. | 896. | 873. | 851. | 831. | 811. | 793. | 774. |
| ENERGY (FT-LB) | 537. | 492. | 455. | 425. | 399. | 376. | 356. | 338. | 322. | 306. | 292. | 279. | 266. |
| DROP (IN) | .00 | -.90 | -3.76 | -8.74 | -15.97 | -25.58 | -37.75 | -52.61 | -70.31 | -91.00 | -114.83 | -141.91 | -172.35 |
| MID-RANGE (IN) | .00 | .23 | .98 | 2.29 | 4.22 | 6.82 | 10.14 | 14.24 | 19.18 | 25.03 | 31.83 | 39.62 | 48.43 |
| BULLET PATH (IN) | -.63 | .66 | .00 | -2.78 | -7.81 | -15.23 | -25.20 | -37.86 | -53.36 | -71.85 | -93.49 | -118.38 | -146.61 |
| TIME OF FLIGHT (SEC) | .000000 | .069752 | .142442 | .217831 | .295746 | .376059 | .458680 | .543543 | .630604 | .719840 | .811240 | .904812 | 1.000635 |
| WIND DEFLECTION (IN) | | | | | | | | | | | | | |
| 5 MPH | .00 | .14 | .53 | 1.17 | 2.03 | 3.09 | 4.36 | 5.83 | 7.49 | 9.35 | 11.39 | 13.62 | 16.06 |
| 10 MPH | .00 | .28 | 1.07 | 2.34 | 4.05 | 6.19 | 8.73 | 11.66 | 14.99 | 18.69 | 22.78 | 27.25 | 32.11 |
| 20 MPH | .00 | .55 | 2.14 | 4.68 | 8.10 | 12.37 | 17.46 | 23.33 | 29.97 | 37.38 | 45.56 | 54.49 | 64.22 |
| 30 MPH | .00 | .83 | 3.21 | 7.01 | 12.15 | 18.56 | 26.18 | 34.99 | 44.96 | 56.08 | 68.33 | 81.74 | 96.34 |
| VELOCITY (FPS) | 1000. | 967. | 938. | 912. | 887. | 865. | 844. | 824. | 805. | 786. | 768. | 750. | 733. |
| ENERGY (FT-LB) | 444. | 415. | 391. | 369. | 350. | 332. | 316. | 301. | 287. | 274. | 262. | 250. | 239. |
| DROP (IN) | .00 | -1.08 | -4.46 | -10.28 | -18.70 | -29.87 | -43.93 | -61.02 | -81.32 | -104.87 | -131.87 | -162.56 | -197.09 |
| MID-RANGE (IN) | .00 | .28 | 1.15 | 2.66 | 4.89 | 7.88 | 11.69 | 16.36 | 21.96 | 28.51 | 36.07 | 44.75 | 54.62 |
| BULLET PATH (IN) | -.63 | .84 | .00 | -3.27 | -9.15 | -17.77 | -29.29 | -43.84 | -61.59 | -82.60 | -107.06 | -135.20 | -167.19 |
| TIME OF FLIGHT (SEC) | .000000 | .076286 | .155049 | .236174 | .319581 | .405213 | .493032 | .583020 | .675172 | .769514 | .866127 | .965033 | 1.066253 |
| WIND DEFLECTION (IN) | | | | | | | | | | | | | |
| 5 MPH | .00 | .11 | .44 | .98 | 1.72 | 2.66 | 3.79 | 5.11 | 6.62 | 8.32 | 10.22 | 12.32 | 14.63 |
| 10 MPH | .00 | .23 | .89 | 1.97 | 3.45 | 5.32 | 7.57 | 10.21 | 13.23 | 16.63 | 20.44 | 24.65 | 29.26 |
| 20 MPH | .00 | .45 | 1.78 | 3.93 | 6.89 | 10.63 | 15.15 | 20.42 | 26.46 | 33.27 | 40.88 | 49.29 | 58.52 |
| 30 MPH | .00 | .68 | 2.67 | 5.90 | 10.34 | 15.95 | 22.72 | 30.63 | 39.69 | 49.90 | 61.32 | 73.94 | 87.78 |

## Bullet: Hornady .410 Dia. 210 Gr. JHP
## Ballistic Coefficient: .168

| RANGE (YDS) | MUZZLE | 25 | 50 | 75 | 100 | 125 | 150 | 175 | 200 | 225 | 250 | 275 | 300 |
|---|---|---|---|---|---|---|---|---|---|---|---|---|---|
| VELOCITY (FPS) | 1500. | 1412. | 1331. | 1258. | 1193. | 1137. | 1089. | 1049. | 1015. | 985. | 958. | 933. | 911. |
| ENERGY (FT-LB) | 1049. | 930. | 826. | 738. | 664. | 603. | 553. | 513. | 480. | 452. | 428. | 406. | 387. |
| DROP (IN) | .00 | -.48 | -2.06 | -4.85 | -9.03 | -14.73 | -22.10 | -31.30 | -42.47 | -55.74 | -71.25 | -89.08 | -109.39 |
| MID-RANGE (IN) | .00 | .13 | .54 | 1.30 | 2.46 | 4.07 | 6.20 | 8.89 | 12.21 | 16.19 | 20.89 | 26.34 | 32.59 |
| BULLET PATH (IN) | -.63 | .23 | .00 | -1.45 | -4.28 | -8.64 | -14.67 | -22.53 | -32.35 | -44.28 | -58.44 | -74.93 | -93.89 |
| TIME OF FLIGHT (SEC) | .000000 | .051538 | .106247 | .164220 | .225461 | .289886 | .357314 | .427495 | .500197 | .575241 | .652493 | .731852 | .813238 |
| WIND DEFLECTION (IN) | | | | | | | | | | | | | |
| 5 MPH | .00 | .14 | .55 | 1.25 | 2.24 | 3.51 | 5.04 | 6.82 | 8.82 | 11.02 | 13.42 | 16.00 | 18.76 |
| 10 MPH | .00 | .27 | 1.10 | 2.50 | 4.48 | 7.02 | 10.09 | 13.64 | 17.63 | 22.04 | 26.84 | 32.01 | 37.53 |
| 20 MPH | .00 | .54 | 2.20 | 5.01 | 8.96 | 14.04 | 20.17 | 27.28 | 35.27 | 44.08 | 53.68 | 64.01 | 75.06 |
| 30 MPH | .00 | .81 | 3.30 | 7.51 | 13.44 | 21.06 | 30.26 | 40.92 | 52.90 | 66.13 | 80.52 | 96.02 | 112.59 |
| VELOCITY (FPS) | 1400. | 1320. | 1248. | 1184. | 1129. | 1083. | 1044. | 1010. | 981. | 954. | 930. | 907. | 887. |
| ENERGY (FT-LB) | 914. | 812. | 726. | 654. | 595. | 547. | 508. | 476. | 448. | 424. | 403. | 384. | 367. |
| DROP (IN) | .00 | -.56 | -2.36 | -5.56 | -10.31 | -16.76 | -25.05 | -35.33 | -47.74 | -62.39 | -79.40 | -98.89 | -121.01 |
| MID-RANGE (IN) | .00 | .15 | .62 | 1.49 | 2.80 | 4.60 | 6.96 | 9.93 | 13.56 | 17.88 | 22.94 | 28.78 | 35.45 |
| BULLET PATH (IN) | -.63 | .31 | .00 | -1.71 | -4.97 | -9.92 | -16.72 | -25.51 | -36.42 | -49.58 | -65.09 | -83.09 | -103.72 |
| TIME OF FLIGHT (SEC) | .000000 | .055186 | .113644 | .175362 | .240244 | .308094 | .378661 | .451717 | .527095 | .604663 | .684324 | .766003 | .849644 |
| WIND DEFLECTION (IN) | | | | | | | | | | | | | |
| 5 MPH | .00 | .14 | .57 | 1.29 | 2.28 | 3.54 | 5.04 | 6.75 | 8.67 | 10.78 | 13.08 | 15.55 | 18.20 |
| 10 MPH | .00 | .28 | 1.14 | 2.58 | 4.57 | 7.08 | 10.07 | 13.50 | 17.34 | 21.56 | 26.16 | 31.10 | 36.39 |
| 20 MPH | .00 | .57 | 2.29 | 5.16 | 9.14 | 14.16 | 20.15 | 27.00 | 34.68 | 43.13 | 52.31 | 62.20 | 72.79 |
| 30 MPH | .00 | .85 | 3.43 | 7.73 | 13.71 | 21.25 | 30.22 | 40.51 | 52.02 | 64.69 | 78.47 | 93.31 | 109.18 |
| VELOCITY (FPS) | 1300. | 1230. | 1169. | 1116. | 1072. | 1035. | 1002. | 973. | 947. | 924. | 902. | 881. | 862. |
| ENERGY (FT-LB) | 788. | 706. | 637. | 581. | 536. | 499. | 468. | 442. | 418. | 398. | 379. | 362. | 347. |
| DROP (IN) | .00 | -.65 | -2.73 | -6.41 | -11.82 | -19.12 | -28.44 | -39.91 | -53.68 | -69.81 | -88.48 | -109.81 | -133.93 |
| MID-RANGE (IN) | .00 | .17 | .72 | 1.70 | 3.18 | 5.20 | 7.81 | 11.07 | 15.02 | 19.68 | 25.12 | 31.39 | 38.53 |
| BULLET PATH (IN) | -.63 | .41 | .00 | -1.99 | -5.72 | -11.34 | -18.97 | -28.77 | -40.85 | -55.31 | -72.29 | -91.94 | -114.38 |
| TIME OF FLIGHT (SEC) | .000000 | .059330 | .121905 | .187599 | .256198 | .327447 | .401136 | .477107 | .555240 | .635444 | .717650 | .801804 | .887871 |
| WIND DEFLECTION (IN) | | | | | | | | | | | | | |
| 5 MPH | .00 | .14 | .57 | 1.28 | 2.24 | 3.43 | 4.84 | 6.45 | 8.25 | 10.23 | 12.38 | 14.71 | 17.21 |
| 10 MPH | .00 | .29 | 1.15 | 2.56 | 4.48 | 6.86 | 9.68 | 12.89 | 16.49 | 20.45 | 24.77 | 29.43 | 34.42 |
| 20 MPH | .00 | .58 | 2.30 | 5.11 | 8.95 | 13.72 | 19.35 | 25.79 | 32.98 | 40.91 | 49.54 | 58.85 | 68.84 |
| 30 MPH | .00 | .86 | 3.44 | 7.67 | 13.43 | 20.58 | 29.03 | 38.68 | 49.47 | 61.36 | 74.30 | 88.28 | 103.26 |
| VELOCITY (FPS) | 1200. | 1143. | 1094. | 1053. | 1019. | 988. | 961. | 936. | 913. | 892. | 872. | 854. | 836. |
| ENERGY (FT-LB) | 671. | 609. | 558. | 517. | 484. | 455. | 430. | 408. | 389. | 371. | 355. | 340. | 326. |
| DROP (IN) | .00 | -.76 | -3.19 | -7.42 | -13.60 | -21.87 | -32.36 | -45.18 | -60.44 | -78.30 | -98.88 | -122.31 | -148.71 |
| MID-RANGE (IN) | .00 | .20 | .83 | 1.95 | 3.61 | 5.87 | 8.76 | 12.33 | 16.62 | 21.68 | 27.57 | 34.32 | 42.00 |
| BULLET PATH (IN) | -.63 | .51 | .00 | -2.32 | -6.60 | -12.96 | -21.54 | -32.45 | -45.80 | -61.75 | -80.43 | -101.95 | -126.44 |
| TIME OF FLIGHT (SEC) | .000000 | .064078 | .131184 | .201072 | .273502 | .348292 | .425303 | .504430 | .585591 | .668726 | .753791 | .840756 | .929602 |
| WIND DEFLECTION (IN) | | | | | | | | | | | | | |
| 5 MPH | .00 | .14 | .54 | 1.19 | 2.07 | 3.15 | 4.43 | 5.89 | 7.53 | 9.35 | 11.33 | 13.49 | 15.80 |
| 10 MPH | .00 | .28 | 1.09 | 2.39 | 4.14 | 6.30 | 8.85 | 11.78 | 15.06 | 18.70 | 22.67 | 26.97 | 31.61 |
| 20 MPH | .00 | .56 | 2.18 | 4.78 | 8.27 | 12.60 | 17.71 | 23.56 | 30.13 | 37.39 | 45.33 | 53.95 | 63.22 |
| 30 MPH | .00 | .83 | 3.27 | 7.17 | 12.41 | 18.90 | 26.56 | 35.34 | 45.19 | 56.09 | 68.00 | 80.92 | 94.83 |

## Bullet: Hornady .410 Dia. 210 Gr. JHP
## Ballistic Coefficient: .168

| RANGE (YDS) | MUZZLE | 25 | 50 | 75 | 100 | 125 | 150 | 175 | 200 | 225 | 250 | 275 | 300 |
|---|---|---|---|---|---|---|---|---|---|---|---|---|---|
| VELOCITY (FPS) | 1100. | 1058. | 1023. | 992. | 964. | 939. | 916. | 895. | 875. | 856. | 838. | 821. | 805. |
| ENERGY (FT-LB) | 564. | 522. | 488. | 458. | 433. | 411. | 391. | 373. | 357. | 341. | 327. | 314. | 302. |
| DROP (IN) | .00 | -.90 | -3.73 | -8.63 | -15.74 | -25.16 | -37.00 | -51.43 | -68.56 | -88.52 | -111.44 | -137.45 | -166.66 |
| MID-RANGE (IN) | .00 | .23 | .97 | 2.25 | 4.14 | 6.66 | 9.87 | 13.81 | 18.54 | 24.11 | 30.57 | 37.96 | 46.33 |
| BULLET PATH (IN) | -.63 | .65 | .00 | -2.72 | -7.65 | -14.89 | -24.55 | -36.80 | -51.75 | -69.54 | -90.28 | -114.10 | -141.14 |
| TIME OF FLIGHT (SEC) | .000000 | .069545 | .141660 | .216155 | .292887 | .371745 | .452648 | .535531 | .620348 | .707068 | .795670 | .886144 | .978488 |
| WIND DEFLECTION (IN) | | | | | | | | | | | | | |
| 5 MPH | .00 | .12 | .47 | 1.02 | 1.77 | 2.71 | 3.83 | 5.13 | 6.59 | 8.22 | 10.02 | 11.98 | 14.11 |
| 10 MPH | .00 | .24 | .93 | 2.04 | 3.55 | 5.43 | 7.67 | 10.25 | 13.18 | 16.44 | 20.04 | 23.96 | 28.21 |
| 20 MPH | .00 | .48 | 1.86 | 4.09 | 7.10 | 10.85 | 15.33 | 20.51 | 26.36 | 32.89 | 40.08 | 47.92 | 56.43 |
| 30 MPH | .00 | .72 | 2.80 | 6.13 | 10.64 | 16.28 | 23.00 | 30.76 | 39.54 | 49.33 | 60.11 | 71.88 | 84.64 |
| VELOCITY (FPS) | 1000. | 971. | 946. | 922. | 900. | 880. | 861. | 843. | 826. | 809. | 793. | 776. | 761. |
| ENERGY (FT-LB) | 466. | 440. | 417. | 396. | 378. | 361. | 346. | 331. | 318. | 305. | 293. | 281. | 270. |
| DROP (IN) | .00 | -1.07 | -4.43 | -10.17 | -18.46 | -29.41 | -43.17 | -59.84 | -79.56 | -102.46 | -128.61 | -158.14 | -191.26 |
| MID-RANGE (IN) | .00 | .28 | 1.15 | 2.63 | 4.80 | 7.71 | 11.41 | 15.93 | 21.32 | 27.64 | 34.89 | 43.14 | 52.46 |
| BULLET PATH (IN) | -.63 | .83 | .00 | -3.21 | -8.96 | -17.39 | -28.61 | -42.75 | -59.94 | -80.31 | -103.93 | -130.93 | -161.52 |
| TIME OF FLIGHT (SEC) | .000000 | .076114 | .154383 | .234716 | .317046 | .401321 | .487506 | .575577 | .665521 | .757334 | .851029 | .946693 | 1.044356 |
| WIND DEFLECTION (IN) | | | | | | | | | | | | | |
| 5 MPH | .00 | .10 | .39 | .86 | 1.50 | 2.32 | 3.30 | 4.45 | 5.77 | 7.25 | 8.89 | 10.71 | 12.70 |
| 10 MPH | .00 | .20 | .77 | 1.71 | 3.00 | 4.63 | 6.60 | 8.90 | 11.53 | 14.49 | 17.78 | 21.42 | 25.41 |
| 20 MPH | .00 | .39 | 1.54 | 3.42 | 6.00 | 9.26 | 13.20 | 17.80 | 23.06 | 28.98 | 35.56 | 42.84 | 50.81 |
| 30 MPH | .00 | .59 | 2.31 | 5.13 | 9.00 | 13.90 | 19.80 | 26.70 | 34.59 | 43.47 | 53.34 | 64.25 | 76.22 |

# Bullet: Speer .410 Dia. 220 Gr. S.P. (SWC)
## Ballistic Coefficient: .170

| RANGE (YDS) | MUZZLE | 25 | 50 | 75 | 100 | 125 | 150 | 175 | 200 | 225 | 250 | 275 | 300 |
|---|---|---|---|---|---|---|---|---|---|---|---|---|---|
| VELOCITY (FPS) | 1400. | 1321. | 1257. | 1201. | 1152. | 1109. | 1073. | 1040. | 1012. | 987. | 963. | 942. | 922. |
| ENERGY (FT-LB) | 957. | 852. | 771. | 705. | 649. | 601. | 562. | 529. | 500. | 475. | 453. | 434. | 416. |
| DROP (IN) | .00 | -.56 | -2.35 | -5.53 | -10.21 | -16.52 | -24.60 | -34.55 | -46.51 | -60.58 | -76.88 | -95.49 | -116.51 |
| MID-RANGE (IN) | .00 | .15 | .62 | 1.47 | 2.75 | 4.50 | 6.77 | 9.60 | 13.04 | 17.14 | 21.92 | 27.41 | 33.66 |
| BULLET PATH (IN) | -.63 | .30 | .00 | -1.69 | -4.88 | -9.70 | -16.28 | -24.75 | -35.22 | -47.80 | -62.60 | -79.73 | -99.26 |
| TIME OF FLIGHT (SEC) | .000000 | .055167 | .113446 | .174505 | .238271 | .304628 | .373409 | .444432 | .517541 | .592618 | .669567 | .748312 | .828792 |
| WIND DEFLECTION (IN) | | | | | | | | | | | | | |
| 5 MPH | .00 | .14 | .55 | 1.21 | 2.11 | 3.24 | 4.57 | 6.11 | 7.83 | 9.72 | 11.78 | 13.99 | 16.36 |
| 10 MPH | .00 | .28 | 1.11 | 2.43 | 4.22 | 6.47 | 9.15 | 12.22 | 15.66 | 19.44 | 23.56 | 27.99 | 32.72 |
| 20 MPH | .00 | .56 | 2.22 | 4.85 | 8.44 | 12.94 | 18.30 | 24.44 | 31.32 | 38.89 | 47.12 | 55.98 | 65.45 |
| 30 MPH | .00 | .84 | 3.33 | 7.28 | 12.66 | 19.41 | 27.45 | 36.66 | 46.98 | 58.33 | 70.67 | 83.97 | 98.17 |
| VELOCITY (FPS) | 1300. | 1240. | 1187. | 1140. | 1099. | 1063. | 1032. | 1005. | 980. | 957. | 936. | 917. | 899. |
| ENERGY (FT-LB) | 825. | 752. | 688. | 634. | 589. | 552. | 520. | 493. | 469. | 448. | 428. | 411. | 395. |
| DROP (IN) | .00 | -.64 | -2.70 | -6.29 | -11.56 | -18.62 | -27.59 | -38.60 | -51.76 | -67.17 | -84.92 | -105.11 | -127.87 |
| MID-RANGE (IN) | .00 | .17 | .71 | 1.66 | 3.08 | 5.01 | 7.50 | 10.59 | 14.32 | 18.73 | 23.84 | 29.70 | 36.34 |
| BULLET PATH (IN) | -.63 | .40 | .00 | -1.93 | -5.54 | -10.93 | -18.25 | -27.59 | -39.09 | -52.84 | -68.92 | -87.45 | -108.55 |
| TIME OF FLIGHT (SEC) | .000000 | .059077 | .120906 | .185414 | .252472 | .321901 | .393524 | .467198 | .542809 | .620269 | .699507 | .780465 | .863100 |
| WIND DEFLECTION (IN) | | | | | | | | | | | | | |
| 5 MPH | .00 | .12 | .49 | 1.09 | 1.91 | 2.94 | 4.17 | 5.57 | 7.15 | 8.89 | 10.79 | 12.83 | 15.03 |
| 10 MPH | .00 | .24 | .97 | 2.17 | 3.82 | 5.89 | 8.34 | 11.15 | 14.30 | 17.78 | 21.57 | 25.67 | 30.06 |
| 20 MPH | .00 | .49 | 1.94 | 4.34 | 7.64 | 11.77 | 16.67 | 22.30 | 28.61 | 35.57 | 43.15 | 51.34 | 60.12 |
| 30 MPH | .00 | .73 | 2.92 | 6.51 | 11.46 | 17.66 | 25.01 | 33.45 | 42.91 | 53.35 | 64.72 | 77.01 | 90.18 |
| VELOCITY (FPS) | 1200. | 1151. | 1108. | 1072. | 1040. | 1011. | 986. | 963. | 942. | 922. | 903. | 886. | 870. |
| ENERGY (FT-LB) | 703. | 647. | 600. | 561. | 528. | 500. | 475. | 453. | 433. | 415. | 399. | 383. | 369. |
| DROP (IN) | .00 | -.76 | -3.15 | -7.31 | -13.35 | -21.39 | -31.56 | -43.95 | -58.66 | -75.78 | -95.44 | -117.75 | -142.83 |
| MID-RANGE (IN) | .00 | .20 | .82 | 1.91 | 3.52 | 5.70 | 8.47 | 11.89 | 15.98 | 20.78 | 26.33 | 32.67 | 39.86 |
| BULLET PATH (IN) | -.63 | .50 | .00 | -2.27 | -6.42 | -12.57 | -20.85 | -31.35 | -44.17 | -59.40 | -77.17 | -97.59 | -120.78 |
| TIME OF FLIGHT (SEC) | .000000 | .063837 | .130262 | .199106 | .270186 | .343349 | .418476 | .495473 | .574263 | .654787 | .736998 | .820858 | .906341 |
| WIND DEFLECTION (IN) | | | | | | | | | | | | | |
| 5 MPH | .00 | .12 | .46 | 1.02 | 1.78 | 2.71 | 3.83 | 5.10 | 6.54 | 8.12 | 9.86 | 11.74 | 13.76 |
| 10 MPH | .00 | .24 | .93 | 2.04 | 3.55 | 5.43 | 7.65 | 10.20 | 13.07 | 16.24 | 19.71 | 23.47 | 27.52 |
| 20 MPH | .00 | .47 | 1.85 | 4.09 | 7.11 | 10.86 | 15.30 | 20.41 | 26.14 | 32.49 | 39.42 | 46.94 | 55.03 |
| 30 MPH | .00 | .71 | 2.78 | 6.13 | 10.66 | 16.29 | 22.96 | 30.61 | 39.21 | 48.73 | 59.13 | 70.41 | 82.55 |
| VELOCITY (FPS) | 1100. | 1064. | 1033. | 1006. | 981. | 958. | 937. | 918. | 900. | 882. | 866. | 851. | 836. |
| ENERGY (FT-LB) | 591. | 553. | 521. | 494. | 470. | 448. | 429. | 411. | 395. | 380. | 366. | 353. | 341. |
| DROP (IN) | .00 | -.89 | -3.69 | -8.52 | -15.50 | -24.72 | -36.28 | -50.28 | -66.84 | -86.08 | -108.10 | -133.01 | -160.92 |
| MID-RANGE (IN) | .00 | .23 | .96 | 2.22 | 4.06 | 6.51 | 9.62 | 13.40 | 17.92 | 23.22 | 29.33 | 36.29 | 44.18 |
| BULLET PATH (IN) | -.63 | .64 | .00 | -2.67 | -7.48 | -14.55 | -23.95 | -35.78 | -50.18 | -67.26 | -87.12 | -109.87 | -135.62 |
| TIME OF FLIGHT (SEC) | .000000 | .069339 | .140878 | .214472 | .290007 | .367394 | .446560 | .527448 | .610014 | .694223 | .780049 | .867474 | .956486 |
| WIND DEFLECTION (IN) | | | | | | | | | | | | | |
| 5 MPH | .00 | .10 | .40 | .87 | 1.52 | 2.33 | 3.30 | 4.42 | 5.68 | 7.09 | 8.64 | 10.34 | 12.17 |
| 10 MPH | .00 | .20 | .79 | 1.75 | 3.04 | 4.66 | 6.59 | 8.83 | 11.36 | 14.18 | 17.29 | 20.68 | 24.34 |
| 20 MPH | .00 | .41 | 1.59 | 3.49 | 6.08 | 9.32 | 13.19 | 17.66 | 22.73 | 28.37 | 34.58 | 41.35 | 48.68 |
| 30 MPH | .00 | .61 | 2.38 | 5.24 | 9.12 | 13.98 | 19.78 | 26.49 | 34.09 | 42.55 | 51.87 | 62.03 | 73.02 |

## Bullet: Speer .410 Dia. 220 Gr. S.P. (SWC)
## Ballistic Coefficient: .170

| RANGE (YDS) | MUZZLE | 25 | 50 | 75 | 100 | 125 | 150 | 175 | 200 | 225 | 250 | 275 | 300 |
|---|---|---|---|---|---|---|---|---|---|---|---|---|---|
| VELOCITY (FPS) | 1000. | 976. | 953. | 933. | 914. | 896. | 879. | 863. | 847. | 832. | 818. | 804. | 789. |
| ENERGY (FT-LB) | 488. | 465. | 444. | 425. | 408. | 392. | 377. | 363. | 350. | 338. | 327. | 316. | 304. |
| DROP (IN) | .00 | -1.06 | -4.40 | -10.08 | -18.24 | -28.99 | -42.44 | -58.69 | -77.85 | -100.03 | -125.33 | -153.88 | -185.73 |
| MID-RANGE (IN) | .00 | .28 | 1.14 | 2.60 | 4.72 | 7.56 | 11.14 | 15.50 | 20.69 | 26.73 | 33.68 | 41.57 | 50.42 |
| BULLET PATH (IN) | -.63 | .82 | .00 | -3.17 | -8.81 | -17.05 | -27.99 | -41.73 | -58.37 | -78.04 | -100.83 | -126.86 | -156.20 |
| TIME OF FLIGHT (SEC) | .000000 | .075944 | .153722 | .233265 | .314520 | .397444 | .482003 | .568174 | .655941 | .745292 | .836223 | .928735 | 1.022887 |
| WIND DEFLECTION (IN) | | | | | | | | | | | | | |
| 5 MPH | .00 | .08 | .33 | .73 | 1.28 | 1.98 | 2.82 | 3.80 | 4.92 | 6.19 | 7.59 | 9.13 | 10.81 |
| 10 MPH | .00 | .17 | .66 | 1.45 | 2.56 | 3.95 | 5.63 | 7.60 | 9.85 | 12.37 | 15.18 | 18.26 | 21.63 |
| 20 MPH | .00 | .33 | 1.31 | 2.91 | 5.11 | 7.90 | 11.27 | 15.20 | 19.69 | 24.74 | 30.35 | 36.51 | 43.26 |
| 30 MPH | .00 | .50 | 1.97 | 4.36 | 7.67 | 11.85 | 16.90 | 22.80 | 29.54 | 37.11 | 45.53 | 54.77 | 64.88 |
| VELOCITY (FPS) | 900. | 883. | 866. | 851. | 836. | 821. | 807. | 793. | 777. | 761. | 746. | 731. | 717. |
| ENERGY (FT-LB) | 396. | 381. | 367. | 354. | 341. | 330. | 318. | 307. | 295. | 283. | 272. | 261. | 251. |
| DROP (IN) | .00 | -1.23 | -5.24 | -12.13 | -22.01 | -34.99 | -51.17 | -70.66 | -93.51 | -119.94 | -150.12 | -184.18 | -222.30 |
| MID-RANGE (IN) | .00 | .34 | 1.39 | 3.18 | 5.77 | 9.18 | 13.46 | 18.64 | 24.75 | 31.87 | 40.07 | 49.42 | 59.97 |
| BULLET PATH (IN) | -.63 | 1.07 | -.00 | -3.96 | -10.90 | -20.95 | -34.20 | -50.75 | -70.67 | -94.17 | -121.41 | -152.54 | -187.72 |
| TIME OF FLIGHT (SEC) | .000000 | .084150 | .169916 | .257281 | .346231 | .436761 | .528871 | .622580 | .718177 | .815771 | .915378 | 1.017015 | 1.120705 |
| WIND DEFLECTION (IN) | | | | | | | | | | | | | |
| 5 MPH | .00 | .07 | .29 | .64 | 1.14 | 1.77 | 2.54 | 3.45 | 4.53 | 5.79 | 7.22 | 8.83 | 10.62 |
| 10 MPH | .00 | .14 | .57 | 1.28 | 2.27 | 3.54 | 5.08 | 6.91 | 9.07 | 11.58 | 14.44 | 17.66 | 21.24 |
| 20 MPH | .00 | .29 | 1.14 | 2.56 | 4.54 | 7.07 | 10.16 | 13.81 | 18.13 | 23.15 | 28.88 | 35.32 | 42.49 |
| 30 MPH | .00 | .43 | 1.72 | 3.84 | 6.81 | 10.61 | 15.24 | 20.72 | 27.20 | 34.73 | 43.32 | 52.98 | 63.73 |
| VELOCITY (FPS) | 800. | 783. | 767. | 752. | 737. | 723. | 708. | 695. | 681. | 668. | 655. | 643. | 631. |
| ENERGY (FT-LB) | 313. | 300. | 288. | 276. | 265. | 255. | 245. | 236. | 227. | 218. | 210. | 202. | 194. |
| DROP (IN) | .00 | -1.46 | -6.43 | -15.06 | -27.52 | -43.96 | -64.54 | -89.41 | -118.78 | -152.81 | -191.69 | -235.62 | -284.80 |
| MID-RANGE (IN) | .00 | .42 | 1.75 | 4.03 | 7.33 | 11.70 | 17.21 | 23.91 | 31.87 | 41.18 | 51.89 | 64.09 | 77.86 |
| BULLET PATH (IN) | -.63 | 1.43 | -.00 | -5.10 | -14.03 | -26.94 | -44.00 | -65.34 | -91.18 | -121.68 | -157.04 | -197.44 | -243.09 |
| TIME OF FLIGHT (SEC) | .000000 | .094743 | .191477 | .290215 | .390974 | .493774 | .598639 | .705598 | .814680 | .925919 | 1.039352 | 1.155018 | 1.272957 |
| WIND DEFLECTION (IN) | | | | | | | | | | | | | |
| 5 MPH | .00 | .09 | .35 | .79 | 1.41 | 2.20 | 3.18 | 4.34 | 5.69 | 7.23 | 8.96 | 10.89 | 13.02 |
| 10 MPH | .00 | .17 | .70 | 1.58 | 2.81 | 4.40 | 6.36 | 8.69 | 11.38 | 14.46 | 17.93 | 21.78 | 26.04 |
| 20 MPH | .00 | .35 | 1.40 | 3.16 | 5.62 | 8.81 | 12.72 | 17.37 | 22.77 | 28.92 | 35.85 | 43.57 | 52.08 |
| 30 MPH | .00 | .52 | 2.10 | 4.73 | 8.43 | 13.21 | 19.08 | 26.06 | 34.15 | 43.39 | 53.78 | 65.35 | 78.12 |

# Bullet: Lyman #41028 206 Gr.
## Ballistic Coefficient: .160

| RANGE (YDS) | MUZZLE | 25 | 50 | 75 | 100 | 125 | 150 | 175 | 200 | 225 | 250 | 275 | 300 |
|---|---|---|---|---|---|---|---|---|---|---|---|---|---|
| VELOCITY (FPS) | 1500. | 1408. | 1324. | 1250. | 1187. | 1132. | 1085. | 1046. | 1012. | 983. | 956. | 932. | 909. |
| ENERGY (FT-LB) | 1029. | 907. | 801. | 715. | 644. | 586. | 539. | 500. | 469. | 442. | 418. | 397. | 378. |
| DROP (IN) | .00 | -.49 | -2.07 | -4.88 | -9.10 | -14.85 | -22.30 | -31.58 | -42.85 | -56.22 | -71.85 | -89.81 | -110.25 |
| MID-RANGE (IN) | .00 | .13 | .55 | 1.31 | 2.48 | 4.11 | 6.27 | 8.98 | 12.32 | 16.34 | 21.07 | 26.55 | 32.82 |
| BULLET PATH (IN) | -.63 | .23 | .00 | -1.47 | -4.34 | -8.74 | -14.84 | -22.77 | -32.69 | -44.72 | -58.99 | -75.60 | -94.69 |
| TIME OF FLIGHT (SEC) | .000000 | .051617 | .106568 | .164911 | .226508 | .291255 | .358965 | .429388 | .502298 | .577525 | .654937 | .734437 | .815949 |
| WIND DEFLECTION (IN) | | | | | | | | | | | | | |
| 5 MPH | .00 | .14 | .58 | 1.31 | 2.33 | 3.63 | 5.19 | 6.99 | 9.00 | 11.22 | 13.63 | 16.23 | 19.00 |
| 10 MPH | .00 | .28 | 1.16 | 2.62 | 4.67 | 7.26 | 10.38 | 13.97 | 18.00 | 22.44 | 27.27 | 32.46 | 38.01 |
| 20 MPH | .00 | .57 | 2.31 | 5.25 | 9.33 | 14.52 | 20.76 | 27.94 | 36.01 | 44.89 | 54.54 | 64.92 | 76.01 |
| 30 MPH | .00 | .85 | 3.47 | 7.87 | 14.00 | 21.78 | 31.13 | 41.92 | 54.01 | 67.33 | 81.81 | 97.38 | 114.02 |
| VELOCITY (FPS) | 1400. | 1316. | 1244. | 1181. | 1127. | 1081. | 1043. | 1009. | 980. | 954. | 929. | 907. | 887. |
| ENERGY (FT-LB) | 896. | 792. | 708. | 638. | 581. | 535. | 497. | 466. | 439. | 416. | 395. | 377. | 360. |
| DROP (IN) | .00 | -.56 | -2.37 | -5.59 | -10.36 | -16.84 | -25.18 | -35.50 | -47.95 | -62.66 | -79.71 | -99.26 | -121.43 |
| MID-RANGE (IN) | .00 | .15 | .63 | 1.50 | 2.81 | 4.63 | 7.00 | 9.98 | 13.61 | 17.95 | 23.01 | 28.86 | 35.54 |
| BULLET PATH (IN) | -.63 | .31 | .00 | -1.72 | -5.00 | -9.98 | -16.81 | -25.64 | -36.59 | -49.80 | -65.35 | -83.40 | -104.07 |
| TIME OF FLIGHT (SEC) | .000000 | .055268 | .113923 | .175823 | .240860 | .308837 | .379504 | .452639 | .528077 | .605691 | .685383 | .767081 | .850728 |
| WIND DEFLECTION (IN) | | | | | | | | | | | | | |
| 5 MPH | .00 | .15 | .60 | 1.33 | 2.34 | 3.61 | 5.11 | 6.83 | 8.76 | 10.87 | 13.17 | 15.65 | 18.29 |
| 10 MPH | .00 | .30 | 1.19 | 2.66 | 4.68 | 7.21 | 10.22 | 13.66 | 17.51 | 21.74 | 26.34 | 31.29 | 36.59 |
| 20 MPH | .00 | .60 | 2.39 | 5.32 | 9.35 | 14.43 | 20.44 | 27.33 | 35.03 | 43.49 | 52.68 | 62.58 | 73.17 |
| 30 MPH | .00 | .90 | 3.58 | 7.98 | 14.03 | 21.64 | 30.66 | 40.99 | 52.54 | 65.23 | 79.03 | 93.88 | 109.76 |
| VELOCITY (FPS) | 1300. | 1231. | 1170. | 1117. | 1073. | 1036. | 1003. | 974. | 948. | 925. | 903. | 883. | 864. |
| ENERGY (FT-LB) | 773. | 693. | 626. | 571. | 527. | 490. | 460. | 434. | 411. | 391. | 373. | 356. | 341. |
| DROP (IN) | .00 | -.65 | -2.73 | -6.40 | -11.81 | -19.10 | -28.41 | -39.87 | -53.61 | -69.71 | -88.34 | -109.63 | -133.70 |
| MID-RANGE (IN) | .00 | .17 | .72 | 1.70 | 3.17 | 5.19 | 7.80 | 11.05 | 14.99 | 19.65 | 25.07 | 31.33 | 38.45 |
| BULLET PATH (IN) | -.63 | .41 | .00 | -1.99 | -5.72 | -11.32 | -18.95 | -28.73 | -40.79 | -55.21 | -72.16 | -91.77 | -114.16 |
| TIME OF FLIGHT (SEC) | .000000 | .059321 | .121867 | .187517 | .256059 | .327240 | .400853 | .476740 | .554780 | .634882 | .716977 | .801013 | .886951 |
| WIND DEFLECTION (IN) | | | | | | | | | | | | | |
| 5 MPH | .00 | .14 | .57 | 1.27 | 2.23 | 3.41 | 4.81 | 6.41 | 8.21 | 10.18 | 12.32 | 14.64 | 17.13 |
| 10 MPH | .00 | .29 | 1.14 | 2.54 | 4.45 | 6.83 | 9.63 | 12.83 | 16.41 | 20.35 | 24.65 | 29.29 | 34.26 |
| 20 MPH | .00 | .57 | 2.28 | 5.08 | 8.90 | 13.65 | 19.25 | 25.66 | 32.82 | 40.71 | 49.30 | 58.57 | 68.51 |
| 30 MPH | .00 | .86 | 3.42 | 7.62 | 13.35 | 20.48 | 28.88 | 38.49 | 49.23 | 61.06 | 73.95 | 87.86 | 102.77 |
| VELOCITY (FPS) | 1200. | 1143. | 1095. | 1054. | 1019. | 989. | 962. | 937. | 914. | 893. | 873. | 855. | 837. |
| ENERGY (FT-LB) | 659. | 598. | 548. | 508. | 475. | 447. | 423. | 401. | 382. | 365. | 349. | 334. | 320. |
| DROP (IN) | .00 | -.76 | -3.19 | -7.41 | -13.59 | -21.85 | -32.33 | -45.13 | -60.37 | -78.20 | -98.75 | -122.14 | -148.49 |
| MID-RANGE (IN) | .00 | .20 | .83 | 1.95 | 3.61 | 5.86 | 8.75 | 12.31 | 16.59 | 21.65 | 27.52 | 34.26 | 41.91 |
| BULLET PATH (IN) | -.63 | .51 | .00 | -2.32 | -6.59 | -12.94 | -21.51 | -32.41 | -45.74 | -61.66 | -80.30 | -101.78 | -126.23 |
| TIME OF FLIGHT (SEC) | .000000 | .064069 | .131150 | .200998 | .273378 | .348108 | .425049 | .504096 | .585169 | .668207 | .753165 | .840014 | .928732 |
| WIND DEFLECTION (IN) | | | | | | | | | | | | | |
| 5 MPH | .00 | .14 | .54 | 1.19 | 2.06 | 3.13 | 4.40 | 5.86 | 7.49 | 9.30 | 11.28 | 13.42 | 15.73 |
| 10 MPH | .00 | .28 | 1.08 | 2.38 | 4.11 | 6.27 | 8.81 | 11.72 | 14.99 | 18.60 | 22.56 | 26.84 | 31.46 |
| 20 MPH | .00 | .55 | 2.16 | 4.75 | 8.23 | 12.53 | 17.62 | 23.44 | 29.98 | 37.21 | 45.11 | 53.68 | 62.91 |
| 30 MPH | .00 | .83 | 3.25 | 7.13 | 12.34 | 18.80 | 26.43 | 35.16 | 44.97 | 55.81 | 67.67 | 80.53 | 94.37 |

## Bullet: Lyman #41028 206 Gr.
### Ballistic Coefficient: .160

| RANGE (YDS) | MUZZLE | 25 | 50 | 75 | 100 | 125 | 150 | 175 | 200 | 225 | 250 | 275 | 300 |
|---|---|---|---|---|---|---|---|---|---|---|---|---|---|
| VELOCITY (FPS) | 1100. | 1059. | 1023. | 992. | 965. | 940. | 917. | 895. | 875. | 857. | 839. | 822. | 806. |
| ENERGY (FT-LB) | 553. | 512. | 479. | 450. | 425. | 404. | 384. | 367. | 351. | 336. | 322. | 309. | 297. |
| DROP (IN) | .00 | -.90 | -3.73 | -8.63 | -15.73 | -25.14 | -36.97 | -51.38 | -68.49 | -88.43 | -111.31 | -137.28 | -166.44 |
| MID-RANGE (IN) | .00 | .23 | .97 | 2.25 | 4.14 | 6.66 | 9.86 | 13.80 | 18.52 | 24.07 | 30.52 | 37.90 | 46.25 |
| BULLET PATH (IN) | -.63 | .65 | .00 | -2.72 | -7.64 | -14.87 | -24.53 | -36.76 | -51.69 | -69.45 | -90.16 | -113.94 | -140.93 |
| TIME OF FLIGHT (SEC) | .000000 | .069537 | .141630 | .216092 | .292778 | .371582 | .452419 | .535228 | .619961 | .706587 | .795084 | .885442 | .977659 |
| WIND DEFLECTION (IN) | | | | | | | | | | | | | |
| 5 MPH | .00 | .12 | .46 | 1.02 | 1.76 | 2.70 | 3.81 | 5.10 | 6.56 | 8.18 | 9.97 | 11.92 | 14.03 |
| 10 MPH | .00 | .24 | .93 | 2.03 | 3.53 | 5.40 | 7.63 | 10.20 | 13.11 | 16.36 | 19.93 | 23.84 | 28.07 |
| 20 MPH | .00 | .48 | 1.85 | 4.06 | 7.06 | 10.80 | 15.25 | 20.40 | 26.23 | 32.72 | 39.87 | 47.68 | 56.14 |
| 30 MPH | .00 | .72 | 2.78 | 6.10 | 10.59 | 16.20 | 22.88 | 30.60 | 39.34 | 49.08 | 59.80 | 71.51 | 84.20 |
| VELOCITY (FPS) | 1000. | 972. | 946. | 922. | 901. | 881. | 862. | 844. | 826. | 810. | 793. | 776. | 759. |
| ENERGY (FT-LB) | 457. | 432. | 409. | 389. | 371. | 355. | 339. | 325. | 312. | 300. | 288. | 275. | 264. |
| DROP (IN) | .00 | -1.07 | -4.43 | -10.17 | -18.45 | -29.40 | -43.14 | -59.80 | -79.49 | -102.36 | -128.49 | -157.99 | -191.09 |
| MID-RANGE (IN) | .00 | .28 | 1.15 | 2.63 | 4.79 | 7.71 | 11.40 | 15.92 | 21.30 | 27.60 | 34.85 | 43.09 | 52.41 |
| BULLET PATH (IN) | -.63 | .83 | .00 | -3.21 | -8.96 | -17.38 | -28.59 | -42.71 | -59.88 | -80.22 | -103.82 | -130.79 | -161.36 |
| TIME OF FLIGHT (SEC) | .000000 | .076108 | .154358 | .234661 | .316951 | .401175 | .487298 | .575298 | .665159 | .756878 | .850480 | .946157 | 1.043967 |
| WIND DEFLECTION (IN) | | | | | | | | | | | | | |
| 5 MPH | .00 | .10 | .38 | .85 | 1.49 | 2.30 | 3.28 | 4.43 | 5.73 | 7.21 | 8.84 | 10.66 | 12.67 |
| 10 MPH | .00 | .19 | .77 | 1.70 | 2.98 | 4.61 | 6.56 | 8.85 | 11.47 | 14.41 | 17.68 | 21.32 | 25.34 |
| 20 MPH | .00 | .39 | 1.53 | 3.40 | 5.97 | 9.21 | 13.13 | 17.70 | 22.94 | 28.82 | 35.37 | 42.65 | 50.68 |
| 30 MPH | .00 | .58 | 2.30 | 5.10 | 8.95 | 13.82 | 19.69 | 26.56 | 34.40 | 43.23 | 53.05 | 63.97 | 76.01 |

# Bullet: Lyman #410610 204 Gr.
## Ballistic Coefficient: .158

| RANGE (YDS) | MUZZLE | 25 | 50 | 75 | 100 | 125 | 150 | 175 | 200 | 225 | 250 | 275 | 300 |
|---|---|---|---|---|---|---|---|---|---|---|---|---|---|
| VELOCITY (FPS) | 1500. | 1407. | 1322. | 1248. | 1185. | 1130. | 1084. | 1045. | 1011. | 982. | 955. | 931. | 909. |
| ENERGY (FT-LB) | 1019. | 897. | 791. | 706. | 636. | 579. | 532. | 495. | 463. | 436. | 413. | 393. | 374. |
| DROP (IN) | .00 | -.49 | -2.07 | -4.89 | -9.12 | -14.89 | -22.35 | -31.66 | -42.95 | -56.36 | -72.02 | -90.02 | -110.50 |
| MID-RANGE (IN) | .00 | .13 | .55 | 1.31 | 2.49 | 4.13 | 6.28 | 9.01 | 12.36 | 16.38 | 21.12 | 26.61 | 32.90 |
| BULLET PATH (IN) | -.63 | .23 | .00 | -1.47 | -4.35 | -8.77 | -14.89 | -22.84 | -32.79 | -44.85 | -59.15 | -75.80 | -94.93 |
| TIME OF FLIGHT (SEC) | .000000 | .051638 | .106654 | .165096 | .226792 | .291635 | .359433 | .429936 | .502921 | .578217 | .655696 | .735260 | .816832 |
| WIND DEFLECTION (IN) | | | | | | | | | | | | | |
| 5 MPH | .00 | .14 | .59 | 1.33 | 2.36 | 3.66 | 5.23 | 7.03 | 9.06 | 11.28 | 13.70 | 16.30 | 19.08 |
| 10 MPH | .00 | .29 | 1.17 | 2.66 | 4.72 | 7.33 | 10.46 | 14.07 | 18.11 | 22.57 | 27.40 | 32.61 | 38.16 |
| 20 MPH | .00 | .58 | 2.34 | 5.31 | 9.43 | 14.66 | 20.92 | 28.14 | 36.23 | 45.13 | 54.80 | 65.21 | 76.32 |
| 30 MPH | .00 | .86 | 3.51 | 7.97 | 14.15 | 21.98 | 31.38 | 42.21 | 54.34 | 67.70 | 82.21 | 97.82 | 114.49 |
| | | | | | | | | | | | | | |
| VELOCITY (FPS) | 1400. | 1315. | 1243. | 1181. | 1126. | 1081. | 1042. | 1009. | 980. | 953. | 929. | 907. | 887. |
| ENERGY (FT-LB) | 888. | 783. | 700. | 631. | 575. | 529. | 492. | 461. | 435. | 412. | 391. | 373. | 356. |
| DROP (IN) | .00 | -.56 | -2.37 | -5.60 | -10.38 | -16.86 | -25.20 | -35.54 | -48.00 | -62.72 | -79.79 | -99.35 | -121.55 |
| MID-RANGE (IN) | .00 | .15 | .63 | 1.50 | 2.82 | 4.63 | 7.01 | 9.99 | 13.62 | 17.96 | 23.03 | 28.88 | 35.57 |
| BULLET PATH (IN) | -.63 | .31 | .00 | -1.72 | -5.00 | -9.99 | -16.83 | -25.67 | -36.63 | -49.85 | -65.42 | -83.48 | -104.17 |
| TIME OF FLIGHT (SEC) | .000000 | .055289 | .113985 | .175923 | .240996 | .309007 | .379703 | .452867 | .528331 | .605969 | .685686 | .767406 | .851077 |
| WIND DEFLECTION (IN) | | | | | | | | | | | | | |
| 5 MPH | .00 | .15 | .60 | 1.34 | 2.35 | 3.62 | 5.13 | 6.85 | 8.78 | 10.90 | 13.20 | 15.67 | 18.32 |
| 10 MPH | .00 | .30 | 1.20 | 2.68 | 4.70 | 7.24 | 10.26 | 13.70 | 17.56 | 21.79 | 26.39 | 31.35 | 36.65 |
| 20 MPH | .00 | .60 | 2.41 | 5.35 | 9.40 | 14.48 | 20.51 | 27.41 | 35.12 | 43.59 | 52.79 | 62.70 | 73.29 |
| 30 MPH | .00 | .91 | 3.61 | 8.03 | 14.10 | 21.73 | 30.77 | 41.11 | 52.67 | 65.38 | 79.18 | 94.05 | 109.94 |
| | | | | | | | | | | | | | |
| VELOCITY (FPS) | 1300. | 1231. | 1170. | 1117. | 1073. | 1036. | 1003. | 974. | 948. | 925. | 903. | 883. | 864. |
| ENERGY (FT-LB) | 765. | 686. | 620. | 565. | 521. | 486. | 456. | 430. | 407. | 387. | 369. | 353. | 338. |
| DROP (IN) | .00 | -.65 | -2.73 | -6.40 | -11.81 | -19.10 | -28.41 | -39.87 | -53.61 | -69.71 | -88.34 | -109.63 | -133.70 |
| MID-RANGE (IN) | .00 | .17 | .72 | 1.70 | 3.17 | 5.19 | 7.80 | 11.05 | 14.99 | 19.65 | 25.07 | 31.33 | 38.45 |
| BULLET PATH (IN) | -.63 | .41 | .00 | -1.99 | -5.72 | -11.32 | -18.95 | -28.73 | -40.79 | -55.21 | -72.16 | -91.77 | -114.16 |
| TIME OF FLIGHT (SEC) | .000000 | .059321 | .121867 | .187517 | .256059 | .327240 | .400853 | .476740 | .554780 | .634882 | .716977 | .801013 | .886951 |
| WIND DEFLECTION (IN) | | | | | | | | | | | | | |
| 5 MPH | .00 | .14 | .57 | 1.27 | 2.23 | 3.41 | 4.81 | 6.41 | 8.21 | 10.18 | 12.32 | 14.64 | 17.13 |
| 10 MPH | .00 | .29 | 1.14 | 2.54 | 4.45 | 6.83 | 9.63 | 12.83 | 16.41 | 20.35 | 24.65 | 29.29 | 34.26 |
| 20 MPH | .00 | .57 | 2.28 | 5.08 | 8.90 | 13.65 | 19.25 | 25.66 | 32.82 | 40.71 | 49.30 | 58.57 | 68.51 |
| 30 MPH | .00 | .86 | 3.42 | 7.62 | 13.35 | 20.48 | 28.88 | 38.49 | 49.23 | 61.06 | 73.95 | 87.86 | 102.77 |
| | | | | | | | | | | | | | |
| VELOCITY (FPS) | 1200. | 1143. | 1095. | 1054. | 1019. | 989. | 962. | 937. | 914. | 893. | 873. | 855. | 837. |
| ENERGY (FT-LB) | 652. | 592. | 543. | 503. | 471. | 443. | 419. | 397. | 378. | 361. | 345. | 331. | 317. |
| DROP (IN) | .00 | -.76 | -3.19 | -7.41 | -13.59 | -21.85 | -32.33 | -45.13 | -60.37 | -78.20 | -98.75 | -122.14 | -148.49 |
| MID-RANGE (IN) | .00 | .20 | .83 | 1.95 | 3.61 | 5.86 | 8.75 | 12.31 | 16.59 | 21.65 | 27.52 | 34.26 | 41.91 |
| BULLET PATH (IN) | -.63 | .51 | .00 | -2.32 | -6.59 | -12.94 | -21.51 | -32.41 | -45.74 | -61.66 | -80.30 | -101.78 | -126.23 |
| TIME OF FLIGHT (SEC) | .000000 | .064069 | .131150 | .200998 | .273378 | .348108 | .425049 | .504096 | .585169 | .668207 | .753165 | .840014 | .928732 |
| WIND DEFLECTION (IN) | | | | | | | | | | | | | |
| 5 MPH | .00 | .14 | .54 | 1.19 | 2.06 | 3.13 | 4.40 | 5.86 | 7.49 | 9.30 | 11.28 | 13.42 | 15.73 |
| 10 MPH | .00 | .28 | 1.08 | 2.38 | 4.11 | 6.27 | 8.81 | 11.72 | 14.99 | 18.60 | 22.56 | 26.84 | 31.46 |
| 20 MPH | .00 | .55 | 2.16 | 4.75 | 8.23 | 12.53 | 17.62 | 23.44 | 29.98 | 37.21 | 45.11 | 53.68 | 62.91 |
| 30 MPH | .00 | .83 | 3.25 | 7.13 | 12.34 | 18.80 | 26.43 | 35.16 | 44.97 | 55.81 | 67.67 | 80.53 | 94.37 |

# Bullet: Lyman #410610 204 Gr.
## Ballistic Coefficient: .158

| RANGE (YDS) | MUZZLE | 25 | 50 | 75 | 100 | 125 | 150 | 175 | 200 | 225 | 250 | 275 | 300 |
|---|---|---|---|---|---|---|---|---|---|---|---|---|---|
| VELOCITY (FPS) | 1100. | 1059. | 1023. | 992. | 965. | 940. | 917. | 895. | 875. | 857. | 839. | 822. | 806. |
| ENERGY (FT-LB) | 548. | 507. | 474. | 446. | 421. | 400. | 381. | 363. | 347. | 332. | 319. | 306. | 294. |
| DROP (IN) | .00 | -.90 | -3.73 | -8.63 | -15.73 | -25.14 | -36.97 | -51.38 | -68.49 | -88.43 | -111.31 | -137.28 | -166.44 |
| MID-RANGE (IN) | .00 | .23 | .97 | 2.25 | 4.14 | 6.66 | 9.86 | 13.80 | 18.52 | 24.07 | 30.52 | 37.90 | 46.25 |
| BULLET PATH (IN) | -.63 | .65 | .00 | -2.72 | -7.64 | -14.87 | -24.53 | -36.76 | -51.69 | -69.45 | -90.16 | -113.94 | -140.93 |
| TIME OF FLIGHT (SEC) | .000000 | .069537 | .141630 | .216092 | .292778 | .371582 | .452419 | .535228 | .619961 | .706587 | .795084 | .885442 | .977659 |
| WIND DEFLECTION (IN) | | | | | | | | | | | | | |
| 5 MPH | .00 | .12 | .46 | 1.02 | 1.76 | 2.70 | 3.81 | 5.10 | 6.56 | 8.18 | 9.97 | 11.92 | 14.03 |
| 10 MPH | .00 | .24 | .93 | 2.03 | 3.53 | 5.40 | 7.63 | 10.20 | 13.11 | 16.36 | 19.93 | 23.84 | 28.07 |
| 20 MPH | .00 | .48 | 1.85 | 4.06 | 7.06 | 10.80 | 15.25 | 20.40 | 26.23 | 32.72 | 39.87 | 47.68 | 56.14 |
| 30 MPH | .00 | .72 | 2.78 | 6.10 | 10.59 | 16.20 | 22.88 | 30.60 | 39.34 | 49.08 | 59.80 | 71.51 | 84.20 |
| VELOCITY (FPS) | 1000. | 972. | 946. | 922. | 901. | 881. | 862. | 844. | 826. | 810. | 793. | 776. | 759. |
| ENERGY (FT-LB) | 453. | 427. | 405. | 385. | 368. | 351. | 336. | 322. | 309. | 297. | 285. | 273. | 261. |
| DROP (IN) | .00 | -1.07 | -4.43 | -10.17 | -18.45 | -29.40 | -43.14 | -59.80 | -79.49 | -102.36 | -128.49 | -157.99 | -191.09 |
| MID-RANGE (IN) | .00 | .28 | 1.15 | 2.63 | 4.79 | 7.71 | 11.40 | 15.92 | 21.30 | 27.60 | 34.85 | 43.09 | 52.41 |
| BULLET PATH (IN) | -.63 | .83 | .00 | -3.21 | -8.96 | -17.38 | -28.59 | -42.71 | -59.88 | -80.22 | -103.82 | -130.79 | -161.36 |
| TIME OF FLIGHT (SEC) | .000000 | .076108 | .154358 | .234661 | .316951 | .401175 | .487298 | .575298 | .665159 | .756878 | .850480 | .946157 | 1.043967 |
| WIND DEFLECTION (IN) | | | | | | | | | | | | | |
| 5 MPH | .00 | .10 | .38 | .85 | 1.49 | 2.30 | 3.28 | 4.43 | 5.73 | 7.21 | 8.84 | 10.66 | 12.67 |
| 10 MPH | .00 | .19 | .77 | 1.70 | 2.98 | 4.61 | 6.56 | 8.85 | 11.47 | 14.41 | 17.68 | 21.32 | 25.34 |
| 20 MPH | .00 | .39 | 1.53 | 3.40 | 5.97 | 9.21 | 13.13 | 17.70 | 22.94 | 28.82 | 35.37 | 42.65 | 50.68 |
| 30 MPH | .00 | .58 | 2.30 | 5.10 | 8.95 | 13.82 | 19.69 | 26.56 | 34.40 | 43.23 | 53.05 | 63.97 | 76.01 |

# Bullet: Lyman #410459 203 Gr.
## Ballistic Coefficient: .150

| RANGE (YDS) | MUZZLE | 25 | 50 | 75 | 100 | 125 | 150 | 175 | 200 | 225 | 250 | 275 | 300 |
|---|---|---|---|---|---|---|---|---|---|---|---|---|---|
| VELOCITY (FPS) | 1400. | 1311. | 1246. | 1190. | 1141. | 1099. | 1063. | 1031. | 1003. | 978. | 955. | 934. | 914. |
| ENERGY (FT-LB) | 883. | 775. | 699. | 639. | 587. | 544. | 509. | 479. | 453. | 431. | 411. | 393. | 377. |
| DROP (IN) | .00 | -.56 | -2.38 | -5.60 | -10.36 | -16.77 | -24.98 | -35.11 | -47.28 | -61.60 | -78.18 | -97.10 | -118.50 |
| MID-RANGE (IN) | .00 | .15 | .63 | 1.50 | 2.80 | 4.58 | 6.89 | 9.78 | 13.28 | 17.45 | 22.32 | 27.91 | 34.27 |
| BULLET PATH (IN) | -.63 | .31 | .00 | -1.72 | -4.97 | -9.89 | -16.60 | -25.22 | -35.88 | -48.70 | -63.78 | -81.20 | -101.09 |
| TIME OF FLIGHT (SEC) | .000000 | .055382 | .114157 | .175772 | .240143 | .307134 | .376561 | .448236 | .522009 | .597764 | .675407 | .754867 | .836085 |
| WIND DEFLECTION (IN) | | | | | | | | | | | | | |
| 5 MPH | .00 | .16 | .62 | 1.33 | 2.28 | 3.46 | 4.85 | 6.44 | 8.22 | 10.17 | 12.29 | 14.57 | 17.00 |
| 10 MPH | .00 | .32 | 1.23 | 2.65 | 4.55 | 6.91 | 9.70 | 12.89 | 16.44 | 20.35 | 24.59 | 29.14 | 34.01 |
| 20 MPH | .00 | .64 | 2.47 | 5.30 | 9.10 | 13.83 | 19.41 | 25.78 | 32.89 | 40.70 | 49.17 | 58.28 | 68.02 |
| 30 MPH | .00 | .96 | 3.70 | 7.95 | 13.65 | 20.74 | 29.11 | 38.67 | 49.33 | 61.05 | 73.76 | 87.43 | 102.02 |
| VELOCITY (FPS) | 1300. | 1239. | 1184. | 1136. | 1095. | 1059. | 1028. | 1000. | 975. | 952. | 931. | 912. | 894. |
| ENERGY (FT-LB) | 762. | 692. | 632. | 582. | 540. | 505. | 476. | 451. | 428. | 409. | 391. | 375. | 360. |
| DROP (IN) | .00 | -.64 | -2.70 | -6.31 | -11.59 | -18.69 | -27.71 | -38.79 | -52.03 | -67.55 | -85.42 | -105.77 | -128.73 |
| MID-RANGE (IN) | .00 | .17 | .71 | 1.67 | 3.10 | 5.04 | 7.55 | 10.66 | 14.42 | 18.87 | 24.02 | 29.94 | 36.65 |
| BULLET PATH (IN) | -.63 | .40 | .00 | -1.94 | -5.56 | -10.99 | -18.35 | -27.76 | -39.34 | -53.19 | -69.40 | -88.08 | -109.37 |
| TIME OF FLIGHT (SEC) | .000000 | .059112 | .121046 | .185722 | .253000 | .322690 | .394610 | .468613 | .544586 | .622439 | .702100 | .783514 | .866636 |
| WIND DEFLECTION (IN) | | | | | | | | | | | | | |
| 5 MPH | .00 | .12 | .50 | 1.11 | 1.96 | 3.01 | 4.26 | 5.70 | 7.31 | 9.08 | 11.02 | 13.10 | 15.34 |
| 10 MPH | .00 | .25 | 1.00 | 2.23 | 3.91 | 6.02 | 8.53 | 11.40 | 14.62 | 18.16 | 22.03 | 26.21 | 30.68 |
| 20 MPH | .00 | .50 | 1.99 | 4.45 | 7.83 | 12.05 | 17.06 | 22.80 | 29.23 | 36.33 | 44.06 | 52.41 | 61.36 |
| 30 MPH | .00 | .75 | 2.99 | 6.68 | 11.74 | 18.07 | 25.58 | 34.20 | 43.85 | 54.49 | 66.09 | 78.62 | 92.05 |
| VELOCITY (FPS) | 1200. | 1150. | 1106. | 1069. | 1037. | 1008. | 982. | 959. | 937. | 917. | 899. | 881. | 865. |
| ENERGY (FT-LB) | 649. | 596. | 552. | 515. | 484. | 458. | 435. | 414. | 396. | 379. | 364. | 350. | 337. |
| DROP (IN) | .00 | -.76 | -3.16 | -7.32 | -13.38 | -21.46 | -31.67 | -44.12 | -58.91 | -76.13 | -95.92 | -118.39 | -143.65 |
| MID-RANGE (IN) | .00 | .20 | .82 | 1.92 | 3.54 | 5.72 | 8.51 | 11.95 | 16.07 | 20.90 | 26.50 | 32.91 | 40.16 |
| BULLET PATH (IN) | -.63 | .50 | .00 | -2.27 | -6.44 | -12.63 | -20.94 | -31.51 | -44.40 | -59.73 | -77.63 | -98.20 | -121.57 |
| TIME OF FLIGHT (SEC) | .000000 | .063871 | .130392 | .199384 | .270656 | .344051 | .419447 | .496747 | .575875 | .656770 | .739385 | .823685 | .909641 |
| WIND DEFLECTION (IN) | | | | | | | | | | | | | |
| 5 MPH | .00 | .12 | .47 | 1.05 | 1.82 | 2.78 | 3.91 | 5.21 | 6.68 | 8.30 | 10.07 | 11.98 | 14.05 |
| 10 MPH | .00 | .24 | .95 | 2.09 | 3.64 | 5.55 | 7.82 | 10.43 | 13.35 | 16.59 | 20.13 | 23.97 | 28.10 |
| 20 MPH | .00 | .48 | 1.90 | 4.18 | 7.27 | 11.11 | 15.65 | 20.85 | 26.71 | 33.18 | 40.26 | 47.94 | 56.19 |
| 30 MPH | .00 | .72 | 2.85 | 6.27 | 10.91 | 16.66 | 23.47 | 31.28 | 40.06 | 49.77 | 60.40 | 71.91 | 84.29 |
| VELOCITY (FPS) | 1100. | 1063. | 1032. | 1004. | 978. | 955. | 934. | 914. | 896. | 879. | 862. | 846. | 831. |
| ENERGY (FT-LB) | 545. | 510. | 480. | 454. | 431. | 411. | 393. | 377. | 362. | 348. | 335. | 323. | 311. |
| DROP (IN) | .00 | -.89 | -3.70 | -8.54 | -15.53 | -24.79 | -36.38 | -50.44 | -67.08 | -86.42 | -108.57 | -133.63 | -161.72 |
| MID-RANGE (IN) | .00 | .23 | .96 | 2.22 | 4.07 | 6.54 | 9.65 | 13.46 | 18.01 | 23.34 | 29.50 | 36.53 | 44.48 |
| BULLET PATH (IN) | -.63 | .64 | .00 | -2.68 | -7.50 | -14.60 | -24.03 | -35.92 | -50.40 | -67.57 | -87.56 | -110.46 | -136.38 |
| TIME OF FLIGHT (SEC) | .000000 | .069368 | .140988 | .214709 | .290414 | .368009 | .447421 | .528592 | .611476 | .696039 | .782255 | .870108 | .959586 |
| WIND DEFLECTION (IN) | | | | | | | | | | | | | |
| 5 MPH | .00 | .10 | .41 | .89 | 1.56 | 2.38 | 3.37 | 4.52 | 5.81 | 7.25 | 8.84 | 10.57 | 12.44 |
| 10 MPH | .00 | .21 | .81 | 1.79 | 3.11 | 4.77 | 6.75 | 9.03 | 11.62 | 14.50 | 17.68 | 21.14 | 24.89 |
| 20 MPH | .00 | .42 | 1.63 | 3.58 | 6.23 | 9.54 | 13.49 | 18.06 | 23.24 | 29.01 | 35.35 | 42.28 | 49.77 |
| 30 MPH | .00 | .63 | 2.44 | 5.37 | 9.34 | 14.31 | 20.24 | 27.10 | 34.86 | 43.51 | 53.03 | 63.42 | 74.66 |

# Bullet: Lyman #410459 203 Gr.
## Ballistic Coefficient: .150

| RANGE (YDS) | MUZZLE | 25 | 50 | 75 | 100 | 125 | 150 | 175 | 200 | 225 | 250 | 275 | 300 |
|---|---|---|---|---|---|---|---|---|---|---|---|---|---|
| VELOCITY (FPS) | 1000. | 975. | 952. | 931. | 912. | 893. | 876. | 860. | 844. | 829. | 814. | 800. | 784. |
| ENERGY (FT-LB) | 451. | 428. | 409. | 391. | 375. | 360. | 346. | 333. | 321. | 310. | 299. | 289. | 277. |
| DROP (IN) | .00 | -1.06 | -4.40 | -10.09 | -18.27 | -29.05 | -42.55 | -58.85 | -78.09 | -100.37 | -125.80 | -154.50 | -186.48 |
| MID-RANGE (IN) | .00 | .28 | 1.14 | 2.61 | 4.73 | 7.58 | 11.18 | 15.56 | 20.78 | 26.86 | 33.85 | 41.80 | 50.70 |
| BULLET PATH (IN) | -.63 | .82 | .00 | -3.17 | -8.83 | -17.10 | -28.08 | -41.87 | -58.59 | -78.36 | -101.27 | -127.46 | -156.92 |
| TIME OF FLIGHT (SEC) | .000000 | .075968 | .153815 | .233469 | .314875 | .397989 | .482777 | .569215 | .657287 | .746982 | .838297 | .931235 | 1.025979 |
| WIND DEFLECTION (IN) | | | | | | | | | | | | | |
| 5 MPH | .00 | .09 | .34 | .75 | 1.31 | 2.02 | 2.88 | 3.89 | 5.04 | 6.33 | 7.77 | 9.35 | 11.09 |
| 10 MPH | .00 | .17 | .67 | 1.49 | 2.62 | 4.05 | 5.77 | 7.78 | 10.08 | 12.67 | 15.54 | 18.70 | 22.17 |
| 20 MPH | .00 | .34 | 1.34 | 2.98 | 5.24 | 8.09 | 11.54 | 15.56 | 20.16 | 25.34 | 31.08 | 37.39 | 44.34 |
| 30 MPH | .00 | .51 | 2.01 | 4.47 | 7.85 | 12.14 | 17.31 | 23.35 | 30.25 | 38.01 | 46.62 | 56.09 | 66.52 |
| VELOCITY (FPS) | 900. | 882. | 866. | 850. | 834. | 820. | 805. | 790. | 774. | 758. | 743. | 729. | 714. |
| ENERGY (FT-LB) | 365. | 351. | 338. | 325. | 314. | 303. | 292. | 281. | 270. | 259. | 249. | 239. | 230. |
| DROP (IN) | .00 | -1.23 | -5.25 | -12.15 | -22.05 | -35.07 | -51.30 | -70.83 | -93.76 | -120.30 | -150.61 | -184.84 | -223.16 |
| MID-RANGE (IN) | .00 | .34 | 1.39 | 3.19 | 5.78 | 9.20 | 13.50 | 18.69 | 24.83 | 31.99 | 40.24 | 49.65 | 60.28 |
| BULLET PATH (IN) | -.63 | 1.08 | -.00 | -3.96 | -10.93 | -21.00 | -34.30 | -50.88 | -70.87 | -94.48 | -121.85 | -153.14 | -188.51 |
| TIME OF FLIGHT (SEC) | .000000 | .084171 | .169999 | .257465 | .346558 | .437270 | .529602 | .623611 | .719570 | .817530 | .917504 | 1.019513 | 1.123579 |
| WIND DEFLECTION (IN) | | | | | | | | | | | | | |
| 5 MPH | .00 | .07 | .29 | .66 | 1.16 | 1.81 | 2.60 | 3.54 | 4.66 | 5.94 | 7.41 | 9.05 | 10.87 |
| 10 MPH | .00 | .15 | .59 | 1.31 | 2.33 | 3.63 | 5.21 | 7.09 | 9.31 | 11.89 | 14.81 | 18.10 | 21.75 |
| 20 MPH | .00 | .29 | 1.17 | 2.63 | 4.65 | 7.25 | 10.42 | 14.18 | 18.62 | 23.77 | 29.63 | 36.20 | 43.50 |
| 30 MPH | .00 | .44 | 1.76 | 3.94 | 6.98 | 10.88 | 15.63 | 21.27 | 27.93 | 35.66 | 44.44 | 54.30 | 65.25 |

# Bullet: Lyman #410426 234 Gr.
## Ballistic Coefficient: .235

| RANGE (YDS) | MUZZLE | 25 | 50 | 75 | 100 | 125 | 150 | 175 | 200 | 225 | 250 | 275 | 300 |
|---|---|---|---|---|---|---|---|---|---|---|---|---|---|
| VELOCITY (FPS) | 1400. | 1342. | 1288. | 1238. | 1192. | 1151. | 1114. | 1082. | 1053. | 1028. | 1005. | 983. | 964. |
| ENERGY (FT-LB) | 1018. | 936. | 862. | 796. | 738. | 688. | 645. | 608. | 576. | 549. | 524. | 502. | 483. |
| DROP (IN) | .00 | -.54 | -2.29 | -5.34 | -9.83 | -15.84 | -23.48 | -32.87 | -44.11 | -57.30 | -72.54 | -89.93 | -109.56 |
| MID-RANGE (IN) | .00 | .14 | .60 | 1.41 | 2.62 | 4.27 | 6.40 | 9.03 | 12.23 | 16.02 | 20.43 | 25.52 | 31.30 |
| BULLET PATH (IN) | -.63 | .29 | .00 | -1.59 | -4.62 | -9.17 | -15.35 | -23.28 | -33.06 | -44.79 | -58.57 | -74.50 | -92.67 |
| TIME OF FLIGHT (SEC) | .000000 | .054722 | .111777 | .171186 | .232937 | .296979 | .363224 | .431551 | .501826 | .573927 | .647757 | .723236 | .800299 |
| WIND DEFLECTION (IN) | | | | | | | | | | | | | |
| 5 MPH | .00 | .10 | .41 | .92 | 1.64 | 2.56 | 3.68 | 4.98 | 6.45 | 8.08 | 9.86 | 11.79 | 13.85 |
| 10 MPH | .00 | .20 | .82 | 1.84 | 3.28 | 5.13 | 7.36 | 9.95 | 12.89 | 16.15 | 19.72 | 23.58 | 27.71 |
| 20 MPH | .00 | .41 | 1.63 | 3.69 | 6.57 | 10.25 | 14.71 | 19.91 | 25.79 | 32.31 | 39.44 | 47.15 | 55.42 |
| 30 MPH | .00 | .61 | 2.45 | 5.53 | 9.85 | 15.38 | 22.07 | 29.86 | 38.68 | 48.46 | 59.16 | 70.73 | 83.13 |
| VELOCITY (FPS) | 1300. | 1249. | 1202. | 1160. | 1122. | 1089. | 1060. | 1033. | 1010. | 988. | 968. | 950. | 932. |
| ENERGY (FT-LB) | 878. | 810. | 751. | 699. | 654. | 616. | 583. | 555. | 530. | 507. | 487. | 468. | 452. |
| DROP (IN) | .00 | -.63 | -2.66 | -6.19 | -11.34 | -18.20 | -26.89 | -37.51 | -50.16 | -64.93 | -81.93 | -101.23 | -122.88 |
| MID-RANGE (IN) | .00 | .17 | .70 | 1.63 | 3.01 | 4.87 | 7.25 | 10.20 | 13.74 | 17.92 | 22.76 | 28.30 | 34.55 |
| BULLET PATH (IN) | -.63 | .39 | .00 | -1.89 | -5.39 | -10.61 | -17.65 | -26.62 | -37.63 | -50.75 | -66.10 | -83.76 | -103.76 |
| TIME OF FLIGHT (SEC) | .000000 | .058871 | .120090 | .183614 | .249364 | .317227 | .387068 | .458760 | .532202 | .607309 | .684014 | .762259 | .841998 |
| WIND DEFLECTION (IN) | | | | | | | | | | | | | |
| 5 MPH | .00 | .10 | .41 | .93 | 1.64 | 2.53 | 3.60 | 4.83 | 6.22 | 7.75 | 9.42 | 11.23 | 13.17 |
| 10 MPH | .00 | .21 | .83 | 1.85 | 3.27 | 5.06 | 7.20 | 9.66 | 12.44 | 15.50 | 18.85 | 22.47 | 26.35 |
| 20 MPH | .00 | .41 | 1.66 | 3.71 | 6.55 | 10.13 | 14.40 | 19.33 | 24.87 | 31.00 | 37.70 | 44.93 | 52.69 |
| 30 MPH | .00 | .62 | 2.48 | 5.56 | 9.82 | 15.19 | 21.60 | 28.99 | 37.31 | 46.51 | 56.54 | 67.40 | 79.04 |
| VELOCITY (FPS) | 1200. | 1158. | 1120. | 1087. | 1058. | 1032. | 1009. | 987. | 967. | 949. | 931. | 915. | 900. |
| ENERGY (FT-LB) | 748. | 697. | 652. | 614. | 582. | 553. | 528. | 506. | 486. | 468. | 451. | 435. | 421. |
| DROP (IN) | .00 | -.75 | -3.12 | -7.21 | -13.13 | -20.99 | -30.89 | -42.91 | -57.15 | -73.72 | -92.63 | -114.03 | -138.02 |
| MID-RANGE (IN) | .00 | .20 | .81 | 1.88 | 3.45 | 5.55 | 8.23 | 11.52 | 15.44 | 20.04 | 25.32 | 31.34 | 38.13 |
| BULLET PATH (IN) | -.63 | .49 | .00 | -2.22 | -6.27 | -12.25 | -20.27 | -30.42 | -42.79 | -57.48 | -74.52 | -94.05 | -116.16 |
| TIME OF FLIGHT (SEC) | .000000 | .063641 | .129502 | .197470 | .267408 | .339191 | .412719 | .487909 | .564692 | .643013 | .722826 | .804092 | .886782 |
| WIND DEFLECTION (IN) | | | | | | | | | | | | | |
| 5 MPH | .00 | .10 | .40 | .88 | 1.53 | 2.35 | 3.32 | 4.44 | 5.69 | 7.09 | 8.61 | 10.26 | 12.04 |
| 10 MPH | .00 | .20 | .79 | 1.75 | 3.06 | 4.70 | 6.64 | 8.87 | 11.39 | 14.17 | 17.22 | 20.52 | 24.07 |
| 20 MPH | .00 | .40 | 1.58 | 3.51 | 6.13 | 9.40 | 13.28 | 17.74 | 22.77 | 28.34 | 34.43 | 41.04 | 48.15 |
| 30 MPH | .00 | .60 | 2.38 | 5.26 | 9.19 | 14.09 | 19.92 | 26.62 | 34.16 | 42.51 | 51.65 | 61.56 | 72.22 |
| VELOCITY (FPS) | 1100. | 1069. | 1042. | 1018. | 995. | 975. | 956. | 938. | 921. | 906. | 891. | 876. | 863. |
| ENERGY (FT-LB) | 629. | 594. | 564. | 538. | 515. | 494. | 475. | 457. | 441. | 426. | 412. | 399. | 387. |
| DROP (IN) | .00 | -.88 | -3.66 | -8.43 | -15.30 | -24.35 | -35.68 | -49.34 | -65.44 | -84.09 | -105.38 | -129.39 | -156.22 |
| MID-RANGE (IN) | .00 | .23 | .95 | 2.19 | 3.99 | 6.38 | 9.41 | 13.07 | 17.42 | 22.50 | 28.34 | 34.97 | 42.43 |
| BULLET PATH (IN) | -.63 | .63 | .00 | -2.63 | -7.35 | -14.25 | -23.44 | -34.95 | -48.91 | -65.41 | -84.55 | -106.42 | -131.11 |
| TIME OF FLIGHT (SEC) | .000000 | .069171 | .140234 | .213078 | .287616 | .363772 | .441486 | .520709 | .601399 | .683524 | .767056 | .851975 | .938263 |
| WIND DEFLECTION (IN) | | | | | | | | | | | | | |
| 5 MPH | .00 | .09 | .34 | .75 | 1.31 | 2.01 | 2.85 | 3.82 | 4.92 | 6.15 | 7.50 | 8.97 | 10.57 |
| 10 MPH | .00 | .17 | .68 | 1.50 | 2.62 | 4.02 | 5.70 | 7.64 | 9.85 | 12.30 | 15.00 | 17.95 | 21.13 |
| 20 MPH | .00 | .35 | 1.36 | 3.00 | 5.24 | 8.05 | 11.40 | 15.29 | 19.69 | 24.60 | 30.00 | 35.90 | 42.27 |
| 30 MPH | .00 | .52 | 2.04 | 4.51 | 7.86 | 12.07 | 17.10 | 22.93 | 29.54 | 36.90 | 45.01 | 53.84 | 63.40 |

## Bullet: Lyman #410426 234 Gr.
### Ballistic Coefficient: .235

| RANGE (YDS) | MUZZLE | 25 | 50 | 75 | 100 | 125 | 150 | 175 | 200 | 225 | 250 | 275 | 300 |
|---|---|---|---|---|---|---|---|---|---|---|---|---|---|
| VELOCITY (FPS) | 1000. | 979. | 960. | 942. | 925. | 909. | 894. | 879. | 866. | 852. | 840. | 827. | 815. |
| ENERGY (FT-LB) | 519. | 498. | 479. | 461. | 444. | 429. | 415. | 402. | 389. | 377. | 366. | 355. | 345. |
| DROP (IN) | .00 | -1.05 | -4.36 | -10.00 | -18.05 | -28.62 | -41.81 | -57.70 | -76.39 | -97.97 | -122.54 | -150.19 | -181.00 |
| MID-RANGE (IN) | .00 | .28 | 1.13 | 2.58 | 4.66 | 7.42 | 10.90 | 15.14 | 20.15 | 25.98 | 32.65 | 40.21 | 48.69 |
| BULLET PATH (IN) | -.63 | .81 | .00 | -3.14 | -8.70 | -16.77 | -27.47 | -40.86 | -57.06 | -76.15 | -98.22 | -123.37 | -151.69 |
| TIME OF FLIGHT (SEC) | .000000 | .075805 | .153181 | .232075 | .312444 | .394254 | .477476 | .562088 | .648072 | .735417 | .824112 | .914154 | 1.005540 |
| WIND DEFLECTION (IN) | | | | | | | | | | | | | |
| 5 MPH | .00 | .07 | .28 | .62 | 1.10 | 1.69 | 2.42 | 3.26 | 4.23 | 5.32 | 6.52 | 7.85 | 9.29 |
| 10 MPH | .00 | .14 | .56 | 1.25 | 2.19 | 3.39 | 4.84 | 6.53 | 8.46 | 10.63 | 13.04 | 15.69 | 18.58 |
| 20 MPH | .00 | .28 | 1.12 | 2.49 | 4.38 | 6.78 | 9.67 | 13.05 | 16.92 | 21.27 | 26.09 | 31.38 | 37.15 |
| 30 MPH | .00 | .43 | 1.68 | 3.74 | 6.57 | 10.17 | 14.51 | 19.58 | 25.38 | 31.90 | 39.13 | 47.07 | 55.73 |
| VELOCITY (FPS) | 900. | 885. | 871. | 858. | 845. | 832. | 820. | 808. | 797. | 785. | 775. | 764. | 753. |
| ENERGY (FT-LB) | 421. | 407. | 394. | 382. | 371. | 360. | 349. | 339. | 330. | 320. | 312. | 303. | 295. |
| DROP (IN) | .00 | -1.19 | -5.15 | -11.95 | -21.70 | -34.49 | -50.42 | -69.56 | -92.02 | -117.69 | -146.86 | -179.64 | -216.14 |
| MID-RANGE (IN) | .00 | .34 | 1.38 | 3.16 | 5.71 | 9.06 | 13.26 | 18.33 | 24.31 | 31.13 | 38.94 | 47.76 | 57.65 |
| BULLET PATH (IN) | -.63 | 1.06 | -.00 | -3.92 | -10.78 | -20.69 | -33.72 | -49.98 | -69.55 | -92.33 | -118.61 | -148.51 | -182.12 |
| TIME OF FLIGHT (SEC) | .000000 | .084029 | .169436 | .256208 | .344333 | .433804 | .524620 | .616781 | .710288 | .805147 | .901362 | .998939 | 1.097885 |
| WIND DEFLECTION (IN) | | | | | | | | | | | | | |
| 5 MPH | .00 | .06 | .24 | .55 | .97 | 1.51 | 2.17 | 2.94 | 3.84 | 4.85 | 5.99 | 7.24 | 8.61 |
| 10 MPH | .00 | .12 | .49 | 1.09 | 1.94 | 3.02 | 4.33 | 5.89 | 7.68 | 9.71 | 11.97 | 14.48 | 17.23 |
| 20 MPH | .00 | .24 | .97 | 2.19 | 3.87 | 6.03 | 8.67 | 11.77 | 15.35 | 19.41 | 23.95 | 28.96 | 34.46 |
| 30 MPH | .00 | .37 | 1.46 | 3.28 | 5.81 | 9.05 | 13.00 | 17.66 | 23.03 | 29.12 | 35.92 | 43.44 | 51.68 |

# Bullet: Sierra .4295 Dia. 180 Gr. JHC
## Ballistic Coefficient: .133

| RANGE (YDS) | MUZZLE | 25 | 50 | 75 | 100 | 125 | 150 | 175 | 200 | 225 | 250 | 275 | 300 |
|---|---|---|---|---|---|---|---|---|---|---|---|---|---|
| VELOCITY (FPS) | 1800. | 1669. | 1546. | 1433. | 1330. | 1238. | 1161. | 1097. | 1046. | 1003. | 967. | 935. | 907. |
| ENERGY (FT-LB) | 1295. | 1113. | 956. | 821. | 707. | 613. | 538. | 481. | 437. | 402. | 374. | 350. | 328. |
| DROP (IN) | .00 | -.33 | -1.45 | -3.47 | -6.56 | -10.88 | -16.61 | -23.96 | -33.11 | -44.23 | -57.51 | -73.09 | -91.13 |
| MID-RANGE (IN) | .00 | .09 | .39 | .95 | 1.83 | 3.10 | 4.83 | 7.11 | 9.99 | 13.56 | 17.88 | 22.99 | 28.95 |
| BULLET PATH (IN) | -.63 | .07 | .00 | -.99 | -3.04 | -6.31 | -11.01 | -17.32 | -25.43 | -35.52 | -47.76 | -62.30 | -79.30 |
| TIME OF FLIGHT (SEC) | .000000 | .043274 | .089963 | .140360 | .194719 | .253202 | .315812 | .382347 | .452437 | .525706 | .601885 | .680783 | .762267 |
| WIND DEFLECTION (IN) | | | | | | | | | | | | | |
| 5 MPH | .00 | .14 | .58 | 1.35 | 2.47 | 3.95 | 5.79 | 7.98 | 10.48 | 13.26 | 16.30 | 19.58 | 23.08 |
| 10 MPH | .00 | .28 | 1.17 | 2.70 | 4.94 | 7.90 | 11.58 | 15.96 | 20.96 | 26.52 | 32.60 | 39.15 | 46.16 |
| 20 MPH | .00 | .57 | 2.33 | 5.41 | 9.87 | 15.79 | 23.17 | 31.92 | 41.92 | 53.05 | 65.20 | 78.30 | 92.32 |
| 30 MPH | .00 | .85 | 3.50 | 8.11 | 14.81 | 23.69 | 34.75 | 47.88 | 62.89 | 79.57 | 97.80 | 117.45 | 138.48 |
| VELOCITY (FPS) | 1700. | 1575. | 1459. | 1354. | 1259. | 1178. | 1111. | 1057. | 1013. | 975. | 943. | 913. | 887. |
| ENERGY (FT-LB) | 1155. | 992. | 851. | 732. | 634. | 555. | 493. | 446. | 410. | 380. | 355. | 333. | 314. |
| DROP (IN) | .00 | -.37 | -1.63 | -3.91 | -7.37 | -12.21 | -18.61 | -26.77 | -36.86 | -49.07 | -63.55 | -80.43 | -99.92 |
| MID-RANGE (IN) | .00 | .10 | .44 | 1.07 | 2.06 | 3.48 | 5.40 | 7.91 | 11.06 | 14.93 | 19.57 | 25.01 | 31.35 |
| BULLET PATH (IN) | -.63 | .12 | .00 | -1.15 | -3.49 | -7.20 | -12.47 | -19.50 | -28.47 | -39.54 | -52.90 | -68.65 | -87.00 |
| TIME OF FLIGHT (SEC) | .000000 | .045840 | .095320 | .148706 | .206189 | .267817 | .333436 | .402708 | .475240 | .550738 | .628996 | .709867 | .793249 |
| WIND DEFLECTION (IN) | | | | | | | | | | | | | |
| 5 MPH | .00 | .15 | .62 | 1.44 | 2.62 | 4.16 | 6.05 | 8.26 | 10.76 | 13.52 | 16.53 | 19.76 | 23.22 |
| 10 MPH | .00 | .30 | 1.25 | 2.88 | 5.23 | 8.31 | 12.10 | 16.52 | 21.52 | 27.05 | 33.06 | 39.52 | 46.44 |
| 20 MPH | .00 | .61 | 2.49 | 5.76 | 10.46 | 16.62 | 24.19 | 33.05 | 43.05 | 54.10 | 66.11 | 79.05 | 92.87 |
| 30 MPH | .00 | .91 | 3.74 | 8.63 | 15.69 | 24.94 | 36.29 | 49.57 | 64.57 | 81.14 | 99.17 | 118.57 | 139.31 |
| VELOCITY (FPS) | 1600. | 1482. | 1374. | 1277. | 1193. | 1123. | 1067. | 1021. | 983. | 949. | 919. | 892. | 867. |
| ENERGY (FT-LB) | 1023. | 878. | 755. | 652. | 569. | 504. | 455. | 417. | 386. | 360. | 337. | 318. | 300. |
| DROP (IN) | .00 | -.42 | -1.85 | -4.42 | -8.33 | -13.76 | -20.91 | -29.96 | -41.08 | -54.45 | -70.19 | -88.49 | -109.52 |
| MID-RANGE (IN) | .00 | .11 | .50 | 1.21 | 2.32 | 3.90 | 6.03 | 8.79 | 12.22 | 16.39 | 21.34 | 27.14 | 33.85 |
| BULLET PATH (IN) | -.63 | .18 | .00 | -1.33 | -4.00 | -8.20 | -14.10 | -21.92 | -31.80 | -43.94 | -58.43 | -75.50 | -95.28 |
| TIME OF FLIGHT (SEC) | .000000 | .048715 | .101286 | .157924 | .218712 | .283540 | .352100 | .423995 | .498907 | .576616 | .656963 | .739839 | .825168 |
| WIND DEFLECTION (IN) | | | | | | | | | | | | | |
| 5 MPH | .00 | .16 | .66 | 1.52 | 2.75 | 4.33 | 6.23 | 8.44 | 10.90 | 13.62 | 16.56 | 19.73 | 23.11 |
| 10 MPH | .00 | .32 | 1.33 | 3.04 | 5.49 | 8.65 | 12.47 | 16.87 | 21.81 | 27.23 | 33.13 | 39.46 | 46.23 |
| 20 MPH | .00 | .65 | 2.65 | 6.09 | 10.99 | 17.31 | 24.94 | 33.75 | 43.62 | 54.47 | 66.25 | 78.92 | 92.46 |
| 30 MPH | .00 | .97 | 3.98 | 9.13 | 16.48 | 25.96 | 37.41 | 50.62 | 65.42 | 81.70 | 99.38 | 118.38 | 138.69 |
| VELOCITY (FPS) | 1500. | 1390. | 1292. | 1206. | 1133. | 1075. | 1028. | 988. | 954. | 923. | 896. | 871. | 847. |
| ENERGY (FT-LB) | 899. | 773. | 667. | 581. | 513. | 462. | 422. | 390. | 364. | 341. | 321. | 303. | 287. |
| DROP (IN) | .00 | -.49 | -2.11 | -5.03 | -9.45 | -15.55 | -23.53 | -33.56 | -45.80 | -60.39 | -77.52 | -97.34 | -120.02 |
| MID-RANGE (IN) | .00 | .13 | .56 | 1.36 | 2.61 | 4.38 | 6.73 | 9.74 | 13.45 | 17.92 | 23.21 | 29.38 | 36.48 |
| BULLET PATH (IN) | -.63 | .25 | .00 | -1.55 | -4.60 | -9.33 | -15.94 | -24.59 | -35.47 | -48.69 | -64.45 | -82.90 | -104.21 |
| TIME OF FLIGHT (SEC) | .000000 | .051951 | .107944 | .168085 | .232296 | .300299 | .371696 | .446152 | .523435 | .603377 | .685862 | .770811 | .858171 |
| WIND DEFLECTION (IN) | | | | | | | | | | | | | |
| 5 MPH | .00 | .17 | .70 | 1.59 | 2.84 | 4.43 | 6.31 | 8.46 | 10.86 | 13.50 | 16.36 | 19.43 | 22.72 |
| 10 MPH | .00 | .34 | 1.40 | 3.18 | 5.68 | 8.85 | 12.62 | 16.92 | 21.72 | 26.99 | 32.71 | 38.86 | 45.44 |
| 20 MPH | .00 | .69 | 2.80 | 6.37 | 11.37 | 17.71 | 25.24 | 33.85 | 43.45 | 53.99 | 65.42 | 77.73 | 90.88 |
| 30 MPH | .00 | 1.03 | 4.19 | 9.55 | 17.05 | 26.56 | 37.86 | 50.77 | 65.17 | 80.98 | 98.14 | 116.59 | 136.31 |

## Bullet: Sierra .4295 Dia. 180 Gr. JHC
## Ballistic Coefficient: .133

| RANGE (YDS) | MUZZLE | 25 | 50 | 75 | 100 | 125 | 150 | 175 | 200 | 225 | 250 | 275 | 300 |
|---|---|---|---|---|---|---|---|---|---|---|---|---|---|
| VELOCITY (FPS) | 1400. | 1300. | 1213. | 1140. | 1080. | 1032. | 992. | 957. | 926. | 898. | 873. | 849. | 827. |
| ENERGY (FT-LB) | 783. | 676. | 588. | 519. | 466. | 425. | 393. | 366. | 343. | 323. | 305. | 288. | 273. |
| DROP (IN) | .00 | -.57 | -2.42 | -5.75 | -10.75 | -17.61 | -26.50 | -37.60 | -51.03 | -66.98 | -85.61 | -107.08 | -131.55 |
| MID-RANGE (IN) | .00 | .15 | .64 | 1.56 | 2.96 | 4.92 | 7.50 | 10.77 | 14.76 | 19.55 | 25.19 | 31.75 | 39.27 |
| BULLET PATH (IN) | -.63 | .33 | .00 | -1.81 | -5.28 | -10.61 | -17.98 | -27.55 | -39.46 | -53.88 | -70.99 | -90.93 | -113.88 |
| TIME OF FLIGHT (SEC) | .000000 | .055616 | .115377 | .179225 | .246897 | .317998 | .392185 | .469216 | .548920 | .631176 | .715901 | .803041 | .892567 |
| WIND DEFLECTION (IN) | | | | | | | | | | | | | |
| 5 MPH | .00 | .18 | .72 | 1.63 | 2.87 | 4.41 | 6.23 | 8.29 | 10.59 | 13.11 | 15.86 | 18.81 | 21.97 |
| 10 MPH | .00 | .36 | 1.45 | 3.26 | 5.74 | 8.82 | 12.45 | 16.58 | 21.18 | 26.23 | 31.71 | 37.62 | 43.95 |
| 20 MPH | .00 | .72 | 2.90 | 6.52 | 11.48 | 17.65 | 24.91 | 33.16 | 42.36 | 52.46 | 63.43 | 75.24 | 87.90 |
| 30 MPH | .00 | 1.08 | 4.35 | 9.77 | 17.22 | 26.47 | 37.36 | 49.75 | 63.54 | 78.69 | 95.14 | 112.86 | 131.85 |
| VELOCITY (FPS) | 1300. | 1213. | 1139. | 1080. | 1032. | 991. | 957. | 926. | 898. | 873. | 849. | 827. | 806. |
| ENERGY (FT-LB) | 675. | 588. | 519. | 466. | 425. | 393. | 366. | 343. | 322. | 304. | 288. | 273. | 260. |
| DROP (IN) | .00 | -.66 | -2.80 | -6.60 | -12.27 | -19.97 | -29.87 | -42.11 | -56.87 | -74.30 | -94.58 | -117.87 | -144.32 |
| MID-RANGE (IN) | .00 | .17 | .74 | 1.77 | 3.34 | 5.50 | 8.33 | 11.86 | 16.17 | 21.30 | 27.32 | 34.30 | 42.29 |
| BULLET PATH (IN) | -.63 | .42 | .00 | -2.09 | -6.04 | -12.03 | -20.22 | -30.75 | -43.79 | -59.51 | -78.08 | -99.65 | -124.39 |
| TIME OF FLIGHT (SEC) | .000000 | .059772 | .123630 | .191311 | .262420 | .336614 | .413652 | .493362 | .575624 | .660354 | .747499 | .837029 | .928936 |
| WIND DEFLECTION (IN) | | | | | | | | | | | | | |
| 5 MPH | .00 | .18 | .73 | 1.60 | 2.79 | 4.24 | 5.94 | 7.88 | 10.04 | 12.42 | 15.01 | 17.81 | 20.82 |
| 10 MPH | .00 | .37 | 1.45 | 3.21 | 5.57 | 8.47 | 11.88 | 15.75 | 20.08 | 24.84 | 30.02 | 35.62 | 41.65 |
| 20 MPH | .00 | .73 | 2.90 | 6.42 | 11.14 | 16.95 | 23.76 | 31.51 | 40.16 | 49.68 | 60.04 | 71.25 | 83.29 |
| 30 MPH | .00 | 1.10 | 4.35 | 9.63 | 16.71 | 25.42 | 35.64 | 47.26 | 60.24 | 74.51 | 90.06 | 106.87 | 124.94 |
| VELOCITY (FPS) | 1200. | 1129. | 1071. | 1025. | 986. | 952. | 921. | 894. | 869. | 846. | 824. | 803. | 782. |
| ENERGY (FT-LB) | 575. | 509. | 459. | 420. | 388. | 362. | 339. | 319. | 302. | 286. | 271. | 257. | 245. |
| DROP (IN) | .00 | -.77 | -3.25 | -7.61 | -14.03 | -22.68 | -33.69 | -47.25 | -63.51 | -82.64 | -104.81 | -130.17 | -158.79 |
| MID-RANGE (IN) | .00 | .20 | .85 | 2.02 | 3.77 | 6.16 | 9.24 | 13.08 | 17.73 | 23.26 | 29.72 | 37.20 | 45.70 |
| BULLET PATH (IN) | -.63 | .53 | .00 | -2.42 | -6.90 | -13.62 | -22.69 | -34.31 | -48.63 | -65.83 | -86.05 | -109.47 | -136.16 |
| TIME OF FLIGHT (SEC) | .000000 | .064495 | .132756 | .204381 | .279047 | .356524 | .436651 | .519314 | .604434 | .691963 | .781875 | .874163 | .968855 |
| WIND DEFLECTION (IN) | | | | | | | | | | | | | |
| 5 MPH | .00 | .18 | .68 | 1.49 | 2.56 | 3.87 | 5.43 | 7.20 | 9.19 | 11.39 | 13.81 | 16.43 | 19.26 |
| 10 MPH | .00 | .35 | 1.36 | 2.97 | 5.11 | 7.75 | 10.85 | 14.40 | 18.38 | 22.79 | 27.61 | 32.85 | 38.52 |
| 20 MPH | .00 | .70 | 2.73 | 5.94 | 10.22 | 15.50 | 21.70 | 28.80 | 36.76 | 45.57 | 55.22 | 65.71 | 77.04 |
| 30 MPH | .00 | 1.05 | 4.09 | 8.91 | 15.34 | 23.24 | 32.55 | 43.20 | 55.14 | 68.36 | 82.83 | 98.56 | 115.56 |
| VELOCITY (FPS) | 1100. | 1048. | 1005. | 969. | 937. | 908. | 882. | 858. | 835. | 814. | 793. | 773. | 754. |
| ENERGY (FT-LB) | 484. | 439. | 404. | 375. | 351. | 330. | 311. | 294. | 279. | 264. | 251. | 239. | 227. |
| DROP (IN) | .00 | -.91 | -3.79 | -8.81 | -16.13 | -25.89 | -38.27 | -53.44 | -71.55 | -92.77 | -117.24 | -145.06 | -176.49 |
| MID-RANGE (IN) | .00 | .24 | .99 | 2.31 | 4.28 | 6.92 | 10.33 | 14.54 | 19.65 | 25.70 | 32.74 | 40.79 | 49.97 |
| BULLET PATH (IN) | -.63 | .67 | .00 | -2.81 | -7.92 | -15.47 | -25.65 | -38.60 | -54.50 | -73.51 | -95.78 | -121.39 | -150.61 |
| TIME OF FLIGHT (SEC) | .000000 | .069900 | .142997 | .219015 | .297762 | .379099 | .462931 | .549194 | .637850 | .728883 | .822296 | .918139 | 1.016444 |
| WIND DEFLECTION (IN) | | | | | | | | | | | | | |
| 5 MPH | .00 | .15 | .58 | 1.27 | 2.20 | 3.36 | 4.74 | 6.33 | 8.13 | 10.14 | 12.36 | 14.80 | 17.45 |
| 10 MPH | .00 | .30 | 1.17 | 2.55 | 4.41 | 6.72 | 9.48 | 12.66 | 16.26 | 20.28 | 24.72 | 29.59 | 34.89 |
| 20 MPH | .00 | .60 | 2.33 | 5.09 | 8.81 | 13.44 | 18.95 | 25.32 | 32.52 | 40.57 | 49.45 | 59.18 | 69.79 |
| 30 MPH | .00 | .91 | 3.50 | 7.64 | 13.22 | 20.16 | 28.43 | 37.97 | 48.78 | 60.85 | 74.17 | 88.78 | 104.68 |

## Bullet: Sierra .4295 Dia. 180 Gr. JHC
## Ballistic Coefficient: .133

| RANGE (YDS) | MUZZLE | 25 | 50 | 75 | 100 | 125 | 150 | 175 | 200 | 225 | 250 | 275 | 300 |
|---|---|---|---|---|---|---|---|---|---|---|---|---|---|
| VELOCITY (FPS) | 900. | 874. | 851. | 828. | 807. | 787. | 767. | 748. | 730. | 712. | 695. | 679. | 663. |
| ENERGY (FT-LB) | 324. | 306. | 289. | 274. | 260. | 247. | 235. | 224. | 213. | 203. | 193. | 184. | 176. |
| DROP (IN) | .00 | -1.28 | -5.39 | -12.49 | -22.74 | -36.27 | -53.25 | -73.87 | -98.35 | -126.88 | -159.68 | -196.95 | -238.92 |
| MID-RANGE (IN) | .00 | .34 | 1.42 | 3.28 | 5.99 | 9.58 | 14.14 | 19.72 | 26.43 | 34.33 | 43.57 | 54.16 | 66.21 |
| BULLET PATH (IN) | -.63 | 1.10 | -.00 | -4.09 | -11.34 | -21.86 | -35.83 | -53.44 | -74.91 | -100.44 | -130.23 | -164.49 | -203.46 |
| TIME OF FLIGHT (SEC) | .000000 | .084558 | .171533 | .260893 | .352627 | .446751 | .543317 | .642351 | .743880 | .847936 | .954560 | 1.063797 | 1.175698 |
| WIND DEFLECTION (IN) | | | | | | | | | | | | | |
| 5 MPH | .00 | .11 | .43 | .96 | 1.70 | 2.65 | 3.81 | 5.19 | 6.79 | 8.62 | 10.67 | 12.95 | 15.46 |
| 10 MPH | .00 | .22 | .86 | 1.92 | 3.40 | 5.29 | 7.62 | 10.39 | 13.59 | 17.24 | 21.34 | 25.89 | 30.92 |
| 20 MPH | .00 | .43 | 1.71 | 3.83 | 6.79 | 10.59 | 15.25 | 20.77 | 27.18 | 34.47 | 42.67 | 51.79 | 61.85 |
| 30 MPH | .00 | .65 | 2.57 | 5.75 | 10.19 | 15.88 | 22.87 | 31.16 | 40.77 | 51.71 | 64.01 | 77.68 | 92.77 |
| VELOCITY (FPS) | 800. | 780. | 760. | 742. | 724. | 706. | 689. | 673. | 657. | 642. | 627. | 612. | 598. |
| ENERGY (FT-LB) | 256. | 243. | 231. | 220. | 209. | 199. | 190. | 181. | 173. | 165. | 157. | 150. | 143. |
| DROP (IN) | .00 | -1.50 | -6.56 | -15.35 | -28.08 | -44.95 | -66.16 | -91.85 | -122.33 | -157.83 | -198.58 | -244.86 | -296.92 |
| MID-RANGE (IN) | .00 | .43 | 1.77 | 4.10 | 7.49 | 12.00 | 17.73 | 24.71 | 33.08 | 42.93 | 54.34 | 67.43 | 82.31 |
| BULLET PATH (IN) | -.63 | 1.46 | -.00 | -5.20 | -14.34 | -27.61 | -45.23 | -67.33 | -94.22 | -126.12 | -163.28 | -205.96 | -254.44 |
| TIME OF FLIGHT (SEC) | .000000 | .094972 | .192399 | .292300 | .394706 | .499652 | .607179 | .717335 | .830172 | .945748 | 1.064127 | 1.185373 | 1.309560 |
| WIND DEFLECTION (IN) | | | | | | | | | | | | | |
| 5 MPH | .00 | .11 | .43 | .97 | 1.73 | 2.72 | 3.93 | 5.38 | 7.06 | 8.98 | 11.14 | 13.56 | 16.24 |
| 10 MPH | .00 | .22 | .86 | 1.94 | 3.47 | 5.44 | 7.86 | 10.75 | 14.11 | 17.95 | 22.29 | 27.13 | 32.48 |
| 20 MPH | .00 | .43 | 1.72 | 3.89 | 6.94 | 10.88 | 15.73 | 21.50 | 28.22 | 35.90 | 44.57 | 54.25 | 64.97 |
| 30 MPH | .00 | .65 | 2.59 | 5.83 | 10.40 | 16.32 | 23.59 | 32.25 | 42.33 | 53.86 | 66.86 | 81.38 | 97.45 |
| VELOCITY (FPS) | 700. | 683. | 667. | 651. | 636. | 621. | 606. | 592. | 578. | 565. | 552. | 539. | 526. |
| ENERGY (FT-LB) | 196. | 187. | 178. | 170. | 162. | 154. | 147. | 140. | 134. | 128. | 122. | 116. | 111. |
| DROP (IN) | .00 | -1.94 | -8.51 | -19.95 | -36.49 | -58.38 | -85.89 | -119.21 | -158.71 | -204.69 | -257.48 | -317.40 | -384.82 |
| MID-RANGE (IN) | .00 | .57 | 2.32 | 5.34 | 9.73 | 15.59 | 23.00 | 32.05 | 42.87 | 55.60 | 70.36 | 87.28 | 106.52 |
| BULLET PATH (IN) | -.63 | 2.00 | -.00 | -6.87 | -18.84 | -36.16 | -59.11 | -87.86 | -122.79 | -164.20 | -212.41 | -267.77 | -330.62 |
| TIME OF FLIGHT (SEC) | .000000 | .108449 | .219543 | .333338 | .449891 | .569266 | .691532 | .816762 | .945030 | 1.076419 | 1.211013 | 1.348900 | 1.490174 |
| WIND DEFLECTION (IN) | | | | | | | | | | | | | |
| 5 MPH | .00 | .11 | .46 | 1.05 | 1.88 | 2.95 | 4.28 | 5.88 | 7.73 | 9.87 | 12.28 | 14.99 | 17.99 |
| 10 MPH | .00 | .23 | .93 | 2.10 | 3.75 | 5.91 | 8.57 | 11.75 | 15.47 | 19.74 | 24.57 | 29.98 | 35.98 |
| 20 MPH | .00 | .46 | 1.85 | 4.19 | 7.50 | 11.81 | 17.13 | 23.50 | 30.94 | 39.47 | 49.13 | 59.96 | 71.97 |
| 30 MPH | .00 | .69 | 2.78 | 6.29 | 11.26 | 17.72 | 25.70 | 35.25 | 46.40 | 59.21 | 73.70 | 89.93 | 107.95 |

## Bullet: Hornady .430 Dia. 200 Gr. JHP
## Ballistic Coefficient: .185

| RANGE (YDS) | MUZZLE | 25 | 50 | 75 | 100 | 125 | 150 | 175 | 200 | 225 | 250 | 275 | 300 |
|---|---|---|---|---|---|---|---|---|---|---|---|---|---|
| VELOCITY (FPS) | 1700. | 1609. | 1523. | 1442. | 1366. | 1295. | 1210. | 1138. | 1079. | 1032. | 992. | 958. | 927. |
| ENERGY (FT-LB) | 1283. | 1150. | 1030. | 923. | 829. | 745. | 650. | 575. | 517. | 473. | 437. | 407. | 382. |
| DROP (IN) | .00 | -.36 | -1.56 | -3.70 | -6.90 | -11.25 | -16.91 | -24.06 | -32.89 | -43.57 | -56.29 | -71.21 | -88.46 |
| MID-RANGE (IN) | .00 | .10 | .42 | 1.00 | 1.89 | 3.12 | 4.75 | 6.87 | 9.55 | 12.87 | 16.90 | 21.70 | 27.32 |
| BULLET PATH (IN) | -.63 | .10 | .00 | -1.05 | -3.14 | -6.40 | -10.97 | -17.02 | -24.75 | -34.34 | -45.96 | -59.78 | -75.94 |
| TIME OF FLIGHT (SEC) | .000000 | .045345 | .093250 | .143861 | .197308 | .253693 | .313661 | .377648 | .445394 | .516514 | .590675 | .667642 | .747246 |
| WIND DEFLECTION (IN) | | | | | | | | | | | | | |
| 5 MPH | .00 | .11 | .44 | 1.01 | 1.83 | 2.91 | 4.31 | 6.06 | 8.14 | 10.51 | 13.16 | 16.05 | 19.17 |
| 10 MPH | .00 | .22 | .88 | 2.03 | 3.67 | 5.83 | 8.62 | 12.11 | 16.27 | 21.02 | 26.31 | 32.09 | 38.34 |
| 20 MPH | .00 | .43 | 1.77 | 4.05 | 7.33 | 11.65 | 17.23 | 24.23 | 32.54 | 42.05 | 52.62 | 64.19 | 76.68 |
| 30 MPH | .00 | .65 | 2.65 | 6.08 | 11.00 | 17.48 | 25.85 | 36.34 | 48.82 | 63.07 | 78.94 | 96.28 | 115.02 |
| VELOCITY (FPS) | 1600. | 1514. | 1434. | 1358. | 1286. | 1202. | 1131. | 1074. | 1028. | 989. | 955. | 925. | 897. |
| ENERGY (FT-LB) | 1137. | 1018. | 913. | 819. | 734. | 641. | 568. | 512. | 469. | 434. | 405. | 380. | 358. |
| DROP (IN) | .00 | -.41 | -1.77 | -4.20 | -7.80 | -12.73 | -19.16 | -27.29 | -37.29 | -49.35 | -63.62 | -80.24 | -99.38 |
| MID-RANGE (IN) | .00 | .11 | .48 | 1.14 | 2.13 | 3.52 | 5.38 | 7.80 | 10.85 | 14.59 | 19.08 | 24.37 | 30.53 |
| BULLET PATH (IN) | -.63 | .16 | .00 | -1.22 | -3.62 | -7.35 | -12.58 | -19.50 | -28.31 | -39.16 | -52.23 | -67.64 | -85.58 |
| TIME OF FLIGHT (SEC) | .000000 | .048188 | .099098 | .152856 | .209567 | .269939 | .334312 | .402409 | .473842 | .548288 | .625522 | .705379 | .787746 |
| WIND DEFLECTION (IN) | | | | | | | | | | | | | |
| 5 MPH | .00 | .12 | .47 | 1.08 | 1.94 | 3.13 | 4.67 | 6.54 | 8.70 | 11.12 | 13.80 | 16.70 | 19.82 |
| 10 MPH | .00 | .23 | .94 | 2.15 | 3.88 | 6.26 | 9.34 | 13.07 | 17.40 | 22.25 | 27.59 | 33.40 | 39.64 |
| 20 MPH | .00 | .46 | 1.88 | 4.31 | 7.77 | 12.52 | 18.68 | 26.15 | 34.79 | 44.50 | 55.18 | 66.79 | 79.29 |
| 30 MPH | .00 | .69 | 2.82 | 6.46 | 11.65 | 18.78 | 28.02 | 39.22 | 52.19 | 66.75 | 82.78 | 100.19 | 118.93 |
| VELOCITY (FPS) | 1500. | 1420. | 1346. | 1271. | 1189. | 1121. | 1066. | 1021. | 983. | 949. | 920. | 893. | 868. |
| ENERGY (FT-LB) | 999. | 896. | 804. | 717. | 628. | 558. | 504. | 463. | 429. | 400. | 376. | 354. | 335. |
| DROP (IN) | .00 | -.48 | -2.04 | -4.79 | -8.91 | -14.56 | -21.93 | -31.21 | -42.57 | -56.17 | -72.14 | -90.67 | -111.90 |
| MID-RANGE (IN) | .00 | .13 | .54 | 1.28 | 2.42 | 4.02 | 6.17 | 8.94 | 12.38 | 16.56 | 21.51 | 27.32 | 34.02 |
| BULLET PATH (IN) | -.63 | .22 | .00 | -1.42 | -4.20 | -8.52 | -14.56 | -22.50 | -32.53 | -44.80 | -59.43 | -76.63 | -96.53 |
| TIME OF FLIGHT (SEC) | .000000 | .051395 | .105658 | .162919 | .223973 | .288995 | .357679 | .429637 | .504564 | .582247 | .662532 | .745310 | .830508 |
| WIND DEFLECTION (IN) | | | | | | | | | | | | | |
| 5 MPH | .00 | .12 | .50 | 1.14 | 2.11 | 3.43 | 5.08 | 7.01 | 9.20 | 11.64 | 14.30 | 17.19 | 20.28 |
| 10 MPH | .00 | .25 | 1.00 | 2.27 | 4.22 | 6.86 | 10.15 | 14.02 | 18.40 | 23.28 | 28.61 | 34.37 | 40.57 |
| 20 MPH | .00 | .49 | 1.99 | 4.55 | 8.44 | 13.73 | 20.30 | 28.03 | 36.81 | 46.55 | 57.21 | 68.75 | 81.14 |
| 30 MPH | .00 | .74 | 2.99 | 6.82 | 12.66 | 20.59 | 30.45 | 42.05 | 55.21 | 69.83 | 85.82 | 103.12 | 121.71 |
| VELOCITY (FPS) | 1400. | 1327. | 1248. | 1170. | 1105. | 1053. | 1010. | 973. | 941. | 913. | 886. | 862. | 840. |
| ENERGY (FT-LB) | 870. | 782. | 691. | 607. | 542. | 492. | 453. | 421. | 393. | 370. | 349. | 330. | 313. |
| DROP (IN) | .00 | -.55 | -2.34 | -5.54 | -10.31 | -16.87 | -25.37 | -36.00 | -48.91 | -64.23 | -82.15 | -102.83 | -126.42 |
| MID-RANGE (IN) | .00 | .15 | .62 | 1.48 | 2.82 | 4.68 | 7.15 | 10.28 | 14.14 | 18.76 | 24.21 | 30.55 | 37.84 |
| BULLET PATH (IN) | -.63 | .30 | .00 | -1.71 | -5.00 | -10.07 | -17.09 | -26.23 | -37.65 | -51.49 | -67.92 | -87.11 | -109.22 |
| TIME OF FLIGHT (SEC) | .000000 | .055036 | .113206 | .175334 | .241365 | .310954 | .383726 | .459403 | .537789 | .618744 | .702170 | .788001 | .876195 |
| WIND DEFLECTION (IN) | | | | | | | | | | | | | |
| 5 MPH | .00 | .13 | .53 | 1.29 | 2.38 | 3.79 | 5.48 | 7.43 | 9.61 | 12.02 | 14.65 | 17.49 | 20.53 |
| 10 MPH | .00 | .26 | 1.07 | 2.57 | 4.77 | 7.59 | 10.96 | 14.85 | 19.22 | 24.04 | 29.30 | 34.97 | 41.07 |
| 20 MPH | .00 | .52 | 2.13 | 5.15 | 9.53 | 15.17 | 21.93 | 29.71 | 38.44 | 48.08 | 58.59 | 69.95 | 82.13 |
| 30 MPH | .00 | .77 | 3.20 | 7.72 | 14.30 | 22.76 | 32.89 | 44.56 | 57.67 | 72.13 | 87.89 | 104.92 | 123.20 |

# Bullet: Hornady .430 Dia. 200 Gr. JHP
## Ballistic Coefficient: .185

| RANGE (YDS) | MUZZLE | 25 | 50 | 75 | 100 | 125 | 150 | 175 | 200 | 225 | 250 | 275 | 300 |
|---|---|---|---|---|---|---|---|---|---|---|---|---|---|
| VELOCITY (FPS) | 1300. | 1214. | 1141. | 1082. | 1034. | 994. | 960. | 929. | 901. | 876. | 853. | 831. | 810. |
| ENERGY (FT-LB) | 750. | 654. | 578. | 520. | 475. | 439. | 409. | 383. | 361. | 341. | 323. | 306. | 291. |
| DROP (IN) | .00 | -.66 | -2.79 | -6.59 | -12.24 | -19.91 | -29.77 | -41.96 | -56.65 | -74.00 | -94.16 | -117.31 | -143.61 |
| MID-RANGE (IN) | .00 | .17 | .74 | 1.77 | 3.33 | 5.48 | 8.30 | 11.81 | 16.09 | 21.19 | 27.17 | 34.10 | 42.03 |
| BULLET PATH (IN) | -.63 | .42 | .00 | -2.09 | -6.02 | -11.98 | -20.13 | -30.61 | -43.59 | -59.22 | -77.68 | -99.12 | -123.70 |
| TIME OF FLIGHT (SEC) | .000000 | .059741 | .123511 | .191057 | .261999 | .335996 | .412810 | .492269 | .574251 | .658674 | .745482 | .834643 | .926145 |
| WIND DEFLECTION (IN) | | | | | | | | | | | | | |
| 5 MPH | .00 | .18 | .72 | 1.58 | 2.75 | 4.18 | 5.87 | 7.78 | 9.92 | 12.27 | 14.83 | 17.60 | 20.58 |
| 10 MPH | .00 | .36 | 1.43 | 3.16 | 5.50 | 8.37 | 11.73 | 15.56 | 19.84 | 24.54 | 29.67 | 35.20 | 41.16 |
| 20 MPH | .00 | .72 | 2.86 | 6.33 | 10.99 | 16.73 | 23.46 | 31.12 | 39.67 | 49.08 | 59.33 | 70.41 | 82.31 |
| 30 MPH | .00 | 1.08 | 4.29 | 9.49 | 16.49 | 25.10 | 35.19 | 46.69 | 59.51 | 73.63 | 89.00 | 105.61 | 123.47 |
| VELOCITY (FPS) | 1200. | 1130. | 1073. | 1027. | 988. | 954. | 924. | 897. | 872. | 849. | 827. | 806. | 779. |
| ENERGY (FT-LB) | 639. | 567. | 511. | 468. | 433. | 404. | 379. | 357. | 337. | 320. | 303. | 288. | 269. |
| DROP (IN) | .00 | -.77 | -3.24 | -7.59 | -14.00 | -22.63 | -33.60 | -47.11 | -63.30 | -82.35 | -104.40 | -129.62 | -158.17 |
| MID-RANGE (IN) | .00 | .20 | .85 | 2.01 | 3.76 | 6.14 | 9.21 | 13.02 | 17.65 | 23.15 | 29.57 | 37.00 | 45.48 |
| BULLET PATH (IN) | -.63 | .53 | .00 | -2.41 | -6.88 | -13.57 | -22.61 | -34.18 | -48.44 | -65.55 | -85.66 | -108.95 | -135.56 |
| TIME OF FLIGHT (SEC) | .000000 | .064467 | .132648 | .204156 | .278671 | .355966 | .435882 | .518305 | .603155 | .690383 | .779960 | .871879 | .966466 |
| WIND DEFLECTION (IN) | | | | | | | | | | | | | |
| 5 MPH | .00 | .17 | .67 | 1.47 | 2.52 | 3.83 | 5.36 | 7.11 | 9.08 | 11.25 | 13.64 | 16.23 | 19.05 |
| 10 MPH | .00 | .35 | 1.35 | 2.93 | 5.05 | 7.65 | 10.72 | 14.22 | 18.16 | 22.51 | 27.27 | 32.45 | 38.10 |
| 20 MPH | .00 | .69 | 2.69 | 5.86 | 10.09 | 15.30 | 21.43 | 28.44 | 36.31 | 45.01 | 54.55 | 64.90 | 76.20 |
| 30 MPH | .00 | 1.04 | 4.04 | 8.79 | 15.14 | 22.95 | 32.15 | 42.66 | 54.47 | 67.52 | 81.82 | 97.35 | 114.29 |
| VELOCITY (FPS) | 1100. | 1049. | 1007. | 970. | 939. | 910. | 884. | 860. | 838. | 816. | 794. | 763. | 735. |
| ENERGY (FT-LB) | 537. | 488. | 450. | 418. | 391. | 368. | 347. | 328. | 311. | 296. | 280. | 259. | 240. |
| DROP (IN) | .00 | -.91 | -3.78 | -8.80 | -16.10 | -25.84 | -38.19 | -53.30 | -71.34 | -92.48 | -116.85 | -144.64 | -176.15 |
| MID-RANGE (IN) | .00 | .24 | .98 | 2.31 | 4.27 | 6.90 | 10.30 | 14.49 | 19.57 | 25.58 | 32.59 | 40.64 | 49.89 |
| BULLET PATH (IN) | -.63 | .67 | .00 | -2.81 | -7.90 | -15.43 | -25.57 | -38.48 | -54.32 | -73.24 | -95.41 | -120.99 | -150.29 |
| TIME OF FLIGHT (SEC) | .000000 | .069875 | .142905 | .218820 | .297429 | .378597 | .462228 | .548259 | .636650 | .727385 | .820491 | .916904 | 1.017103 |
| WIND DEFLECTION (IN) | | | | | | | | | | | | | |
| 5 MPH | .00 | .15 | .58 | 1.26 | 2.17 | 3.32 | 4.68 | 6.25 | 8.03 | 10.01 | 12.20 | 14.69 | 17.51 |
| 10 MPH | .00 | .30 | 1.15 | 2.51 | 4.35 | 6.63 | 9.35 | 12.49 | 16.05 | 20.02 | 24.41 | 29.38 | 35.01 |
| 20 MPH | .00 | .60 | 2.30 | 5.02 | 8.69 | 13.27 | 18.70 | 24.99 | 32.10 | 40.04 | 48.81 | 58.75 | 70.02 |
| 30 MPH | .00 | .89 | 3.45 | 7.54 | 13.04 | 19.90 | 28.06 | 37.48 | 48.15 | 60.06 | 73.22 | 88.13 | 105.03 |
| VELOCITY (FPS) | 1000. | 965. | 934. | 905. | 880. | 856. | 834. | 813. | 788. | 758. | 730. | 704. | 678. |
| ENERGY (FT-LB) | 444. | 413. | 387. | 364. | 344. | 325. | 309. | 293. | 276. | 255. | 237. | 220. | 204. |
| DROP (IN) | .00 | -1.08 | -4.47 | -10.33 | -18.83 | -30.12 | -44.37 | -61.74 | -82.37 | -106.46 | -134.31 | -166.23 | -202.52 |
| MID-RANGE (IN) | .00 | .28 | 1.15 | 2.68 | 4.94 | 7.98 | 11.86 | 16.63 | 22.35 | 29.11 | 37.03 | 46.24 | 56.88 |
| BULLET PATH (IN) | -.63 | .84 | -.00 | -3.31 | -9.25 | -18.00 | -29.69 | -44.51 | -62.58 | -84.12 | -109.43 | -138.79 | -172.53 |
| TIME OF FLIGHT (SEC) | .000000 | .076389 | .155445 | .237040 | .321086 | .407523 | .496315 | .587448 | .681017 | .778053 | .878882 | .983578 | 1.092237 |
| WIND DEFLECTION (IN) | | | | | | | | | | | | | |
| 5 MPH | .00 | .12 | .48 | 1.06 | 1.86 | 2.86 | 4.08 | 5.50 | 7.13 | 9.07 | 11.34 | 13.95 | 16.92 |
| 10 MPH | .00 | .24 | .96 | 2.12 | 3.71 | 5.72 | 8.15 | 10.99 | 14.26 | 18.14 | 22.68 | 27.91 | 33.83 |
| 20 MPH | .00 | .49 | 1.92 | 4.24 | 7.42 | 11.45 | 16.30 | 21.98 | 28.52 | 36.27 | 45.37 | 55.82 | 67.67 |
| 30 MPH | .00 | .73 | 2.87 | 6.36 | 11.13 | 17.17 | 24.45 | 32.97 | 42.78 | 54.41 | 68.05 | 83.73 | 101.50 |

# Bullet: Hornady .430 Dia. 200 Gr. JHP
## Ballistic Coefficient: .185

| RANGE (YDS) | MUZZLE | 25 | 50 | 75 | 100 | 125 | 150 | 175 | 200 | 225 | 250 | 275 | 300 |
|---|---|---|---|---|---|---|---|---|---|---|---|---|---|
| VELOCITY (FPS) | 900. | 875. | 851. | 829. | 808. | 783. | 753. | 725. | 699. | 673. | 649. | 626. | 604. |
| ENERGY (FT-LB) | 360. | 340. | 322. | 305. | 290. | 272. | 252. | 233. | 217. | 201. | 187. | 174. | 162. |
| DROP (IN) | .00 | -1.28 | -5.38 | -12.47 | -22.71 | -36.25 | -53.32 | -74.18 | -99.17 | -128.59 | -162.79 | -202.13 | -247.01 |
| MID-RANGE (IN) | .00 | .34 | 1.41 | 3.27 | 5.97 | 9.58 | 14.19 | 19.90 | 26.87 | 35.23 | 45.14 | 56.74 | 70.19 |
| BULLET PATH (IN) | -.63 | 1.10 | -.00 | -4.09 | -11.32 | -21.86 | -35.91 | -53.78 | -75.76 | -102.17 | -133.37 | -169.70 | -211.57 |
| TIME OF FLIGHT (SEC) | .000000 | .084540 | .171461 | .260733 | .352345 | .446490 | .544230 | .645773 | .751198 | .860605 | .974113 | 1.091859 | 1.213996 |
| WIND DEFLECTION (IN) | | | | | | | | | | | | | |
| 5 MPH | .00 | .11 | .42 | .94 | 1.67 | 2.62 | 3.89 | 5.49 | 7.44 | 9.73 | 12.39 | 15.42 | 18.83 |
| 10 MPH | .00 | .21 | .84 | 1.89 | 3.35 | 5.25 | 7.78 | 10.99 | 14.88 | 19.47 | 24.78 | 30.83 | 37.66 |
| 20 MPH | .00 | .42 | 1.69 | 3.78 | 6.69 | 10.50 | 15.57 | 21.98 | 29.76 | 38.93 | 49.55 | 61.67 | 75.33 |
| 30 MPH | .00 | .64 | 2.53 | 5.67 | 10.04 | 15.75 | 23.35 | 32.97 | 44.63 | 58.40 | 74.33 | 92.50 | 112.99 |
| VELOCITY (FPS) | 800. | 769. | 740. | 713. | 687. | 663. | 639. | 616. | 594. | 573. | 552. | 532. | 513. |
| ENERGY (FT-LB) | 284. | 263. | 243. | 226. | 210. | 195. | 181. | 168. | 157. | 146. | 135. | 126. | 117. |
| DROP (IN) | .00 | -1.58 | -6.81 | -15.99 | -29.42 | -47.43 | -70.37 | -98.62 | -132.57 | -172.67 | -219.39 | -273.23 | -334.74 |
| MID-RANGE (IN) | .00 | .44 | 1.83 | 4.27 | 7.90 | 12.83 | 19.20 | 27.16 | 36.87 | 48.51 | 62.27 | 78.35 | 97.00 |
| BULLET PATH (IN) | -.63 | 1.51 | -.00 | -5.46 | -15.16 | -29.45 | -48.68 | -73.20 | -103.43 | -139.81 | -182.81 | -232.93 | -290.72 |
| TIME OF FLIGHT (SEC) | .000000 | .095621 | .195013 | .298238 | .405384 | .516560 | .631892 | .751525 | .875618 | 1.004347 | 1.137902 | 1.276484 | 1.420311 |
| WIND DEFLECTION (IN) | | | | | | | | | | | | | |
| 5 MPH | .00 | .16 | .66 | 1.49 | 2.67 | 4.21 | 6.11 | 8.38 | 11.05 | 14.13 | 17.64 | 21.58 | 25.99 |
| 10 MPH | .00 | .33 | 1.32 | 2.99 | 5.35 | 8.41 | 12.21 | 16.77 | 22.11 | 28.27 | 35.27 | 43.16 | 51.97 |
| 20 MPH | .00 | .66 | 2.64 | 5.98 | 10.70 | 16.83 | 24.43 | 33.54 | 44.22 | 56.53 | 70.54 | 86.32 | 103.95 |
| 30 MPH | .00 | .99 | 3.97 | 8.97 | 16.04 | 25.24 | 36.64 | 50.30 | 66.33 | 84.80 | 105.81 | 129.48 | 155.92 |
| VELOCITY (FPS) | 700. | 675. | 650. | 627. | 605. | 583. | 562. | 542. | 522. | 503. | 485. | 468. | 451. |
| ENERGY (FT-LB) | 218. | 202. | 188. | 175. | 162. | 151. | 140. | 130. | 121. | 113. | 105. | 97. | 90. |
| DROP (IN) | .00 | -2.04 | -8.84 | -20.76 | -38.18 | -61.53 | -91.25 | -127.83 | -171.79 | -223.71 | -284.21 | -353.96 | -433.67 |
| MID-RANGE (IN) | .00 | .57 | 2.38 | 5.55 | 10.25 | 16.63 | 24.87 | 35.16 | 47.72 | 62.77 | 80.58 | 101.42 | 125.58 |
| BULLET PATH (IN) | -.63 | 2.06 | .00 | -7.18 | -19.87 | -38.48 | -63.47 | -95.31 | -134.54 | -181.72 | -237.49 | -302.50 | -377.47 |
| TIME OF FLIGHT (SEC) | .000000 | .109147 | .222384 | .339846 | .461685 | .588069 | .719178 | .855209 | .996371 | 1.142887 | 1.294993 | 1.452941 | 1.616997 |
| WIND DEFLECTION (IN) | | | | | | | | | | | | | |
| 5 MPH | .00 | .18 | .71 | 1.62 | 2.91 | 4.61 | 6.72 | 9.26 | 12.25 | 15.72 | 19.67 | 24.14 | 29.15 |
| 10 MPH | .00 | .35 | 1.43 | 3.24 | 5.83 | 9.21 | 13.43 | 18.52 | 24.50 | 31.43 | 39.35 | 48.29 | 58.31 |
| 20 MPH | .00 | .71 | 2.85 | 6.48 | 11.66 | 18.43 | 26.87 | 37.03 | 49.01 | 62.87 | 78.69 | 96.58 | 116.61 |
| 30 MPH | .00 | 1.06 | 4.28 | 9.72 | 17.48 | 27.64 | 40.30 | 55.55 | 73.51 | 94.30 | 118.04 | 144.87 | 174.92 |

## Bullet: Speer .429 Dia. 225 Gr. HJHP (SWC)
## Ballistic Coefficient: .140

| RANGE (YDS) | MUZZLE | 25 | 50 | 75 | 100 | 125 | 150 | 175 | 200 | 225 | 250 | 275 | 300 |
|---|---|---|---|---|---|---|---|---|---|---|---|---|---|
| VELOCITY (FPS) | 1700. | 1581. | 1470. | 1369. | 1274. | 1182. | 1108. | 1049. | 1002. | 963. | 928. | 897. | 869. |
| ENERGY (FT-LB) | 1444. | 1249. | 1080. | 936. | 811. | 698. | 613. | 550. | 502. | 463. | 430. | 402. | 377. |
| DROP (IN) | .00 | -.37 | -1.61 | -3.87 | -7.28 | -12.05 | -18.36 | -26.44 | -36.48 | -48.68 | -63.20 | -80.22 | -99.92 |
| MID-RANGE (IN) | .00 | .10 | .43 | 1.06 | 2.03 | 3.41 | 5.31 | 7.80 | 10.96 | 14.86 | 19.55 | 25.11 | 31.60 |
| BULLET PATH (IN) | -.63 | .12 | .00 | -1.13 | -3.42 | -7.06 | -12.25 | -19.21 | -28.13 | -39.21 | -52.60 | -68.50 | -87.08 |
| TIME OF FLIGHT (SEC) | .000000 | .045751 | .094947 | .147832 | .204604 | .265767 | .331366 | .400990 | .474170 | .550570 | .629963 | .712192 | .797154 |
| WIND DEFLECTION (IN) | | | | | | | | | | | | | |
| 5 MPH | .00 | .14 | .59 | 1.36 | 2.48 | 3.98 | 5.87 | 8.11 | 10.67 | 13.51 | 16.61 | 19.97 | 23.56 |
| 10 MPH | .00 | .29 | 1.18 | 2.72 | 4.95 | 7.95 | 11.73 | 16.22 | 21.34 | 27.02 | 33.23 | 39.93 | 47.12 |
| 20 MPH | .00 | .57 | 2.36 | 5.45 | 9.90 | 15.90 | 23.46 | 32.44 | 42.67 | 54.04 | 66.45 | 79.87 | 94.25 |
| 30 MPH | .00 | .86 | 3.54 | 8.17 | 14.85 | 23.85 | 35.20 | 48.66 | 64.01 | 81.05 | 99.68 | 119.80 | 141.37 |
| VELOCITY (FPS) | 1600. | 1488. | 1385. | 1290. | 1196. | 1118. | 1058. | 1009. | 968. | 933. | 902. | 873. | 847. |
| ENERGY (FT-LB) | 1279. | 1106. | 958. | 831. | 714. | 625. | 559. | 509. | 468. | 435. | 406. | 381. | 359. |
| DROP (IN) | .00 | -.42 | -1.84 | -4.38 | -8.24 | -13.62 | -20.73 | -29.78 | -40.96 | -54.43 | -70.37 | -88.96 | -110.40 |
| MID-RANGE (IN) | .00 | .11 | .50 | 1.19 | 2.29 | 3.85 | 5.99 | 8.76 | 12.24 | 16.48 | 21.56 | 27.53 | 34.47 |
| BULLET PATH (IN) | -.63 | .18 | .00 | -1.31 | -3.94 | -8.09 | -13.97 | -21.78 | -31.73 | -43.96 | -58.67 | -76.03 | -96.24 |
| TIME OF FLIGHT (SEC) | .000000 | .048621 | .100893 | .157017 | .217471 | .282403 | .351435 | .424088 | .500006 | .578945 | .660742 | .745285 | .832506 |
| WIND DEFLECTION (IN) | | | | | | | | | | | | | |
| 5 MPH | .00 | .15 | .63 | 1.44 | 2.64 | 4.23 | 6.18 | 8.44 | 11.00 | 13.82 | 16.90 | 20.21 | 23.76 |
| 10 MPH | .00 | .31 | 1.26 | 2.89 | 5.27 | 8.45 | 12.35 | 16.89 | 22.00 | 27.64 | 33.79 | 40.42 | 47.52 |
| 20 MPH | .00 | .61 | 2.51 | 5.77 | 10.55 | 16.91 | 24.71 | 33.78 | 44.00 | 55.29 | 67.58 | 80.84 | 95.04 |
| 30 MPH | .00 | .92 | 3.77 | 8.66 | 15.82 | 25.36 | 37.06 | 50.67 | 66.00 | 82.93 | 101.37 | 121.26 | 142.56 |
| VELOCITY (FPS) | 1500. | 1396. | 1301. | 1206. | 1127. | 1064. | 1014. | 973. | 937. | 905. | 877. | 850. | 826. |
| ENERGY (FT-LB) | 1124. | 973. | 846. | 726. | 634. | 566. | 514. | 473. | 439. | 409. | 384. | 361. | 341. |
| DROP (IN) | .00 | -.49 | -2.10 | -5.00 | -9.39 | -15.49 | -23.50 | -33.62 | -46.02 | -60.86 | -78.33 | -98.63 | -121.93 |
| MID-RANGE (IN) | .00 | .13 | .56 | 1.35 | 2.60 | 4.37 | 6.76 | 9.82 | 13.62 | 18.21 | 23.68 | 30.07 | 37.46 |
| BULLET PATH (IN) | -.63 | .24 | .00 | -1.53 | -4.56 | -9.30 | -15.95 | -24.70 | -35.74 | -49.21 | -65.33 | -84.26 | -106.19 |
| TIME OF FLIGHT (SEC) | .000000 | .051852 | .107535 | .167448 | .231867 | .300442 | .372689 | .448236 | .526828 | .608294 | .692516 | .779424 | .868976 |
| WIND DEFLECTION (IN) | | | | | | | | | | | | | |
| 5 MPH | .00 | .16 | .66 | 1.54 | 2.80 | 4.44 | 6.40 | 8.64 | 11.16 | 13.93 | 16.94 | 20.19 | 23.67 |
| 10 MPH | .00 | .33 | 1.33 | 3.07 | 5.61 | 8.88 | 12.79 | 17.29 | 22.32 | 27.86 | 33.88 | 40.38 | 47.34 |
| 20 MPH | .00 | .65 | 2.65 | 6.14 | 11.22 | 17.76 | 25.59 | 34.58 | 44.64 | 55.72 | 67.77 | 80.76 | 94.68 |
| 30 MPH | .00 | .98 | 3.98 | 9.21 | 16.83 | 26.63 | 38.38 | 51.87 | 66.97 | 83.58 | 101.65 | 121.14 | 142.02 |
| VELOCITY (FPS) | 1400. | 1305. | 1209. | 1129. | 1066. | 1016. | 974. | 938. | 906. | 878. | 851. | 827. | 803. |
| ENERGY (FT-LB) | 979. | 851. | 730. | 637. | 568. | 516. | 474. | 440. | 410. | 385. | 362. | 341. | 322. |
| DROP (IN) | .00 | -.56 | -2.41 | -5.75 | -10.78 | -17.72 | -26.76 | -38.07 | -51.82 | -68.19 | -87.39 | -109.57 | -134.92 |
| MID-RANGE (IN) | .00 | .15 | .64 | 1.56 | 2.98 | 4.98 | 7.63 | 10.99 | 15.12 | 20.09 | 25.97 | 32.82 | 40.70 |
| BULLET PATH (IN) | -.63 | .33 | .00 | -1.82 | -5.33 | -10.75 | -18.26 | -28.05 | -40.28 | -55.13 | -72.80 | -93.46 | -117.29 |
| TIME OF FLIGHT (SEC) | .000000 | .055513 | .115254 | .179508 | .247936 | .320053 | .395481 | .473963 | .555322 | .639443 | .726249 | .815702 | .907789 |
| WIND DEFLECTION (IN) | | | | | | | | | | | | | |
| 5 MPH | .00 | .17 | .71 | 1.65 | 2.96 | 4.59 | 6.52 | 8.71 | 11.15 | 13.84 | 16.77 | 19.92 | 23.31 |
| 10 MPH | .00 | .34 | 1.43 | 3.31 | 5.92 | 9.19 | 13.03 | 17.42 | 22.31 | 27.68 | 33.53 | 39.85 | 46.63 |
| 20 MPH | .00 | .68 | 2.85 | 6.62 | 11.84 | 18.37 | 26.07 | 34.83 | 44.62 | 55.37 | 67.07 | 79.70 | 93.26 |
| 30 MPH | .00 | 1.02 | 4.28 | 9.92 | 17.77 | 27.56 | 39.10 | 52.25 | 66.92 | 83.05 | 100.60 | 119.55 | 139.88 |

## Bullet: Speer .429 Dia. 225 Gr. HJHP (SWC)
### Ballistic Coefficient: .140

| RANGE (YDS) | MUZZLE | 25 | 50 | 75 | 100 | 125 | 150 | 175 | 200 | 225 | 250 | 275 | 300 |
|---|---|---|---|---|---|---|---|---|---|---|---|---|---|
| VELOCITY (FPS) | 1300. | 1204. | 1125. | 1063. | 1013. | 972. | 936. | 905. | 876. | 850. | 825. | 802. | 774. |
| ENERGY (FT-LB) | 844. | 724. | 632. | 564. | 513. | 472. | 438. | 409. | 383. | 361. | 340. | 321. | 299. |
| DROP (IN) | .00 | -.66 | -2.83 | -6.70 | -12.49 | -20.39 | -30.57 | -43.20 | -58.46 | -76.56 | -97.65 | -121.92 | -149.51 |
| MID-RANGE (IN) | .00 | .18 | .75 | 1.81 | 3.42 | 5.66 | 8.58 | 12.25 | 16.74 | 22.11 | 28.43 | 35.77 | 44.19 |
| BULLET PATH (IN) | -.63 | .44 | .00 | -2.15 | -6.21 | -12.38 | -20.83 | -31.72 | -45.26 | -61.63 | -80.99 | -103.53 | -129.39 |
| TIME OF FLIGHT (SEC) | .000000 | .059996 | .124495 | .193141 | .265451 | .341055 | .419700 | .501216 | .585486 | .672440 | .762039 | .854271 | .949473 |
| WIND DEFLECTION (IN) | | | | | | | | | | | | | |
| 5 MPH | .00 | .20 | .80 | 1.77 | 3.05 | 4.63 | 6.47 | 8.57 | 10.91 | 13.48 | 16.29 | 19.33 | 22.63 |
| 10 MPH | .00 | .41 | 1.60 | 3.53 | 6.10 | 9.26 | 12.94 | 17.14 | 21.81 | 26.96 | 32.58 | 38.66 | 45.26 |
| 20 MPH | .00 | .81 | 3.21 | 7.06 | 12.21 | 18.51 | 25.89 | 34.27 | 43.63 | 53.93 | 65.16 | 77.32 | 90.52 |
| 30 MPH | .00 | 1.22 | 4.81 | 10.59 | 18.31 | 27.77 | 38.83 | 51.41 | 65.44 | 80.89 | 97.74 | 115.98 | 135.78 |
| VELOCITY (FPS) | 1200. | 1122. | 1060. | 1011. | 970. | 935. | 903. | 875. | 849. | 824. | 801. | 773. | 745. |
| ENERGY (FT-LB) | 719. | 629. | 562. | 511. | 470. | 436. | 408. | 382. | 360. | 339. | 320. | 298. | 278. |
| DROP (IN) | .00 | -.78 | -3.28 | -7.70 | -14.24 | -23.07 | -34.36 | -48.29 | -65.06 | -84.84 | -107.80 | -134.10 | -164.01 |
| MID-RANGE (IN) | .00 | .20 | .86 | 2.05 | 3.84 | 6.30 | 9.48 | 13.45 | 18.29 | 24.05 | 30.83 | 38.65 | 47.65 |
| BULLET PATH (IN) | -.63 | .55 | .00 | -2.47 | -7.06 | -13.93 | -23.27 | -35.25 | -50.06 | -67.88 | -88.90 | -113.24 | -141.20 |
| TIME OF FLIGHT (SEC) | .000000 | .064705 | .133536 | .206009 | .281762 | .360546 | .442194 | .526593 | .613672 | .703394 | .795750 | .891077 | .989953 |
| WIND DEFLECTION (IN) | | | | | | | | | | | | | |
| 5 MPH | .00 | .19 | .75 | 1.63 | 2.80 | 4.23 | 5.91 | 7.84 | 10.00 | 12.40 | 15.03 | 17.91 | 21.12 |
| 10 MPH | .00 | .39 | 1.50 | 3.26 | 5.59 | 8.46 | 11.83 | 15.68 | 20.01 | 24.80 | 30.05 | 35.83 | 42.23 |
| 20 MPH | .00 | .78 | 3.00 | 6.52 | 11.18 | 16.91 | 23.65 | 31.36 | 40.01 | 49.59 | 60.10 | 71.66 | 84.46 |
| 30 MPH | .00 | 1.16 | 4.51 | 9.77 | 16.77 | 25.37 | 35.48 | 47.04 | 60.02 | 74.39 | 90.16 | 107.49 | 126.70 |
| VELOCITY (FPS) | 1100. | 1043. | 997. | 958. | 924. | 894. | 866. | 840. | 816. | 791. | 763. | 736. | 711. |
| ENERGY (FT-LB) | 604. | 544. | 497. | 458. | 426. | 399. | 375. | 353. | 333. | 313. | 291. | 271. | 252. |
| DROP (IN) | .00 | -.91 | -3.82 | -8.90 | -16.32 | -26.26 | -38.91 | -54.45 | -73.06 | -94.91 | -120.18 | -149.16 | -182.14 |
| MID-RANGE (IN) | .00 | .24 | .99 | 2.34 | 4.34 | 7.06 | 10.56 | 14.92 | 20.21 | 26.49 | 33.83 | 42.34 | 52.15 |
| BULLET PATH (IN) | -.63 | .68 | .00 | -2.86 | -8.05 | -15.77 | -26.20 | -39.52 | -55.90 | -75.53 | -98.57 | -125.33 | -156.08 |
| TIME OF FLIGHT (SEC) | .000000 | .070078 | .143664 | .220436 | .300177 | .382739 | .468025 | .555975 | .646562 | .739822 | .836400 | .936551 | 1.040339 |
| WIND DEFLECTION (IN) | | | | | | | | | | | | | |
| 5 MPH | .00 | 17 | .64 | 1.40 | 2.42 | 3.68 | 5.19 | 6.93 | 8.90 | 11.10 | 13.60 | 16.42 | 19.55 |
| 10 MPH | .00 | 33 | 1.28 | 2.80 | 4.83 | 7.36 | 10.37 | 13.85 | 17.79 | 22.21 | 27.21 | 32.83 | 39.10 |
| 20 MPH | .00 | 67 | 2.57 | 5.59 | 9.66 | 14.72 | 20.74 | 27.70 | 35.59 | 44.42 | 54.41 | 65.67 | 78.20 |
| 30 MPH | .00 | 1 00 | 3.85 | 8.39 | 14.49 | 22.09 | 31.12 | 41.55 | 53.38 | 66.63 | 81.62 | 98.50 | 117.30 |
| VELOCITY (FPS) | 1000. | 961. | 926. | 896. | 868. | 842. | 818. | 794. | 765. | 738. | 713. | 688. | 665. |
| ENERGY (FT-LB) | 500. | 461. | 429. | 401. | 376. | 354. | 334. | 315. | 293. | 272. | 254. | 237. | 221. |
| DROP (IN) | .00 | -1 09 | -4.50 | -10.43 | -19.05 | -30.55 | -45.11 | -62.89 | -84.07 | -108.94 | -137.78 | -170.89 | -208.56 |
| MID-RANGE (IN) | .00 | .28 | 1.16 | 2.72 | 5.02 | 8.13 | 12.12 | 17.05 | 22.98 | 30.04 | 38.34 | 48.01 | 59.17 |
| BULLET PATH (IN) | -.63 | .85 | .00 | -3.36 | -9.41 | -18.34 | -30.33 | -45.55 | -64.16 | -86.46 | -112.74 | -143.27 | -178.38 |
| TIME OF FLIGHT (SEC) | .000000 | .076558 | .156097 | .238466 | .323563 | .411327 | .501729 | .594775 | .691039 | .790872 | .894334 | 1.001505 | 1.112482 |
| WIND DEFLECTION (IN) | | | | | | | | | | | | | |
| 5 MPH | .00 | .14 | .54 | 1.18 | 2.07 | 3.20 | 4.55 | 6.14 | 8.01 | 10.20 | 12.70 | 15.53 | 18.70 |
| 10 MPH | .00 | .27 | 1.07 | 2.37 | 4.15 | 6.39 | 9.10 | 12.28 | 16.02 | 20.39 | 25.40 | 31.06 | 37.40 |
| 20 MPH | .00 | .55 | 2.15 | 4.74 | 8.29 | 12.79 | 18.21 | 24.56 | 32.05 | 40.79 | 50.81 | 62.13 | 74.79 |
| 30 MPH | .00 | .82 | 3.22 | 7.11 | 12.44 | 19.18 | 27.31 | 36.84 | 48.07 | 61.18 | 76.21 | 93.19 | 112.19 |

# Bullet: Speer .429 Dia. 225 Gr. HJHP (SWC)
## Ballistic Coefficient: .140

| RANGE (YDS) | MUZZLE | 25 | 50 | 75 | 100 | 125 | 150 | 175 | 200 | 225 | 250 | 275 | 300 |
|---|---|---|---|---|---|---|---|---|---|---|---|---|---|
| VELOCITY (FPS) | 900. | 872. | 846. | 821. | 798. | 769. | 742. | 716. | 691. | 668. | 645. | 623. | 602. |
| ENERGY (FT-LB) | 405. | 380. | 357. | 337. | 318. | 295. | 275. | 256. | 239. | 223. | 208. | 194. | 181. |
| DROP (IN) | .00 | -1.28 | -5.42 | -12.58 | -22.96 | -36.69 | -54.08 | -75.39 | -100.93 | -131.00 | -165.93 | -206.07 | -251.78 |
| MID-RANGE (IN) | .00 | .35 | 1.43 | 3.31 | 6.06 | 9.73 | 14.46 | 20.34 | 27.51 | 36.12 | 46.27 | 58.12 | 71.81 |
| BULLET PATH (IN) | -.63 | 1.11 | -.00 | -4.14 | -11.49 | -22.20 | -36.56 | -54.86 | -77.37 | -104.41 | -136.32 | -173.43 | -216.12 |
| TIME OF FLIGHT (SEC) | .000000 | .084689 | .172053 | .262056 | .354695 | .450486 | .549838 | .652810 | .759477 | .869935 | .984295 | 1.102682 | 1.225235 |
| WIND DEFLECTION (IN) | | | | | | | | | | | | | |
| 5 MPH | .00 | .12 | .47 | 1.06 | 1.88 | 2.98 | 4.39 | 6.11 | 8.17 | 10.55 | 13.28 | 16.37 | 19.82 |
| 10 MPH | .00 | .24 | .95 | 2.12 | 3.76 | 5.95 | 8.77 | 12.23 | 16.33 | 21.11 | 26.57 | 32.74 | 39.64 |
| 20 MPH | .00 | .48 | 1.90 | 4.24 | 7.52 | 11.90 | 17.54 | 24.46 | 32.67 | 42.22 | 53.14 | 65.48 | 79.28 |
| 30 MPH | .00 | .72 | 2.84 | 6.37 | 11.28 | 17.86 | 26.31 | 36.68 | 49.00 | 63.33 | 79.71 | 98.22 | 118.92 |
| VELOCITY (FPS) | 800. | 771. | 744. | 718. | 693. | 669. | 647. | 625. | 604. | 583. | 564. | 545. | 526. |
| ENERGY (FT-LB) | 320. | 297. | 276. | 257. | 240. | 224. | 209. | 195. | 182. | 170. | 159. | 148. | 138. |
| DROP (IN) | .00 | -1.57 | -6.77 | -15.88 | -29.19 | -47.01 | -69.66 | -97.48 | -130.85 | -170.18 | -215.88 | -268.42 | -328.29 |
| MID-RANGE (IN) | .00 | .43 | 1.81 | 4.24 | 7.82 | 12.68 | 18.94 | 26.74 | 36.23 | 47.57 | 60.93 | 76.50 | 94.49 |
| BULLET PATH (IN) | -.63 | 1.50 | -.00 | -5.41 | -15.02 | -29.13 | -48.08 | -72.20 | -101.87 | -137.49 | -179.49 | -228.33 | -284.49 |
| TIME OF FLIGHT (SEC) | .000000 | .095517 | .194592 | .297279 | .403655 | .513812 | .627859 | .745921 | .868135 | .994653 | 1.125637 | 1.261262 | 1.401715 |
| WIND DEFLECTION (IN) | | | | | | | | | | | | | |
| 5 MPH | .00 | .16 | .62 | 1.41 | 2.52 | 3.97 | 5.75 | 7.89 | 10.40 | 13.28 | 16.56 | 20.24 | 24.35 |
| 10 MPH | .00 | .31 | 1.25 | 2.82 | 5.04 | 7.93 | 11.50 | 15.78 | 20.79 | 26.56 | 33.11 | 40.48 | 48.70 |
| 20 MPH | .00 | .62 | 2.50 | 5.64 | 10.09 | 15.86 | 23.01 | 31.56 | 41.58 | 53.12 | 66.22 | 80.96 | 97.40 |
| 30 MPH | .00 | .93 | 3.74 | 8.46 | 15.13 | 23.79 | 34.51 | 47.35 | 62.38 | 79.68 | 99.34 | 121.45 | 146.11 |
| VELOCITY (FPS) | 700. | 676. | 653. | 631. | 610. | 589. | 569. | 550. | 531. | 513. | 495. | 478. | 462. |
| ENERGY (FT-LB) | 245. | 228. | 213. | 199. | 186. | 173. | 162. | 151. | 141. | 131. | 123. | 114. | 107. |
| DROP (IN) | .00 | -2.02 | -8.77 | -20.60 | -37.87 | -60.97 | -90.31 | -126.35 | -169.57 | -220.49 | -279.67 | -347.72 | -425.29 |
| MID-RANGE (IN) | .00 | .57 | 2.37 | 5.52 | 10.16 | 16.45 | 24.55 | 34.64 | 46.91 | 61.58 | 78.87 | 99.03 | 122.33 |
| BULLET PATH (IN) | -.63 | 2.05 | .00 | -7.13 | -19.69 | -38.09 | -62.73 | -94.07 | -132.58 | -178.80 | -233.28 | -296.63 | -369.50 |
| TIME OF FLIGHT (SEC) | .000000 | .109035 | .221925 | .338791 | .459766 | .584996 | .714641 | .848871 | .987869 | 1.131829 | 1.280956 | 1.435468 | 1.595595 |
| WIND DEFLECTION (IN) | | | | | | | | | | | | | |
| 5 MPH | .00 | .17 | .67 | 1.53 | 2.75 | 4.34 | 6.32 | 8.70 | 11.50 | 14.74 | 18.44 | 22.61 | 27.27 |
| 10 MPH | .00 | .33 | 1.34 | 3.06 | 5.49 | 8.67 | 12.63 | 17.40 | 23.01 | 29.49 | 36.88 | 45.21 | 54.54 |
| 20 MPH | .00 | .67 | 2.69 | 6.11 | 10.98 | 17.35 | 25.27 | 34.80 | 46.02 | 58.98 | 73.75 | 90.43 | 109.08 |
| 30 MPH | .00 | 1.00 | 4.03 | 9.17 | 16.47 | 26.02 | 37.90 | 52.20 | 69.02 | 88.46 | 110.63 | 135.64 | 163.62 |

## Bullet: Sierra .4295 Dia. 240 Gr. JHC
## Bullet Coefficient: .183

| RANGE (YDS) | MUZZLE | 25 | 50 | 75 | 100 | 125 | 150 | 175 | 200 | 225 | 250 | 275 | 300 |
|---|---|---|---|---|---|---|---|---|---|---|---|---|---|
| VELOCITY (FPS) | 1600. | 1513. | 1432. | 1356. | 1285. | 1218. | 1159. | 1109. | 1067. | 1031. | 999. | 971. | 946. |
| ENERGY (FT-LB) | 1364. | 1220. | 1093. | 980. | 880. | 791. | 716. | 655. | 606. | 566. | 532. | 502. | 477. |
| DROP (IN) | .00 | -.41 | -1.78 | -4.21 | -7.82 | -12.74 | -19.14 | -27.15 | -36.92 | -48.59 | -62.31 | -78.19 | -96.37 |
| MID-RANGE (IN) | .00 | .11 | .48 | 1.14 | 2.13 | 3.52 | 5.36 | 7.72 | 10.64 | 14.19 | 18.41 | 23.35 | 29.04 |
| BULLET PATH (IN) | -.63 | .16 | .00 | -1.23 | -3.63 | -7.36 | -12.55 | -19.36 | -27.92 | -38.39 | -50.90 | -65.59 | -82.56 |
| TIME OF FLIGHT (SEC) | .000000 | .048203 | .099158 | .152996 | .209818 | .269791 | .332937 | .399119 | .468117 | .539687 | .613634 | .689811 | .768105 |
| WIND DEFLECTION (IN) | | | | | | | | | | | | | |
| 5 MPH | .00 | .12 | .48 | 1.09 | 1.96 | 3.12 | 4.55 | 6.25 | 8.19 | 10.37 | 12.75 | 15.33 | 18.09 |
| 10 MPH | .00 | .23 | .95 | 2.18 | 3.93 | 6.23 | 9.10 | 12.49 | 16.39 | 20.73 | 25.50 | 30.66 | 36.19 |
| 20 MPH | .00 | .47 | 1.90 | 4.35 | 7.86 | 12.47 | 18.19 | 24.99 | 32.78 | 41.47 | 51.00 | 61.31 | 72.37 |
| 30 MPH | .00 | .70 | 2.86 | 6.53 | 11.78 | 18.70 | 27.29 | 37.48 | 49.17 | 62.20 | 76.50 | 91.97 | 108.56 |
| VELOCITY (FPS) | 1500. | 1419. | 1344. | 1274. | 1208. | 1151. | 1102. | 1060. | 1025. | 994. | 967. | 942. | 919. |
| ENERGY (FT-LB) | 1199. | 1073. | 963. | 864. | 778. | 706. | 647. | 599. | 560. | 527. | 498. | 473. | 450. |
| DROP (IN) | .00 | -.48 | -2.04 | -4.80 | -8.90 | -14.50 | -21.73 | -30.74 | -41.69 | -54.69 | -69.88 | -87.38 | -107.28 |
| MID-RANGE (IN) | .00 | .13 | .54 | 1.28 | 2.41 | 3.99 | 6.07 | 8.70 | 11.94 | 15.84 | 20.44 | 25.79 | 31.91 |
| BULLET PATH (IN) | -.63 | .22 | .00 | -1.43 | -4.19 | -8.45 | -14.35 | -22.03 | -31.64 | -43.31 | -57.16 | -73.33 | -91.90 |
| TIME OF FLIGHT (SEC) | .000000 | .051411 | .105722 | .163036 | .223523 | .287166 | .353815 | .423240 | .495202 | .569515 | .646037 | .724660 | .805301 |
| WIND DEFLECTION (IN) | | | | | | | | | | | | | |
| 5 MPH | .00 | .12 | .50 | 1.15 | 2.07 | 3.27 | 4.74 | 6.45 | 8.38 | 10.52 | 12.85 | 15.37 | 18.07 |
| 10 MPH | .00 | .25 | 1.01 | 2.29 | 4.14 | 6.54 | 9.47 | 12.89 | 16.76 | 21.03 | 25.70 | 30.74 | 36.13 |
| 20 MPH | .00 | .50 | 2.01 | 4.59 | 8.28 | 13.08 | 18.94 | 25.78 | 33.51 | 42.07 | 51.41 | 61.48 | 72.27 |
| 30 MPH | .00 | .74 | 3.02 | 6.88 | 12.42 | 19.62 | 28.41 | 38.67 | 50.27 | 63.10 | 77.11 | 92.22 | 108.40 |
| VELOCITY (FPS) | 1400. | 1326. | 1257. | 1193. | 1138. | 1091. | 1051. | 1017. | 987. | 960. | 936. | 914. | 893. |
| ENERGY (FT-LB) | 1044. | 937. | 842. | 758. | 690. | 634. | 589. | 551. | 519. | 491. | 467. | 445. | 425. |
| DROP (IN) | .00 | -.55 | -2.34 | -5.51 | -10.21 | -16.57 | -24.76 | -34.91 | -47.15 | -61.61 | -78.40 | -97.64 | -119.47 |
| MID-RANGE (IN) | .00 | .15 | .62 | 1.47 | 2.76 | 4.54 | 6.87 | 9.80 | 13.37 | 17.64 | 22.63 | 28.39 | 34.97 |
| BULLET PATH (IN) | -.63 | .30 | .00 | -1.69 | -4.89 | -9.78 | -16.48 | -25.14 | -35.90 | -48.88 | -64.18 | -81.93 | -102.28 |
| TIME OF FLIGHT (SEC) | .000000 | .055053 | .113139 | .174414 | .238817 | .306174 | .376245 | .448803 | .523672 | .600721 | .679848 | .760975 | .844044 |
| WIND DEFLECTION (IN) | | | | | | | | | | | | | |
| 5 MPH | .00 | .13 | .53 | 1.21 | 2.16 | 3.37 | 4.82 | 6.49 | 8.37 | 10.43 | 12.68 | 15.11 | 17.70 |
| 10 MPH | .00 | .26 | 1.06 | 2.41 | 4.32 | 6.74 | 9.65 | 12.99 | 16.74 | 20.87 | 25.37 | 30.22 | 35.41 |
| 20 MPH | .00 | .52 | 2.11 | 4.82 | 8.63 | 13.49 | 19.30 | 25.98 | 33.48 | 41.74 | 50.73 | 60.43 | 70.82 |
| 30 MPH | .00 | .78 | 3.17 | 7.23 | 12.95 | 20.23 | 28.94 | 38.97 | 50.21 | 62.61 | 76.10 | 90.65 | 106.23 |
| VELOCITY (FPS) | 1300. | 1231. | 1171. | 1119. | 1075. | 1038. | 1005. | 977. | 951. | 927. | 906. | 885. | 866. |
| ENERGY (FT-LB) | 900. | 808. | 730. | 667. | 616. | 574. | 538. | 508. | 482. | 458. | 437. | 418. | 400. |
| DROP (IN) | .00 | -.65 | -2.73 | -6.40 | -11.79 | -19.06 | -28.34 | -39.77 | -53.47 | -69.52 | -88.08 | -109.28 | -133.25 |
| MID-RANGE (IN) | .00 | .17 | .72 | 1.70 | 3.17 | 5.17 | 7.78 | 11.02 | 14.94 | 19.58 | 24.98 | 31.20 | 38.28 |
| BULLET PATH (IN) | -.63 | .40 | .00 | -1.99 | -5.70 | -11.29 | -18.89 | -28.64 | -40.66 | -55.03 | -71.91 | -91.43 | -113.72 |
| TIME OF FLIGHT (SEC) | .000000 | .059302 | .121793 | .187355 | .255785 | .326834 | .400297 | .476016 | .553873 | .633775 | .715653 | .799453 | .885140 |
| WIND DEFLECTION (IN) | | | | | | | | | | | | | |
| 5 MPH | .00 | .14 | .56 | 1.26 | 2.20 | 3.38 | 4.76 | 6.35 | 8.13 | 10.08 | 12.21 | 14.51 | 16.97 |
| 10 MPH | .00 | .28 | 1.13 | 2.51 | 4.40 | 6.75 | 9.53 | 12.70 | 16.25 | 20.16 | 24.42 | 29.01 | 33.94 |
| 20 MPH | .00 | .57 | 2.26 | 5.03 | 8.81 | 13.51 | 19.06 | 25.40 | 32.50 | 40.32 | 48.83 | 58.02 | 67.88 |
| 30 MPH | .00 | .85 | 3.38 | 7.54 | 13.21 | 20.26 | 28.59 | 38.11 | 48.75 | 60.48 | 73.25 | 87.03 | 101.82 |

## Bullet: Sierra .4295 Dia. 240 Gr. JHC
## Bullet Coefficient: .183

| RANGE (YDS) | MUZZLE | 25 | 50 | 75 | 100 | 125 | 150 | 175 | 200 | 225 | 250 | 275 | 300 |
|---|---|---|---|---|---|---|---|---|---|---|---|---|---|
| VELOCITY (FPS) | 1200. | 1144. | 1096. | 1055. | 1021. | 990. | 963. | 939. | 916. | 895. | 876. | 857. | 839. |
| ENERGY (FT-LB) | 767. | 697. | 640. | 594. | 555. | 523. | 494. | 470. | 447. | 427. | 408. | 391. | 375. |
| DROP (IN) | .00 | -.76 | -3.18 | -7.40 | -13.57 | -21.82 | -32.27 | -45.04 | -60.23 | -78.01 | -98.49 | -121.79 | -148.05 |
| MID-RANGE (IN) | .00 | .20 | .83 | 1.95 | 3.60 | 5.85 | 8.73 | 12.28 | 16.54 | 21.58 | 27.43 | 34.14 | 41.75 |
| BULLET PATH (IN) | -.63 | .51 | .00 | -2.32 | -6.58 | -12.92 | -21.46 | -32.32 | -45.61 | -61.48 | -80.06 | -101.46 | -125.80 |
| TIME OF FLIGHT (SEC) | .000000 | .064051 | .131081 | .200853 | .273135 | .347745 | .424548 | .503439 | .584338 | .667184 | .751932 | .838552 | .927021 |
| WIND DEFLECTION (IN) | | | | | | | | | | | | | |
| 5 MPH | .00 | .14 | .54 | 1.18 | 2.04 | 3.10 | 4.36 | 5.80 | 7.42 | 9.21 | 11.17 | 13.29 | 15.58 |
| 10 MPH | .00 | .27 | 1.07 | 2.35 | 4.07 | 6.20 | 8.72 | 11.61 | 14.84 | 18.42 | 22.34 | 26.59 | 31.16 |
| 20 MPH | .00 | .55 | 2.14 | 4.70 | 8.14 | 12.41 | 17.44 | 23.21 | 29.69 | 36.85 | 44.68 | 53.17 | 62.31 |
| 30 MPH | .00 | .82 | 3.21 | 7.05 | 12.22 | 18.61 | 26.16 | 34.82 | 44.53 | 55.27 | 67.02 | 79.76 | 93.47 |
| VELOCITY (FPS) | 1100. | 1059. | 1024. | 993. | 966. | 941. | 918. | 897. | 877. | 859. | 841. | 824. | 808. |
| ENERGY (FT-LB) | 645. | 598. | 559. | 526. | 497. | 472. | 449. | 429. | 410. | 393. | 377. | 362. | 348. |
| DROP (IN) | .00 | -.90 | -3.73 | -8.62 | -15.71 | -25.11 | -36.92 | -51.30 | -68.37 | -88.25 | -111.07 | -136.94 | -166.01 |
| MID-RANGE (IN) | .00 | .23 | .97 | 2.25 | 4.13 | 6.65 | 9.84 | 13.77 | 18.47 | 24.01 | 30.42 | 37.77 | 46.09 |
| BULLET PATH (IN) | -.63 | .65 | .00 | -2.72 | -7.63 | -14.85 | -24.48 | -36.68 | -51.57 | -69.28 | -89.92 | -113.62 | -140.51 |
| TIME OF FLIGHT (SEC) | .000000 | .069522 | .141572 | .215967 | .292566 | .371261 | .451971 | .534632 | .619199 | .705639 | .793930 | .884061 | .976029 |
| WIND DEFLECTION (IN) | | | | | | | | | | | | | |
| 5 MPH | .00 | .12 | .46 | 1.01 | 1.75 | 2.67 | 3.77 | 5.05 | 6.49 | 8.10 | 9.87 | 11.80 | 13.89 |
| 10 MPH | .00 | .24 | .92 | 2.01 | 3.49 | 5.34 | 7.55 | 10.10 | 12.98 | 16.19 | 19.73 | 23.59 | 27.78 |
| 20 MPH | .00 | .47 | 1.83 | 4.02 | 6.98 | 10.68 | 15.09 | 20.19 | 25.96 | 32.38 | 39.46 | 47.19 | 55.56 |
| 30 MPH | .00 | .71 | 2.75 | 6.03 | 10.47 | 16.03 | 22.64 | 30.29 | 38.94 | 48.58 | 59.19 | 70.78 | 83.34 |
| VELOCITY (FPS) | 1000. | 972. | 946. | 923. | 902. | 882. | 863. | 845. | 828. | 812. | 796. | 780. | 764. |
| ENERGY (FT-LB) | 533. | 503. | 477. | 454. | 433. | 414. | 397. | 380. | 365. | 351. | 337. | 324. | 311. |
| DROP (IN) | .00 | -1.07 | -4.43 | -10.16 | -18.43 | -29.36 | -43.09 | -59.71 | -79.36 | -102.18 | -128.28 | -157.68 | -190.64 |
| MID-RANGE (IN) | .00 | .28 | 1.15 | 2.62 | 4.79 | 7.70 | 11.38 | 15.88 | 21.25 | 27.53 | 34.78 | 42.97 | 52.23 |
| BULLET PATH (IN) | -.63 | .83 | .00 | -3.20 | -8.94 | -17.35 | -28.54 | -42.64 | -59.76 | -80.05 | -103.62 | -130.49 | -160.92 |
| TIME OF FLIGHT (SEC) | .000000 | .076095 | .154309 | .234554 | .316763 | .400887 | .486890 | .574748 | .664448 | .755984 | .849361 | .944638 | 1.041852 |
| WIND DEFLECTION (IN) | | | | | | | | | | | | | |
| 5 MPH | .00 | .10 | .38 | .84 | 1.48 | 2.28 | 3.25 | 4.38 | 5.67 | 7.13 | 8.74 | 10.53 | 12.48 |
| 10 MPH | .00 | .19 | .76 | 1.68 | 2.95 | 4.56 | 6.49 | 8.76 | 11.34 | 14.25 | 17.49 | 21.06 | 24.97 |
| 20 MPH | .00 | .39 | 1.52 | 3.36 | 5.90 | 9.11 | 12.99 | 17.51 | 22.69 | 28.51 | 34.97 | 42.11 | 49.93 |
| 30 MPH | .00 | .58 | 2.28 | 5.04 | 8.85 | 13.67 | 19.48 | 26.27 | 34.03 | 42.76 | 52.46 | 63.17 | 74.90 |
| VELOCITY (FPS) | 900. | 880. | 861. | 843. | 826. | 810. | 794. | 778. | 763. | 748. | 734. | 720. | 706. |
| ENERGY (FT-LB) | 432. | 413. | 395. | 379. | 364. | 350. | 336. | 323. | 310. | 298. | 287. | 276. | 266. |
| DROP (IN) | .00 | -1.25 | -5.29 | -12.25 | -22.26 | -35.44 | -51.90 | -71.67 | -95.01 | -122.07 | -152.99 | -187.94 | -227.07 |
| MID-RANGE (IN) | .00 | .34 | 1.40 | 3.22 | 5.84 | 9.32 | 13.70 | 18.97 | 25.24 | 32.59 | 41.05 | 50.71 | 61.64 |
| BULLET PATH (IN) | -.63 | 1.08 | -.00 | -4.00 | -11.05 | -21.28 | -34.77 | -51.58 | -71.97 | -96.07 | -124.03 | -156.02 | -192.19 |
| TIME OF FLIGHT (SEC) | .000000 | .084282 | .170440 | .258451 | .348301 | .439988 | .533518 | .628954 | .726326 | .825646 | .926933 | 1.030206 | 1.135489 |
| WIND DEFLECTION (IN) | | | | | | | | | | | | | |
| 5 MPH | .00 | .08 | .33 | .74 | 1.32 | 2.05 | 2.95 | 4.01 | 5.25 | 6.66 | 8.24 | 9.99 | 11.92 |
| 10 MPH | .00 | .17 | .66 | 1.49 | 2.63 | 4.10 | 5.90 | 8.03 | 10.50 | 13.31 | 16.47 | 19.98 | 23.85 |
| 20 MPH | .00 | .33 | 1.33 | 2.97 | 5.27 | 8.21 | 11.80 | 16.06 | 21.00 | 26.63 | 32.95 | 39.97 | 47.69 |
| 30 MPH | .00 | .50 | 1.99 | 4.46 | 7.90 | 12.31 | 17.70 | 24.09 | 31.50 | 39.94 | 49.42 | 59.95 | 71.54 |

## Bullet: Sierra .4295 Dia. 240 Gr. JHC
## Bullet Coefficient: .183

| RANGE (YDS) | MUZZLE | 25 | 50 | 75 | 100 | 125 | 150 | 175 | 200 | 225 | 250 | 275 | 300 |
|---|---|---|---|---|---|---|---|---|---|---|---|---|---|
| VELOCITY (FPS) | 800. | 734. | 768. | 753. | 739. | 725. | 711. | 698. | 685. | 672. | 659. | 647. | 635. |
| ENERGY (FT-LB) | 341. | 327. | 315. | 302. | 291. | 280. | 269. | 259. | 250. | 240. | 232. | 223. | 215. |
| DROP (IN) | .00 | -1.46 | -6.41 | -15.01 | -27.43 | -43.80 | -64.30 | -89.06 | -118.27 | -152.11 | -190.74 | -234.37 | -283.17 |
| MID-RANGE (IN) | .00 | .42 | 1.74 | 4.02 | 7.31 | 11.66 | 17.14 | 23.80 | 31.71 | 40.94 | 51.57 | 63.66 | 77.29 |
| BULLET PATH (IN) | -.63 | 1 43 | -.00 | -5.09 | -13.99 | -26.84 | -43.82 | -65.07 | -90.76 | -121.08 | -156.20 | -196.30 | -241.59 |
| TIME OF FLIGHT (SEC) | .000000 | .094713 | .191356 | .289942 | .390485 | .493005 | .597525 | .704069 | .812666 | .923346 | 1.036144 | 1.151095 | 1.268238 |
| WIND DEFLECTION (IN) | | | | | | | | | | | | | |
| 5 MPH | .00 | 08 | .34 | .76 | 1.36 | 2.13 | 3.08 | 4.21 | 5.51 | 7.00 | 8.68 | 10.55 | 12.60 |
| 10 MPH | .00 | 17 | .68 | 1.53 | 2.73 | 4.27 | 6.16 | 8.42 | 11.03 | 14.01 | 17.36 | 21.09 | 25.21 |
| 20 MPH | .00 | 34 | 1.36 | 3.06 | 5.45 | 8.54 | 12.33 | 16.83 | 22.06 | 28.02 | 34.72 | 42.19 | 50.42 |
| 30 MPH | .00 | 51 | 2.04 | 4.59 | 8.18 | 12.81 | 18.49 | 25.25 | 33.09 | 42.03 | 52.08 | 63.28 | 75.63 |
| VELOCITY (FPS) | 700. | 687. | 674. | 661. | 649. | 637. | 625. | 614. | 602. | 591. | 580. | 570. | 559. |
| ENERGY (FT-LB) | 261. | 251. | 242. | 233. | 224. | 216. | 208. | 201. | 193. | 186. | 180. | 173. | 167. |
| DROP (IN) | .00 | -.83 | -8.24 | -19.42 | -35.54 | -56.81 | -83.41 | -115.54 | -153.42 | -197.29 | -247.36 | -303.88 | -367.10 |
| MID-RANGE (IN) | .00 | .56 | 2.29 | 5.26 | 9.53 | 15.18 | 22.29 | 30.92 | 41.17 | 53.12 | 66.87 | 82.51 | 100.14 |
| BULLET PATH (IN) | -.63 | 1.98 | -.00 | -6.74 | -18.43 | -35.26 | -57.43 | -85.12 | -118.57 | -158.00 | -203.64 | -255.72 | -314.50 |
| TIME OF FLIGHT (SEC) | .000000 | .108170 | .218417 | .330771 | .445270 | .561949 | .680849 | .802012 | .925482 | 1.051305 | 1.179530 | 1.310206 | 1.443388 |
| WIND DEFLECTION (IN) | | | | | | | | | | | | | |
| 5 MPH | .00 | .09 | .36 | .82 | 1.47 | 2.31 | 3.34 | 4.58 | 6.01 | 7.66 | 9.51 | 11.58 | 13.88 |
| 10 MPH | .00 | .18 | .73 | 1.64 | 2.94 | 4.62 | 6.69 | 9.15 | 12.03 | 15.32 | 19.03 | 23.17 | 27.75 |
| 20 MPH | .00 | .36 | 1.45 | 3.29 | 5.88 | 9.23 | 13.37 | 18.31 | 24.06 | 30.63 | 38.05 | 46.34 | 55.50 |
| 30 MPH | .00 | .54 | 2.18 | 4.93 | 8.82 | 13.85 | 20.06 | 27.46 | 36.08 | 45.95 | 57.08 | 69.50 | 83.25 |

## Bullet: Speer .429 Dia. 240 Gr. JHP
## Ballistic Coefficient: .210

| RANGE (YDS) | MUZZLE | 25 | 50 | 75 | 100 | 125 | 150 | 175 | 200 | 225 | 250 | 275 | 300 |
|---|---|---|---|---|---|---|---|---|---|---|---|---|---|
| VELOCITY (FPS) | 1600. | 1524. | 1452. | 1385. | 1321. | 1262. | 1209. | 1161. | 1119. | 1083. | 1051. | 1023. | 997. |
| ENERGY (FT-LB) | 1364. | 1238. | 1124. | 1021. | 930. | 849. | 779. | 719. | 668. | 625. | 588. | 557. | 530. |
| DROP (IN) | .00 | -.40 | -1.74 | -4.13 | -7.64 | -12.39 | -18.52 | -26.13 | -35.35 | -46.30 | -59.09 | -73.85 | -90.68 |
| MID-RANGE (IN) | .00 | .11 | .47 | 1.12 | 2.07 | 3.40 | 5.13 | 7.33 | 10.04 | 13.29 | 17.15 | 21.65 | 26.82 |
| BULLET PATH (IN) | -.63 | .16 | .00 | -1.19 | -3.52 | -7.09 | -12.02 | -18.45 | -26.48 | -36.24 | -47.85 | -61.42 | -77.06 |
| TIME OF FLIGHT (SEC) | .000000 | .048029 | .098442 | .151339 | .206806 | .264896 | .325619 | .388930 | .454736 | .522893 | .593234 | .665607 | .739895 |
| WIND DEFLECTION (IN) | | | | | | | | | | | | | |
| 5 MPH | .00 | .10 | .41 | .94 | 1.70 | 2.69 | 3.90 | 5.35 | 7.02 | 8.89 | 10.95 | 13.20 | 15.61 |
| 10 MPH | .00 | .20 | .83 | 1.89 | 3.40 | 5.37 | 7.81 | 10.70 | 14.03 | 17.78 | 21.91 | 26.40 | 31.22 |
| 20 MPH | .00 | .41 | 1.65 | 3.77 | 6.80 | 10.74 | 15.62 | 21.40 | 28.07 | 35.56 | 43.82 | 52.79 | 62.44 |
| 30 MPH | .00 | .61 | 2.48 | 5.66 | 10.19 | 16.12 | 23.43 | 32.11 | 42.10 | 53.34 | 65.73 | 79.19 | 93.66 |
| VELOCITY (FPS) | 1500. | 1429. | 1363. | 1301. | 1244. | 1193. | 1147. | 1107. | 1072. | 1041. | 1014. | 989. | 967. |
| ENERGY (FT-LB) | 1199. | 1089. | 990. | 902. | 825. | 758. | 701. | 652. | 612. | 578. | 548. | 522. | 498. |
| DROP (IN) | .00 | -.47 | -2.01 | -4.72 | -8.71 | -14.11 | -21.03 | -29.61 | -39.95 | -52.18 | -66.40 | -82.74 | -101.28 |
| MID-RANGE (IN) | .00 | .13 | .53 | 1.26 | 2.34 | 3.84 | 5.80 | 8.26 | 11.27 | 14.86 | 19.09 | 24.00 | 29.61 |
| BULLET PATH (IN) | -.63 | .22 | .00 | -1.38 | -4.05 | -8.13 | -13.73 | -20.99 | -30.01 | -40.91 | -53.82 | -68.83 | -86.06 |
| TIME OF FLIGHT (SEC) | .000000 | .051227 | .104969 | .161303 | .220268 | .281857 | .346009 | .412613 | .481513 | .552543 | .625563 | .700464 | .777158 |
| WIND DEFLECTION (IN) | | | | | | | | | | | | | |
| 5 MPH | .00 | .11 | .44 | .99 | 1.78 | 2.80 | 4.05 | 5.51 | 7.17 | 9.02 | 11.05 | 13.24 | 15.59 |
| 10 MPH | .00 | .22 | .87 | 1.99 | 3.57 | 5.61 | 8.10 | 11.02 | 14.35 | 18.05 | 22.10 | 26.48 | 31.18 |
| 20 MPH | .00 | .43 | 1.75 | 3.98 | 7.13 | 11.21 | 16.20 | 22.04 | 28.69 | 36.10 | 44.20 | 52.96 | 62.36 |
| 30 MPH | .00 | .65 | 2.62 | 5.97 | 10.70 | 16.82 | 24.29 | 33.06 | 43.04 | 54.14 | 66.30 | 79.45 | 93.54 |
| VELOCITY (FPS) | 1400. | 1335. | 1276. | 1221. | 1172. | 1129. | 1091. | 1058. | 1029. | 1003. | 980. | 958. | 938. |
| ENERGY (FT-LB) | 1044. | 950. | 867. | 794. | 732. | 679. | 634. | 596. | 564. | 536. | 511. | 489. | 469. |
| DROP (IN) | .00 | -.55 | -2.31 | -5.41 | -9.98 | -16.12 | -23.96 | -33.62 | -45.22 | -58.86 | -74.66 | -92.72 | -113.11 |
| MID-RANGE (IN) | .00 | .14 | .61 | 1.44 | 2.68 | 4.37 | 6.57 | 9.31 | 12.63 | 16.59 | 21.21 | 26.54 | 32.59 |
| BULLET PATH (IN) | -.63 | .29 | .00 | -1.63 | -4.73 | -9.40 | -15.77 | -23.96 | -34.09 | -46.26 | -60.59 | -77.17 | -96.09 |
| TIME OF FLIGHT (SEC) | .000000 | .054861 | .112335 | .172444 | .235157 | .300391 | .368013 | .437857 | .509766 | .583615 | .659304 | .736752 | .815895 |
| WIND DEFLECTION (IN) | | | | | | | | | | | | | |
| 5 MPH | .00 | .11 | .46 | 1.03 | 1.84 | 2.86 | 4.10 | 5.53 | 7.15 | 8.93 | 10.88 | 12.98 | 15.23 |
| 10 MPH | .00 | .23 | .91 | 2.06 | 3.67 | 5.73 | 8.20 | 11.06 | 14.29 | 17.86 | 21.75 | 25.95 | 30.45 |
| 20 MPH | .00 | .45 | 1.83 | 4.13 | 7.35 | 11.45 | 16.40 | 22.13 | 28.58 | 35.72 | 43.50 | 51.91 | 60.91 |
| 30 MPH | .00 | .68 | 2.74 | 6.19 | 11.02 | 17.18 | 24.60 | 33.19 | 42.87 | 53.58 | 65.26 | 77.86 | 91.36 |
| VELOCITY (FPS) | 1300. | 1243. | 1192. | 1146. | 1106. | 1071. | 1041. | 1013. | 989. | 967. | 946. | 927. | 909. |
| ENERGY (FT-LB) | 900. | 823. | 757. | 700. | 652. | 611. | 577. | 547. | 521. | 498. | 477. | 458. | 440. |
| DROP (IN) | .00 | -.63 | -2.68 | -6.26 | -11.49 | -18.48 | -27.37 | -38.26 | -51.25 | -66.46 | -83.98 | -103.87 | -126.28 |
| MID-RANGE (IN) | .00 | .17 | .71 | 1.65 | 3.06 | 4.97 | 7.42 | 10.47 | 14.14 | 18.47 | 23.51 | 29.25 | 35.77 |
| BULLET PATH (IN) | -.63 | .39 | .00 | -1.92 | -5.49 | -10.83 | -18.06 | -27.29 | -38.62 | -52.18 | -68.04 | -86.27 | -107.02 |
| TIME OF FLIGHT (SEC) | .000000 | .059011 | .120645 | .184841 | .251486 | .320424 | .391488 | .464541 | .539472 | .616194 | .694636 | .774740 | .856463 |
| WIND DEFLECTION (IN) | | | | | | | | | | | | | |
| 5 MPH | .00 | .12 | .46 | 1.04 | 1.82 | 2.81 | 3.99 | 5.34 | 6.86 | 8.53 | 10.36 | 12.33 | 14.45 |
| 10 MPH | .00 | .23 | .93 | 2.07 | 3.65 | 5.63 | 7.98 | 10.68 | 13.72 | 17.07 | 20.72 | 24.66 | 28.89 |
| 20 MPH | .00 | .46 | 1.85 | 4.14 | 7.29 | 11.25 | 15.96 | 21.36 | 27.43 | 34.13 | 41.43 | 49.32 | 57.78 |
| 30 MPH | .00 | .70 | 2.78 | 6.21 | 10.94 | 16.88 | 23.94 | 32.05 | 41.15 | 51.20 | 62.15 | 73.99 | 86.67 |

## Bullet: Speer .429 Dia. 240 Gr. JHP
### Ballistic Coefficient: .210

| RANGE (YDS) | MUZZLE | 25 | 50 | 75 | 100 | 125 | 150 | 175 | 200 | 225 | 250 | 275 | 300 |
|---|---|---|---|---|---|---|---|---|---|---|---|---|---|
| VELOCITY (FPS) | 1200. | 1153. | 1112. | 1077. | 1045. | 1018. | 993. | 970. | 949. | 930. | 912. | 895. | 879. |
| ENERGY (FT-LB) | 767. | 709. | 659. | 617. | 582. | 552. | 525. | 502. | 480. | 461. | 443. | 427. | 412. |
| DROP (IN) | .00 | -.76 | -3.14 | -7.28 | -13.28 | -21.27 | -31.34 | -43.62 | -58.19 | -75.12 | -94.54 | -116.56 | -141.30 |
| MID-RANGE (IN) | .00 | .20 | .81 | 1.90 | 3.50 | 5.65 | 8.40 | 11.77 | 15.82 | 20.54 | 26.00 | 32.25 | 39.31 |
| BULLET PATH (IN) | -.63 | .50 | .00 | -2.25 | -6.37 | -12.47 | -20.66 | -31.05 | -43.74 | -58.78 | -76.32 | -96.46 | -119.31 |
| TIME OF FLIGHT (SEC) | .000000 | .063775 | .130020 | .198587 | .269306 | .342035 | .416658 | .493085 | .571243 | .651072 | .732525 | .815565 | .900164 |
| WIND DEFLECTION (IN) | | | | | | | | | | | | | |
| 5 MPH | .00 | .11 | .44 | .98 | 1.70 | 2.60 | 3.67 | 4.89 | 6.27 | 7.79 | 9.46 | 11.27 | 13.21 |
| 10 MPH | .00 | .22 | .88 | 1.95 | 3.40 | 5.20 | 7.33 | 9.78 | 12.54 | 15.59 | 18.92 | 22.54 | 26.43 |
| 20 MPH | .00 | .45 | 1.77 | 3.90 | 6.80 | 10.40 | 14.66 | 19.57 | 25.08 | 31.18 | 37.85 | 45.08 | 52.86 |
| 30 MPH | .00 | .67 | 2.65 | 5.85 | 10.19 | 15.59 | 22.00 | 29.35 | 37.62 | 46.77 | 56.77 | 67.62 | 79.29 |
| VELOCITY (FPS) | 1100. | 1066. | 1036. | 1009. | 985. | 963. | 943. | 924. | 906. | 890. | 874. | 859. | 844. |
| ENERGY (FT-LB) | 645. | 605. | 572. | 543. | 517. | 494. | 474. | 455. | 438. | 422. | 407. | 393. | 380. |
| DROP (IN) | .00 | -.89 | -3.68 | -8.49 | -15.43 | -24.60 | -36.09 | -49.98 | -66.39 | -85.44 | -107.23 | -131.86 | -159.43 |
| MID-RANGE (IN) | .00 | .23 | .95 | 2.21 | 4.03 | 6.47 | 9.55 | 13.30 | 17.76 | 22.99 | 29.01 | 35.87 | 43.62 |
| BULLET PATH (IN) | -.63 | .64 | .00 | -2.66 | -7.44 | -14.45 | -23.79 | -35.51 | -49.77 | -66.67 | -86.30 | -108.77 | -134.18 |
| TIME OF FLIGHT (SEC) | .000000 | .069286 | .140673 | .214029 | .289248 | .366245 | .444951 | .525312 | .607284 | .690832 | .775929 | .862557 | .950702 |
| WIND DEFLECTION (IN) | | | | | | | | | | | | | |
| 5 MPH | .00 | .10 | .38 | .83 | 1.45 | 2.23 | 3.16 | 4.23 | 5.44 | 6.79 | 8.28 | 9.91 | 11.66 |
| 10 MPH | .00 | .19 | .76 | 1.67 | 2.91 | 4.46 | 6.31 | 8.45 | 10.88 | 13.59 | 16.56 | 19.81 | 23.32 |
| 20 MPH | .00 | .39 | 1.52 | 3.34 | 5.82 | 8.92 | 12.62 | 16.91 | 21.76 | 27.17 | 33.13 | 39.62 | 46.65 |
| 30 MPH | .00 | .58 | 2.28 | 5.01 | 8.72 | 13.38 | 18.93 | 25.36 | 32.65 | 40.76 | 49.69 | 59.43 | 69.97 |
| VELOCITY (FPS) | 1000. | 977. | 955. | 936. | 917. | 900. | 883. | 868. | 853. | 839. | 825. | 811. | 798. |
| ENERGY (FT-LB) | 533. | 508. | 486. | 466. | 448. | 431. | 416. | 401. | 388. | 375. | 362. | 351. | 340. |
| DROP (IN) | .00 | -1.06 | -4.39 | -10.05 | -18.18 | -28.87 | -42.24 | -58.38 | -77.39 | -99.38 | -124.45 | -152.70 | -184.24 |
| MID-RANGE (IN) | .00 | .28 | 1.14 | 2.60 | 4.70 | 7.51 | 11.07 | 15.39 | 20.52 | 26.49 | 33.35 | 41.13 | 49.88 |
| BULLET PATH (IN) | -.63 | .32 | .00 | -3.16 | -8.78 | -16.96 | -27.83 | -41.46 | -57.96 | -77.44 | -100.00 | -125.74 | -154.77 |
| TIME OF FLIGHT (SEC) | .000000 | .075930 | .153549 | .232886 | .313859 | .396428 | .480562 | .566237 | .653435 | .742146 | .832363 | .924084 | 1.017313 |
| WIND DEFLECTION (IN) | | | | | | | | | | | | | |
| 5 MPH | .00 | .08 | .31 | .69 | 1.22 | 1.89 | 2.69 | 3.63 | 4.70 | 5.91 | 7.25 | 8.72 | 10.32 |
| 10 MPH | .00 | .16 | .62 | 1.39 | 2.44 | 3.77 | 5.38 | 7.26 | 9.40 | 11.82 | 14.50 | 17.44 | 20.65 |
| 20 MPH | .00 | .32 | 1.25 | 2.78 | 4.88 | 7.54 | 10.76 | 14.52 | 18.81 | 23.64 | 28.99 | 34.88 | 41.29 |
| 30 MPH | .00 | .48 | 1.87 | 4.16 | 7.32 | 11.31 | 16.14 | 21.77 | 28.21 | 35.45 | 43.49 | 52.32 | 61.94 |
| VELOCITY (FPS) | 900. | 884. | 868. | 853. | 839. | 825. | 811. | 798. | 779. | 760. | 742. | 725. | 708. |
| ENERGY (FT-LB) | 432. | 416. | 401. | 388. | 375. | 362. | 351. | 340. | 323. | 308. | 294. | 280. | 267. |
| DROP (IN) | .00 | -1.22 | -5.21 | -12.08 | -21.92 | -34.83 | -50.93 | -70.31 | -93.04 | -119.33 | -149.36 | -183.32 | -221.40 |
| MID-RANGE (IN) | .00 | .34 | 1.39 | 3.17 | 5.74 | 9.13 | 13.38 | 18.54 | 24.60 | 31.68 | 39.85 | 49.19 | 59.78 |
| BULLET PATH (IN) | -.63 | 1.07 | -.00 | -3.94 | -10.86 | -20.85 | -34.02 | -50.48 | -70.29 | -93.66 | -120.77 | -151.80 | -186.96 |
| TIME OF FLIGHT (SEC) | .000000 | .084111 | .169763 | .256938 | .345625 | .435817 | .527512 | .620714 | .715857 | .813366 | .913260 | 1.015568 | 1.120320 |
| WIND DEFLECTION (IN) | | | | | | | | | | | | | |
| 5 MPH | .00 | .07 | .27 | .61 | 1.08 | 1.69 | 2.42 | 3.29 | 4.33 | 5.58 | 7.03 | 8.70 | 10.59 |
| 10 MPH | .00 | .14 | .54 | 1.22 | 2.16 | 3.37 | 4.84 | 6.58 | 8.66 | 11.15 | 14.07 | 17.41 | 21.18 |
| 20 MPH | .00 | .27 | 1.09 | 2.44 | 4.33 | 6.74 | 9.68 | 13.16 | 17.32 | 22.30 | 28.13 | 34.81 | 42.35 |
| 30 MPH | .00 | .41 | 1.63 | 3.66 | 6.49 | 10.11 | 14.53 | 19.74 | 25.97 | 33.46 | 42.20 | 52.22 | 63.53 |

# Bullet: Speer .429 Dia. 240 Gr. JHP
## Ballistic Coefficient: .210

| RANGE (YDS) | MUZZLE | 25 | 50 | 75 | 100 | 125 | 150 | 175 | 200 | 225 | 250 | 275 | 300 |
|---|---|---|---|---|---|---|---|---|---|---|---|---|---|
| VELOCITY (FPS) | 800. | 780. | 762. | 744. | 726. | 709. | 693. | 677. | 662. | 647. | 632. | | |
| ENERGY (FT-LB) | 341. | 324. | 309. | 295. | 281. | 268. | 256. | 244. | 233. | 223. | 213. | 204. | 195. |
| DROP (IN) | .00 | -1.49 | -6.53 | -15.30 | -27.98 | -44.77 | -65.85 | -91.39 | -121.66 | -156.86 | -197.25 | -243.06 | -294.55 |
| MID-RANGE (IN) | .00 | .43 | 1.77 | 4.09 | 7.46 | 11.95 | 17.63 | 24.56 | 32.85 | 42.58 | 53.86 | 66.77 | 81.42 |
| BULLET PATH (IN) | -.63 | 1.46 | -.00 | -5.19 | -14.29 | -27.49 | -45.00 | -66.96 | -93.64 | -125.27 | -162.07 | -204.30 | -252.21 |
| TIME OF FLIGHT (SEC) | .000000 | .094927 | .192216 | .291888 | .393967 | .498487 | .605484 | .715002 | .827089 | .941797 | 1.059183 | 1.179306 | 1.302233 |
| WIND DEFLECTION (IN) | | | | | | | | | | | | | |
| 5 MPH | .00 | .10 | .42 | .94 | 1.67 | 2.62 | 3.78 | 5.17 | 6.78 | 8.63 | 10.71 | 13.03 | 15.60 |
| 10 MPH | .00 | .21 | .83 | 1.87 | 3.34 | 5.23 | 7.57 | 10.34 | 13.57 | 17.26 | 21.42 | 26.06 | 31.19 |
| 20 MPH | .00 | .41 | 1.66 | 3.74 | 6.68 | 10.47 | 15.13 | 20.68 | 27.14 | 34.51 | 42.83 | 52.12 | 62.39 |
| 30 MPH | .00 | .62 | 2.49 | 5.62 | 10.01 | 15.70 | 22.70 | 31.02 | 40.70 | 51.77 | 64.25 | 78.17 | 93.58 |
| VELOCITY (FPS) | 700. | 684. | 668. | 653. | 638. | 624. | 610. | 596. | 583. | 569. | 557. | 544. | 532. |
| ENERGY (FT-LB) | 261. | 249. | 238. | 227. | 217. | 207. | 198. | 189. | 181. | 173. | 165. | 158. | 151. |
| DROP (IN) | .00 | -1.93 | -8.48 | -19.88 | -36.35 | -58.15 | -85.51 | -118.63 | -157.85 | -203.46 | -255.77 | -315.09 | -381.76 |
| MID-RANGE (IN) | .00 | .57 | 2.31 | 5.33 | 9.70 | 15.52 | 22.88 | 31.85 | 42.57 | 55.16 | 69.74 | 86.43 | 105.37 |
| BULLET PATH (IN) | -.63 | 2.00 | -.00 | -6.85 | -18.77 | -36.01 | -58.81 | -87.38 | -122.05 | -163.11 | -210.86 | -265.63 | -327.74 |
| TIME OF FLIGHT (SEC) | .000000 | .108400 | .219346 | .332888 | .449080 | .567980 | .689652 | .814161 | .941577 | 1.071976 | 1.205433 | 1.342030 | 1.481851 |
| WIND DEFLECTION (IN) | | | | | | | | | | | | | |
| 5 MPH | .00 | .11 | .45 | 1.01 | 1.80 | 2.84 | 4.12 | 5.65 | 7.43 | 9.48 | 11.79 | 14.38 | 17.26 |
| 10 MPH | .00 | .22 | .89 | 2.02 | 3.61 | 5.68 | 8.24 | 11.29 | 14.86 | 18.95 | 23.58 | 28.77 | 34.52 |
| 20 MPH | .00 | .44 | 1.78 | 4.03 | 7.22 | 11.36 | 16.47 | 22.58 | 29.72 | 37.91 | 47.17 | 57.54 | 69.04 |
| 30 MPH | .00 | .66 | 2.67 | 6.05 | 10.83 | 17.04 | 24.71 | 33.88 | 44.58 | 56.86 | 70.75 | 86.31 | 103.56 |

## Bullet: Hornady .430 Dia. 265 Gr. FN
### Ballistic Coefficient: .267

| RANGE (YDS) | MUZZLE | 25 | 50 | 75 | 100 | 125 | 150 | 175 | 200 | 225 | 250 | 275 | 300 |
|---|---|---|---|---|---|---|---|---|---|---|---|---|---|
| VELOCITY (FPS) | 1500. | 1444. | 1391. | 1340. | 1292. | 1247. | 1206. | 1168. | 1134. | 1103. | 1075. | 1050. | 1028. |
| ENERGY (FT-LB) | 1324. | 1227. | 1138. | 1057. | 983. | 916. | 856. | 803. | 756. | 715. | 680. | 649. | 621. |
| DROP (IN) | .00 | -.46 | -1.97 | -4.60 | -8.44 | -13.57 | -20.12 | -28.16 | -37.78 | -49.09 | -62.18 | -77.15 | -94.08 |
| MID-RANGE (IN) | .00 | .13 | .52 | 1.22 | 2.25 | 3.64 | 5.46 | 7.71 | 10.45 | 13.71 | 17.52 | 21.91 | 26.92 |
| BULLET PATH (IN) | -.63 | .21 | .00 | -1.33 | -3.86 | -7.70 | -12.94 | -19.67 | -28.00 | -38.00 | -49.80 | -63.46 | -79.09 |
| TIME OF FLIGHT (SEC) | .000000 | .050952 | .103888 | .158830 | .215827 | .274904 | .336064 | .399274 | .464474 | .531583 | .600498 | .671111 | .743333 |
| WIND DEFLECTION (IN) | | | | | | | | | | | | | |
| 5 MPH | .00 | .08 | .34 | .78 | 1.39 | 2.19 | 3.17 | 4.34 | 5.67 | 7.18 | 8.84 | 10.66 | 12.61 |
| 10 MPH | .00 | .17 | .68 | 1.55 | 2.79 | 4.38 | 6.35 | 8.67 | 11.35 | 14.36 | 17.69 | 21.32 | 25.23 |
| 20 MPH | .00 | .34 | 1.37 | 3.11 | 5.57 | 8.77 | 12.69 | 17.34 | 22.69 | 28.72 | 35.38 | 42.63 | 50.45 |
| 30 MPH | .00 | .51 | 2.05 | 4.66 | 8.36 | 13.15 | 19.04 | 26.02 | 34.04 | 43.08 | 53.06 | 63.95 | 75.68 |
| VELOCITY (FPS) | 1400. | 1349. | 1301. | 1255. | 1213. | 1174. | 1139. | 1108. | 1080. | 1054. | 1031. | 1010. | 991. |
| ENERGY (FT-LB) | 1153. | 1070. | 995. | 927. | 866. | 811. | 764. | 722. | 686. | 654. | 626. | 601. | 578. |
| DROP (IN) | .00 | -.54 | -2.27 | -5.28 | -9.68 | -15.56 | -23.01 | -32.13 | -43.01 | -55.75 | -70.44 | -87.17 | -106.02 |
| MID-RANGE (IN) | .00 | .14 | .60 | 1.39 | 2.57 | 4.17 | 6.22 | 8.76 | 11.83 | 15.45 | 19.66 | 24.51 | 30.00 |
| BULLET PATH (IN) | -.63 | .28 | .00 | -1.56 | -4.51 | -8.94 | -14.94 | -22.61 | -32.05 | -43.34 | -56.58 | -71.86 | -89.26 |
| TIME OF FLIGHT (SEC) | .000000 | .054584 | .111216 | .169926 | .230720 | .293572 | .358427 | .425206 | .493810 | .564131 | .636074 | .709562 | .784529 |
| WIND DEFLECTION (IN) | | | | | | | | | | | | | |
| 5 MPH | .00 | .09 | .36 | .81 | 1.45 | 2.26 | 3.26 | 4.42 | 5.74 | 7.21 | 8.83 | 10.58 | 12.47 |
| 10 MPH | .00 | .18 | .72 | 1.62 | 2.89 | 4.53 | 6.51 | 8.84 | 11.48 | 14.43 | 17.66 | 21.17 | 24.93 |
| 20 MPH | .00 | .36 | 1.43 | 3.24 | 5.78 | 9.05 | 13.02 | 17.67 | 22.96 | 28.86 | 35.33 | 42.34 | 49.87 |
| 30 MPH | .00 | .53 | 2.15 | 4.86 | 8.68 | 13.58 | 19.54 | 26.51 | 34.45 | 43.29 | 52.99 | 63.51 | 74.80 |
| VELOCITY (FPS) | 1300. | 1255. | 1212. | 1174. | 1139. | 1107. | 1079. | 1054. | 1031. | 1010. | 991. | 973. | 956. |
| ENERGY (FT-LB) | 994. | 926. | 865. | 811. | 763. | 721. | 685. | 654. | 625. | 600. | 578. | 557. | 538. |
| DROP (IN) | .00 | -.62 | -2.64 | -6.13 | -11.20 | -17.93 | -26.43 | -36.80 | -49.11 | -63.46 | -79.93 | -98.61 | -119.58 |
| MID-RANGE (IN) | .00 | .17 | .70 | 1.61 | 2.96 | 4.77 | 7.09 | 9.94 | 13.36 | 17.38 | 22.03 | 27.35 | 33.36 |
| BULLET PATH (IN) | -.63 | .38 | .00 | -1.86 | -5.29 | -10.39 | -17.26 | -25.99 | -36.67 | -49.38 | -64.22 | -81.27 | -100.60 |
| TIME OF FLIGHT (SEC) | .000000 | .058737 | .119560 | .182439 | .247320 | .314124 | .382751 | .453093 | .525056 | .598563 | .673549 | .749958 | .827745 |
| WIND DEFLECTION (IN) | | | | | | | | | | | | | |
| 5 MPH | .00 | .09 | .37 | .82 | 1.46 | 2.26 | 3.22 | 4.33 | 5.59 | 6.98 | 8.50 | 10.15 | 11.92 |
| 10 MPH | .00 | .18 | .73 | 1.65 | 2.91 | 4.52 | 6.44 | 8.67 | 11.18 | 13.96 | 17.01 | 20.30 | 23.84 |
| 20 MPH | .00 | .37 | 1.47 | 3.30 | 5.83 | 9.03 | 12.88 | 17.33 | 22.36 | 27.93 | 34.01 | 40.60 | 47.67 |
| 30 MPH | .00 | .55 | 2.20 | 4.94 | 8.74 | 13.55 | 19.32 | 26.00 | 33.54 | 41.89 | 51.02 | 60.90 | 71.51 |
| VELOCITY (FPS) | 1200. | 1163. | 1129. | 1098. | 1071. | 1047. | 1024. | 1004. | 985. | 968. | 951. | 936. | 921. |
| ENERGY (FT-LB) | 847. | 795. | 749. | 709. | 675. | 644. | 617. | 593. | 571. | 551. | 532. | 515. | 499. |
| DROP (IN) | .00 | -.75 | -3.10 | -7.15 | -12.99 | -20.73 | -30.44 | -42.22 | -56.15 | -72.31 | -90.79 | -111.60 | -134.87 |
| MID-RANGE (IN) | .00 | .19 | .80 | 1.86 | 3.40 | 5.46 | 8.07 | 11.27 | 15.08 | 19.53 | 24.66 | 30.47 | 36.99 |
| BULLET PATH (IN) | -.63 | .49 | .00 | -2.19 | -6.17 | -12.04 | -19.89 | -29.81 | -41.87 | -56.17 | -72.78 | -91.73 | -113.14 |
| TIME OF FLIGHT (SEC) | .000000 | .063513 | .129005 | .196391 | .265566 | .336425 | .408878 | .482853 | .558288 | .635132 | .713339 | .792875 | .873708 |
| WIND DEFLECTION (IN) | | | | | | | | | | | | | |
| 5 MPH | .00 | .09 | .35 | .78 | 1.37 | 2.11 | 2.98 | 3.99 | 5.13 | 6.39 | 7.77 | 9.27 | 10.89 |
| 10 MPH | .00 | .18 | .70 | 1.56 | 2.74 | 4.21 | 5.96 | 7.98 | 10.26 | 12.78 | 15.55 | 18.55 | 21.77 |
| 20 MPH | .00 | .36 | 1.41 | 3.13 | 5.48 | 8.42 | 11.93 | 15.96 | 20.52 | 25.57 | 31.10 | 37.09 | 43.55 |
| 30 MPH | .00 | .53 | 2.11 | 4.69 | 8.22 | 12.63 | 17.89 | 23.95 | 30.78 | 38.35 | 46.64 | 55.64 | 65.32 |

# Bullet: Hornady .430 Dia. 265 Gr. FN
## Ballistic Coefficient: .267

| RANGE (YDS) | MUZZLE | 25 | 50 | 75 | 100 | 125 | 150 | 175 | 200 | 225 | 250 | 275 | 300 |
|---|---|---|---|---|---|---|---|---|---|---|---|---|---|
| VELOCITY (FPS) | 1100. | 1073. | 1048. | 1026. | 1005. | 986. | 969. | 952. | 936. | 922. | 908. | 894. | 882. |
| ENERGY (FT-LB) | 712. | 677. | 646. | 619. | 594. | 572. | 552. | 533. | 516. | 500. | 485. | 471. | 457. |
| DROP (IN) | .00 | -.88 | -3.64 | -8.37 | -15.16 | -24.10 | -35.26 | -48.73 | -64.54 | -82.80 | -103.61 | -127.04 | -153.19 |
| MID-RANGE (IN) | .00 | .23 | .94 | 2.17 | 3.94 | 6.30 | 9.26 | 12.86 | 17.11 | 22.05 | 27.71 | 34.13 | 41.33 |
| BULLET PATH (IN) | -.63 | .63 | .00 | -2.60 | -7.26 | -14.06 | -23.09 | -34.43 | -48.10 | -64.23 | -82.90 | -104.20 | -128.22 |
| TIME OF FLIGHT (SEC) | .000000 | .069061 | .139812 | .212163 | .286040 | .361381 | .438132 | .516250 | .595697 | .676443 | .758462 | .841731 | .926235 |
| WIND DEFLECTION (IN) | | | | | | | | | | | | | |
| 5 MPH | .00 | .08 | .30 | .67 | 1.17 | 1.80 | 2.56 | 3.43 | 4.42 | 5.53 | 6.74 | 8.07 | 9.51 |
| 10 MPH | .00 | .15 | .61 | 1.34 | 2.34 | 3.60 | 5.11 | 6.86 | 8.84 | 11.05 | 13.49 | 16.14 | 19.02 |
| 20 MPH | .00 | .31 | 1.21 | 2.68 | 4.69 | 7.21 | 10.22 | 13.72 | 17.69 | 22.11 | 26.98 | 32.29 | 38.03 |
| 30 MPH | .00 | .46 | 1.82 | 4.02 | 7.03 | 10.81 | 15.33 | 20.58 | 26.53 | 33.16 | 40.47 | 48.43 | 57.05 |
| VELOCITY (FPS) | 1000. | 981. | 964. | 948. | 932. | 918. | 904. | 891. | 878. | 866. | 854. | 843. | 832. |
| ENERGY (FT-LB) | 588. | 567. | 547. | 528. | 511. | 496. | 481. | 467. | 454. | 441. | 429. | 418. | 407. |
| DROP (IN) | .00 | -1.04 | -4.34 | -9.96 | -17.92 | -28.37 | -41.39 | -57.06 | -75.46 | -96.67 | -120.77 | -147.84 | -177.96 |
| MID-RANGE (IN) | .00 | .28 | 1.13 | 2.58 | 4.62 | 7.33 | 10.73 | 14.89 | 19.81 | 25.51 | 32.02 | 39.37 | 47.59 |
| BULLET PATH (IN) | -.63 | .81 | .00 | -3.14 | -8.62 | -16.58 | -27.12 | -40.31 | -56.22 | -74.95 | -96.56 | -121.15 | -148.78 |
| TIME OF FLIGHT (SEC) | .000000 | .075715 | .152828 | .231298 | .311087 | .392168 | .474514 | .558105 | .642926 | .728963 | .816206 | .904648 | .994285 |
| WIND DEFLECTION (IN) | | | | | | | | | | | | | |
| 5 MPH | .00 | .06 | .25 | .55 | .98 | 1.51 | 2.16 | 2.91 | 3.78 | 4.75 | 5.83 | 7.01 | 8.30 |
| 10 MPH | .00 | .13 | .50 | 1.11 | 1.95 | 3.02 | 4.31 | 5.83 | 7.55 | 9.50 | 11.65 | 14.02 | 16.59 |
| 20 MPH | .00 | .25 | 1.00 | 2.22 | 3.90 | 6.04 | 8.63 | 11.65 | 15.11 | 18.99 | 23.30 | 28.04 | 33.19 |
| 30 MPH | .00 | .38 | 1.49 | 3.33 | 5.85 | 9.06 | 12.94 | 17.48 | 22.66 | 28.49 | 34.96 | 42.05 | 49.78 |
| VELOCITY (FPS) | 900. | 887. | 874. | 862. | 851. | 839. | 828. | 818. | 807. | 797. | 787. | 777. | 768. |
| ENERGY (FT-LB) | 477. | 463. | 450. | 437. | 426. | 414. | 404. | 393. | 383. | 374. | 364. | 355. | 347. |
| DROP (IN) | .00 | -1.18 | -5.12 | -11.89 | -21.58 | -34.26 | -50.02 | -68.92 | -91.06 | -116.53 | -145.21 | -177.38 | -213.14 |
| MID-RANGE (IN) | .00 | .34 | 1.38 | 3.14 | 5.67 | 8.98 | 13.12 | 18.10 | 23.95 | 30.73 | 38.34 | 46.94 | 56.55 |
| BULLET PATH (IN) | -.63 | 1.06 | -.00 | -3.90 | -10.71 | -20.52 | -33.40 | -49.43 | -68.70 | -91.29 | -117.10 | -146.40 | -179.28 |
| TIME OF FLIGHT (SEC) | .000000 | .083950 | .169125 | .255512 | .343101 | .431888 | .521867 | .613039 | .705402 | .798961 | .893719 | .989679 | 1.086845 |
| WIND DEFLECTION (IN) | | | | | | | | | | | | | |
| 5 MPH | .00 | .05 | .22 | .49 | .86 | 1.34 | 1.92 | 2.61 | 3.41 | 4.31 | 5.31 | 6.43 | 7.64 |
| 10 MPH | .00 | .11 | .43 | .97 | 1.72 | 2.68 | 3.85 | 5.23 | 6.82 | 8.62 | 10.63 | 12.85 | 15.28 |
| 20 MPH | .00 | .22 | .87 | 1.94 | 3.44 | 5.36 | 7.70 | 10.46 | 13.63 | 17.23 | 21.26 | 25.70 | 30.57 |
| 30 MPH | .00 | .33 | 1.30 | 2.91 | 5.16 | 8.04 | 11.55 | 15.68 | 20.45 | 25.85 | 31.88 | 38.55 | 45.85 |
| VELOCITY (FPS) | 800. | 790. | 780. | 770. | 761. | 752. | 743. | 734. | 725. | 717. | 708. | 700. | 692. |
| ENERGY (FT-LB) | 377. | 367. | 358. | 349. | 341. | 332. | 325. | 317. | 309. | 302. | 295. | 288. | 282. |
| DROP (IN) | .00 | -1.38 | -6.13 | -14.44 | -26.39 | -42.09 | -61.61 | -85.02 | -112.44 | -143.98 | -179.74 | -219.82 | -264.33 |
| MID-RANGE (IN) | .00 | .38 | 1.68 | 3.90 | 7.07 | 11.21 | 16.37 | 22.56 | 29.83 | 38.22 | 47.78 | 58.55 | 70.55 |
| BULLET PATH (IN) | -.63 | 1.37 | -.00 | -4.93 | -13.50 | -25.82 | -41.97 | -61.99 | -86.03 | -114.20 | -146.58 | -183.28 | -224.41 |
| TIME OF FLIGHT (SEC) | .000000 | .094349 | .189898 | .286651 | .384612 | .483786 | .584181 | .685804 | .788663 | .892767 | .998125 | 1.104749 | 1.212649 |
| WIND DEFLECTION (IN) | | | | | | | | | | | | | |
| 5 MPH | .00 | .05 | .21 | .48 | .85 | 1.32 | 1.91 | 2.60 | 3.40 | 4.31 | 5.34 | 6.47 | 7.71 |
| 10 MPH | .00 | .11 | .42 | .95 | 1.69 | 2.65 | 3.82 | 5.20 | 6.80 | 8.63 | 10.67 | 12.94 | 15.43 |
| 20 MPH | .00 | .21 | .84 | 1.90 | 3.38 | 5.29 | 7.63 | 10.40 | 13.61 | 17.25 | 21.34 | 25.87 | 30.85 |
| 30 MPH | .00 | .32 | 1.27 | 2.85 | 5.08 | 7.94 | 11.45 | 15.60 | 20.41 | 25.88 | 32.01 | 38.81 | 46.28 |

## Bullet: Hornady .430 Dia. 265 Gr. FN
### Ballistic Coefficient: .267

| RANGE (YDS) | MUZZLE | 25 | 50 | 75 | 100 | 125 | 150 | 175 | 200 | 225 | 250 | 275 | 300 |
|---|---|---|---|---|---|---|---|---|---|---|---|---|---|
| VELOCITY (FPS) | 700. | 692. | 684. | 676. | 668. | 660. | 652. | 645. | 638. | 630. | 623. | 616. | 609. |
| ENERGY (FT-LB) | 288. | 282. | 275. | 269. | 262. | 256. | 250. | 245. | 239. | 234. | 228. | 223. | 218. |
| DROP (IN) | .00 | -1.67 | -7.87 | -18.70 | -34.27 | -54.70 | -80.11 | -110.52 | -146.14 | -187.10 | -233.52 | -285.52 | -343.26 |
| MID-RANGE (IN) | .00 | .56 | 2.26 | 5.15 | 9.27 | 14.66 | 21.36 | 29.38 | 38.80 | 49.68 | 62.05 | 75.99 | 91.52 |
| BULLET PATH (IN) | -.63 | 1.95 | -.00 | -6.58 | -17.91 | -34.09 | -55.25 | -81.41 | -112.78 | -149.49 | -191.66 | -239.42 | -292.91 |
| TIME OF FLIGHT (SEC) | .000000 | .107781 | .216848 | .327212 | .438886 | .551883 | .666218 | .781905 | .898959 | 1.017397 | 1.137233 | 1.258486 | 1.381173 |
| WIND DEFLECTION (IN) | | | | | | | | | | | | | |
| 5 MPH | .00 | .06 | .23 | .51 | .91 | 1.42 | 2.06 | 2.81 | 3.68 | 4.67 | 5.79 | 7.03 | 8.40 |
| 10 MPH | .00 | .11 | .45 | 1.02 | 1.82 | 2.85 | 4.11 | 5.62 | 7.36 | 9.35 | 11.58 | 14.07 | 16.80 |
| 20 MPH | .00 | .22 | .90 | 2.04 | 3.63 | 5.69 | 8.22 | 11.23 | 14.72 | 18.70 | 23.16 | 28.13 | 33.60 |
| 30 MPH | .00 | .34 | 1.35 | 3.05 | 5.45 | 8.54 | 12.33 | 16.85 | 22.08 | 28.04 | 34.74 | 42.20 | 50.40 |
| VELOCITY (FPS) | 600. | 593. | 586. | 579. | 573. | 566. | 560. | 553. | 547. | 541. | 535. | 529. | 523. |
| ENERGY (FT-LB) | 212. | 207. | 202. | 197. | 193. | 189. | 184. | 180. | 176. | 172. | 168. | 164. | 161. |
| DROP (IN) | .00 | -2.27 | -10.71 | -25.44 | -46.64 | -74.44 | -109.00 | -150.39 | -198.86 | -254.60 | -317.78 | -388.58 | -467.18 |
| MID-RANGE (IN) | .00 | .76 | 3.08 | 7.01 | 12.61 | 19.94 | 29.06 | 39.97 | 52.80 | 67.60 | 84.46 | 103.43 | 124.58 |
| BULLET PATH (IN) | -.63 | 2.76 | .00 | -9.07 | -24.59 | -46.72 | -75.62 | -111.34 | -154.15 | -204.22 | -261.73 | -326.86 | -399.79 |
| TIME OF FLIGHT (SEC) | .000000 | .125737 | .252961 | .381691 | .511945 | .643745 | .777110 | .912061 | 1.048622 | 1.186812 | 1.326656 | 1.468176 | 1.611397 |
| WIND DEFLECTION (IN) | | | | | | | | | | | | | |
| 5 MPH | .00 | .06 | .26 | .59 | 1.05 | 1.65 | 2.39 | 3.26 | 4.28 | 5.44 | 6.75 | 8.20 | 9.80 |
| 10 MPH | .00 | .13 | .52 | 1.18 | 2.10 | 3.30 | 4.77 | 6.52 | 8.56 | 10.88 | 13.49 | 16.40 | 19.61 |
| 20 MPH | .00 | .26 | 1.04 | 2.35 | 4.20 | 6.60 | 9.54 | 13.05 | 17.11 | 21.76 | 26.98 | 32.80 | 39.21 |
| 30 MPH | .00 | .39 | 1.56 | 3.53 | 6.31 | 9.90 | 14.31 | 19.57 | 25.67 | 32.64 | 40.47 | 49.20 | 58.82 |

# Bullet: Lyman #429348 175 Gr.
## Bullet Coefficient: .065

| RANGE (YDS) | MUZZLE | 25 | 50 | 75 | 100 | 125 | 150 | 175 | 200 | 225 | 250 | 275 | 300 |
|---|---|---|---|---|---|---|---|---|---|---|---|---|---|
| VELOCITY (FPS) | 1800. | 1541. | 1321. | 1100. | 970. | 884. | 816. | 753. | 695. | 643. | 595. | 550. | 509. |
| ENERGY (FT-LB) | 1259. | 923. | 678. | 470. | 366. | 304. | 259. | 220. | 188. | 161. | 138. | 118. | 101. |
| DROP (IN) | .00 | -.36 | -1.65 | -4.21 | -8.56 | -15.19 | -24.59 | -37.23 | -53.67 | -74.61 | -100.78 | -133.09 | -172.57 |
| MID-RANGE (IN) | .00 | .10 | .46 | 1.23 | 2.63 | 4.86 | 8.09 | 12.49 | 18.28 | 25.74 | 35.20 | 47.03 | 61.69 |
| BULLET PATH (IN) | -.63 | .15 | .00 | -1.42 | -4.63 | -10.12 | -18.39 | -29.88 | -45.19 | -64.98 | -90.02 | -121.19 | -159.53 |
| TIME OF FLIGHT (SEC) | .000000 | .045063 | .097717 | .160171 | .233129 | .314280 | .402678 | .498363 | .602116 | .714370 | .835715 | .966895 | 1.108792 |
| WIND DEFLECTION (IN) | | | | | | | | | | | | | |
| 5 MPH | .00 | .30 | 1.27 | 3.10 | 5.85 | 9.32 | 13.44 | 18.19 | 23.65 | 29.86 | 36.88 | 44.75 | 53.57 |
| 10 MPH | .00 | .60 | 2.53 | 6.19 | 11.70 | 18.65 | 26.87 | 36.38 | 47.31 | 59.73 | 73.75 | 89.51 | 107.15 |
| 20 MPH | .00 | 1.20 | 5.06 | 12.38 | 23.39 | 37.29 | 53.74 | 72.76 | 94.61 | 119.46 | 147.51 | 179.01 | 214.29 |
| 30 MPH | .00 | 1.79 | 7.59 | 18.57 | 35.09 | 55.94 | 80.61 | 109.14 | 141.92 | 179.19 | 221.26 | 268.52 | 321.44 |
| VELOCITY (FPS) | 1700. | 1454. | 1231. | 1044. | 935. | 857. | 793. | 731. | 675. | 625. | 578. | 535. | 494. |
| ENERGY (FT-LB) | 1123. | 822. | 589. | 423. | 340. | 286. | 244. | 208. | 177. | 152. | 130. | 111. | 95. |
| DROP (IN) | .00 | -.41 | -1.85 | -4.75 | -9.63 | -16.97 | -27.25 | -40.95 | -58.71 | -81.22 | -109.29 | -143.84 | -185.99 |
| MID-RANGE (IN) | .00 | .11 | .52 | 1.39 | 2.96 | 5.40 | 8.88 | 13.57 | 19.73 | 27.65 | 37.68 | 50.21 | 65.75 |
| BULLET PATH (IN) | -.63 | .20 | .00 | -1.65 | -5.29 | -11.39 | -20.43 | -32.89 | -49.41 | -70.67 | -97.50 | -130.81 | -171.72 |
| TIME OF FLIGHT (SEC) | .000000 | .047746 | .103591 | .170247 | .246419 | .330309 | .421313 | .519905 | .626717 | .742228 | .867086 | 1.002087 | 1.148159 |
| WIND DEFLECTION (IN) | | | | | | | | | | | | | |
| 5 MPH | .00 | .32 | 1.35 | 3.33 | 6.16 | 9.66 | 13.78 | 18.58 | 24.09 | 30.37 | 37.48 | 45.48 | 54.45 |
| 10 MPH | .00 | .64 | 2.70 | 6.67 | 12.31 | 19.31 | 27.56 | 37.15 | 48.18 | 60.75 | 74.96 | 90.96 | 108.90 |
| 20 MPH | .00 | 1.28 | 5.41 | 13.34 | 24.62 | 38.62 | 55.13 | 74.30 | 96.37 | 121.50 | 149.92 | 181.91 | 217.80 |
| 30 MPH | .00 | 1.92 | 8.11 | 20.01 | 36.93 | 57.93 | 82.69 | 111.45 | 144.55 | 182.25 | 224.88 | 272.87 | 326.70 |
| VELOCITY (FPS) | 1600. | 1369. | 1144. | 996. | 902. | 831. | 768. | 709. | 655. | 606. | 561. | 519. | 479. |
| ENERGY (FT-LB) | 995. | 729. | 508. | 385. | 316. | 268. | 229. | 195. | 167. | 143. | 122. | 105. | 89. |
| DROP (IN) | .00 | -.46 | -2.10 | -5.40 | -10.87 | -18.99 | -30.23 | -45.13 | -64.34 | -88.60 | -118.76 | -155.82 | -200.95 |
| MID-RANGE (IN) | .00 | .13 | .59 | 1.59 | 3.33 | 5.99 | 9.72 | 14.73 | 21.30 | 29.73 | 40.39 | 53.72 | 70.25 |
| BULLET PATH (IN) | -.63 | .27 | .00 | -1.93 | -6.03 | -12.79 | -22.67 | -36.19 | -54.04 | -76.93 | -105.73 | -141.42 | -185.19 |
| TIME OF FLIGHT (SEC) | .000000 | .050734 | .110523 | .181225 | .260541 | .347242 | .441083 | .542823 | .652946 | .772001 | .900695 | 1.039877 | 1.190527 |
| WIND DEFLECTION (IN) | | | | | | | | | | | | | |
| 5 MPH | .00 | .34 | 1.48 | 3.57 | 6.43 | 9.93 | 14.07 | 18.89 | 24.46 | 30.81 | 38.01 | 46.13 | 55.27 |
| 10 MPH | .00 | .68 | 2.95 | 7.15 | 12.86 | 19.86 | 28.13 | 37.79 | 48.92 | 61.62 | 76.02 | 92.27 | 110.53 |
| 20 MPH | .00 | 1.36 | 5.90 | 14.29 | 25.71 | 39.73 | 56.26 | 75.57 | 97.84 | 123.24 | 152.04 | 184.54 | 221.07 |
| 30 MPH | .00 | 2.04 | 8.86 | 21.44 | 38.57 | 59.59 | 84.39 | 113.36 | 146.76 | 184.87 | 228.07 | 276.81 | 331.60 |
| VELOCITY (FPS) | 1500. | 1283. | 1073. | 954. | 872. | 806. | 743. | 686. | 635. | 587. | 543. | 502. | 464. |
| ENERGY (FT-LB) | 874. | 640. | 447. | 353. | 295. | 252. | 214. | 183. | 157. | 134. | 115. | 98. | 84. |
| DROP (IN) | .00 | -.53 | -2.41 | -6.17 | -12.29 | -21.26 | -33.54 | -49.75 | -70.56 | -96.76 | -129.24 | -169.09 | -217.55 |
| MID-RANGE (IN) | .00 | .14 | .68 | 1.81 | 3.73 | 6.61 | 10.61 | 15.97 | 22.99 | 31.99 | 43.36 | 57.59 | 75.23 |
| BULLET PATH (IN) | -.63 | .36 | .00 | -2.24 | -6.84 | -14.29 | -25.06 | -39.74 | -59.03 | -83.71 | -114.68 | -153.00 | -199.95 |
| TIME OF FLIGHT (SEC) | .000000 | .054071 | .118456 | .192904 | .275307 | .364885 | .461872 | .567006 | .680726 | .803650 | .936547 | 1.080319 | 1.235997 |
| WIND DEFLECTION (IN) | | | | | | | | | | | | | |
| 5 MPH | .00 | .36 | 1.62 | 3.78 | 6.63 | 10.11 | 14.24 | 19.10 | 24.70 | 31.12 | 38.42 | 46.67 | 55.97 |
| 10 MPH | .00 | .72 | 3.25 | 7.55 | 13.25 | 20.22 | 28.49 | 38.19 | 49.41 | 62.24 | 76.83 | 93.34 | 111.94 |
| 20 MPH | .00 | 1.43 | 6.50 | 15.10 | 26.51 | 40.44 | 56.98 | 76.39 | 98.82 | 124.48 | 153.66 | 186.67 | 223.87 |
| 30 MPH | .00 | 2.15 | 9.74 | 22.65 | 39.76 | 60.66 | 85.47 | 114.58 | 148.22 | 186.73 | 230.50 | 280.01 | 335.81 |

## Bullet: Lyman #429348 175 Gr.
### Ballistic Coefficient: .065

| RANGE (YDS) | MUZZLE | 25 | 50 | 75 | 100 | 125 | 150 | 175 | 200 | 225 | 250 | 275 | 300 |
|---|---|---|---|---|---|---|---|---|---|---|---|---|---|
| VELOCITY (FPS) | 1400. | 1174. | 1013. | 914. | 841. | 777. | 717. | 663. | 613. | 567. | 525. | 485. | 447. |
| ENERGY (FT-LB) | 761. | 536. | 398. | 325. | 275. | 234. | 200. | 171. | 146. | 125. | 107. | 91. | 78. |
| DROP (IN) | .00 | -.61 | -2.80 | -7.09 | -13.96 | -23.89 | -37.38 | -55.09 | -77.74 | -106.16 | -141.33 | -184.39 | -236.71 |
| MID-RANGE (IN) | .00 | .16 | .79 | 2.07 | 4.18 | 7.29 | 11.60 | 17.37 | 24.91 | 34.56 | 46.77 | 62.04 | 80.97 |
| BULLET PATH (IN) | -.63 | .47 | .00 | -2.58 | -7.73 | -15.95 | -27.72 | -43.72 | -64.65 | -91.35 | -124.81 | -166.15 | -216.76 |
| TIME OF FLIGHT (SEC) | .000000 | .058252 | .127489 | .205649 | .291302 | .384071 | .484652 | .593553 | .711296 | .838569 | .976202 | 1.125153 | 1.286512 |
| WIND DEFLECTION (IN) | | | | | | | | | | | | | |
| 5 MPH | .00 | .41 | 1.79 | 3.95 | 6.78 | 10.23 | 14.36 | 19.23 | 24.88 | 31.37 | 38.76 | 47.16 | 56.64 |
| 10 MPH | .00 | .82 | 3.58 | 7.91 | 13.55 | 20.45 | 28.73 | 38.47 | 49.76 | 62.73 | 77.53 | 94.31 | 113.28 |
| 20 MPH | .00 | 1.65 | 7.16 | 15.82 | 27.11 | 40.91 | 57.45 | 76.93 | 99.52 | 125.46 | 155.05 | 188.63 | 226.57 |
| 30 MPH | .00 | 2.47 | 10.74 | 23.73 | 40.66 | 61.36 | 86.18 | 115.40 | 149.28 | 188.19 | 232.58 | 282.94 | 339.85 |
| VELOCITY (FPS) | 1300. | 1082. | 960. | 876. | 809. | 746. | 689. | 638. | 590. | 546. | 504. | 466. | 430. |
| ENERGY (FT-LB) | 657. | 455. | 358. | 298. | 255. | 216. | 185. | 158. | 135. | 116. | 99. | 84. | 72. |
| DROP (IN) | .00 | -.73 | -3.31 | -8.22 | -15.95 | -26.97 | -41.87 | -61.33 | -86.13 | -117.15 | -155.47 | -202.33 | -259.20 |
| MID-RANGE (IN) | .00 | .20 | .92 | 2.35 | 4.67 | 8.04 | 12.71 | 18.96 | 27.11 | 37.56 | 50.76 | 67.27 | 87.73 |
| BULLET PATH (IN) | -.63 | .61 | .00 | -2.94 | -8.71 | -17.76 | -30.69 | -48.18 | -71.00 | -100.06 | -136.41 | -181.30 | -236.21 |
| TIME OF FLIGHT (SEC) | .000000 | .063686 | .137603 | .219555 | .308706 | .405222 | .509858 | .623049 | .745402 | .877678 | 1.020771 | 1.175703 | 1.343630 |
| WIND DEFLECTION (IN) | | | | | | | | | | | | | |
| 5 MPH | .00 | .53 | 1.96 | 4.09 | 6.86 | 10.27 | 14.41 | 19.29 | 24.98 | 31.54 | 39.06 | 47.62 | 57.32 |
| 10 MPH | .00 | 1.05 | 3.91 | 8.18 | 13.72 | 20.55 | 28.81 | 38.58 | 49.96 | 63.09 | 78.12 | 95.23 | 114.63 |
| 20 MPH | .00 | 2.11 | 7.82 | 16.36 | 27.43 | 41.10 | 57.62 | 77.16 | 99.92 | 126.17 | 156.23 | 190.46 | 229.27 |
| 30 MPH | .00 | 3.16 | 11.73 | 24.54 | 41.15 | 61.65 | 86.44 | 115.74 | 149.88 | 189.26 | 234.35 | 285.69 | 343.90 |
| VELOCITY (FPS) | 1200. | 1027. | 924. | 848. | 784. | 723. | 668. | 618. | 572. | 529. | 489. | 452. | 417. |
| ENERGY (FT-LB) | 559. | 409. | 332. | 280. | 239. | 203. | 174. | 149. | 127. | 109. | 93. | 79. | 67. |
| DROP (IN) | .00 | -.84 | -3.73 | -9.15 | -17.57 | -29.48 | -45.54 | -66.45 | -93.03 | -126.24 | -167.20 | -217.25 | -277.99 |
| MID-RANGE (IN) | .00 | .22 | 1.02 | 2.57 | 5.05 | 8.65 | 13.62 | 20.27 | 28.95 | 40.07 | 54.12 | 71.68 | 93.45 |
| BULLET PATH (IN) | -.63 | .71 | .00 | -3.23 | -9.47 | -19.20 | -33.08 | -51.81 | -76.20 | -107.23 | -146.01 | -193.88 | -252.44 |
| TIME OF FLIGHT (SEC) | .000000 | .068052 | .145293 | .230124 | .322063 | .421726 | .529662 | .646372 | .772526 | .908938 | 1.056553 | 1.216445 | 1.389822 |
| WIND DEFLECTION (IN) | | | | | | | | | | | | | |
| 5 MPH | .00 | .49 | 1.79 | 3.75 | 6.34 | 9.61 | 13.61 | 18.38 | 23.98 | 30.49 | 37.98 | 46.55 | 56.30 |
| 10 MPH | .00 | .98 | 3.57 | 7.50 | 12.68 | 19.22 | 27.22 | 36.76 | 47.96 | 60.97 | 75.95 | 93.09 | 112.61 |
| 20 MPH | .00 | 1.95 | 7.14 | 15.00 | 25.37 | 38.45 | 54.44 | 73.52 | 95.93 | 121.95 | 151.91 | 186.19 | 225.22 |
| 30 MPH | .00 | 2.93 | 10.71 | 22.51 | 38.05 | 57.67 | 81.66 | 110.28 | 143.89 | 182.92 | 227.86 | 279.28 | 337.83 |
| VELOCITY (FPS) | 1100. | 970. | 884. | 816. | 753. | 695. | 643. | 595. | 550. | 509. | 470. | 434. | 400. |
| ENERGY (FT-LB) | 470. | 366. | 304. | 259. | 220. | 188. | 161. | 138. | 118. | 101. | 86. | 73. | 62. |
| DROP (IN) | .00 | -.98 | -4.24 | -10.27 | -19.54 | -32.61 | -50.17 | -72.97 | -101.90 | -137.99 | -182.48 | -236.82 | -302.74 |
| MID-RANGE (IN) | .00 | .26 | 1.14 | 2.84 | 5.53 | 9.42 | 14.81 | 22.01 | 31.41 | 43.44 | 58.63 | 77.63 | 101.20 |
| BULLET PATH (IN) | -.63 | .83 | -.00 | -3.60 | -10.42 | -21.06 | -36.18 | -56.55 | -83.04 | -116.70 | -158.75 | -210.65 | -274.14 |
| TIME OF FLIGHT (SEC) | .000000 | .072940 | .154075 | .242458 | .338127 | .441863 | .554099 | .675423 | .806579 | .948448 | 1.102039 | 1.268491 | 1.449085 |
| WIND DEFLECTION (IN) | | | | | | | | | | | | | |
| 5 MPH | .00 | .42 | 1.56 | 3.34 | 5.76 | 8.88 | 12.76 | 17.44 | 22.98 | 29.46 | 36.98 | 45.63 | 55.52 |
| 10 MPH | .00 | .84 | 3.12 | 6.67 | 11.51 | 17.77 | 25.52 | 34.87 | 45.96 | 58.93 | 73.96 | 91.25 | 111.04 |
| 20 MPH | .00 | 1 67 | 6.23 | 13.35 | 23.02 | 35.54 | 51.04 | 69.75 | 91.92 | 117.85 | 147.92 | 182.51 | 222.08 |
| 30 MPH | .00 | 2 51 | 9.35 | 20.02 | 34.53 | 53.30 | 76.56 | 104.62 | 137.87 | 176.78 | 221.88 | 273.76 | 333.12 |

# Bullet: Lyman #429348 175 Gr.
## Bullet Coefficient: .065

| RANGE (YDS) | MUZZLE | 25 | 50 | 75 | 100 | 125 | 150 | 175 | 200 | 225 | 250 | 275 | 300 |
|---|---|---|---|---|---|---|---|---|---|---|---|---|---|
| VELOCITY (FPS) | 1000. | 905. | 834. | 770. | 711. | 657. | 608. | 562. | 520. | 480. | 444. | 409. | 377. |
| ENERGY (FT-LB) | 389. | 319. | 270. | 230. | 196. | 168. | 144. | 123. | 105. | 90. | 76. | 65. | 55. |
| DROP (IN) | .00 | -1.15 | -4.93 | -11.81 | -22.33 | -37.14 | -56.97 | -82.66 | -115.22 | -155.81 | -205.82 | -266.90 | -341.01 |
| MID-RANGE (IN) | .00 | .30 | 1.32 | 3.23 | 6.24 | 10.61 | 16.67 | 24.76 | 35.28 | 48.75 | 65.77 | 87.08 | 113.54 |
| BULLET PATH (IN) | -.63 | 1.00 | -.00 | -4.11 | -11.84 | -23.87 | -40.92 | -63.84 | -93.62 | -131.43 | -178.66 | -236.96 | -308.29 |
| TIME OF FLIGHT (SEC) | .000000 | .079013 | .165437 | .259004 | .360461 | .470284 | .589016 | .717359 | .856157 | 1.006383 | 1.169139 | 1.345664 | 1.537340 |
| WIND DEFLECTION (IN) | | | | | | | | | | | | | |
| 5 MPH | .00 | .35 | 1.36 | 2.99 | 5.32 | 8.39 | 12.23 | 16.93 | 22.54 | 29.16 | 36.88 | 45.82 | 56.09 |
| 10 MPH | .00 | .71 | 2.72 | 5.98 | 10.64 | 16.77 | 24.47 | 33.86 | 45.08 | 58.32 | 73.77 | 91.64 | 112.17 |
| 20 MPH | .00 | 1.41 | 5.43 | 11.97 | 21.28 | 33.54 | 48.93 | 67.71 | 90.17 | 116.65 | 147.54 | 183.27 | 224.34 |
| 30 MPH | .00 | 2.12 | 8.15 | 17.95 | 31.92 | 50.31 | 73.40 | 101.57 | 135.25 | 174.97 | 221.31 | 274.91 | 336.52 |
| VELOCITY (FPS) | 900. | 829. | 766. | 707. | 653. | 605. | 559. | 517. | 478. | 441. | 407. | 375. | 346. |
| ENERGY (FT-LB) | 315. | 267. | 228. | 194. | 166. | 142. | 122. | 104. | 89. | 76. | 64. | 55. | 46. |
| DROP (IN) | .00 | -1.38 | -5.90 | -14.10 | -26.63 | -44.24 | -67.78 | -98.26 | -136.86 | -184.98 | -244.30 | -316.80 | -404.86 |
| MID-RANGE (IN) | .00 | .36 | 1.57 | 3.82 | 7.41 | 12.62 | 19.80 | 29.35 | 41.80 | 57.74 | 77.91 | 103.19 | 134.64 |
| BULLET PATH (IN) | -.63 | 1.25 | -.00 | -4.93 | -14.19 | -28.54 | -48.82 | -76.03 | -111.36 | -156.21 | -212.27 | -281.50 | -366.29 |
| TIME OF FLIGHT (SEC) | .000000 | .086911 | .180994 | .283020 | .393445 | .512820 | .641861 | .781419 | .932476 | 1:096143 | 1.273669 | 1.466440 | 1.675858 |
| WIND DEFLECTION (IN) | | | | | | | | | | | | | |
| 5 MPH | .00 | .31 | 1.26 | 2.91 | 5.29 | 8.46 | 12.48 | 17.43 | 23.39 | 30.46 | 38.75 | 48.38 | 59.48 |
| 10 MPH | .00 | .63 | 2.52 | 5.81 | 10.58 | 16.92 | 24.97 | 34.86 | 46.78 | 60.92 | 77.50 | 96.76 | 118.95 |
| 20 MPH | .00 | 1.26 | 5.04 | 11.62 | 21.16 | 33.85 | 49.94 | 69.73 | 93.56 | 121.84 | 155.00 | 193.52 | 237.90 |
| 30 MPH | .00 | 1.89 | 7.56 | 17.43 | 31.74 | 50.77 | 74.90 | 104.59 | 140.35 | 182.76 | 232.50 | 290.28 | 356.85 |
| VELOCITY (FPS) | 800. | 737. | 681. | 630. | 583. | 539. | 498. | 460. | 424. | 391. | 361. | 333. | 307. |
| ENERGY (FT-LB) | 249. | 211. | 180. | 154. | 132. | 113. | 96. | 82. | 70. | 59. | 51. | 43. | 37. |
| DROP (IN) | .00 | -1.71 | -7.40 | -17.77 | -33.60 | -55.81 | -85.49 | -123.92 | -172.63 | -233.42 | -308.47 | -400.32 | -512.01 |
| MID-RANGE (IN) | .00 | .46 | 1.99 | 4.87 | 9.40 | 15.95 | 24.98 | 37.02 | 52.72 | 72.87 | 98.42 | 130.51 | 170.51 |
| BULLET PATH (IN) | -.63 | 1.68 | -.00 | -6.35 | -18.17 | -36.37 | -62.03 | -96.45 | -141.14 | -197.92 | -268.95 | -356.78 | -464.46 |
| TIME OF FLIGHT (SEC) | .000000 | .097743 | .203668 | .318229 | .442059 | .575937 | .720777 | .877621 | 1.047641 | 1.232156 | 1.432571 | 1.650299 | 1.886876 |
| WIND DEFLECTION (IN) | | | | | | | | | | | | | |
| 5 MPH | .00 | .35 | 1.42 | 3.25 | 5.90 | 9.43 | 13.93 | 19.48 | 26.19 | 34.18 | 43.57 | 54.48 | 67.05 |
| 10 MPH | .00 | .70 | 2.85 | 6.51 | 11.80 | 18.86 | 27.86 | 38.96 | 52.38 | 68.36 | 87.13 | 108.95 | 134.09 |
| 20 MPH | .00 | 1.41 | 5.69 | 13.02 | 23.60 | 37.73 | 55.71 | 77.92 | 104.77 | 136.72 | 174.26 | 217.91 | 268.18 |
| 30 MPH | .00 | 2.11 | 8.54 | 19.52 | 35.41 | 56.59 | 83.57 | 116.88 | 157.15 | 205.08 | 261.40 | 326.86 | 402.27 |
| VELOCITY (FPS) | 700. | 647. | 599. | 554. | 512. | 473. | 437. | 403. | 371. | 342. | 316. | 291. | 269. |
| ENERGY (FT-LB) | 190. | 163. | 139. | 119. | 102. | 87. | 74. | 63. | 53. | 45. | 39. | 33. | 28. |
| DROP (IN) | .00 | -2.22 | -9.60 | -23.03 | -43.53 | -72.30 | -110.77 | -160.65 | -223.97 | -303.14 | -401.02 | -520.99 | -667.04 |
| MID-RANGE (IN) | .00 | .60 | 2.59 | 6.31 | 12.16 | 20.64 | 32.35 | 48.00 | 68.46 | 94.78 | 128.21 | 170.28 | 222.75 |
| BULLET PATH (IN) | -.63 | 2.27 | .00 | -8.31 | -23.69 | -47.34 | -80.70 | -125.46 | -183.67 | -257.72 | -350.48 | -465.33 | -606.27 |
| TIME OF FLIGHT (SEC) | .000000 | .111451 | .231928 | .362162 | .503020 | .655497 | .820716 | .999942 | 1.194566 | 1.405982 | 1.635672 | 1.885265 | 2.156567 |
| WIND DEFLECTION (IN) | | | | | | | | | | | | | |
| 5 MPH | .00 | .38 | 1.55 | 3.58 | 6.55 | 10.54 | 15.65 | 21.99 | 29.69 | 38.87 | 49.65 | 62.19 | 76.64 |
| 10 MPH | .00 | .76 | 3.11 | 7.17 | 13.10 | 21.08 | 31.30 | 43.99 | 59.39 | 77.74 | 99.31 | 124.38 | 153.27 |
| 20 MPH | .00 | 1.52 | 6.21 | 14.34 | 26.21 | 42.16 | 62.61 | 87.98 | 118.77 | 155.48 | 198.61 | 248.76 | 306.54 |
| 30 MPH | .00 | 2.27 | 9.32 | 21.51 | 39.31 | 63.24 | 93.91 | 131.97 | 178.16 | 233.22 | 297.92 | 373.13 | 459.81 |

# Bullet: Lyman #429303 189 Gr.
## Ballistic Coefficient: .105

| RANGE (YDS) | MUZZLE | 25 | 50 | 75 | 100 | 125 | 150 | 175 | 200 | 225 | 250 | 275 | 300 |
|---|---|---|---|---|---|---|---|---|---|---|---|---|---|
| VELOCITY (FPS) | 1600. | 1452. | 1321. | 1201. | 1105. | 1034. | 979. | 933. | 894. | 860. | 828. | 799. | 770. |
| ENERGY (FT-LB) | 1074. | 885. | 732. | 605. | 512. | 449. | 402. | 366. | 336. | 310. | 288. | 268. | 249. |
| DROP (IN) | .00 | -.44 | -1.91 | -4.63 | -8.86 | -14.87 | -22.89 | -33.17 | -45.92 | -61.38 | -79.76 | -101.30 | -126.16 |
| MID-RANGE (IN) | .00 | .12 | .52 | 1.28 | 2.52 | 4.33 | 6.82 | 10.04 | 14.10 | 19.06 | 25.01 | 32.04 | 40.19 |
| BULLET PATH (IN) | -.63 | .20 | .00 | -1.45 | -4.41 | -9.15 | -15.91 | -24.92 | -36.39 | -50.58 | -67.69 | -87.96 | -111.55 |
| TIME OF FLIGHT (SEC) | .000000 | .049221 | .103403 | .162978 | .228198 | .298471 | .373096 | .451614 | .533746 | .619323 | .708256 | .800517 | .896214 |
| WIND DEFLECTION (IN) | | | | | | | | | | | | | |
| 5 MPH | .00 | .21 | .85 | 1.97 | 3.58 | 5.64 | 8.08 | 10.87 | 13.97 | 17.38 | 21.08 | 25.07 | 29.37 |
| 10 MPH | .00 | .41 | 1.70 | 3.93 | 7.16 | 11.28 | 16.16 | 21.73 | 27.94 | 34.75 | 42.15 | 50.14 | 58.73 |
| 20 MPH | .00 | .83 | 3.40 | 7.87 | 14.33 | 22.56 | 32.33 | 43.47 | 55.88 | 69.50 | 84.31 | 100.28 | 117.47 |
| 30 MPH | .00 | 1.24 | 5.10 | 11.80 | 21.49 | 33.84 | 48.49 | 65.20 | 83.82 | 104.25 | 126.46 | 150.42 | 176.20 |
| VELOCITY (FPS) | 1500. | 1363. | 1239. | 1134. | 1055. | 996. | 948. | 907. | 871. | 838. | 808. | 779. | 751. |
| ENERGY (FT-LB) | 944. | 779. | 644. | 539. | 467. | 416. | 377. | 345. | 318. | 295. | 274. | 255. | 237. |
| DROP (IN) | .00 | -.50 | -2.18 | -5.27 | -10.05 | -16.78 | -25.68 | -36.97 | -50.89 | -67.67 | -87.53 | -110.65 | -137.31 |
| MID-RANGE (IN) | .00 | .13 | .59 | 1.46 | 2.85 | 4.86 | 7.57 | 11.05 | 15.39 | 20.68 | 26.99 | 34.39 | 42.98 |
| BULLET PATH (IN) | -.63 | .27 | .00 | -1.69 | -5.07 | -10.39 | -17.89 | -27.77 | -40.29 | -55.67 | -74.12 | -95.84 | -121.10 |
| TIME OF FLIGHT (SEC) | .000000 | .052485 | .110192 | .173572 | .242248 | .315484 | .392743 | .473694 | .558136 | .645956 | .737108 | .831634 | .929700 |
| WIND DEFLECTION (IN) | | | | | | | | | | | | | |
| 5 MPH | .00 | .22 | .90 | 2.07 | 3.72 | 5.76 | 8.16 | 10.89 | 13.92 | 17.24 | 20.87 | 24.78 | 29.01 |
| 10 MPH | .00 | .44 | 1.79 | 4.15 | 7.44 | 11.53 | 16.32 | 21.77 | 27.83 | 34.49 | 41.73 | 49.57 | 58.03 |
| 20 MPH | .00 | .87 | 3.59 | 8.30 | 14.87 | 23.05 | 32.65 | 43.54 | 55.66 | 68.98 | 83.46 | 99.14 | 116.05 |
| 30 MPH | .00 | 1.31 | 5.38 | 12.45 | 22.31 | 34.58 | 48.97 | 65.31 | 83.50 | 103.46 | 125.19 | 148.70 | 174.08 |
| VELOCITY (FPS) | 1400. | 1273. | 1161. | 1075. | 1011. | 960. | 918. | 880. | 847. | 816. | 787. | 759. | 733. |
| ENERGY (FT-LB) | 822. | 680. | 565. | 485. | 429. | 387. | 353. | 325. | 301. | 280. | 260. | 242. | 225. |
| DROP (IN) | .00 | -.58 | -2.50 | -6.04 | -11.44 | -18.97 | -28.82 | -41.24 | -56.45 | -74.68 | -96.13 | -121.02 | -149.67 |
| MID-RANGE (IN) | .00 | .15 | .67 | 1.66 | 3.22 | 5.43 | 8.38 | 12.13 | 16.78 | 22.41 | 29.10 | 36.92 | 46.01 |
| BULLET PATH (IN) | -.63 | .36 | .00 | -1.97 | -5.81 | -11.77 | -20.06 | -30.91 | -44.55 | -61.21 | -81.10 | -104.42 | -131.50 |
| TIME OF FLIGHT (SEC) | .000000 | .056178 | .117967 | .185230 | .257253 | .333423 | .413363 | .496839 | .583718 | .673936 | .767493 | .864548 | .965184 |
| WIND DEFLECTION (IN) | | | | | | | | | | | | | |
| 5 MPH | .00 | .23 | .95 | 2.16 | 3.78 | 5.77 | 8.09 | 10.72 | 13.65 | 16.88 | 20.40 | 24.22 | 28.36 |
| 10 MPH | .00 | .46 | 1.91 | 4.31 | 7.56 | 11.54 | 16.18 | 21.44 | 27.31 | 33.76 | 40.79 | 48.45 | 56.73 |
| 20 MPH | .00 | .92 | 3.81 | 8.63 | 15.12 | 23.08 | 32.36 | 42.89 | 54.61 | 67.51 | 81.59 | 96.89 | 113.46 |
| 30 MPH | .00 | 1.38 | 5.72 | 12.94 | 22.69 | 34.62 | 48.54 | 64.33 | 81.92 | 101.27 | 122.38 | 145.34 | 170.19 |
| VELOCITY (FPS) | 1300. | 1182. | 1091. | 1023. | 970. | 926. | 888. | 854. | 823. | 793. | 765. | 738. | 712. |
| ENERGY (FT-LB) | 709. | 586. | 499. | 439. | 395. | 360. | 331. | 306. | 284. | 264. | 245. | 228. | 213. |
| DROP (IN) | .00 | -.68 | -2.91 | -6.96 | -13.08 | -21.49 | -32.41 | -46.07 | -62.70 | -82.52 | -105.72 | -132.61 | -153.49 |
| MID-RANGE (IN) | .00 | .18 | .78 | 1.90 | 3.63 | 6.05 | 9.24 | 13.29 | 18.27 | 24.27 | 31.37 | 39.69 | 49.34 |
| BULLET PATH (IN) | -.63 | .46 | .00 | -2.28 | -6.63 | -13.27 | -22.42 | -34.32 | -49.18 | -67.23 | -88.66 | -113.79 | -142.89 |
| TIME OF FLIGHT (SEC) | .000000 | .060597 | .126776 | .197872 | .273219 | .352400 | .435157 | .521337 | .610864 | .703723 | .800059 | .899965 | 1.003502 |
| WIND DEFLECTION (IN) | | | | | | | | | | | | | |
| 5 MPH | .00 | .26 | 1.00 | 2.18 | 3.74 | 5.63 | 7.83 | 10.34 | 13.14 | 16.24 | 19.64 | 23.35 | 27.39 |
| 10 MPH | .00 | .51 | 2.00 | 4.36 | 7.47 | 11.25 | 15.66 | 20.68 | 26.28 | 32.47 | 39.27 | 46.70 | 54.77 |
| 20 MPH | .00 | 1.02 | 4.01 | 8.73 | 14.94 | 22.51 | 31.33 | 41.36 | 52.56 | 64.94 | 78.54 | 93.40 | 109.54 |
| 30 MPH | .00 | 1.53 | 6.01 | 13.09 | 22.41 | 33.76 | 46.99 | 62.03 | 78.84 | 97.41 | 117.82 | 140.10 | 164.31 |

# Bullet: Lyman #429303 189 Gr.
## Ballistic Coefficient: .105

| RANGE (YDS) | MUZZLE | 25 | 50 | 75 | 100 | 125 | 150 | 175 | 200 | 225 | 250 | 275 | 300 |
|---|---|---|---|---|---|---|---|---|---|---|---|---|---|
| VELOCITY (FPS) | 1200. | 1104. | 1033. | 978. | 933. | 894. | 859. | 828. | 798. | 770. | 742. | 717. | 692. |
| ENERGY (FT-LB) | 604. | 512. | 448. | 402. | 365. | 335. | 310. | 287. | 268. | 248. | 231. | 215. | 201. |
| DROP (IN) | .00 | -.79 | -3.35 | -7.95 | -14.79 | -24.11 | -36.13 | -51.08 | -69.20 | -90.63 | -115.72 | -144.73 | -177.97 |
| MID-RANGE (IN) | .00 | .21 | .89 | 2.14 | 4.04 | 6.67 | 10.12 | 14.46 | 19.81 | 26.19 | 33.75 | 42.60 | 52.85 |
| BULLET PATH (IN) | -.63 | .57 | .00 | -2.60 | -7.45 | -14.78 | -24.80 | -37.77 | -53.89 | -73.33 | -96.43 | -123.45 | -154.69 |
| TIME OF FLIGHT (SEC) | .000000 | .065264 | .135574 | .210231 | .288778 | .370937 | .456539 | .545497 | .637782 | .733501 | .832784 | .935685 | 1.042281 |
| WIND DEFLECTION (IN) | | | | | | | | | | | | | |
| 5 MPH | .00 | .24 | .93 | 2.00 | 3.41 | 5.14 | 7.18 | 9.50 | 12.12 | 15.05 | 18.28 | 21.84 | 25.72 |
| 10 MPH | .00 | .49 | 1.86 | 4.00 | 6.83 | 10.28 | 14.35 | 19.01 | 24.25 | 30.10 | 36.57 | 43.68 | 51.44 |
| 20 MPH | .00 | .97 | 3.72 | 8.00 | 13.65 | 20.57 | 28.70 | 38.01 | 48.50 | 60.19 | 73.14 | 87.36 | 102.88 |
| 30 MPH | .00 | 1.46 | 5.58 | 12.00 | 20.48 | 30.85 | 43.05 | 57.02 | 72.75 | 90.29 | 109.71 | 131.04 | 154.32 |
| VELOCITY (FPS) | 1100. | 1030. | 976. | 931. | 892. | 858. | 826. | 797. | 768. | 741. | 715. | 691. | 667. |
| ENERGY (FT-LB) | 508. | 445. | 400. | 364. | 334. | 309. | 286. | 266. | 247. | 230. | 215. | 200. | 187. |
| DROP (IN) | .00 | -.93 | -3.89 | -9.12 | -16.83 | -27.26 | -40.64 | -57.19 | -77.07 | -100.61 | -128.10 | -159.83 | -196.09 |
| MID-RANGE (IN) | .00 | .24 | 1.02 | 2.42 | 4.53 | 7.42 | 11.20 | 15.94 | 21.70 | 28.61 | 36.79 | 46.34 | 57.41 |
| BULLET PATH (IN) | -.63 | .70 | .00 | -2.97 | -8.42 | -16.59 | -27.71 | -41.99 | -59.61 | -80.90 | -106.13 | -135.59 | -169.60 |
| TIME OF FLIGHT (SEC) | .000000 | .070552 | .145420 | .224161 | .306502 | .392280 | .481410 | .573868 | .669782 | .769261 | .872362 | .979162 | 1.089758 |
| WIND DEFLECTION (IN) | | | | | | | | | | | | | |
| 5 MPH | .00 | .21 | .80 | 1.73 | 2.97 | 4.52 | 6.36 | 8.50 | 10.94 | 13.69 | 16.77 | 20.17 | 23.90 |
| 10 MPH | .00 | .42 | 1.59 | 3.45 | 5.94 | 9.04 | 12.73 | 17.00 | 21.88 | 27.39 | 33.54 | 40.33 | 47.80 |
| 20 MPH | .00 | .83 | 3.19 | 6.90 | 11.89 | 18.08 | 25.46 | 34.00 | 43.76 | 54.78 | 67.07 | 80.67 | 95.59 |
| 30 MPH | .00 | 1.25 | 4.78 | 10.36 | 17.83 | 27.12 | 38.18 | 51.00 | 65.64 | 82.17 | 100.61 | 121.00 | 143.39 |
| VELOCITY (FPS) | 1000. | 951. | 910. | 873. | 841. | 810. | 781. | 754. | 727. | 702. | 678. | 655. | 633. |
| ENERGY (FT-LB) | 420. | 380. | 347. | 320. | 297. | 276. | 256. | 238. | 222. | 207. | 193. | 180. | 168. |
| DROP (IN) | .00 | -1.10 | -4.57 | -10.66 | -19.58 | -31.56 | -46.81 | -65.56 | -88.11 | -114.76 | -145.80 | -181.55 | -222.35 |
| MID-RANGE (IN) | .00 | .29 | 1.18 | 2.80 | 5.22 | 8.51 | 12.75 | 18.03 | 24.48 | 32.21 | 41.34 | 51.99 | 64.36 |
| BULLET PATH (IN) | -.63 | .87 | .00 | -3.48 | -9.80 | -19.19 | -31.83 | -47.98 | -67.93 | -91.98 | -120.42 | -153.57 | -191.77 |
| TIME OF FLIGHT (SEC) | .000000 | .076956 | .157624 | .241794 | .329348 | .420234 | .514480 | .612253 | .713617 | .818641 | .927412 | 1.040034 | 1.156627 |
| WIND DEFLECTION (IN) | | | | | | | | | | | | | |
| 5 MPH | .00 | .17 | .67 | 1.48 | 2.58 | 3.98 | 5.67 | 7.68 | 10.00 | 12.64 | 15.61 | 18.92 | 22.58 |
| 10 MPH | .00 | .34 | 1.34 | 2.96 | 5.17 | 7.96 | 11.35 | 15.36 | 20.00 | 25.28 | 31.22 | 37.85 | 45.17 |
| 20 MPH | .00 | .69 | 2.68 | 5.91 | 10.33 | 15.92 | 22.70 | 30.71 | 39.99 | 50.56 | 62.45 | 75.69 | 90.33 |
| 30 MPH | .00 | 1.03 | 4.03 | 8.87 | 15.50 | 23.88 | 34.05 | 46.07 | 59.99 | 75.84 | 93.67 | 113.54 | 135.50 |
| VELOCITY (FPS) | 900. | 865. | 833. | 803. | 774. | 746. | 720. | 696. | 672. | 649. | 627. | 606. | 586. |
| ENERGY (FT-LB) | 340. | 314. | 291. | 271. | 251. | 234. | 218. | 203. | 189. | 177. | 165. | 154. | 144. |
| DROP (IN) | .00 | -1.31 | -5.51 | -12.85 | -23.47 | -37.69 | -55.79 | -78.06 | -104.79 | -136.33 | -173.01 | -215.20 | -263.29 |
| MID-RANGE (IN) | .00 | .35 | 1.45 | 3.39 | 6.22 | 10.07 | 15.05 | 21.30 | 28.93 | 38.06 | 48.82 | 61.36 | 75.86 |
| BULLET PATH (IN) | -.63 | 1.13 | -.00 | -4.26 | -11.81 | -22.96 | -37.99 | -57.18 | -80.85 | -109.31 | -142.92 | -182.03 | -227.06 |
| TIME OF FLIGHT (SEC) | .000000 | .085043 | .173450 | .265185 | .360330 | .459028 | .561332 | .667317 | .777073 | .890709 | 1.008347 | 1.130125 | 1.256192 |
| WIND DEFLECTION (IN) | | | | | | | | | | | | | |
| 5 MPH | .00 | .15 | .60 | 1.34 | 2.38 | 3.73 | 5.40 | 7.39 | 9.72 | 12.38 | 15.40 | 18.78 | 22.54 |
| 10 MPH | .00 | .30 | 1.19 | 2.67 | 4.75 | 7.46 | 10.79 | 14.78 | 19.43 | 24.76 | 30.80 | 37.57 | 45.09 |
| 20 MPH | .00 | .60 | 2.39 | 5.35 | 9.50 | 14.91 | 21.59 | 29.56 | 38.86 | 49.53 | 61.60 | 75.14 | 90.18 |
| 30 MPH | .00 | .90 | 3.58 | 8.02 | 14.25 | 22.37 | 32.38 | 44.34 | 58.29 | 74.29 | 92.41 | 112.71 | 135.27 |

## Bullet: Lyman #429303 189 Gr.
### Ballistic Coefficient: .105

| RANGE (YDS) | MUZZLE | 25 | 50 | 75 | 100 | 125 | 150 | 175 | 200 | 225 | 250 | 275 | 300 |
|---|---|---|---|---|---|---|---|---|---|---|---|---|---|
| VELOCITY (FPS) | 800. | 771. | 744. | 718. | 693. | 669. | 647. | 625. | 604. | 583. | 564. | 545. | 526. |
| ENERGY (FT-LB) | 269. | 249. | 232. | 216. | 201. | 188. | 175. | 164. | 153. | 143. | 133. | 124. | 116. |
| DROP (IN) | .00 | -1.57 | -6.77 | -15.88 | -29.19 | -47.01 | -69.66 | -97.48 | -130.85 | -170.18 | -215.88 | -268.42 | -328.29 |
| MID-RANGE (IN) | .00 | .43 | 1.81 | 4.24 | 7.82 | 12.68 | 18.94 | 26.74 | 36.23 | 47.57 | 60.93 | 76.50 | 94.49 |
| BULLET PATH (IN) | -.63 | 1.50 | -.00 | -5.41 | -15.02 | -29.13 | -48.08 | -72.20 | -101.87 | -137.49 | -179.49 | -228.33 | -284.49 |
| TIME OF FLIGHT (SEC) | .000000 | .C95517 | .194592 | .297279 | .403655 | .513812 | .627859 | .745921 | .868135 | .994653 | 1.125637 | 1.261262 | 1.401715 |
| WIND DEFLECTION (IN) | | | | | | | | | | | | | |
| 5 MPH | .00 | .16 | .62 | 1.41 | 2.52 | 3.97 | 5.75 | 7.89 | 10.40 | 13.28 | 16.56 | 20.24 | 24.35 |
| 10 MPH | .00 | .31 | 1.25 | 2.82 | 5.04 | 7.93 | 11.50 | 15.78 | 20.79 | 26.56 | 33.11 | 40.48 | 48.70 |
| 20 MPH | .00 | .62 | 2.50 | 5.64 | 10.09 | 15.86 | 23.01 | 31.56 | 41.58 | 53.12 | 66.22 | 80.96 | 97.40 |
| 30 MPH | .00 | .93 | 3.74 | 8.46 | 15.13 | 23.79 | 34.51 | 47.35 | 62.38 | 79.68 | 99.34 | 121.45 | 146.11 |
| VELOCITY (FPS) | 700. | 676. | 653. | 631. | 610. | 589. | 569. | 550. | 531. | 513. | 495. | 478. | 462. |
| ENERGY (FT-LB) | 206. | 192. | 179. | 167. | 156. | 146. | 136. | 127. | 118. | 110. | 103. | 96. | 90. |
| DROP (IN) | .00 | -2.02 | -8.77 | -20.60 | -37.87 | -60.97 | -90.31 | -126.35 | -169.57 | -220.49 | -279.67 | -347.72 | -425.29 |
| MID-RANGE (IN) | .00 | .57 | 2.37 | 5.52 | 10.16 | 16.45 | 24.55 | 34.64 | 46.91 | 61.58 | 78.87 | 99.03 | 122.33 |
| BULLET PATH (IN) | -.63 | 2.05 | .00 | -7.13 | -19.69 | -38.09 | -62.73 | -94.07 | -132.58 | -178.80 | -233.28 | -296.63 | -369.50 |
| TIME OF FLIGHT (SEC) | .000000 | .109035 | .221925 | .338791 | .459766 | .584996 | .714641 | .848871 | .987869 | 1.131829 | 1.280956 | 1.435468 | 1.595595 |
| WIND DEFLECTION (IN) | | | | | | | | | | | | | |
| 5 MPH | .00 | .17 | .67 | 1.53 | 2.75 | 4.34 | 6.32 | 8.70 | 11.50 | 14.74 | 18.44 | 22.61 | 27.27 |
| 10 MPH | .00 | .33 | 1.34 | 3.06 | 5.49 | 8.67 | 12.63 | 17.40 | 23.01 | 29.49 | 36.88 | 45.21 | 54.54 |
| 20 MPH | .00 | .67 | 2.69 | 6.11 | 10.98 | 17.35 | 25.27 | 34.80 | 46.02 | 58.98 | 73.75 | 90.43 | 109.08 |
| 30 MPH | .00 | 1.00 | 4.03 | 9.17 | 16.47 | 26.02 | 37.90 | 52.20 | 69.02 | 88.46 | 110.63 | 135.64 | 163.62 |

# Bullet: Lyman #429215 203 Gr.
## Ballistic Coefficient: .188

| RANGE (YDS) | MUZZLE | 25 | 50 | 75 | 100 | 125 | 150 | 175 | 200 | 225 | 250 | 275 | 300 |
|---|---|---|---|---|---|---|---|---|---|---|---|---|---|
| VELOCITY (FPS) | 1600. | 1516. | 1436. | 1362. | 1293. | 1238. | 1189. | 1144. | 1105. | 1071. | 1041. | 1015. | 991. |
| ENERGY (FT-LB) | 1154. | 1035. | 930. | 836. | 754. | 691. | 637. | 590. | 550. | 517. | 489. | 464. | 442. |
| DROP (IN) | .00 | -.41 | -1.77 | -4.19 | -7.78 | -12.66 | -18.97 | -26.81 | -36.31 | -47.58 | -60.73 | -75.89 | -93.15 |
| MID-RANGE (IN) | .00 | .11 | .48 | 1.13 | 2.12 | 3.49 | 5.29 | 7.57 | 10.37 | 13.74 | 17.71 | 22.32 | 27.61 |
| BULLET PATH (IN) | -.63 | .16 | .00 | -1.22 | -3.61 | -7.29 | -12.39 | -19.03 | -27.33 | -37.40 | -49.36 | -63.31 | -79.37 |
| TIME OF FLIGHT (SEC) | .000000 | .048167 | .099010 | .152652 | .209189 | .268471 | .330312 | .394649 | .461373 | .530333 | .601371 | .674354 | .749178 |
| WIND DEFLECTION (IN) | | | | | | | | | | | | | |
| 5 MPH | .00 | .11 | .46 | 1.06 | 1.91 | 3.00 | 4.32 | 5.85 | 7.60 | 9.54 | 11.67 | 13.97 | 16.43 |
| 10 MPH | .00 | .23 | .93 | 2.12 | 3.82 | 6.00 | 8.63 | 11.71 | 15.20 | 19.09 | 23.34 | 27.94 | 32.86 |
| 20 MPH | .00 | .45 | 1.85 | 4.23 | 7.63 | 12.00 | 17.27 | 23.42 | 30.40 | 38.18 | 46.68 | 55.87 | 65.71 |
| 30 MPH | .00 | .68 | 2.78 | 6.35 | 11.45 | 18.00 | 25.90 | 35.12 | 45.60 | 57.27 | 70.02 | 83.81 | 98.57 |
| VELOCITY (FPS) | 1500. | 1421. | 1348. | 1283. | 1229. | 1180. | 1136. | 1098. | 1065. | 1036. | 1010. | 987. | 965. |
| ENERGY (FT-LB) | 1014. | 911. | 819. | 741. | 680. | 627. | 582. | 544. | 511. | 484. | 460. | 439. | 420. |
| DROP (IN) | .00 | -.48 | -2.03 | -4.78 | -8.85 | -14.36 | -21.43 | -30.18 | -40.72 | -53.17 | -67.64 | -84.23 | -103.05 |
| MID-RANGE (IN) | .00 | .13 | .54 | 1.28 | 2.39 | 3.93 | 5.93 | 8.45 | 11.51 | 15.17 | 19.46 | 24.42 | 30.09 |
| BULLET PATH (IN) | -.63 | .22 | .00 | -1.41 | -4.15 | -8.33 | -14.07 | -21.48 | -30.69 | -41.81 | -54.94 | -70.20 | -87.69 |
| TIME OF FLIGHT (SEC) | .000000 | .051372 | .105566 | .162654 | .222416 | .284728 | .349520 | .416673 | .486032 | .557442 | .630775 | .705931 | .782828 |
| WIND DEFLECTION (IN) | | | | | | | | | | | | | |
| 5 MPH | .00 | .12 | .49 | 1.11 | 1.97 | 3.06 | 4.36 | 5.87 | 7.57 | 9.45 | 11.51 | 13.72 | 16.09 |
| 10 MPH | .00 | .24 | .98 | 2.23 | 3.95 | 6.11 | 8.72 | 11.73 | 15.14 | 18.91 | 23.02 | 27.44 | 32.18 |
| 20 MPH | .00 | .48 | 1.96 | 4.45 | 7.89 | 12.22 | 17.43 | 23.47 | 30.28 | 37.82 | 46.03 | 54.89 | 64.36 |
| 30 MPH | .00 | .72 | 2.94 | 6.68 | 11.84 | 18.34 | 26.15 | 35.20 | 45.43 | 56.73 | 69.05 | 82.33 | 96.53 |
| VELOCITY (FPS) | 1400. | 1328. | 1267. | 1214. | 1167. | 1125. | 1088. | 1057. | 1028. | 1003. | 980. | 959. | 940. |
| ENERGY (FT-LB) | 883. | 795. | 723. | 664. | 614. | 570. | 534. | 503. | 477. | 453. | 433. | 415. | 398. |
| DROP (IN) | .00 | -.55 | -2.33 | -5.47 | -10.08 | -16.29 | -24.20 | -33.95 | -45.63 | -59.37 | -75.26 | -93.40 | -113.87 |
| MID-RANGE (IN) | .00 | .15 | .61 | 1.45 | 2.71 | 4.42 | 6.63 | 9.39 | 12.73 | 16.71 | 21.34 | 26.67 | 32.73 |
| BULLET PATH (IN) | -.63 | .30 | .00 | -1.66 | -4.79 | -9.51 | -15.95 | -24.21 | -34.42 | -46.67 | -61.08 | -77.74 | -96.73 |
| TIME OF FLIGHT (SEC) | .000000 | .055013 | .112897 | .173394 | .236427 | .301910 | .369712 | .439673 | .511645 | .585509 | .661171 | .738553 | .817593 |
| WIND DEFLECTION (IN) | | | | | | | | | | | | | |
| 5 MPH | .00 | .13 | .51 | 1.12 | 1.95 | 3.00 | 4.25 | 5.69 | 7.31 | 9.10 | 11.04 | 13.14 | 15.38 |
| 10 MPH | .00 | .25 | 1.01 | 2.23 | 3.90 | 5.99 | 8.50 | 11.38 | 14.62 | 18.19 | 22.08 | 26.27 | 30.75 |
| 20 MPH | .00 | .51 | 2.03 | 4.46 | 7.79 | 11.99 | 17.00 | 22.76 | 29.24 | 36.38 | 44.16 | 52.54 | 61.51 |
| 30 MPH | .00 | .76 | 3.04 | 6.70 | 11.69 | 17.98 | 25.49 | 34.15 | 43.86 | 54.58 | 66.24 | 78.81 | 92.26 |
| VELOCITY (FPS) | 1300. | 1244. | 1194. | 1149. | 1109. | 1075. | 1045. | 1018. | 993. | 971. | 951. | 932. | 914. |
| ENERGY (FT-LB) | 762. | 698. | 643. | 595. | 555. | 521. | 492. | 467. | 445. | 425. | 407. | 391. | 377. |
| DROP (IN) | .00 | -.63 | -2.68 | -6.25 | -11.46 | -18.42 | -27.27 | -38.09 | -51.01 | -66.13 | -83.54 | -103.30 | -125.54 |
| MID-RANGE (IN) | .00 | .17 | .71 | 1.65 | 3.05 | 4.95 | 7.39 | 10.41 | 14.05 | 18.35 | 23.35 | 29.05 | 35.51 |
| BULLET PATH (IN) | -.63 | .39 | .00 | -1.91 | -5.47 | -10.78 | -17.97 | -27.14 | -38.40 | -51.87 | -67.62 | -85.73 | -106.32 |
| TIME OF FLIGHT (SEC) | .000000 | .058980 | .120524 | .184574 | .251025 | .319731 | .390532 | .463292 | .537903 | .614277 | .692344 | .772048 | .853342 |
| WIND DEFLECTION (IN) | | | | | | | | | | | | | |
| 5 MPH | .00 | .11 | .45 | 1.01 | 1.78 | 2.75 | 3.91 | 5.23 | 6.72 | 8.36 | 10.16 | 12.09 | 14.17 |
| 10 MPH | .00 | .23 | .90 | 2.02 | 3.56 | 5.50 | 7.81 | 10.46 | 13.44 | 16.73 | 20.31 | 24.19 | 28.34 |
| 20 MPH | .00 | .45 | 1.81 | 4.05 | 7.13 | 11.01 | 15.62 | 20.92 | 26.88 | 33.46 | 40.63 | 48.38 | 56.68 |
| 30 MPH | .00 | .68 | 2.71 | 6.07 | 10.69 | 16.51 | 23.43 | 31.39 | 40.32 | 50.18 | 60.94 | 72.56 | 85.03 |

## Bullet: Lyman #429215 203 Gr.
### Ballistic Coefficient: .188

| RANGE (YDS) | MUZZLE | 25 | 50 | 75 | 100 | 125 | 150 | 175 | 200 | 225 | 250 | 275 | 300 |
|---|---|---|---|---|---|---|---|---|---|---|---|---|---|
| VELOCITY (FPS) | 1200. | 1154. | 1114. | 1079. | 1048. | 1021. | 996. | 974. | 953. | 934. | 916. | 899. | 883. |
| ENERGY (FT-LB) | 649. | 600. | 559. | 525. | 495. | 470. | 447. | 427. | 410. | 393. | 378. | 364. | 352. |
| DROP (IN) | .00 | -.75 | -3.14 | -7.26 | -13.25 | -21.21 | -31.24 | -43.46 | -57.97 | -74.81 | -94.12 | -116.01 | -140.58 |
| MID-RANGE (IN) | .00 | .20 | .81 | 1.90 | 3.49 | 5.63 | 8.36 | 11.72 | 15.74 | 20.43 | 25.85 | 32.05 | 39.05 |
| BULLET PATH (IN) | -.63 | .50 | .00 | -2.24 | -6.35 | -12.42 | -20.58 | -30.91 | -43.54 | -58.50 | -75.92 | -95.93 | -118.62 |
| TIME OF FLIGHT (SEC) | .000000 | .063745 | .129907 | .198344 | .268894 | .341419 | .415805 | .491965 | .569826 | .649329 | .730427 | .813083 | .897267 |
| WIND DEFLECTION (IN) | | | | | | | | | | | | | |
| 5 MPH | .00 | .11 | .43 | .95 | 1.66 | 2.54 | 3.59 | 4.79 | 6.14 | 7.64 | 9.28 | 11.05 | 12.96 |
| 10 MPH | .00 | .22 | .86 | 1.91 | 3.33 | 5.09 | 7.18 | 9.59 | 12.29 | 15.28 | 18.56 | 22.10 | 25.92 |
| 20 MPH | .00 | .44 | 1.73 | 3.82 | 6.65 | 10.18 | 14.36 | 19.17 | 24.58 | 30.56 | 37.11 | 44.21 | 51.84 |
| 30 MPH | .00 | .66 | 2.59 | 5.73 | 9.98 | 15.27 | 21.55 | 28.76 | 36.87 | 45.85 | 55.67 | 66.31 | 77.76 |
| VELOCITY (FPS) | 1100. | 1067. | 1037. | 1011. | 987. | 966. | 946. | 927. | 910. | 893. | 877. | 862. | 848. |
| ENERGY (FT-LB) | 545. | 513. | 485. | 461. | 439. | 420. | 403. | 387. | 373. | 359. | 347. | 335. | 324. |
| DROP (IN) | .00 | -.89 | -3.68 | -8.48 | -15.40 | -24.55 | -36.00 | -49.83 | -66.18 | -85.14 | -106.83 | -131.32 | -158.72 |
| MID-RANGE (IN) | .00 | .23 | .95 | 2.20 | 4.02 | 6.45 | 9.52 | 13.24 | 17.69 | 22.88 | 28.87 | 35.67 | 43.36 |
| BULLET PATH (IN) | -.63 | .64 | .00 | -2.65 | -7.42 | -14.41 | -23.71 | -35.38 | -49.58 | -66.39 | -85.92 | -108.26 | -133.51 |
| TIME OF FLIGHT (SEC) | .000000 | .069260 | .140577 | .213822 | .288893 | .365708 | .444199 | .524313 | .606007 | .689245 | .774003 | .860259 | .947999 |
| WIND DEFLECTION (IN) | | | | | | | | | | | | | |
| 5 MPH | .00 | .09 | .37 | .82 | 1.42 | 2.18 | 3.09 | 4.14 | 5.33 | 6.65 | 8.11 | 9.70 | 11.42 |
| 10 MPH | .00 | .19 | .74 | 1.63 | 2.85 | 4.36 | 6.18 | 8.28 | 10.66 | 13.31 | 16.22 | 19.41 | 22.85 |
| 20 MPH | .00 | .38 | 1.48 | 3.27 | 5.69 | 8.73 | 12.36 | 16.56 | 21.31 | 26.61 | 32.45 | 38.81 | 45.70 |
| 30 MPH | .00 | .57 | 2.22 | 4.90 | 8.54 | 13.09 | 18.54 | 24.84 | 31.97 | 39.92 | 48.67 | 58.22 | 68.54 |
| VELOCITY (FPS) | 1000. | 977. | 956. | 937. | 919. | 902. | 886. | 870. | 856. | 842. | 828. | 815. | 802. |
| ENERGY (FT-LB) | 451. | 430. | 412. | 396. | 380. | 367. | 354. | 341. | 330. | 319. | 309. | 299. | 290. |
| DROP (IN) | .00 | -1.06 | -4.38 | -10.05 | -18.16 | -28.83 | -42.16 | -58.24 | -77.19 | -99.08 | -124.04 | -152.15 | -183.53 |
| MID-RANGE (IN) | .00 | .28 | 1.13 | 2.59 | 4.70 | 7.49 | 11.03 | 15.33 | 20.44 | 26.38 | 33.19 | 40.92 | 49.61 |
| BULLET PATH (IN) | -.63 | .82 | .00 | -3.16 | -8.76 | -16.93 | -27.75 | -41.33 | -57.77 | -77.16 | -99.61 | -125.22 | -154.09 |
| TIME OF FLIGHT (SEC) | .000000 | .075879 | .153469 | .232709 | .313551 | .395954 | .479890 | .565333 | .652267 | .740679 | .830564 | .921918 | 1.014743 |
| WIND DEFLECTION (IN) | | | | | | | | | | | | | |
| 5 MPH | .00 | .08 | .31 | .68 | 1.19 | 1.84 | 2.63 | 3.55 | 4.60 | 5.78 | 7.09 | 8.53 | 10.10 |
| 10 MPH | .00 | .15 | .61 | 1.36 | 2.38 | 3.69 | 5.26 | 7.10 | 9.20 | 11.56 | 14.18 | 17.06 | 20.19 |
| 20 MPH | .00 | .31 | 1.22 | 2.71 | 4.77 | 7.38 | 10.52 | 14.20 | 18.40 | 23.12 | 28.36 | 34.12 | 40.39 |
| 30 MPH | .00 | .46 | 1.83 | 4.07 | 7.15 | 11.06 | 15.78 | 21.30 | 27.60 | 34.68 | 42.54 | 51.17 | 60.58 |
| VELOCITY (FPS) | 900. | 884. | 869. | 854. | 840. | 826. | 813. | 801. | 787. | 773. | 759. | 746. | 734. |
| ENERGY (FT-LB) | 365. | 352. | 340. | 329. | 318. | 308. | 298. | 289. | 279. | 269. | 260. | 251. | 243. |
| DROP (IN) | .00 | -1.22 | -5.20 | -12.05 | -21.87 | -34.75 | -50.80 | -70.13 | -92.76 | -118.83 | -148.52 | -181.96 | -219.29 |
| MID-RANGE (IN) | .00 | .34 | 1.38 | 3.17 | 5.73 | 9.11 | 13.35 | 18.48 | 24.51 | 31.49 | 39.51 | 48.61 | 58.84 |
| BULLET PATH (IN) | -.63 | 1.07 | -.00 | -3.93 | -10.83 | -20.80 | -33.94 | -50.35 | -70.06 | -93.21 | -119.99 | -150.52 | -184.93 |
| TIME OF FLIGHT (SEC) | .000000 | .034093 | .169692 | .256778 | .345343 | .435377 | .526880 | .619854 | .714371 | .810601 | .908562 | 1.008265 | 1.109724 |
| WIND DEFLECTION (IN) | | | | | | | | | | | | | |
| 5 MPH | .00 | .07 | .27 | .60 | 1.06 | 1.65 | 2.37 | 3.21 | 4.20 | 5.33 | 6.62 | 8.06 | 9.66 |
| 10 MPH | .00 | .13 | .53 | 1.19 | 2.11 | 3.29 | 4.73 | 6.43 | 8.40 | 10.67 | 13.24 | 16.12 | 19.31 |
| 20 MPH | .00 | .27 | 1.06 | 2.39 | 4.23 | 6.59 | 9.46 | 12.86 | 16.79 | 21.33 | 26.48 | 32.24 | 38.62 |
| 30 MPH | .00 | .40 | 1.60 | 3.58 | 6.34 | 9.88 | 14.19 | 19.28 | 25.19 | 32.00 | 39.72 | 48.36 | 57.93 |

# Bullet: Lyman #429215 203 Gr.
## Ballistic Coefficient: .188

| RANGE (YDS) | MUZZLE | 25 | 50 | 75 | 100 | 125 | 150 | 175 | 200 | 225 | 250 | 275 | 300 |
|---|---|---|---|---|---|---|---|---|---|---|---|---|---|
| VELOCITY (FPS) | 800. | 786. | 772. | 758. | 745. | 732. | 720. | 708. | 696. | 684. | 673. | 662. | 651. |
| ENERGY (FT-LB) | 288. | 278. | 268. | 259. | 250. | 242. | 234. | 226. | 218. | 211. | 204. | 197. | 191. |
| DROP (IN) | .00 | -1.43 | -6.30 | -14.80 | -27.05 | -43.19 | -63.37 | -87.72 | -116.38 | -149.52 | -187.28 | -229.83 | -277.32 |
| MID-RANGE (IN) | .00 | .40 | 1.72 | 3.98 | 7.22 | 11.51 | 16.89 | 23.41 | 31.14 | 40.13 | 50.45 | 62.14 | 75.29 |
| BULLET PATH (IN) | -.63 | 1.40 | -.00 | -5.03 | -13.82 | -26.50 | -43.20 | -64.09 | -89.29 | -118.96 | -153.25 | -192.33 | -236.36 |
| TIME OF FLIGHT (SEC) | .000000 | .094608 | .190938 | .288997 | .388798 | .490354 | .593681 | .698800 | .805731 | .914496 | 1.025121 | 1.137632 | 1.252058 |
| WIND DEFLECTION (IN) | | | | | | | | | | | | | |
| 5 MPH | .00 | .08 | .30 | .68 | 1.21 | 1.90 | 2.74 | 3.74 | 4.90 | 6.23 | 7.71 | 9.36 | 11.18 |
| 10 MPH | .00 | .15 | .61 | 1.36 | 2.43 | 3.80 | 5.49 | 7.49 | 9.81 | 12.45 | 15.42 | 18.72 | 22.36 |
| 20 MPH | .00 | .30 | 1.21 | 2.73 | 4.86 | 7.60 | 10.98 | 14.98 | 19.62 | 24.90 | 30.84 | 37.45 | 44.72 |
| 30 MPH | .00 | .45 | 1.82 | 4.09 | 7.29 | 11.41 | 16.46 | 22.47 | 29.43 | 37.35 | 46.26 | 56.17 | 67.09 |
| VELOCITY (FPS) | 700. | 688. | 677. | 665. | 654. | 644. | 633. | 622. | 612. | 602. | 592. | 583. | 573. |
| ENERGY (FT-LB) | 221. | 213. | 206. | 200. | 193. | 187. | 181. | 175. | 169. | 163. | 158. | 153. | 148. |
| DROP (IN) | .00 | -1.73 | -8.03 | -19.05 | -34.96 | -55.92 | -82.11 | -113.70 | -150.89 | -193.86 | -242.81 | -297.95 | -359.49 |
| MID-RANGE (IN) | .00 | .56 | 2.28 | 5.22 | 9.45 | 15.02 | 22.00 | 30.47 | 40.48 | 52.13 | 65.49 | 80.63 | 97.64 |
| BULLET PATH (IN) | -.63 | 1.97 | .00 | -6.69 | -18.27 | -34.90 | -56.75 | -84.02 | -116.87 | -155.51 | -200.13 | -250.94 | -308.16 |
| TIME OF FLIGHT (SEC) | .000000 | .108059 | .217966 | .329747 | .443430 | .559043 | .676618 | .796186 | .917783 | 1.041443 | 1.167203 | 1.295102 | 1.425181 |
| WIND DEFLECTION (IN) | | | | | | | | | | | | | |
| 5 MPH | .00 | .08 | .32 | .73 | 1.31 | 2.05 | 2.97 | 4.06 | 5.34 | 6.79 | 8.43 | 10.25 | 12.27 |
| 10 MPH | .00 | .16 | .65 | 1.46 | 2.62 | 4.11 | 5.94 | 8.13 | 10.67 | 13.58 | 16.86 | 20.51 | 24.55 |
| 20 MPH | .00 | .32 | 1.30 | 2.93 | 5.23 | 8.21 | 11.88 | 16.26 | 21.35 | 27.16 | 33.71 | 41.02 | 49.09 |
| 30 MPH | .00 | .48 | 1.94 | 4.39 | 7.85 | 12.32 | 17.83 | 24.39 | 32.02 | 40.74 | 50.57 | 61.53 | 73.64 |

# Bullet: Lyman #429360 225 Gr.
## Ballistic Coefficient: .169

| RANGE (YDS) | MUZZLE | 25 | 50 | 75 | 100 | 125 | 150 | 175 | 200 | 225 | 250 | 275 | 300 |
|---|---|---|---|---|---|---|---|---|---|---|---|---|---|
| VELOCITY (FPS) | 1600. | 1507. | 1419. | 1338. | 1268. | 1210. | 1159. | 1114. | 1076. | 1043. | 1013. | 987. | 963. |
| ENERGY (FT-LB) | 1279. | 1134. | 1006. | 894. | 803. | 732. | 671. | 620. | 578. | 543. | 513. | 487. | 463. |
| DROP (IN) | .00 | -.41 | -1.79 | -4.25 | -7.92 | -12.95 | -19.46 | -27.58 | -37.45 | -49.18 | -62.91 | -78.76 | -96.82 |
| MID-RANGE (IN) | .00 | .11 | .48 | 1.15 | 2.17 | 3.59 | 5.47 | 7.86 | 10.80 | 14.33 | 18.51 | 23.37 | 28.95 |
| BULLET PATH (IN) | -.63 | .17 | .00 | -1.25 | -3.71 | -7.52 | -12.81 | -19.72 | -28.38 | -38.91 | -51.42 | -66.05 | -82.91 |
| TIME OF FLIGHT (SEC) | .000000 | .048315 | .099623 | .154072 | .211724 | .272283 | .335631 | .401657 | .470193 | .541043 | .614044 | .689068 | .766013 |
| WIND DEFLECTION (IN) | | | | | | | | | | | | | |
| 5 MPH | .00 | .13 | .52 | 1.18 | 2.13 | 3.34 | 4.79 | 6.47 | 8.38 | 10.49 | 12.79 | 15.26 | 17.91 |
| 10 MPH | .00 | .25 | 1.03 | 2.37 | 4.26 | 6.67 | 9.57 | 12.94 | 16.75 | 20.97 | 25.57 | 30.53 | 35.82 |
| 20 MPH | .00 | .51 | 2.07 | 4.73 | 8.53 | 13.34 | 19.14 | 25.88 | 33.51 | 41.95 | 51.14 | 61.05 | 71.64 |
| 30 MPH | .00 | .76 | 3.10 | 7.10 | 12.79 | 20.02 | 28.71 | 38.83 | 50.26 | 62.92 | 76.72 | 91.58 | 107.45 |
| VELOCITY (FPS) | 1500. | 1413. | 1332. | 1264. | 1207. | 1156. | 1111. | 1073. | 1040. | 1011. | 985. | 962. | 940. |
| ENERGY (FT-LB) | 1124. | 997. | 887. | 798. | 727. | 667. | 617. | 575. | 541. | 511. | 485. | 462. | 441. |
| DROP (IN) | .00 | -.48 | -2.06 | -4.85 | -9.00 | -14.65 | -21.92 | -30.95 | -41.86 | -54.76 | -69.79 | -87.05 | -106.63 |
| MID-RANGE (IN) | .00 | .13 | .54 | 1.30 | 2.45 | 4.04 | 6.12 | 8.73 | 11.92 | 15.75 | 20.24 | 25.45 | 31.39 |
| BULLET PATH (IN) | -.63 | .23 | .00 | -1.45 | -4.26 | -8.57 | -14.50 | -22.18 | -31.74 | -43.31 | -56.99 | -72.90 | -91.14 |
| TIME OF FLIGHT (SEC) | .000000 | .05[illegible]529 | .106209 | .164080 | .224838 | .288379 | .354589 | .423294 | .494300 | .567448 | .642609 | .719686 | .798599 |
| WIND DEFLECTION (IN) | | | | | | | | | | | | | |
| 5 MPH | .00 | .13 | .55 | 1.24 | 2.19 | 3.38 | 4.80 | 6.45 | 8.30 | 10.34 | 12.55 | 14.93 | 17.48 |
| 10 MPH | .00 | .27 | 1.09 | 2.48 | 4.37 | 6.75 | 9.61 | 12.90 | 16.60 | 20.67 | 25.10 | 29.86 | 34.95 |
| 20 MPH | .00 | .54 | 2.19 | 4.96 | 8.74 | 13.51 | 19.22 | 25.80 | 33.19 | 41.34 | 50.20 | 59.73 | 69.91 |
| 30 MPH | .00 | .81 | 3.28 | 7.43 | 13.11 | 20.26 | 28.82 | 38.70 | 49.79 | 62.01 | 75.30 | 89.59 | 104.86 |
| VELOCITY (FPS) | 1400. | 1321. | 1255. | 1199. | 1149. | 1105. | 1068. | 1036. | 1007. | 982. | 958. | 937. | 917. |
| ENERGY (FT-LB) | 979. | 871. | 787. | 718. | 659. | 610. | 570. | 536. | 507. | 481. | 459. | 439. | 420. |
| DROP (IN) | .00 | -.56 | -2.35 | -5.54 | -10.23 | -16.56 | -24.67 | -34.68 | -46.70 | -60.86 | -77.27 | -96.01 | -117.20 |
| MID-RANGE (IN) | .00 | .15 | .62 | 1.48 | 2.76 | 4.52 | 6.80 | 9.65 | 13.12 | 17.25 | 22.07 | 27.62 | 33.93 |
| BULLET PATH (IN) | -.63 | .30 | .00 | -1.69 | -4.89 | -9.74 | -16.36 | -24.87 | -35.40 | -48.07 | -62.99 | -80.24 | -99.93 |
| TIME OF FLIGHT (SEC) | .000000 | .055176 | .113497 | .174666 | .238605 | .305190 | .374242 | .445570 | .519018 | .594464 | .671812 | .750987 | .831928 |
| WIND DEFLECTION (IN) | | | | | | | | | | | | | |
| 5 MPH | .00 | .14 | .56 | 1.23 | 2.14 | 3.29 | 4.65 | 6.21 | 7.96 | 9.88 | 11.98 | 14.23 | 16.64 |
| 10 MPH | .00 | .28 | 1.12 | 2.46 | 4.28 | 6.57 | 9.30 | 12.42 | 15.92 | 19.77 | 23.95 | 28.46 | 33.28 |
| 20 MPH | .00 | .56 | 2.24 | 4.91 | 8.56 | 13.14 | 18.59 | 24.84 | 31.84 | 39.54 | 47.91 | 56.92 | 66.55 |
| 30 MPH | .00 | .85 | 3.35 | 7.37 | 12.84 | 19.71 | 27.89 | 37.26 | 47.76 | 59.31 | 71.86 | 85.38 | 99.83 |
| VELOCITY (FPS) | 1300. | 1239. | 1184. | 1136. | 1095. | 1059. | 1028. | 1000. | 975. | 952. | 931. | 912. | 894. |
| ENERGY (FT-LB) | 844. | 767. | 701. | 645. | 599. | 560. | 528. | 500. | 475. | 453. | 433. | 415. | 399. |
| DROP (IN) | .00 | -.64 | -2.70 | -6.31 | -11.59 | -18.69 | -27.71 | -38.79 | -52.03 | -67.55 | -85.42 | -105.77 | -128.73 |
| MID-RANGE (IN) | .00 | .17 | .71 | 1.67 | 3.10 | 5.04 | 7.55 | 10.66 | 14.42 | 18.87 | 24.02 | 29.94 | 36.65 |
| BULLET PATH (IN) | -.63 | .40 | .00 | -1.94 | -5.56 | -10.99 | -18.35 | -27.76 | -39.34 | -53.19 | -69.40 | -88.08 | -109.37 |
| TIME OF FLIGHT (SEC) | .000000 | .059112 | .121046 | .185722 | .253000 | .322690 | .394610 | .468613 | .544586 | .622439 | .702100 | .783514 | .866636 |
| WIND DEFLECTION (IN) | | | | | | | | | | | | | |
| 5 MPH | .00 | .12 | .50 | 1.11 | 1.96 | 3.01 | 4.26 | 5.70 | 7.31 | 9.08 | 11.02 | 13.10 | 15.34 |
| 10 MPH | .00 | .25 | 1.00 | 2.23 | 3.91 | 6.02 | 8.53 | 11.40 | 14.62 | 18.16 | 22.03 | 26.21 | 30.68 |
| 20 MPH | .00 | .50 | 1.99 | 4.45 | 7.83 | 12.05 | 17.06 | 22.80 | 29.23 | 36.33 | 44.06 | 52.41 | 61.36 |
| 30 MPH | .00 | .75 | 2.99 | 6.68 | 11.74 | 18.07 | 25.58 | 34.20 | 43.85 | 54.49 | 66.09 | 78.62 | 92.05 |

# Bullet: Lyman #429360 225 Gr.
## Ballistic Coefficient: .169

| RANGE (YDS) | MUZZLE | 25 | 50 | 75 | 100 | 125 | 150 | 175 | 200 | 225 | 250 | 275 | 300 |
|---|---|---|---|---|---|---|---|---|---|---|---|---|---|
| VELOCITY (FPS) | 1200. | 1150. | 1106. | 1069. | 1037. | 1008. | 982. | 959. | 937. | 917. | 899. | 881. | 865. |
| ENERGY (FT-LB) | 719. | 661. | 611. | 571. | 537. | 507. | 482. | 459. | 439. | 420. | 404. | 388. | 373. |
| DROP (IN) | .00 | -.76 | -3.16 | -7.32 | -13.38 | -21.46 | -31.67 | -44.12 | -58.91 | -76.13 | -95.92 | -118.39 | -143.65 |
| MID-RANGE (IN) | .00 | .20 | .82 | 1.92 | 3.54 | 5.72 | 8.51 | 11.95 | 16.07 | 20.90 | 26.50 | 32.91 | 40.16 |
| BULLET PATH (IN) | -.63 | .50 | .00 | -2.27 | -6.44 | -12.63 | -20.94 | -31.51 | -44.40 | -59.73 | -77.63 | -98.20 | -121.57 |
| TIME OF FLIGHT (SEC) | .000000 | .063871 | .130392 | .199384 | .270656 | .344051 | .419447 | .496747 | .575875 | .656770 | .739385 | .823685 | .909641 |
| WIND DEFLECTION (IN) | | | | | | | | | | | | | |
| 5 MPH | .00 | .12 | .47 | 1.05 | 1.82 | 2.78 | 3.91 | 5.21 | 6.68 | 8.30 | 10.07 | 11.98 | 14.05 |
| 10 MPH | .00 | .24 | .95 | 2.09 | 3.64 | 5.55 | 7.82 | 10.43 | 13.35 | 16.59 | 20.13 | 23.97 | 28.10 |
| 20 MPH | .00 | .48 | 1.90 | 4.18 | 7.27 | 11.11 | 15.65 | 20.85 | 26.71 | 33.18 | 40.26 | 47.94 | 56.19 |
| 30 MPH | .00 | .72 | 2.85 | 6.27 | 10.91 | 16.66 | 23.47 | 31.28 | 40.06 | 49.77 | 60.40 | 71.91 | 84.29 |
| VELOCITY (FPS) | 1100. | 1063. | 1032. | 1004. | 978. | 955. | 934. | 914. | 896. | 879. | 862. | 846. | 831. |
| ENERGY (FT-LB) | 604. | 565. | 532. | 503. | 478. | 456. | 436. | 418. | 401. | 386. | 371. | 358. | 345. |
| DROP (IN) | .00 | -.89 | -3.70 | -8.54 | -15.53 | -24.79 | -36.38 | -50.44 | -67.08 | -86.42 | -108.57 | -133.63 | -161.72 |
| MID-RANGE (IN) | .00 | .23 | .96 | 2.22 | 4.07 | 6.54 | 9.65 | 13.46 | 18.01 | 23.34 | 29.50 | 36.53 | 44.48 |
| BULLET PATH (IN) | -.63 | .64 | .00 | -2.68 | -7.50 | -14.60 | -24.03 | -35.92 | -50.40 | -67.57 | -87.56 | -110.46 | -136.38 |
| TIME OF FLIGHT (SEC) | .000000 | .069368 | .140988 | .214709 | .290414 | .368009 | .447421 | .528592 | .611476 | .696039 | .782255 | .870108 | .959586 |
| WIND DEFLECTION (IN) | | | | | | | | | | | | | |
| 5 MPH | .00 | .10 | .41 | .89 | 1.56 | 2.38 | 3.37 | 4.52 | 5.81 | 7.25 | 8.84 | 10.57 | 12.44 |
| 10 MPH | .00 | .21 | .81 | 1.79 | 3.11 | 4.77 | 6.75 | 9.03 | 11.62 | 14.50 | 17.68 | 21.14 | 24.89 |
| 20 MPH | .00 | .42 | 1.63 | 3.58 | 6.23 | 9.54 | 13.49 | 18.06 | 23.24 | 29.01 | 35.35 | 42.28 | 49.77 |
| 30 MPH | .00 | .63 | 2.44 | 5.37 | 9.34 | 14.31 | 20.24 | 27.10 | 34.86 | 43.51 | 53.03 | 63.42 | 74.66 |
| VELOCITY (FPS) | 1000. | 975. | 952. | 931. | 912. | 893. | 876. | 860. | 844. | 829. | 814. | 800. | 785. |
| ENERGY (FT-LB) | 500. | 475. | 453. | 433. | 415. | 399. | 383. | 369. | 356. | 343. | 331. | 320. | 308. |
| DROP (IN) | .00 | -1.06 | -4.40 | -10.09 | -18.27 | -29.05 | -42.55 | -58.85 | -78.09 | -100.37 | -125.80 | -154.50 | -186.47 |
| MID-RANGE (IN) | .00 | .28 | 1.14 | 2.61 | 4.73 | 7.58 | 11.18 | 15.56 | 20.78 | 26.86 | 33.85 | 41.80 | 50.69 |
| BULLET PATH (IN) | -.63 | .82 | .00 | -3.17 | -8.83 | -17.10 | -28.08 | -41.87 | -58.59 | -78.36 | -101.27 | -127.46 | -156.90 |
| TIME OF FLIGHT (SEC) | .000000 | .075968 | .153815 | .233469 | .314875 | .397989 | .482777 | .569215 | .657287 | .746982 | .838297 | .931235 | 1.025921 |
| WIND DEFLECTION (IN) | | | | | | | | | | | | | |
| 5 MPH | .00 | .09 | .34 | .75 | 1.31 | 2.02 | 2.88 | 3.89 | 5.04 | 6.33 | 7.77 | 9.35 | 11.08 |
| 10 MPH | .00 | .17 | .67 | 1.49 | 2.62 | 4.05 | 5.77 | 7.78 | 10.08 | 12.67 | 15.54 | 18.70 | 22.16 |
| 20 MPH | .00 | .34 | 1.34 | 2.98 | 5.24 | 8.09 | 11.54 | 15.56 | 20.16 | 25.34 | 31.08 | 37.39 | 44.32 |
| 30 MPH | .00 | .51 | 2.01 | 4.47 | 7.85 | 12.14 | 17.31 | 23.35 | 30.25 | 38.01 | 46.62 | 56.09 | 66.49 |
| VELOCITY (FPS) | 900. | 882. | 866. | 850. | 834. | 820. | 805. | 791. | 775. | 761. | 746. | 732. | 719. |
| ENERGY (FT-LB) | 405. | 389. | 374. | 361. | 348. | 336. | 324. | 312. | 300. | 289. | 278. | 268. | 258. |
| DROP (IN) | .00 | -1.23 | -5.25 | -12.15 | -22.05 | -35.07 | -51.30 | -70.82 | -93.73 | -120.24 | -150.49 | -184.62 | -222.80 |
| MID-RANGE (IN) | .00 | .34 | 1.39 | 3.19 | 5.78 | 9.20 | 13.50 | 18.69 | 24.81 | 31.95 | 40.18 | 49.54 | 60.10 |
| BULLET PATH (IN) | -.63 | 1.08 | -.00 | -3.96 | -10.93 | -21.00 | -34.30 | -50.87 | -70.84 | -94.41 | -121.72 | -152.92 | -188.16 |
| TIME OF FLIGHT (SEC) | .000000 | .084171 | .169999 | .257465 | .346558 | .437270 | .529602 | .623594 | .719432 | .817150 | .916763 | 1.018287 | 1.121741 |
| WIND DEFLECTION (IN) | | | | | | | | | | | | | |
| 5 MPH | .00 | .07 | .29 | .66 | 1.16 | 1.81 | 2.60 | 3.54 | 4.64 | 5.91 | 7.34 | 8.94 | 10.71 |
| 10 MPH | .00 | .15 | .59 | 1.31 | 2.33 | 3.63 | 5.21 | 7.09 | 9.29 | 11.82 | 14.68 | 17.89 | 21.43 |
| 20 MPH | .00 | .29 | 1.17 | 2.63 | 4.65 | 7.25 | 10.42 | 14.17 | 18.57 | 23.64 | 29.37 | 35.77 | 42.85 |
| 30 MPH | .00 | .44 | 1.76 | 3.94 | 6.98 | 10.88 | 15.63 | 21.26 | 27.86 | 35.46 | 44.05 | 53.66 | 64.28 |

## Bullet: Lyman #429360 225 Gr.
### Ballistic Coefficient: .169

| RANGE (YDS) | MUZZLE | 25 | 50 | 75 | 100 | 125 | 150 | 175 | 200 | 225 | 250 | 275 | 300 |
|---|---|---|---|---|---|---|---|---|---|---|---|---|---|
| VELOCITY (FPS) | 800. | 784. | 769. | 755. | 741. | 727. | 713. | 700. | 688. | 675. | 663. | 651. | 640. |
| ENERGY (FT-LB) | 320. | 307. | 296. | 284. | 274. | 264. | 254. | 245. | 236. | 228. | 220. | 212. | 204. |
| DROP (IN) | .00 | -1.45 | -6.38 | -14.96 | -27.33 | -43.65 | -64.06 | -88.71 | -117.78 | -151.43 | -189.83 | -233.16 | -281.61 |
| MID-RANGE (IN) | .00 | .41 | 1.74 | 4.01 | 7.29 | 11.62 | 17.07 | 23.69 | 31.56 | 40.72 | 51.26 | 63.25 | 76.74 |
| BULLET PATH (IN) | -.63 | 1.42 | -.00 | -5.07 | -13.94 | -26.75 | -43.66 | -64.81 | -90.37 | -120.51 | -155.40 | -195.23 | -240.17 |
| TIME OF FLIGHT (SEC) | .000000 | .094684 | .191242 | .289685 | .390026 | .492283 | .596477 | .702632 | .810774 | .920931 | 1.033134 | 1.147416 | 1.263814 |
| WIND DEFLECTION (IN) | | | | | | | | | | | | | |
| 5 MPH | .00 | .08 | .33 | .74 | 1.32 | 2.07 | 2.99 | 4.08 | 5.35 | 6.79 | 8.42 | 10.22 | 12.22 |
| 10 MPH | .00 | .16 | .66 | 1.48 | 2.64 | 4.14 | 5.98 | 8.16 | 10.70 | 13.58 | 16.83 | 20.45 | 24.43 |
| 20 MPH | .00 | .33 | 1.32 | 2.97 | 5.29 | 8.28 | 11.96 | 16.33 | 21.39 | 27.17 | 33.66 | 40.89 | 48.86 |
| 30 MPH | .00 | .49 | 1.98 | 4.45 | 7.93 | 12.43 | 17.94 | 24.49 | 32.09 | 40.75 | 50.49 | 61.34 | 73.29 |
| VELOCITY (FPS) | 700. | 687. | 675. | 662. | 650. | 639. | 627. | 616. | 605. | 594. | 584. | 573. | 563. |
| ENERGY (FT-LB) | 245. | 236. | 227. | 219. | 211. | 204. | 197. | 190. | 183. | 176. | 170. | 164. | 158. |
| DROP (IN) | .00 | -1.81 | -8.19 | -19.33 | -35.40 | -56.59 | -83.08 | -115.07 | -152.77 | -196.40 | -246.17 | -302.31 | -365.08 |
| MID-RANGE (IN) | .00 | .56 | 2.29 | 5.25 | 9.51 | 15.14 | 22.21 | 30.80 | 40.98 | 52.86 | 66.50 | 82.00 | 99.46 |
| BULLET PATH (IN) | -.63 | 1.98 | -.00 | -6.73 | -18.39 | -35.16 | -57.25 | -84.82 | -118.11 | -157.33 | -202.69 | -254.42 | -312.78 |
| TIME OF FLIGHT (SEC) | .000000 | .108140 | .218294 | .330492 | .444768 | .561157 | .679695 | .800422 | .923379 | 1.048610 | 1.176159 | 1.306073 | 1.438403 |
| WIND DEFLECTION (IN) | | | | | | | | | | | | | |
| 5 MPH | .00 | .09 | .35 | .80 | 1.43 | 2.24 | 3.24 | 4.44 | 5.83 | 7.42 | 9.22 | 11.22 | 13.44 |
| 10 MPH | .00 | .18 | .71 | 1.60 | 2.85 | 4.48 | 6.48 | 8.87 | 11.66 | 14.84 | 18.43 | 22.44 | 26.87 |
| 20 MPH | .00 | .35 | 1.41 | 3.19 | 5.70 | 8.96 | 12.97 | 17.75 | 23.32 | 29.68 | 36.87 | 44.88 | 53.75 |
| 30 MPH | .00 | .53 | 2.12 | 4.79 | 8.55 | 13.43 | 19.45 | 26.62 | 34.97 | 44.52 | 55.30 | 67.32 | 80.62 |

## Bullet: Lyman #429421 235 Gr.
### Ballistic Coefficient: .209

| RANGE (YDS) | MUZZLE | 25 | 50 | 75 | 100 | 125 | 150 | 175 | 200 | 225 | 250 | 275 | 300 |
|---|---|---|---|---|---|---|---|---|---|---|---|---|---|
| VELOCITY (FPS) | 1500. | 1429. | 1362. | 1300. | 1251. | 1207. | 1167. | 1131. | 1098. | 1069. | 1044. | 1021. | 999. |
| ENERGY (FT-LB) | 1174. | 1065. | 968. | 882. | 817. | 760. | 710. | 667. | 629. | 597. | 568. | 543. | 521. |
| DROP (IN) | .00 | -.47 | -2.01 | -4.72 | -8.70 | -14.09 | -20.96 | -29.42 | -39.58 | -51.53 | -65.38 | -81.22 | -99.14 |
| MID-RANGE (IN) | .00 | .13 | .53 | 1.26 | 2.34 | 3.83 | 5.76 | 8.16 | 11.09 | 14.55 | 18.61 | 23.28 | 28.61 |
| BULLET PATH (IN) | -.63 | .22 | .00 | -1.39 | -4.05 | -8.11 | -13.66 | -20.80 | -29.63 | -40.27 | -52.79 | -67.31 | -83.91 |
| TIME OF FLIGHT (SEC) | .000000 | .051233 | .104994 | .161359 | .220187 | .281226 | .344438 | .409758 | .477091 | .546321 | .617329 | .690019 | .764310 |
| WIND DEFLECTION (IN) | | | | | | | | | | | | | |
| 5 MPH | .00 | .11 | .44 | 1.00 | 1.78 | 2.75 | 3.91 | 5.26 | 6.78 | 8.48 | 10.32 | 12.32 | 14.46 |
| 10 MPH | .00 | .22 | .88 | 2.00 | 3.55 | 5.50 | 7.82 | 10.52 | 13.57 | 16.95 | 20.65 | 24.64 | 28.92 |
| 20 MPH | .00 | .43 | 1.76 | 4.00 | 7.11 | 10.99 | 15.64 | 21.03 | 27.14 | 33.90 | 41.30 | 49.29 | 57.84 |
| 30 MPH | .00 | .65 | 2.64 | 6.00 | 10.66 | 16.49 | 23.46 | 31.55 | 40.70 | 50.86 | 61.95 | 73.93 | 86.76 |
| VELOCITY (FPS) | 1400. | 1335. | 1279. | 1233. | 1190. | 1151. | 1117. | 1086. | 1058. | 1034. | 1012. | 991. | 972. |
| ENERGY (FT-LB) | 1023. | 930. | 853. | 793. | 739. | 692. | 651. | 615. | 585. | 558. | 534. | 512. | 493. |
| DROP (IN) | .00 | -.55 | -2.31 | -5.40 | -9.92 | -15.98 | -23.67 | -33.10 | -44.37 | -57.57 | -72.80 | -90.14 | -109.69 |
| MID-RANGE (IN) | .00 | .14 | .61 | 1.43 | 2.65 | 4.31 | 6.44 | 9.08 | 12.26 | 16.03 | 20.42 | 25.45 | 31.17 |
| BULLET PATH (IN) | -.63 | .29 | .00 | -1.62 | -4.67 | -9.26 | -15.49 | -23.45 | -33.24 | -44.97 | -58.73 | -74.61 | -92.69 |
| TIME OF FLIGHT (SEC) | .000000 | .054867 | .112323 | .172068 | .234012 | .298105 | .364270 | .432401 | .502379 | .574093 | .647452 | .722384 | .798825 |
| WIND DEFLECTION (IN) | | | | | | | | | | | | | |
| 5 MPH | .00 | .11 | .46 | 1.00 | 1.74 | 2.66 | 3.77 | 5.05 | 6.50 | 8.09 | 9.83 | 11.71 | 13.73 |
| 10 MPH | .00 | .23 | .91 | 2.00 | 3.47 | 5.32 | 7.54 | 10.10 | 12.99 | 16.18 | 19.67 | 23.43 | 27.45 |
| 20 MPH | .00 | .46 | 1.82 | 4.00 | 6.94 | 10.65 | 15.08 | 20.21 | 25.98 | 32.37 | 39.33 | 46.85 | 54.90 |
| 30 MPH | .00 | .68 | 2.74 | 5.99 | 10.42 | 15.97 | 22.62 | 30.31 | 38.97 | 48.55 | 59.00 | 70.28 | 82.35 |
| VELOCITY (FPS) | 1300. | 1252. | 1208. | 1167. | 1131. | 1099. | 1070. | 1044. | 1021. | 1000. | 980. | 962. | 945. |
| ENERGY (FT-LB) | 882. | 818. | 761. | 711. | 667. | 630. | 597. | 569. | 544. | 521. | 501. | 483. | 466. |
| DROP (IN) | .00 | -.63 | -2.65 | -6.16 | -11.26 | -18.06 | -26.65 | -37.13 | -49.60 | -64.15 | -80.87 | -99.85 | -121.15 |
| MID-RANGE (IN) | .00 | .17 | .70 | 1.62 | 2.98 | 4.82 | 7.16 | 10.06 | 13.54 | 17.63 | 22.38 | 27.80 | 33.93 |
| BULLET PATH (IN) | -.63 | .38 | .00 | -1.87 | -5.34 | -10.49 | -17.45 | -26.29 | -37.12 | -50.03 | -65.11 | -82.45 | -102.11 |
| TIME OF FLIGHT (SEC) | .000000 | .058800 | .119809 | .182992 | .248284 | .315590 | .384794 | .455778 | .528444 | .602712 | .678514 | .755796 | .834509 |
| WIND DEFLECTION (IN) | | | | | | | | | | | | | |
| 5 MPH | .00 | .10 | .39 | .87 | 1.54 | 2.39 | 3.40 | 4.57 | 5.89 | 7.35 | 8.94 | 10.66 | 12.51 |
| 10 MPH | .00 | .19 | .78 | 1.75 | 3.08 | 4.77 | 6.80 | 9.14 | 11.78 | 14.69 | 17.88 | 21.33 | 25.03 |
| 20 MPH | .00 | .39 | 1.56 | 3.49 | 6.17 | 9.55 | 13.60 | 18.28 | 23.55 | 29.39 | 35.76 | 42.66 | 50.05 |
| 30 MPH | .00 | .58 | 2.34 | 5.24 | 9.25 | 14.32 | 20.40 | 27.42 | 35.33 | 44.08 | 53.64 | 63.98 | 75.08 |
| VELOCITY (FPS) | 1200. | 1160. | 1125. | 1093. | 1065. | 1040. | 1017. | 996. | 976. | 958. | 942. | 926. | 911. |
| ENERGY (FT-LB) | 751. | 702. | 660. | 623. | 592. | 564. | 539. | 517. | 497. | 479. | 463. | 447. | 433. |
| DROP (IN) | .00 | -.75 | -3.11 | -7.18 | -13.06 | -20.85 | -30.65 | -42.55 | -56.62 | -72.97 | -91.65 | -112.74 | -136.35 |
| MID-RANGE (IN) | .00 | .19 | .80 | 1.87 | 3.42 | 5.51 | 8.15 | 11.39 | 15.25 | 19.77 | 24.97 | 30.87 | 37.52 |
| BULLET PATH (IN) | -.63 | .49 | .00 | -2.20 | -6.22 | -12.14 | -20.07 | -30.10 | -42.31 | -56.79 | -73.60 | -92.82 | -114.56 |
| TIME OF FLIGHT (SEC) | .000000 | .063573 | .129239 | .196900 | .266436 | .337733 | .410695 | .485246 | .561321 | .638864 | .717832 | .798187 | .879899 |
| WIND DEFLECTION (IN) | | | | | | | | | | | | | |
| 5 MPH | .00 | .09 | .37 | .83 | 1.45 | 2.22 | 3.14 | 4.20 | 5.40 | 6.72 | 8.17 | 9.74 | 11.43 |
| 10 MPH | .00 | .19 | .75 | 1.65 | 2.89 | 4.44 | 6.28 | 8.40 | 10.79 | 13.44 | 16.34 | 19.48 | 22.86 |
| 20 MPH | .00 | .38 | 1.49 | 3.31 | 5.79 | 8.88 | 12.56 | 16.81 | 21.58 | 26.88 | 32.68 | 38.96 | 45.72 |
| 30 MPH | .00 | .57 | 2.24 | 4.96 | 8.68 | 13.32 | 18.85 | 25.21 | 32.38 | 40.32 | 49.02 | 58.44 | 68.59 |

## Bullet: Lyman #429421 235 Gr.
### Ballistic Coefficient: .209

| RANGE (YDS) | MUZZLE | 25 | 50 | 75 | 100 | 125 | 150 | 175 | 200 | 225 | 250 | 275 | 300 |
|---|---|---|---|---|---|---|---|---|---|---|---|---|---|
| VELOCITY (FPS) | 1100. | 1071. | 1045. | 1022. | 1000. | 981. | 962. | 945. | 929. | 914. | 900. | 886. | 873. |
| ENERGY (FT-LB) | 631. | 599. | 570. | 545. | 522. | 502. | 483. | 466. | 451. | 436. | 422. | 409. | 397. |
| DROP (IN) | .00 | -.88 | -3.65 | -8.40 | -15.22 | -24.21 | -35.46 | -49.02 | -64.96 | -83.40 | -104.43 | -128.14 | -154.61 |
| MID-RANGE (IN) | .00 | .23 | .94 | 2.18 | 3.96 | 6.34 | 9.33 | 12.96 | 17.26 | 22.26 | 28.00 | 34.52 | 41.85 |
| BULLET PATH (IN) | -.63 | .63 | .00 | -2.61 | -7.30 | -14.15 | -23.25 | -34.68 | -48.48 | -64.78 | -83.67 | -105.24 | -129.57 |
| TIME OF FLIGHT (SEC) | .000000 | .069113 | .140011 | .212595 | .286784 | .362509 | .439716 | .518356 | .598391 | .679788 | .762522 | .846570 | .931915 |
| WIND DEFLECTION (IN) | | | | | | | | | | | | | |
| 5 MPH | .00 | .08 | .32 | .71 | 1.24 | 1.90 | 2.69 | 3.62 | 4.66 | 5.82 | 7.10 | 8.50 | 10.01 |
| 10 MPH | .00 | .16 | .64 | 1.42 | 2.47 | 3.80 | 5.39 | 7.23 | 9.32 | 11.64 | 14.20 | 17.00 | 20.02 |
| 20 MPH | .00 | .33 | 1.28 | 2.83 | 4.95 | 7.60 | 10.78 | 14.46 | 18.63 | 23.29 | 28.41 | 33.99 | 40.03 |
| 30 MPH | .00 | .49 | 1.93 | 4.25 | 7.42 | 11.40 | 16.17 | 21.69 | 27.95 | 34.93 | 42.61 | 50.99 | 60.05 |
| VELOCITY (FPS) | 1000. | 980. | 962. | 945. | 929. | 914. | 899. | 885. | 872. | 859. | 847. | 835. | 824. |
| ENERGY (FT-LB) | 522. | 501. | 483. | 466. | 450. | 436. | 422. | 409. | 397. | 385. | 374. | 364. | 354. |
| DROP (IN) | .00 | -1.05 | -4.35 | -9.98 | -17.98 | -28.48 | -41.58 | -57.35 | -75.89 | -97.27 | -121.60 | -148.94 | -179.38 |
| MID-RANGE (IN) | .00 | .28 | 1.13 | 2.58 | 4.64 | 7.37 | 10.81 | 15.01 | 19.97 | 25.73 | 32.32 | 39.77 | 48.12 |
| BULLET PATH (IN) | -.63 | .81 | .00 | -3.14 | -8.65 | -16.67 | -27.27 | -40.56 | -56.61 | -75.51 | -97.34 | -122.19 | -150.15 |
| TIME OF FLIGHT (SEC) | .000000 | .075758 | .152994 | .231663 | .311726 | .393150 | .475909 | .559982 | .645351 | .732003 | .819930 | .909125 | .999584 |
| WIND DEFLECTION (IN) | | | | | | | | | | | | | |
| 5 MPH | .00 | .07 | .26 | .59 | 1.03 | 1.60 | 2.28 | 3.08 | 3.99 | 5.02 | 6.15 | 7.40 | 8.76 |
| 10 MPH | .00 | .13 | .53 | 1.17 | 2.06 | 3.19 | 4.56 | 6.16 | 7.98 | 10.03 | 12.31 | 14.81 | 17.53 |
| 20 MPH | .00 | .27 | 1.05 | 2.35 | 4.13 | 6.39 | 9.12 | 12.31 | 15.96 | 20.07 | 24.62 | 29.61 | 35.05 |
| 30 MPH | .00 | .40 | 1.58 | 3.52 | 6.19 | 9.58 | 13.68 | 18.47 | 23.95 | 30.10 | 36.92 | 44.42 | 52.58 |
| VELOCITY (FPS) | 900. | 886. | 873. | 860. | 848. | 836. | 824. | 813. | 802. | 790. | 777. | 764. | 752. |
| ENERGY (FT-LB) | 423. | 410. | 397. | 386. | 375. | 364. | 354. | 345. | 336. | 325. | 315. | 305. | 295. |
| DROP (IN) | .00 | -1.18 | -5.13 | -11.92 | -21.63 | -34.37 | -50.21 | -69.22 | -91.52 | -117.11 | -146.08 | -178.63 | -214.89 |
| MID-RANGE (IN) | .00 | .34 | 1.38 | 3.15 | 5.69 | 9.02 | 13.19 | 18.21 | 24.13 | 30.93 | 38.67 | 47.42 | 57.24 |
| BULLET PATH (IN) | -.63 | 1.07 | -.00 | -3.91 | -10.75 | -20.61 | -33.56 | -49.70 | -69.12 | -91.83 | -117.92 | -147.60 | -180.98 |
| TIME OF FLIGHT (SEC) | .000000 | .083987 | .169271 | .255839 | .343680 | .432789 | .523162 | .614798 | .707698 | .801947 | .897773 | .995198 | 1.094230 |
| WIND DEFLECTION (IN) | | | | | | | | | | | | | |
| 5 MPH | .00 | .06 | .23 | .51 | .91 | 1.42 | 2.04 | 2.77 | 3.61 | 4.57 | 5.67 | 6.91 | 8.29 |
| 10 MPH | .00 | .12 | .46 | 1.03 | 1.82 | 2.84 | 4.08 | 5.54 | 7.22 | 9.14 | 11.34 | 13.82 | 16.58 |
| 20 MPH | .00 | .23 | .92 | 2.06 | 3.64 | 5.68 | 8.15 | 11.08 | 14.44 | 18.29 | 22.68 | 27.64 | 33.17 |
| 30 MPH | .00 | .35 | 1.38 | 3.08 | 5.46 | 8.51 | 12.23 | 16.61 | 21.66 | 27.43 | 34.02 | 41.46 | 49.75 |
| VELOCITY (FPS) | 800. | 787. | 774. | 761. | 749. | 737. | 725. | 714. | 703. | 692. | 682. | 671. | 661. |
| ENERGY (FT-LB) | 334. | 323. | 312. | 302. | 293. | 283. | 275. | 266. | 258. | 250. | 242. | 235. | 228. |
| DROP (IN) | .00 | -1.41 | -6.21 | -14.62 | -26.75 | -42.73 | -62.70 | -86.77 | -115.10 | -147.80 | -185.03 | -226.92 | -273.63 |
| MID-RANGE (IN) | .00 | .38 | 1.70 | 3.94 | 7.17 | 11.41 | 16.73 | 23.18 | 30.80 | 39.65 | 49.78 | 61.25 | 74.12 |
| BULLET PATH (IN) | -.63 | 1.38 | -.00 | -4.99 | -13.70 | -26.26 | -42.81 | -63.46 | -88.36 | -117.65 | -151.46 | -189.93 | -233.22 |
| TIME OF FLIGHT (SEC) | .000000 | .094544 | .190679 | .288414 | .387756 | .488717 | .591312 | .695555 | .801464 | .909057 | 1.018355 | 1.129380 | 1.242154 |
| WIND DEFLECTION (IN) | | | | | | | | | | | | | |
| 5 MPH | .00 | .07 | .28 | .63 | 1.12 | 1.76 | 2.54 | 3.46 | 4.53 | 5.75 | 7.12 | 8.64 | 10.31 |
| 10 MPH | .00 | .14 | .56 | 1.26 | 2.24 | 3.51 | 5.07 | 6.92 | 9.06 | 11.49 | 14.23 | 17.27 | 20.62 |
| 20 MPH | .00 | .28 | 1.12 | 2.52 | 4.49 | 7.03 | 10.14 | 13.84 | 18.12 | 22.99 | 28.46 | 34.54 | 41.24 |
| 30 MPH | .00 | .42 | 1.68 | 3.78 | 6.73 | 10.54 | 15.21 | 20.75 | 27.17 | 34.48 | 42.69 | 51.81 | 61.86 |

## Bullet: Lyman #429421 235 Gr.
### Ballistic Coefficient: .209

| RANGE (YDS) | MUZZLE | 25 | 50 | 75 | 100 | 125 | 150 | 175 | 200 | 225 | 250 | 275 | 300 |
|---|---|---|---|---|---|---|---|---|---|---|---|---|---|
| VELOCITY (FPS) | 700. | 689. | 678. | 668. | 658. | 648. | 638. | 628. | 618. | 609. | 600. | 591. | 582. |
| ENERGY (FT-LB) | 256. | 248. | 240. | 233. | 226. | 219. | 212. | 206. | 200. | 194. | 188. | 182. | 177. |
| DROP (IN) | .00 | -1.67 | -7.90 | -18.83 | -34.61 | -55.40 | -81.35 | -112.62 | -149.39 | -191.82 | -240.10 | -294.41 | -354.95 |
| MID-RANGE (IN) | .00 | .56 | 2.28 | 5.21 | 9.40 | 14.93 | 21.84 | 30.20 | 40.08 | 51.54 | 64.65 | 79.49 | 96.13 |
| BULLET PATH (IN) | -.63 | 1.96 | -.00 | -6.66 | -18.18 | -34.70 | -56.38 | -83.38 | -115.89 | -154.05 | -198.07 | -248.11 | -304.38 |
| TIME OF FLIGHT (SEC) | .000000 | .107990 | .217688 | .329115 | .442296 | .557254 | .674016 | .792609 | .913060 | 1.035401 | 1.159662 | 1.285875 | 1.414074 |
| WIND DEFLECTION (IN) | | | | | | | | | | | | | |
| 5 MPH | .00 | .07 | .30 | .68 | 1.21 | 1.90 | 2.74 | 3.75 | 4.92 | 6.26 | 7.76 | 9.44 | 11.30 |
| 10 MPH | .00 | .15 | .60 | 1.35 | 2.42 | 3.79 | 5.48 | 7.50 | 9.84 | 12.52 | 15.53 | 18.89 | 22.59 |
| 20 MPH | .00 | .30 | 1.20 | 2.71 | 4.83 | 7.58 | 10.97 | 15.00 | 19.68 | 25.03 | 31.06 | 37.77 | 45.18 |
| 30 MPH | .00 | .45 | 1.80 | 4.06 | 7.25 | 11.37 | 16.45 | 22.50 | 29.52 | 37.55 | 46.59 | 56.66 | 67.77 |
| VELOCITY (FPS) | 600. | 591. | 582. | 573. | 564. | 555. | 547. | 539. | 530. | 522. | 514. | 507. | 499. |
| ENERGY (FT-LB) | 188. | 182. | 177. | 171. | 166. | 161. | 156. | 151. | 147. | 142. | 138. | 134. | 130. |
| DROP (IN) | .00 | -2.28 | -10.76 | -25.63 | -47.10 | -75.38 | -110.69 | -153.24 | -203.28 | -261.03 | -326.75 | -400.70 | -483.14 |
| MID-RANGE (IN) | .00 | .76 | 3.10 | 7.08 | 12.80 | 20.31 | 29.71 | 41.09 | 54.54 | 70.14 | 87.99 | 108.21 | 130.88 |
| BULLET PATH (IN) | -.63 | 2.78 | .00 | -9.18 | -24.96 | -47.55 | -77.16 | -114.02 | -158.36 | -210.43 | -270.46 | -338.71 | -415.46 |
| TIME OF FLIGHT (SEC) | .000000 | .125979 | .253934 | .383902 | .515916 | .650012 | .786227 | .924600 | 1.065170 | 1.207977 | 1.353064 | 1.500473 | 1.650249 |
| WIND DEFLECTION (IN) | | | | | | | | | | | | | |
| 5 MPH | .00 | .09 | .35 | .78 | 1.40 | 2.20 | 3.19 | 4.36 | 5.73 | 7.30 | 9.07 | 11.04 | 13.22 |
| 10 MPH | .00 | .17 | .69 | 1.57 | 2.80 | 4.40 | 6.38 | 8.73 | 11.47 | 14.60 | 18.14 | 22.08 | 26.44 |
| 20 MPH | .00 | .34 | 1.38 | 3.13 | 5.60 | 8.80 | 12.75 | 17.46 | 22.94 | 29.21 | 36.28 | 44.17 | 52.89 |
| 30 MPH | .00 | .52 | 2.08 | 4.70 | 8.40 | 13.21 | 19.13 | 26.19 | 34.41 | 43.81 | 54.42 | 66.25 | 79.33 |

# Bullet: Lyman #429244 237 Gr.
## Ballistic Coefficient: .201

| RANGE (YDS) | MUZZLE | 25 | 50 | 75 | 100 | 125 | 150 | 175 | 200 | 225 | 250 | 275 | 300 |
|---|---|---|---|---|---|---|---|---|---|---|---|---|---|
| VELOCITY (FPS) | 1500. | 1426. | 1357. | 1295. | 1252. | 1212. | 1176. | 1142. | 1112. | 1085. | 1060. | 1038. | 1018. |
| ENERGY (FT-LB) | 1184. | 1070. | 969. | 882. | 825. | 773. | 727. | 686. | 650. | 619. | 591. | 567. | 545. |
| DROP (IN) | .00 | -.48 | -2.02 | -4.74 | -8.74 | -14.14 | -21.02 | -29.47 | -39.58 | -51.44 | -65.14 | -80.77 | -98.41 |
| MID-RANGE (IN) | .00 | .13 | .53 | 1.26 | 2.35 | 3.84 | 5.77 | 8.16 | 11.05 | 14.46 | 18.43 | 22.99 | 28.18 |
| BULLET PATH (IN) | -.63 | .22 | .00 | -1.40 | -4.07 | -8.14 | -13.70 | -20.82 | -29.60 | -40.14 | -52.51 | -66.82 | -83.13 |
| TIME OF FLIGHT (SEC) | .000000 | .051282 | .105197 | .161822 | .220747 | .281645 | .344489 | .409231 | .475802 | .544114 | .614069 | .685581 | .758575 |
| WIND DEFLECTION (IN) | | | | | | | | | | | | | |
| 5 MPH | .00 | .11 | .46 | 1.04 | 1.83 | 2.78 | 3.92 | 5.21 | 6.67 | 8.28 | 10.04 | 11.93 | 13.95 |
| 10 MPH | .00 | .23 | .91 | 2.08 | 3.65 | 5.57 | 7.83 | 10.42 | 13.34 | 16.56 | 20.08 | 23.86 | 27.91 |
| 20 MPH | .00 | .45 | 1.83 | 4.16 | 7.30 | 11.14 | 15.66 | 20.85 | 26.68 | 33.13 | 40.15 | 47.72 | 55.82 |
| 30 MPH | .00 | .68 | 2.74 | 6.24 | 10.95 | 16.71 | 23.49 | 31.27 | 40.02 | 49.69 | 60.23 | 71.59 | 83.73 |
| VELOCITY (FPS) | 1400. | 1333. | 1278. | 1236. | 1198. | 1162. | 1130. | 1101. | 1075. | 1051. | 1030. | 1010. | 992. |
| ENERGY (FT-LB) | 1031. | 934. | 859. | 804. | 755. | 711. | 672. | 638. | 608. | 582. | 558. | 537. | 518. |
| DROP (IN) | .00 | -.55 | -2.32 | -5.41 | -9.93 | -15.97 | -23.61 | -32.94 | -44.07 | -57.07 | -72.03 | -89.03 | -108.15 |
| MID-RANGE (IN) | .00 | .15 | .61 | 1.43 | 2.65 | 4.30 | 6.40 | 8.99 | 12.10 | 15.78 | 20.05 | 24.93 | 30.47 |
| BULLET PATH (IN) | -.63 | .29 | .00 | -1.62 | -4.67 | -9.23 | -15.40 | -23.27 | -32.92 | -44.44 | -57.93 | -73.46 | -91.11 |
| TIME OF FLIGHT (SEC) | .000000 | .054919 | .112503 | .172197 | .233855 | .297444 | .362906 | .430165 | .499127 | .569696 | .641791 | .715343 | .790295 |
| WIND DEFLECTION (IN) | | | | | | | | | | | | | |
| 5 MPH | .00 | .12 | .47 | 1.01 | 1.72 | 2.60 | 3.65 | 4.85 | 6.21 | 7.70 | 9.33 | 11.09 | 12.97 |
| 10 MPH | .00 | .24 | .94 | 2.02 | 3.44 | 5.21 | 7.30 | 9.71 | 12.42 | 15.41 | 18.67 | 22.19 | 25.95 |
| 20 MPH | .00 | .47 | 1.89 | 4.04 | 6.89 | 10.41 | 14.60 | 19.42 | 24.84 | 30.82 | 37.34 | 44.37 | 51.90 |
| 30 MPH | .00 | .71 | 2.83 | 6.06 | 10.33 | 15.62 | 21.90 | 29.13 | 37.25 | 46.23 | 56.01 | 66.56 | 77.85 |
| VELOCITY (FPS) | 1300. | 1257. | 1217. | 1180. | 1146. | 1115. | 1088. | 1063. | 1041. | 1020. | 1001. | 983. | 967. |
| ENERGY (FT-LB) | 889. | 831. | 779. | 732. | 691. | 655. | 623. | 595. | 570. | 547. | 527. | 509. | 492. |
| DROP (IN) | .00 | -.62 | -2.62 | -6.10 | -11.13 | -17.81 | -26.24 | -36.49 | -48.66 | -62.83 | -79.08 | -97.50 | -118.15 |
| MID-RANGE (IN) | .00 | .17 | .69 | 1.61 | 2.94 | 4.73 | 7.02 | 9.83 | 13.20 | 17.15 | 21.73 | 26.95 | 32.84 |
| BULLET PATH (IN) | -.63 | .38 | .00 | -1.85 | -5.25 | -10.31 | -17.10 | -25.73 | -36.27 | -48.81 | -63.44 | -80.23 | -99.26 |
| TIME OF FLIGHT (SEC) | .000000 | .058681 | .119337 | .181943 | .246455 | .312804 | .380906 | .450663 | .521985 | .594799 | .669040 | .744655 | .821598 |
| WIND DEFLECTION (IN) | | | | | | | | | | | | | |
| 5 MPH | .00 | .09 | .35 | .78 | 1.38 | 2.14 | 3.06 | 4.12 | 5.32 | 6.65 | 8.11 | 9.68 | 11.38 |
| 10 MPH | .00 | .17 | .70 | 1.56 | 2.76 | 4.28 | 6.12 | 8.24 | 10.64 | 13.30 | 16.21 | 19.37 | 22.76 |
| 20 MPH | .00 | .35 | 1.39 | 3.12 | 5.52 | 8.57 | 12.23 | 16.48 | 21.28 | 26.60 | 32.43 | 38.73 | 45.51 |
| 30 MPH | .00 | .52 | 2.09 | 4.68 | 8.28 | 12.85 | 18.35 | 24.72 | 31.92 | 39.90 | 48.64 | 58.10 | 68.27 |
| VELOCITY (FPS) | 1200. | 1164. | 1132. | 1103. | 1077. | 1053. | 1031. | 1011. | 993. | 976. | 960. | 945. | 930. |
| ENERGY (FT-LB) | 758. | 713. | 674. | 640. | 610. | 583. | 560. | 538. | 519. | 501. | 485. | 469. | 455. |
| DROP (IN) | .00 | -.74 | -3.09 | -7.12 | -12.93 | -20.62 | -30.26 | -41.93 | -55.72 | -71.71 | -89.98 | -110.58 | -133.55 |
| MID-RANGE (IN) | .00 | .19 | .80 | 1.85 | 3.38 | 5.42 | 8.01 | 11.17 | 14.93 | 19.32 | 24.37 | 30.10 | 36.52 |
| BULLET PATH (IN) | -.63 | .48 | .00 | -2.18 | -6.13 | -11.96 | -19.73 | -29.55 | -41.48 | -55.62 | -72.02 | -90.76 | -111.88 |
| TIME OF FLIGHT (SEC) | .000000 | .063459 | .128796 | .195935 | .264783 | .335244 | .407237 | .480690 | .555546 | .631754 | .709273 | .788066 | .868104 |
| WIND DEFLECTION (IN) | | | | | | | | | | | | | |
| 5 MPH | .00 | .08 | .33 | .74 | 1.30 | 2.00 | 2.84 | 3.80 | 4.89 | 6.09 | 7.42 | 8.85 | 10.39 |
| 10 MPH | .00 | .17 | .67 | 1.48 | 2.60 | 4.00 | 5.67 | 7.60 | 9.78 | 12.19 | 14.83 | 17.70 | 20.79 |
| 20 MPH | .00 | .34 | 1.34 | 2.97 | 5.20 | 8.01 | 11.35 | 15.20 | 19.55 | 24.38 | 29.66 | 35.40 | 41.57 |
| 30 MPH | .00 | .51 | 2.00 | 4.45 | 7.81 | 12.01 | 17.02 | 22.80 | 29.33 | 36.57 | 44.50 | 53.10 | 62.36 |

# Bullet: Lyman #429244 237 Gr.
## Ballistic Coefficient: .201

| RANGE (YDS) | MUZZLE | 25 | 50 | 75 | 100 | 125 | 150 | 175 | 200 | 225 | 250 | 275 | 300 |
|---|---|---|---|---|---|---|---|---|---|---|---|---|---|
| VELOCITY (FPS) | 1100. | 1074. | 1051. | 1029. | 1009. | 991. | 974. | 958. | 943. | 929. | 915. | 902. | 890. |
| ENERGY (FT-LB) | 637. | 607. | 581. | 557. | 536. | 517. | 499. | 483. | 468. | 454. | 441. | 428. | 417. |
| DROP (IN) | .00 | -.87 | -3.63 | -8.34 | -15.10 | -23.99 | -35.08 | -48.46 | -64.17 | -82.27 | -102.88 | -126.07 | -151.93 |
| MID-RANGE (IN) | .00 | .23 | .94 | 2.16 | 3.92 | 6.26 | 9.20 | 12.77 | 16.99 | 21.86 | 27.45 | 33.78 | 40.88 |
| BULLET PATH (IN) | -.63 | .62 | .00 | -2.59 | -7.22 | -13.98 | -22.94 | -34.19 | -47.78 | -63.74 | -82.22 | -103.29 | -127.02 |
| TIME OF FLIGHT (SEC) | .000000 | .069015 | .139634 | .211776 | .285372 | .360365 | .436706 | .514353 | .593270 | .673429 | .754803 | .837372 | .921118 |
| WIND DEFLECTION (IN) | | | | | | | | | | | | | |
| 5 MPH | .00 | .07 | .29 | .64 | 1.11 | 1.71 | 2.43 | 3.26 | 4.21 | 5.26 | 6.42 | 7.69 | 9.06 |
| 10 MPH | .00 | .15 | .58 | 1.27 | 2.23 | 3.42 | 4.86 | 6.53 | 8.42 | 10.52 | 12.85 | 15.38 | 18.12 |
| 20 MPH | .00 | .29 | 1.15 | 2.55 | 4.45 | 6.85 | 9.72 | 13.05 | 16.83 | 21.05 | 25.69 | 30.75 | 36.23 |
| 30 MPH | .00 | .44 | 1.73 | 3.82 | 6.68 | 10.27 | 14.58 | 19.58 | 25.25 | 31.57 | 38.54 | 46.13 | 54.35 |
| VELOCITY (FPS) | 1000. | 982. | 966. | 950. | 936. | 922. | 908. | 896. | 884. | 872. | 860. | 849. | 839. |
| ENERGY (FT-LB) | 526. | 508. | 491. | 475. | 461. | 447. | 434. | 422. | 411. | 400. | 390. | 380. | 370. |
| DROP (IN) | .00 | -1.04 | -4.32 | -9.93 | -17.88 | -28.28 | -41.24 | -56.81 | -75.09 | -96.13 | -120.03 | -146.85 | -176.67 |
| MID-RANGE (IN) | .00 | .28 | 1.12 | 2.57 | 4.61 | 7.31 | 10.69 | 14.80 | 19.67 | 25.30 | 31.73 | 38.99 | 47.10 |
| BULLET PATH (IN) | -.63 | .81 | .00 | -3.13 | -8.60 | -16.52 | -27.00 | -40.10 | -55.90 | -74.47 | -95.89 | -120.23 | -147.57 |
| TIME OF FLIGHT (SEC) | .000000 | .075677 | .152680 | .230970 | .310515 | .391287 | .473262 | .556422 | .640751 | .726236 | .812866 | .900635 | .989537 |
| WIND DEFLECTION (IN) | | | | | | | | | | | | | |
| 5 MPH | .00 | .06 | .24 | .53 | .93 | 1.43 | 2.05 | 2.77 | 3.59 | 4.51 | 5.53 | 6.66 | 7.88 |
| 10 MPH | .00 | .12 | .47 | 1.05 | 1.85 | 2.87 | 4.09 | 5.53 | 7.17 | 9.02 | 11.06 | 13.31 | 15.76 |
| 20 MPH | .00 | .24 | .94 | 2.10 | 3.70 | 5.73 | 8.19 | 11.06 | 14.34 | 18.03 | 22.13 | 26.62 | 31.52 |
| 30 MPH | .00 | .36 | 1.42 | 3.15 | 5.55 | 8.60 | 12.28 | 16.59 | 21.52 | 27.05 | 33.19 | 39.94 | 47.28 |
| VELOCITY (FPS) | 900. | 888. | 876. | 864. | 853. | 842. | 832. | 822. | 812. | 802. | 790. | 777. | 764. |
| ENERGY (FT-LB) | 426. | 415. | 403. | 393. | 383. | 373. | 364. | 355. | 347. | 338. | 328. | 318. | 307. |
| DROP (IN) | .00 | -1.18 | -5.11 | -11.86 | -21.52 | -34.15 | -49.84 | -68.63 | -90.63 | -115.92 | -144.54 | -176.55 | -212.14 |
| MID-RANGE (IN) | .00 | .34 | 1.38 | 3.14 | 5.65 | 8.95 | 13.06 | 17.99 | 23.80 | 30.50 | 38.12 | 46.66 | 56.23 |
| BULLET PATH (IN) | -.63 | 1.06 | -.00 | -3.89 | -10.67 | -20.44 | -33.26 | -49.18 | -68.31 | -90.73 | -116.48 | -145.62 | -178.35 |
| TIME OF FLIGHT (SEC) | .000000 | .083917 | .168994 | .255219 | .342584 | .431083 | .520711 | .611467 | .703352 | .796366 | .890615 | .986431 | 1.083846 |
| WIND DEFLECTION (IN) | | | | | | | | | | | | | |
| 5 MPH | .00 | .05 | .20 | .46 | .81 | 1.27 | 1.82 | 2.48 | 3.23 | 4.08 | 5.04 | 6.14 | 7.38 |
| 10 MPH | .00 | .10 | .41 | .92 | 1.63 | 2.54 | 3.65 | 4.95 | 6.46 | 8.16 | 10.08 | 12.28 | 14.76 |
| 20 MPH | .00 | .21 | .82 | 1.84 | 3.26 | 5.07 | 7.29 | 9.90 | 12.91 | 16.32 | 20.16 | 24.56 | 29.51 |
| 30 MPH | .00 | .31 | 1.23 | 2.76 | 4.88 | 7.61 | 10.94 | 14.85 | 19.37 | 24.48 | 30.24 | 36.84 | 44.27 |
| VELOCITY (FPS) | 800. | 787. | 774. | 761. | 749. | 737. | 725. | 714. | 703. | 692. | 682. | 671. | 661. |
| ENERGY (FT-LB) | 337. | 326. | 315. | 305. | 295. | 286. | 277. | 268. | 260. | 252. | 244. | 237. | 230. |
| DROP (IN) | .00 | -1.41 | -6.21 | -14.62 | -26.75 | -42.73 | -62.70 | -86.77 | -115.10 | -147.80 | -185.03 | -226.92 | -273.63 |
| MID-RANGE (IN) | .00 | .38 | 1.70 | 3.94 | 7.17 | 11.41 | 16.73 | 23.18 | 30.80 | 39.65 | 49.78 | 61.25 | 74.12 |
| BULLET PATH (IN) | -.63 | 1.38 | -.00 | -4.99 | -13.70 | -26.26 | -42.81 | -63.46 | -88.36 | -117.65 | -151.46 | -189.93 | -233.22 |
| TIME OF FLIGHT (SEC) | .000000 | .094544 | .190679 | .288414 | .387756 | .488717 | .591312 | .695555 | .801464 | .909057 | 1.018355 | 1.129380 | 1.242154 |
| WIND DEFLECTION (IN) | | | | | | | | | | | | | |
| 5 MPH | .00 | .07 | .28 | .63 | 1.12 | 1.76 | 2.54 | 3.46 | 4.53 | 5.75 | 7.12 | 8.64 | 10.31 |
| 10 MPH | .00 | .14 | .56 | 1.26 | 2.24 | 3.51 | 5.07 | 6.92 | 9.06 | 11.49 | 14.23 | 17.27 | 20.62 |
| 20 MPH | .00 | .28 | 1.12 | 2.52 | 4.49 | 7.03 | 10.14 | 13.84 | 18.12 | 22.99 | 28.46 | 34.54 | 41.24 |
| 30 MPH | .00 | .42 | 1.68 | 3.78 | 6.73 | 10.54 | 15.21 | 20.75 | 27.17 | 34.48 | 42.69 | 51.81 | 61.86 |

## Bullet: Lyman #429244 237 Gr.
### Ballistic Coefficient: .201

| RANGE (YDS) | MUZZLE | 25 | 50 | 75 | 100 | 125 | 150 | 175 | 200 | 225 | 250 | 275 | 300 |
|---|---|---|---|---|---|---|---|---|---|---|---|---|---|
| VELOCITY (FPS) | 700. | 689. | 678. | 668. | 658. | 648. | 638. | 628. | 618. | 609. | 600. | 591. | 582. |
| ENERGY (FT-LB) | 258. | 250. | 242. | 235. | 228. | 221. | 214. | 208. | 201. | 195. | 189. | 184. | 178. |
| DROP (IN) | .00 | -1.67 | -7.90 | -18.83 | -34.61 | -55.40 | -81.35 | -112.62 | -149.39 | -191.82 | -240.10 | -294.41 | -354.95 |
| MID-RANGE (IN) | .00 | .56 | 2.28 | 5.21 | 9.40 | 14.93 | 21.84 | 30.20 | 40.08 | 51.54 | 64.65 | 79.49 | 96.13 |
| BULLET PATH (IN) | -.63 | 1.96 | -.00 | -6.66 | -18.18 | -34.70 | -56.38 | -83.38 | -115.89 | -154.05 | -198.07 | -248.11 | -304.38 |
| TIME OF FLIGHT (SEC) | .000000 | .107990 | .217688 | .329115 | .442296 | .557254 | .674016 | .792609 | .913060 | 1.035401 | 1.159662 | 1.285875 | 1.414074 |
| WIND DEFLECTION (IN) | | | | | | | | | | | | | |
| 5 MPH | .00 | .07 | .30 | .68 | 1.21 | 1.90 | 2.74 | 3.75 | 4.92 | 6.26 | 7.76 | 9.44 | 11.30 |
| 10 MPH | .00 | .15 | .60 | 1.35 | 2.42 | 3.79 | 5.48 | 7.50 | 9.84 | 12.52 | 15.53 | 18.89 | 22.59 |
| 20 MPH | .00 | .30 | 1.20 | 2.71 | 4.83 | 7.58 | 10.97 | 15.00 | 19.68 | 25.03 | 31.06 | 37.77 | 45.18 |
| 30 MPH | .00 | .45 | 1.80 | 4.06 | 7.25 | 11.37 | 16.45 | 22.50 | 29.52 | 37.55 | 46.59 | 56.66 | 67.77 |
| VELOCITY (FPS) | 600. | 591. | 582. | 573. | 564. | 555. | 547. | 539. | 530. | 522. | 514. | 507. | 499. |
| ENERGY (FT-LB) | 189. | 184. | 178. | 173. | 167. | 162. | 157. | 153. | 148. | 144. | 139. | 135. | 131. |
| DROP (IN) | .00 | -2.28 | -10.76 | -25.63 | -47.10 | -75.38 | -110.69 | -153.24 | -203.28 | -261.03 | -326.75 | -400.70 | -483.14 |
| MID-RANGE (IN) | .00 | .76 | 3.10 | 7.08 | 12.80 | 20.31 | 29.71 | 41.09 | 54.54 | 70.14 | 87.99 | 108.21 | 130.88 |
| BULLET PATH (IN) | -.63 | 2.78 | .00 | -9.18 | -24.96 | -47.55 | -77.16 | -114.02 | -158.36 | -210.43 | -270.46 | -338.71 | -415.46 |
| TIME OF FLIGHT (SEC) | .000000 | .125979 | .253934 | .383902 | .515916 | .650012 | .786227 | .924600 | 1.065170 | 1.207977 | 1.353064 | 1.500473 | 1.650249 |
| WIND DEFLECTION (IN) | | | | | | | | | | | | | |
| 5 MPH | .00 | .09 | .35 | .78 | 1.40 | 2.20 | 3.19 | 4.36 | 5.73 | 7.30 | 9.07 | 11.04 | 13.22 |
| 10 MPH | .00 | .17 | .69 | 1.57 | 2.80 | 4.40 | 6.38 | 8.73 | 11.47 | 14.60 | 18.14 | 22.08 | 26.44 |
| 20 MPH | .00 | .34 | 1.38 | 3.13 | 5.60 | 8.80 | 12.75 | 17.46 | 22.94 | 29.21 | 36.28 | 44.17 | 52.89 |
| 30 MPH | .00 | .52 | 2.08 | 4.70 | 8.40 | 13.21 | 19.13 | 26.19 | 34.41 | 43.81 | 54.42 | 66.25 | 79.33 |

## Bullet: Sierra .4515 Dia. 185 Gr. JHC
## Ballistic Coefficient: .120

| RANGE (YDS) | MUZZLE | 25 | 50 | 75 | 100 | 125 | 150 | 175 | 200 | 225 | 250 | 275 | 300 |
|---|---|---|---|---|---|---|---|---|---|---|---|---|---|
| VELOCITY (FPS) | 1400. | 1290. | 1196. | 1119. | 1058. | 1009. | 968. | 933. | 902. | 874. | 847. | 823. | 800. |
| ENERGY (FT-LB) | 805. | 684. | 587. | 514. | 460. | 418. | 385. | 358. | 334. | 313. | 295. | 278. | 263. |
| DROP (IN) | .00 | -.57 | -2.45 | -5.85 | -10.98 | -18.04 | -27.23 | -38.72 | -52.67 | -69.28 | -88.73 | -111.20 | -136.87 |
| MID-RANGE (IN) | .00 | .15 | .65 | 1.59 | 3.04 | 5.07 | 7.77 | 11.17 | 15.36 | 20.38 | 26.32 | 33.24 | 41.20 |
| BULLET PATH (IN) | -.63 | .34 | .00 | -1.86 | -5.45 | -10.97 | -18.62 | -28.57 | -40.98 | -56.05 | -73.96 | -94.89 | -119.01 |
| TIME OF FLIGHT (SEC) | .000000 | .055840 | .116274 | .181188 | .250204 | .322842 | .398746 | .477672 | .559455 | .643985 | .731194 | .821045 | .913530 |
| WIND DEFLECTION (IN) | | | | | | | | | | | | | |
| 5 MPH | .00 | .20 | .80 | 1.80 | 3.16 | 4.84 | 6.80 | 9.04 | 11.52 | 14.24 | 17.20 | 20.39 | 23.82 |
| 10 MPH | .00 | .40 | 1.61 | 3.60 | 6.32 | 9.68 | 13.61 | 18.07 | 23.04 | 28.48 | 34.40 | 40.79 | 47.64 |
| 20 MPH | .00 | .80 | 3.21 | 7.21 | 12.64 | 19.35 | 27.22 | 36.14 | 46.07 | 56.97 | 68.81 | 81.58 | 95.28 |
| 30 MPH | .00 | 1.20 | 4.82 | 10.81 | 18.96 | 29.03 | 40.82 | 54.21 | 69.11 | 85.45 | 103.21 | 122.37 | 142.92 |
| VELOCITY (FPS) | 1300. | 1204. | 1125. | 1063. | 1013. | 972. | 936. | 905. | 876. | 850. | 825. | 802. | 781. |
| ENERGY (FT-LB) | 694. | 596. | 520. | 464. | 422. | 388. | 360. | 336. | 315. | 297. | 280. | 264. | 251. |
| DROP (IN) | .00 | -.66 | -2.83 | -6.70 | -12.49 | -20.39 | -30.57 | -43.20 | -58.46 | -76.56 | -97.65 | -121.92 | -149.45 |
| MID-RANGE (IN) | .00 | .18 | .75 | 1.81 | 3.42 | 5.66 | 8.58 | 12.25 | 16.74 | 22.11 | 28.43 | 35.77 | 44.15 |
| BULLET PATH (IN) | -.63 | .44 | .00 | -2.15 | -6.21 | -12.38 | -20.83 | -31.72 | -45.26 | -61.63 | -80.99 | -103.53 | -129.33 |
| TIME OF FLIGHT (SEC) | .000000 | .059996 | .124495 | .193141 | .265451 | .341055 | .419700 | .501216 | .585486 | .672440 | .762039 | .854271 | .949072 |
| WIND DEFLECTION (IN) | | | | | | | | | | | | | |
| 5 MPH | .00 | .20 | .80 | 1.77 | 3.05 | 4.63 | 6.47 | 8.57 | 10.91 | 13.48 | 16.29 | 19.33 | 22.60 |
| 10 MPH | .00 | .41 | 1.60 | 3.53 | 6.10 | 9.26 | 12.94 | 17.14 | 21.81 | 26.96 | 32.58 | 38.66 | 45.19 |
| 20 MPH | .00 | .81 | 3.21 | 7.06 | 12.21 | 18.51 | 25.89 | 34.27 | 43.63 | 53.93 | 65.16 | 77.32 | 90.38 |
| 30 MPH | .00 | 1.22 | 4.81 | 10.59 | 18.31 | 27.77 | 38.83 | 51.41 | 65.44 | 80.89 | 97.74 | 115.98 | 135.57 |
| VELOCITY (FPS) | 1200. | 1122. | 1060. | 1011. | 970. | 935. | 903. | 875. | 849. | 824. | 801. | 780. | 761. |
| ENERGY (FT-LB) | 591. | 517. | 462. | 420. | 387. | 359. | 335. | 314. | 296. | 279. | 263. | 250. | 238. |
| DROP (IN) | .00 | -.78 | -3.28 | -7.70 | -14.24 | -23.07 | -34.36 | -48.29 | -65.06 | -84.84 | -107.80 | -134.03 | -163.79 |
| MID-RANGE (IN) | .00 | .20 | .86 | 2.05 | 3.84 | 6.30 | 9.48 | 13.45 | 18.29 | 24.05 | 30.83 | 38.62 | 47.54 |
| BULLET PATH (IN) | -.63 | .55 | .00 | -2.47 | -7.06 | -13.93 | -23.27 | -35.25 | -50.06 | -67.88 | -88.90 | -113.17 | -140.97 |
| TIME OF FLIGHT (SEC) | .000000 | .064705 | .133536 | .206009 | .281762 | .360546 | .442194 | .526593 | .613672 | .703394 | .795750 | .890675 | .988059 |
| WIND DEFLECTION (IN) | | | | | | | | | | | | | |
| 5 MPH | .00 | .19 | .75 | 1.63 | 2.80 | 4.23 | 5.91 | 7.84 | 10.00 | 12.40 | 15.03 | 17.88 | 20.95 |
| 10 MPH | .00 | .39 | 1.50 | 3.26 | 5.59 | 8.46 | 11.83 | 15.68 | 20.01 | 24.80 | 30.05 | 35.76 | 41.90 |
| 20 MPH | .00 | .78 | 3.00 | 6.52 | 11.18 | 16.91 | 23.65 | 31.36 | 40.01 | 49.59 | 60.10 | 71.52 | 83.80 |
| 30 MPH | .00 | 1.16 | 4.51 | 9.77 | 16.77 | 25.37 | 35.48 | 47.04 | 60.02 | 74.39 | 90.16 | 107.28 | 125.69 |
| VELOCITY (FPS) | 1100. | 1043. | 997. | 958. | 924. | 894. | 866. | 840. | 816. | 794. | 774. | 755. | 737. |
| ENERGY (FT-LB) | 497. | 447. | 408. | 377. | 351. | 328. | 308. | 290. | 274. | 259. | 246. | 234. | 223. |
| DROP (IN) | .00 | -.91 | -3.82 | -8.90 | -16.32 | -26.26 | -38.91 | -54.45 | -73.06 | -94.88 | -120.04 | -148.80 | -181.36 |
| MID-RANGE (IN) | .00 | .24 | .99 | 2.34 | 4.34 | 7.06 | 10.56 | 14.92 | 20.21 | 26.48 | 33.76 | 42.16 | 51.77 |
| BULLET PATH (IN) | -.63 | .68 | .00 | -2.86 | -8.05 | -15.77 | -26.20 | -39.52 | -55.90 | -75.50 | -98.43 | -124.97 | -155.31 |
| TIME OF FLIGHT (SEC) | .000000 | .070078 | .143664 | .220436 | .300177 | .382739 | .468025 | .555975 | .646562 | .739775 | .835484 | .933655 | 1.034310 |
| WIND DEFLECTION (IN) | | | | | | | | | | | | | |
| 5 MPH | .00 | .17 | .64 | 1.40 | 2.42 | 3.68 | 5.19 | 6.93 | 8.90 | 11.10 | 13.52 | 16.16 | 19.02 |
| 10 MPH | .00 | .33 | 1.28 | 2.80 | 4.83 | 7.36 | 10.37 | 13.85 | 17.79 | 22.20 | 27.05 | 32.32 | 38.04 |
| 20 MPH | .00 | .67 | 2.57 | 5.59 | 9.66 | 14.72 | 20.74 | 27.70 | 35.59 | 44.40 | 54.09 | 64.65 | 76.08 |
| 30 MPH | .00 | 1.00 | 3.85 | 8.39 | 14.49 | 22.09 | 31.12 | 41.55 | 53.38 | 66.60 | 81.14 | 96.97 | 114.12 |

## Bullet: Sierra .4515 Dia. 185 Gr. JHC
### Ballistic Coefficient: .120

| RANGE (YDS) | MUZZLE | 25 | 50 | 75 | 100 | 125 | 150 | 175 | 200 | 225 | 250 | 275 | 300 |
|---|---|---|---|---|---|---|---|---|---|---|---|---|---|
| VELOCITY (FPS) | 1000. | 961. | 926. | 896. | 868. | 842. | 818. | 795. | 775. | 756. | 738. | 720. | 703. |
| ENERGY (FT-LB) | 411. | 379. | 352. | 329. | 309. | 291. | 275. | 260. | 247. | 235. | 223. | 213. | 203. |
| DROP (IN) | .00 | -1.09 | -4.50 | -10.43 | -19.05 | -30.55 | -45.11 | -62.89 | -83.97 | -108.64 | -137.10 | -169.52 | -206.13 |
| MID-RANGE (IN) | .00 | .28 | 1.16 | 2.72 | 5.02 | 8.13 | 12.12 | 17.05 | 22.93 | 29.89 | 38.00 | 47.33 | 57.96 |
| BULLET PATH (IN) | -.63 | .85 | .00 | -3.36 | -9.41 | -18.34 | -30.33 | -45.55 | -64.06 | -86.17 | -112.05 | -141.91 | -175.95 |
| TIME OF FLIGHT (SEC) | .000000 | .076558 | .156097 | .238466 | .323563 | .411327 | .501729 | .594762 | .690312 | .788322 | .888815 | .991821 | 1.097379 |
| WIND DEFLECTION (IN) | | | | | | | | | | | | | |
| 5 MPH | .00 | .14 | .54 | 1.18 | 2.07 | 3.20 | 4.55 | 6.14 | 7.95 | 9.97 | 12.22 | 14.68 | 17.37 |
| 10 MPH | .00 | .27 | 1.07 | 2.37 | 4.15 | 6.39 | 9.10 | 12.28 | 15.89 | 19.94 | 24.43 | 29.36 | 34.74 |
| 20 MPH | .00 | .55 | 2.15 | 4.74 | 8.29 | 12.79 | 18.21 | 24.56 | 31.79 | 39.89 | 48.86 | 58.72 | 69.48 |
| 30 MPH | .00 | .82 | 3.22 | 7.11 | 12.44 | 19.18 | 27.31 | 36.83 | 47.68 | 59.83 | 73.29 | 88.08 | 104.22 |
| VELOCITY (FPS) | 900. | 872. | 846. | 821. | 798. | 778. | 759. | 740. | 722. | 705. | 688. | 672. | 656. |
| ENERGY (FT-LB) | 333. | 312. | 294. | 277. | 262. | 249. | 237. | 225. | 214. | 204. | 195. | 186. | 177. |
| DROP (IN) | .00 | -1.28 | -5.42 | -12.58 | -22.96 | -36.62 | -53.85 | -74.79 | -99.69 | -128.72 | -162.11 | -200.06 | -242.82 |
| MID-RANGE (IN) | .00 | .35 | 1.43 | 3.31 | 6.06 | 9.70 | 14.34 | 20.04 | 26.88 | 35.01 | 44.44 | 55.26 | 67.56 |
| BULLET PATH (IN) | -.63 | 1.11 | -.00 | -4.14 | -11.49 | -22.13 | -36.33 | -54.25 | -76.12 | -102.13 | -132.49 | -167.43 | -207.16 |
| TIME OF FLIGHT (SEC) | .000000 | .084689 | .172053 | .262056 | .354692 | .449855 | .547474 | .647572 | .750177 | .855326 | .963061 | 1.073430 | 1.186487 |
| WIND DEFLECTION (IN) | | | | | | | | | | | | | |
| 5 MPH | .00 | .12 | .47 | 1.06 | 1.88 | 2.92 | 4.18 | 5.65 | 7.35 | 9.27 | 11.42 | 13.80 | 16.41 |
| 10 MPH | .00 | .24 | .95 | 2.12 | 3.76 | 5.84 | 8.36 | 11.31 | 14.70 | 18.54 | 22.83 | 27.59 | 32.82 |
| 20 MPH | .00 | .48 | 1.90 | 4.24 | 7.52 | 11.68 | 16.71 | 22.61 | 29.40 | 37.07 | 45.66 | 55.18 | 65.64 |
| 30 MPH | .00 | .72 | 2.84 | 6.37 | 11.28 | 17.52 | 25.07 | 33.92 | 44.09 | 55.61 | 68.50 | 82.77 | 98.46 |
| VELOCITY (FPS) | 800. | 780. | 760. | 742. | 724. | 706. | 689. | 673. | 657. | 642. | 627. | 612. | 598. |
| ENERGY (FT-LB) | 263. | 250. | 237. | 226. | 215. | 205. | 195. | 186. | 177. | 169. | 161. | 154. | 147. |
| DROP (IN) | .00 | -1.50 | -6.56 | -15.35 | -28.08 | -44.95 | -66.16 | -91.85 | -122.33 | -157.83 | -198.58 | -244.86 | -296.92 |
| MID-RANGE (IN) | .00 | .43 | 1.77 | 4.10 | 7.49 | 12.00 | 17.73 | 24.71 | 33.08 | 42.93 | 54.34 | 67.43 | 82.31 |
| BULLET PATH (IN) | -.63 | 1.46 | -.00 | -5.20 | -14.34 | -27.61 | -45.23 | -67.33 | -94.22 | -126.12 | -163.28 | -205.96 | -254.44 |
| TIME OF FLIGHT (SEC) | .000000 | .094972 | .192399 | .292300 | .394706 | .499652 | .607179 | .717335 | .830172 | .945748 | 1.064127 | 1.185373 | 1.309560 |
| WIND DEFLECTION (IN) | | | | | | | | | | | | | |
| 5 MPH | .00 | .11 | .43 | .97 | 1.73 | 2.72 | 3.93 | 5.38 | 7.06 | 8.98 | 11.14 | 13.56 | 16.24 |
| 10 MPH | .00 | .22 | .86 | 1.94 | 3.47 | 5.44 | 7.86 | 10.75 | 14.11 | 17.95 | 22.29 | 27.13 | 32.48 |
| 20 MPH | .00 | .43 | 1.72 | 3.89 | 6.94 | 10.88 | 15.73 | 21.50 | 28.22 | 35.90 | 44.57 | 54.25 | 64.97 |
| 30 MPH | .00 | 65 | 2.59 | 5.83 | 10.40 | 16.32 | 23.59 | 32.25 | 42.33 | 53.86 | 66.86 | 81.38 | 97.45 |

## Bullet: Speer .451 Dia. 200 Gr. FJHP
### Ballistic Coefficient: .165

| RANGE (YDS) | MUZZLE | 25 | 50 | 75 | 100 | 125 | 150 | 175 | 200 | 225 | 250 | 275 | 300 |
|---|---|---|---|---|---|---|---|---|---|---|---|---|---|
| VELOCITY (FPS) | 1300. | 1229. | 1166. | 1113. | 1068. | 1030. | 998. | 969. | 943. | 919. | 897. | 876. | 857. |
| ENERGY (FT-LB) | 750. | 670. | 604. | 550. | 507. | 471. | 442. | 417. | 394. | 375. | 357. | 341. | 326. |
| DROP (IN) | .00 | -.65 | -2.74 | -6.43 | -11.86 | -19.19 | -28.57 | -40.12 | -53.97 | -70.22 | -89.03 | -110.54 | -134.87 |
| MID-RANGE (IN) | .00 | .17 | .72 | 1.71 | 3.19 | 5.22 | 7.86 | 11.14 | 15.12 | 19.83 | 25.32 | 31.65 | 38.87 |
| BULLET PATH (IN) | -.63 | .41 | .00 | -2.00 | -5.75 | -11.40 | -19.09 | -28.95 | -41.11 | -55.68 | -72.81 | -92.63 | -115.28 |
| TIME OF FLIGHT (SEC) | .000000 | .059370 | .122060 | .187937 | .256770 | .328294 | .402295 | .478614 | .557129 | .637750 | .720408 | .805052 | .891647 |
| WIND DEFLECTION (IN) | | | | | | | | | | | | | |
| 5 MPH | .00 | .15 | .59 | 1.31 | 2.29 | 3.51 | 4.94 | 6.58 | 8.41 | 10.43 | 12.63 | 15.00 | 17.54 |
| 10 MPH | .00 | .30 | 1.17 | 2.62 | 4.58 | 7.01 | 9.88 | 13.16 | 16.82 | 20.86 | 25.25 | 30.00 | 35.08 |
| 20 MPH | .00 | .59 | 2.35 | 5.23 | 9.15 | 14.02 | 19.76 | 26.32 | 33.65 | 41.72 | 50.51 | 59.99 | 70.17 |
| 30 MPH | .00 | .89 | 3.52 | 7.85 | 13.73 | 21.03 | 29.64 | 39.48 | 50.47 | 62.58 | 75.76 | 89.99 | 105.25 |
| VELOCITY (FPS) | 1200. | 1142. | 1092. | 1051. | 1015. | 984. | 957. | 932. | 909. | 888. | 868. | 849. | 831. |
| ENERGY (FT-LB) | 639. | 579. | 530. | 490. | 458. | 430. | 407. | 386. | 367. | 350. | 334. | 320. | 306. |
| DROP (IN) | .00 | -.76 | -3.19 | -7.43 | -13.64 | -21.94 | -32.48 | -45.36 | -60.72 | -78.69 | -99.41 | -123.01 | -149.62 |
| MID-RANGE (IN) | .00 | .20 | .83 | 1.96 | 3.63 | 5.89 | 8.81 | 12.40 | 16.72 | 21.82 | 27.76 | 34.58 | 42.33 |
| BULLET PATH (IN) | -.63 | .52 | .00 | -2.33 | -6.62 | -13.02 | -21.65 | -32.61 | -46.06 | -62.12 | -80.93 | -102.62 | -127.32 |
| TIME OF FLIGHT (SEC) | .000000 | .064116 | .131327 | .201375 | .274011 | .349049 | .426347 | .505799 | .587323 | .670859 | .756364 | .843808 | .933175 |
| WIND DEFLECTION (IN) | | | | | | | | | | | | | |
| 5 MPH | .00 | .14 | .56 | 1.22 | 2.11 | 3.22 | 4.52 | 6.01 | 7.68 | 9.54 | 11.56 | 13.76 | 16.12 |
| 10 MPH | .00 | .28 | 1.11 | 2.44 | 4.23 | 6.43 | 9.04 | 12.02 | 15.37 | 19.07 | 23.12 | 27.51 | 32.24 |
| 20 MPH | .00 | .57 | 2.23 | 4.88 | 8.45 | 12.87 | 18.07 | 24.04 | 30.74 | 38.14 | 46.24 | 55.02 | 64.48 |
| 30 MPH | .00 | .85 | 3.34 | 7.33 | 12.68 | 19.30 | 27.11 | 36.06 | 46.11 | 57.21 | 69.36 | 82.53 | 96.72 |
| VELOCITY (FPS) | 1100. | 1057. | 1021. | 990. | 961. | 936. | 913. | 891. | 871. | 852. | 834. | 816. | 800. |
| ENERGY (FT-LB) | 537. | 496. | 463. | 435. | 410. | 389. | 370. | 353. | 337. | 322. | 309. | 296. | 284. |
| DROP (IN) | .00 | -.90 | -3.73 | -8.65 | -15.77 | -25.22 | -37.12 | -51.61 | -68.83 | -88.90 | -111.96 | -138.14 | -167.57 |
| MID-RANGE (IN) | .00 | .23 | .97 | 2.26 | 4.15 | 6.69 | 9.91 | 13.88 | 18.64 | 24.25 | 30.76 | 38.22 | 46.67 |
| BULLET PATH (IN) | -.63 | .65 | .00 | -2.73 | -7.67 | -14.94 | -24.65 | -36.96 | -52.00 | -69.89 | -90.77 | -114.77 | -142.01 |
| TIME OF FLIGHT (SEC) | .000000 | .069577 | .141781 | .216415 | .293331 | .372416 | .453585 | .536776 | .621941 | .709050 | .798083 | .889033 | .981899 |
| WIND DEFLECTION (IN) | | | | | | | | | | | | | |
| 5 MPH | .00 | .12 | .48 | 1.04 | 1.81 | 2.77 | 3.92 | 5.24 | 6.73 | 8.40 | 10.23 | 12.23 | 14.41 |
| 10 MPH | .00 | .25 | .95 | 2.09 | 3.63 | 5.55 | 7.83 | 10.47 | 13.46 | 16.79 | 20.46 | 24.47 | 28.81 |
| 20 MPH | .00 | .49 | 1.91 | 4.18 | 7.25 | 11.09 | 15.66 | 20.95 | 26.92 | 33.59 | 40.93 | 48.94 | 57.63 |
| 30 MPH | .00 | .74 | 2.86 | 6.27 | 10.88 | 16.64 | 23.49 | 31.42 | 40.38 | 50.38 | 61.39 | 73.41 | 86.44 |
| VELOCITY (FPS) | 1000. | 971. | 944. | 920. | 898. | 878. | 858. | 840. | 822. | 805. | 783. | 758. | 734. |
| ENERGY (FT-LB) | 444. | 418. | 396. | 376. | 358. | 342. | 327. | 313. | 300. | 288. | 272. | 255. | 239. |
| DROP (IN) | .00 | -1.07 | -4.43 | -10.19 | -18.49 | -29.48 | -43.28 | -60.02 | -79.83 | -102.85 | -129.19 | -159.04 | -192.65 |
| MID-RANGE (IN) | .00 | .28 | 1.15 | 2.63 | 4.81 | 7.74 | 11.45 | 16.00 | 21.42 | 27.78 | 35.11 | 43.50 | 53.05 |
| BULLET PATH (IN) | -.63 | .83 | .00 | -3.22 | -8.99 | -17.45 | -28.72 | -42.92 | -60.20 | -80.68 | -104.49 | -131.81 | -162.90 |
| TIME OF FLIGHT (SEC) | .000000 | .076141 | .154486 | .234942 | .317438 | .401923 | .488360 | .576727 | .667010 | .759209 | .853561 | .951009 | 1.051684 |
| WIND DEFLECTION (IN) | | | | | | | | | | | | | |
| 5 MPH | .00 | .10 | .39 | .87 | 1.53 | 2.37 | 3.38 | 4.55 | 5.90 | 7.41 | 9.11 | 11.09 | 13.35 |
| 10 MPH | .00 | .20 | .79 | 1.75 | 3.07 | 4.74 | 6.75 | 9.10 | 11.79 | 14.82 | 18.23 | 22.18 | 26.70 |
| 20 MPH | .00 | .40 | 1.58 | 3.50 | 6.14 | 9.48 | 13.50 | 18.21 | 23.59 | 29.64 | 36.45 | 44.36 | 53.39 |
| 30 MPH | .00 | .60 | 2.37 | 5.25 | 9.21 | 14.22 | 20.25 | 27.31 | 35.38 | 44.46 | 54.68 | 66.53 | 80.09 |

# Bullet: Speer .451 Dia. 200 Gr. FJHP
## Ballistic Coefficient: .165

| RANGE (YDS) | MUZZLE | 25 | 50 | 75 | 100 | 125 | 150 | 175 | 200 | 225 | 250 | 275 | 300 |
|---|---|---|---|---|---|---|---|---|---|---|---|---|---|
| VELOCITY (FPS) | 900. | 879. | 860. | 841. | 824. | 807. | 785. | 760. | 736. | 713. | 691. | 670. | 649. |
| ENERGY (FT-LB) | 360. | 343. | 328. | 314. | 301. | 289. | 274. | 256. | 240. | 226. | 212. | 199. | 187. |
| DROP (IN) | .00 | -1.25 | -5.30 | -12.28 | -22.32 | -35.56 | -52.11 | -72.15 | -95.92 | -123.70 | -155.73 | -192.32 | -233.73 |
| MID-RANGE (IN) | .00 | .34 | 1.40 | 3.23 | 5.86 | 9.37 | 13.78 | 19.16 | 25.64 | 33.31 | 42.30 | 52.73 | 64.75 |
| BULLET PATH (IN) | -.63 | 1.09 | -.00 | -4.02 | -11.10 | -21.38 | -34.96 | -52.03 | -72.84 | -97.65 | -126.73 | -160.35 | -198.80 |
| TIME OF FLIGHT (SEC) | .000000 | .084322 | .170598 | .258805 | .348928 | .440965 | .535078 | .632222 | .732587 | .836223 | .943197 | 1.053586 | 1.167481 |
| WIND DEFLECTION (IN) | | | | | | | | | | | | | |
| 5 MPH | .00 | .09 | .35 | .77 | 1.37 | 2.14 | 3.09 | 4.30 | 5.80 | 7.59 | 9.67 | 12.05 | 14.74 |
| 10 MPH | .00 | .17 | .69 | 1.55 | 2.74 | 4.28 | 6.17 | 8.60 | 11.60 | 15.18 | 19.34 | 24.10 | 29.48 |
| 20 MPH | .00 | .35 | 1.38 | 3.10 | 5.49 | 8.55 | 12.35 | 17.21 | 23.20 | 30.35 | 38.67 | 48.20 | 58.95 |
| 30 MPH | .00 | .52 | 2.08 | 4.65 | 8.23 | 12.83 | 18.52 | 25.81 | 34.81 | 45.53 | 58.01 | 72.29 | 88.43 |
| VELOCITY (FPS) | 800. | 774. | 749. | 725. | 703. | 681. | 660. | 640. | 621. | 602. | 583. | 566. | 549. |
| ENERGY (FT-LB) | 284. | 266. | 249. | 234. | 219. | 206. | 194. | 182. | 171. | 161. | 151. | 142. | 134. |
| DROP (IN) | .00 | -1.55 | -6.69 | -15.67 | -28.77 | -46.25 | -68.40 | -95.52 | -127.93 | -165.96 | -209.99 | -260.38 | -317.57 |
| MID-RANGE (IN) | .00 | .43 | 1.79 | 4.18 | 7.70 | 12.45 | 18.53 | 26.07 | 35.19 | 46.03 | 58.74 | 73.47 | 90.38 |
| BULLET PATH (IN) | -.63 | 1.48 | -.00 | -5.33 | -14.77 | -28.59 | -47.09 | -70.55 | -99.30 | -133.67 | -174.04 | -220.78 | -274.31 |
| TIME OF FLIGHT (SEC) | .000000 | .095340 | .193878 | .295655 | .400730 | .509172 | .621064 | .736501 | .855588 | .978439 | 1.105177 | 1.235936 | 1.370857 |
| WIND DEFLECTION (IN) | | | | | | | | | | | | | |
| 5 MPH | .00 | .14 | .56 | 1.27 | 2.26 | 3.56 | 5.15 | 7.06 | 9.29 | 11.85 | 14.76 | 18.01 | 21.64 |
| 10 MPH | .00 | .28 | 1.12 | 2.54 | 4.53 | 7.11 | 10.31 | 14.12 | 18.58 | 23.71 | 29.51 | 36.02 | 43.27 |
| 20 MPH | .00 | .56 | 2.24 | 5.07 | 9.06 | 14.23 | 20.61 | 28.25 | 37.17 | 47.41 | 59.02 | 72.05 | 86.54 |
| 30 MPH | .00 | .84 | 3.37 | 7.61 | 13.59 | 21.34 | 30.92 | 42.37 | 55.75 | 71.12 | 88.53 | 108.07 | 129.81 |
| VELOCITY (FPS) | 700. | 678. | 658. | 637. | 618. | 599. | 581. | 563. | 546. | 529. | 513. | 497. | 482. |
| ENERGY (FT-LB) | 218. | 204. | 192. | 180. | 170. | 159. | 150. | 141. | 132. | 124. | 117. | 110. | 103. |
| DROP (IN) | .00 | -1.98 | -8.65 | -20.34 | -37.35 | -60.02 | -88.74 | -123.87 | -165.83 | -215.08 | -272.07 | -337.33 | -411.38 |
| MID-RANGE (IN) | .00 | .57 | 2.35 | 5.46 | 10.02 | 16.16 | 24.03 | 33.78 | 45.57 | 59.58 | 76.01 | 95.06 | 116.95 |
| BULLET PATH (IN) | -.63 | 2.04 | .00 | -7.04 | -19.41 | -37.45 | -61.52 | -92.00 | -129.33 | -173.93 | -226.29 | -286.90 | -356.31 |
| TIME OF FLIGHT (SEC) | .000000 | .108844 | .221148 | .337007 | .456527 | .579823 | .707019 | .838248 | .973653 | 1.113383 | 1.257599 | 1.406467 | 1.560164 |
| WIND DEFLECTION (IN) | | | | | | | | | | | | | |
| 5 MPH | .00 | .15 | .60 | 1.37 | 2.46 | 3.88 | 5.65 | 7.77 | 10.25 | 13.12 | 16.38 | 20.05 | 24.15 |
| 10 MPH | .00 | .30 | 1.21 | 2.74 | 4.92 | 7.76 | 11.29 | 15.53 | 20.51 | 26.24 | 32.77 | 40.11 | 48.30 |
| 20 MPH | .00 | .60 | 2.42 | 5.48 | 9.84 | 15.53 | 22.58 | 31.06 | 41.01 | 52.48 | 65.53 | 80.22 | 96.61 |
| 30 MPH | .00 | .90 | 3.62 | 8.23 | 14.76 | 23.29 | 33.88 | 46.59 | 61.52 | 78.72 | 98.30 | 120.33 | 144.91 |

## Bullet: Speer .451 Dia. 225 Gr. JHP
## Ballistic Coefficient: .175

| RANGE (YDS) | MUZZLE | 25 | 50 | 75 | 100 | 125 | 150 | 175 | 200 | 225 | 250 | 275 | 300 |
|---|---|---|---|---|---|---|---|---|---|---|---|---|---|
| VELOCITY (FPS) | 1300. | 1232. | 1173. | 1121. | 1078. | 1041. | 1008. | 980. | 954. | 931. | 909. | 889. | 870. |
| ENERGY (FT-LB) | 844. | 759. | 687. | 628. | 580. | 541. | 508. | 480. | 455. | 433. | 413. | 395. | 378. |
| DROP (IN) | .00 | -.64 | -2.73 | -6.38 | -11.76 | -19.01 | -28.25 | -39.63 | -53.26 | -69.23 | -87.70 | -108.78 | -132.60 |
| MID-RANGE (IN) | .00 | .17 | .72 | 1.69 | 3.16 | 5.16 | 7.74 | 10.97 | 14.87 | 19.47 | 24.84 | 31.02 | 38.05 |
| BULLET PATH (IN) | -.63 | .40 | .00 | -1.98 | -5.68 | -11.25 | -18.82 | -28.51 | -40.47 | -54.77 | -71.55 | -90.95 | -113.10 |
| TIME OF FLIGHT (SEC) | .000000 | .059274 | .121684 | .187119 | .255384 | .326240 | .399482 | .474957 | .552545 | .632154 | .713714 | .797172 | .882489 |
| WIND DEFLECTION (IN) | | | | | | | | | | | | | |
| 5 MPH | .00 | .14 | .55 | 1.24 | 2.17 | 3.32 | 4.69 | 6.26 | 8.01 | 9.94 | 12.04 | 14.30 | 16.74 |
| 10 MPH | .00 | .28 | 1.11 | 2.47 | 4.33 | 6.65 | 9.39 | 12.52 | 16.02 | 19.87 | 24.08 | 28.61 | 33.47 |
| 20 MPH | .00 | .56 | 2.22 | 4.94 | 8.66 | 13.30 | 18.77 | 25.03 | 32.03 | 39.75 | 48.15 | 57.22 | 66.94 |
| 30 MPH | .00 | .84 | 3.33 | 7.41 | 13.00 | 19.95 | 28.16 | 37.55 | 48.05 | 59.62 | 72.23 | 85.83 | 100.42 |
| VELOCITY (FPS) | 1200. | 1145. | 1097. | 1057. | 1023. | 993. | 966. | 942. | 919. | 898. | 879. | 860. | 843. |
| ENERGY (FT-LB) | 719. | 654. | 601. | 558. | 523. | 492. | 466. | 443. | 422. | 403. | 386. | 370. | 355. |
| DROP (IN) | .00 | -.76 | -3.18 | -7.39 | -13.54 | -21.77 | -32.18 | -44.91 | -60.04 | -77.74 | -98.12 | -121.30 | -147.41 |
| MID-RANGE (IN) | .00 | .20 | .83 | 1.94 | 3.59 | 5.83 | 8.70 | 12.24 | 16.48 | 21.48 | 27.29 | 33.96 | 41.52 |
| BULLET PATH (IN) | -.63 | .51 | .00 | -2.31 | -6.56 | -12.87 | -21.39 | -32.21 | -45.44 | -61.23 | -79.71 | -100.99 | -125.19 |
| TIME OF FLIGHT (SEC) | .000000 | .064025 | .130982 | .200642 | .272778 | .347215 | .423816 | .502479 | .583124 | .665689 | .750130 | .836415 | .924521 |
| WIND DEFLECTION (IN) | | | | | | | | | | | | | |
| 5 MPH | .00 | .13 | .53 | 1.16 | 2.00 | 3.05 | 4.30 | 5.72 | 7.31 | 9.08 | 11.01 | 13.10 | 15.36 |
| 10 MPH | .00 | .27 | 1.05 | 2.31 | 4.01 | 6.11 | 8.59 | 11.44 | 14.63 | 18.16 | 22.02 | 26.21 | 30.72 |
| 20 MPH | .00 | .54 | 2.11 | 4.63 | 8.02 | 12.22 | 17.18 | 22.87 | 29.26 | 36.32 | 44.05 | 52.42 | 61.43 |
| 30 MPH | .00 | .81 | 3.16 | 6.94 | 12.03 | 18.33 | 25.77 | 34.31 | 43.89 | 54.48 | 66.07 | 78.63 | 92.15 |
| VELOCITY (FPS) | 1100. | 1060. | 1025. | 995. | 968. | 943. | 920. | 900. | 880. | 861. | 844. | 827. | 811. |
| ENERGY (FT-LB) | 604. | 561. | 525. | 494. | 468. | 444. | 423. | 404. | 387. | 371. | 356. | 342. | 329. |
| DROP (IN) | .00 | -.90 | -3.72 | -8.61 | -15.68 | -25.06 | -36.84 | -51.17 | -68.18 | -87.98 | -110.70 | -136.46 | -165.39 |
| MID-RANGE (IN) | .00 | .23 | .96 | 2.24 | 4.12 | 6.63 | 9.81 | 13.72 | 18.41 | 23.91 | 30.29 | 37.59 | 45.85 |
| BULLET PATH (IN) | -.63 | .65 | .00 | -2.71 | -7.61 | -14.82 | -24.42 | -36.57 | -51.41 | -69.03 | -89.58 | -113.16 | -139.91 |
| TIME OF FLIGHT (SEC) | .000000 | .069500 | .141488 | .215786 | .292256 | .370792 | .451315 | .533761 | .618086 | .704254 | .792245 | .882046 | .973652 |
| WIND DEFLECTION (IN) | | | | | | | | | | | | | |
| 5 MPH | .00 | .12 | .45 | .99 | 1.72 | 2.63 | 3.72 | 4.97 | 6.39 | 7.97 | 9.72 | 11.62 | 13.68 |
| 10 MPH | .00 | .23 | .90 | 1.98 | 3.44 | 5.26 | 7.43 | 9.94 | 12.78 | 15.95 | 19.44 | 23.24 | 27.36 |
| 20 MPH | .00 | .46 | 1.80 | 3.96 | 6.87 | 10.52 | 14.86 | 19.88 | 25.57 | 31.90 | 38.87 | 46.48 | 54.73 |
| 30 MPH | .00 | .70 | 2.71 | 5.93 | 10.31 | 15.78 | 22.29 | 29.83 | 38.35 | 47.85 | 58.31 | 69.72 | 82.09 |
| VELOCITY (FPS) | 1000. | 972. | 947. | 924. | 903. | 883. | 865. | 847. | 830. | 814. | 798. | 774. | 750. |
| ENERGY (FT-LB) | 500. | 472. | 448. | 427. | 407. | 390. | 374. | 358. | 344. | 331. | 318. | 299. | 281. |
| DROP (IN) | .00 | -1.07 | -4.42 | -10.15 | -18.40 | -29.32 | -43.01 | -59.58 | -79.17 | -101.91 | -127.92 | -157.30 | -190.30 |
| MID-RANGE (IN) | .00 | .28 | 1.14 | 2.62 | 4.78 | 7.68 | 11.35 | 15.84 | 21.18 | 27.43 | 34.64 | 42.84 | 52.15 |
| BULLET PATH (IN) | -.63 | .83 | .00 | -3.20 | -8.92 | -17.31 | -28.47 | -42.52 | -59.59 | -79.79 | -103.28 | -130.13 | -160.60 |
| TIME OF FLIGHT (SEC) | .000000 | .076077 | .154237 | .234397 | .316490 | .400468 | .486295 | .573947 | .663410 | .754678 | .847754 | .943252 | 1.041800 |
| WIND DEFLECTION (IN) | | | | | | | | | | | | | |
| 5 MPH | .00 | .09 | .37 | .83 | 1.45 | 2.24 | 3.19 | 4.31 | 5.58 | 7.01 | 8.60 | 10.41 | 12.48 |
| 10 MPH | .00 | .19 | .75 | 1.65 | 2.90 | 4.48 | 6.39 | 8.61 | 11.16 | 14.02 | 17.20 | 20.81 | 24.96 |
| 20 MPH | .00 | .38 | 1.49 | 3.31 | 5.80 | 8.96 | 12.78 | 17.23 | 22.32 | 28.05 | 34.41 | 41.62 | 49.91 |
| 30 MPH | .00 | .57 | 2.24 | 4.96 | 8.71 | 13.45 | 19.16 | 25.84 | 33.48 | 42.07 | 51.61 | 62.44 | 74.87 |

**Bullet: Speer .451 Dia. 225 Gr. JHP**
**Ballistic Coefficient: .175**

| RANGE (YDS) | MUZZLE | 25 | 50 | 75 | 100 | 125 | 150 | 175 | 200 | 225 | 250 | 275 | 300 |
|---|---|---|---|---|---|---|---|---|---|---|---|---|---|
| VELOCITY (FPS) | 900. | 880. | 862. | 844. | 827. | 811. | 795. | 770. | 747. | 724. | 703. | 682. | 663. |
| ENERGY (FT-LB) | 405. | 387. | 371. | 356. | 342. | 329. | 316. | 296. | 279. | 262. | 247. | 233. | 219. |
| DROP (IN) | .00 | -1.24 | -5.28 | -12.24 | -22.23 | -35.39 | -51.84 | -71.67 | -95.16 | -122.52 | -154.01 | -189.88 | -230.41 |
| MID-RANGE (IN) | .00 | .34 | 1.40 | 3.21 | 5.83 | 9.31 | 13.68 | 18.99 | 25.34 | 32.85 | 41.61 | 51.74 | 63.36 |
| BULLET PATH (IN) | -.63 | 1.08 | -.00 | -4.00 | -11.04 | -21.24 | -34.73 | -51.61 | -72.13 | -96.54 | -125.07 | -157.99 | -195.56 |
| TIME OF FLIGHT (SEC) | .000000 | .084266 | .170376 | .258308 | .348048 | .439593 | .532950 | .628829 | .727760 | .829783 | .934950 | 1.043328 | 1.154990 |
| WIND DEFLECTION (IN) | | | | | | | | | | | | | |
| 5 MPH | .00 | .08 | .33 | .73 | 1.29 | 2.02 | 2.90 | 4.00 | 5.38 | 7.02 | 8.94 | 11.15 | 13.64 |
| 10 MPH | .00 | .16 | .65 | 1.46 | 2.59 | 4.04 | 5.80 | 8.01 | 10.75 | 14.04 | 17.88 | 22.29 | 27.28 |
| 20 MPH | .00 | .33 | 1.31 | 2.92 | 5.18 | 8.07 | 11.60 | 16.01 | 21.50 | 28.08 | 35.77 | 44.58 | 54.56 |
| 30 MPH | .00 | .49 | 1.96 | 4.39 | 7.77 | 12.11 | 17.40 | 24.02 | 32.26 | 42.13 | 53.65 | 66.88 | 81.83 |
| VELOCITY (FPS) | 800. | 775. | 751. | 729. | 707. | 686. | 666. | 647. | 628. | 610. | 592. | 575. | 558. |
| ENERGY (FT-LB) | 320. | 300. | 282. | 265. | 250. | 235. | 222. | 209. | 197. | 186. | 175. | 165. | 156. |
| DROP (IN) | .00 | -1.54 | -6.66 | -15.61 | -28.63 | -45.99 | -67.95 | -94.78 | -126.79 | -164.30 | -207.65 | -257.18 | -313.28 |
| MID-RANGE (IN) | .00 | .42 | 1.79 | 4.17 | 7.66 | 12.36 | 18.37 | 25.80 | 34.76 | 45.40 | 57.83 | 72.21 | 88.69 |
| BULLET PATH (IN) | -.63 | 1.47 | -.00 | -5.31 | -14.69 | -28.41 | -46.72 | -69.91 | -98.28 | -132.15 | -171.85 | -217.74 | -270.20 |
| TIME OF FLIGHT (SEC) | .000000 | .095264 | .193572 | .294961 | .399481 | .507195 | .618175 | .732504 | .850275 | .971588 | 1.096552 | 1.225285 | 1.357910 |
| WIND DEFLECTION (IN) | | | | | | | | | | | | | |
| 5 MPH | .00 | .13 | .53 | 1.21 | 2.15 | 3.38 | 4.90 | 6.71 | 8.82 | 11.25 | 14.00 | 17.08 | 20.50 |
| 10 MPH | .00 | .27 | 1.07 | 2.41 | 4.31 | 6.77 | 9.80 | 13.42 | 17.65 | 22.50 | 27.99 | 34.15 | 40.99 |
| 20 MPH | .00 | .53 | 2.14 | 4.83 | 8.62 | 13.53 | 19.60 | 26.84 | 35.30 | 45.00 | 55.99 | 68.30 | 81.98 |
| 30 MPH | .00 | .80 | 3.21 | 7.24 | 12.93 | 20.30 | 29.40 | 40.26 | 52.95 | 67.50 | 83.98 | 102.45 | 122.98 |
| VELOCITY (FPS) | 700. | 679. | 660. | 640. | 622. | 604. | 586. | 569. | 553. | 536. | 521. | 506. | 491. |
| ENERGY (FT-LB) | 245. | 231. | 217. | 205. | 193. | 182. | 172. | 162. | 153. | 144. | 136. | 128. | 120. |
| DROP (IN) | .00 | -1.97 | -8.63 | -20.27 | -37.19 | -59.71 | -88.17 | -122.94 | -164.41 | -212.98 | -269.09 | -333.22 | -405.87 |
| MID-RANGE (IN) | .00 | .57 | 2.34 | 5.44 | 9.96 | 16.04 | 23.82 | 33.42 | 45.01 | 58.76 | 74.84 | 93.44 | 114.76 |
| BULLET PATH (IN) | -.63 | 2.03 | .00 | -7.01 | -19.29 | -37.18 | -61.01 | -91.15 | -127.98 | -171.92 | -223.40 | -282.90 | -350.92 |
| TIME OF FLIGHT (SEC) | .000000 | .108762 | .220816 | .336246 | .455148 | .577624 | .703785 | .833751 | .967648 | 1.105610 | 1.247778 | 1.394300 | 1.545334 |
| WIND DEFLECTION (IN) | | | | | | | | | | | | | |
| 5 MPH | .00 | .14 | .57 | 1.30 | 2.34 | 3.69 | 5.36 | 7.37 | 9.72 | 12.44 | 15.52 | 18.98 | 22.85 |
| 10 MPH | .00 | .29 | 1.15 | 2.61 | 4.68 | 7.38 | 10.72 | 14.74 | 19.45 | 24.87 | 31.04 | 37.97 | 45.69 |
| 20 MPH | .00 | .57 | 2.30 | 5.22 | 9.35 | 14.75 | 21.45 | 29.48 | 38.90 | 49.75 | 62.07 | 75.94 | 91.39 |
| 30 MPH | .00 | .86 | 3.45 | 7.82 | 14.03 | 22.13 | 32.17 | 44.22 | 58.35 | 74.62 | 93.11 | 113.90 | 137.08 |

## Bullet: Sierra .4515 Dia. 240 Gr. JHC
## Ballistic Coefficient: .168

| RANGE (YDS) | MUZZLE | 25 | 50 | 75 | 100 | 125 | 150 | 175 | 200 | 225 | 250 | 275 | 300 |
|---|---|---|---|---|---|---|---|---|---|---|---|---|---|
| VELOCITY (FPS) | 1300. | 1230. | 1168. | 1115. | 1071. | 1034. | 1001. | 972. | 946. | 922. | 901. | 880. | 861. |
| ENERGY (FT-LB) | 900. | 806. | 727. | 663. | 611. | 569. | 534. | 504. | 477. | 453. | 432. | 413. | 395. |
| DROP (IN) | .00 | -.65 | -2.74 | -6.41 | -11.83 | -19.14 | -28.47 | -39.96 | -53.75 | -69.91 | -88.61 | -109.99 | -134.16 |
| MID-RANGE (IN) | .00 | .17 | .72 | 1.70 | 3.18 | 5.20 | 7.82 | 11.09 | 15.04 | 19.72 | 25.17 | 31.45 | 38.61 |
| BULLET PATH (IN) | -.63 | .41 | .00 | -1.99 | -5.73 | -11.35 | -19.00 | -28.81 | -40.92 | -55.40 | -72.42 | -92.11 | -114.60 |
| TIME OF FLIGHT (SEC) | .000000 | .059340 | .121943 | .187682 | .256339 | .327655 | .401421 | .477478 | .555706 | .636012 | .718329 | .802603 | .888801 |
| WIND DEFLECTION (IN) | | | | | | | | | | | | | |
| 5 MPH | .00 | .15 | .58 | 1.29 | 2.25 | 3.45 | 4.86 | 6.48 | 8.29 | 10.28 | 12.44 | 14.78 | 17.29 |
| 10 MPH | .00 | .29 | 1.15 | 2.57 | 4.50 | 6.90 | 9.73 | 12.96 | 16.57 | 20.55 | 24.89 | 29.57 | 34.58 |
| 20 MPH | .00 | .58 | 2.31 | 5.14 | 9.00 | 13.80 | 19.45 | 25.92 | 33.15 | 41.11 | 49.77 | 59.13 | 69.17 |
| 30 MPH | .00 | .87 | 3.46 | 7.71 | 13.50 | 20.69 | 29.18 | 38.88 | 49.72 | 61.66 | 74.66 | 88.70 | 103.75 |
| VELOCITY (FPS) | 1200. | 1142. | 1094. | 1053. | 1018. | 987. | 960. | 935. | 912. | 891. | 871. | 852. | 835. |
| ENERGY (FT-LB) | 767. | 695. | 637. | 591. | 552. | 519. | 491. | 466. | 443. | 423. | 404. | 387. | 371. |
| DROP (IN) | .00 | -.76 | -3.19 | -7.42 | -13.61 | -21.89 | -32.39 | -45.22 | -60.51 | -78.39 | -99.01 | -122.48 | -148.94 |
| MID-RANGE (IN) | .00 | .20 | .83 | 1.95 | 3.62 | 5.87 | 8.77 | 12.35 | 16.64 | 21.72 | 27.62 | 34.39 | 42.08 |
| BULLET PATH (IN) | -.63 | .51 | .00 | -2.32 | -6.60 | -12.97 | -21.57 | -32.49 | -45.87 | -61.84 | -80.55 | -102.12 | -126.66 |
| TIME OF FLIGHT (SEC) | .000000 | .064088 | .131219 | .201146 | .273627 | .348478 | .425560 | .504766 | .586017 | .669251 | .754424 | .841507 | .930480 |
| WIND DEFLECTION (IN) | | | | | | | | | | | | | |
| 5 MPH | .00 | .14 | .55 | 1.20 | 2.08 | 3.17 | 4.45 | 5.92 | 7.57 | 9.39 | 11.39 | 13.55 | 15.88 |
| 10 MPH | .00 | .28 | 1.09 | 2.40 | 4.16 | 6.33 | 8.90 | 11.84 | 15.14 | 18.79 | 22.78 | 27.11 | 31.76 |
| 20 MPH | .00 | .56 | 2.19 | 4.80 | 8.32 | 12.66 | 17.80 | 23.68 | 30.28 | 37.58 | 45.56 | 54.21 | 63.53 |
| 30 MPH | .00 | .84 | 3.28 | 7.21 | 12.48 | 19.00 | 26.70 | 35.52 | 45.42 | 56.36 | 68.34 | 81.32 | 95.29 |
| VELOCITY (FPS) | 1100. | 1058. | 1022. | 991. | 963. | 938. | 915. | 894. | 874. | 855. | 837. | 820. | 803. |
| ENERGY (FT-LB) | 645. | 597. | 557. | 523. | 494. | 469. | 446. | 426. | 407. | 389. | 373. | 358. | 344. |
| DROP (IN) | .00 | -.90 | -3.73 | -8.64 | -15.74 | -25.17 | -37.03 | -51.47 | -68.62 | -88.62 | -111.57 | -137.62 | -166.89 |
| MID-RANGE (IN) | .00 | .23 | .97 | 2.25 | 4.14 | 6.67 | 9.88 | 13.83 | 18.57 | 24.14 | 30.61 | 38.02 | 46.41 |
| BULLET PATH (IN) | -.63 | .65 | .00 | -2.72 | -7.65 | -14.90 | -24.58 | -36.84 | -51.81 | -69.62 | -90.40 | -114.27 | -141.36 |
| TIME OF FLIGHT (SEC) | .000000 | .069553 | .141690 | .216219 | .292996 | .371910 | .452878 | .535837 | .620740 | .707556 | .796264 | .886854 | .979326 |
| WIND DEFLECTION (IN) | | | | | | | | | | | | | |
| 5 MPH | .00 | .12 | .47 | 1.03 | 1.78 | 2.73 | 3.85 | 5.15 | 6.63 | 8.26 | 10.07 | 12.04 | 14.18 |
| 10 MPH | .00 | .24 | .94 | 2.05 | 3.57 | 5.46 | 7.71 | 10.31 | 13.25 | 16.53 | 20.14 | 24.09 | 28.36 |
| 20 MPH | .00 | .48 | 1.87 | 4.11 | 7.13 | 10.91 | 15.41 | 20.61 | 26.50 | 33.06 | 40.28 | 48.17 | 56.72 |
| 30 MPH | .00 | .72 | 2.81 | 6.16 | 10.70 | 16.37 | 23.12 | 30.92 | 39.75 | 49.59 | 60.43 | 72.26 | 85.08 |
| VELOCITY (FPS) | 1000. | 971. | 945. | 922. | 900. | 879. | 860. | 842. | 825. | 808. | 792. | 776. | 761. |
| ENERGY (FT-LB) | 533. | 503. | 476. | 453. | 431. | 412. | 394. | 378. | 362. | 348. | 334. | 321. | 309. |
| DROP (IN) | .00 | -1.07 | -4.43 | -10.18 | -18.47 | -29.43 | -43.20 | -59.89 | -79.63 | -102.55 | -128.73 | -158.29 | -191.45 |
| MID-RANGE (IN) | .00 | .28 | 1.15 | 2.63 | 4.80 | 7.72 | 11.42 | 15.95 | 21.35 | 27.67 | 34.94 | 43.19 | 52.53 |
| BULLET PATH (IN) | -.63 | .83 | .00 | -3.21 | -8.97 | -17.40 | -28.64 | -42.80 | -60.01 | -80.40 | -104.05 | -131.08 | -161.70 |
| TIME OF FLIGHT (SEC) | .000000 | .076121 | .154408 | .234772 | .317142 | .401469 | .487716 | .575859 | .665887 | .757794 | .851590 | .947308 | 1.044964 |
| WIND DEFLECTION (IN) | | | | | | | | | | | | | |
| 5 MPH | .00 | .10 | .39 | .86 | 1.51 | 2.33 | 3.32 | 4.48 | 5.80 | 7.29 | 8.94 | 10.76 | 12.76 |
| 10 MPH | .00 | .20 | .78 | 1.72 | 3.02 | 4.66 | 6.64 | 8.95 | 11.60 | 14.57 | 17.88 | 21.53 | 25.51 |
| 20 MPH | .00 | .39 | 1.55 | 3.44 | 6.03 | 9.32 | 13.28 | 17.90 | 23.19 | 29.14 | 35.76 | 43.05 | 51.03 |
| 30 MPH | .00 | .59 | 2.33 | 5.16 | 9.05 | 13.98 | 19.91 | 26.85 | 34.79 | 43.72 | 53.64 | 64.58 | 76.54 |

## Bullet: Sierra .4515 Dia. 240 Gr. JHC
### Ballistic Coefficient: .168

| RANGE (YDS) | MUZZLE | 25 | 50 | 75 | 100 | 125 | 150 | 175 | 200 | 225 | 250 | 275 | 300 |
|---|---|---|---|---|---|---|---|---|---|---|---|---|---|
| VELOCITY (FPS) | 900. | 880. | 860. | 842. | 825. | 808. | 792. | 776. | 761. | 746. | 732. | 718. | 705. |
| ENERGY (FT-LB) | 432. | 412. | 394. | 378. | 362. | 348. | 334. | 321. | 309. | 297. | 285. | 275. | 264. |
| DROP (IN) | .00 | -1.25 | -5.29 | -12.27 | -22.30 | -35.51 | -51.98 | -71.81 | -95.24 | -122.40 | -153.44 | -188.53 | -227.83 |
| MID-RANGE (IN) | .00 | .34 | 1.40 | 3.22 | 5.86 | 9.35 | 13.72 | 19.01 | 25.32 | 32.70 | 41.21 | 50.91 | 61.93 |
| BULLET PATH (IN) | -.63 | 1.09 | -.00 | -4.01 | -11.08 | -21.33 | -34.84 | -51.71 | -72.17 | -96.37 | -124.46 | -156.59 | -192.92 |
| TIME OF FLIGHT (SEC) | .000000 | .084304 | .170529 | .258649 | .348653 | .440536 | .534309 | .630003 | .727635 | .827218 | .928769 | 1.032309 | 1.137862 |
| WIND DEFLECTION (IN) | | | | | | | | | | | | | |
| 5 MPH | .00 | .09 | .34 | .76 | 1.35 | 2.10 | 3.02 | 4.11 | 5.37 | 6.80 | 8.40 | 10.18 | 12.13 |
| 10 MPH | .00 | .17 | .68 | 1.52 | 2.70 | 4.20 | 6.04 | 8.21 | 10.73 | 13.59 | 16.80 | 20.35 | 24.26 |
| 20 MPH | .00 | .34 | 1.36 | 3.04 | 5.39 | 8.40 | 12.08 | 16.43 | 21.46 | 27.18 | 33.59 | 40.71 | 48.53 |
| 30 MPH | .00 | .51 | 2.04 | 4.57 | 8.09 | 12.60 | 18.11 | 24.64 | 32.19 | 40.77 | 50.39 | 61.06 | 72.79 |
| VELOCITY (FPS) | 800. | 784. | 768. | 753. | 739. | 725. | 711. | 698. | 685. | 672. | 659. | 647. | 635. |
| ENERGY (FT-LB) | 341. | 327. | 315. | 302. | 291. | 280. | 269. | 259. | 250. | 240. | 232. | 223. | 215. |
| DROP (IN) | .00 | -1.46 | -6.41 | -15.01 | -27.43 | -43.80 | -64.30 | -89.06 | -118.27 | -152.11 | -190.74 | -234.37 | -283.17 |
| MID-RANGE (IN) | .00 | 42 | 1.74 | 4.02 | 7.31 | 11.66 | 17.14 | 23.80 | 31.71 | 40.94 | 51.57 | 63.66 | 77.29 |
| BULLET PATH (IN) | -.63 | 1.43 | -.00 | -5.09 | -13.99 | -26.84 | -43.82 | -65.07 | -90.76 | -121.08 | -156.20 | -196.30 | -241.59 |
| TIME OF FLIGHT (SEC) | .000000 | .094713 | .191356 | .289942 | .390485 | .493005 | .597525 | .704069 | .812666 | .923346 | 1.036144 | 1.151095 | 1.268238 |
| WIND DEFLECTION (IN) | | | | | | | | | | | | | |
| 5 MPH | .00 | .08 | .34 | .76 | 1.36 | 2.13 | 3.08 | 4.21 | 5.51 | 7.00 | 8.68 | 10.55 | 12.60 |
| 10 MPH | .00 | .17 | .68 | 1.53 | 2.73 | 4.27 | 6.16 | 8.42 | 11.03 | 14.01 | 17.36 | 21.09 | 25.21 |
| 20 MPH | .00 | .34 | 1.36 | 3.06 | 5.45 | 8.54 | 12.33 | 16.83 | 22.06 | 28.02 | 34.72 | 42.19 | 50.42 |
| 30 MPH | .00 | .51 | 2.04 | 4.59 | 8.18 | 12.81 | 18.49 | 25.25 | 33.09 | 42.03 | 52.08 | 63.28 | 75.63 |
| VELOCITY (FPS) | 700. | 687. | 674. | 661. | 649. | 637. | 625. | 614. | 602. | 591. | 580. | 570. | 559. |
| ENERGY (FT-LB) | 261. | 251. | 242. | 233. | 224. | 216. | 208. | 201. | 193. | 186. | 180. | 173. | 167. |
| DROP (IN) | .00 | -1.83 | -8.24 | -19.42 | -35.54 | -56.81 | -83.41 | -115.54 | -153.42 | -197.29 | -247.36 | -303.88 | -367.10 |
| MID-RANGE (IN) | .00 | .56 | 2.29 | 5.26 | 9.53 | 15.18 | 22.29 | 30.92 | 41.17 | 53.12 | 66.87 | 82.51 | 100.14 |
| BULLET PATH (IN) | -.63 | 1.98 | -.00 | -6.74 | -18.43 | -35.26 | -57.43 | -85.12 | -118.57 | -158.00 | -203.64 | -255.72 | -314.50 |
| TIME OF FLIGHT (SEC) | .000000 | .108170 | .218417 | .330771 | .445270 | .561949 | .680849 | .802012 | .925482 | 1.051305 | 1.179530 | 1.310206 | 1.443388 |
| WIND DEFLECTION (IN) | | | | | | | | | | | | | |
| 5 MPH | .00 | .09 | .36 | .82 | 1.47 | 2.31 | 3.34 | 4.58 | 6.01 | 7.66 | 9.51 | 11.58 | 13.88 |
| 10 MPH | .00 | .18 | .73 | 1.64 | 2.94 | 4.62 | 6.69 | 9.15 | 12.03 | 15.32 | 19.03 | 23.17 | 27.75 |
| 20 MPH | .00 | .36 | 1.45 | 3.29 | 5.88 | 9.23 | 13.37 | 18.31 | 24.06 | 30.63 | 38.05 | 46.34 | 55.50 |
| 30 MPH | .00 | .54 | 2.18 | 4.93 | 8.82 | 13.85 | 20.06 | 27.46 | 36.08 | 45.95 | 57.08 | 69.50 | 83.25 |

## Bullet: Hornady .452 Dia. 250 Gr. H.P.
## Ballistic Coefficient: .237

| RANGE (YDS) | MUZZLE | 25 | 50 | 75 | 100 | 125 | 150 | 175 | 200 | 225 | 250 | 275 | 300 |
|---|---|---|---|---|---|---|---|---|---|---|---|---|---|
| VELOCITY (FPS) | 1300. | 1249. | 1203. | 1161. | 1124. | 1090. | 1061. | 1035. | 1011. | 990. | 970. | 951. | 934. |
| ENERGY (FT-LB) | 938. | 866. | 803. | 748. | 701. | 660. | 625. | 594. | 568. | 544. | 522. | 502. | 484. |
| DROP (IN) | .00 | -.63 | -2.66 | -6.19 | -11.33 | -18.18 | -26.86 | -37.46 | -50.08 | -64.82 | -81.78 | -101.04 | -122.63 |
| MID-RANGE (IN) | .00 | .17 | .70 | 1.63 | 3.00 | 4.86 | 7.24 | 10.18 | 13.71 | 17.88 | 22.71 | 28.23 | 34.46 |
| BULLET PATH (IN) | -.63 | .39 | .00 | -1.89 | -5.38 | -10.59 | -17.62 | -26.58 | -37.56 | -50.65 | -65.96 | -83.58 | -103.53 |
| TIME OF FLIGHT (SEC) | .000000 | .058861 | .120050 | .183527 | .249213 | .316998 | .386750 | .458344 | .531677 | .606668 | .683246 | .761357 | .840953 |
| WIND DEFLECTION (IN) | | | | | | | | | | | | | |
| 5 MPH | .00 | .10 | .41 | .92 | 1.62 | 2.51 | 3.57 | 4.80 | 6.17 | 7.69 | 9.36 | 11.15 | 13.08 |
| 10 MPH | .00 | .21 | .82 | 1.84 | 3.25 | 5.02 | 7.14 | 9.59 | 12.34 | 15.39 | 18.71 | 22.31 | 26.16 |
| 20 MPH | .00 | .41 | 1.64 | 3.68 | 6.49 | 10.04 | 14.29 | 19.18 | 24.69 | 30.78 | 37.43 | 44.61 | 52.32 |
| 30 MPH | .00 | .62 | 2.46 | 5.52 | 9.74 | 15.07 | 21.43 | 28.77 | 37.03 | 46.17 | 56.14 | 66.92 | 78.48 |
| VELOCITY (FPS) | 1200. | 1158. | 1121. | 1088. | 1059. | 1033. | 1010. | 988. | 968. | 950. | 933. | 917. | 901. |
| ENERGY (FT-LB) | 799. | 745. | 698. | 657. | 623. | 592. | 566. | 542. | 521. | 501. | 483. | 466. | 451. |
| DROP (IN) | .00 | -.75 | -3.12 | -7.21 | -13.12 | -20.97 | -30.85 | -42.86 | -57.08 | -73.61 | -92.49 | -113.85 | -137.78 |
| MID-RANGE (IN) | .00 | .20 | .81 | 1.88 | 3.45 | 5.55 | 8.22 | 11.50 | 15.42 | 20.00 | 25.27 | 31.27 | 38.04 |
| BULLET PATH (IN) | -.63 | .49 | .00 | -2.22 | -6.26 | -12.24 | -20.24 | -30.37 | -42.72 | -57.38 | -74.39 | -93.87 | -115.94 |
| TIME OF FLIGHT (SEC) | .000000 | .063631 | .129465 | .197390 | .267271 | .338987 | .412436 | .487536 | .564221 | .642433 | .722128 | .803267 | .885820 |
| WIND DEFLECTION (IN) | | | | | | | | | | | | | |
| 5 MPH | .00 | .10 | .39 | .87 | 1.52 | 2.33 | 3.29 | 4.40 | 5.65 | 7.03 | 8.55 | 10.19 | 11.95 |
| 10 MPH | .00 | .20 | .79 | 1.74 | 3.04 | 4.66 | 6.59 | 8.81 | 11.30 | 14.07 | 17.09 | 20.37 | 23.90 |
| 20 MPH | .00 | .40 | 1.57 | 3.48 | 6.08 | 9.32 | 13.18 | 17.61 | 22.61 | 28.14 | 34.19 | 40.75 | 47.81 |
| 30 MPH | .00 | .60 | 2.36 | 5.22 | 9.12 | 13.99 | 19.77 | 26.42 | 33.91 | 42.20 | 51.28 | 61.12 | 71.71 |
| VELOCITY (FPS) | 1100. | 1070. | 1043. | 1018. | 996. | 976. | 957. | 939. | 923. | 907. | 892. | 878. | 864. |
| ENERGY (FT-LB) | 672. | 635. | 603. | 575. | 551. | 528. | 508. | 489. | 472. | 456. | 442. | 428. | 414. |
| DROP (IN) | .00 | -.88 | -3.66 | -8.43 | -15.29 | -24.33 | -35.65 | -49.30 | -65.38 | -84.00 | -105.25 | -129.22 | -156.00 |
| MID-RANGE (IN) | .00 | .23 | .95 | 2.19 | 3.98 | 6.38 | 9.40 | 13.06 | 17.40 | 22.47 | 28.30 | 34.91 | 42.35 |
| BULLET PATH (IN) | -.63 | .63 | .00 | -2.62 | -7.34 | -14.24 | -23.41 | -34.92 | -48.85 | -65.33 | -84.44 | -106.26 | -130.90 |
| TIME OF FLIGHT (SEC) | .000000 | .069163 | .140202 | .213011 | .287499 | .363595 | .441238 | .520379 | .600978 | .683001 | .766421 | .851218 | .937374 |
| WIND DEFLECTION (IN) | | | | | | | | | | | | | |
| 5 MPH | .00 | .09 | .34 | .74 | 1.30 | 2.00 | 2.83 | 3.79 | 4.89 | 6.10 | 7.45 | 8.91 | 10.49 |
| 10 MPH | .00 | .17 | .68 | 1.49 | 2.60 | 3.99 | 5.66 | 7.59 | 9.77 | 12.21 | 14.89 | 17.81 | 20.98 |
| 20 MPH | .00 | .35 | 1.35 | 2.98 | 5.20 | 7.99 | 11.32 | 15.17 | 19.54 | 24.42 | 29.78 | 35.63 | 41.96 |
| 30 MPH | .00 | .52 | 2.03 | 4.47 | 7.80 | 11.98 | 16.97 | 22.76 | 29.32 | 36.62 | 44.67 | 53.44 | 62.93 |
| VELOCITY (FPS) | 1000. | 979. | 960. | 942. | 925. | 910. | 895. | 880. | 867. | 853. | 841. | 828. | 816. |
| ENERGY (FT-LB) | 555. | 532. | 512. | 493. | 475. | 459. | 444. | 430. | 417. | 404. | 392. | 381. | 370. |
| DROP (IN) | .00 | -1.05 | -4.36 | -10.00 | -18.04 | -28.60 | -41.77 | -57.65 | -76.32 | -97.87 | -122.41 | -150.01 | -180.77 |
| MID-RANGE (IN) | .00 | .28 | 1.13 | 2.58 | 4.66 | 7.41 | 10.89 | 15.12 | 20.12 | 25.94 | 32.60 | 40.15 | 48.61 |
| BULLET PATH (IN) | -.63 | .81 | .00 | -3.14 | -8.69 | -16.76 | -27.44 | -40.82 | -56.99 | -76.06 | -98.10 | -123.21 | -151.48 |
| TIME OF FLIGHT (SEC) | .000000 | .075798 | .153155 | .232017 | .312343 | .394099 | .477256 | .561792 | .647690 | .734937 | .823525 | .913448 | 1.004704 |
| WIND DEFLECTION (IN) | | | | | | | | | | | | | |
| 5 MPH | .00 | .07 | .28 | .62 | 1.09 | 1.68 | 2.40 | 3.24 | 4.20 | 5.27 | 6.47 | 7.78 | 9.21 |
| 10 MPH | .00 | .14 | .56 | 1.23 | 2.17 | 3.36 | 4.80 | 6.48 | 8.39 | 10.55 | 12.94 | 15.57 | 18.43 |
| 20 MPH | .00 | .28 | 1.11 | 2.47 | 4.34 | 6.72 | 9.59 | 12.95 | 16.79 | 21.10 | 25.88 | 31.13 | 36.86 |
| 30 MPH | .00 | .42 | 1.67 | 3.70 | 6.52 | 10.08 | 14.39 | 19.43 | 25.18 | 31.65 | 38.82 | 46.70 | 55.28 |

## Bullet: Hornady .452 Dia. 250 Gr. H.P.
### Ballistic Coefficient: .237

| RANGE (YDS) | MUZZLE | 25 | 50 | 75 | 100 | 125 | 150 | 175 | 200 | 225 | 250 | 275 | 300 |
|---|---|---|---|---|---|---|---|---|---|---|---|---|---|
| VELOCITY (FPS) | 900. | 885. | 871. | 858. | 845. | 833. | 821. | 809. | 796. | 775. | 755. | 736. | 717. |
| ENERGY (FT-LB) | 450. | 435. | 421. | 409. | 396. | 385. | 374. | 363. | 351. | 333. | 316. | 301. | 286. |
| DROP (IN) | .00 | -1.19 | -5.14 | -11.94 | -21.69 | -34.47 | -50.38 | -69.51 | -91.95 | -117.80 | -147.26 | -180.50 | -217.75 |
| MID-RANGE (IN) | .00 | .34 | 1.38 | 3.16 | 5.71 | 9.06 | 13.25 | 18.31 | 24.28 | 31.20 | 39.15 | 48.22 | 58.49 |
| BULLET PATH (IN) | -.63 | 1.06 | -.00 | -3.92 | -10.78 | -20.68 | -33.70 | -49.94 | -69.50 | -92.47 | -119.03 | -149.40 | -183.76 |
| TIME OF FLIGHT (SEC) | .000000 | .084023 | .169413 | .256156 | .344241 | .433662 | .524415 | .616502 | .709939 | .805498 | .903616 | 1.004318 | 1.107640 |
| WIND DEFLECTION (IN) | | | | | | | | | | | | | |
| 5 MPH | .00 | .06 | .24 | .54 | .96 | 1.50 | 2.15 | 2.92 | 3.81 | 4.88 | 6.18 | 7.71 | 9.47 |
| 10 MPH | .00 | .12 | .48 | 1.08 | 1.92 | 2.99 | 4.30 | 5.84 | 7.62 | 9.77 | 12.37 | 15.43 | 18.94 |
| 20 MPH | .00 | 24 | .97 | 2.17 | 3.84 | 5.98 | 8.59 | 11.68 | 15.23 | 19.54 | 24.74 | 30.85 | 37.89 |
| 30 MPH | .00 | .36 | 1.45 | 3.25 | 5.76 | 8.97 | 12.89 | 17.51 | 22.85 | 29.30 | 37.11 | 46.28 | 56.83 |
| VELOCITY (FPS) | 800. | 779. | 759. | 739. | 721. | 703. | 685. | 669. | 652. | 636. | 621. | 606. | 591. |
| ENERGY (FT-LB) | 355. | 337. | 319. | 303. | 288. | 274. | 261. | 248. | 236. | 225. | 214. | 204. | 194. |
| DROP (IN) | .00 | -1.51 | -6.58 | -15.40 | -28.19 | -45.14 | -66.47 | -92.37 | -123.11 | -158.94 | -200.13 | -246.94 | -299.67 |
| MID-RANGE (IN) | .00 | .43 | 1.78 | 4.11 | 7.51 | 12.06 | 17.83 | 24.90 | 33.37 | 43.34 | 54.93 | 68.23 | 83.36 |
| BULLET PATH (IN) | -.63 | 1.46 | -.00 | -5.22 | -14.40 | -27.75 | -45.47 | -67.77 | -94.90 | -127.13 | -164.71 | -207.92 | -257.05 |
| TIME OF FLIGHT (SEC) | .000000 | .095021 | .192595 | .292746 | .395506 | .500913 | .609014 | .719861 | .833514 | .950035 | 1.069495 | 1.191967 | 1.317530 |
| WIND DEFLECTION (IN) | | | | | | | | | | | | | |
| 5 MPH | .00 | .11 | .45 | 1.01 | 1.80 | 2.83 | 4.09 | 5.60 | 7.35 | 9.35 | 11.62 | 14.14 | 16.94 |
| 10 MPH | .00 | .22 | .90 | 2.02 | 3.61 | 5.66 | 8.19 | 11.20 | 14.70 | 18.71 | 23.23 | 28.29 | 33.89 |
| 20 MPH | .00 | .45 | 1.79 | 4.05 | 7.22 | 11.32 | 16.37 | 22.39 | 29.40 | 37.41 | 46.46 | 56.57 | 67.77 |
| 30 MPH | .00 | .67 | 2.69 | 6.07 | 10.83 | 16.98 | 24.56 | 33.59 | 44.10 | 56.12 | 69.69 | 84.86 | 101.66 |
| VELOCITY (FPS) | 700. | 683. | 666. | 649. | 634. | 618. | 603. | 588. | 574. | 560. | 546. | 533. | 520. |
| ENERGY (FT-LB) | 272. | 259. | 246. | 234. | 223. | 212. | 202. | 192. | 183. | 174. | 166. | 158. | 150. |
| DROP (IN) | .00 | -1.95 | -8.54 | -20.01 | -36.62 | -58.62 | -86.30 | -119.87 | -159.70 | -206.12 | -259.47 | -320.09 | -388.37 |
| MID-RANGE (IN) | .00 | .57 | 2.32 | 5.36 | 9.77 | 15.66 | 23.14 | 32.28 | 43.23 | 56.13 | 71.11 | 88.31 | 107.89 |
| BULLET PATH (IN) | -.63 | 2.01 | -.00 | -6.89 | -18.91 | -36.33 | -59.42 | -88.41 | -123.66 | -165.50 | -214.26 | -270.30 | -334.00 |
| TIME OF FLIGHT (SEC) | .000000 | .108501 | .219757 | .333824 | .450769 | .570659 | .693571 | .819583 | .948779 | 1.081247 | 1.217081 | 1.356377 | 1.499239 |
| WIND DEFLECTION (IN) | | | | | | | | | | | | | |
| 5 MPH | .00 | .12 | .48 | 1.09 | 1.95 | 3.08 | 4.46 | 6.12 | 8.06 | 10.29 | 12.82 | 15.65 | 18.79 |
| 10 MPH | .00 | .24 | .96 | 2.18 | 3.91 | 6.15 | 8.93 | 12.25 | 16.13 | 20.59 | 25.63 | 31.29 | 37.58 |
| 20 MPH | .00 | .48 | 1.93 | 4.36 | 7.81 | 12.30 | 17.85 | 24.49 | 32.26 | 41.17 | 51.27 | 62.59 | 75.16 |
| 30 MPH | .00 | .72 | 2.89 | 6.54 | 11.72 | 18.45 | 26.78 | 36.74 | 48.38 | 61.76 | 76.90 | 93.88 | 112.74 |

# Bullet: Lyman #45468 161 Gr.
## Ballistic Coefficient: .105

| RANGE (YDS) | MUZZLE | 25 | 50 | 75 | 100 | 125 | 150 | 175 | 200 | 225 | 250 | 275 | 300 |
|---|---|---|---|---|---|---|---|---|---|---|---|---|---|
| VELOCITY (FPS) | 1400. | 1276. | 1172. | 1091. | 1029. | 979. | 938. | 902. | 870. | 840. | 813. | 790. | 770. |
| ENERGY (FT-LB) | 701. | 582. | 491. | 425. | 378. | 343. | 314. | 291. | 270. | 252. | 236. | 223. | 212. |
| DROP (IN) | .00 | -.58 | -2.49 | -5.99 | -11.30 | -18.66 | -28.28 | -40.32 | -55.03 | -72.61 | -93.25 | -117.11 | -144.35 |
| MID-RANGE (IN) | .00 | .15 | .67 | 1.64 | 3.16 | 5.30 | 8.15 | 11.76 | 16.21 | 21.59 | 27.96 | 35.37 | 43.90 |
| BULLET PATH (IN) | -.63 | .35 | .00 | -1.93 | -5.69 | -11.49 | -19.53 | -30.02 | -43.17 | -59.18 | -78.26 | -100.56 | -126.23 |
| TIME OF FLIGHT (SEC) | .000000 | .056168 | .117579 | .184006 | .254882 | .329649 | .407941 | .489522 | .574244 | .662019 | .752809 | .846538 | .942787 |
| WIND DEFLECTION (IN) | | | | | | | | | | | | | |
| 5 MPH | .00 | .23 | .92 | 2.05 | 3.57 | 5.44 | 7.61 | 10.08 | 12.82 | 15.83 | 19.10 | 22.64 | 26.39 |
| 10 MPH | .00 | .46 | 1.84 | 4.10 | 7.14 | 10.88 | 15.23 | 20.16 | 25.64 | 31.66 | 38.21 | 45.28 | 52.79 |
| 20 MPH | .00 | .91 | 3.67 | 8.20 | 14.29 | 21.75 | 30.45 | 40.31 | 51.28 | 63.32 | 76.42 | 90.55 | 105.58 |
| 30 MPH | .00 | 1.37 | 5.51 | 12.30 | 21.43 | 32.63 | 45.68 | 60.47 | 76.91 | 94.97 | 114.63 | 135.83 | 158.36 |
| VELOCITY (FPS) | 1300. | 1192. | 1106. | 1041. | 989. | 946. | 909. | 876. | 846. | 818. | 794. | 774. | 755. |
| ENERGY (FT-LB) | 604. | 508. | 437. | 387. | 350. | 320. | 295. | 274. | 256. | 239. | 225. | 214. | 203. |
| DROP (IN) | .00 | -.67 | -2.87 | -6.84 | -12.81 | -20.99 | -31.57 | -44.76 | -60.77 | -79.81 | -102.06 | -127.61 | -156.78 |
| MID-RANGE (IN) | .00 | .18 | .76 | 1.86 | 3.53 | 5.88 | 8.94 | 12.82 | 17.58 | 23.29 | 30.03 | 37.84 | 46.82 |
| BULLET PATH (IN) | -.63 | .45 | .00 | -2.22 | -6.44 | -12.87 | -21.69 | -33.14 | -47.40 | -64.69 | -85.18 | -108.98 | -136.40 |
| TIME OF FLIGHT (SEC) | .000000 | .060323 | .125743 | .195744 | .269734 | .347314 | .428224 | .512299 | .599442 | .689603 | .782757 | .878514 | .976734 |
| WIND DEFLECTION (IN) | | | | | | | | | | | | | |
| 5 MPH | .00 | .23 | .91 | 1.99 | 3.43 | 5.18 | 7.22 | 9.54 | 12.14 | 14.99 | 18.11 | 21.46 | 25.03 |
| 10 MPH | .00 | .46 | 1.82 | 3.99 | 6.86 | 10.36 | 14.44 | 19.09 | 24.27 | 29.99 | 36.23 | 42.93 | 50.06 |
| 20 MPH | .00 | .93 | 3.65 | 7.98 | 13.72 | 20.72 | 28.89 | 38.18 | 48.54 | 59.97 | 72.45 | 85.85 | 100.12 |
| 30 MPH | .00 | 1.39 | 5.47 | 11.97 | 20.57 | 31.07 | 43.33 | 57.26 | 72.81 | 89.96 | 108.68 | 128.78 | 150.18 |
| VELOCITY (FPS) | 1200. | 1112. | 1045. | 993. | 949. | 912. | 879. | 849. | 821. | 795. | 775. | 756. | 737. |
| ENERGY (FT-LB) | 515. | 442. | 391. | 352. | 322. | 297. | 276. | 257. | 241. | 226. | 215. | 204. | 194. |
| DROP (IN) | .00 | -.78 | -3.32 | -7.84 | -14.55 | -23.63 | -35.32 | -49.81 | -67.31 | -88.03 | -112.01 | -139.60 | -170.96 |
| MID-RANGE (IN) | .00 | .20 | .87 | 2.10 | 3.95 | 6.50 | 9.83 | 14.00 | 19.11 | 25.23 | 32.37 | 40.66 | 50.16 |
| BULLET PATH (IN) | -.63 | .56 | .00 | -2.54 | -7.28 | -14.39 | -24.10 | -36.62 | -52.14 | -70.88 | -92.89 | -118.50 | -147.89 |
| TIME OF FLIGHT (SEC) | .000000 | .065010 | .134653 | .208328 | .285620 | .366258 | .450073 | .536961 | .626871 | .719785 | .815360 | .913395 | 1.013914 |
| WIND DEFLECTION (IN) | | | | | | | | | | | | | |
| 5 MPH | .00 | .22 | .85 | 1.83 | 3.13 | 4.73 | 6.61 | 8.75 | 11.16 | 13.84 | 16.75 | 19.88 | 23.22 |
| 10 MPH | .00 | .44 | 1.70 | 3.67 | 6.27 | 9.46 | 13.21 | 17.51 | 22.33 | 27.68 | 33.50 | 39.76 | 46.45 |
| 20 MPH | .00 | .88 | 3.40 | 7.33 | 12.54 | 18.92 | 26.43 | 35.01 | 44.66 | 55.36 | 67.01 | 79.52 | 92.90 |
| 30 MPH | .00 | 1.33 | 5.10 | 11.00 | 18.81 | 28.38 | 39.64 | 52.52 | 66.99 | 83.05 | 100.51 | 119.27 | 139.35 |
| VELOCITY (FPS) | 1100. | 1036. | 985. | 943. | 906. | 874. | 844. | 816. | 792. | 772. | 753. | 735. | 717. |
| ENERGY (FT-LB) | 432. | 384. | 347. | 318. | 294. | 273. | 254. | 238. | 224. | 213. | 203. | 193. | 184. |
| DROP (IN) | .00 | -.92 | -3.86 | -9.02 | -16.60 | -26.81 | -39.85 | -55.94 | -75.23 | -97.87 | -124.12 | -154.19 | -188.26 |
| MID-RANGE (IN) | .00 | .24 | 1.01 | 2.39 | 4.44 | 7.25 | 10.91 | 15.48 | 21.02 | 27.57 | 35.26 | 44.13 | 54.28 |
| BULLET PATH (IN) | -.63 | .69 | .00 | -2.92 | -8.25 | -16.22 | -27.02 | -40.86 | -57.91 | -78.30 | -102.32 | -130.14 | -161.97 |
| TIME OF FLIGHT (SEC) | .000000 | .070336 | .144624 | .222475 | .303640 | .387960 | .475342 | .565740 | .659107 | .755031 | .853418 | .954292 | 1.057685 |
| WIND DEFLECTION (IN) | | | | | | | | | | | | | |
| 5 MPH | .00 | .19 | .73 | 1.58 | 2.72 | 4.14 | 5.83 | 7.79 | 10.00 | 12.44 | 15.10 | 17.98 | 21.08 |
| 10 MPH | .00 | .38 | 1.45 | 3.16 | 5.44 | 8.28 | 11.66 | 15.57 | 20.00 | 24.89 | 30.20 | 35.96 | 42.15 |
| 20 MPH | .00 | .76 | 2.91 | 6.31 | 10.88 | 16.56 | 23.32 | 31.14 | 40.01 | 49.77 | 60.40 | 71.91 | 84.31 |
| 30 MPH | .00 | 1.14 | 4.36 | 9.47 | 16.32 | 24.84 | 34.98 | 46.71 | 60.01 | 74.66 | 90.60 | 107.87 | 126.46 |

## Bullet: Lyman #45468 161 Gr.
### Ballistic Coefficient: .105

| RANGE (YDS) | MUZZLE | 25 | 50 | 75 | 100 | 125 | 150 | 175 | 200 | 225 | 250 | 275 | 300 |
|---|---|---|---|---|---|---|---|---|---|---|---|---|---|
| VELOCITY (FPS) | 1000. | 955. | 917. | 883. | 853. | 825. | 798. | 778. | 759. | 740. | 722. | 705. | 688. |
| ENERGY (FT-LB) | 357. | 326. | 301. | 279. | 260. | 243. | 228. | 216. | 206. | 196. | 186. | 178. | 169. |
| DROP (IN) | .00 | -1.10 | -4.54 | -10.55 | -19.34 | -31.10 | -46.05 | -64.24 | -86.00 | -111.51 | -140.97 | -174.58 | -212.54 |
| MID-RANGE (IN) | .00 | .28 | 1.17 | 2.76 | 5.13 | 8.34 | 12.47 | 17.53 | 23.66 | 30.92 | 39.38 | 49.12 | 60.22 |
| BULLET PATH (IN) | -.63 | .86 | .00 | -3.43 | -9.63 | -18.81 | -31.17 | -46.78 | -65.95 | -88.88 | -115.75 | -146.77 | -182.15 |
| TIME OF FLIGHT (SEC) | .000000 | .076775 | .156929 | .240280 | .326716 | .416177 | .508647 | .603846 | .701501 | .801636 | .904279 | 1.009467 | 1.117243 |
| WIND DEFLECTION (IN) | | | | | | | | | | | | | |
| 5 MPH | .00 | .16 | .61 | 1.34 | 2.35 | 3.62 | 5.16 | 6.94 | 8.93 | 11.14 | 13.58 | 16.23 | 19.12 |
| 10 MPH | .00 | .31 | 1.22 | 2.69 | 4.70 | 7.25 | 10.32 | 13.88 | 17.86 | 22.29 | 27.15 | 32.47 | 38.23 |
| 20 MPH | .00 | .62 | 2.44 | 5.38 | 9.40 | 14.49 | 20.64 | 27.75 | 35.73 | 44.58 | 54.31 | 64.93 | 76.47 |
| 30 MPH | .00 | .94 | 3.66 | 8.07 | 14.11 | 21.74 | 30.97 | 41.63 | 53.59 | 66.86 | 81.46 | 97.40 | 114.70 |
| VELOCITY (FPS) | 900. | 868. | 839. | 811. | 788. | 768. | 749. | 731. | 714. | 697. | 680. | 664. | 648. |
| ENERGY (FT-LB) | 290. | 269. | 251. | 235. | 222. | 211. | 201. | 191. | 182. | 173. | 165. | 158. | 150. |
| DROP (IN) | .00 | -1.30 | -5.48 | -12.73 | -23.21 | -37.11 | -54.68 | -76.06 | -101.48 | -131.15 | -165.28 | -204.09 | -247.82 |
| MID-RANGE (IN) | .00 | .35 | 1.44 | 3.35 | 6.13 | 9.85 | 14.61 | 20.45 | 27.53 | 35.86 | 45.53 | 56.62 | 69.23 |
| BULLET PATH (IN) | -.63 | 1.12 | -.00 | -4.20 | -11.63 | -22.48 | -36.99 | -55.32 | -77.68 | -104.30 | -135.38 | -171.14 | -211.81 |
| TIME OF FLIGHT (SEC) | .000000 | .084881 | .172812 | .263755 | .357638 | .454039 | .552907 | .654266 | .758151 | .864599 | .973658 | 1.085376 | 1.199811 |
| WIND DEFLECTION (IN) | | | | | | | | | | | | | |
| 5 MPH | .00 | .14 | .54 | 1.21 | 2.14 | 3.29 | 4.66 | 6.24 | 8.05 | 10.08 | 12.35 | 14.85 | 17.58 |
| 10 MPH | .00 | .27 | 1.08 | 2.42 | 4.28 | 6.58 | 9.31 | 12.48 | 16.10 | 20.17 | 24.70 | 29.69 | 35.17 |
| 20 MPH | .00 | 54 | 2.16 | 4.84 | 8.56 | 13.16 | 18.62 | 24.97 | 32.20 | 40.34 | 49.39 | 59.39 | 70.33 |
| 30 MPH | .00 | .82 | 3.24 | 7.26 | 12.83 | 19.73 | 27.93 | 37.45 | 48.30 | 60.51 | 74.09 | 89.08 | 105.50 |
| VELOCITY (FPS) | 800. | 780. | 760. | 742. | 724. | 706. | 689. | 673. | 657. | 642. | 627. | 612. | 598. |
| ENERGY (FT-LB) | 229. | 217. | 207. | 197. | 187. | 178. | 170. | 162. | 154. | 147. | 140. | 134. | 128. |
| DROP (IN) | .00 | -1.50 | -6.56 | -15.35 | -28.08 | -44.95 | -66.16 | -91.85 | -122.33 | -157.83 | -198.58 | -244.86 | -296.92 |
| MID-RANGE (IN) | .00 | .43 | 1.77 | 4.10 | 7.49 | 12.00 | 17.73 | 24.71 | 33.08 | 42.93 | 54.34 | 67.43 | 82.31 |
| BULLET PATH (IN) | -.63 | 1.46 | -.00 | -5.20 | -14.34 | -27.61 | -45.23 | -67.33 | -94.22 | -126.12 | -163.28 | -205.96 | -254.44 |
| TIME OF FLIGHT (SEC) | .000000 | .094972 | .192399 | .292300 | .394706 | .499652 | .607179 | .717335 | .830172 | .945748 | 1.064127 | 1.185373 | 1.309560 |
| WIND DEFLECTION (IN) | | | | | | | | | | | | | |
| 5 MPH | .00 | .11 | .43 | .97 | 1.73 | 2.72 | 3.93 | 5.38 | 7.06 | 8.98 | 11.14 | 13.56 | 16.24 |
| 10 MPH | .00 | .22 | .86 | 1.94 | 3.47 | 5.44 | 7.86 | 10.75 | 14.11 | 17.95 | 22.29 | 27.13 | 32.48 |
| 20 MPH | .00 | .43 | 1.72 | 3.89 | 6.94 | 10.88 | 15.73 | 21.50 | 28.22 | 35.90 | 44.57 | 54.25 | 64.97 |
| 30 MPH | .00 | .65 | 2.59 | 5.83 | 10.40 | 16.32 | 23.59 | 32.25 | 42.33 | 53.86 | 66.86 | 81.38 | 97.45 |

## Bullet: Lyman #454190 238 Gr.
### Ballistic Coefficient: .269

| RANGE (YDS) | MUZZLE | 25 | 50 | 75 | 100 | 125 | 150 | 175 | 200 | 225 | 250 | 275 | 300 |
|---|---|---|---|---|---|---|---|---|---|---|---|---|---|
| VELOCITY (FPS) | 1300. | 1255. | 1214. | 1176. | 1141. | 1110. | 1082. | 1056. | 1034. | 1013. | 994. | 976. | 959. |
| ENERGY (FT-LB) | 893. | 832. | 778. | 730. | 688. | 650. | 618. | 590. | 565. | 542. | 522. | 503. | 486. |
| DROP (IN) | .00 | -.62 | -2.63 | -6.12 | -11.18 | -17.90 | -26.38 | -36.71 | -48.98 | -63.28 | -79.70 | -98.30 | -119.19 |
| MID-RANGE (IN) | .00 | .17 | .69 | 1.61 | 2.96 | 4.76 | 7.07 | 9.91 | 13.31 | 17.32 | 21.95 | 27.24 | 33.21 |
| BULLET PATH (IN) | -.63 | .38 | .00 | -1.86 | -5.28 | -10.37 | -17.22 | -25.92 | -36.56 | -49.23 | -64.01 | -80.99 | -100.24 |
| TIME OF FLIGHT (SEC) | .000000 | .058722 | .119498 | .182301 | .247080 | .313759 | .382241 | .452421 | .524208 | .597524 | .672304 | .748495 | .826049 |
| WIND DEFLECTION (IN) | | | | | | | | | | | | | |
| 5 MPH | .00 | .09 | .36 | .81 | 1.44 | 2.23 | 3.18 | 4.27 | 5.51 | 6.89 | 8.39 | 10.02 | 11.77 |
| 10 MPH | .00 | .18 | .72 | 1.62 | 2.87 | 4.45 | 6.35 | 8.55 | 11.03 | 13.78 | 16.79 | 20.04 | 23.54 |
| 20 MPH | .00 | .36 | 1.45 | 3.25 | 5.74 | 8.90 | 12.70 | 17.10 | 22.06 | 27.56 | 33.57 | 40.09 | 47.08 |
| 30 MPH | .00 | .54 | 2.17 | 4.87 | 8.61 | 13.36 | 19.05 | 25.65 | 33.09 | 41.34 | 50.36 | 60.13 | 70.62 |
| VELOCITY (FPS) | 1200. | 1163. | 1130. | 1099. | 1072. | 1048. | 1026. | 1006. | 987. | 970. | 953. | 938. | 923. |
| ENERGY (FT-LB) | 761. | 715. | 674. | 639. | 608. | 581. | 556. | 535. | 515. | 497. | 480. | 465. | 451. |
| DROP (IN) | .00 | -.75 | -3.09 | -7.14 | -12.98 | -20.70 | -30.39 | -42.14 | -56.03 | -72.14 | -90.56 | -111.31 | -134.50 |
| MID-RANGE (IN) | .00 | .19 | .80 | 1.86 | 3.40 | 5.45 | 8.06 | 11.24 | 15.04 | 19.47 | 24.58 | 30.36 | 36.86 |
| BULLET PATH (IN) | -.63 | .49 | .00 | -2.18 | -6.16 | -12.02 | -19.85 | -29.74 | -41.76 | -56.02 | -72.57 | -91.46 | -112.78 |
| TIME OF FLIGHT (SEC) | .000000 | .063498 | .128947 | .196265 | .265349 | .336098 | .408424 | .482255 | .557530 | .634198 | .712216 | .791546 | .872160 |
| WIND DEFLECTION (IN) | | | | | | | | | | | | | |
| 5 MPH | .00 | .09 | .35 | .77 | 1.35 | 2.08 | 2.94 | 3.94 | 5.06 | 6.31 | 7.68 | 9.16 | 10.75 |
| 10 MPH | .00 | .18 | .69 | 1.54 | 2.70 | 4.15 | 5.88 | 7.88 | 10.13 | 12.62 | 15.35 | 18.31 | 21.50 |
| 20 MPH | .00 | .35 | 1.39 | 3.09 | 5.40 | 8.31 | 11.77 | 15.75 | 20.25 | 25.24 | 30.70 | 36.62 | 43.00 |
| 30 MPH | .00 | .53 | 2.08 | 4.63 | 8.10 | 12.46 | 17.65 | 23.63 | 30.38 | 37.86 | 46.05 | 54.94 | 64.50 |
| VELOCITY (FPS) | 1100. | 1073. | 1049. | 1027. | 1006. | 988. | 970. | 954. | 938. | 924. | 910. | 897. | 884. |
| ENERGY (FT-LB) | 639. | 608. | 581. | 557. | 535. | 515. | 497. | 481. | 465. | 451. | 437. | 425. | 413. |
| DROP (IN) | .00 | -.88 | -3.63 | -8.36 | -15.14 | -24.07 | -35.21 | -48.66 | -64.43 | -82.64 | -103.39 | -126.77 | -152.83 |
| MID-RANGE (IN) | .00 | .23 | .94 | 2.17 | 3.94 | 6.29 | 9.24 | 12.84 | 17.07 | 21.99 | 27.63 | 34.03 | 41.20 |
| BULLET PATH (IN) | -.63 | .63 | .00 | -2.60 | -7.25 | -14.04 | -23.05 | -34.36 | -48.00 | -64.08 | -82.70 | -103.94 | -127.88 |
| TIME OF FLIGHT (SEC) | .000000 | .069048 | .139763 | .212056 | .285855 | .361099 | .437737 | .515725 | .595026 | .675609 | .757449 | .840525 | .924819 |
| WIND DEFLECTION (IN) | | | | | | | | | | | | | |
| 5 MPH | .00 | .08 | .30 | .66 | 1.16 | 1.78 | 2.52 | 3.38 | 4.36 | 5.45 | 6.66 | 7.97 | 9.38 |
| 10 MPH | .00 | .15 | .60 | 1.32 | 2.31 | 3.55 | 5.04 | 6.77 | 8.72 | 10.91 | 13.31 | 15.93 | 18.77 |
| 20 MPH | .00 | .30 | 1.20 | 2.64 | 4.62 | 7.11 | 10.08 | 13.54 | 17.45 | 21.81 | 26.62 | 31.86 | 37.54 |
| 30 MPH | .00 | .46 | 1.79 | 3.97 | 6.93 | 10.66 | 15.13 | 20.30 | 26.17 | 32.72 | 39.93 | 47.80 | 56.30 |
| VELOCITY (FPS) | 1000. | 982. | 965. | 948. | 933. | 919. | 905. | 892. | 880. | 868. | 856. | 845. | 834. |
| ENERGY (FT-LB) | 528. | 509. | 492. | 475. | 460. | 446. | 433. | 421. | 409. | 398. | 387. | 377. | 367. |
| DROP (IN) | .00 | -1.04 | -4.33 | -9.95 | -17.90 | -28.34 | -41.34 | -56.99 | -75.35 | -96.52 | -120.56 | -147.56 | -177.60 |
| MID-RANGE (IN) | .00 | .28 | 1.12 | 2.57 | 4.62 | 7.32 | 10.72 | 14.87 | 19.77 | 25.45 | 31.94 | 39.27 | 47.46 |
| BULLET PATH (IN) | -.63 | .81 | .00 | -3.14 | -8.61 | -16.56 | -27.08 | -40.24 | -56.13 | -74.81 | -96.37 | -120.89 | -148.44 |
| TIME OF FLIGHT (SEC) | .000000 | .075705 | .152787 | .231207 | .310929 | .391923 | .474167 | .557639 | .642323 | .728207 | .815280 | .903536 | .992969 |
| WIND DEFLECTION (IN) | | | | | | | | | | | | | |
| 5 MPH | .00 | .06 | .25 | .55 | .96 | 1.49 | 2.13 | 2.87 | 3.72 | 4.68 | 5.74 | 6.91 | 8.18 |
| 10 MPH | .00 | .12 | .49 | 1.09 | 1.92 | 2.98 | 4.25 | 5.74 | 7.45 | 9.36 | 11.49 | 13.82 | 16.36 |
| 20 MPH | .00 | .25 | .98 | 2.18 | 3.85 | 5.96 | 8.51 | 11.49 | 14.90 | 18.73 | 22.98 | 27.64 | 32.72 |
| 30 MPH | .00 | .37 | 1.47 | 3.28 | 5.77 | 8.94 | 12.76 | 17.23 | 22.35 | 28.09 | 34.47 | 41.47 | 49.09 |

## Bullet: Lyman #454190 238 Gr.
**Ballistic Coefficient: .269**

| RANGE (YDS) | MUZZLE | 25 | 50 | 75 | 100 | 125 | 150 | 175 | 200 | 225 | 250 | 275 | 300 |
|---|---|---|---|---|---|---|---|---|---|---|---|---|---|
| VELOCITY (FPS) | 900. | 887. | 875. | 863. | 851. | 840. | 829. | 819. | 808. | 798. | 789. | 780. | 771. |
| ENERGY (FT-LB) | 428. | 416. | 404. | 393. | 383. | 373. | 363. | 354. | 345. | 337. | 329. | 321. | 314. |
| DROP (IN) | .00 | -1.18 | -5.12 | -11.88 | -21.56 | -34.23 | -49.97 | -68.84 | -90.95 | -116.36 | -144.97 | -177.05 | -212.70 |
| MID-RANGE (IN) | .00 | .34 | 1.38 | 3.14 | 5.66 | 8.98 | 13.10 | 18.07 | 23.91 | 30.66 | 38.25 | 46.81 | 56.38 |
| BULLET PATH (IN) | -.63 | 1.06 | -.00 | -3.89 | -10.70 | -20.50 | -33.37 | -49.36 | -68.59 | -91.14 | -116.87 | -146.08 | -178.86 |
| TIME OF FLIGHT (SEC) | .000000 | .083941 | .169089 | .255431 | .342958 | .431665 | .521547 | .612603 | .704833 | .798241 | .892806 | .988512 | 1.085363 |
| WIND DEFLECTION (IN) | | | | | | | | | | | | | |
| 5 MPH | .00 | .05 | .21 | .48 | .85 | 1.32 | 1.90 | 2.58 | 3.36 | 4.25 | 5.23 | 6.32 | 7.51 |
| 10 MPH | .00 | .11 | .43 | .96 | 1.69 | 2.64 | 3.79 | 5.15 | 6.72 | 8.49 | 10.47 | 12.64 | 15.02 |
| 20 MPH | .00 | .21 | .85 | 1.91 | 3.39 | 5.28 | 7.58 | 10.30 | 13.43 | 16.98 | 20.93 | 25.29 | 30.05 |
| 30 MPH | .00 | .32 | 1.28 | 2.87 | 5.08 | 7.92 | 11.38 | 15.45 | 20.15 | 25.47 | 31.40 | 37.93 | 45.07 |
| VELOCITY (FPS) | 800. | 790. | 781. | 772. | 763. | 754. | 745. | 737. | 729. | 721. | 713. | 705. | 697. |
| ENERGY (FT-LB) | 338. | 330. | 322. | 315. | 308. | 300. | 294. | 287. | 281. | 274. | 268. | 262. | 257. |
| DROP (IN) | .00 | -1.38 | -6.12 | -14.41 | -26.34 | -42.00 | -61.46 | -84.74 | -112.01 | -143.36 | -178.88 | -218.67 | -262.83 |
| MID-RANGE (IN) | .00 | .38 | 1.68 | 3.90 | 7.05 | 11.18 | 16.31 | 22.45 | 29.66 | 37.98 | 47.44 | 58.09 | 69.96 |
| BULLET PATH (IN) | -.63 | 1.37 | -.00 | -4.92 | -13.47 | -25.75 | -41.84 | -61.75 | -85.64 | -113.61 | -145.76 | -182.18 | -222.96 |
| TIME OF FLIGHT (SEC) | .000000 | .094319 | .189778 | .286379 | .384127 | .483027 | .583085 | .684307 | .786699 | .890271 | .995031 | 1.100987 | 1.208150 |
| WIND DEFLECTION (IN) | | | | | | | | | | | | | |
| 5 MPH | .00 | .05 | .20 | .45 | .80 | 1.26 | 1.81 | 2.47 | 3.23 | 4.09 | 5.06 | 6.14 | 7.32 |
| 10 MPH | .00 | .10 | .40 | .90 | 1.61 | 2.51 | 3.62 | 4.94 | 6.46 | 8.19 | 10.13 | 12.27 | 14.63 |
| 20 MPH | .00 | .20 | .80 | 1.81 | 3.21 | 5.03 | 7.25 | 9.88 | 12.92 | 16.38 | 20.25 | 24.55 | 29.27 |
| 30 MPH | .00 | .30 | 1.20 | 2.71 | 4.82 | 7.54 | 10.87 | 14.81 | 19.38 | 24.56 | 30.38 | 36.82 | 43.90 |
| VELOCITY (FPS) | 700. | 692. | 684. | 677. | 669. | 662. | 655. | 648. | 641. | 634. | 627. | 620. | 613. |
| ENERGY (FT-LB) | 259. | 253. | 248. | 242. | 237. | 232. | 227. | 222. | 217. | 212. | 208. | 203. | 199. |
| DROP (IN) | .00 | -1.67 | -7.86 | -18.68 | -34.22 | -54.60 | -79.92 | -110.18 | -145.60 | -186.31 | -232.42 | -284.06 | -341.35 |
| MID-RANGE (IN) | .00 | .56 | 2.26 | 5.14 | 9.25 | 14.61 | 21.28 | 29.24 | 38.58 | 49.36 | 61.61 | 75.40 | 90.76 |
| BULLET PATH (IN) | -.63 | 1.95 | -.00 | -6.57 | -17.87 | -34.00 | -55.08 | -81.09 | -112.27 | -148.73 | -190.60 | -237.99 | -291.04 |
| TIME OF FLIGHT (SEC) | .000000 | .107749 | .216719 | .326919 | .438362 | .551060 | .665024 | .780268 | .896806 | 1.014652 | 1.133820 | 1.254325 | 1.376183 |
| WIND DEFLECTION (IN) | | | | | | | | | | | | | |
| 5 MPH | .00 | .05 | .21 | .48 | .86 | 1.35 | 1.95 | 2.66 | 3.49 | 4.43 | 5.49 | 6.67 | 7.96 |
| 10 MPH | .00 | .11 | .43 | .97 | 1.72 | 2.70 | 3.90 | 5.33 | 6.98 | 8.86 | 10.98 | 13.33 | 15.92 |
| 20 MPH | .00 | .21 | .86 | 1.93 | 3.45 | 5.40 | 7.80 | 10.65 | 13.96 | 17.73 | 21.96 | 26.67 | 31.85 |
| 30 MPH | .00 | .32 | 1.28 | 2.90 | 5.17 | 8.10 | 11.70 | 15.98 | 20.94 | 26.59 | 32.94 | 40.00 | 47.77 |

## Bullet: Lyman #454424 245 Gr.
### Ballistic Coefficient: .210

| RANGE (YDS) | MUZZLE | 25 | 50 | 75 | 100 | 125 | 150 | 175 | 200 | 225 | 250 | 275 | 300 |
|---|---|---|---|---|---|---|---|---|---|---|---|---|---|
| VELOCITY (FPS) | 1300. | 1243. | 1192. | 1146. | 1106. | 1071. | 1041. | 1013. | 989. | 967. | 946. | 927. | 909. |
| ENERGY (FT-LB) | 919. | 841. | 773. | 714. | 665. | 624. | 589. | 559. | 532. | 508. | 487. | 467. | 450. |
| DROP (IN) | .00 | -.63 | -2.68 | -6.26 | -11.49 | -18.48 | -27.37 | -38.26 | -51.25 | -66.46 | -83.98 | -103.87 | -126.28 |
| MID-RANGE (IN) | .00 | .17 | .71 | 1.65 | 3.06 | 4.97 | 7.42 | 10.47 | 14.14 | 18.47 | 23.51 | 29.25 | 35.77 |
| BULLET PATH (IN) | -.63 | .39 | .00 | -1.92 | -5.49 | -10.83 | -18.06 | -27.29 | -38.62 | -52.18 | -68.04 | -86.27 | -107.02 |
| TIME OF FLIGHT (SEC) | .000000 | .059011 | .120645 | .184841 | .251486 | .320424 | .391488 | .464541 | .539472 | .616194 | .694636 | .774740 | .856463 |
| WIND DEFLECTION (IN) | | | | | | | | | | | | | |
| 5 MPH | .00 | .12 | .46 | 1.04 | 1.82 | 2.81 | 3.99 | 5.34 | 6.86 | 8.53 | 10.36 | 12.33 | 14.45 |
| 10 MPH | .00 | .23 | .93 | 2.07 | 3.65 | 5.63 | 7.98 | 10.68 | 13.72 | 17.07 | 20.72 | 24.66 | 28.89 |
| 20 MPH | .00 | .46 | 1.85 | 4.14 | 7.29 | 11.25 | 15.96 | 21.36 | 27.43 | 34.13 | 41.43 | 49.32 | 57.78 |
| 30 MPH | .00 | .70 | 2.78 | 6.21 | 10.94 | 16.88 | 23.94 | 32.05 | 41.15 | 51.20 | 62.15 | 73.99 | 86.67 |
| VELOCITY (FPS) | 1200. | 1153. | 1112. | 1077. | 1045. | 1018. | 993. | 970. | 949. | 930. | 912. | 895. | 879. |
| ENERGY (FT-LB) | 783. | 723. | 673. | 630. | 594. | 563. | 536. | 512. | 490. | 471. | 452. | 436. | 420. |
| DROP (IN) | .00 | -.76 | -3.14 | -7.28 | -13.28 | -21.27 | -31.34 | -43.62 | -58.19 | -75.12 | -94.54 | -116.56 | -141.30 |
| MID-RANGE (IN) | .00 | .20 | .81 | 1.90 | 3.50 | 5.65 | 8.40 | 11.77 | 15.82 | 20.54 | 26.00 | 32.25 | 39.31 |
| BULLET PATH (IN) | -.63 | .50 | .00 | -2.25 | -6.37 | -12.47 | -20.66 | -31.05 | -43.74 | -58.78 | -76.32 | -96.46 | -119.31 |
| TIME OF FLIGHT (SEC) | .000000 | .063775 | .130020 | .198587 | .269306 | .342035 | .416658 | .493085 | .571243 | .651072 | .732525 | .815565 | .900164 |
| WIND DEFLECTION (IN) | | | | | | | | | | | | | |
| 5 MPH | .00 | .11 | .44 | .98 | 1.70 | 2.60 | 3.67 | 4.89 | 6.27 | 7.79 | 9.46 | 11.27 | 13.21 |
| 10 MPH | .00 | .22 | .88 | 1.95 | 3.40 | 5.20 | 7.33 | 9.78 | 12.54 | 15.59 | 18.92 | 22.54 | 26.43 |
| 20 MPH | .00 | .45 | 1.77 | 3.90 | 6.80 | 10.40 | 14.66 | 19.57 | 25.08 | 31.18 | 37.85 | 45.08 | 52.86 |
| 30 MPH | .00 | .67 | 2.65 | 5.85 | 10.19 | 15.59 | 22.00 | 29.35 | 37.62 | 46.77 | 56.77 | 67.62 | 79.29 |
| VELOCITY (FPS) | 1100. | 1066. | 1036. | 1009. | 985. | 963. | 943. | 924. | 906. | 890. | 874. | 859. | 844. |
| ENERGY (FT-LB) | 658. | 618. | 584. | 554. | 528. | 505. | 484. | 464. | 447. | 430. | 415. | 401. | 388. |
| DROP (IN) | .00 | -.89 | -3.68 | -8.49 | -15.43 | -24.60 | -36.09 | -49.98 | -66.39 | -85.44 | -107.23 | -131.86 | -159.43 |
| MID-RANGE (IN) | .00 | .23 | .95 | 2.21 | 4.03 | 6.47 | 9.55 | 13.30 | 17.76 | 22.99 | 29.01 | 35.87 | 43.62 |
| BULLET PATH (IN) | -.63 | .64 | .00 | -2.66 | -7.44 | -14.45 | -23.79 | -35.51 | -49.77 | -66.67 | -86.30 | -108.77 | -134.18 |
| TIME OF FLIGHT (SEC) | .000000 | .069286 | .140673 | .214029 | .289248 | .366245 | .444951 | .525312 | .607284 | .690832 | .775929 | .862557 | .950702 |
| WIND DEFLECTION (IN) | | | | | | | | | | | | | |
| 5 MPH | .00 | .10 | .38 | .83 | 1.45 | 2.23 | 3.16 | 4.23 | 5.44 | 6.79 | 8.28 | 9.91 | 11.66 |
| 10 MPH | .00 | .19 | .76 | 1.67 | 2.91 | 4.46 | 6.31 | 8.45 | 10.88 | 13.59 | 16.56 | 19.81 | 23.32 |
| 20 MPH | .00 | .39 | 1.52 | 3.34 | 5.82 | 8.92 | 12.62 | 16.91 | 21.76 | 27.17 | 33.13 | 39.62 | 46.65 |
| 30 MPH | .00 | .58 | 2.28 | 5.01 | 8.72 | 13.38 | 18.93 | 25.36 | 32.65 | 40.76 | 49.69 | 59.43 | 69.97 |
| VELOCITY (FPS) | 1000. | 977. | 955. | 936. | 917. | 900. | 883. | 868. | 853. | 839. | 825. | 811. | 799. |
| ENERGY (FT-LB) | 544. | 519. | 496. | 476. | 458. | 440. | 425. | 410. | 396. | 383. | 370. | 358. | 347. |
| DROP (IN) | .00 | -1.06 | -4.39 | -10.05 | -18.18 | -28.87 | -42.24 | -58.38 | -77.39 | -99.38 | -124.45 | -152.70 | -184.24 |
| MID-RANGE (IN) | .00 | .28 | 1.14 | 2.60 | 4.70 | 7.51 | 11.07 | 15.39 | 20.52 | 26.49 | 33.35 | 41.13 | 49.88 |
| BULLET PATH (IN) | -.63 | .82 | .00 | -3.16 | -8.78 | -16.96 | -27.83 | -41.46 | -57.96 | -77.44 | -100.00 | -125.74 | -154.77 |
| TIME OF FLIGHT (SEC) | .000000 | .075900 | .153549 | .232886 | .313859 | .396428 | .480562 | .566237 | .653435 | .742146 | .832363 | .924084 | 1.017313 |
| WIND DEFLECTION (IN) | | | | | | | | | | | | | |
| 5 MPH | .00 | .08 | .31 | .69 | 1.22 | 1.89 | 2.69 | 3.63 | 4.70 | 5.91 | 7.25 | 8.72 | 10.32 |
| 10 MPH | .00 | .16 | .62 | 1.39 | 2.44 | 3.77 | 5.38 | 7.26 | 9.40 | 11.82 | 14.50 | 17.44 | 20.65 |
| 20 MPH | .00 | .32 | 1.25 | 2.78 | 4.88 | 7.54 | 10.76 | 14.52 | 18.31 | 23.64 | 28.99 | 34.88 | 41.29 |
| 30 MPH | .00 | .48 | 1.87 | 4.16 | 7.32 | 11.31 | 16.14 | 21.77 | 28.21 | 35.45 | 43.49 | 52.32 | 61.94 |

## Bullet: Lyman #454424 245 Gr.
### Ballistic Coefficient: .210

| RANGE (YDS) | MUZZLE | 25 | 50 | 75 | 100 | 125 | 150 | 175 | 200 | 225 | 250 | 275 | 300 |
|---|---|---|---|---|---|---|---|---|---|---|---|---|---|
| VELOCITY (FPS) | 900. | 884. | 868. | 853. | 839. | 825. | 811. | 798. | 782. | 766. | 751. | 736. | 721. |
| ENERGY (FT-LB) | 441. | 425. | 410. | 396. | 383. | 370. | 358. | 347. | 333. | 319. | 307. | 295. | 283. |
| DROP (IN) | .00 | -1.22 | -5.21 | -12.08 | -21.92 | -34.83 | -50.93 | -70.31 | -93.01 | -119.25 | -149.17 | -182.94 | -220.71 |
| MID-RANGE (IN) | .00 | .34 | 1.39 | 3.17 | 5.74 | 9.13 | 13.38 | 18.54 | 24.59 | 31.64 | 39.76 | 49.00 | 59.43 |
| BULLET PATH (IN) | -.63 | 1.07 | -.00 | -3.94 | -10.86 | -20.85 | -34.02 | -50.48 | -70.26 | -93.58 | -120.58 | -151.42 | -186.27 |
| TIME OF FLIGHT (SEC) | .000000 | .084111 | .169763 | .256938 | .345625 | .435817 | .527512 | .620714 | .715669 | .812618 | .911574 | 1.012555 | 1.115581 |
| WIND DEFLECTION (IN) | | | | | | | | | | | | | |
| 5 MPH | .00 | .07 | .27 | .61 | 1.08 | 1.69 | 2.42 | 3.29 | 4.31 | 5.51 | 6.89 | 8.44 | 10.17 |
| 10 MPH | .00 | .14 | .54 | 1.22 | 2.16 | 3.37 | 4.84 | 6.58 | 8.62 | 11.02 | 13.77 | 16.88 | 20.34 |
| 20 MPH | .00 | .27 | 1.09 | 2.44 | 4.33 | 6.74 | 9.68 | 13.16 | 17.25 | 22.04 | 27.54 | 33.75 | 40.68 |
| 30 MPH | .00 | .41 | 1.63 | 3.66 | 6.49 | 10.11 | 14.53 | 19.74 | 25.87 | 33.06 | 41.31 | 50.63 | 61.03 |
| VELOCITY (FPS) | 800. | 783. | 767. | 752. | 737. | 723. | 708. | 695. | 681. | 668. | 655. | 643. | 631. |
| ENERGY (FT-LB) | 348. | 334. | 320. | 308. | 295. | 284. | 273. | 262. | 252. | 243. | 234. | 225. | 216. |
| DROP (IN) | .00 | -1.46 | -6.43 | -15.06 | -27.52 | -43.96 | -64.54 | -89.41 | -118.78 | -152.81 | -191.69 | -235.62 | -284.80 |
| MID-RANGE (IN) | .00 | .42 | 1.75 | 4.03 | 7.33 | 11.70 | 17.21 | 23.91 | 31.87 | 41.18 | 51.89 | 64.09 | 77.86 |
| BULLET PATH (IN) | -.63 | 1.43 | -.00 | -5.10 | -14.03 | -26.94 | -44.00 | -65.34 | -91.18 | -121.68 | -157.04 | -197.44 | -243.09 |
| TIME OF FLIGHT (SEC) | .000000 | .094743 | .191477 | .290215 | .390974 | .493774 | .598639 | .705598 | .814680 | .925919 | 1.039352 | 1.155018 | 1.272957 |
| WIND DEFLECTION (IN) | | | | | | | | | | | | | |
| 5 MPH | .00 | .09 | .35 | .79 | 1.41 | 2.20 | 3.18 | 4.34 | 5.69 | 7.23 | 8.96 | 10.89 | 13.02 |
| 10 MPH | .00 | .17 | .70 | 1.58 | 2.81 | 4.40 | 6.36 | 8.69 | 11.38 | 14.46 | 17.93 | 21.78 | 26.04 |
| 20 MPH | .00 | .35 | 1.40 | 3.16 | 5.62 | 8.81 | 12.72 | 17.37 | 22.77 | 28.92 | 35.85 | 43.57 | 52.08 |
| 30 MPH | .00 | .52 | 2.10 | 4.73 | 8.43 | 13.21 | 19.08 | 26.06 | 34.15 | 43.39 | 53.78 | 65.35 | 78.12 |
| VELOCITY (FPS) | 700. | 686. | 673. | 660. | 648. | 635. | 623. | 611. | 600. | 588. | 577. | 566. | 555. |
| ENERGY (FT-LB) | 267. | 256. | 246. | 237. | 228. | 219. | 211. | 203. | 196. | 188. | 181. | 174. | 168. |
| DROP (IN) | .00 | -1.85 | -8.29 | -19.50 | -35.68 | -57.03 | -83.74 | -116.02 | -154.10 | -198.21 | -248.60 | -305.51 | -369.21 |
| MID-RANGE (IN) | .00 | .56 | 2.30 | 5.27 | 9.55 | 15.23 | 22.37 | 31.05 | 41.37 | 53.41 | 67.27 | 83.06 | 100.86 |
| BULLET PATH (IN) | -.63 | 1.98 | -.00 | -6.76 | -18.48 | -35.37 | -57.62 | -85.44 | -119.06 | -158.72 | -204.65 | -257.10 | -316.34 |
| TIME OF FLIGHT (SEC) | .000000 | .108203 | .218547 | .331068 | .445803 | .562792 | .682078 | .803706 | .927723 | 1.054179 | 1.183126 | 1.314618 | 1.448711 |
| WIND DEFLECTION (IN) | | | | | | | | | | | | | |
| 5 MPH | .00 | .09 | .38 | .85 | 1.52 | 2.38 | 3.45 | 4.73 | 6.21 | 7.91 | 9.83 | 11.97 | 14.34 |
| 10 MPH | .00 | .19 | .75 | 1.70 | 3.03 | 4.77 | 6.90 | 9.45 | 12.42 | 15.82 | 19.66 | 23.94 | 28.69 |
| 20 MPH | .00 | .37 | 1.50 | 3.39 | 6.07 | 9.53 | 13.81 | 18.90 | 24.84 | 31.64 | 39.32 | 47.89 | 57.37 |
| 30 MPH | .00 | .56 | 2.25 | 5.09 | 9.10 | 14.30 | 20.71 | 28.36 | 37.27 | 47.46 | 58.98 | 71.83 | 86.06 |

# Bullet: .319 Round Ball
## Ballistic Coefficient: .045

| RANGE (YARDS) | MUZZLE | 25 | 50 | 75 | 100 | 125 | 150 |
|---|---|---|---|---|---|---|---|
| VELOCITY (FPS) | 800. | 720. | 651. | 589. | 532. | 480. | 433. |
| ENERGY (FT-LB) | 70. | 56. | 46. | 38. | 31. | 25. | 20. |
| DROP (IN) | 0.0 | -1.76 | -7.69 | -18.75 | -36.08 | -61.08 | -95.50 |
| MID-RANGE (IN) | 0.0 | 0.47 | 2.09 | 5.23 | 10.35 | 18.03 | 29.00 |
| BULLET PATH (IN) | -0.85 | -0.00 | -3.33 | -11.79 | -26.51 | -48.90 | -80.72 |
| TIME OF FLIGHT | 0.0 | 0.098918 | 0.208552 | 0.329788 | 0.463859 | 0.612287 | 0.776869 |
| WIND DEFLECTION (IN) | | | | | | | |
| 5 MPH | 0.0 | 0.45 | 1.85 | 4.27 | 7.82 | 12.63 | 18.86 |
| 10 MPH | 0.0 | 0.91 | 3.71 | 8.54 | 15.64 | 25.26 | 37.73 |
| 20 MPH | 0.0 | 1.82 | 7.41 | 17.09 | 31.28 | 50.53 | 75.46 |
| 30 MPH | 0.0 | 2.73 | 11.12 | 25.63 | 46.92 | 75.79 | 113.19 |
| VELOCITY (FPS) | 700. | 633. | 572. | 517. | 467. | 421. | 379. |
| ENERGY (FT-LB) | 53. | 44. | 36. | 29. | 24. | 19. | 16. |
| DROP (IN) | 0.0 | -2.28 | -9.97 | -24.30 | -46.74 | -79.16 | -123.87 |
| MID-RANGE (IN) | 0.0 | 0.61 | 2.71 | 6.77 | 13.40 | 23.36 | 37.64 |
| BULLET PATH (IN) | -0.85 | 0.00 | -4.57 | -15.76 | -35.08 | -64.37 | -105.96 |
| TIME OF FLIGHT | 0.0 | 0.112738 | 0.237388 | 0.375265 | 0.527966 | 0.697365 | 0.885630 |
| WIND DEFLECTION (IN) | | | | | | | |
| 5 MPH | 0.0 | 0.49 | 2.03 | 4.74 | 8.75 | 14.23 | 21.36 |
| 10 MPH | 0.0 | 0.98 | 4.07 | 9.48 | 17.49 | 28.45 | 42.73 |
| 20 MPH | 0.0 | 1.97 | 8.13 | 18.95 | 34.99 | 56.90 | 85.46 |
| 30 MPH | 0.0 | 2.95 | 12.20 | 28.43 | 52.48 | 85.35 | 128.18 |
| VELOCITY (FPS) | 600. | 542. | 490. | 442. | 398. | 358. | 322. |
| ENERGY (FT-LB) | 39. | 32. | 26. | 21. | 17. | 14. | 11. |
| DROP (IN) | 0.0 | -3.10 | -13.58 | -33.12 | -63.80 | -108.25 | -169.73 |
| MID-RANGE (IN) | 0.0 | 0.83 | 3.69 | 9.24 | 18.32 | 32.03 | 51.75 |
| BULLET PATH (IN) | -0.85 | 0.00 | -6.53 | -22.12 | -48.85 | -89.35 | -146.88 |
| TIME OF FLIGHT | 0.0 | 0.131508 | 0.277059 | 0.438392 | 0.617537 | 0.816769 | 1.038422 |
| WIND DEFLECTION (IN) | | | | | | | |
| 5 MPH | 0.0 | 0.57 | 2.38 | 5.58 | 10.34 | 16.88 | 25.38 |
| 10 MPH | 0.0 | 1.15 | 4.76 | 11.16 | 20.69 | 33.75 | 50.76 |
| 20 MPH | 0.0 | 2.29 | 9.52 | 22.31 | 41.37 | 67.50 | 101.52 |
| 30 MPH | 0.0 | 3.44 | 14.29 | 33.47 | 62.06 | 101.25 | 152.29 |
| VELOCITY (FPS) | 500. | 451. | 406. | 365. | 329. | 296. | 267. |
| ENERGY (FT-LB) | 27. | 22. | 18. | 15. | 12. | 10. | 8. |
| DROP (IN) | 0.0 | -4.48 | -19.64 | -47.99 | -92.66 | -157.52 | -247.37 |
| MID-RANGE (IN) | 0.0 | 1.20 | 5.34 | 13.42 | 26.69 | 46.77 | 75.69 |
| BULLET PATH (IN) | -0.85 | 0.00 | -9.84 | -32.86 | -72.20 | -131.73 | -216.26 |
| TIME OF FLIGHT | 0.0 | 0.158008 | 0.333392 | 0.528389 | 0.745309 | 0.986643 | 1.255190 |
| WIND DEFLECTION (IN) | | | | | | | |
| 5 MPH | 0.0 | 0.70 | 2.94 | 6.90 | 12.79 | 20.82 | 31.26 |
| 10 MPH | 0.0 | 1.41 | 5.88 | 13.80 | 25.57 | 41.65 | 62.51 |
| 20 MPH | 0.0 | 2.82 | 11.75 | 27.59 | 51.15 | 83.30 | 125.03 |
| 30 MPH | 0.0 | 4.23 | 17.63 | 41.39 | 76.72 | 124.95 | 187.54 |
| VELOCITY (FPS) | 400. | 360. | 324. | 291. | 262. | 237. | 214. |
| ENERGY (FT-LB) | 17. | 14. | 11. | 9. | 7. | 6. | 5. |
| DROP (IN) | 0.0 | -7.02 | -30.83 | -75.42 | -145.73 | -247.88 | -389.48 |
| MID-RANGE (IN) | 0.0 | 1.88 | 8.40 | 21.11 | 42.03 | 73.69 | 119.32 |
| BULLET PATH (IN) | -0.85 | 0.00 | -15.94 | -52.66 | -115.10 | -209.38 | -343.11 |
| TIME OF FLIGHT | 0.0 | 0.197845 | 0.417917 | 0.662728 | 0.935097 | 1.238211 | 1.575702 |
| WIND DEFLECTION (IN) | | | | | | | |
| 5 MPH | 0.0 | 0.91 | 3.78 | 8.82 | 16.29 | 26.46 | 39.66 |
| 10 MPH | 0.0 | 1.82 | 7.55 | 17.64 | 32.58 | 52.93 | 79.32 |
| 20 MPH | 0.0 | 3.64 | 15.11 | 35.28 | 65.15 | 105.85 | 158.65 |
| 30 MPH | 0.0 | 5.46 | 22.66 | 52.92 | 97.73 | 158.78 | 237.97 |

# Bullet: .375 Round Ball
## Ballistic Coefficient: .053

| RANGE (YARDS) | MUZZLE | 25 | 50 | 75 | 100 | 125 | 150 |
|---|---|---|---|---|---|---|---|
| VELOCITY (FPS) | 1100. | 963. | 863. | 785. | 720. | 661. | 608. |
| ENERGY (FT-LB) | 212. | 163. | 131. | 108. | 91. | 77. | 65. |
| DROP (IN) | 0.0 | -0.98 | -4.29 | -10.50 | -20.20 | -34.08 | -52.93 |
| MID-RANGE (IN) | 0.0 | 0.26 | 1.17 | 2.94 | 5.81 | 10.04 | 15.96 |
| BULLET PATH (IN) | -0.85 | 0.00 | -1.48 | -5.86 | -13.73 | -25.78 | -42.79 |
| TIME OF FLIGHT | 0.0 | 0.073102 | 0.155612 | 0.246825 | 0.346667 | 0.455483 | 0.573879 |
| WIND DEFLECTION (IN) | | | | | | | |
| 5 MPH | 0.0 | 0.43 | 1.69 | 3.72 | 6.51 | 10.08 | 14.50 |
| 10 MPH | 0.0 | 0.87 | 3.39 | 7.44 | 13.01 | 20.16 | 29.00 |
| 20 MPH | 0.0 | 1.73 | 6.78 | 14.88 | 26.03 | 40.33 | 58.01 |
| 30 MPH | 0.0 | 2.60 | 10.16 | 22.32 | 39.04 | 60.49 | 87.01 |
| VELOCITY (FPS) | 1000. | 889. | 806. | 738. | 677. | 623. | 573. |
| ENERGY (FT-LB) | 175. | 138. | 114. | 95. | 80. | 68. | 57. |
| DROP (IN) | 0.0 | -1.17 | -5.07 | -12.27 | -23.45 | -39.36 | -60.85 |
| MID-RANGE (IN) | 0.0 | 0.30 | 1.37 | 3.40 | 6.66 | 11.46 | 18.15 |
| BULLET PATH (IN) | -0.85 | -0.00 | -1.89 | -7.08 | -16.24 | -30.13 | -49.61 |
| TIME OF FLIGHT | 0.0 | 0.079833 | 0.168552 | 0.265861 | 0.372013 | 0.487545 | 0.613217 |
| WIND DEFLECTION (IN) | | | | | | | |
| 5 MPH | 0.0 | 0.43 | 1.63 | 3.60 | 6.34 | 9.90 | 14.36 |
| 10 MPH | 0.0 | 0.85 | 3.27 | 7.19 | 12.67 | 19.81 | 28.73 |
| 20 MPH | 0.0 | 1.70 | 6.53 | 14.38 | 25.35 | 39.62 | 57.45 |
| 30 MPH | 0.0 | 2.55 | 9.80 | 21.57 | 38.02 | 59.42 | 86.18 |
| VELOCITY (FPS) | 900. | 816. | 746. | 684. | 629. | 578. | 532. |
| ENERGY (FT-LB) | 142. | 117. | 98. | 82. | 69. | 59. | 50. |
| DROP (IN) | 0.0 | -1.41 | -6.04 | -14.57 | -27.72 | -46.35 | -71.47 |
| MID-RANGE (IN) | 0.0 | 0.37 | 1.62 | 4.00 | 7.82 | 13.40 | 21.17 |
| BULLET PATH (IN) | -0.85 | -0.00 | -2.38 | -8.65 | -19.54 | -35.92 | -58.78 |
| TIME OF FLIGHT | 0.0 | 0.087674 | 0.183931 | 0.288984 | 0.403338 | 0.527729 | 0.663076 |
| WIND DEFLECTION (IN) | | | | | | | |
| 5 MPH | 0.0 | 0.38 | 1.52 | 3.43 | 6.16 | 9.77 | 14.35 |
| 10 MPH | 0.0 | 0.76 | 3.04 | 6.86 | 12.32 | 19.55 | 28.70 |
| 20 MPH | 0.0 | 1.53 | 6.08 | 13.72 | 24.64 | 39.09 | 57.40 |
| 30 MPH | 0.0 | 2.29 | 9.12 | 20.58 | 36.96 | 58.64 | 86.10 |
| VELOCITY (FPS) | 800. | 732. | 672. | 618. | 568. | 522. | 479. |
| ENERGY (FT-LB) | 112. | 94. | 79. | 67. | 57. | 48. | 40. |
| DROP (IN) | 0.0 | -1.72 | -7.48 | -18.04 | -34.28 | -57.23 | -88.15 |
| MID-RANGE (IN) | 0.0 | 0.46 | 2.02 | 4.97 | 9.65 | 16.50 | 26.03 |
| BULLET PATH (IN) | -0.85 | -0.00 | -3.19 | -11.17 | -24.83 | -45.22 | -73.57 |
| TIME OF FLIGHT | 0.0 | 0.098071 | 0.205021 | 0.321409 | 0.448010 | 0.585782 | 0.735845 |
| WIND DEFLECTION (IN) | | | | | | | |
| 5 MPH | 0.0 | 0.38 | 1.54 | 3.53 | 6.42 | 10.30 | 15.25 |
| 10 MPH | 0.0 | 0.76 | 3.08 | 7.07 | 12.85 | 20.60 | 30.51 |
| 20 MPH | 0.0 | 1.52 | 6.17 | 14.14 | 25.70 | 41.20 | 61.02 |
| 30 MPH | 0.0 | 2.28 | 9.25 | 21.20 | 38.55 | 61.79 | 91.53 |
| VELOCITY (FPS) | 700. | 643. | 591. | 544. | 499. | 458. | 420. |
| ENERGY (FT-LB) | 86. | 73. | 61. | 52. | 44. | 37. | 31. |
| DROP (IN) | 0.0 | -2.24 | -9.71 | -23.39 | -44.41 | -74.14 | -114.25 |
| MID-RANGE (IN) | 0.0 | 0.60 | 2.62 | 6.43 | 12.49 | 21.37 | 33.74 |
| BULLET PATH (IN) | -0.85 | 0.00 | -4.39 | -14.98 | -32.91 | -59.57 | -96.58 |
| TIME OF FLIGHT | 0.0 | 0.111808 | 0.233436 | 0.365755 | 0.509810 | 0.666806 | 0.838113 |
| WIND DEFLECTION (IN) | | | | | | | |
| 5 MPH | 0.0 | 0.41 | 1.69 | 3.90 | 7.15 | 11.54 | 17.18 |
| 10 MPH | 0.0 | 0.82 | 3.37 | 7.80 | 14.30 | 23.07 | 34.36 |
| 20 MPH | 0.0 | 1.64 | 6.74 | 15.60 | 28.60 | 46.14 | 68.73 |
| 30 MPH | 0.0 | 2.46 | 10.11 | 23.40 | 42.89 | 69.22 | 103.09 |
| VELOCITY (FPS) | 600. | 551. | 506. | 465. | 426. | 390. | 357. |
| ENERGY (FT-LB) | 63. | 53. | 45. | 38. | 32. | 27. | 22. |
| DROP (IN) | 0.0 | -3.04 | -13.22 | -31.86 | -60.56 | -101.24 | -156.24 |
| MID-RANGE (IN) | 0.0 | 0.82 | 3.57 | 8.77 | 17.06 | 29.23 | 46.26 |
| BULLET PATH (IN) | -0.85 | 0.00 | -6.28 | -21.03 | -45.84 | -82.63 | -133.74 |
| TIME OF FLIGHT | 0.0 | 0.130420 | 0.272384 | 0.427068 | 0.595812 | 0.780134 | 0.981661 |
| WIND DEFLECTION (IN) | | | | | | | |
| 5 MPH | 0.0 | 0.48 | 1.97 | 4.58 | 8.43 | 13.65 | 20.39 |
| 10 MPH | 0.0 | 0.95 | 3.94 | 9.16 | 16.86 | 27.30 | 40.77 |
| 20 MPH | 0.0 | 1.91 | 7.88 | 18.33 | 33.73 | 54.61 | 81.54 |
| 30 MPH | 0.0 | 2.86 | 11.82 | 27.49 | 50.59 | 81.91 | 122.32 |

| RANGE (YARDS) | MUZZLE | 25 | 50 | 75 | 100 | 125 | 150 |
|---|---|---|---|---|---|---|---|
| VELOCITY (FPS) | 500. | 459. | 420. | 385. | 352. | 322. | 295. |
| ENERGY (FT-LB) | 44. | 37. | 31. | 26. | 22. | 18. | 15. |
| DROP (IN) | 0.0 | -4.39 | -19.11 | -46.12 | -87.83 | -147.10 | -227.38 |
| MID-RANGE (IN) | 0.0 | 1.18 | 5.16 | 12.72 | 24.81 | 42.61 | 67.57 |
| BULLET PATH (IN) | -0.85 | 0.00 | -9.47 | -31.24 | -67.71 | -121.74 | -196.77 |
| TIME OF FLIGHT | 0.0 | 0.156666 | 0.327599 | 0.514339 | 0.718505 | 0.941749 | 1.185886 |
| WIND DEFLECTION (IN) | | | | | | | |
| 5 MPH | 0.0 | 0.59 | 2.43 | 5.66 | 10.43 | 16.87 | 25.16 |
| 10 MPH | 0.0 | 1.17 | 4.86 | 11.32 | 20.86 | 33.75 | 50.32 |
| 20 MPH | 0.0 | 2.35 | 9.71 | 22.65 | 41.71 | 67.50 | 100.63 |
| 30 MPH | 0.0 | 3.52 | 14.57 | 33.97 | 62.57 | 101.24 | 150.95 |

## Bullet: No. 37583 Solid Base
### Ballistic Coefficient: .095

| RANGE (YARDS) | MUZZLE | 25 | 50 | 75 | 100 | 125 | 150 |
|---|---|---|---|---|---|---|---|
| VELOCITY (FPS) | 800. | 772. | 746. | 722. | 698. | 675. | 654. |
| ENERGY (FT-LB) | 206. | 192. | 179. | 168. | 157. | 147. | 138. |
| DROP (IN) | 0.0 | -1.56 | -6.73 | -15.78 | -28.98 | -46.61 | -68.99 |
| MID-RANGE (IN) | 0.0 | 0.43 | 1.80 | 4.21 | 7.76 | 12.55 | 18.72 |
| BULLET PATH (IN) | -0.85 | -0.00 | -2.76 | -9.39 | -20.18 | -35.40 | -55.37 |
| TIME OF FLIGHT | 0.0 | 0.095424 | 0.194216 | 0.296424 | 0.402113 | 0.511365 | 0.624273 |
| WIND DEFLECTION (IN) | | | | | | | |
| 5 MPH | 0.0 | 0.15 | 0.59 | 1.34 | 2.39 | 3.75 | 5.44 |
| 10 MPH | 0.0 | 0.29 | 1.18 | 2.67 | 4.77 | 7.50 | 10.87 |
| 20 MPH | 0.0 | 0.59 | 2.36 | 5.34 | 9.54 | 15.00 | 21.74 |
| 30 MPH | 0.0 | 0.88 | 3.55 | 8.01 | 14.32 | 22.50 | 32.62 |
| VELOCITY (FPS) | 700. | 677. | 655. | 634. | 614. | 594. | 575. |
| ENERGY (FT-LB) | 158. | 148. | 138. | 130. | 121. | 114. | 107. |
| DROP (IN) | 0.0 | -1.99 | -8.70 | -20.44 | -37.57 | -60.43 | -89.44 |
| MID-RANGE (IN) | 0.0 | 0.57 | 2.36 | 5.48 | 10.08 | 16.30 | 24.27 |
| BULLET PATH (IN) | -0.85 | 0.00 | -3.86 | -12.77 | -27.05 | -47.07 | -73.23 |
| TIME OF FLIGHT | 0.0 | 0.108934 | 0.221516 | 0.337851 | 0.458058 | 0.582266 | 0.710615 |
| WIND DEFLECTION (IN) | | | | | | | |
| 5 MPH | 0.0 | 0.16 | 0.64 | 1.45 | 2.59 | 4.10 | 5.96 |
| 10 MPH | 0.0 | 0.32 | 1.27 | 2.89 | 5.19 | 8.19 | 11.93 |
| 20 MPH | 0.0 | 0.63 | 2.54 | 5.78 | 10.38 | 16.39 | 23.85 |
| 30 MPH | 0.0 | 0.95 | 3.82 | 8.67 | 15.57 | 24.58 | 35.78 |
| VELOCITY (FPS) | 600. | 581. | 562. | 544. | 526. | 509. | 492. |
| ENERGY (FT-LB) | 116. | 109. | 102. | 95. | 89. | 83. | 78. |
| DROP (IN) | 0.0 | -2.71 | -11.84 | -27.82 | -51.13 | -82.28 | -121.79 |
| MID-RANGE (IN) | 0.0 | 0.78 | 3.21 | 7.47 | 13.73 | 22.20 | 33.07 |
| BULLET PATH (IN) | -0.85 | 0.00 | -5.57 | -18.00 | -37.76 | -65.34 | -101.30 |
| TIME OF FLIGHT | 0.0 | 0.127072 | 0.258386 | 0.394100 | 0.534379 | 0.679399 | 0.829346 |
| WIND DEFLECTION (IN) | | | | | | | |
| 5 MPH | 0.0 | 0.18 | 0.74 | 1.68 | 3.03 | 4.79 | 6.98 |
| 10 MPH | 0.0 | 0.36 | 1.48 | 3.36 | 6.05 | 9.57 | 13.96 |
| 20 MPH | 0.0 | 0.73 | 2.95 | 6.72 | 12.10 | 19.15 | 27.93 |
| 30 MPH | 0.0 | 1.09 | 4.43 | 10.08 | 18.15 | 28.72 | 41.89 |
| VELOCITY (FPS) | 500. | 484. | 468. | 452. | 437. | 423. | 409. |
| ENERGY (FT-LB) | 80. | 75. | 70. | 66. | 62. | 58. | 54. |
| DROP (IN) | 0.0 | -3.92 | -17.11 | -40.21 | -73.91 | -118.97 | -176.19 |
| MID-RANGE (IN) | 0.0 | 1.12 | 4.63 | 10.78 | 19.85 | 32.11 | 47.89 |
| BULLET PATH (IN) | -0.85 | 0.00 | -8.41 | -26.74 | -55.67 | -95.95 | -148.40 |
| TIME OF FLIGHT | 0.0 | 0.152543 | 0.310307 | 0.473501 | 0.642344 | 0.817069 | 0.997920 |
| WIND DEFLECTION (IN) | | | | | | | |
| 5 MPH | 0.0 | 0.22 | 0.91 | 2.07 | 3.73 | 5.90 | 8.62 |
| 10 MPH | 0.0 | 0.45 | 1.81 | 4.14 | 7.45 | 11.80 | 17.23 |
| 20 MPH | 0.0 | 0.90 | 3.63 | 8.27 | 14.91 | 23.61 | 34.47 |
| 30 MPH | 0.0 | 1.34 | 5.44 | 12.41 | 22.36 | 35.41 | 51.70 |
| VELOCITY (FPS) | 400. | 386. | 373. | 361. | 349. | 338. | 327. |
| ENERGY (FT-LB) | 52. | 48. | 45. | 42. | 39. | 37. | 34. |
| DROP (IN) | 0.0 | -6.18 | -26.87 | -63.09 | -115.99 | -186.75 | -276.67 |
| MID-RANGE (IN) | 0.0 | 1.75 | 7.25 | 16.90 | 31.13 | 50.41 | 75.24 |
| BULLET PATH (IN) | -0.85 | 0.00 | -13.65 | -42.85 | -88.71 | -152.44 | -235.33 |
| TIME OF FLIGHT | 0.0 | 0.190796 | 0.388336 | 0.592863 | 0.804629 | 1.023898 | 1.250950 |
| WIND DEFLECTION (IN) | | | | | | | |
| 5 MPH | 0.0 | 0.29 | 1.17 | 2.67 | 4.81 | 7.60 | 11.08 |
| 10 MPH | 0.0 | 0.58 | 2.35 | 5.34 | 9.61 | 15.21 | 22.17 |
| 20 MPH | 0.0 | 1.16 | 4.69 | 10.69 | 19.23 | 30.41 | 44.33 |
| 30 MPH | 0.0 | 1.74 | 7.04 | 16.03 | 28.84 | 45.62 | 66.50 |

## Bullet: .451 Round Ball
### Ballistic Coefficient: .064

| RANGE (YARDS) | MUZZLE | 25 | 50 | 75 | 100 | 125 | 150 |
|---|---|---|---|---|---|---|---|
| VELOCITY (FPS) | 1100. | 982. | 893. | 823. | 764. | 711. | 664. |
| ENERGY (FT-LB) | 371. | 296. | 244. | 208. | 179. | 155. | 135. |
| DROP (IN) | 0.0 | -0.97 | -4.17 | -10.09 | -19.18 | -31.96 | -49.02 |
| MID-RANGE (IN) | 0.0 | 0.25 | 1.12 | 2.78 | 5.42 | 9.22 | 14.42 |
| BULLET PATH (IN) | -0.85 | 0.00 | -1.39 | -5.49 | -12.77 | -23.74 | -38.98 |
| TIME OF FLIGHT | 0.0 | 0.072361 | 0.152629 | 0.240204 | 0.334825 | 0.436597 | 0.545796 |
| WIND DEFLECTION (IN) | | | | | | | |
| 5 MPH | 0.0 | 0.37 | 1.43 | 3.14 | 5.46 | 8.42 | 12.03 |
| 10 MPH | 0.0 | 0.74 | 2.86 | 6.28 | 10.93 | 16.84 | 24.06 |
| 20 MPH | 0.0 | 1.47 | 5.73 | 12.55 | 21.86 | 33.68 | 48.12 |
| 30 MPH | 0.0 | 2.21 | 8.59 | 18.83 | 32.79 | 50.52 | 72.18 |
| VELOCITY (FPS) | 1000. | 905. | 834. | 773. | 720. | 671. | 626. |
| ENERGY (FT-LB) | 306. | 251. | 213. | 183. | 159. | 138. | 120. |
| DROP (IN) | 0.0 | -1.15 | -4.93 | -11.81 | -22.28 | -36.94 | -56.41 |
| MID-RANGE (IN) | 0.0 | 0.30 | 1.32 | 3.22 | 6.21 | 10.51 | 16.40 |
| BULLET PATH (IN) | -0.85 | 0.00 | -1.78 | -6.66 | -15.14 | -27.80 | -45.27 |
| TIME OF FLIGHT | 0.0 | 0.079013 | 0.165437 | 0.258915 | 0.359513 | 0.467483 | 0.583223 |
| WIND DEFLECTION (IN) | | | | | | | |
| 5 MPH | 0.0 | 0.35 | 1.36 | 2.98 | 5.24 | 8.14 | 11.72 |
| 10 MPH | 0.0 | 0.71 | 2.72 | 5.97 | 10.47 | 16.28 | 23.45 |
| 20 MPH | 0.0 | 1.41 | 5.43 | 11.94 | 20.95 | 32.55 | 46.89 |
| 30 MPH | 0.0 | 2.12 | 8.15 | 17.91 | 31.42 | 48.83 | 70.34 |
| VELOCITY (FPS) | 900. | 829. | 769. | 716. | 668. | 623. | 582. |
| ENERGY (FT-LB) | 248. | 211. | 181. | 157. | 137. | 119. | 104. |
| DROP (IN) | 0.0 | -1.38 | -5.90 | -14.05 | -26.42 | -43.65 | -66.45 |
| MID-RANGE (IN) | 0.0 | 0.36 | 1.56 | 3.80 | 7.31 | 12.34 | 19.18 |
| BULLET PATH (IN) | -0.85 | -0.00 | -2.28 | -8.20 | -18.33 | -33.33 | -53.90 |
| TIME OF FLIGHT | 0.0 | 0.086911 | 0.180871 | 0.281963 | 0.390449 | 0.506736 | 0.631348 |
| WIND DEFLECTION (IN) | | | | | | | |
| 5 MPH | 0.0 | 0.31 | 1.25 | 2.81 | 5.03 | 7.93 | 11.56 |
| 10 MPH | 0.0 | 0.63 | 2.50 | 5.63 | 10.05 | 15.85 | 23.12 |
| 20 MPH | 0.0 | 1.26 | 5.00 | 11.25 | 20.10 | 31.70 | 46.23 |
| 30 MPH | 0.0 | 1.89 | 7.50 | 16.88 | 30.16 | 47.56 | 69.35 |
| VELOCITY (FPS) | 800. | 744. | 693. | 646. | 603. | 563. | 525. |
| ENERGY (FT-LB) | 196. | 169. | 147. | 128. | 112. | 97. | 84. |
| DROP (IN) | 0.0 | -1.69 | -7.29 | -17.41 | -32.71 | -53.96 | -82.06 |
| MID-RANGE (IN) | 0.0 | 0.45 | 1.96 | 4.74 | 9.06 | 15.23 | 23.62 |
| BULLET PATH (IN) | -0.85 | -0.00 | -3.06 | -10.64 | -23.40 | -42.11 | -67.67 |
| TIME OF FLIGHT | 0.0 | 0.097296 | 0.201827 | 0.313928 | 0.434062 | 0.562803 | 0.700821 |
| WIND DEFLECTION (IN) | | | | | | | |
| 5 MPH | 0.0 | 0.31 | 1.26 | 2.88 | 5.20 | 8.28 | 12.17 |
| 10 MPH | 0.0 | 0.62 | 2.52 | 5.75 | 10.39 | 16.55 | 24.34 |
| 20 MPH | 0.0 | 1.25 | 5.04 | 11.50 | 20.79 | 33.11 | 48.69 |
| 30 MPH | 0.0 | 1.87 | 7.56 | 17.25 | 31.18 | 49.66 | 73.03 |
| VELOCITY (FPS) | 700. | 653. | 609. | 569. | 530. | 495. | 461. |
| ENERGY (FT-LB) | 150. | 131. | 114. | 99. | 86. | 75. | 65. |
| DROP (IN) | 0.0 | -2.19 | -9.46 | -22.57 | -42.39 | -69.91 | -106.31 |
| MID-RANGE (IN) | 0.0 | 0.59 | 2.54 | 6.14 | 11.73 | 19.72 | 30.58 |
| BULLET PATH (IN) | -0.85 | 0.00 | -4.23 | -14.30 | -31.07 | -55.55 | -88.91 |
| TIME OF FLIGHT | 0.0 | 0.110963 | 0.229883 | 0.357318 | 0.493925 | 0.640452 | 0.797735 |
| WIND DEFLECTION (IN) | | | | | | | |
| 5 MPH | 0.0 | 0.34 | 1.37 | 3.16 | 5.75 | 9.22 | 13.63 |
| 10 MPH | 0.0 | 0.67 | 2.75 | 6.32 | 11.50 | 18.43 | 27.26 |
| 20 MPH | 0.0 | 1.34 | 5.49 | 12.63 | 23.00 | 36.87 | 54.52 |
| 30 MPH | 0.0 | 2.02 | 8.24 | 18.95 | 34.51 | 55.30 | 81.78 |
| VELOCITY (FPS) | 600. | 560. | 522. | 487. | 453. | 422. | 392. |
| ENERGY (FT-LB) | 110. | 96. | 84. | 73. | 63. | 55. | 47. |
| DROP (IN) | 0.0 | -2.98 | -12.88 | -30.74 | -57.76 | -95.34 | -145.15 |
| MID-RANGE (IN) | 0.0 | 0.81 | 3.46 | 8.36 | 16.00 | 26.93 | 41.84 |
| BULLET PATH (IN) | -0.85 | 0.00 | -6.07 | -20.09 | -43.28 | -77.03 | -123.01 |
| TIME OF FLIGHT | 0.0 | 0.129432 | 0.268194 | 0.417053 | 0.576864 | 0.748579 | 0.933252 |
| WIND DEFLECTION (IN) | | | | | | | |
| 5 MPH | 0.0 | 0.39 | 1.60 | 3.70 | 6.76 | 10.87 | 16.13 |
| 10 MPH | 0.0 | 0.78 | 3.20 | 7.40 | 13.53 | 21.75 | 32.25 |
| 20 MPH | 0.0 | 1.56 | 6.40 | 14.80 | 27.06 | 43.50 | 64.50 |
| 30 MPH | 0.0 | 2.34 | 9.61 | 22.20 | 40.58 | 65.25 | 96.76 |

| RANGE (YARDS) | MUZZLE | 25 | 50 | 75 | 100 | 125 | 150 |
|---|---|---|---|---|---|---|---|
| VELOCITY (FPS) | 500. | 466. | 434. | 403. | 375. | 349. | 25. |
| ENERGY (FT-LB) | 77. | 66. | 58. | 50. | 43. | 37. | 32. |
| DROP (IN) | 0.0 | -4.31 | -18.61 | -44.46 | -83.65 | -138.31 | -210.87 |
| MID-RANGE (IN) | 0.0 | 1.16 | 5.00 | 12.11 | 23.22 | 39.17 | 60.98 |
| BULLET PATH (IN) | -0.85 | 0.00 | -9.15 | -29.84 | -63.88 | -113.38 | -180.78 |
| TIME OF FLIGHT | 0.0 | 0.155448 | 0.322413 | 0.501905 | 0.695024 | 0.902858 | 1.12547 |
| WIND DEFLECTION (IN) | | | | | | | |
| 5 MPH | 0.0 | 0.48 | 1.97 | 4.57 | 8.36 | 13.45 | 19.94 |
| 10 MPH | 0.0 | 0.96 | 3.94 | 9.14 | 16.72 | 26.90 | 39.87 |
| 20 MPH | 0.0 | 1.92 | 7.89 | 18.27 | 33.45 | 53.81 | 79.74 |
| 30 MPH | 0.0 | 2.88 | 11.83 | 27.41 | 50.17 | 80.71 | 119.62 |

## Bullet: No. 450229 Hollow Base
### Ballistic Coefficient: .050

| RANGE (YARDS) | MUZZLE | 25 | 50 | 75 | 100 | 125 | 150 |
|---|---|---|---|---|---|---|---|
| VELOCITY (FPS) | 900. | 836. | 781. | 732. | 687. | 645. | 606. |
| ENERGY (FT-LB) | 279. | 240. | 210. | 184. | 162. | 143. | 127. |
| DROP (IN) | 0.0 | -1.37 | -5.83 | -13.81 | -25.82 | -42.43 | -64.25 |
| MID-RANGE (IN) | 0.0 | 0.36 | 1.54 | 3.71 | 7.08 | 11.87 | 18.32 |
| BULLET PATH (IN) | -0.85 | -0.00 | -2.24 | -7.99 | -17.78 | -32.17 | -51.76 |
| TIME OF FLIGHT | 0.0 | 0.086558 | 0.179460 | 0.278749 | 0.384605 | 0.497320 | 0.617279 |
| WIND DEFLECTION (IN) | | | | | | | |
| 5 MPH | 0.0 | 0.28 | 1.13 | 2.53 | 4.51 | 7.10 | 10.32 |
| 10 MPH | 0.0 | 0.57 | 2.25 | 5.06 | 9.02 | 14.19 | 20.64 |
| 20 MPH | 0.0 | 1.13 | 4.50 | 10.12 | 18.05 | 28.39 | 41.28 |
| 30 MPH | 0.0 | 1.70 | 6.75 | 15.18 | 27.07 | 42.58 | 61.92 |
| VELOCITY (FPS) | 800. | 749. | 703. | 660. | 620. | 583. | 548. |
| ENERGY (FT-LB) | 220. | 193. | 170. | 150. | 132. | 117. | 103. |
| DROP (IN) | 0.0 | -1.67 | -7.19 | -17.10 | -31.98 | -52.49 | -79.39 |
| MID-RANGE (IN) | 0.0 | 0.45 | 1.93 | 4.63 | 8.80 | 14.68 | 22.59 |
| BULLET PATH (IN) | -0.85 | -0.00 | -3.00 | -10.39 | -22.75 | -40.74 | -65.11 |
| TIME OF FLIGHT | 0.0 | 0.096939 | 0.200365 | 0.310530 | 0.427789 | 0.552575 | 0.685398 |
| WIND DEFLECTION (IN) | | | | | | | |
| 5 MPH | 0.0 | 0.28 | 1.13 | 2.58 | 4.65 | 7.38 | 10.81 |
| 10 MPH | 0.0 | 0.56 | 2.26 | 5.15 | 9.29 | 14.75 | 21.63 |
| 20 MPH | 0.0 | 1.12 | 4.53 | 10.31 | 18.58 | 29.51 | 43.26 |
| 30 MPH | 0.0 | 1.68 | 6.79 | 15.46 | 27.87 | 44.26 | 64.89 |
| VELOCITY (FPS) | 700. | 658. | 618. | 581. | 546. | 512. | 481. |
| ENERGY (FT-LB) | 169. | 149. | 131. | 116. | 102. | 90. | 80. |
| DROP (IN) | 0.0 | -2.16 | -9.34 | -22.18 | -41.46 | -68.02 | -102.84 |
| MID-RANGE (IN) | 0.0 | 0.59 | 2.50 | 6.01 | 11.39 | 19.00 | 29.24 |
| BULLET PATH (IN) | -0.85 | 0.00 | -4.16 | -13.99 | -30.25 | -53.80 | -85.61 |
| TIME OF FLIGHT | 0.0 | 0.110574 | 0.228263 | 0.353507 | 0.486818 | 0.628778 | 0.780031 |
| WIND DEFLECTION (IN) | | | | | | | |
| 5 MPH | 0.0 | 0.30 | 1.23 | 2.82 | 5.13 | 8.19 | 12.07 |
| 10 MPH | 0.0 | 0.60 | 2.46 | 5.65 | 10.25 | 16.38 | 24.14 |
| 20 MPH | 0.0 | 1.21 | 4.92 | 11.29 | 20.50 | 32.76 | 48.29 |
| 30 MPH | 0.0 | 1.81 | 7.38 | 16.94 | 30.75 | 49.14 | 72.43 |
| VELOCITY (FPS) | 600. | 564. | 529. | 497. | 466. | 437. | 410. |
| ENERGY (FT-LB) | 124. | 109. | 96. | 85. | 75. | 66. | 58. |
| DROP (IN) | 0.0 | -2.94 | -12.70 | -30.19 | -56.46 | -92.70 | -140.30 |
| MID-RANGE (IN) | 0.0 | 0.80 | 3.41 | 8.18 | 15.53 | 25.93 | 39.96 |
| BULLET PATH (IN) | -0.85 | 0.00 | -5.97 | -19.66 | -42.14 | -74.58 | -118.39 |
| TIME OF FLIGHT | 0.0 | 0.128979 | 0.266289 | 0.412540 | 0.568409 | 0.734637 | 0.912041 |
| WIND DEFLECTION (IN) | | | | | | | |
| 5 MPH | 0.0 | 0.35 | 1.43 | 3.30 | 6.02 | 9.65 | 14.26 |
| 10 MPH | 0.0 | 0.70 | 2.87 | 6.61 | 12.04 | 19.30 | 28.52 |
| 20 MPH | 0.0 | 1.40 | 5.73 | 13.21 | 24.08 | 38.59 | 57.04 |
| 30 MPH | 0.0 | 2.10 | 8.60 | 19.82 | 36.12 | 57.89 | 85.56 |
| VELOCITY (FPS) | 500. | 469. | 440. | 412. | 386. | 362. | 339. |
| ENERGY (FT-LB) | 86. | 76. | 67. | 58. | 51. | 45. | 40. |
| DROP (IN) | 0.0 | -4.24 | -18.34 | -43.64 | -81.72 | -134.38 | -203.66 |
| MID-RANGE (IN) | 0.0 | 1.15 | 4.92 | 11.84 | 22.53 | 37.69 | 58.19 |
| BULLET PATH (IN) | -0.85 | 0.00 | -9.00 | -29.21 | -62.20 | -109.76 | -173.95 |
| TIME OF FLIGHT | 0.0 | 0.154890 | 0.320058 | 0.496310 | 0.684526 | 0.885599 | 1.100424 |
| WIND DEFLECTION (IN) | | | | | | | |
| 5 MPH | 0.0 | 0.43 | 1.77 | 4.08 | 7.44 | 11.93 | 17.64 |
| 10 MPH | 0.0 | 0.86 | 3.53 | 8.15 | 14.88 | 23.87 | 35.27 |
| 20 MPH | 0.0 | 1.72 | 7.06 | 16.30 | 29.75 | 47.73 | 70.55 |
| 30 MPH | 0.0 | 2.58 | 10.59 | 24.45 | 44.63 | 71.60 | 105.82 |

## Bullet: .457 Round Ball
### Ballistic Coefficient: .064

| RANGE (YARDS) | MUZZLE | 25 | 50 | 75 | 100 | 125 | 150 |
|---|---|---|---|---|---|---|---|
| VELOCITY (FPS) | 1100. | 984. | 894. | 824. | 765. | 712. | 664. |
| ENERGY (FT-LB) | 384. | 307. | 254. | 216. | 186. | 161. | 140. |
| DROP (IN) | 0.0 | -0.96 | -4.16 | -10.07 | -19.14 | -31.89 | -48.91 |
| MID-RANGE (IN) | 0.0 | 0.25 | 1.12 | 2.78 | 5.41 | 9.20 | 14.39 |
| BULLET PATH (IN) | -0.85 | 0.00 | -1.39 | -5.48 | -12.73 | -23.67 | -38.87 |
| TIME OF FLIGHT | 0.0 | 0.072289 | 0.152444 | 0.239915 | 0.334433 | 0.436097 | 0.545185 |
| WIND DEFLECTION (IN) | | | | | | | |
| 5 MPH | 0.0 | 0.36 | 1.42 | 3.11 | 5.43 | 8.38 | 11.98 |
| 10 MPH | 0.0 | 0.72 | 2.83 | 6.23 | 10.86 | 16.75 | 23.95 |
| 20 MPH | 0.0 | 1.45 | 5.66 | 12.45 | 21.72 | 33.51 | 47.90 |
| 30 MPH | 0.0 | 2.17 | 8.49 | 18.68 | 32.58 | 50.26 | 71.86 |
| VELOCITY (FPS) | 1000. | 905. | 834. | 773. | 720. | 671. | 626. |
| ENERGY (FT-LB) | 317. | 260. | 221. | 190. | 164. | 143. | 125. |
| DROP (IN) | 0.0 | -1.15 | -4.93 | -11.81 | -22.28 | -36.94 | -56.41 |
| MID-RANGE (IN) | 0.0 | 0.30 | 1.32 | 3.22 | 6.21 | 10.51 | 16.40 |
| BULLET PATH (IN) | -0.85 | 0.00 | -1.78 | -6.66 | -15.14 | -27.80 | -45.27 |
| TIME OF FLIGHT | 0.0 | 0.079013 | 0.165437 | 0.258915 | 0.359513 | 0.467483 | 0.583223 |
| WIND DEFLECTION (IN) | | | | | | | |
| 5 MPH | 0.0 | 0.35 | 1.36 | 2.98 | 5.24 | 8.14 | 11.72 |
| 10 MPH | 0.0 | 0.71 | 2.72 | 5.97 | 10.47 | 16.28 | 23.45 |
| 20 MPH | 0.0 | 1.41 | 5.43 | 11.94 | 20.95 | 32.55 | 46.89 |
| 30 MPH | 0.0 | 2.12 | 8.15 | 17.91 | 31.42 | 48.83 | 70.34 |
| VELOCITY (FPS) | 900. | 829. | 769. | 716. | 668. | 623. | 582. |
| ENERGY (FT-LB) | 257. | 218. | 188. | 163. | 142. | 123. | 107. |
| DROP (IN) | 0.0 | -1.38 | -5.90 | -14.05 | -26.42 | -43.65 | -66.45 |
| MID-RANGE (IN) | 0.0 | 0.36 | 1.56 | 3.80 | 7.31 | 12.34 | 19.18 |
| BULLET PATH (IN) | -0.85 | -0.00 | -2.28 | -8.20 | -18.33 | -33.33 | -53.90 |
| TIME OF FLIGHT | 0.0 | 0.086911 | 0.180871 | 0.281963 | 0.390449 | 0.506736 | 0.631348 |
| WIND DEFLECTION (IN) | | | | | | | |
| 5 MPH | 0.0 | 0.31 | 1.25 | 2.81 | 5.03 | 7.93 | 11.56 |
| 10 MPH | 0.0 | 0.63 | 2.50 | 5.63 | 10.05 | 15.85 | 23.12 |
| 20 MPH | 0.0 | 1.26 | 5.00 | 11.25 | 20.10 | 31.70 | 46.23 |
| 30 MPH | 0.0 | 1.89 | 7.50 | 16.88 | 30.16 | 47.56 | 69.35 |
| VELOCITY (FPS) | 800. | 744. | 693. | 646. | 603. | 563. | 525. |
| ENERGY (FT-LB) | 203. | 176. | 152. | 133. | 116. | 101. | 88. |
| DROP (IN) | 0.0 | -1.69 | -7.29 | -17.41 | -32.71 | -53.96 | -82.06 |
| MID-RANGE (IN) | 0.0 | 0.45 | 1.96 | 4.74 | 9.06 | 15.23 | 23.62 |
| BULLET PATH (IN) | -0.85 | -0.00 | -3.06 | -10.64 | -23.40 | -42.11 | -67.67 |
| TIME OF FLIGHT | 0.0 | 0.097296 | 0.201827 | 0.313928 | 0.434062 | 0.562803 | 0.700821 |
| WIND DEFLECTION (IN) | | | | | | | |
| 5 MPH | 0.0 | 0.31 | 1.26 | 2.88 | 5.20 | 8.28 | 12.17 |
| 10 MPH | 0.0 | 0.62 | 2.52 | 5.75 | 10.39 | 16.55 | 24.34 |
| 20 MPH | 0.0 | 1.25 | 5.04 | 11.50 | 20.79 | 33.11 | 48.69 |
| 30 MPH | 0.0 | 1.87 | 7.56 | 17.25 | 31.18 | 49.66 | 73.03 |
| VELOCITY (FPS) | 700. | 653. | 609. | 569. | 530. | 495. | 461. |
| ENERGY (FT-LB) | 156. | 135. | 118. | 103. | 89. | 78. | 67. |
| DROP (IN) | 0.0 | -2.19 | -9.46 | -22.57 | -42.39 | -69.91 | -106.31 |
| MID-RANGE (IN) | 0.0 | 0.59 | 2.54 | 6.14 | 11.73 | 19.72 | 30.58 |
| BULLET PATH (IN) | -0.85 | 0.00 | -4.23 | -14.30 | -31.07 | -55.55 | -88.91 |
| TIME OF FLIGHT | 0.0 | 0.110963 | 0.229883 | 0.357318 | 0.493925 | 0.640452 | 0.797735 |
| WIND DEFLECTION (IN) | | | | | | | |
| 5 MPH | 0.0 | 0.34 | 1.37 | 3.16 | 5.75 | 9.22 | 13.63 |
| 10 MPH | 0.0 | 0.67 | 2.75 | 6.32 | 11.50 | 18.43 | 27.26 |
| 20 MPH | 0.0 | 1.34 | 5.49 | 12.63 | 23.00 | 36.87 | 54.52 |
| 30 MPH | 0.0 | 2.02 | 8.24 | 18.95 | 34.51 | 55.30 | 81.78 |
| VELOCITY (FPS) | 600. | 560. | 522. | 487. | 453. | 422. | 392. |
| ENERGY (FT-LB) | 114. | 99. | 87. | 75. | 65. | 56. | 49. |
| DROP (IN) | 0.0 | -2.98 | -12.88 | -30.74 | -57.76 | -95.34 | -145.15 |
| MID-RANGE (IN) | 0.0 | 0.81 | 3.46 | 8.36 | 16.00 | 26.93 | 41.84 |
| BULLET PATH (IN) | -0.85 | 0.00 | -6.07 | -20.09 | -43.28 | -77.03 | -123.01 |
| TIME OF FLIGHT | 0.0 | 0.129432 | 0.268194 | 0.417053 | 0.576864 | 0.748579 | 0.933252 |
| WIND DEFLECTION (IN) | | | | | | | |
| 5 MPH | 0.0 | 0.39 | 1.60 | 3.70 | 6.76 | 10.87 | 16.13 |
| 10 MPH | 0.0 | 0.78 | 3.20 | 7.40 | 13.53 | 21.75 | 32.25 |
| 20 MPH | 0.0 | 1.56 | 6.40 | 14.80 | 27.06 | 43.50 | 64.50 |
| 30 MPH | 0.0 | 2.34 | 9.61 | 22.20 | 40.58 | 65.25 | 96.76 |

| RANGE (YARDS) | MUZZLE | 25 | 50 | 75 | 100 | 125 | 150 |
|---|---|---|---|---|---|---|---|
| VELOCITY (FPS) | 500. | 466. | 434. | 403. | 375. | 349. | 325. |
| ENERGY (FT-LB) | 79. | 69. | 60. | 52. | 45. | 39. | 33. |
| DROP (IN) | 0.0 | -4.31 | -18.61 | -44.46 | -83.65 | -138.31 | -210.87 |
| MID-RANGE (IN) | 0.0 | 1.16 | 5.00 | 12.11 | 23.22 | 39.17 | 60.98 |
| BULLET PATH (IN) | -0.85 | 0.00 | -9.15 | -29.84 | -63.88 | -113.38 | -180.78 |
| TIME OF FLIGHT | 0.0 | 0.155448 | 0.322413 | 0.501905 | 0.695024 | 0.902858 | 1.126547 |
| WIND DEFLECTION (IN) | | | | | | | |
| 5 MPH | 0.0 | 0.48 | 1.97 | 4.57 | 8.36 | 13.45 | 19.94 |
| 10 MPH | 0.0 | 0.96 | 3.94 | 9.14 | 16.72 | 26.90 | 39.87 |
| 20 MPH | 0.0 | 1.92 | 7.89 | 18.27 | 33.45 | 53.81 | 79.74 |
| 30 MPH | 0.0 | 2.88 | 11.83 | 27.41 | 50.17 | 80.71 | 119.62 |

## Bullet: No. 45467 Solid Base
### Ballistic Coefficient: .112

| RANGE (YARDS) | MUZZLE | 25 | 50 | 75 | 100 | 125 | 150 |
|---|---|---|---|---|---|---|---|
| VELOCITY (FPS) | 900. | 870. | 842. | 816. | 792. | 769. | 747. |
| ENERGY (FT-LB) | 360. | 336. | 315. | 296. | 278. | 262. | 248. |
| DROP (IN) | 0.0 | -1.29 | -5.45 | -12.65 | -23.09 | -36.89 | -54.36 |
| MID-RANGE (IN) | 0.0 | 0.35 | 1.43 | 3.33 | 6.10 | 9.79 | 14.53 |
| BULLET PATH (IN) | -0.85 | -0.00 | -2.01 | -7.07 | -15.36 | -27.02 | -42.34 |
| TIME OF FLIGHT | 0.0 | 0.084786 | 0.172433 | 0.262906 | 0.356202 | 0.452336 | 0.551332 |
| WIND DEFLECTION (IN) | | | | | | | |
| 5 MPH | 0.0 | 0.13 | 0.51 | 1.14 | 2.01 | 3.14 | 4.52 |
| 10 MPH | 0.0 | 0.26 | 1.01 | 2.27 | 4.02 | 6.28 | 9.03 |
| 20 MPH | 0.0 | 0.51 | 2.03 | 4.54 | 8.05 | 12.56 | 18.07 |
| 30 MPH | 0.0 | 0.77 | 3.04 | 6.81 | 12.07 | 18.83 | 27.10 |
| VELOCITY (FPS) | 800. | 776. | 754. | 733. | 712. | 693. | 674. |
| ENERGY (FT-LB) | 284. | 268. | 253. | 238. | 225. | 213. | 201. |
| DROP (IN) | 0.0 | -1.53 | -6.63 | -15.53 | -28.47 | -45.67 | -67.38 |
| MID-RANGE (IN) | 0.0 | 0.42 | 1.78 | 4.15 | 7.60 | 12.24 | 18.16 |
| BULLET PATH (IN) | -0.85 | -0.00 | -2.72 | -9.24 | -19.79 | -34.61 | -53.94 |
| TIME OF FLIGHT | 0.0 | 0.095169 | 0.193190 | 0.294095 | 0.397925 | 0.504733 | 0.614581 |
| WIND DEFLECTION (IN) | | | | | | | |
| 5 MPH | 0.0 | 0.12 | 0.50 | 1.13 | 2.02 | 3.17 | 4.58 |
| 10 MPH | 0.0 | 0.25 | 1.00 | 2.26 | 4.03 | 6.33 | 9.17 |
| 20 MPH | 0.0 | 0.50 | 2.00 | 4.52 | 8.07 | 12.67 | 18.33 |
| 30 MPH | 0.0 | 0.75 | 3.00 | 6.78 | 12.10 | 19.00 | 27.50 |
| VELOCITY (FPS) | 700. | 681. | 662. | 644. | 626. | 609. | 593. |
| ENERGY (FT-LB) | 218. | 206. | 195. | 184. | 174. | 165. | 156. |
| DROP (IN) | 0.0 | -1.97 | -8.60 | -20.18 | -36.98 | -59.30 | -87.46 |
| MID-RANGE (IN) | 0.0 | 0.57 | 2.34 | 5.41 | 9.89 | 15.89 | 23.55 |
| BULLET PATH (IN) | -0.85 | -0.00 | -3.82 | -12.58 | -26.57 | -46.07 | -71.41 |
| TIME OF FLIGHT | 0.0 | 0.108660 | 0.220401 | 0.335297 | 0.453431 | 0.574890 | 0.699772 |
| WIND DEFLECTION (IN) | | | | | | | |
| 5 MPH | 0.0 | 0.13 | 0.54 | 1.22 | 2.19 | 3.45 | 5.01 |
| 10 MPH | 0.0 | 0.27 | 1.08 | 2.44 | 4.38 | 6.89 | 10.02 |
| 20 MPH | 0.0 | 0.53 | 2.15 | 4.88 | 8.75 | 13.79 | 20.03 |
| 30 MPH | 0.0 | 0.80 | 3.23 | 7.32 | 13.13 | 20.68 | 30.05 |
| VELOCITY (FPS) | 600. | 584. | 568. | 552. | 537. | 522. | 508. |
| ENERGY (FT-LB) | 160. | 151. | 143. | 135. | 128. | 121. | 114. |
| DROP (IN) | 0.0 | -2.67 | -11.71 | -27.46 | -50.32 | -80.70 | -119.03 |
| MID-RANGE (IN) | 0.0 | 0.77 | 3.18 | 7.36 | 13.45 | 21.63 | 32.06 |
| BULLET PATH (IN) | -0.85 | 0.00 | -5.51 | -17.74 | -37.07 | -63.93 | -98.74 |
| TIME OF FLIGHT | 0.0 | 0.126754 | 0.257088 | 0.391111 | 0.528942 | 0.670703 | 0.816523 |
| WIND DEFLECTION (IN) | | | | | | | |
| 5 MPH | 0.0 | 0.15 | 0.62 | 1.42 | 2.55 | 4.02 | 5.85 |
| 10 MPH | 0.0 | 0.31 | 1.25 | 2.84 | 5.09 | 8.04 | 11.71 |
| 20 MPH | 0.0 | 0.62 | 2.49 | 5.67 | 10.19 | 16.09 | 23.42 |
| 30 MPH | 0.0 | 0.93 | 3.74 | 8.51 | 15.28 | 24.13 | 35.12 |
| VELOCITY (FPS) | 500. | 486. | 472. | 459. | 446. | 434. | 422. |
| ENERGY (FT-LB) | 111. | 105. | 99. | 94. | 88. | 84. | 79. |
| DROP (IN) | 0.0 | -3.86 | -16.88 | -39.61 | -72.61 | -116.51 | -171.94 |
| MID-RANGE (IN) | 0.0 | 1.11 | 4.58 | 10.61 | 19.43 | 31.26 | 46.36 |
| BULLET PATH (IN) | -0.85 | 0.00 | -8.32 | -26.34 | -54.64 | -93.82 | -144.55 |
| TIME OF FLIGHT | 0.0 | 0.152153 | 0.308708 | 0.469814 | 0.635626 | 0.806307 | 0.982028 |
| WIND DEFLECTION (IN) | | | | | | | |
| 5 MPH | 0.0 | 0.19 | 0.77 | 1.74 | 3.14 | 4.96 | 7.22 |
| 10 MPH | 0.0 | 0.38 | 1.53 | 3.49 | 6.27 | 9.91 | 14.44 |
| 20 MPH | 0.0 | 0.76 | 3.07 | 6.97 | 12.54 | 19.82 | 28.87 |
| 30 MPH | 0.0 | 1.14 | 4.60 | 10.46 | 18.81 | 29.73 | 43.31 |

# Reader Service Page

**CARD #1**

**SEND TO COMPONENT MANUFACTURER**

Gentlemen:

I have read about your products in Lyman's PISTOL & REVOLVER HANDBOOK. Please send me a current catalog showing your reloading components.

NAME: ____________________

ADDRESS: ____________________

____________________

ZIP CODE: __________

P&RHB-2

---

**CARD #2**

**SEND TO LYMAN**

Please send me your free annual catalog showing the complete Lyman line.

NAME: ____________________

ADDRESS: ____________________

____________________ZIP: __________

Please help by answering the following questions:

1. Do you use any of the following handbooks?
   Speer ☐ Sierra ☐ Hornady ☐ Lyman 45th ☐
   Lyman Shotshell ☐ Lyman Blackpowder ☐
2. How many metallic cartridges do you load each year? __________
3. What Lyman products do you use?
   Metallic Reloading ☐ Shotshell Reloading ☐ Telescopic Sights ☐
   Metallic Sights ☐ Bullet Casting ☐
   Black Powder Guns ☐ Publications ☐
4. Do you shoot black powder? __________

   Cast your own bullets? __________

Publisher's Offer: We have a limited stock of Shotshell Handbooks, 1st Edition, available.

Send ______ copies @ $5.95 postpaid.

Send ______ copies COD @ $5.95 plus COD charges.

P&RHB-2

---

**CARD #3**

**SEND TO LYMAN**

Gentlemen:

I'd like to order additional copies of the PISTOL & REVOLVER HANDBOOK. I'll send check or money order for postpaid delivery. Orders without check will be sent COD.

______ copies of the P&RHB, @ $10.95 postpaid

NAME: ____________________

ADDRESS: ____________________

ZIP: ____________________

P&RHB-2

PLACE

STAMP

HERE

PRINT NAME AND ADDRESS OF MANUFACTURER BELOW.

PLACE

STAMP

HERE

**LYMAN PUBLICATIONS**

**ROUTE 147**

**MIDDLEFIELD, CT 06455**

PLACE

STAMP

HERE

**LYMAN PUBLICATIONS**

**ROUTE 147**

**MIDDLEFIELD, CT 06455**